Natural Toxicants in Feeds, Forages, and Poisonous Plants

Second Edition

Natural Toxicants in Feeds, Forages, and Poisonous Plants

by

Peter R. Cheeke

Department of Animal Sciences
Oregon State University

Interstate Publishers, Inc.
Danville, Illinois

Natural Toxicants in Feeds, Forages, and Poisonous Plants

Second Edition

First edition published under the title *Natural Toxicants in Feeds and Poisonous Plants*, 1985.

Library of Congress Catalog Card No. 96-80185

ISBN 0-8134-3128-X

2 3 4 5 6 7 8 9 10   03 02 01 00 99

Order from

Interstate Publishers, Inc.

510 North Vermilion Street
P.O. Box 50
Danville, IL 61834-0050

Phone: (800) 843-4774
Fax: (217) 446-9706
Email: info-ipp@IPPINC.com

World Wide Web: http://www.IPPINC.com

Preface

There is widespread societal concern about environmental toxins and their potential adverse effects on humans and animals. Many people fail to understand that we are constantly exposed to toxic substances, not produced by mad scientists in obscure industrial laboratories, but by the greatest chemist of all, Mother Nature. Virtually all plants that humans and our domestic animals consume are potentially toxic. If plants were not toxic, they would long ago have been eaten to extinction. Plants defend themselves against being eaten (herbivory) with an incredible diversity of chemical toxins. Thus, we are constantly exposed to "Nature's pesticides," or natural toxicants. Humans and animals have evolved feeding and metabolic strategies to overcome the chemical defenses of plants. This book provides a comprehensive treatment of natural toxicants in plants consumed by domestic animals and in crop plants used in human nutrition. It emphasizes the occurrence of toxins in plants, their metabolism and toxicologic effects in animals, and means of overcoming their effects. The book will be useful as a text and reference to upper division undergraduate students, graduate students, and professional resource managers in disciplines such as Animal Science, Poultry Science, Agronomy and Crop Science, Range Management, Forestry, Wildlife Science, Veterinary Medicine, Toxicology, Food Science and Technology, and Nutrition.

The book is organized to facilitate its use as a text. Animal and Poultry Science courses might emphasize Parts II and III, which deal with natural toxins in feeds and forages. In Range Management courses, Part IV on poisonous plants will be highly relevant. For veterinary students and veterinarians, the entire text will be a useful treatment of naturally occurring toxicoses in animals. Thus, the organization of topics will allow individual instructors to tailor courses relevant to their discipline's priorities and needs. The book will also be a valuable resource for extension specialists, who are often the first to be contacted when problems with poisonous plants occur.

It is difficult, if not impossible, for one person to be fully knowledgeable in all the areas covered in a book of this scope. In numerous instances, I have had sections reviewed by leading authorities in particular subject areas to eliminate inaccuracies and minimize misinterpretations. I believe single authorship has some advantages for the reader, including cohesiveness of style and format. It is my intention that this book be readable and useful to a wide, diverse audience, with adequate referencing of current literature to allow further specialization if necessary.

I have endeavored to cover the recent literature thoroughly and have selected post-1990 references for citation whenever possible. Because a toxicant rarely has an exclusive effect on one organ or tissue, a certain amount of cross-referencing between chapters is necessary. This has been accomplished by making reference within the chapters to other chapters where a particular toxicant or plant may be mentioned and by having a complete, thorough index. Key words have been highlighted to make it easier to locate information.

It has been a pleasure to work with the staff at Interstate Publishers, Inc., and especially Ronald L. McDaniel, Vice President–Editorial. He facilitated rapid publication after receipt of the manuscript and showed admirable good humor when I proposed last-minute changes and additions.

Two people have made particularly valuable contributions to the preparation of this edition. Keyboarding and layout of the manuscript were largely done by Helen Chesbrough, who worked long, hard, and cheerfully to see the project through. Her competence, enthusiasm, and grammatical skills are greatly appreciated. I also wish to acknowledge with pleasure the work of Dr. Jianya Huan, who completed his Ph.D. under me in 1995. He prepared the figures of chemical structure and metabolic pathways, using current methodology that is beyond my computer capabilities. His assistance was very useful and much appreciated. I also acknowledge Dr. Lee Shull for the use of some of the material that he prepared for the First Edition.

I began preparation of this edition in 1992. It has been a long struggle, nurtured by the numerous people who kept asking, "When's the revised edition of your book coming out?" We finally made it!

Peter R. Cheeke

Table of Contents

PART III

Forage-Induced Toxicoses

PART IV

Plants Poisonous to Livestock

PART I

Metabolic Effects, Metabolism, and Ecological Roles of Plant Toxins

CHAPTER 1

Natural Toxicants and Their General Biological Effects

"All substances are poisons; there is none which is not a poison.
The right dose differentiates a poison and a remedy."

Paracelsus (1493–1541)

As the above quotation suggests, it has been known for centuries that "the dose makes the poison". Despite this, the common reaction is one of apprehension and fear upon learning that a toxin is present in food or the environment, even though the level present may be so low as to present no discernible risk. Another relevant point in the quotation is that all drugs and medicines are poisons. At a particular dose level, their pharmacological properties may be useful, but at higher levels, they are poisonous. Also of interest is that many natural toxins in poisonous plants are in fact used medicinally, particularly in herbal preparations.

Toxicants, also known as toxins, poisons and **xenobiotics** (xeno = Gr: strange or foreign) are substances which under practical circumstances can impair some aspect of animal metabolism and produce adverse biological or economic effects in animal production. This is a broad definition, but encompasses those aspects that are relevant in livestock production. Virtually everything is toxic, including oxygen, water, and all nutrients, if given in a large enough dose. Thus, the term "toxicant" refers only to those substances which might normally be encountered at toxic levels.

The term **natural toxicant** is used to refer to toxins that occur in nature as opposed to those that are synthesized by humans (e.g. agricultural chemicals—insecticides, herbicides, etc.). Natural might be loosely defined as not being produced or influenced by the activities of humans. On the other hand, humans are obviously part of the natural world, so it is considered by some that "natural vs. human-made" is an illogical distinction. In any case, natural toxicants in this book refers to those that occur naturally in plants, animals and microbes, as contrasted with those chemically synthesized by humans. There is a common perception that "human-made" chemicals are more dangerous than "natural" substances. Interestingly, the most toxic compounds known are natural. Aflatoxin is the most potent carcinogen, while the botulism toxin is the most poisonous organic substance ever encountered. Both of these extremely poisonous substances commonly occur in food (aflatoxin in peanut products, botulism toxin in honey) but are not generally involved in poisoning of humans because of the low levels present (honey should not be fed to human infants because of their high susceptibility to botulism).

Toxicants can be classified in various ways, including by chemical structure. An outline of the major chemical categories of natural toxins is presented in this chapter, with a brief description of their general biological effects. The remainder of the book is concerned with a more detailed treatment of each toxicant, including historical and contemporary perspectives on livestock problems, the chemical nature of the toxins, their biochemistry and mode of action, the pathological signs of toxicity, and the treatment and prevention of toxicity.

Toxicants discussed have been selected on the basis of their importance in livestock production in North America and to some extent in the rest of the world. Also, consideration has been given to including representative examples of toxicants of particular chemical types or unique metabolism in order to illustrate the diversity of structure and biological effects of toxicants. The compounds discussed are overwhelmingly of plant (**phytotoxin**) or microbial (**mycotoxin**) origin, with only a few toxins of animal (**zootoxin**) origin considered. This is not a book on poisonous plants per se, as many of the important toxins affecting livestock production occur in common feedstuffs. In fact, very few feedstuffs do not contain one or more substances that have deleterious effects.

Toxicants can influence animal agriculture in several ways. They can directly intoxicate animals, resulting in mortality or decreased production. Toxicants may be implicated in reducing the wholesomeness of meat, poultry, and dairy products due to the presence of hazardous **residues** in animal products. Natural toxins may reduce the availability or usability of nutritious feedstuffs, or may necessitate the use of costly feed processing techniques to eliminate their effects.

In the past few years, the focus of toxicology has widened dramatically. The importance of chronic toxicoses has been realized, with greater emphasis placed on delayed responses such as mutations, cancer, birth defects, and neurological and immunological effects. Thus large animal toxicology is an extremely broad discipline, incorporating pharmacology, botany, physiology, pathology, chemistry, biochemistry, immunology, animal nutrition, range and animal management, veterinary medicine, and economics.

The advances in the study of toxicants have in large part been due to developments in analytical techniques and instrumentation. Sophisticated methodology allows detection of toxicants at extremely low levels and the identification of short-lived and reactive metabolites. In spite of these advances, there are still important natural toxicoses for which the toxic agents have not been conclusively identified. Many opportunities exist for further advances in the understanding of the roles of toxicants in animal production.

ECOLOGICAL ROLES OF PLANT TOXINS

Plants contain toxic substances to protect themselves from **herbivory** (i.e. being consumed by herbivores). Plants and animals are mutually dependent inhabitants of the planet, and neither can survive without the other. The photosynthetic capacity of plants allows them to utilize solar energy, carbon dioxide and water to synthesize organic compounds, releasing oxygen as a waste product. Animals utilize the oxygen and organic compounds (nutrients) and excrete carbon dioxide which is used by plants. This plant–animal interrelationship has maintained the **biosphere** for several billion years. It is not in the evolutionary best interest of plants to be eaten to extinction by animals, or for plants to be totally resistant to animal herbivory, leading to extinction of animals. Thus plants and animals exist in a mutually beneficial adversarial relationship. Plant toxins have an intimate role in this relationship.

Since plants are immobile and unable to resist herbivory by moving away from herbivores, they must have other means of defending themselves. Their defenses are primarily physical and chemical. **Physical defenses** include spines, thorns, leaf hairs, leaf wax, and highly lignified or silica-loaded tissue. Spines and thorns deter large mammalian herbivores by their physical effects on the mouth, while leaf hairs and wax may help to prevent insect herbivory. The **chemical defenses** of plants consist of chemicals which in some way deter herbivory. They may cause the plant to be unpalatable through effects on taste or smell. Many plant toxins (e.g. alkaloids, glucosinolates, saponins) are bitter or otherwise unpleasant tasting, while terpenes and essential oils (e.g. in sagebrush) have unpleasant odors which reduce palatability. Phenolic compounds (tannins) have an astringent effect, because they react with taste receptors in the mouth. Many other plant toxins exert their effects elsewhere in the body, producing characteristic pathologies.

Much of the selection pressure on plants for the synthesis of toxins comes from insect herbivores and microbes (plant diseases) rather than from large animals. Because of their short generation intervals, insects and microbes can quickly evolve resistance to plant toxins, exerting selection pressure on the plant to respond with greater toxin synthesis or evolution of new defensive chemicals. Thus plants and herbivores coevolve. **Coevolution** implies that chemical changes in plants in response to herbivory lead to compensatory metabolic changes in herbivores to adapt to the new plant chemistry, ultimately leading to further alterations in plant chemistry. Thus coevolution is a continuum of changes in production of defensive chemicals in plants and their metabolism by herbivores. In many cases, the poisoning of large animals by poisonous plants is probably incidental to the real role of the toxins in protecting the plant against insects. Large herbivores play a greater role in stimulating the evolution of physical defenses such as spines and thorns. Some toxins, such as the phenolic compounds, probably evolved as protection against ultra-violet radiation, which was much more intense when Angiosperms first evolved than it is now. Phenolics are excellent antioxidants.

Animals have evolved various strategies for overcoming the effects of plant toxins. They can surmount physical defenses such as spines and thorns by developing **feeding strategies** that allow them to daintily browse on leaves while avoiding spines (e.g. goat), or developing tough mouth parts that allow them to consume and macerate spines with no ill effect (e.g. rhinoceros). Chemical defenses can be overcome in several ways. Some animals such as sheep and goats are tolerant of bitter compounds, and may consume unpalatable plants which cattle avoid. **Detoxification** may occur in the digestive tract, especially in ruminants. Ruminants are often more resistant to plant toxins than nonruminants, because of the degradation or inactivation of toxins by the rumen microbes. The liver is a major site of detoxification. It is strategically located for interception of absorbed toxins to prevent their entry into the general circulation. The liver contains a number of enzymes, such as the cytochrome P450 system (see Chapter 3), which are effective in detoxification. **Liver detoxifying enzyme systems** are quite substrate non-specific, giving the animal the capacity to withstand a wide variety of challenges from natural and synthetic toxins (herbicides, pesticides and other agricultural chemicals as well as natural toxins in plants). These hepatic enzymes are readily and quickly inducible, allowing a rapid detoxification response to ingested toxins. There is a high degree of variability in hepatic enzyme activities among individuals, providing the genetic basis for selection pressure for the evolution of species differences in susceptibility to poisonous plants (e.g. sheep vs. cattle). Thus the toxins in poisonous plants play an important ecological role as chemical defenses. They are **Nature's pesticides.**

HISTORICAL ASPECTS OF NATURAL TOXICANTS IN LIVESTOCK PRODUCTION IN NORTH AMERICA

Natural toxicants have had significant effects on livestock production in North America. In the frontier days, extensive stock losses occurred due to consumption of poisonous plants. In the settlement of new areas, both the livestock and the ranchers were inexperienced as to the toxicity of native plants, and some spectacular losses occurred as this experience was gained. In the late 1800s and early 1900s, some of the major toxicity problems in the U.S. included milk sickness in humans caused by milk transfer of a toxin from white snakeroot, and livestock poisonings from consumption of such plants as locoweed, larkspur, water hemlock, lupin, death camas, bitterweed, and sleepy grass. An early function of the United States Department of Agriculture (USDA), which was organized in 1862, was the investigation of toxic plants. Numerous field stations in the western U.S. were established by the USDA in the early 1900s for the study of specific local problems (Fig. 1–1). Some of these included stations in Hugo, Colorado, to study locoweed, in Gunnison National forest, Colorado, to study lark-

FIGURE 1–1 USDA poisonous plant investigators performing an autopsy—1906. USDA field stations were consolidated into one entity, the USDA Poisonous Plants Research Laboratory, Logan, Utah. (Courtesy of R. F. Keeler)

spur, lupin, and water hemlock poisoning, in Greycliff, Montana, to investigate lupin and death camas, and in Salina, Utah, to study oak, sneezewood, locoweed, and milkweed problems. In 1955, these stations were consolidated with the establishment of the **USDA Poisonous Plants Research Laboratory** in Logan, Utah. In cooperation with Utah State University in Logan, scientists at this laboratory have conducted in-depth investigations of many plant toxins and poisonous plants, including locoweed, larkspur, lupin, halogeton, ponderosa pine, pyrrolizidine alkaloid–containing plants, and numerous others. Considerable effort has been placed on the study of teratogens in plants, such as those in lupin and poison hemlock responsible for crooked calf disease, and *Veratrum* alkaloids which cause birth defects in sheep.

The days of massive outbreaks of plant poisonings of livestock in North America are probably over. Agriculture has become more intensive, and range and pasture management has improved. Herbicides are important tools in controlling poisonous plants, as are biological control agents (usually insects). Nevertheless, localized outbreaks of plant poisonings still occur. In addition, as animal production has become more intensive, the importance of natural toxins in cultivated forages and in feed ingredients has increased. Tall fescue toxicosis, ryegrass staggers, and gossypol toxicosis from cottonseed products are examples. Mycotoxins in grains and other feed ingredients are of major economic importance.

TOXICANT PROBLEMS AROUND THE WORLD

Natural toxicants are and have been responsible for livestock problems in many parts of the world. Australia might be regarded as the land of poisonous plants. Its arid, hostile environment and extensive pastoral industries contribute to the scope of the problems. In times of drought, stockmen might regard the availability of poisonous plants as being better than nothing at all. This could account for the contradictory common names of Paterson's curse and Salvation Jane for *Echium plantagineum*, a pyrrolizidine alkaloid–containing plant. Among the significant Australian toxicity situations are disorders in all livestock species caused by pyrrolizidine alkaloids, and fluoroacetate toxicity, lupinosis, ryegrass staggers, annual ryegrass toxicity, clover disease caused by phytoestrogens, *Swainsona* poisoning, oxalate poisoning, and numerous others. In New Zealand, some of the major problems have involved **photosensitization** in which severe skin lesions occur with exposure to sunlight. Probably the most important of these is facial eczema in dairy cattle and sheep, caused by a mycotoxin, sporidesmin. Another mycotoxin problem of significance in New Zealand is perennial ryegrass staggers. Brassica poisoning occurs in those areas where *Brassica* spp. such as rape and kale are used as forage. In the past, *Senecio jacobaea* poisoning has been a significant problem in parts of New Zealand. In Southeast Asian countries such as Indonesia, the Philippines, Papua New Guinea, and tropical areas of Australia, toxins in tropical forages such as *Leucaena leucocephala* are of concern. In China and Central Asia, poisonous plants on rangelands are important, including *Astragalus* spp. (locoweeds) and *Senecio* spp. In South Africa, numerous poisonous plant problems have affected the extensive livestock industries of that country; many of these problems are similar to those seen in Australia. Some of the conditions reported from South Africa include annual ryegrass toxicity, poisoning from various *Senecio* spp., and mycotoxin contamination of feedstuffs. Photosensitization caused by consumption of a variety of plants is a particular problem in South Africa. Cardiac glycoside–containing plants are widespread and cause considerable losses. Vomiting disease caused by sesquiterpene lactones in *Geigeria* spp. is significant. Fluoroacetate toxicity is one of the major problems. Mycotoxins in grains and protein supplements are major problems in equatorial Africa because of high temperature and humidity, and poor crop storage facilities. Aflatoxin contamination of foodstuffs is believed to be a significant contributing factor to liver cancer in humans in tropical African countries. In northern Africa, tannins in sorghum and millets have important implications in livestock production and human nutrition. In Great Britain and western Europe, some of the toxicants of interest include glucosinolates in rapeseed, the brassica anemia factor, pyrrolizidine alkaloids, and a variety of mycotoxins in feeds. Acute bovine pulmonary emphysema and bracken poisoning are other important European problems. In Latin American countries, a number of toxicants and poisonous plant problems have been reported, including bracken poisoning, mycotoxicoses, *Senecio* poisoning, and toxins such as fluoroacetate, oxalates, and thiaminases.

While this is a very brief overview of the situation, it is apparent that many toxicants affect livestock production in all parts of the world.

CLASSIFICATION OF NATURAL TOXICANTS BY CHEMICAL STRUCTURE

Natural toxicants can be classified on the basis of their chemical structures. The following categories describe many of the common plant toxins. In numerous instances, a neat, concise categorization is not possible, so arbitrary groupings may be made. For example, solanine in green potato tissue is both an alkaloid and a glycoside, and may be referred to as a glycoalkaloid.

1. Alkaloids

Alkaloids ("alkali-like") are compounds that contain nitrogen, usually in a heterocyclic ring, and are generally basic substances. They are usually bitter, and most are toxic. They are subclassified on the basis of the chemical type of the nitrogen-containing heterocyclic ring. They are synthesized in plants from amino acids. Many alkaloids are used medicinally (e.g. atropine, reserpine), some are common substances of drug abuse (e.g. cocaine, LSD), and some are used as stimulants (e.g. caffeine, nicotine). The following categories illustrate some of the most important alkaloids in poisonous plants.

A. Pyrrolizidine Alkaloids

N

The nucleus consists of two five-membered rings. Pyrrolizidine alkaloids are responsible for the toxicity of various *Senecio* spp. (tansy ragwort, common groundsel), crotalaria, *E. plantagineum* (Paterson's curse), and *Heliotropium europaeum*. The latter two are particularly important in Australia. Pyrrolizidine alkaloids cause irreversible liver damage. Examples are monocrotaline, heliotrine, lasiocarpine, and senecionine.

B. Piperidine Alkaloids

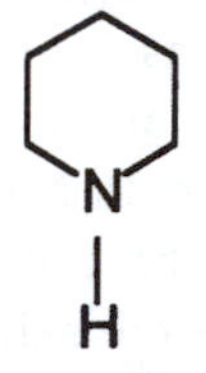

The most important piperidine alkaloids in animal production are coniine and related alkaloids found in *Conium maculatum* (poison hemlock). These alkaloids affect the central nervous system and are also teratogens.

C. Pyridine Alkaloids

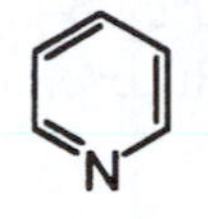

An example of a pyridine alkaloid is nicotine, in *Nicotiana* spp. (cultivated and wild tobacco).

D. Indole Alkaloids

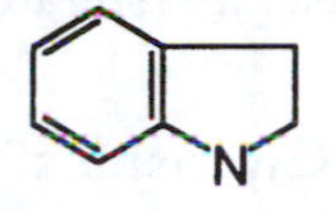

Indoles are derivatives of the amino acid tryptophan. Examples are the ergot alkaloids, the alkaloids such as perloline in tall fescue, tryptamine alkaloids found in *Phalaris* grasses, and 3-methylindole, an alkaloid implicated in bovine pulmonary emphysema.

E. Quinolizidine Alkaloids

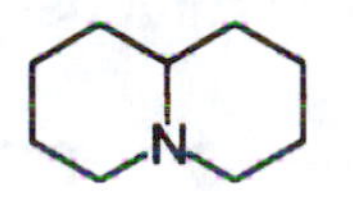

The quinolizidine nucleus consists of two six-membered rings. Lupins contain these alkaloids, which cause acute poisoning in sheep and teratogenic effects in calves (crooked calf disease).

F. Indolizidine Alkaloids

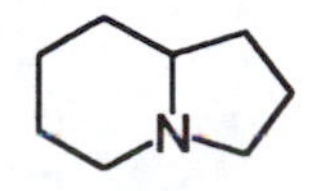

Indolizidine alkaloids, such as swainsonine, have been identified as the toxic components of *Swainsona* spp. in Australia and *Astragalus* spp. (locoweeds) in the U.S. They are inhibitors of α-mannosidase, resulting in the accumulation of mannose deposits in lysosomes of nerve cells. Slaframine is an indolizidine alkaloid mycotoxin produced on red clover forage by a fungus; it causes profuse salivation.

G. Steroid Alkaloids

i. Solanum type (e.g., solanidine)

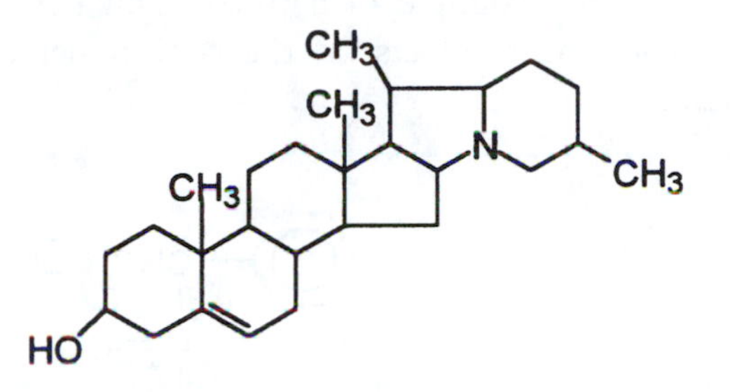

These are found in green potatoes, tomatoes, and nightshade. They are central nervous system poisons and cholinesterase inhibitors. Solanidine in potatoes has human health implications; an upper limit for alkaloid in new potato cultivars in the U.S. has been established by regulatory agencies.

ii. Veratrum type (e.g., veratramine)

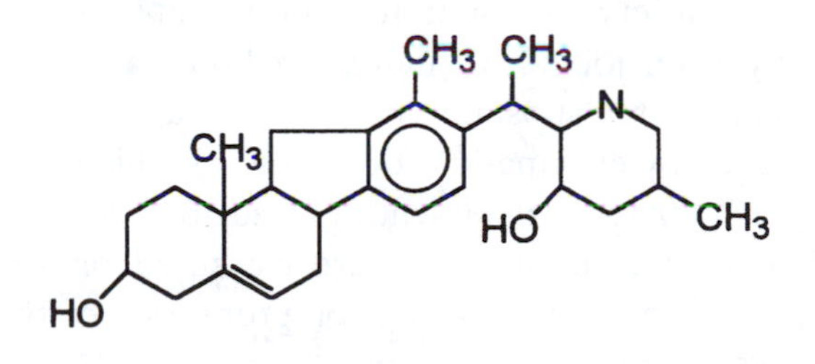

False hellebore (*Veratrum californicum*) contains veratrum alkaloids that produce teratogenic effects in lambs (cyclops lamb) and prolonged gestation in ewes. Death camas, a source of extensive sheep losses on western American rangelands in the past, contains alkaloids of this type.

H. Polycyclic Diterpene Alkaloids

These complex alkaloids are found in *Delphinium* spp., commonly known as larkspurs. Larkspurs are responsible for more cattle losses in the U.S. than all other toxic plants combined. The larkspur alkaloids cause acute central nervous system effects.

I. Tropane Alkaloids

Atropine, found in *Datura* spp. (Jimsonweed), is an example of a tropane alkaloid. It has pronounced effects on the central nervous system.

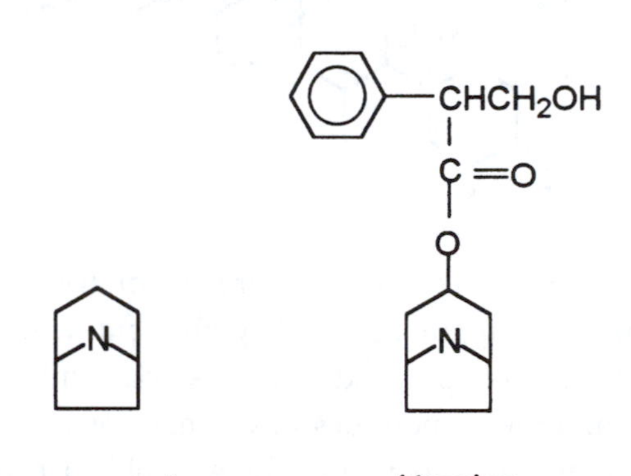

Tropane nucleus Atropine

2. Glycosides

Glycosides are ethers containing a carbohydrate moiety and a non-carbohydrate moiety (**aglycone**) joined with an ether bond. They are usually bitter substances. Often the aglycone is released by enzymatic action when the plant tissue is damaged, as by wilting, freezing, mastication, or trampling. They are classified on the basis of the structure and/or properties of the aglycone.

A. Cyanogenic Glycosides

These yield hydrocyanic (prussic) acid when hydrolyzed. An example is laetrile (amygdalin) found in the kernels of almonds, apricots, peaches, apples, and the leaves of chokecherries. It is hydrolyzed by β-glycosidase to release hydrogen cyanide, glucose, and benzaldehyde as follows:

$$C_6H_5\text{–}CH(CN)\text{–}O\text{–}C_6H_{10}O_4\text{–}C_6H_{11}O_5 \longrightarrow C_6H_5\text{–}CHO + HCN + 2C_6H_{12}O_6$$

The hydrogen cyanide is a potent inhibitor of cytochrome oxidase, a respiratory enzyme.

B. Goitrogenic Glycosides

Goitrogens decrease production of the thyroid hormones by inhibiting their synthesis by the thyroid gland. As a result, the thyroid enlarges to compensate for reduced thyroxin output, producing a goiter.

Goitrogenic glycosides are thioethers, containing an organic aglycone:

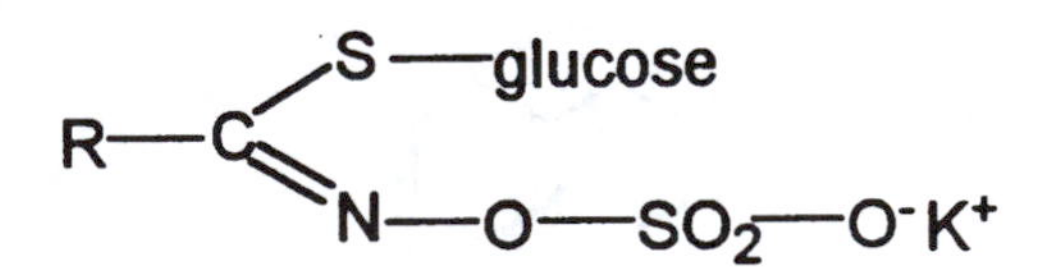

where R (the aglycone) is an alkyl group. These compounds, called **glucosinolates** (formerly called thioglucosides) are commonly found in *Brassica* spp. such as cabbage, kale, and rape. They are hydrolyzed by glucosinolases to β-D-glucose, HSO^-_4, and derivatives of the aglycone. These include isothiocyanates, nitriles, and thiocyanates. The glucosinolates are anions, usually found as potassium salts.

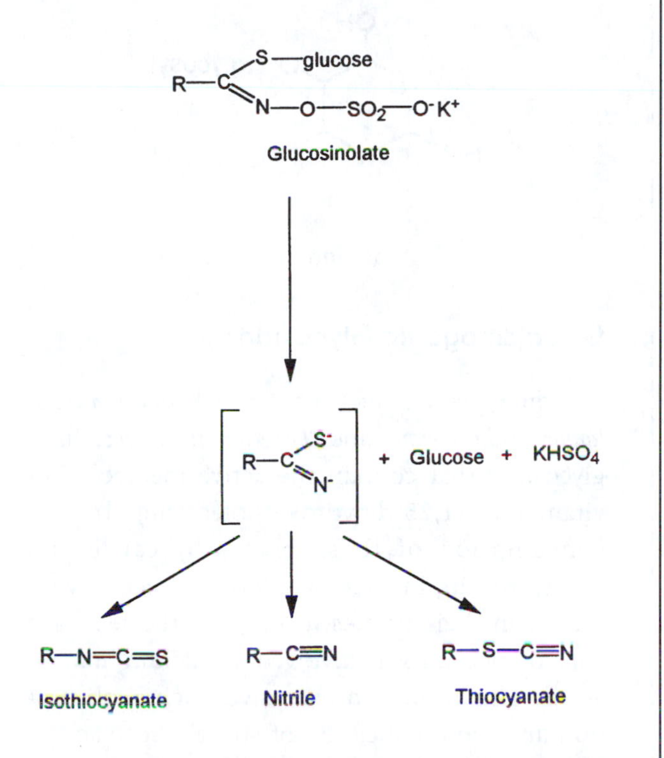

The glucosinolase is released from plant tissue by crushing (mastication) and is also produced by rumen microorganisms. Isothiocyanates may form ring structures such as goitrin (5-vinyl-2-oxazolidine-thione):

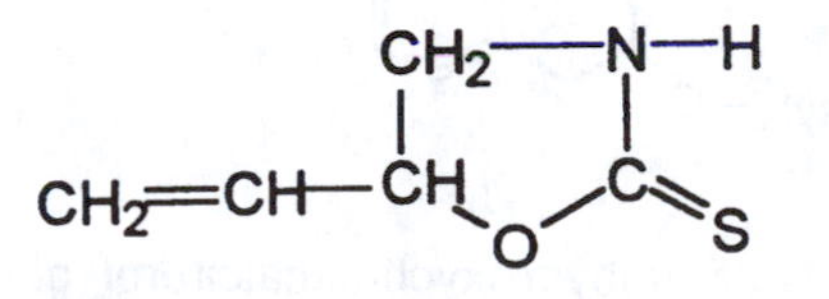

C. Coumarin Glycosides

Coumarin is found in sweet clover (*Melilotus* spp.) as melilotoside.

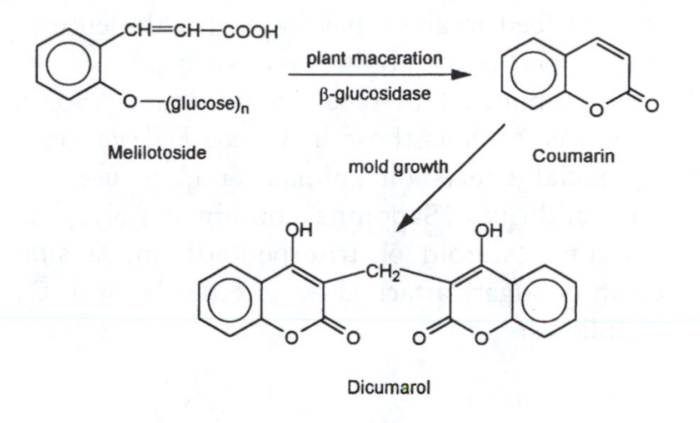

Coumarin is converted by mold growth to dicumarol, an antagonist of vitamin K. Sweet clover poisoning, caused by feeding moldy sweet clover hay, is therefore an induced vitamin K deficiency.

D. Steroid and Triterpenoid Glycosides

i. Cardiac glycosides

The best known cardiac glycoside is digitonin, contained in foxgloves (*Digitalis* spp.). These glycosides contain a sterol group in their structure. Physiologically, they are potent stimulators of heart rate and are used medicinally. Milkweeds (*Asclepias* spp.) contain cardiac glycosides (cardenolides) and are extremely toxic to livestock.

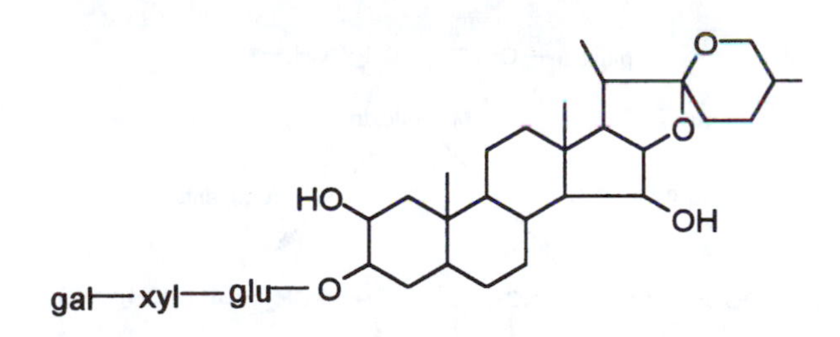

ii. Saponins

Saponins are glycosides which have profuse foaming properties, producing a distinctive honey-combed stable foam when an aqueous solution is shaken. They are widely distributed in plants and, in animal nutrition, are particularly important in temperate legume forages. They are bitter compounds, affecting palatability and feed intake. They have growth-depressing properties in poultry and swine, and have been implicated in bloat in ruminants. Some saponins, such as those in *Yucca schidigera*, have beneficial effects on animals and are used as feed additives. Saponins contain a polycyclic aglycone (steroid or triterpenoid) and a side chain of sugars attached by an ether bond to C3 as follows:

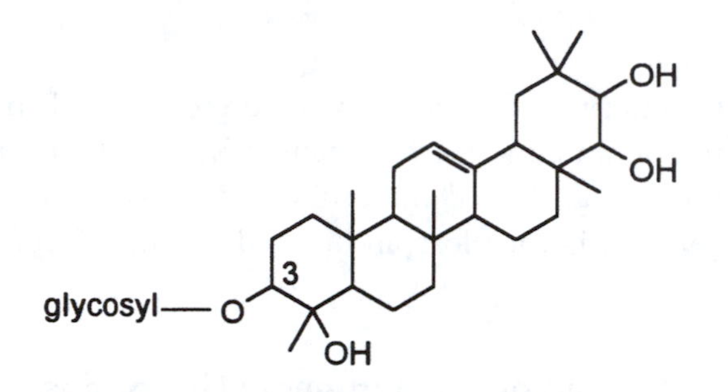

E. Nitropropanol Glycosides

Many *Astragalus* species, such as timber milk vetch (*Astragalus miser*), owe their toxicity to this class of glycoside. They are metabolized to 3-nitro-1-propanol (3NPOH) in ruminants, and 3-nitropropionic acid (3-NPA) in nonruminants. These compounds, especially 3NPOH, are acutely toxic, producing methemoglobinemia. They also produce chronic toxicity symptoms, involving permanent nerve damage.

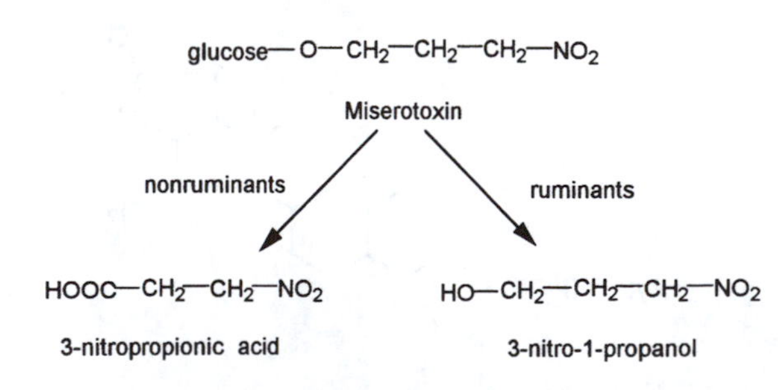

This class of glycosides is also found in crown vetch (*Coronilla* spp.).

F. Vicine

Vicine is a glycoside in fava beans (*Vicia faba*). It causes hemolytic anemia (favism) in people who have a genetic deficiency of glucose-6-phosphate dehydrogenase activity in their red blood cells. Fava beans are grown to some extent as a protein supplement for livestock.

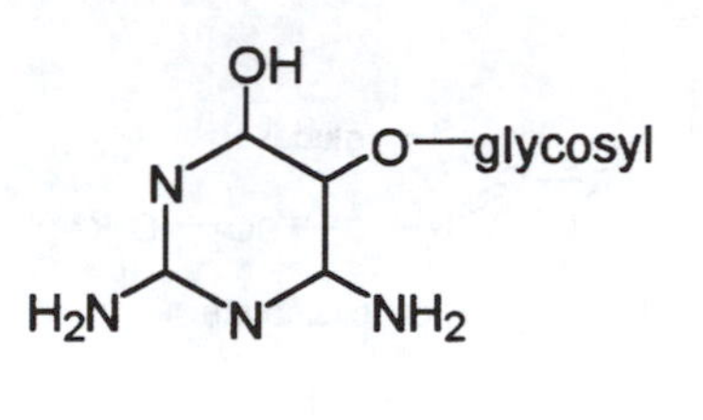

Vicine

G. Calcinogenic Glycosides

Some plants, such as *Cestrum diurnum*, *Solanum malacoxylon*, and *Trisetum flavescens*, have glycosides that contain the active metabolite of vitamin D (1,25-dihydroxycholecalciferol). The consumption of these plants by cattle and horses results in excessive levels of active vitamin D in their tissues, overriding the feedback control mechanisms involved in calcium homeostasis. This results in excessive calcium absorption and the calcification of soft tissues such as arteries and kidney.

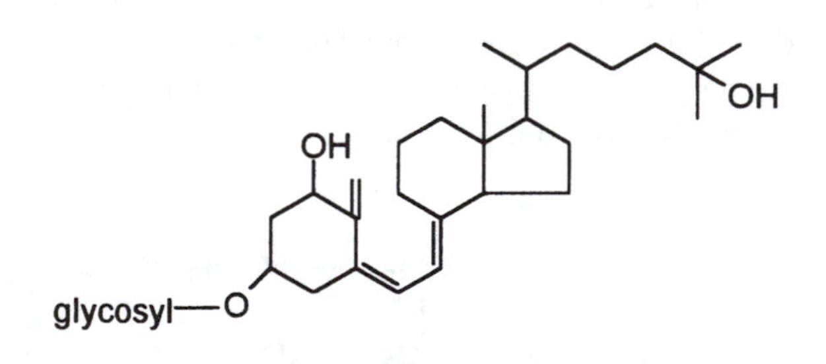

1,25 - dihydroxycholecalciferol glycoside

H. Carboxyatractyloside

A glycoside called carboxyatractyloside has been identified as the toxicant in cocklebur (*Xanthium strumarium*). It produces hepatic lesions, convulsions, and severe hypoglycemia.

I. Isoflavones

These compounds contain a flavone nucleus. Examples of isoflavones are genistein, formononetin, and coumestrol. They have estrogenic activity in mammals, and are referred to as plant estrogens or **phytoestrogens**.

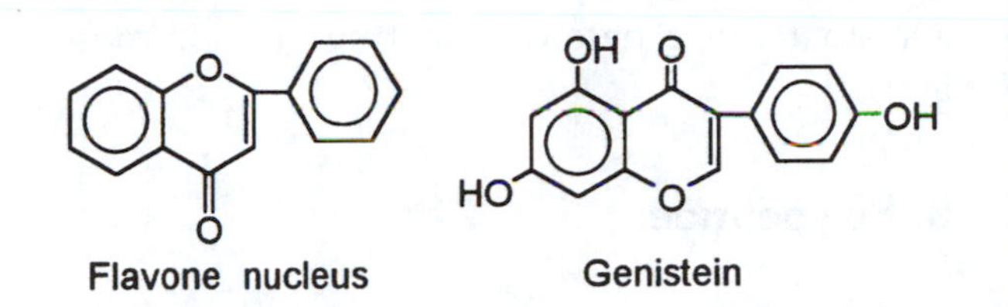

Flavone nucleus Genistein

Isoflavones are particularly important in subterranean clover and have caused extensive reproductive problems in sheep in Australia. Soybean isoflavones are also of much current interest in human nutrition because of their possible beneficial effects on estrogen-dependent cancers.

3. Proteins

Several important inhibitors in plants are proteins. Of interest is that in some cases the effect of these is to inhibit the utilization of other proteins by animals.

A. Protease (Trypsin) and Amylase Inhibitors

Soybeans, most other legume seeds, potatoes, and some grains (e.g., rye and triticale) contain **trypsin inhibitors**. These are small protein molecules which combine with and inactivate the digestive enzymes trypsin and chymotrypsin in the small intestine. They cause reduced growth and pancreatic hypertrophy. Amylase inhibitors also occur in beans and reduce the digestibility of starch in nonruminants.

B. Lectins (Hemagglutinins)

These are glycoproteins of 60,000–100,000 MW that cause agglutination (clumping) of red blood cells in vitro. They are found in most types of beans, including soybeans. They cause reduced growth, diarrhea, severe damage to the intestinal mucosa and interfere with nutrient absorption.

C. Enzymes

An example of an enzyme toxin is thiaminase, found in bracken fern (*Pteridium aquilinum*) and the tissues of certain fish such as carp. It is also produced by some ruminal microbes. The enzyme cleaves thiamin (vitamin B_1), thereby inactivating it. Consumption of bracken fern causes thiamin deficiency in some animals. Use of carp and other types of thiaminase-containing fish in mink diets has produced thiamin deficiency (Chastek's paralysis).

Other enzymes in feeds which produce deleterious effects in livestock include lipoxidases in soybeans and alfalfa, which degrade fat-soluble vitamins.

D. Plant Cytoplasmic Proteins

Many leguminous forages such as alfalfa and numerous clovers may cause bloat in ruminants. This is primarily due to formation of a stable foam in the rumen, involving cytoplasmic proteins in the plant cell contents. These proteins are not specifically toxic, but because they induce a pathological condition (bloat) they are considered in this text as natural toxicants.

4. Amino Acids and Amino Acid Derivatives

A. Amino Acids

Over 600 different amino acids have been identified in the plant kingdom (Rosenthal, 1991). Not surprisingly, some of these are toxic. One of the best known relative to livestock poisoning is **mimosine**, which is structurally similar to tyrosine:

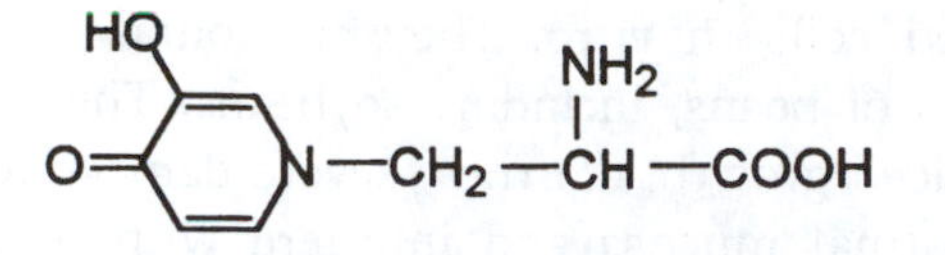

Mimosine occurs in the tropical forage *L. leucocephala*, which produces protein-rich leaves that have considerable potential as livestock feed. Mimosine causes reduced growth and alopecia (loss of hair) in nonruminants and is metabolized in the rumen to a goitrogenic compound, producing goiter in ruminants.

Tryptophan is a dietary essential amino acid. It can also be regarded as a toxicant because under some conditions it is metabolized in cattle to **3-methylindole**, a compound responsible for acute bovine pulmonary emphysema.

Dihydroxyphenylalanine (dopa) occurs in fava beans and has pharmacological effects.

Selenoamino acids, such as selenocystine, methylselenocysteine, selenocystathionine, and selenomethionine, which contain selenium in place of sulfur, are implicated in selenium toxicity due to consumption of selenium accumulator plants such as *Astragalaus* spp.

There are several **lathyrogenic amino acids**, including β-cyano-L-alanine in *Vicia* spp. and β-amino propionitrile in *Lathyrus* spp. Lathyrus seeds are consumed as food in India, causing a major public health problem, as permanent paralysis and skeletal deformity may occur. In livestock, consumption of *Lathyrus odoratus* seeds by cattle and horses causes paralysis, aortic aneurysm, and skeletal deformity. **Lathyrism** is the term used to describe the symptoms; both neurolathyrism (paralysis) and osteolathyrism (skeletal deformity) are observed.

Another toxic amino acid is 1-amino-D-proline. It is produced from **linatine**, found in linseed meal. 1-Amino-D-proline is an antagonist of pyridoxine (vitamin B_6).

Indospecine is a toxic amino acid found in *Indigofera spicata*, a potentially useful tropical pasture legume. It is an antagonist of arginine, resulting in depressed incorporation of arginine into liver-synthesized proteins.

The brassica anemia factor, **S-methylcysteine sulfoxide**, is an amino acid derivative. It leads to red blood cell hemolysis and anemia. It is found in forage brassicas such as kale and turnips.

Most of the toxic amino acids could also be classified as alkaloids, because their structures satisfy the definition of an alkaloid if they have a N-containing heterocyclic ring (e.g. mimosine).

B. Polypeptides

Amanita mushrooms are extremely poisonous and owe their toxicity to cyclic polypeptides. An unusual livestock poisoning problem in Australia is caused by the larvae of the sawfly, which contain an octapeptide called lophyrotomin. The sawfly larvae congregate in mounds beneath the silver-leaf ironbark tree (*Eucalyptus melanophloia*) and are avidly consumed by cattle. Affected animals show incoordination, trembling, and liver damage.

5. Carbohydrates

There are few toxicity problems due to carbohydrates. **Xylose**, a hexose sugar, causes growth retardation and eye cataracts in pigs and poultry. Certain **oligosaccharides**, such as raffinose, are not digested in the small intestine, and so promote bacterial growth in the hindgut. These are the flatulence factors in beans. The β-glucans in certain barley varieties sometimes cause nutritional problems in poultry. These effects meet the broad definition of a toxicant (pg. 4).

6. Lipids

Several fatty acids are toxic. These include **erucic acid** in rapeseed, which may result in myocardial lesions in rats. **Cyclopropenoid fatty acids**, such as sterculic and malvalic acids in cottonseed, have toxic properties and cause pink albumins to develop in stored eggs. They are also cocarcinogens, increasing the carcinogenicity of aflatoxins.

7. Glycoproteins

Lectins (discussed under Proteins) are glycoproteins. **Avidin**, a glycoprotein in egg albumin, is an antagonist of the B vitamin biotin. Raw eggs can be used to induce biotin deficiency in experimental animals. Biotin deficiency has occurred in fur animals (foxes and mink) fed raw eggs or poultry-processing plant wastes containing unlaid eggs.

8. Glycolipids

The cause of annual ryegrass toxicity (ARGT) has been identified as a glycolipid(s) called **corynetoxin**. This toxin is synthesized by bacteria which colonize nematode galls in the ryegrass seed head. The toxin affects the brain, leading to incoordination, staggering and death.

9. Metal-Binding Substances

A. Oxalates

Oxalic acid is a chelating agent which chelates calcium very effectively. Plants with a high oxalate content, such as the U.S. range weed *Halogeton glomeratus*, may produce acute metabolic calcium deficiency (hypocalcemia) when consumed by livestock.

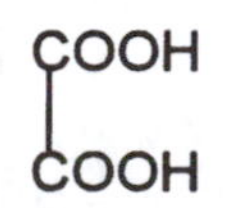

Oxalic acid

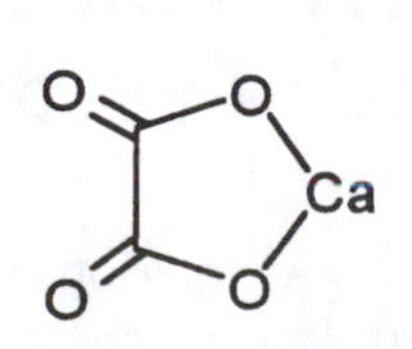

Calcium oxalate

B. Phytates

Phytic acid in cereal grains and soybean meal causes reduced mineral availability, particularly of zinc, through the formation of unabsorbable phytates. Organic phosphorus (phytin phosphorus) is of low availability to nonruminant animals.

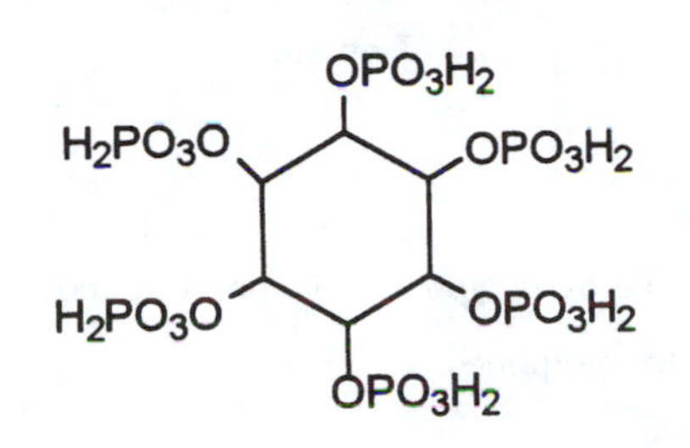

C. Mimosine

Mimosine, the toxic amino acid in *L. leucocephala,* has metal-binding properties.

10. Phenolic Compounds

Phenolics contain an aromatic ring with one or more hydroxyl groups. Polyphenolics are those with an abundance of hydroxyl groups. Examples of toxic phenolics include hypericin, gossypol and tannins.

Hypericin is a phenolic compound in *Hypericum perforatum* (St. Johnswort). It is a primary photosensitizing agent, producing dermatitis and skin lesions in light-skinned animals by reacting with ultraviolet light at the skin surface to produce a photodynamic reaction.

Cottonseed meal contains a toxic polyphenol called **gossypol**. Gossypol causes reduced growth and feed intake, cardiac lesions, and male infertility. Gossypol can also be classified as a sesquiterpene (see Terpenes and Terpenoids).

Tannins are phenolic compounds that react with proteins. The term was originally used for plant extracts that were used in tanning leather. They are astringent and adversely affect feed intake. Tannins are important in oak poisoning and in nutritional problems with sorghum grain (milo).

11. Terpenes and Terpenoids

Terpenes and terpenoids are plant substances derived from the 5-carbon isoprene unit:

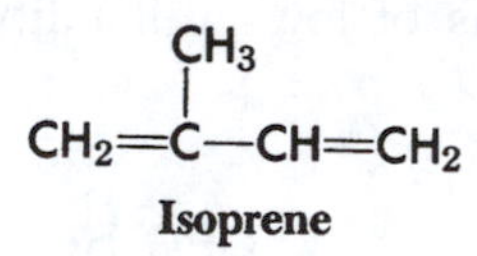

Terpenes contain an integral number of isoprene units, while **terpenoids** are compounds with varying numbers of carbon atoms that are clearly derived from isoprene units. They are classified according to the number of 5-carbon units in their structures:

Terpene type	No. of isoprene units	No. of C atoms	Examples
Monoterpenes	2	10	Essential oils—e.g. in mint and sagebrush; monoterpenes in Ponderosa pine (abortion-inducing agents)
Sesquiterpenes	3	15	Sesquiterpene lactones in sneezeweeds (*Helenium* spp.) and bitterweeds (*Hymenoxys* spp.). Gossypol.
Diterpenes	4	20	Polycyclic diterpene alkaloids in larkspurs

The biochemistry of terpenes is complex. Examples of other polyterpenes include rubber and other latexes (e.g. chicle used to make chewing gum) and carotenoid pigments. Most gums and resins in plants, such as the pitch (resin) of pines and other conifers, are terpenes. Conifer resins are mixtures of monoterpenes, sesquiterpenes, diterpene acids and phenolic compounds. Some of these terpenes are the causative agents of pine needle abortion in cattle.

12. Other Toxins

There are many other plant toxins of miscellaneous chemical structures. Examples include **fluoroacetate (1080)** and other organofluorine compounds which occur in a number of Australian poisonous plants (e.g. *Gastrolobium* spp.). Fluoroacetate is a potent metabolic toxin inhibiting the conversion of citrate to isocitrate in the tricarboxylic acid cycle in cellular metabolism. Water hemlock (*Cicuta* spp.) contains an extremely potent resin called cicutotoxin, which acts directly on the central nervous sytem to produce violent convulsions.

Onions contain n-propyl disulfide, an inhibitor of the red blood cell enzyme glucose-6-phosphate dehydrogenase. Onion poisoning induces anemia as a result of red blood cell hemolysis caused by the enzyme inhibition.

CLASSIFICATION OF NATURAL TOXICANTS BY THEIR OCCURRENCE IN FEEDS

Most feedstuffs and forages used in the feeding of livestock contain potentially deleterious factors. Indeed, it is rare to find a feedstuff in which a deleterious factor cannot be identified. Plants lacking chemical defenses would be unlikely to survive in nature. Crop plants with low toxin levels, such as corn (maize) and wheat, require human intervention (use of pesticides) to protect them from pests and diseases. Humans have by plant breeding produced crops with low levels of Nature's pesticides, so we must replace them with our own pesticides (agricultural chemicals). Some of the plant toxins in feeds are of academic interest only, while others are of considerable concern in animal production. Even the ubiquitous chlorophyll can have adverse effects on animals, causing photosensitization reactions under certain conditions.

A listing of some of the more common and important toxicants in feedstuffs is provided to illustrate their widespread distribution and to provide ready access to potential deleterious factors in common feeds (Table 1–1).

TABLE 1–1 Natural Toxicants in Common Feedstuffs

Feedstuff	Toxicant
Grains	
All	Phytates, mycotoxins
Rye, triticale	Trypsin inhibitors, ergot
Milo	Tannins
Grain amaranth	Oxalates, saponins
Buckwheat	Fagopyrin
Tubers	
Potatoes	Solanum alkaloids
Cassava	Cyanogenic glycosides
Protein supplements	
Soybeans	Trypsin inhibitors, lectins, goitrogens, saponins, phytates, mycotoxins
Cottonseed	Gossypol, tannins, cyclopropenoid fatty acids, mycotoxins
Rapeseed	Glucosinolates, tannins, erucic acid, sinapine
Linseed meal	Linatine, linamarin
Fava beans	Trypsin inhibitors, vicine, lectins
Field beans	Trypsin inhibitors, lectins
Forages	
Legumes	
Alfalfa	Saponins, phytoestrogens, bloating agents
White clover	Cyanogens, phytoestrogens, bloating agents
Red clover	Slaframine, phytoestrogens, bloating agents
Alsike clover	Photosensitizing agents
Sweet clover	Coumarin
Subterranean clover	Photoestrogens
Crown vetch	β-Nitropropanol glycosides
Leucaena spp.	Mimosine
Indigofera spp.	Indospecine
Grasses	
Forage sorghums	Cyanogens
Tall fescue	Ergot alkaloids
Tropical grasses	Oxalates, saponins
Others	
Forage brassicas	Brassica anemia factor

CLASSIFICATION OF NATURAL TOXICANTS BY THEIR SITE OF ACTION AND/OR METABOLIC EFFECT

It is remarkable that for practically all major metabolic functions in animals, there exists in the plant world inhibitors of these functions. For virtually every organ, endocrine gland, and metabolic pathway in animals, there is a corresponding inhibitor in plants. Considerable speculation has been made as to the reasons why plants contain such a wide variety of deleterious (to animals) compounds. The following is by no means a complete listing, but is intended to give an appreciation for the diversity of effects that plant substances can have on animals.

1. Mouth
   - A. Proteolytic enzymes
     Bromelain (from pineapple) and papain (from papaya) are proteolytic enzymes that digest cellular proteins in the buccal cavity.
   - B. Oxalate crystals
     The common houseplant *Dieffenbachia sequine* (dumb cane) contains calcium oxalate crystals which penetrate the mucous membranes of the mouth and throat and cause intense irritation.
   - C. Phenolics (tannins)
     Tannins have an astringent effect due to their binding with proteins on the tongue and in the oral cavity.
2. Digestive tract
   - A. Rumen
     - i. Nitrate and nitrite
       Nitrates are metabolized to nitrites by rumen microorganisms.
     - ii. Rumen stasis
       Mesquite poisoning, lupinosis, and oxalate toxicity cause rumen stasis.
     - iii. Bloat
       Cytoplasmic proteins and saponins have been implicated in causing bloat. Tannins may have protective effects.
     - iv. Rumenitis
       Oxalate poisoning results in damage to the lining of the rumen.
   - B. Intestine
     - i. Irritation
       Saponins, tannins, and selenoamino acids cause irritation and damage to the intestinal mucosa.
     - ii. Enzyme inhibitors
       Trypsin and amylase inhibitors in feeds exert their inhibitory effects on digestive enzymes in the small intestine.
     - iii. Nutrient absorption
       Lectins alter intestinal permeability, cause extensive damage to microvilli and reduce the absorption of nutrients.
   - C. Diarrhea
     Diarrhea is commonly observed in pyrrolizidine alkaloid toxicity. Nitrates also cause diarrhea.
   - D. Prolapsed rectum
     This is a symptom of pyrrolizidine alkaloid poisoning.
3. Liver
   - A. Hepatotoxins
     Pyrrolizidine alkaloids cause irreversible liver damage (Fig. 1–2). Lupinosis results in cirrhotic, fatty livers. Facial eczema in sheep and cattle, caused by the mycotoxin sporidesmin, is accompanied by liver damage.
   - B. Mineral and vitamin metabolism
     Pyrrolizidine alkaloid poisoning and lupinosis result in elevated liver copper concentrations and alterations in hepatic zinc and iron metabolism. Liver vi-

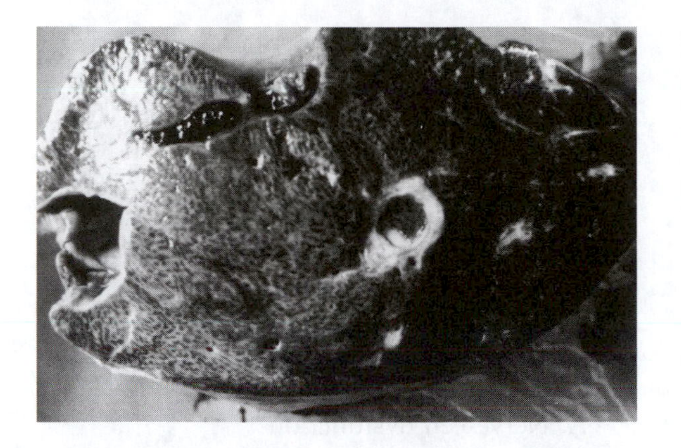

FIGURE 1–2 The liver of a horse poisoned by hepatoxic pyrrolizidine alkaloids in *Echium plantagineum*, showing fibrosis and loss of normal structure. (Courtesy of John Seaman)

tamin A concentrations are depressed with pyrrolizidine alkaloid poisoning.

C. Cholestasis
Lantana poisoning causes cholestasis, or cessation of bile flow, producing an enlarged green liver. *Lantana camara* is the poisonous plant involved.

4. Lung
   A. Pyrrolizidine alkaloids cause pulmonary damage.
   B. Acute bovine pulmonary emphysema is caused by indoles (e.g., 3-methylindole) which are products of rumen metabolism of tryptophan.
5. Kidney
   Kidney damage occurs in poisonings from pyrrolizidine alkaloids, oxalates, tannins, and sesquiterpene lactones.
6. Circulatory system
   A. Aortic aneurysm occurs in lathyrism.
   B. Erythrocyte hemolysis is caused by saponins, brassica anemia factor, favism, copper toxicity, and pyrrolizidine alkaloids.
   C. Hematopoesis is impaired in pyrrolizidine alkaloid poisoning.
   D. Anemia occurs in favism and in poisonings due to pyrrolizidine alkaloids, brassica anemia factor, and copper toxicity.
   E. Delayed blood clotting occurs in sweet clover poisoning because of the properties of dicumarol as a vitamin K antagonist.
   F. Agglutination of red blood cells in vitro is caused by hemagglutinins (lectins) found in beans.
   G. Vasoconstriction is caused by ergot alkaloids, resulting in loss of blood supply to the extremities, and gangrene (fescue foot).
   H. Vasodilation is caused by veratrum alkaloids.
   I. Hypocalcemia is observed with oxalate poisoning.
   J. Hypercalcemia is caused by calcinogenic glycosides.
   K. Hypomagnesemia may be induced by *trans*-aconitic acid in early spring grass, leading to the condition of grass tetany.
   L. Hypoglycemia occurs in cocklebur poisoning due to the action of carboxyatractyloside. Severe hypoglycemia is also noted in "vomiting sickness" induced by the toxic amino acid hypoglycin in ackee (*Blighia sapida*).
   M. Hyperglycemia occurs in fluoroacetate (1080) poisoning. The 1080 compound occurs in several poisonous plants in Australia and South Africa.
   N. Hypercholesterolemia is induced by cyclopropenoid fatty acids in cottonseed oil.
7. Heart
   A. Heart lesions
   Erucic acid in rapeseed oil and gossypol in cottonseed meal are known to produce cardiac lesions.
   B. Increased heart rate is caused by digitonin from foxglove and the piperidine alkaloid coniine in poison hemlock.

C. Decreased heart rate is characteristic of the toxic effects of veratrum alkaloids.

D. Cardiac irregularity is caused by gossypol.

8. Bone

A. Skeletal deformity (Fig. 1–3)

A wide variety of teratogenic agents, such as lupin and *Astragalus* toxins, cause skeletal deformities in fetuses. Skeletal deformity is also observed in lathyrism.

FIGURE 1–3 Bones and joints affected by toxicants. For example, lambs from ewes consuming *Veratrum californicum* may have shortened metatarsal and metacarpal bones. Lambs and calves from dams consuming locoweed may have a permanent flexure of the carpal joints. *Cestrum diurnum* intoxication in horses results in hyperextension of the metacarpophalangeal and metatarsophalangeal joints.

B. Destruction of bone marrow occurs in bracken fern poisoning of cattle.

C. Fibrosis of bone occurs in nutritional secondary hyperparathyroidism. The condition, called "bighead", is seen in horses consuming tropical grasses that contain oxalates.

9. Eye

A. Disturbed vision occurs in animals consuming *Datura* (jimsonweed), which contains atropine.

B. Blindness and impaired vision occur in *Astragalus* poisoning due to degeneration of the optic nerve. Blind grass (*Stypandra imbricata*), a plant in Western Australia, causes degeneration of the optic nerve. Selenium toxicity, from consumption of *Astragalus* spp. and other selenium accumulators, causes "blind staggers." Sheep grazing bracken fern may develop a degeneration of the neuroepithelium of the retina causing blindness.

10. Nervous system

A. Nerve cell dysfunction

The indolizidine alkaloids in *Swainsona* and *Astragalus* spp. cause accumulation of mannose in nerve cell lysosomes, causing axonal dystrophy.

B. Degeneration of the spinal cord occurs in lathyrism.

C. Cholinesterase inhibitors, such as solanine in green potatoes, inhibit breakdown of acetylcholine at neuromuscular junctions.

D. Somnolence effects

Sleepy grass (*Stipa robusta*) induces a profound somnolence in horses; they may remain in deep slumber for several days.

E. Incoordination (ataxia) occurs with locoweed poisoning, in pyrrolizidine alkaloid poisoning in horses, and in ryegrass staggers and annual ryegrass toxicity in sheep.

F. Polyneuritis occurs in thiamin deficiency, induced by thiaminase in bracken fern. It also is observed in mink fed fish that contain thiaminase.

11. Muscle

Nutritional myopathy, similar to white muscle disease in sheep due to selenium deficiency, has been observed associated with lupinosis in Western Australia. Skeletal muscle degeneration is noted in cattle consuming *Thermopsis montana*, a toxic range plant on western U.S. rangelands.

12. Thyroid gland

A variety of goitrogenic agents in plants inhibit the synthesis of thyroxin. These include glucosinolates in *Brassica* spp., thiocyanate produced from detoxification of cyanogens, and dihydropyridone (DHP), a rumen metabolite of mimosine produced in animals grazing *Leucaena leucocephala.*

13. Reproductive system

    A. Various mycotoxins such as zearalenone have estrogenic effects.

    B. Female infertility is caused by isoflavones (phytoestrogens) in subterranean clover.

    C. Male infertility is caused by gossypol in cottonseed meal.

    D. Teratogenic effects have been observed with a number of toxic plants, including lupins, poison hemlock, and veratrum.

    E. Abortifacients or agents that induce abortion are found in ponderosa pine needles.

14. Milk-transferred toxins

    The best known example is white snakeroot toxin, which produced milk sickness in pioneer communities in the U.S. Pyrrolizidine alkaloids and mycotoxins such as aflatoxin may be transferred in milk. Bitterweed (*Helenium amarum*) toxins give a bitter taste to milk, reducing its consumer acceptance.

15. Immune system

    Some toxicants affect the immune system and resistance to disease. Examples are lectins in beans, and impairment of the immune system in chronic toxicoses of aflatoxin and trichothecenes.

16. Hair, skin, and extremities

    A. Photosensitization

    Primary photosensitization is caused by photodynamic agents, such as hypericin in St. Johnswort (*H. perforatum*), which are absorbed and react with light at the skin's surface, producing lesions. **Secondary photosensitization** is caused by phylloerythrin, a metabolite of chlorophyll. It is normally excreted in the bile, but in cases of liver damage, such as with pyrrolizidine alkaloid toxicity, or facial eczema caused by sporidesmin, phylloerythrin enters the general circulation and causes skin lesions when it reacts with ultraviolet light.

    B. Contact dermatitis

    Poison ivy, poison oak and stinging nettles contain irritants which cause allergenic reactions and pruritus (intense itching).

    C. Achromatrichia (loss of hair pigmentation)

    Two examples of this in mink are cotton fur in which a white band develops in the fur due to iron-deficiency anemia induced by trimethylamine oxide and formaldehyde in fish, and a graying of the fur caused by biotin deficiency, induced by avidin in raw eggs (turkey waste graying).

    D. Alopecia (hair loss)

    This occurs with mimosine and selenium toxicities.

    E. Hoof deformities occur with mimosine and selenium toxicities.

    F. Sloughing off of extremities, due to vasoconstriction and gangrene, occurs with ergot poisoning and tall fescue toxicity (fescue foot).

17. Energy and protein metabolism

    A. Protein

    i. Digestion

    Protein digestion is impaired by protease inhibitors, such as trypsin inhibitors in soybeans.

    ii. Amino acid metabolism

    Indospecine, an analog of arginine, induces arginine deficiency in animals grazing *I. spicata* (creeping indigo). Pyrrolizidine alkaloid toxicity causes impairment of liver functions, reducing the liver's ability to deaminate amino acids and synthesize urea from ammonia.

iii. Protein synthesis by cells is inhibited by pyrrolizidine alkaloids, which by crosslinking DNA strands interfere with DNA replication and RNA synthesis.

B. Carbohydrates

i. Starch digestion is inhibited by α-amylase inhibitors in wheat, oats, rye, navy beans, and kidney beans.

ii. Mannosidosis
Accumulation of α-mannose in nerve tissue occurs as a result of an α-mannosidase inhibitor in *Swainsona* spp. and *Astragalus* spp. (locoweed).

C. Lipids

i. Fatty livers occur in lupinosis in sheep.

ii. Fat absorption is inhibited by aflatoxins, which cause steatorrhea (lipid in feces).

D. Energy metabolism

i. Fluoroacetate, found in several Australian and South African toxic plants, inhibits the tricarboxylic acid (TCA) cycle by inhibiting aconitase.

ii. Nitropropionic acid in crown vetch is an inhibitor of the TCA cycle enzyme succinate dehydrogenase.

iii. Hydrogen cyandide (HCN) from cyanogens is a potent inhibitor of cytochrome oxidase, the terminal respiratory enzyme.

18. Cell division
Pyrrolizidine alkaloids inhibit the prophase stage of mitosis.
Metaphase arrest is observed in lupinosis.

19. Mineral metabolism

A. Absorption
Chelating agents such as phytic acid and oxalates impair mineral absorption.

B. Liver storage of copper and zinc is altered by hepatotoxins.
Liver copper is dramatically increased in animals poisoned with pyrrolizidine alkaloids, whereas the concentration of zinc is decreased.

C. Blood levels of minerals are influenced by toxicants. Oxalates cause hypocalcemia, while calcinogenic glycosides cause hypercalcemia.

20. Vitamin metabolism
Various vitamin antagonists occur in feeds. These include avidin in raw eggs, which is an antagonist of biotin, and linatine, a pyridoxine (vitamin B_6) antagonist, in linseed meal. Another plant with a pyridoxine antagonist is the African tree *Albizia tanganyicensis*. Thiaminases in bracken and certain fish cause thiamin deficiency. Calcinogenic glycosides provide the active metabolite of vitamin D, and so interfere with the homeostatic effects of vitamin D on calcium metabolism. Dicumarol causes a vitamin K deficiency. Lipoxidases destroy carotene, the precursor of vitamin A. Pyrrolizidine alkaloids cause markedly depressed liver and blood vitamin A levels, by inhibiting hepatic synthesis of retinol-binding protein.

This list of animal tissues and systems influenced by plant toxicants is by no means complete, but gives an indication of the remarkably diverse effects that feed toxicants have on animal metabolism.

SOURCES OF INFORMATION ON TOXICOLOGY AND POISONOUS PLANTS

The classic book "Poisonous Plants of the United States and Canada" by Kingsbury (1964) provides an excellent historical background to poisonous plant problems in North America. Even though much of the information on chemical identity of toxins, toxin metabolism, etc., is out of date, Kingsbury's book is still an excellent reference for general information on poisonous plants. More current information is available in a series of proceedings of International Symposia on Poisonous Plants: Keeler *et al.*, 1978; Seawright *et al.*, 1985; James *et al.*, 1992; Colegate and Dorling, 1994.

These symposia are held approximately every four years, and are organized and sponsored by an informal international network of scientists conducting poisonous plants research.

Some other useful references on poisonous plants and natural toxins are: Cheeke (1989), Everist (1981), James *et al.* (1988, 1991), Keeler and Tu (1983, 1991), Kellerman *et al.* (1988), and Seawright (1989).

Liener's *Toxic Constituents of Plant Foodstuffs* (1980) is an excellent reference for a detailed treatment of some of the toxins in feeds, such as protease inhibitors, lectins and glucosinolates. The series "Toxicants of Plant Origin" edited by Cheeke (1989) also covers toxins in feeds in depth.

The principles of toxicology are comprehensively covered in *Casarett and Doull's Toxicology* (Amdur *et al.*, 1991). *Clinical and Diagnostic Veterinary Toxicology*, by Osweiler *et al.* (1985) deals with practical aspects of livestock toxicology from a veterinary perspective.

The study of natural toxicants, particularly those in plants poisonous to livestock, has been a less active area of research than is the case for agricultural chemicals, pharmaceuticals, feed additives and other synthetic toxicants of concern to society. This in part reflects the attitude that natural substances are more benign than those developed by humans. However, the widespread distribution of natural toxicants in food plants and the search for new crops to maximize efficient utilization of resources for the ever-increasing human population suggest that the study of natural toxins will be of increasing importance. Ames and coworkers (Ames, 1983; Ames *et al.*, 1990a,b) have been leading proponents of the need for maintaining a realistic perspective on the relative importance of natural vs. human-made chemicals. Ames (1983) estimated that the human dietary intake of natural toxicants (Nature's pesticides) is several grams per day, or at least 10,000 times the dietary intake of human-made pesticides. Expressed another way, about 99.99% of the pesticides in the American diet are natural compounds produced by plants as chemical defenses (Ames *et al.*, 1990a). This doesn't minimize the need to study synthetic chemicals, but does suggest that a full understanding of natural toxicants is important.

REFERENCES

Amdur, M.O., J. Doull, and C.D. Klaassen. 1991. Casarett and Doull's Toxicology. The Basic Science of Poisons. Pergamon Press, New York.

Ames, B.N. 1983. Dietary carcinogens and anticarcinogens. Science 221:1256–1264.

Ames, B.N., M. Proffet, and L.S. Gold. 1990a. Dietary pesticides (99.99% all natural). Proc. Natl. Acad. Sci. USA 87:7777–7781.

Ames, B.N., M. Proffet, and L.S. Gold. 1990b. Nature's chemicals and synthetic chemicals: Comparative toxicology. Proc. Natl. Acad. Sci. USA 87:7782–7786.

Cheeke, P.R. (Ed.). 1989. Toxicants of Plant Origin. Vol. 1. Alkaloids. Vol. II. Glycosides. Vol. III. Proteins and Amino Acids. Vol. IV. Phenolics. CRC Press, Boca Raton, FL.

Colegate, S.M., and P.R. Dorling (Eds.). 1994. Plant-Associated Toxins. Agricultural, Phytochemical and Ecological Aspects. CAB International, Wallingford, UK.

Everest, S.L. 1981. Poisonous Plants of Australia. Angus and Robertson Publishers, Sydney.

James, L.F., J.O. Evans, M.H. Ralphs, and R.D. Child (Eds.). 1991. Noxious Range Weeds. Westview Press, Boulder, CO.

James, L.F., R.F. Keeler, E.M. Bailey, Jr., P.R. Cheeke, and M.P. Hegarty (Eds.). 1992. Poisonous Plants. Proceedings of the Third International Symposium. Iowa State University Press, Ames.

James, L.F., Ralphs, M.H., and D.B. Nielsen (Eds.). 1988. The Ecology and Economic Impact of Poisonous Plants on Livestock Production. Westview Press, Boulder, CO.

Keeler, R.F., and A.T. Tu (Eds.). 1983. Handbook of Natural Toxins. Vol. 1. Plant and Fungal Toxins. Marcel Dekker, Inc., New York.

Keeler, R.F., and A.T. Tu (Eds.). 1991. Handbook of Natural Toxins. Vol. 6. Toxicology of Plant and Fungal Compounds. Marcel Dekker, Inc., New York.

Keeler, R.F., K.R. Van Kampen, and L.F. James (Eds.). 1978. Effects of Poisonous Plants on Livestock. Academic Press, NY.

Kellerman, T.S., J.A.W. Coetzer, and T.W. Naude. 1988. Plant Poisonings and Mycotoxicoses of Livestock in Southern Africa. Oxford University Press, Cape Town.

Kingsbury, J.M. 1964. Poisonous Plants of the United States and Canada. Prentice–Hall, Englewood Cliffs, NJ.

Liener, I.E. (Ed.). 1980. Toxic Constituents of Plant Foodstuffs, 2nd Edition. Academic Press, NY.

Osweiler, G.D., T.L. Carson, W.B. Buck, and G.A. Van Gelder. 1985. Clinical and Diagnostic Veterinary Toxicology. Kendall/Hunt Publishing Co., Dubuque, IA.

Seawright, A.A. 1989. Animal Health in Australia. Vol. 2. Chemical and Plant Poisonings. Australian Government Printing Service, Canberra.

Seawright, A.A., M.P. Hegarty, L.F. James, and R.F. Keeler (Eds.). 1985. Plant Toxicology. Queensland Poisonous Plants Committee, Yeerongpilly, Australia.

CHAPTER 2

Techniques and Calculations in Toxicology

ISOLATION AND IDENTIFICATION OF PLANT TOXINS

In studying the chemical nature of plant toxins and their metabolism in animals, it is necessary to extract and purify them. While specific techniques and methods are published for many toxicants, a few general guidelines are of interest. In cases where the chemical nature of the toxicant is not yet known, it may be necessary to employ a **bioassay** technique to identify the location of the toxic component(s) during the fractionation and purification procedures. For example, in fractionation of tall fescue extracts to attempt to identify the causative agent of fescue foot, fractions produced by extraction and ion-exchange chromatography were tested for biological activity by injection into calves and the subsequent measurement of skin temperature in the area of the coronary band of the hoof (Garner *et al.*, 1982).

The stability of toxicants in plant tissue varies with the chemical nature of the specific compounds involved. In many cases, no special handling is required, and samples of hay or sun-cured or oven-dried plant material may be used. In some cases, rapid freezing in liquid nitrogen followed by freeze-drying may be desirable to prevent postharvest changes from occurring. Enzymatic action, resulting in hydrolysis and oxidation, may occur as breakdown in cell structure occurs. Special techniques, such as homogenization with polyvinyl pyrrolidone to bind tannins, may be used to prevent such action (Loomis and Battaile, 1966).

Extraction of plant tissue can be performed in several ways. Green plant tissue can be macerated in a Waring blender. Dried material may be extracted with an appropriate solvent in a Soxhlet extraction unit. For extraction of large quantities of plant material, the dried plant tissue can be soaked in drums or large vats, with the solvent exchanged several times, followed by evaporation of the solvent. Selection of the appropriate solvent is based upon the chemical characteristics of the specific toxicant. For example, alkaloids normally occur in plant tissue as salts of organic acids. They can be extracted with an acidic, aqueous solvent; the aqueous extract may be made basic and the alkaloids extracted with an organic solvent, leaving neutral and acidic water-soluble compounds behind. Alternatively, the plant material can be made alkaline, and the free alkaloid bases extracted with organic solvents.

Following extraction of the plant tissue, various techniques are employed to further isolate and identify specific plant toxins. Extracts may be passed through cation or anion exchange columns, followed by elution with various solvents. Various chromatographic techniques can be used to separate individual compounds, including paper, thin-layer, column, or **gas-liquid chromatography (GLC)**, and **high-performance liquid chromatography (HPLC)**. The HPLC technique is now widely employed for isolation and identification of plant toxins. Detection of individual compounds is often based on colorimetric procedures. For example, pyrrolizidine alkaloids are commonly detected by the Ehrlich reagent test. Ehrlich reagent (4-dimethylaminobenzaldehyde) reacts with pyrroles to produce a purple chromophore, which may be detected visually with paper or thin-layer chromatography, or measured spectrophotometrically. Other methods of identification include infrared, ultraviolet, and nuclear magnetic resonance (NMR) spectra, and mass spectra. Various derivatives may be made to improve the gas chromatography separation, mass spectra, and detector sensitivity. Trimethylsilyl derivatives are commonly made for gas chromatography–mass spectrometry (GC/MS) analysis. The GC procedure allows separation of extracts into specific peaks based on retention time; the peaks can be subjected to MS analysis for identification of functional groups. Holstege *et al.* (1996, 1995) developed a rapid GC/MS method for screening plants, rumen contents, animal tissues and urine for various alkaloids, to aid in the diagnosis of plant toxicoses.

Specific techniques for analysis and identification of most plant toxicants have been published. A detailed consideration of analytical techniques is beyond the scope of this book. Pertinent current literature should be consulted by those who need to pursue this aspect.

BIOASSAY OF TOXICANTS

Coincident with chemical analysis, or sometimes in place of it, bioassays are often employed in the study of toxicants. A **bioassay** is a test using a living organism for measurement of the response. A few examples will illustrate their application. Saponins in alfalfa have been measured using a soil fungus, *Trichoderma viride*. The fungus is grown on a culture medium containing various levels of alfalfa extract; the fungal growth is inversely proportional to the saponin content (Livingston *et al.*, 1977). Another saponin bioassay has involved the use of guppies or minnows. Fish are extremely sensitive to the detergent properties of saponins, so that fish paralysis can be correlated with saponin content of the solution. Saponins can also be measured by erythrocyte hemolysis. Suspensions of red blood cells are exposed to serial dilutions of alfalfa extracts, and the minimum dilution causing complete hemolysis is determined.

Another bioassay technique, alluded to earlier, is that for the study of the toxin in tall fescue responsible for fescue foot. During the fractionation of tall fescue extracts with ion-exchange chromatography (Garner *et al.*, 1982), the activity of the various fractions was tested to determine if they contained the fescue foot toxin. The fractions were injected intraperitoneally into calves and the temperature of the coronary band of the rear hoof measured by an infrared technique. Those fractions containing the fescue foot toxin produced an elevation in the coronary band temperature.

A final example of a bioassay technique is the mouse or rat uterine weight bioassay for plant estrogens (Fig. 2–1). Plant extracts are injected into immature female mice or rats; 24 hr later the uterine weight is measured. With plant extracts containing phytoestrogen activity, the uterine weight is elevated. A standard curve is prepared using an estrogenic hormone such as diethylstilbestrol or estradiol.

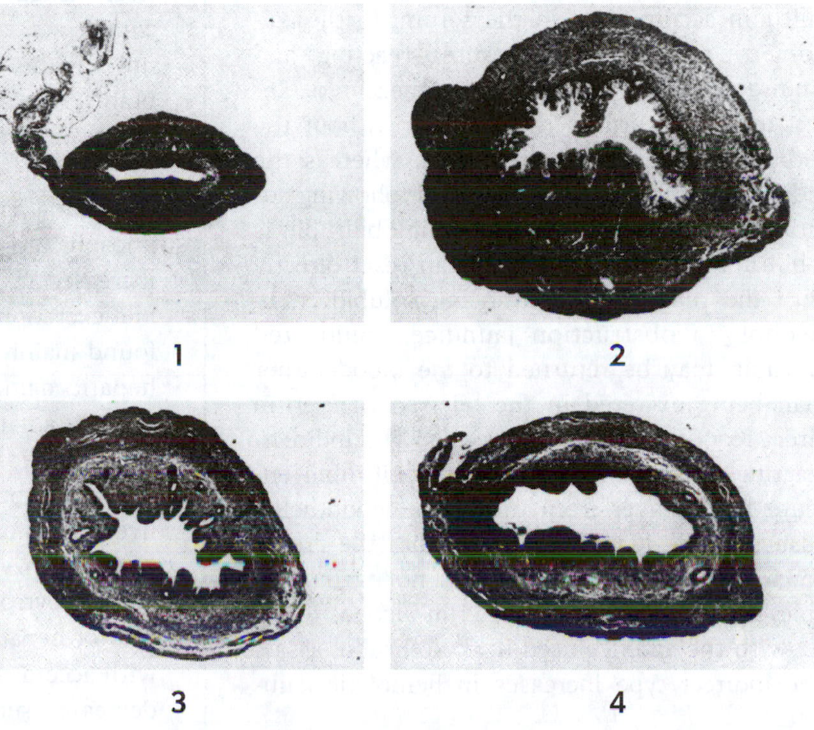

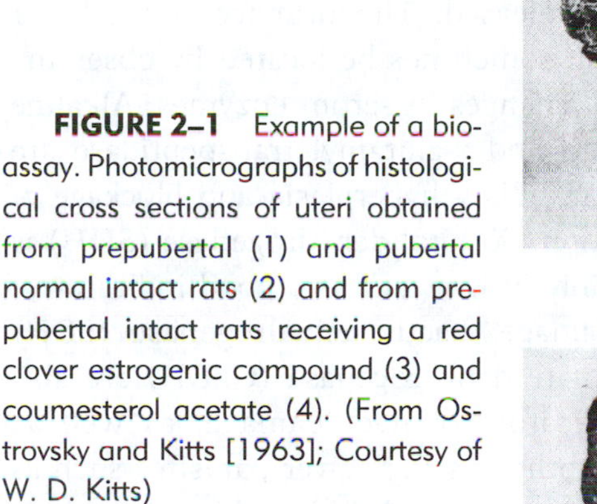

FIGURE 2–1 Example of a bioassay. Photomicrographs of histological cross sections of uteri obtained from prepubertal (1) and pubertal normal intact rats (2) and from prepubertal intact rats receiving a red clover estrogenic compound (3) and coumesterol acetate (4). (From Ostrovsky and Kitts [1963]; Courtesy of W. D. Kitts)

ASSESSMENT OF TISSUE DAMAGE

Various procedures are used to assess pathology when animals are fed toxicants or poisonous plants. The specific techniques used to assess tissue damage depend, of course, on what tissues are affected by the particular toxicant being studied. A necropsy of affected animals should be performed by a veterinarian to assess gross pathology. Tissues with evidence of pathology should then be submitted to a veterinary pathologist for microscopic assessment of tissue lesions. In the case of hepatotoxins, liver biopsies can be used to assess sequential changes in liver histology. A number of tests can be performed to assess liver function (Cornelius, 1989). The **sulfobromophthalein (BSP) clearance rate** is widely used; this test measures the ability of the liver to remove a dye (BSP) from the blood. The BSP is injected intravenously, and sequential blood samples are taken over a period of 2.5–15 min. In hepatic insufficiency, the removal of BSP from the blood is impaired. The BSP is conjugated with glutathione and excreted in the urine as a mercaptide (see Chapter 3). Normal BSP clearance rates for livestock species are $t_{1/2}$ of 2–5 minutes. The $t_{1\backslash 2}$ is the time taken for half of the administered dose to be cleared from the blood. With liver damage, $t_{1/2}$ can be as long as 30–60 minutes. The serum bilirubin level may be measured to assess liver function. **Bilirubin** is a metabolite of hemoglobin that is normally excreted in the bile. With biliary obstruction, bilirubin accumulates in the serum. In the colorimetric procedures used, "direct-reacting" and "indirect-reacting" bilirubin are measured. Direct-reacting bilirubin is measured without the addition of alcohol to the serum, whereas the indirect-reacting form is measured following addition of alcohol. The direct-reacting bilirubin is a bilirubin glucuronide which can react directly with the reagent without being solubilized in alcohol. In obstruction jaundice, conjugated bilirubin may be returned to the blood rather than being excreted in the bile, so the serum direct-reacting bilirubin increases. The indirect-reacting bilirubin is unconjugated bilirubin enroute to the liver from the reticuloendothelial tissues where it was produced from the breakdown of heme porphyrins. It is not water soluble and has to be solubilized in alcohol to react with the azo dye used in the bilirubin assay. The indirect type increases in hemolytic jaundice.

Another serum test used to assess liver function is the measurement of serum albumin. **Albumin** is synthesized in the liver, so hypoalbuminemia reflects impaired liver function. Because serum albumin has a major function in maintaining osmotic balance between the blood and tissues, hypoalbuminemia is often accompanied by edema and ascites (fluid accumulation in body cavities). If the concentration of dissolved substances (solutes) in the blood is lowered because of reduced serum albumin, fluid will move from the dilute solution (blood) across semipermeable membranes to the more concentrated solution (tissue spaces), resulting in edema. A variety of **serum enzymes** is measured in assessment of liver function. These include alkaline phosphatase, glutamic-oxaloacetic transaminase (**GOT**), lactic dehydrogenase (**LDH**), γ-glutamyl transpeptidase (**GGT**), and several others. These enzymes are released from liver tissue during hepatic necrosis. The enzymes are compartmentalized into functional units within liver cells, so when the cell membranes are disrupted by liver damage, the enzymes are released. The primary site of liver damage can sometimes be located by observing differential changes in serum enzymes. Alkaline phosphatase and γ-glutamyl transpeptidase are increased in biliary hyperplasia and blockage of bile excretion. Sorbitol dehydrogenase (**SDH**) is found mainly in liver and is a good indicator of hepatic damage. Lactic dehydrogenase (LDH) and glutamate dehydrogenase (GLDH) are also useful indicators of liver damage, as well as damage to other tissues. Liver parasites such as liver flukes can cause elevations in serum levels of these enzymes.

The pyrrolizidine alkaloids are typical examples of hepatotoxins. They cause liver necrosis, with accompanying elevation in serum bilirubin, decreased serum albumin (resulting in ascites), impaired BSP clearance, and elevations in serum enzymes such as GOT and GGT.

γ-Glutamyl transpeptidase is found mainly in the kidney, but its activity in the liver is relatively high in cattle, horses, sheep, and goats (Braun *et al.*, 1983), and very low in dogs, cats, rabbits and birds. Measurement of this enzyme in serum is useful in hepatobiliary diseases of cattle, sheep, goats, and cholestatic disorders of dogs. The urinary activity is a good test for kidney damage.

TECHNIQUES FOR STUDYING TOXICANT METABOLISM

Isolated Cellular Preparations and Whole Animal Studies

The metabolism of toxicants is often studied in isolated liver tissue preparations. This reflects the importance of the liver in the metabolism of toxicants. Preparations of other tissues are used when appropriate. For example, the metabolism of 3-methylindole in lung tissue has been investigated (see Chapter 9) in studies of acute bovine pulmonary emphysema. The tissue is obtained immediately upon sacrifice of the animal and placed in cold isotonic saline in ice. Homogenization should be conducted as soon as possible. Homogenates are prepared by mincing the tissue in a tissue homogenizer equipped with a Teflon pestle. The tissue is homogenized in a buffer solution containing nutrients and cofactors to allow continued metabolism. Further fractionation of the homogenate to produce fractions of nuclei and debris, mitochondria, lysosomes, microsomes, and the cytosol is accomplished by selective ultracentrifugation. Typical fractionation of homogenates involves the following sequential centrifugation of a homogenate: nuclei and debris, 600 g; mitochondria, 4100 g; lysosomal fraction 24,500 g; and microsomes, 105,500 g (g = 1 gravitational field). The cytosolic fraction is the supernatant remaining. The **microsome** fraction consists primarily of endoplasmic reticulum in which drug-metabolizing enzymes are located. While fractionation of tissue is very useful in isolating enzyme systems and studying the metabolism of toxins without extraneous complications, it is necessary to maintain the perspective that the intact animal is what is ultimately important. Sometimes there is a tendency to regard an animal as a giant liver, or even a microsomal preparation. The use of isolated hepatocytes is a technique that more closely correlates with intact liver metabolism than the study of microsomes or other individual fractions. Studies involving the whole animal employ techniques such as the distribution of isotope-labeled compounds and their metabolism and excretion of metabolites.

For ruminant animals, study of toxin metabolism in the rumen is often desirable. In vitro rumen fermentation techniques are useful. Continuous flow fermentors most closely duplicate rumen fermentation. These techniques allow duplication of the events occurring in the rumen without the expense and complications of utilizing an intact animal.

Mutagenesis

Other techniques sometimes employed in large animal toxicology include the study of mutagenesis, carcinogenicity, and teratogenesis. **Mutagenesis** is the alteration of DNA, involving chromosomal changes such as breaks or rearrangements. One of the common tests for mutagenicity is the **Ames test**. Histidine-dependent *Salmonella* organisms are cultured on a histidine-deficient medium containing the test substance and animal tissue microsomes (Fig. 2–2). The microsomal fraction (usually rat liver microsomes) is used to mimic mammalian metabolism. The test measures the extent of mutagenesis producing a wild-type nonhistidine-dependent *Salmonella* which can grow on the histidine-deficient medium. This test (Maron and Ames, 1983), developed by Ames *et al.* (1975), is widely used as a short-term predictive screening test to detect possible carcinogens. There is a high correlation between mutagenic activity in the Ames test and carcinogenicity (McCann *et al.*, 1975). A number of other tests for mutagenicity are described by Hoffmann (1991).

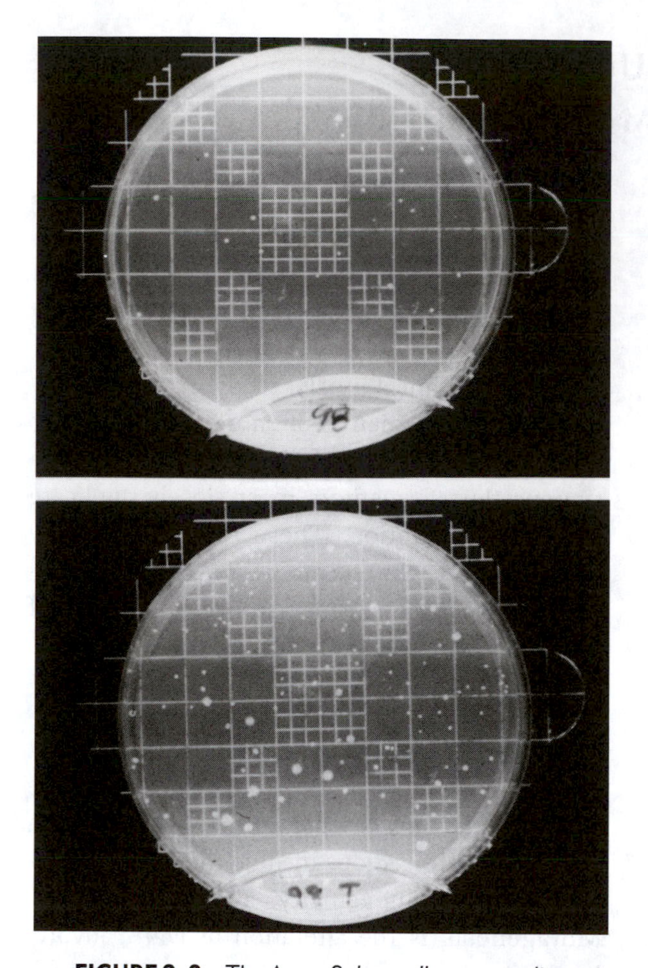

FIGURE 2–2 The Ames *Salmonella* mammalian microsome mutagenicity test. An example of a negative response with the Ames test (top), showing a few bacterial colonies resulting from spontaneous mutation. On the bottom is an example of a postive Ames test response, showing a large number of colonies on the media. The mutagenic agent added to the media caused an increased rate of mutation, resulting in an increased quantity of bacteria capable of synthesizing histidine and thus able to grow on the histidine-free medium. (Courtesy of R. D. White)

Carcinogenesis

A tumor (neoplasm) is an abnormal mass of tissue whose growth exceeds and is uncoordinated with that of normal tissue. Benign tumors are those with cells structurally identical to those of normal tissues, and they are confined to the area of origination, without invasions of neighboring tissue. A malignant tumor (**cancer**) contains cells that are not typical of the structures from which they arise, and they have a tendency to invade neighboring tissue. They also have a tendency to **metastasize**; that is, they form secondary tumors at a site distant to the primary tumor. **Carcinogens** are substances that induce cancer. Primary or direct-acting carcinogens are those that do not require metabolic activation. They generally act at the point of application. An example is mustard gas. Secondary carcinogens require metabolic activation to a reactive (electrophilic) form. This is usually accomplished by cytochrome P450 enzymes (see Chapter 3). They usually affect a specific target organ and may not act at the point of application. Examples of secondary carcinogens are pyrrolizidine alkaloids, aflatoxin B_1, and carbon tetrachloride. Pyrrolizidine alkaloids are bioactivated to reactive pyrrole derivatives (see Chapter 12) while aflatoxin B_1 is converted to the active carcinogen, aflatoxin B_1 2,3-epoxide, by liver enzymes. **Cocarcinogens** (promoters) are substances that potentiate or promote the effects of carcinogens. For example, cyclopropenoid fatty acids (see Chapter 8) potentiate the effects of aflatoxin so that with exposure to both cyclopropenoids and aflatoxin, the tumor incidence is greater than what would be caused by aflatoxin alone.

Carcinogenesis involves two phases: initiation and promotion. In the **initiation phase**, an irreversible genetic change (mutation) of a cell occurs. Initiators are substances that are mutagenic: they react with DNA to cause mutations or chromosomal damage. In the **promotion phase**, the mutated cell acquires a selective growth advantage over surrounding cells and develops into a tumor. Promoters in some way influence the differentiation and proliferation of cells; their action is membrane oriented rather than affecting the DNA.

There is a difference between non-carcinogenic toxins and carcinogens in the dose-response relationship. With other toxins, the dose needs to exceed the capacity of detoxification mechanisms for toxicity signs to occur. For example, cyanide is readily detoxified in the tissues (see Chapter 6), and becomes toxic only when detoxification capacity is exceeded. Carcinogens react with specific tissue receptors and may produce a permanent change. Subsequent doses add to such a change. After a sufficient number of cells have been altered, a visible neoplasm may result. Thus it is difficult to establish if there is a "no-effect" level for carcinogens. There is generally a latent period between exposure and occurrence of neoplasms. Increasing doses of carcinogens increase the number of tumors and reduce the latent period.

Various factors may modify carcinogenesis, including dietary factors. In some cases, different carcinogens that act on the same target organ have either additive or synergistic effects. Factors that modify microsomal drug metabolizing enzyme activity may modify carcinogenesis. For example, feeding phenothiazine, an enzyme inducer, to rats consuming bracken fern reduced tumor incidence by 60% in the studies of Pamukcu *et al.* (1971). The carcinogenicity of aflatoxin is increased by stimulation of microsomal enzymes.

Natural toxicants that have carcinogenic activity include aflatoxins, pyrrolizidine alkaloids, and bracken carcinogens. Except with bracken, cancer is not normally encountered in livestock exposed to these carcinogens because most livestock are marketed or culled before cancer would normally be seen. Probably of greater significance to animal production than direct carcinogenesis is the possibility of transfer of carcinogens to humans through animal products. Aflatoxin and pyrrolizidine alkaloids in milk are examples.

Testing for carcinogenic activity is usually performed with laboratory animals such as rats. Several dosage levels of the test substance are used and fed in diets for a prolonged period, such as 24 months for rats and 21–24 months for mice. At the termination, tissues are examined by a pathologist for tumors. At least 20 animals per dosage level are recommended.

Teratogenesis

The term "teratology" is derived from the Greek work "teras", meaning monsters. In more common terms, teratology is the study of congenital malformations. **Teratogens** are agents that induce fetal abnormalities.

The study of the effect of plant teratogens on livestock has been quite limited, mainly because of the expense of conducting this type of research with large animal species. There are a number of significant teratogenic problems in livestock, which in the U.S. include crooked calf disease, induced by lupins and poison hemlock, abnormalities associated with the consumption of locoweeds, and the bizarre "cyclops lambs" produced when pregnant ewes consume *Veratrum californicum*. Keeler (1984) has reviewed some general principles of teratology. These will be enumerated briefly.

Principle 1. Genotype determines susceptibility.

There is species variation in response to teratogens. For example, consumption of lupins by pregnant cows may cause crooked calf disease, but no teratogenic effects are seen when lupins are consumed by pregnant ewes.

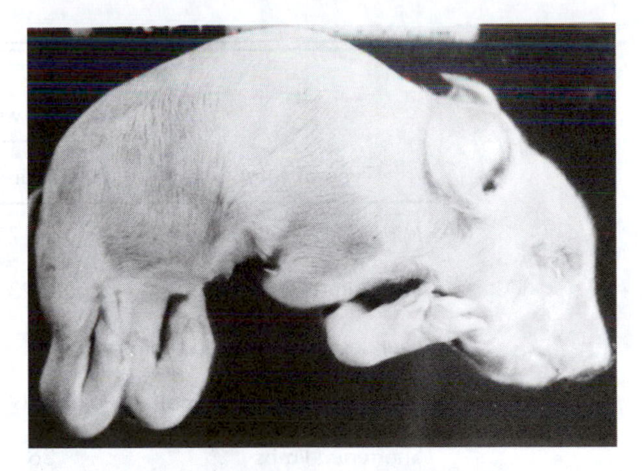

FIGURE 2–3 Arthrogryposis in a baby pig born to a sow that consumed tobacco stalks during gestation. Tobacco contains teratogenic alkaloids. (Courtesy of M. Ward Crowe)

Principle 2. A teratogen must reach the fetus or produce a metabolite which does.

The teratogen exerts its effect within the fetal tissue.

Principle 3. Deformities are dose dependent.

The incidence of malformations varies according to the amount of teratogen reaching the fetus. This ranges from a no-effect level where no observable incidence of teratogenesis occurs to toxic levels where fetal death occurs.

Principle 4. A teratogen can produce death rather than deformities.

In livestock, it is possible that a problem with plant teratogens could be manifested mainly in a high incidence of abortions or fetal resorptions.

Principle 5. The fetus must be exposed at the susceptible period in gestation.

The fetus is susceptible to the effects of a teratogen during the sharply defined period when the particular tissue affected is undergoing development. For example, the cyclops lamb condition occurs only if a ewe consumes *V. californicum* on the fourteenth day of pregnancy. The time in days of gestation when an insult induces a specific defect varies greatly among species because of differences in the length of the gestation period. The approximate periods when teratogens produce specific defects in a number of animal species are shown in Table 2–1.

Principle 6. Teratogens exert their effects by specific mechanisms.

Dissimilar teratogens may induce the same metabolic defect to give rise to similar deformities. For example, crooked calf disease may be induced by the quinolizidine alkaloid anagyrine in lupins and the piperidine alkaloid coniine in poison hemlock, via a sedative effect on the fetus (see Chapters 11 and 13).

Plants implicated in teratogenesis can be grouped into three categories (Keeler, 1984): those with known teratogens, teratogenic plants with as yet unidentified teratogens, and suspected teratogenic plants. Included in the first group with known teratogens are plants in the *Lupinus, Veratrum, Conium,* and *Leucaena* genera. Examples of teratogenic plants for which the

TABLE 2–1 Approximate Gestation Period When Teratogens Produce Defects in Five Animal Species*

	Days of gestation when teratogenic effects are produced				
Type of fetal defect	**Humans**	**Hamsters**	**Sheep**	**Cattle**	**Pigs**
Microphthalmia (small eye)	16–17	6.5	12–13	16–18	10–11
Cyclopia (one eye)	21–23	7	14	21	12
Exencephaly (exposed brain)	26	7.5	16.5	26–27	15.5
Spina bifida (open spine)	28–29	8–8.5	18	29–31	17
Shortened limbs	36	9.5	29–31	42	20
Total gestation period	267	16	147	283	115

*Adapted from Keeler (1978).

specific teratogenic agent has not been identified are *Astragalus* and *Nicotiana* spp. Suspected teratogenic plants, for which definitive feeding trials have not yet been conducted or for which field cases suggest an involvement, include *Datura, Prunus,* and *Sorghum* spp.

In general, the critical period for most birth defects is the first trimester of pregnancy. Chemical attack by a teratogen during early organogenesis is most likely to produce a malformation; during the fetal growth period, after the organs are fully formed, changes induced by exogenous substances are almost always toxic rather than teratologic effects. Effects of plant toxins on reproduction in livestock, including teratogenic effects, have been reviewed by James *et al.* (1992).

EXPRESSION OF BIOLOGICAL RESPONSES TO TOXICANTS

The expression of toxicity is generally made with respect to body size, for example, in milligrams of compound per kilogram body weight. The most commonly used measure of acute toxicity is the **lethal dose (LD_{50})** which is the dosage of toxin that on average will kill 50% of the test animals. Several methods for calculation of the LD_{50} are available. The most widely used are those of Miller and Tainter (1944), Litchfield and Wilcoxon (1949), and Weil (1952). The method of Weil is probably the simplest to use. Its application involves the use of four or more dosage levels, and 2, 3, 4, 5, 6, or 10 animals per dose level. The LD_{50} and a 95% confidence interval are derived from tables published by Weil (1952), using the number of mortalities at each dosage level. Modified LD_{50} techniques, which use fewer animals and avoid lethality as the end-point, are now available (Lipnick *et al.*, 1995).

Acute toxicity is usually used to describe the effects seen following a single dose or doses within a 24-hr period. **Chronic toxicity** refers to effects produced by prolonged exposure to the test substance over a period of 3 months or longer. **Subchronic toxicity** refers to effects of exposure of more than 24 hr but less than 3 months.

In expressing the chronic toxicity of poisonous plants, the lethal dosage as a percentage of body weight is often given and is useful for rough comparisons. For example, the lethal dose of tansy ragwort (*Senecio jacobaea*) is about 5–10% of body weight for cattle and horses, and 100–200% for sheep and goats. A problem with this mode of expression for chronic toxicity is deciding what body weight to use. With growing animals, the body weight may change considerably during a chronic toxicity study. Also, there may be a marked decline in body weight near the terminal stages of toxicity. Methods to compensate for these difficulties include using either initial body weight or average body weight for the period of the study.

Because of increasing concern and attention paid to animal suffering and animal rights issues in biomedical research involving laboratory animals, the use of procedures such as LD_{50} and toxicity responses to poisonous plants and other toxins is becoming less common. Toxins that primarily affect livestock, such as those in poisonous range and pasture plants, are often virtually unstudied, and many of them are likely to remain that way. Priorities of agricultural research have for many years been higher for food crops than for livestock. Current trends in interest in vegetarian diets and in animal rights and animal welfare, such as opposition to "factory farming" of animals, show no sign of diminishing; these attitudes tend to discourage higher priority for livestock-related research. Toxicity studies with poisonous plants, such as determining the lethal dose of plant toxins to livestock, are becoming increasingly expensive and difficult to perform. In the U.S., and in many other countries, research protocols must be submitted to animal use or animal care committees which must approve the research. Studies involving pain or poisoning of animals are scrutinized very carefully. Thus it is likely that with poisonous plants of minor economic significance, further research will be very limited, particularly whole-animal studies and feeding trials.

MANAGEMENT OF TOXICOSES

Toxicoses in livestock may involve spectacular acute effects or insidious conditions of a chronic nature. Detailed consideration of antidotes and therapeutic agents is beyond the scope of this text, and should be conducted with the supervision or collaboration of veterinarians. In the U.S. and Canada, assistance may be obtained from the Animal Poison Control Center, University of Illinois. The Center was established in 1978 to provide telephone consultation to veterinarians seeking assistance with suspected animal poisonings and to provide information on the potential hazard and toxicity of pharmaceutical products, agrichemicals, and environmental pollutants. A description of the activities of the Center is provided by Haliburton and Buck (1983). A rapid-response investigation service is available; a team of toxicology specialists will travel to assist with the identification and containment of poisoning outbreaks. The address and phone number of the Center are

Animal Poison Control Center
University of Illinois
College of Veterinary Medicine
2001 S. Lincoln Avenue
Urbana, IL 61810
(217) 333-3611

Osweiler *et al.* (1985) described a management plan for toxicologic emergencies, as follows:

1. Institute the necessary emergency and supportive therapy to keep the animal alive.
2. Establish a tentative clinical diagnosis on which to base therapy.
3. Institute the appropriate remedial and antidotal procedures.
4. Identify the toxic agent as rapidly as possible.
5. Determine the source of the toxin.
6. Counsel the livestock owner on the hazards of the implicated toxicant, and provide instruction for the avoidance of the problem in the future.

The services of a veterinarian should be sought immediately when a toxicosis is suspected. The livestock owner should provide a description of the clinical signs, and any information on the suspected source of the toxin that might be available. Therapeutic measures should be attempted by the owner only following veterinary advice.

PLANT IDENTIFICATION AND USE OF BOTANICAL NAMES

All plant material used in toxicology studies should be identified by genus and species. Publications arising from experimental work should indicate the authority by which the plant was identified. **Voucher specimens** of the material should be deposited in an herbarium for future verification if necessary. Herbaria can be regarded as libraries, being depositories of materials in good condition which can be recalled at a later date if necessary. The importance of the use of botanical names is obvious when one considers the multiplicity of common names used for the same plant and the multiplicity of plants identified by the same common name. Tansy (*Tanacetum vulgare*) and tansy ragwort (*Senecio jacobaea*) are both commonly referred to as tansy. *Tanacetum vulgare* is an herb used safely in cooking, whereas *Senecio jacobaea* is a dangerous poisonous plant causing irreversible liver damage. "Hemlock" is a common name for water hemlock (*Cicuta maculata*), poison hemlock (*Conium maculatum*), the hemlock tree (*Tsuga canadensis*), and ground hemlock (*Taxus canadensis*). One of these "hemlocks" is deadly poisonous (*Cicuta maculata*), while another is harmless (*Tsuga canadensis*). Thus, confusion as to the identity of a particular plant can be avoided if the botanical names are used.

The **binomial system** (genus, species) of plant taxonomy was developed by the Swedish botanist Linnaeus. Generally the Latin botanical name is descriptive of the plant or its habitat. Sometimes the name reflects use of the plant as an herbal or in folk-lore. For example, the botanical name of soapwort is *Saponaria officinalis*:

saponis: soap
aria: belonging to
officinalis: of the apothecary (pharmacy)

The genus name is thus derived from the foaming property of the crushed plant (due to the saponins it contains) and the species name indicates that it has been used medicinally. Plants containing saponins have been used for soap and shampoo since ancient times.

Information on natural toxins and poisonous plants is available in various computer databases. The Poisonous Plants Information System (PPIS) is probably the largest data base available. Information on and access to PPIS can be obtained from:

Dr. D. Jesse Wagstaff
Food and Drug Administration HFF-108
2000 C St. SW
Washington, DC 20204
(202) 245-3115

REFERENCES

Ames, B.N., J. McCann, and E. Yamasaki. 1975. Methods for detecting carcinogens and mutagens with the *Salmonella* mammalian microsome mutagenicity test. Mutat. Res. 31:347–364.

Beckman, D.A., and R.L. Brent. 1984. Mechanisms of teratogenesis. Annu. Rev. Pharmacol. Toxicol. 24:483–50.

Braun, J.P., P. Benard, V. Burgat, and A.G. Rico. 1983. Gamma glutamyl transferase in domestic animals. Vet. Res. Commun. 6:77–90.

Cornelius, C.E. 1989. Liver function. pp. 364–397 in: J. J. Kaneko (Ed.). Clinical Biochemistry of Domestic Animals. 4th Ed. Academic Press, San Diego.

Garner, G.B., C.N. Cornell, S.G. Yates, R.D. Plattner, J.A. Rothfos, and W.F. Kwolek. 1982. Fescue foot: Assay of extracts of toxic tall fescue herbage. J. Anim. Sci. 55:185–193.

Haliburton, J.C., and W.B. Buck. 1983. Animal Poison Control Center: Summary of telephone inquiries during first three years of service. J. Am. Vet. Med. Assoc. 182:514–515.

Hoffmann, G.R. 1991. Genetic toxicology. pp. 201–225 in: M.O. Amdur, J. Doull and C.D. Klaassen (Eds.). Casarett and Doull's Toxicology. The Basic Science of Poisons. Pergamon Press, New York.

Holstege, D.M., F.D. Galey, B. Johnson, and J.N. Seiber. 1996. Determination of alkaloid exposure in a model ruminant (goat) using a multiresidue screening method. J. Agric. Food Chem. 44:2310–2315.

Holstege, D.M., J.N. Seiber, and F.D. Galey. 1995. A rapid multiresidue screen for alkaloids in plant material and biological samples. J. Agric. Food Chem. 43:691–699.

James, L.F., K.E. Panter, D.B. Nielsen, and R.J. Molyneux. 1992. The effect of natural toxins on reproduction in livestock. J. Anim. Sci. 70:1573–1579.

Johnson, E.M., and D.M. Kochhar (Eds.). 1983. Teratogenesis and reproductive toxicology. Handbook of Experimental Pharmacology, Vol. 65. Springer–Verlag, Berlin, 1983.

Keeler, R.F. 1978. Reducing incidence of plant-caused congenital deformities in livestock by grazing management. J. Range Manage. 31:355–360.

Keeler, R.F. 1984. Teratogens in plants. J. Anim. Sci. 58:1029–1039.

Kirk, R.W. 1983. Current Veterinary Therapy, 8th Edition. W. B. Saunders Co., Philadelphia, PA.

Lipnick, R.L., J.A. Cotruvo, R.N. Hill, R.D. Bruce, K.A. Stitzel, A.P. Walker, I. Chu, M. Goddard, L. Segal, J.A. Springer, and R.C. Myers. 1995. Comparison of the up-and-down, conventional LD_{50}, and fixed-dose acute toxicity procedures. Fd. Chem. Toxic. 33:223–231.

Litchfield, J.T., and F. Wilcoxon. 1949. A simplified method of evaluating dose-effect experiments. J. Pharmacol. Exp. Ther. 96:99–113.

Livingston, A.L., L.C. Whitehand, and G.O. Kohler. 1977. Microbiological assay for saponin in alfalfa products. J. Assoc. Off. Anal. Chem. 60:957–960.

Loomis, W.D., and J. Battaile. 1966. Plant phenolic compounds and isolation of plant enzymes. Phytochemistry 5:423–438.

Maron, D.M., and B.N. Ames. 1983. Revised methods for the *Salmonella* mutagenicity test. Mutat. Res. 113:173–215.

McCann, J., E. Choi, E. Yamasaki, and B.N. Ames. 1975. Detection of carcinogens as mutagens in the *Salmonella* microsome test: Assay of 300 chemicals. Proc. Natl. Acad. Sci. U.S.A. 72:5135–5139.

Miller, L.C., and M.L. Tainter. 1944. Estimation of the ED_{50} and its error by means of logarithmic–Probic graph paper. Proc. Soc. Exp. Biol. Med. 57:261–264.

Ostrovsky, D., and W.D. Kitts. 1963. The effect of estrogenic plant extracts on the uterus of the laboratory rat. Can. J. Anim. Sci. 43:106–112.

Osweiler, G.D., T.L. Carson, W.B. Buck, and G.A. Van Gelder. 1985. Clinical and Diagnostic Veterinary Toxicology. Kendall–Hunt Publishing Co., Dubuque, IA.

Pamukcu, A.M., L.W. Wattenberg, J.M. Price, and G.T. Bryan. 1971. Phenothiazine inhibition of intestinal and urinary bladder tumors induced in rats by bracken fern. J. Natl. Cancer Inst. (U.S.) 47:155–159.

Weil, C. 1952. Tables for convenient calculation of median-effective dose (LD_{50} or ED_{50}) and instructions in their use. Biometrics 8:249–263.

CHAPTER 3

Metabolism and Metabolic Effects of Toxicants

METABOLIC FATE OF TOXICANTS

The principal route of exposure of most natural toxicants to livestock is through the diet. Ingested toxins may be subjected to a number of metabolic processes in the digestive tract and in various tissues prior to excretion. Animals have been exposed to toxic constituents of plants during the long period of their coevolution and have developed numerous biochemical strategies for detoxification of poisonous compounds. This capacity is shared by all plant predators including vertebrate herbivores, insects, and microorganisms. Once a toxicant has been consumed, there are several barriers it must surmount before reaching its critical target. These include chemical and microbiological detoxification mechanisms in the gastrointestinal tract, a host of detoxifying enzymes in the liver, and similar enzymes in all other tissues. Probably the ability to overcome the effects of toxicants is most developed in the insect world, where there are numerous examples of **specialist feeders** which feed with impunity on poisonous plants. The cinnabar moth (*Tyria jacobaea*) larvae feed only on plants such as *Senecio* spp. that contain pyrrolizidine alkaloids; the larvae have evolved a metabolic requirement for these alkaloids. In Australia, numerous species of two butterfly families and one moth family require pyrrolizidine alkaloids for synthesis of a phermone. The *Chrysolina* beetle was introduced into California and the Pacific Northwest to control St. Johnswort (*Hypericum perforatum*), a toxic weed. The larvae of the monarch butterfly feed on milkweeds and accumulate the toxic milkweed cardenolides, making the larvae repellent to birds. Many more examples are known of insects that have evolved mechanisms to detoxify specific toxicants to the point where in some cases they now have a metabolic requirement for them. Among mammals and specifically livestock, there are few examples of specialist feeders. One example of a wild species is

FIGURE 3–1 The koala is a specialist feeder which has evolved metabolic pathways and taste preferences which overcome the chemical defenses (phenolics and terpenoids) of eucalyptus trees.

the **koala**, which specializes in eating eucalyptus leaves that are rich in phenolics and terpenoids. Vultures have developed immunity to the deadly **botulin** toxin, produced in rotting carcasses which they consume. Many mammals, including some livestock species, do have specialized resistance to some toxicants. Sheep and goats can tolerate more than ten times as much pyrrolizidine alkaloid as can cattle and horses. Sheep will avidly consume foliage of plants containing these alkaloids, whereas cattle and horses will avoid them. Sheep are more resistant to larkspur and poison hemlock toxicity than are cattle. Sheep are not affected by the teratogenic effects of quinolizidine alkaloids in lupins which cause birth defects in cattle. Many more examples of these differences in livestock response to poisonous plants are discussed in this book.

After an animal has ingested a toxic plant, numerous metabolic processes occur before the toxicant involved exerts its toxic effect or is excreted. These processes may occur in the gastrointestinal tract, in the liver, or in other tissues. Liver tissue has a very high level of toxin-metabolizing enzymes. This is of obvious significance, as absorbed substances are taken to the liver by the portal circulation before entering the general circulatory system of the body. Thus, the liver is a first line of defense and can detoxify many poisons before other tissues are exposed to them. A consequence of this activity is that the liver is particularly likely to be the target organ of many toxicants.

Some generalizations can be made concerning the overall fate of toxicants. They are absorbed in the lipid-soluble form and are excreted as water-soluble metabolites. Thus metabolism of toxicants involves enzymatic reactions to convert fat-soluble substances to water-soluble compounds. These metabolic processes, known as **biotransformation**, may either increase or decrease the toxicity of the ingested toxicant.

Toxicant Metabolism in the Gastrointestinal Tract

Most toxicants do not produce their toxic effects in the gut, with the exception of irritants such as saponins, *Solanum* alkaloids, selenium-containing amino acids, and oxalates, and toxicants which affect digestive processes, such as trypsin inhibitors, tannins and lectins.

The rate of absorption of toxicants is largely determined by their **lipid solubility**, which is commonly expressed as the lipid/water partition coefficient. Nonionic compounds are more readily absorbed than ionized substances, as they are more lipophilic. Thus, organic acids are more likely to be absorbed in an acid environment (the stomach) while organic bases are absorbed in a basic environment (small intestine). The extent of dissociation of a compound will determine the proportion that is ionized at a given pH. Most toxicants are absorbed by simple diffusion.

An example of the effect of water solubility and base strength on toxicity is shown in Table 3–1, in which the toxicity of numerous pyrrolizidine alkaloids is compared to these characteristics. The lower the pK_a, the lower the base strength. The most toxic alkaloids, such as lasiocarpine, have a low water solubility and low base strength (low degree of ionization) while alkaloids with a low toxicity, such as heliotridine, have a high water solubility and a high base strength (high degree of ionization). Of course, other factors influence the toxicity of these alkaloids, but there is a clear association between toxicity and lipid solubility.

TABLE 3–1 Comparison of Acute Toxicity of Pyrrolizidine Alkaloids with Water Solubility and Base Strength*

Alkaloid	LD_{50} in male rats (mg/kg)	Solubility in water	pK_a
Lasiocarpine	72	Low	7.6
Seneciphylline	77	Low	7.6
Senecionine	85	Low	7.7
Monocrotaline	175	Medium	7.9
Heliotrine	300	Medium	8.5
Echinatine	350	High	8.4
Intermedine + lycopsamine	>1000	High	8.5
Heliotridine	1200	High	9.0
Heliotrine *N*-oxide	5000	High	>9.0

*Adapted from Bull *et al.* (1968).

In ruminant animals, considerable metabolism of toxicants occurs in the rumen. This aspect is discussed separately. In nonruminants, there may be metabolism of toxicants by microorganisms in the hindgut. Because of the anatomical location of these processes in the posterior region of the gut, metabolism of toxins by cecal organisms generally has limited effects in livestock. In humans, this aspect is of interest as there may be microbial metabolism producing carcinogens involved in colon cancer. In some breeds of chickens, cecal metabolism of sinapine in rapeseed meal can result in formation of trimethylamine, causing a fishy flavor of the eggs.

The degree of absorption of toxicants can be influenced by several other factors. Anion exchange resins, bentonite and zeolites (clays with ion-exchange capacity), and alfalfa meal in the diet reduce the absorption of mycotoxins such as zearalenone. Factors affecting gastrointestinal motility affect absorption; in general, reduction in motility results in a longer period during which absorption can occur, and greater absorption.

Ruminal Metabolism of Toxicants

The rumen is a specialized area that may have pronounced effects on ingested toxicants in ruminant animals. It generally has a slightly acid pH and has an immense population of diverse types of microorganisms. Ingested material remains in the rumen for a much longer period than is the case for ingesta in the gut of nonruminants. It is an area where small water-soluble molecules such as the volatile fatty acids are absorbed.

Ruminal metabolism by microorganisms has diverse effects on toxicants. Some of the potential interactions of toxicants and ruminal metabolism are: (1) toxicity may be increased as a result of ruminal metabolism; (2) toxicity may be decreased; (3) rumen microorganisms may produce toxins; and (4) dietary toxicants may inhibit rumen fermentation. Examples of each of these will be discussed briefly here and in more detail later. Examples of toxicants for which an increase in toxicity due to rumen metabolism occurs include nitrates, cyanogens, mimosine, the brassica anemia factor (S-methylcysteine sulfoxide), and phytoestrogens (Fig. 3–2). Nitrate is converted to the more toxic nitrite in the rumen. Cyanogenic glycosides are hydrolyzed more rapidly in ruminants than in nonruminants, and thus are more toxic because the enzymatic hydrolysis of the glycosides is favored by the higher pH of the rumen than the highly acid nonruminant stomach. The toxic amino acid mimosine in *Leucaena leucocephala* is converted by rumen metabolism to dihydroxypyridine, a more toxic metabolite. The brassica anemia factor is metabolized in the rumen to dimethyl disulfide, which when absorbed causes hemolytic anemia. The phytoestrogen formononetin in subterranean clover is metabolized in the sheep rumen to a more po-

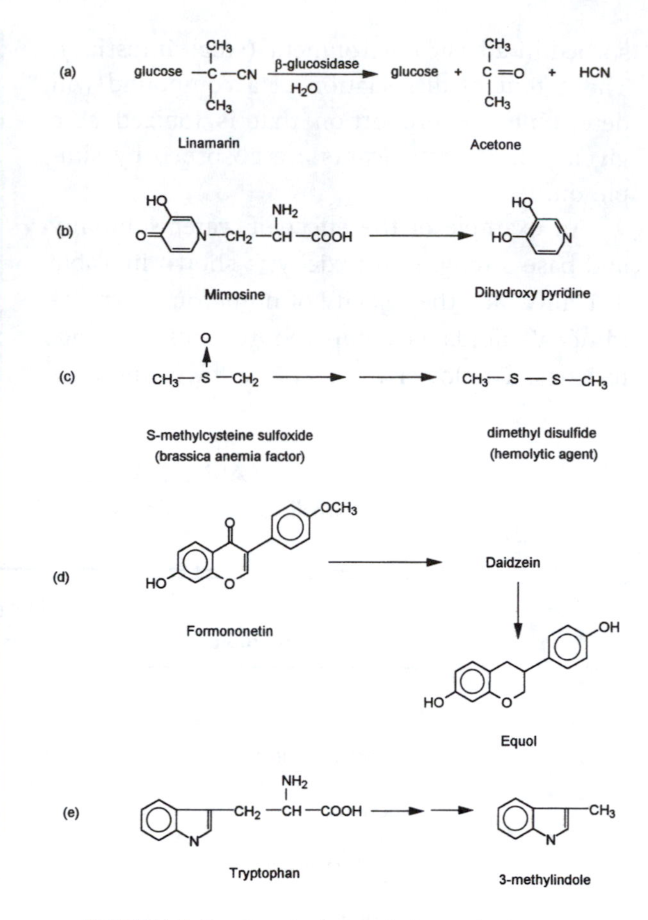

FIGURE 3–2 Ruminal metabolism which increases toxicity of toxicants. (a) Cyanide formation. (b) Mimosine metabolism. (c) Metabolism of brassica anemia factor. (d) Metabolism of formononetin. (e) Formation of 3-methylindole.

tent estrogen. These are a few examples of how rumen metabolism can increase toxicity. Alternatively, detoxification can occur in the rumen (Fig. 3–3). Oxalates can be degraded by rumen microorganisms. There is evidence that pyrrolizidine alkaloids in *Heliotropium europaeum* are detoxified in the sheep rumen with the formation of methylated derivatives. Gossypol, a phenolic substance in cottonseed meal, is detoxified in the rumen, as are trypsin inhibitors in soybeans and glucosinolates in rapeseed meal. Some phytoestrogens, such as biochanin A and genistein in subterranean clover, are converted to *p*-ethylphenol, a nonestrogenic compound, in the sheep rumen. A mycotoxin, ochratoxin A, is de-

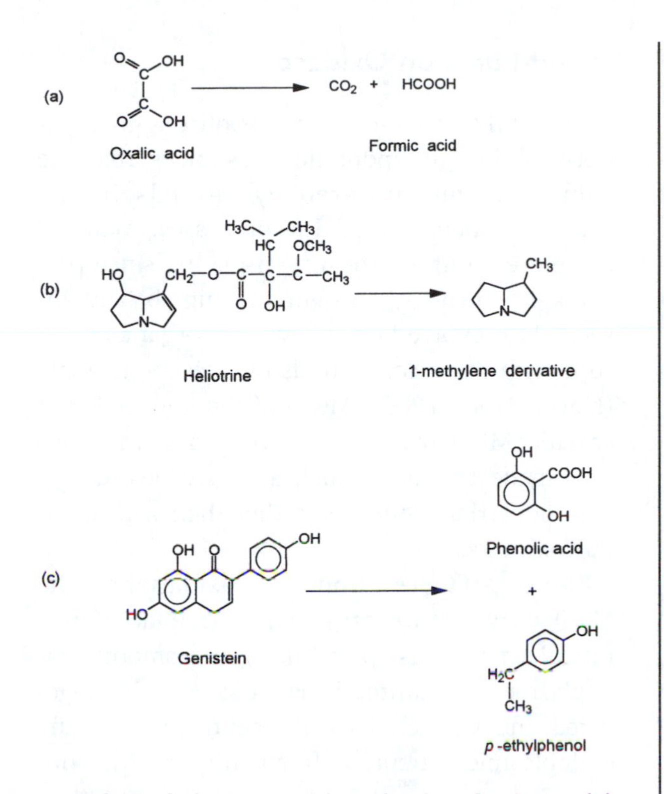

FIGURE 3–3 Ruminal detoxification of toxicants. (a) Oxalate metabolism. (b) Pyrrolizidine alkaloid metabolism. (c) Phytoestrogen metabolism.

toxified in the rumen. There is evidence of ruminal degradation of toxic ergot alkaloids in tall fescue. Thus, it is apparent that in many cases the rumen has a favorable effect in detoxifying many of the toxic substances that ruminant animals are exposed to in their typically diverse types of diets. In some instances, usually associated with a dietary change, rumen microorganisms may produce toxins. An example is the production of lactic acid by *Streptococcus bovis*, causing acute acid indigestion when a sudden shift to a high starch diet is made. The condition of acute bovine pulmonary emphysema is caused by the abnormal rumen metabolism of tryptophan to 3-methylindole, following an abrupt dietary change. Polioencephalomalacia is a thiamin deficiency in ruminants, provoked by a thiamin-degrading enzyme, thiaminase I, produced by rumen microorganisms when high concentrate diets are fed. Ruminants are normally protected against thiamin deficiency due to thiaminase in bracken fern because of the abundant synthesis of thiamin by the rumen bacteria.

In some cases, dietary toxicants may inhibit rumen function. Oxalates and mesquite bean toxicity result in rumen stasis, perloline in tall fescue reduces cellulose digestibility, tannins reduce protein digestibility, aflatoxins may inhibit rumen fermentation, and fungi on soil or herbage may have antimicrobial effects.

It is apparent that there are many interactions between toxicants and rumen microorganisms. In many cases, the result is detoxification, with the ability of ruminants to utilize feedstuffs that are toxic to monogastric animals. Among ruminant species, differences in susceptibility to plant toxins are largely due to postabsorptive differences in tissue enzyme activities. There is little evidence that toxins are degraded in the rumen of some ruminants (e.g. sheep) and not in others (e.g. cattle). Rumen microbes are a reflection of diet and other external factors, rather than being characteristic of animal species. A possible exception was reported by Kronberg and Walker (1993). They hypothesized that leafy spurge (*Euphorbia esula*) is detoxified in the rumen of goats so that its ability to elicit feed aversion responses is reduced. Sheep developed an aversion to a novel feed when its intake was followed by a dose of leafy spurge fermented with sheep rumen fluid, but not when followed by a dose of the plant fermented with goat rumen fluid. This suggests that goat ruminal microbes may modify leafy spurge so that it does not elicit an aversion in sheep or goats. Goats readily consume the plant whereas sheep graze it reluctantly.

Postabsorptive Metabolism of Toxicants

Toxicants are absorbed as lipid-soluble substances and are metabolized in the tissues to water-soluble metabolites that can be excreted in the urine. This metabolism is largely carried out in the liver by enzymes referred to collectively as the **mixed function oxidases (MFO)**. This name is derived from the double role of the oxygen molecule in MFO-catalyzed reactions, both as an oxidizing (to form water) and as an oxygenating agent. The MFO system has a remarkable degree of nonspecificity, in contrast to most enzymes, so it metabolizes many diverse compounds. The MFO enzymes catalyze numerous oxidative reactions, producing more polar, water-soluble metabolites. Another attribute of MFOs in detoxification is that they respond very quickly to the presence of dietary toxicants with a marked increase in level or activity. This process is called **enzyme induction**, whereby the presence of a toxicant induces increased activity of the enzymes that detoxify it. Along with the MFO system, a variety of other enzymes such as esterases, reductases, and transferases are involved in detoxification.

Mixed-Function Oxidases

The MFO system can metabolize a great variety of foreign lipophilic substances that are commonly encountered by animals in the course of their normal life processes. Many of these are natural toxicants in plants, since presumably the MFO system in animals and insects has evolved to allow the organisms to cope with these compounds in their food supply (Nebert *et al.*, 1989). Much of the knowledge of specific MFO reactions is concerned with human-made chemicals such as pesticides, drugs, and industrial chemicals, rather than with natural toxicants.

Most MFO reactions are oxidations. The MFO enzymes are present in the endoplasmic reticulum of cells, particularly the smooth endoplasmic reticulum. When tissue is homogenized and subjected to ultracentrifugation, the endoplasmic reticulum fragments are fraction-

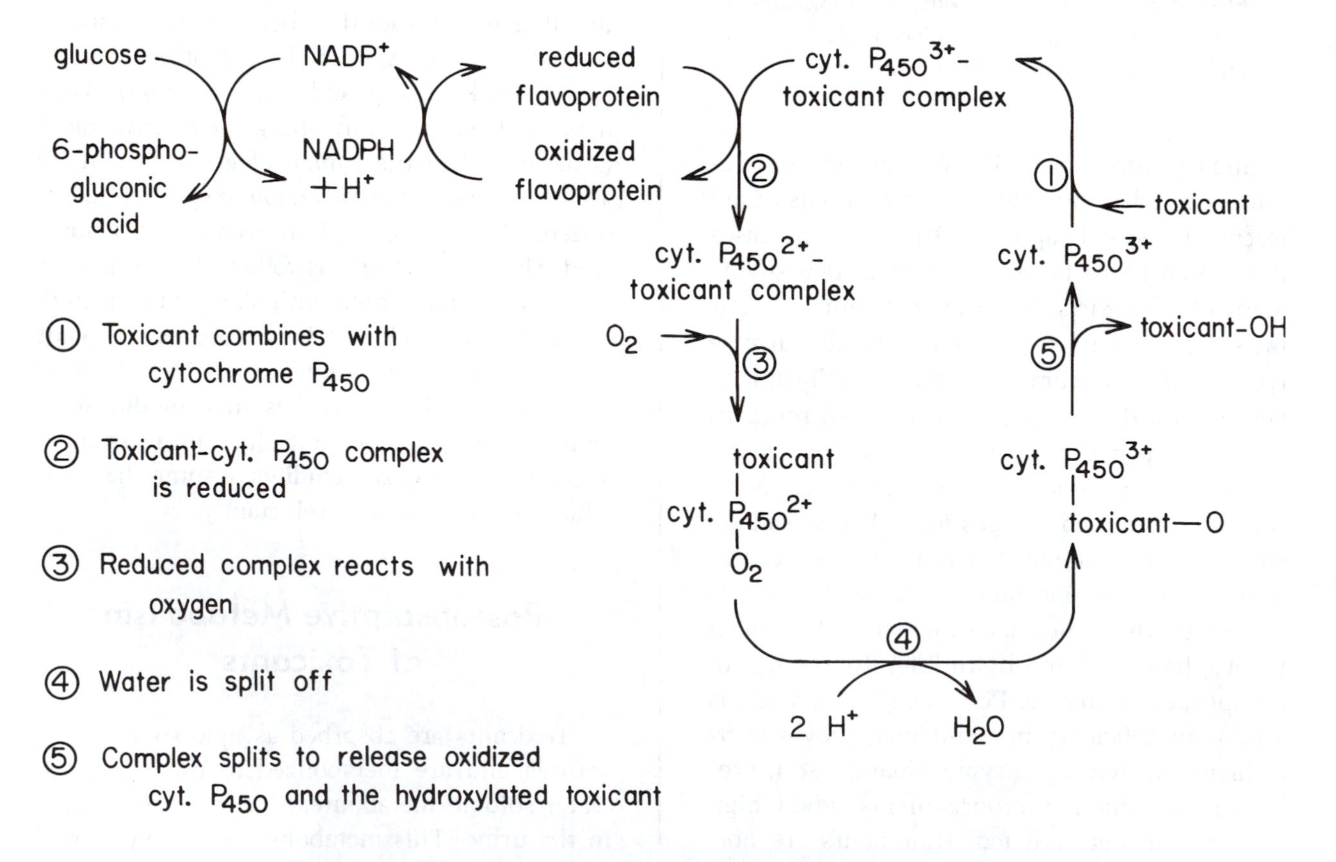

FIGURE 3–4 The mixed function oxidase (MFO) enzyme system.

ated out as a pellet called the microsomal pellet or **microsomes**. The MFO system is often referred to as microsomal enzymes. The MFO system has several components: cytochrome P450, NADPH, a flavoprotein enzyme called NADPH-cytochrome P450 reductase, and phosphatidylcholine.

Cytochrome P450 is the terminal oxidase of the MFO system. It is a b-type cytochrome (a hemoprotein containing heme) which binds carbon monoxide. The reduced cytochrome–carbon monoxide complex has an absorption peak at 450 nm, hence its name. The mechanism of action of the MFO system is that the toxicant reacts with the oxidized cytochrome P450, producing a complex which then is reduced by picking up hydrogen from the reduced flavoprotein, and subsequently reacts with molecular oxygen to produce water, reoxidized cytochrome P450, and the toxicant with a hydroxyl group attached. These reactions are shown in Fig. 3–4. The net result is that an aromatic substrate has been hydroxylated, so it can then be conjugated with polar compounds to produce a water-soluble metabolite (Parke *et al.*, 1991).

There are several isozymes of cytochrome P450 with different substrate specificities. Some of the animal species differences in susceptibility to toxins may be due to differences in cytochrome P450 isozyme activities (Ioannides and Parke, 1990; Smith, 1991). At least eight major gene families for cytochrome P450 have been identified, with the types and amounts varying with species, organ, age, sex, stress and toxicant exposure (Ioannides and Parke, 1990; Sipes and Gandolfi, 1991). Thus individual isozymes are relatively substrate specific, but the full complement of P450 enzymes provides for a remarkably broad capacity for toxin metabolism.

Specific MFO Reactions

The MFO reactions are referred to as **biotransformation** because the substrate is transformed into metabolites. Examples of oxidation reactions include the metabolism of safrole to hydroxysafrole (Fig. 3–5); this is a hydroxylation reaction. Safrole is a component of essential oils such as sassafras oil; it was widely used as a flavoring agent in soft drinks until the discovery that it has carcinogenic activity in rodents. Another type of oxidation is epoxide formation, such as the metabolism of aflatoxin B_1 (Fig. 3–5) to produce aflatoxin 2,3-epoxide. Another group of reactions involves formation of N-oxides from tertiary amines, such as from nicotine and pyrrolizidine alkaloids (Fig. 3–5). Oxidation can also result in oxidation of thioethers to sulfoxides and sulfones, as in the formation of the brassica anemia factor (S-methylcysteine sulfoxide). N-Demethylation is another example of an MFO-catalyzed reaction. The activity of amino-pyrine-N-demethylase is often measured as an indicator of drug-metabolizing ability. This is easily accomplished by measuring the rate of formation of formaldehyde, an end product of the reaction (Fig. 3–5). Reductions are another detoxification reaction, as in the formation of toxic pyrroles from pyrrolizidine alkaloids and conversion of the mycotoxin zearalenone to zearalenol. These types of reactions are often referred to as **phase I reactions**. In the phase I reactions, reactive functional groups are added or exposed (e.g. –OH, –SH, $-NH_2$, –COOH). The MFO system is the primary mediator of phase I reactions. Phase II involves conjugation of the metabolites produced in the MFO-mediated phase I reactions.

Biotransformation in general is a detoxification process, converting toxic lipophilic compounds to nontoxic water-soluble metabolites, although sometimes the metabolites are more toxic than the original compounds. Examples of increased toxicity include the conversion of aflatoxin to the carcinogenic epoxide and the conversion of the relatively nontoxic pyrrolizidine alkaloids to very toxic pyrroles.

Factors Affecting MFO Activity

The MFO enzyme activities are generally much higher in the liver than in other organs, so that the liver is the major site of detoxifica-

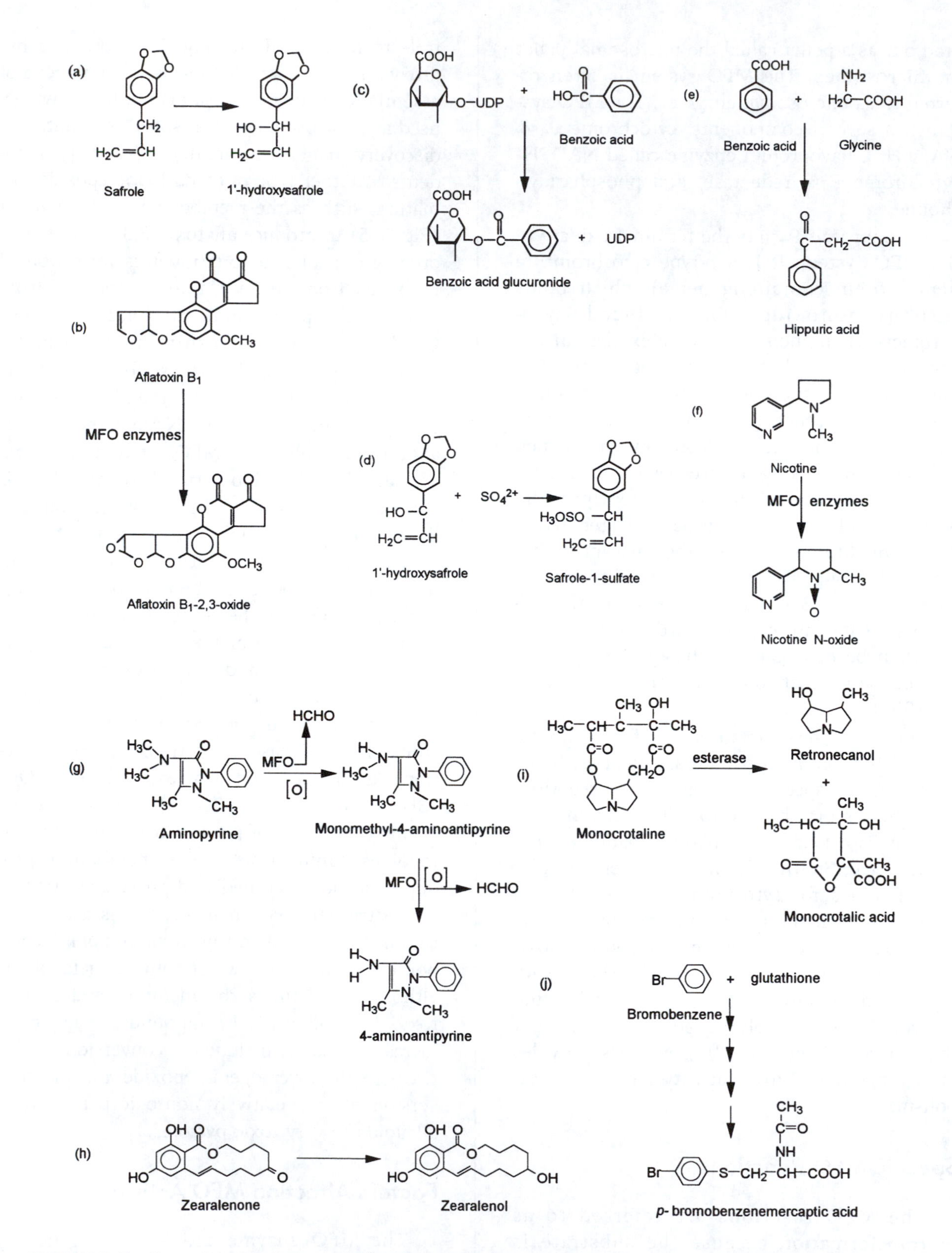

FIGURE 3–5 Examples of typical metabolic reactions in the conversion of toxicants to metabolites. (a) Hydroxylation. (b) Epoxidation. (c) Glucuronide formation. (d) Sulfation. (e) Glycine conjugation. (f) N-oxide formation. (g) N-demethylation. (h) Reduction. (i) Esterase activity. (j) Glutathione conjugation.

tion. All other tissues do possess MFO activity. Thus in acute bovine pulmonary emphysema, for instance, 3-methylindole is metabolized by the lung to an active metabolite that attacks lung tissue. Both the liver and lung are important sites for detoxification enzymes, since the portal vein and the lungs are the major routes for entrance of toxicants into the internal environment. The nasal mucosa contains cytochrome P450 activity, which presumably aids in detoxification of toxins that are inhaled (Longo *et al.*, 1991). This could perhaps have some relevance to the ability of animals to detect poisonous plants, by producing metabolites in the nasal tissue, eliciting an olfactory response. Rumen and intestinal mucosa tissues have cytochrome P450 activity. Smith (1992) speculated that biotransformation of absorbed toxins in gut mucosal tissue may be more important than currently recognized.

Fetal and newborn animals have very low MFO activity, and thus lack the ability to metabolize many toxicants and drugs. Children may be more sensitive to pesticide residues on food than are adults (NRC, 1993), for example. There are numerous sex-related differences in MFO activity, presumably related to steroid hormone balance. Steroid hormones are metabolized by the MFO enzymes.

There are pronounced species differences in microsomal enzyme activities. Ducklings and trout are extremely sensitive to aflatoxins because of a rapid rate of metabolism to the epoxide, whereas sheep and rats metabolize aflatoxin more slowly and are much more resistant to it. Animals that are resistant to the toxic effects of pyrrolizidine alkaloids, such as sheep, guinea pigs, and Japanese quail, have a much lower rate of pyrrole production than susceptible species such as cattle, horses, and rats (Cheeke, 1994).

Nutritional factors influence MFO activity. Low dietary protein levels generally result in reduced MFO function. Thus the toxicity of aflatoxin is reduced on low-protein diets because less of the active metabolite is produced. Vitamin and mineral deficiencies may affect MFO status.

Factors that stimulate MFO activity are called **inducers** or inducing agents. Most toxicants are inducers and therefore exposure to toxicants results in an increased ability of animals to detoxify them. This, of course, is a sound strategy for coping with environmental hazards. Certain drugs are widely used for research purposes to induce MFO activity. The most common is phenobarbital. Pretreatment of animals with barbiturates markedly increases their microsomal enzyme activities. This increases the toxicity if the product of biotransformation is more toxic than the original toxicant. For example, guinea pigs are resistant to the pyrrolizidine alkaloid monocrotaline (Swick *et al.*, 1982), but if pretreated with phenobarbital, they become susceptible (White *et al.*, 1973). Other drugs are used to inhibit MFO activity. These include SKF 525-A (from Smith, Kline, & French Co.), and piperonyl butoxide (Fig. 3–6). Pretreatment of animals with these drugs blocks the formation of active metabolites. Thus by studying the effect of inducers and inhibitors on the toxicity of a compound it is possible to determine if it is bioactivated by the MFO enzymes.

One procedure used to assess MFO activity is phenobarbital sleeping time. Animals are given a dose of barbiturate, and the time taken

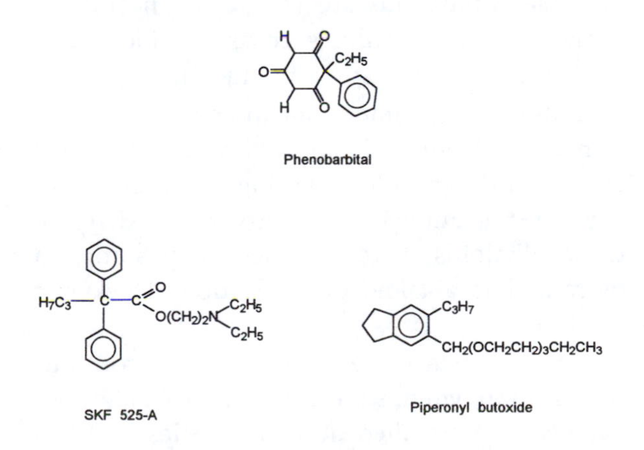

FIGURE 3–6 Structures of drugs commonly used to induce or inhibit MFO activity for experimental purposes.

for them to awaken is measured (sleeping time). The more rapidly the phenobarbital is metabolized, the more rapidly they awaken. Therefore a high MFO activity will result in a short sleeping time.

Other Phase I Detoxifying Enzyme Systems

Various enzymes besides the MFO system are involved in detoxification. **Epoxide hydrolases** are enzymes that convert epoxides to diols. This is a detoxification reaction because diols (adjacent hydroxyls) are less chemically reactive than epoxides. For example, aflatoxin B_1 2,3-epoxide is detoxified by an epoxide hydrolase. The pyrrolizidine alkaloid jacobine in *Senecio jacobaea* (tansy ragwort) is an epoxide. Swick *et al.* (1983) have shown that sheep liver has high epoxide hydrolase activity, which may account in part for the resistance of sheep to tansy ragwort. Phenobarbital induces increased epoxide hydrolase activity.

Mullin and Croft (1984) suggest a role of epoxide hydrolase in plant–herbivore interactions. Plants produce a variety of fatty acids, phenolics, alkaloids, and terpenoids which may be metabolized in mammalian and insect herbivores to epoxides. Some plants counter insect herbivory by synthesizing mimics of insect juvenile hormones or antagonists of the hormone's synthesis or degradation within insects. Some of these compounds are epoxides. The insects evolve countermeasures such as epoxide hydrolase (the coevolutionary arms race!).

Esterases are important in some detoxifications. Pyrrolizidine alkaloids have esterified side chains on the pyrrolizidine ring; the presence of the ester groups is necessary for toxicity of these alkaloids. Esterases may play a role in pyrrolizidine alkaloid detoxification (Dueker *et al.*, 1992).

Another phase I enzyme is **flavin-containing monooxygenase (FMO)**. Pyrrolizidine alkaloids are detoxified in some species by FMO activity to convert the alkaloids to N-oxides (Williams *et al.*, 1989). Cashman (1995) has reviewed the roles of FMO in xenobiotic metabolism.

Phase II: Conjugation Reactions

Metabolites produced by MFO activity are generally excreted in the urine as conjugated compounds; that is, they are conjugated with other substances such as glutathione, glycine, glucuronic acid, and sulfate to increase their water solubility. These reactions are catalyzed by group transferase enzymes, such as **glutathione-S-transferase**. This enzyme results in the reaction of an hydroxylated metabolite with glutathione:

$$\underset{\text{Hydroxylated compound}}{R{-}OH} + \underset{\text{Reduced glutathione}}{GSH} \longrightarrow \underset{\text{Conjugated metabolite}}{R{-}S{-}G + H_2O}$$

UDPglucuronosyltransferase catalyzes the formation of a β-glucuronide:

$$R{-}OH + \text{UDP-glucuronic acid} \longrightarrow R{-}O{-}\beta\text{-glucuronide} + UDP + H_2O$$

Sulfotransferases result in the formation of water-soluble sulfates. Rhodanese is a transferase involved in the detoxification of cyanide. Its function will be further discussed in the section on cyanogenic glycosides. There are species differences in activities of these enzymes. Cats lack glucuronosyltransferase and so cannot form glucuronides. Therefore they are more susceptible than other species to intoxication by benzoic acid and phenol. Guinea pigs lack the ability to synthesize mercaptic acid metabolites from glutathione conjugation, perhaps explaining the lack of protection against pyrrolizidine alkaloid toxicity by dietary cysteine (Swick *et al.*, 1982) as compared to the effectiveness of cysteine in rats (Buckmaster *et al.*, 1976).

The activity of conjugation mechanisms can be altered by various additives. This is useful for experimental purposes and in some cases may present possibilities for diet supplementation programs to reduce toxicities. Dietary cysteine stimulates increased glutathione production, whereas administration of diethyl maleate blocks glutathione conjugation. Salicylamide blocks glucuronide conjugation. These compounds can be given to test animals prior to administration of a toxicant, and the effects on toxicity can be observed. An example of the use of the technique is that (Table 3–2) of Jennings *et al.* (1978), who showed the effect of various pretreatments on mice given a dose of tetradymol. Tetradymol is the toxic agent of *Tetradymia* spp. (horsebrush, rabbit brush) which causes photosensitization in sheep. These data indicate that tetradymol is activated by MFO activity to a more toxic metabolite, and that glutathione conjugates with tetradymol or its metabolites, resulting in reduced toxicity. Blocking glutathione conjugation with diethyl maleate increased toxicity. Piperonyl butoxide and SKF 525-A inhibit MFO activity; because they reduced tetradymol toxicity, it can be concluded that tetradymol is bioactivated by MFO.

TABLE 3–2 Effect of Various Pretreatments on Toxicity of Tetradymol in Mice*

Pretreatment	Survival time (hr)	Interpretation
None	7.5	—
Phenobarbital	5.5	Increased MFO, increased toxicity
SKF 525-A	13.6	Decreased MFO, decreased toxicity
Piperonyl butoxide	17.9	Decreased MFO, decreased toxicity
Cysteine	9.2	Increased glutathione, decreased toxicity
Diethyl maleate	6.3	Decreased glutathione, increased toxicity

*Adapted from Jennings *et al.* (1978).

The activity of various drug-metabolizing enzymes can be influenced by other additives and by exposure to toxicants. Miranda *et al.* (1980a) found that administration of jacobine, a pyrrolizidine alkaloid with an epoxide group, to rats resulted in a significant increase in epoxide hydrolase and glutathione S-transferase activity, and reductions in cytochrome P450 and aminopyrine demethylase activity. Feeding a plant, *S. jacobaea*, which contains jacobine and other pyrrolizidine alkaloids, resulted in similar changes, with elevated epoxide hydrolase and glutathione S-transferase activities (Miranda *et al.* , 1980b). Thus these alkaloids induce detoxification mechanisms. Synthetic antioxidants such as ethoxyquin and butylated hydroxyanisole (BHA) also induce conjugating enzymes. Miranda *et al.* (1981a) found that dietary ethoxyquin caused increases in liver glutathione, glutathione S-transferase, and cytochrome P450 levels, and reduced the toxicity of pyrrolizidine alkaloids in mice. Other studies (Miranda *et al.*, 1981b, 1982) have shown protective effects of synthetic antioxidants against pyrrolizidine alkaloid toxicity, while Kim *et al.* (1981, 1982) have observed similar protection with synthetic antioxidants against bitterweed (*Hymenoxys odorata*) toxins.

Much of the information on the metabolism of toxicants has been obtained using pesticides and other synthetic compounds. This is primarily because of the economic incentives and the potential toxic effects of pesticide residues in humans. For many of the natural plant toxicants affecting livestock, knowledge of their metabolism in large animals is nonexistent or skimpy. The activity of toxin-metabolizing enzymes in various tissues of livestock species have been studied by Smith *et al.* (1984), Watkins and Klaassen (1986) and Watkins *et al.* (1987). Smith (1992) reviewed detoxification of plant toxins by ruminant tissues.

General Biological Effects of Natural Toxins

Toxicants exert their effects on tissues by a variety of mechanisms. Some of the more important of these will be briefly discussed.

1. **Oxidant and Free Radical Damage. Free radicals** (reactive substances containing an unpaired electron) and **oxidants** such as hydrogen peroxide (H_2O_2) are formed continuously during metabolism. Both free radicals and oxidants react with double bonds in unsaturated fatty acids in cell membranes. This damage to the membrane components leads to membrane failure and tissue damage. Because free radicals and oxidants are continuously being generated by metabolic processes, there must be mechanisms for constantly inactivating them, to prevent tissue damage. The mode of action of numerous plant toxins is to over-ride these protective mechanisms, resulting in pathology.

   The formation and disposal of peroxides and other oxidants is complicated. A brief over-view will be presented here, sufficient to illustrate how tissue pathology may occur. In the presence of metal ions, such as Fe^{3+}, molecular oxygen (O_2) may react with unsaturated fatty acids to produce hydroperoxides, which undergo further reactions to produce various aldehydes and ketones. The hydroperoxides, being free radicals, can initiate a **chain reaction** by reacting with another unsaturated fatty acid to produce more hydroperoxide.

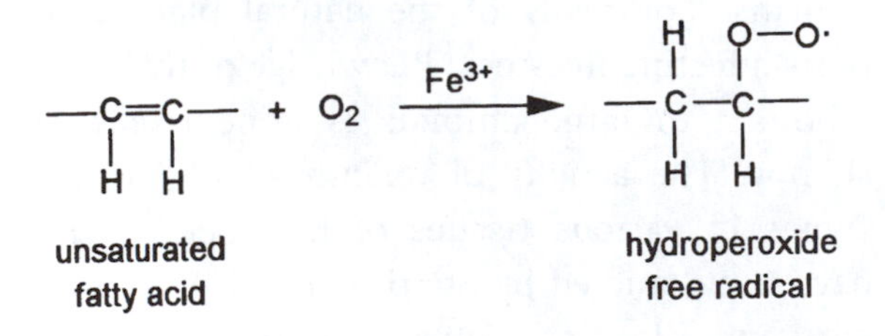

   Vitamin E functions as an **antioxidant** by breaking the free radical chain reaction. It reacts with hydroperoxide free radicals to prevent them from reacting with membrane fatty acids.

   Another free radical formed continuously during metabolism is superoxide (O_2^-). For example, it is generated by the cytochrome P450 enzymes.

   Hydrogen peroxide is an oxidant which attacks cell membranes by reacting with unsaturated fatty acids in membrane phospholipids. Numerous metabolic reactions generate hydrogen peroxide. For example, the enzyme **superoxide dismutase** converts the superoxide anion produced in the cytochrome P450 reactions to hydrogen peroxide:

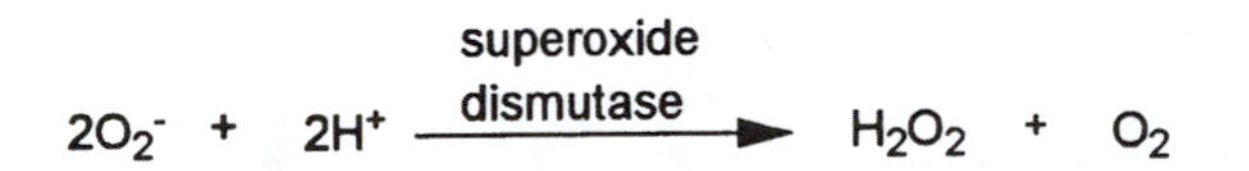

   Superoxide dismutase is a metalloenzyme containing copper and zinc.

   Hydrogen peroxide is also produced when reduced flavin cofactors such as $FMNH_2$ are reoxidized during the electron transport reactions:

$$FMNH_2 + O_2 \longrightarrow FMN + H_2O_2$$

   Another source of hydrogen peroxide is the metabolism of long chain fatty acids in the **peroxisomes**, which are subcellular particles involved in fat metabolism.

   Several types of reactions are involved in the disposal of hydrogen peroxide and other oxidants. They are detoxified by being reduced (biological reduction is the addition of hydrogen to a molecule). Two enzymes that detoxify hydrogen peroxide are catalase and glutathione peroxidase. Both these enzymes catalyze the addition of hydrogen to hydrogen peroxide to form water:

$$H_2O_2 + H_2 \longrightarrow 2H_2O$$

Glutathione peroxidase is a selenium-containing enzyme. It is particularly important in the red blood cell, protecting the erythrocyte membrane from oxidative damage. Several plant toxicities that result in red blood cell hemolysis involve the production of free radicals and the overwhelming of the capacity of glutathione peroxidase to detoxify hydrogen peroxide. Examples include favism (fava bean toxicity), brassica-induced anemia, onion poisoning, and red maple poisoning of horses. The specific mode of action will be discussed under Favism.

Free radicals are involved in other pathologies besides hemolysis. **Photosensitization** is caused by free radicals produced in the epidermis by the action of ultra-violet radiation on photodynamic compounds such as hypericin. These photodynamic substances absorb the energy of a photon, and are raised to a higher energy state. In this excited state, they react with another molecule. Often this molecule is oxygen, producing the superoxide radical and hydrogen peroxide. These reactive products damage proteins and cell membranes in the epidermis, producing severe skin lesions.

Various other toxicity conditions may have a free radical involvement, including pulmonary emphysema, lathyrism and pyrrolizidine alkaloid toxicity. The roles of free radicals in tissue injury and disease have been reviewed by Kehrer (1993).

Some toxins such as phenoxy acid herbicides are **peroxisome proliferators**, producing an increased number of peroxisomes per cell and modifying peroxisomal enzyme activities (Gibson, 1993). Increased free radical and hydrogen peroxide concentrations induced by peroxisome proliferators may produce tissue damage.

2. **Binding to Macromolecules**. Many toxins exert their effects by binding to the active sites of enzymes, to DNA, or to other macromolecules in cells. For example, cyanide binds to the respiratory enzyme cytochrome oxidase and prevents it from serving as an electron acceptor. Nitrite binds to hemoglobin and prevents it from reacting with oxygen. Free radicals, hydroperoxides and other products of lipid peroxidation may bind to membrane components, leading to cell death. Pyrrolizidine alkaloids are bioactivated to pyrroles in the liver. Pyrroles bind to DNA, forming DNA adducts and crosslinks between DNA strands. This impairs cell division and protein synthesis. Electrophilic (electron-seeking) metabolites such as pyrroles may also bind to RNA, directly inhibiting protein synthesis. Tannins and other phenolic compounds bind to proteins, including taste receptors in the mouth, and digestive enzymes.

   Avidin, a glycoprotein in egg albumin, is an inhibitor of the B vitamin biotin. It binds biotin in a tight complex that resists digestion. Another deleterious substance which acts within the digestive tract is trypsin inhibitor, found in raw legume seeds. Trypsin inhibitors bind irreversibly to the digestive enzyme trypsin, thus impairing protein digestion. Lectins in raw beans bind to glycoproteins in the intestinal microvilli, causing mucosal damage.

3. **Binding to Cell Receptors**. **Receptors** are macromolecular components of cell membranes which recognize specific molecules by their shape and/or chemical structure. For example, most hormones affect their target cells by binding to membrane receptors, an event which initiates a cascade of reactions in the cell, to produce a specific response. Some toxins exert their effects by binding to receptors and thus preventing the intended molecule from reacting with the receptor. For example, phytoestrogens bind to estrogen receptors in the female reproductive tract, inhibiting the binding of endogenous estrogens. Some goitrogenic substances may bind to thyroxin receptors. Neurotoxins may bind to receptors such as the cholinergic receptors of the parasympathetic nervous system (e.g. *Solanum* alka-

loids). Glucosinolates may bind to the **Ah receptor** (arylhydrocarbon receptor) that transports certain toxins (e.g. the dioxin TCDD) to the cell nucleus.

4. **Inhibition of Energy Metabolism.** Numerous toxins inhibit some aspect of energy metabolism. For example, cyanide inhibits the electron transport enzyme cytochrome oxidase. Fluoroacetate inhibits metabolism of citric acid in the Kreb's cycle reactions. Another route for toxins to inhibit energy metabolism is via competitive inhibition of vitamin cofactors (e.g. linatine in linseed meal is a competitive inhibitor of pyridoxine).
5. **Structural Antagonism.** Some natural toxins exert their effects as competitive inhibitors of essential nutrients, because of their structural similarity. Dicumarol, the causative agent of sweet clover poisoning, is structurally similar to vitamin K. It creates a vitamin K deficiency by binding to vitamin K-responsive enzymes. Some plants have calcinogenic glycosides which contain the metabolically active form of vitamin D (1,25-dihydroxycholecalciferol). Because of their structural similarity, they cause the pathological state associated with vitamin D toxicity by over-riding the normal regulatory mechanisms controlling activation of vitamin D.
6. **Disruption of Membrane Function.** Toxins can inhibit the movement of ions across cell membranes. Ion channel blockers block the flux of ions across membranes in various tissues. Cardiac glycosides in *Digitalis* spp. and milkweeds inhibit Na^+/K^+-ATPase in heart muscle.

   Perturbed intracellular calcium concentrations due to cell membrane damage is characteristic of many pathologies (Harman and Maxwell, 1995). Calcium plays an important role as a second messenger in the regulation of many intracellular functions. Liver damage in pyrrolizidine alkaloid toxicity is associated with an influx of Ca^{++} into the cells (Griffin and Segall, 1989). The increased intracellular Ca^{++} may play a role in cell death. The neurological disorder of lathyrism (see Chapter 7) may involve an influx of calcium ions into brain cells, causing their death.
7. **Enzymatic Activity.** Thiaminases are enzymes that cleave thiamin (vitamin B_1), thus destroying its activity. Thiaminases occur in diverse sources, including bracken fern, horsetails, rumen microbes and raw fish (carp).
8. **Acid–Base Balance Disturbances and Mineral Depletion.** Phase I and II reactions often result in the formation of organic acids which are excreted as salts of ammonium, sodium or other cations. Foley *et al.* (1995) have proposed that acidosis induced by metabolism of plant toxins may contribute to disturbances in acid–base balance and nutrient retention. For example, generation of calcium ions from the skeletal system and ammonium ions from muscle tissue to buffer organic acids may lead to nutritional and physiological problems. These reactions could be particularly important when animals are consuming nutrient-poor, toxin-rich plants, such as the winter diets of bark and twigs of arctic hares, or the *Eucalyptus* foliage consumed by marsupials such as the koala (Foley *et al.*, 1995). Freeland *et al.* (1985) demonstrated that diets containing hydrolyzable or condensed tannins caused sodium depletion in herbivores, and suggested that chemical defenses of plants exert their effects by causing mineral depletion and deficiency in herbivores varying in size from moose to meadow voles. Iason and Palo (1991) found that dietary birch twig phenolics cause severe sodium losses in European hares (grazers) but not in mountain hares (browsers). They suggested that browsers may have adapted to phenolic-rich diets by metabolic regulation of sodium balance. Some northern rodents and lagomorphs de-

pendent on hindgut digestion of bark and other nutrient-poor fibrous materials are able to maintain high levels of minerals in the hindgut by efficient mechanisms of mineral absorption and retention via coprophagy and cecotrophy (Staaland *et al.*, 1995).

9. **Tissue Repair Mechanisms.** Besides having the ability to biotransform toxins via detoxification mechanisms, animals can also respond by repairing cellular damage caused by toxins. There are numerous enzymes which repair damage to DNA, by removing defective purine bases and reinsertion of the correct bases. Some of the intra-species and inter-species variations in susceptibility to toxins are apparently a reflection of differences in tissue repair capacities (Calabrese and Mehendale, 1996).

TABLE 3–3 General Mechanisms of Toxin Action

Type of action	Examples
1. Oxidant and free radical damage	fava bean glycosides, brassica anemia factor, pyrrolizidine alkaloids
2. Binding to macromolecules	pyrrolizidine alkaloids, aflatoxin, trypsin inhibitors
3. Binding to cell receptors	phytoestrogens, glucosinolates
4. Inhibition of energy metabolism	cyanide, fluoroacetate
5. Structural antagonism	dicumarol, fumonisins
6. Disruption of membrane function	cardiac glycosides, pyrrolizidine alkaloids
7. Enzymatic activity	thiaminases, lipoxidases

REFERENCES

Buckmaster, G.W., P.R. Cheeke, and L.R. Shull. 1976. Pyrrolizidine alkaloid poisoning in rats: Protective effects of dietary cysteine. J. Anim. Sci. 43:464–473.

Bull, L.B., C.C.J. Culvenor, and A.T. Dick. 1968. The Pyrrolizidine Alkaloids. American Elsevier, NY.

Calabrese, E.J., and H.M. Mehendale. 1996. A review of the role of tissue repair as an adaptive strategy: Why low doses are often non-toxic and why high doses can be fatal. Fd. Chem. Toxic. 34:301–311.

Carlson, J.R., and R.G. Breeze. 1984. Ruminal metabolism of plant toxins with emphasis on indolic compounds. J. Anim. Sci. 58:1040–1049.

Cashman, J.R. 1995. Structural and catalytic properties of the mammalian flavin-containing monooxygenase. Chem. Res. Toxicol. 8:165–181.

Cheeke, P.R. 1994. A review of the functional and evolutionary roles of the liver in the detoxification of poisonous plants, with special reference to pyrrolizidine alkaloids. Vet. Human Toxicol. 36:240–247.

Dueker, S.R., M.W. Lame, and H.J. Segall. 1992. Hydrolysis of pyrrolizidine alkaloids by guinea pig hepatic carboxylesterases. Toxicol. Appl. Pharmacol. 117:116–121.

Foley, W.J., S. McLean, and S.J. Cork. 1995. Consequences of biotransformation of plant secondary metabolites on acid–base metabolism in mammals—A final common pathway? J. Chem. Ecol. 21:721–743.

Freeland, W.J., P.H. Calcott, and D.P. Geiss. 1985. Allelochemicals, minerals and herbivore population size. Biochemical Systematics and Ecology 13:195–206.

Gibson, G.G. 1993. Peroxisome proliferators: Paradigms and prospects. Toxicol. Lett. 68:193–201.

Griffin, D.S., and H.J. Segall. 1989. Effects of the pyrrolizidine alkaloid senecionine and the alkenals *trans*-4-OH-hexenal and *trans*-2-hexenal on intracellular calcium compartmentation in isolated hepatocytes. Biochem. Pharmacol. 38:391–397.

Harman, A.W., and M.J. Maxwell. 1995. An evaluation of the role of calcium in cell injury. Annu. Rev. Pharmacol. Toxicol. 35:129–144.

Iason, G.R., and R.T. Palo. 1991. Effects of birch phenolics on a grazing and a browsing mammal: A comparison of hares. J. Chem. Ecol. 17:1733–1743.

Ioannides, C., and D.V. Parke. 1990. The cytochrome P450 I gene family of microsomal hemoproteins and their role in the metabolic activation of chemicals. Drug Met. Rev. 22:1–85.

Jennings, P.W., S.K. Reeder, J.C. Hurley, J.E. Robbins, S.K. Holian, A. Holian, P. Lee, J.A. Pribanic, and M. Hull. 1978. Toxic constituents and hepatotoxicity of the plant *Tetradymia glabrata*. pp. 217–228 in: R. F. Keeler, K. R. Van Kampen, and L. R. James (Eds.). Effects of Poisonous Plants on Livestock. Academic Press, NY.

Kehrer, J.P. 1993. Free radicals as mediators of tissue injury and disease. Critical Rev. Toxicol. 23:21–48.

Kim, H.L., A.C. Anderson, B.W. Herrig, L.P. Jones, and M.C. Calhoun. 1982. Protective effects of antioxidants on bitterweed (*Hymenoxys odorata* DC) toxicity in sheep. Am. J. Vet. Res. 43:1945–1950.

Kim, H.L., A.C. Anderson, M.K. Terry, and E.M. Bailey. 1981. Protective effect of butylated hydroxyanisole on acute hymenoxon and bitterweed poisoning. Res. Comun. Chem. Pathol. Pharmacol. 33:365–368.

Kronberg, S.L., and J.W. Walker. 1993. Ruminal metabolism of leafy spurge in sheep and goats: A potential explanation for differential foraging on spurge by sheep, goats and cattle. J. Chem. Ecol. 19:2007–2017.

Longo, V., A. Mazzaccaro, F. Naldi, and P.G. Gervasi. 1991. Drug-metabolizing enzymes in liver, olfactory, and respiratory epithelium of cattle. J. Biochem. Toxicol. 6:123–128.

Miranda, C.L., D.R. Buhler, H.S. Ramsdell, P.R. Cheeke, and J.A. Schmitz. 1982. Modifications of chronic hepatotoxicity of pyrrolizidine (*Senecio*) alkaloids by butylated hydroxyanisole and cysteine. Toxicol. Lett. 10:177–182.

Miranda, C.L., H.M. Carpenter, P.R. Cheeke, and D.R. Buhler. 1981a. Effect of ethoxyquin on the toxicity of the pyrrolizidine alkaloid monocrotaline and on hepatic drug metabolism in mice. Chem.-Biol. Interact. 37:95–107.

Miranda, C.L., P.R. Cheeke, and D.R. Buhler. 1980a. Effect of pyrrolizidine alkaloids from tansy ragwort (*Senecio jacobaea*) on hepatic drug-metabolizing enzymes in male rats. Biochem. Pharmacol. 29:2645–2649.

Miranda, C.L., P.R. Cheeke, and D.R. Buhler. 1980b. Comparative effects of the pyrrolizidine alkaloids jacobine and monocrotaline on hepatic drug-metabolizing enzymes in the rat. Res. Commun. Chem. Pathol. Pharmacol. 29:573–587.

Miranda, C.L., R.L. Reed, P.R. Cheeke, and D.R. Buhler. 1981b. Protective effects of butylated hydroxyanisole against the acute toxicity of monocrotaline in mice. Toxicol. Appl. Pharmacol. 59:424–430.

Mullin, C.A., and B.A. Croft. 1984. Trans-epoxide hydrolase: A key indicator enzyme for herbivory in arthropods. Experientia 40:176–178.

National Research Council. 1993. Pesticides in the Diets of Infants and Children. National Academy Press, Washington, D.C.

Nebert, D.W., D.R. Nelson, and R. Feyereisen. 1989. Evolution of the cytochrome P450 genes. Xenobiotica 19:1149–1160.

Parke, D.V., C. Ioannides, and D.F.V. Lewis. 1990. The role of the cytochromes P450 in the detoxication and activation of drugs and other chemicals. Can. J. Physiol. Pharmacol. 69:537–549.

Short, C.R. 1994. Consideration of sheep as a minor species: Comparison of drug metabolism and disposition with other domestic ruminants. Vet. Human Toxicol. 36:24–40.

Sipes, I.G., and A.J. Gandolfi. 1991. Biotransformation of toxicants. pp. 88–126 in: M.O. Amdur, J. Doull, and C.D. Klaassen. Casarett and Doull's Toxicology, the Basic Science of Poisons, 4th ed. Pergamon Press, New York.

Smith, D.A. 1991. Species differences in metabolism and pharmacokinetics: Are we close to an understanding? Drug Met. Rev. 23:355–373.

Smith, G.S. 1992. Toxification and detoxification of plant compounds by ruminants: An overview. J. Range Mgmt. 45:25–30.

Smith, G.S., J.B. Watkins, C.D. Klaassen, K. Rozman, and T.N. Thompson. 1984. Oxidative and conjugative metabolism of xenobiotics by livers of cattle, sheep, swine and rats. J. Anim. Sci. 58:386–395.

Staaland, H., R.G. White and H. Kortner. 1995. Mineral concentrations in the alimentary tract of northern rodents and lagomorphs. Comp. Biochem. Physiol. 112A:619–627.

Swick, R.A. 1984. Hepatic metabolism and bioactivation of mycotoxins and plant toxins. J. Anim. Sci. 58:1017–1028.

Swick, R.A., P.R. Cheeke, D.E. Goeger, and D.R. Buhler. 1982. Effect of dietary *Senecio jacobaea* and injected *Senecio* alkaloids and monocrotaline on guinea pigs. J. Anim. Sci. 55:1411–1416.

Swick, R.A., C.L. Miranda, P.R. Cheeke, and D.R. Buhler. 1983. Effect of phenobarbitol on toxicity of pyrrolizidine (*Senecio*) alkaloids in sheep. J. Anim. Sci. 56:887–894.

Watkins, J.B. III, and C.D. Klaassen. 1986. Xenobiotic biotransformation in livestock: Comparison to other species commonly used in toxicity testing. J. Anim. Sci. 63:933–942.

Watkins, J.B. III, G.S. Smith, and D.M. Hallford. 1987. Characterization of xenobiotic biotransformation in hepatic, renal and gut tissues of cattle and sheep. J. Anim. Sci. 65:186–195.

White, I.N.H., A.R. Mattocks, and W.H. Butler. 1973. The conversion of the pyrrolizidine alkaloid retrorsine to pyrrolic derivatives *in vivo* and *in vitro* and its acute toxicity to various animal species. Chem.-Biol. Interact. 6:207–218.

Williams, D.E., R.L. Reed, B. Kedzierski, D.M. Ziegler, and D.R. Buhler. 1989. The role of flavin-containing monooxygenase in the N-oxidation of the pyrrolizidine alkaloid senecionine. Drug Metabol. Disposition 17:380–386.

CHAPTER 4

Ecological Roles of Plant Toxicants and Impacts on Livestock Production and Human Health

WHY DO PLANTS CONTAIN TOXINS?

As discussed in Chapter 1, plants contain a bewildering array of substances that are toxic to animals. For virtually every animal organ, tissue and system, there is, lurking somewhere in the plant kingdom, a chemical which is antagonistic to it. Why do plants contain toxic substances? What purpose do plant toxins serve in the global ecosystem?

Plants and animals are mutually dependent inhabitants of the planet. When looked at from a geologic time perspective, neither one can do without the other. The biosynthetic capacities of plants allow them to utilize solar energy to synthesize organic compounds from carbon dioxide, releasing oxygen as a waste product. Animals feed on plants as their source of nutrients and utilize the oxygen, and excrete carbon dioxide which is used by plants. It is not in the best interests of either plants or animals, from an evolutionary perspective, for plants to be eaten to extinction by animals, or for plants to be totally resistant to animal herbivory. Either choice would ultimately lead to extinction of both classes of organisms. Thus plants and animals exist in a mutually beneficial adversarial relationship and plant toxins have an important role in this relationship.

Since plants are immobile and unable to resist herbivory by moving, they must have other means of defending themselves from being excessively eaten. These defenses are primarily physical and chemical. Plant defenses are particularly important in harsh environments where plant growth is slow, thus inhibiting recovery from excessive herbivory. Kingsbury (1983) aptly states "An old saying about the plants of the inimical Texas plateau country is that they all 'stick, stink or sting'." **Physical defenses** include spines, thorns, leaf hairs, small leaves and highly lignified or silica-loaded tissue. Spines and thorns are particularly effective deterrants of large vertebrate herbivores, especially when combined with small leaves (Fig. 4–1). **Chemical defenses** consist of compounds which in some way deter herbivory. These chemical compounds are often referred to as **secondary compounds**, to distinguish them from the primary substances of plant metabolism. Secondary substances exert their effects in several ways. They may cause the plant to be unpalatable through effects on taste or smell. Many plant toxins (e.g. alkaloids, glucosinolates, saponins) are bitter or otherwise unpleasant tasting, while terpenes and essential oils (e.g. in sagebrush) have unpleasant odors which reduce palatability to herbivores. Phenolic compounds have an astringent effect, because they react with taste receptors in the mouth. These chemical defenses of plants confer resistance to mammalian, invertebrate, fungal and bacterial organisms. From the plant viewpoint, they are chemical defenses; from our viewpoint, they are toxins or poisons.

FIGURE 4–1 (Top): Physical defenses can be formidable, such as these spines on *Acacia* in South Africa. (Bottom): Some animals, such as the giraffe, can overcome this physical defense by delicately plucking leaves between the spines.

Plants and animals have coevolved. As plants have evolved particular defense mechanisms, herbivores have evolved various means to overcome the plant defenses. These processes are known as **coevolution**. If herbivores are able to surmount plant defense mechanisms to a sufficient degree to endanger the plant's survival as a species, the plant must respond by producing new chemical defenses, or probably become extinct. Thus a coevolutionary arms race exists, punctuated by periods of stalemates. These interactions are most obvious with insects, which with their short generation times are able to quickly evolve resistance to toxins, both human-made and natural. Many examples of coevolution, especially involving plants and insects, are provided by Harborne (1993), while chemical mechanisms of coevolutionary processes are described by Spencer (1988). Pirozynski and Hawksworth (1988) provide an especially comprehensive treatise on coevolution of fungi with plants and animals, and the mammalian interactions with plant defenses are reviewed by Palo and Robbins (1991). Rosenthal and Berenbaum (1991, 1992) provide an extensive treatment of the interactions of herbivores with secondary plant metabolites. Much of the ecological literature dealing with plant–herbivore interactions emphasizes insect herbivores. The references cited above do emphasize or consider mammalian herbivores. Laycock (1978) specifically reviewed coevolution of livestock and poisonous plants.

ANIMAL RESPONSES TO PLANT TOXINS

Animals have various ways to overcome the effects of plant toxins. They can surmount physical defenses such as thorns and spines by feeding strategies that allow them to daintily browse leafy material while avoiding spines (e.g. Fig. 4–2) or having a tough mouth which allows them to consume spines (Fig. 4–3) with no ill effect (e.g. rhinoceros). Chemical defenses can be overcome in several ways. Some animals have developed tolerance to bitterness. Sheep and goats are more tolerant of bitterness than cattle, and may consume plants which cattle avoid. In the evolutionary process, some animals may become **specialist feeders**, and acquire a dependency on certain plants. Koalas, for instance, are specialist feeders that can survive only on a few species of *Eucalyptus*. Specialist feeding is most advanced in the insect world. There are numerous poisonous plants for which certain insect species have evolved detoxification mechanisms to allow consumption of the plant material. Often this leads to a chemical dependency of the insect for the plant compounds, which become intimate components of

FIGURE 4–2 Goats have several means of overcoming the physical defenses of plants. These goats in north Africa are able to eat the sharp spines of *Acacia* trees, and even climb the trees to go after the spine-protected foliage!

FIGURE 4–3 Goats consuming gorse (*Ulex europaeus*), a plant consisting entirely of viciously-sharp spines (modified leaves). Goats will consume this plant with impunity and are used as a biological control in gorse-infested areas (Radcliffe, 1986). (Courtesy of J.S. Powley)

its metabolism. For example, cinnabar moth larvae (Fig. 4–4) feed on the pyrrolizidine alkaloid–containing plant tansy ragwort (*Senecio jacobaea*), and use the alkaloids in the synthesis of phermones. The insect cannot survive without consuming these alkaloids. Some insects use plant compounds in their own defenses, such as the Monarch butterfly larvae which sequester milkweed glycosides as protection against avian predators.

FIGURE 4–4 Larvae of the cinnabar moth, showing defoliation of a tansy ragwort plant. The cinnabar moth is used in biological control of tansy ragwort. The larvae are yellow and black.

Many plants are protected against herbivory by polyphenolic compounds (tannins). This is particularly true of browse species (shrubs, small trees). Robbins *et al.* (1987) reported that deer have a salivary protein that binds tannins, thus allowing the animals to utilize tannin-containing browse species. **Tannin-binding salivary proteins** are not present in sheep and cattle (Austin *et al.*, 1989) but perhaps occur in goats, which are able to browse on high-tannin oak foliage with no apparent ill effect (Nastis and Malechek, 1981), although according to McArthur *et al.* (1993), goats do not have salivary tannin-binding proteins. Species differences in salivary tannin-binding proteins have been reviewed by Mole *et al.* (1990), Robbins *et al.* (1991), and Hagerman and Robbins (1993).

As discussed in Chapter 3, the digestive tract is another potential site of detoxification. The rumen, with its abundant microflora, is often involved in metabolism of toxicants. Ruminants are often more resistant to plant toxins than nonruminants, because of the degradation or inactivation of toxins in the rumen. Glucosinolates in rapeseed and other *Brassica* spp. and gossypol in cottonseed are examples of natural toxins detoxified in the rumen. An interesting example of **rumen detoxification** is that of mimosine, a toxic amino acid in the tropical forage legume *Leucaena leucocephala*. In Australia, leucaena has been toxic to ruminants. However, in Hawaii and Indonesia, leucaena is not toxic to ruminants, because of the occurrence of a mimosine-degrading rumen microorganism. This organism has now been introduced into Australian cattle (Quirk *et al.*, 1988), allowing them to utilize leucaena without suffering mimosine toxicity (see Chapter 11).

While some toxins are counteracted or detoxified in the mouth or digestive tract, many plant toxins are metabolized primarily in the liver (Cheeke, 1994). There are several reasons why the liver is strategically located to be selected by evolutionary pressures as a major site of detoxification. In contrast to gut microbes, liver enzyme activities are under genetic control of the animal, whereas gut microbes are subject to external influences and are not heritable. The liver is the first post-absorptive organ encountered by absorbed substances; thus it is in a strategic position to prevent exposure of the rest of the body to toxins. It is energetically more efficient for animals to direct their detoxification activities to absorbed toxins rather than expending metabolic activity on nonabsorbed molecules which may pass harmlessly from the gut. Liver detoxifying enzyme systems such as cytochrome P450 have broad substrate specificity, giving the animal the capacity to withstand a wide variety of challenges from natural and human-made toxins (e.g. herbicides, pesticides and other agricultural chemicals as well as plant toxins). Hepatic enzymes are readily and quickly inducible, allowing a rapid detoxification response to the intake of toxic substances. There is a high degree of variability in hepatic enzyme activities among individuals, providing the genetic basis for selection pressure to induce species resistance to particular toxins which an animal species might be routinely exposed to on an evolutionary basis (e.g. grazers vs. browsers).

With domestic livestock, there are a number of examples of **species differences** in susceptibility to plant toxins. While these will be considered in more detail in the coverage of specific toxicants, it is useful to list some to illustrate the variety of factors that may influence whether livestock are affected by certain plants. These factors may include simple differences in management or feeding programs, digestive tract differences, palatability variations, or differences in metabolism.

There are many situations where a toxicant is important in nonruminants (swine, poultry) and not in ruminants, and vice versa. This may simply be a reflection of the type of diet; nonruminants are fed high-concentrate diets and so are not likely to be exposed to large quantities of some of the poisonous plants that ruminants consume. Many toxicants are degraded or otherwise rendered inactive in the rumen. These include trypsin inhibitors, lectins, gossypol, glucosinolates, and certain pyrrolizidine alkaloids. Some toxins are made more potent or their toxic nature altered as a result of rumen metabolism. For example, *L. leucocephala* is goitrogenic in ruminants but not in nonruminants because the toxicant it contains, mimosine, is converted in the rumen to dihydropyridone, a goitrogen. There are a number of species differences in susceptibility to pyrrolizidine alkaloids. Cattle, horses, chickens, rats and swine are quite susceptible, whereas sheep, goats, Japanese quail, guinea pigs, and rabbits are very resistant. These differences appear to be due to variations in the way that the alkaloids are bioactivated and metabolized in the liver.

Animal management can affect losses. Sheep on western U.S. ranges are usually herded, whereas cattle are free-ranging. Hence, sheep may have less opportunity to avoid poisonous plants, and the extent of losses may depend as much on the skill of the herder as on other factors. Grazing behavior and palatability differences are important. Sheep tend to consume short grasses, forbs, and shrubs, whereas cattle consume more coarse vegetation. Cattle may develop fescue foot on tall fescue pastures; sheep won't readily eat tall fescue because of its coarse texture, whereas it is readily consumed by cattle. Horses can graze without ill effect tall fescue pastures that are toxic to cattle. Sheep are much more likely than cattle to be killed by acute lupin toxicity, because they eat the toxic lupin pods in large quantities while cattle do not. On the other hand, consumption of lupins by cattle can produce teratogenic effects (crooked calf disease) that do not occur in sheep. Sheep are much more resistant than cattle to the polycyclic diterpene alkaloids in larkspur. Cattle are more susceptible to *Conium* alkaloids than are horses, while sheep are highly resistant. The teratogenic effects of *Conium* alkaloids (crooked calf disease) are seen in calves, but not in lambs or foals. *Tetradymia* consumption causes photosensitization in sheep (big head), but cattle are not affected. These examples illustrate that for a variety of reasons there are many species differences among livestock in response to dietary toxicants.

There may be **breed differences** in susceptibility to plant poisoning. In Australia, echium or Paterson's curse seems to be more toxic to British breeds of sheep than to merinos (Culvenor *et al.*, 1984). Seawright (1989) suggests that this is because British breeds and their merino crosses graze plants with pyrrolizidine alkaloids, such as echium and heliotrope, much more readily than do merinos, and so consume a larger dose of the toxins.

Many of the animal species differences in susceptibility to poisonous plants are due to differences in Phase I and Phase II **liver enzyme activities**, and particularly the cytochrome P450 system (see Chapter 3). This enzyme system is ancient, being present in prokaryotes (single cells with no internal organelles—e.g. bacteria) and fungi, functioning in the oxidation of substrates as sources of energy. In higher animals, the cytochrome P450 enzymes also function in energy metabolism (e.g. fatty acid oxidation) and the oxidation of endogenous substrates such as steroid hormones. Nebert and Gonzalez (1987) speculate that with the evolution of flowering plants containing defensive chemicals, cytochrome P450 in animal tissues assumed an additional role in detoxification of plant toxins and products of plant combustion (e.g. benzpyrene produced by forest fires).

Because of the inducibility and genetic inheritance of cytochrome P450 activity (Nebert and Gonzalez, 1987), natural selection for resistance to plant toxins on an evolutionary scale is feasible. Among livestock and small herbivore species, there seems to be a relationship involving toxin resistance, **feeding strategy** and nature of the diet upon which the species evolved (Cheeke, 1994; Cheeke and Palo, 1995). Cattle and horses evolved as grazing animals on grasslands, while sheep and goats and other small herbivores such as rabbits, guinea pigs, gerbils, etc., consume non-grass herbaceous plants as major components of their natural diets. Herbaceous plants (e.g. forbs, shrubs, trees) are generally well equipped with defensive chemicals (toxins). Animals which evolved on diets of herbaceous plants have coevolved adequate detoxification mechanisms; otherwise they could not have survived and evolved on diets rich in secondary compounds. Conversely, grazing animals such as cattle and horses which evolved on grassland may have less well-developed detoxification systems, because of the relative absence of chemical defenses in the grasses in their diets. Grasses are poorly defended with plant toxins, and depend more on growth habit (e.g. prostrate growth, regrowth from protected crown, etc.), physical defenses (lignin and silica deposition), and the periodicity of grazing under natural grassland conditions, for their protection against herbivory (Cheeke, 1995). Thus in general, sheep and goats as browsing animals are more resistant to many toxic plants than grazing animals such as cattle and horses.

On an evolutionary basis, feeding behavior and resistance to plant toxins appear to be linked. The biological purpose of resistance to plant toxins is to allow animals to utilize toxic plants to which they are resistant. It would not be an evolutionarily-sound strategy to acquire enzymatic resistance to a plant toxin without also exploiting the plants as food resources. Browsers such as small herbivores (e.g. sheep, goats, rabbits, guinea pigs) are resistant to many plant toxins, and are cosmopolitan in their feeding behavior. Sheep and goats will willingly or avidly consume many toxic plants that are aversive to cattle and horses. For example, plants containing pyrrolizidine alkaloids are well-accepted by sheep and goats (Fig. 4–5), which are resistant to the alkaloids, but very unpalatable to cattle and horses, which are susceptible. Provenza (1995) has described mechanisms by which **linkage of detoxification ability and feeding behavior** could evolve. Toxins produce unpleasant sensations when consumed (for lack of a better description, animals feel sick after consuming a toxin that has caused pathologic or pharmacologic effects). Animals can associate these sensations with the novel feed recently consumed, and develop **aversion** to it (Pfister *et al.*, 1990). If, on the other hand, toxic reactions do not occur because of detoxifying ability, aversive feed-back is not generated,

FIGURE 4–5 Goats readily consume *Senecio* species such as tansy ragwort, showing a linkage of detoxifying ability and feeding behavior.

and the animal may continue to feed on the novel plant. This learned response can be passed on to the offspring, which acquire much of their feeding behavior in a learning process from their dams (Provenza, 1995). Assuming that taste responses are heritable, a willingness to consume an "unpleasant" toxic plant could on an evolutionary basis lead to selection for a feeding strategy which encompasses plants with particular types of toxins. For example, species such as sheep and goats with the ability to detoxify pyrrolizidine alkaloids do not have aversive taste responses to plants containing these alkaloids. Thus there appears to be a genetic linkage of hepatic detoxification activity and feeding behavior that developed as browsing animals coevolved with herbaceous plants. In insects, this interaction between detoxification and feeding behavior leads to the evolution of **specialist feeders**, which consume only plants containing particular toxic chemicals. This is less common in large animals, but a good example is the koala, which has evolved hepatic activity for detoxifying *Eucalyptus* toxins. Koalas have a specialist feeding behavior, and restrict their diet entirely to a few *Eucalyptus* spp. containing the toxins to which they are resistant. Another example is the pygmy rabbit, which has an obligate dietary relationship with terpene-containing sagebrush (White *et al.*, 1982). Drug-metabolizing enzymes in **respiratory tract** tissue (Dahl and Hadley, 1991; Longo *et al.*, 1991) might have a role in palatability responses to toxic plants. Thus a linkage between toxin susceptibility and feeding behavior might involve detoxification activity at the site of food prehension. This may be the case with the pygmy rabbit, because most of the terpenoids in sagebrush are volatilized during mastication. Characteristic **odors** of poisonous plants may serve a warning function to herbivores (Camazine, 1985). It is possible that nasal tissue cytochrome P450 activity could provide a mechanism for animals resistant to a particular toxic plant to overcome the offensive nature of its odor. For example, *Senecio jacobaea* has a very offensive odor to humans but it is readily consumed by sheep and goats which are resistant to pyrrolizidine alkaloid toxicity.

Another way that herbivores could minimize the effects of plant toxins is by consuming a **generalized diet** rather than feeding exclusively on just a few species. Livestock do consume a generalized diet, in contrast to species such as the koala which feeds exclusively on a few species of eucalyptus. There is evidence that livestock can detect and avoid most plant toxins (Provenza, 1995). Most poisonous plants are bitter and not generally consumed when other forage is available in adequate quantities. This is not invariably true; some toxic plants, such as *Leucaena leucocephala*, *Delphinium* spp., and fluoroacetate-containing plants are highly palatable. For those forage species that do contain toxicants, livestock show a preference for nontoxic or less toxic strains (Laycock, 1978).

ECOLOGICAL ROLES OF PLANT TOXINS

Interactions between plant toxins and herbivores have many interesting consequences. Specialist feeding, as mentioned above for the koala and pygmy rabbit, is an example. Another is the role of plant toxins in population cycles and distribution of herbivores. Many plants respond to herbivory by synthesizing increased concentrations of defensive chemicals. When aspen trees are cut down by beavers, the new shoots emerging from the stumps are rich in phenolic compounds, which the beavers then avoid (Basey *et al.*, 1988). Rather than clear-cutting forests, beavers select trees with low concentrations of phenolics, and trees they have cut can regenerate if they increase their chemical protection. Eventually, when most of the trees are well defended chemically, the animals move to a new habitat, beginning the cycle again. The current annual growth of many woody species contains higher concentrations of defensive chemicals than does older growth, which is reflected in the pattern of herbivory (Bryant *et al.*, 1992).

The snowshoe hare in arctic regions has about a ten year **population cycle**, in which populations build up to a peak, and then "crash". Bryant *et al.* (1991) reviewed extensive studies which demonstrate that browsing by hares on the bark of birches, willows and other woody plants results in the regrowth of shoots having high concentrations of defensive chemicals, including phenolics which reduce protein digestibility and palatability. When hare populations approach their peak, the woody plant material has become highly unpalatable and indigestible, leading to a dramatic winter population crash of hares. With reduced herbivory pressure following the decline in hare numbers, the plants reduce their synthesis of toxins, allowing surviving hares to begin the population cycle anew. The interaction is actually more complicated than this (Krebs *et al.*, 1995), with a three way interaction involving hare populations, predation, and food availability (plant chemistry). Similarly, population cycles in lemmings may be associated with grazing-induced changes in protease inhibitors in plants (Seldahl *et al.*, 1994). Induction of plant chemicals by herbivory has been comprehensively reviewed by Tallamy and Raupp (1991).

Substances in plants may function in regulating reproduction in small herbivorous mammals, such as the meadow vole and the lemming, by providing dietary cues as to when food supplies are adequate for reproduction. Berger *et al.* (1977) found that cinnamic acids and their related vinyl phenols inhibited reproductive function in the meadow vole (*Microtus montanus*). When fed these compounds, the animals exhibited decreased uterine weight, inhibiton of follicular development, and a cessation of breeding activity. The compounds were at highest concentration in native plants at the end of the vegetative growing season, possibly providing a dietary cue to turn off reproduction. Later studies by these workers (Sanders *et al.*, 1981; Berger *et al.*, 1981) demonstrated that a plant-derived cyclic carbamate, **6-methoxybenzoxazolinone (6-MBOA)**, stimulates reproductive activity in meadow voles. They suggest that 6-MBOA may trigger reproductive activity in the spring when food supplies are abundant. The significance, if any, of these compounds in livestock production is not known. However, if late-season forage contains phenolics that inhibit reproductive processes and spring forage has reproduction-stimulating compounds, there would be potential implications for livestock production. 6-MBOA has a structural resemblance to melatonin (Fig. 4–6), a hormone whose activity is related to light exposure and prolactin secretion. Melatonin is produced in the pineal gland and may have a role in the regulation of seasonal breeding. Thus, 6-MBOA may exert its effects through an interaction with melatonin.

A controversial area in plant–herbivore ecology is whether or not there is **chemical communication** among plants. The term "talking

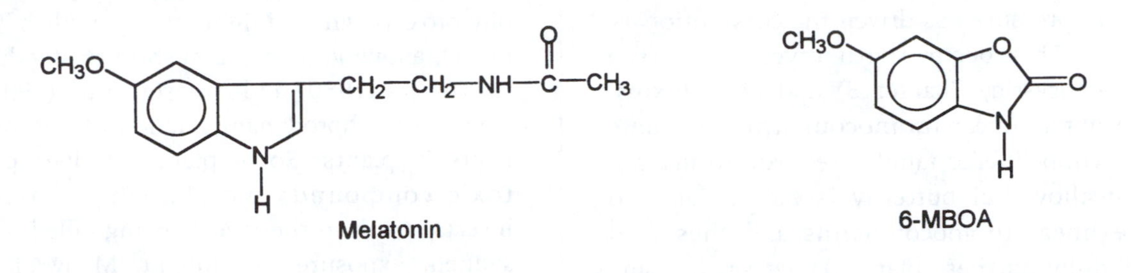

FIGURE 4–6 Structures of melatonin and 6-methoxybenzoxazolinone (6-MBOA), illustrating structural similarity.

trees" has been associated with this controversy (Fowler and Lawton, 1985). Baldwin and Schultz (1983) reported that when trees are experiencing herbivory, the chemical defenses of non-attacked adjacent trees are increased, suggesting that herbivory elicits a chemical cue that warns nearby plants of impending attack. A mode of action for this communication has been reviewed (Karban and Myers, 1989). In essence, it is postulated that herbivore damage to plant tissue may release **ethylene**, a volatile phytohormone. Ethylene stimulates synthesis of phenolic compounds. Thus trees exposed to herbivory may release ethylene, which stimulates adjacent trees to increase their synthesis of defensive phenolic compounds. Furstenburg and Van Hoven (1994) in South Africa reported that browsing of *Acacia* trees by giraffe resulted in an increase in the tannin content of the foliage within 2–10 minutes of initiation of browsing, causing the tree's foliage to become less palatable. A mechanism by which such a rapid change in tannin content could occur might be that the quinones produced by the action of **polyphenoloxidases** on phenolics might be more toxic and of greater feeding deterrence than the original phenolics (Lowry *et al.*, 1996). Tissue maceration caused by browsing would activate polyphenoloxidase activity (see Browning Reactions, Chapter 8). Not only did the trees browsed by giraffes show a rise in tannin content and lowered palatability, adjacent non-browsed *Acacia* trees also showed a marked increase in tannin content in the same time period (2–10 minutes). Further work is necessary to conclusively demonstrate chemical communication among plants. There is also evidence that plants attacked by insect herbivores release volatile chemicals that attract insect predators of the herbivores (Dicke *et al.*, 1993). These are among the many fascinating interactions between plants and herbivores.

Symbiotic relationships between plants and fungi also have ecological implications with herbivores. **Endophytes** are fungi which live entirely within plant tissues, in the intercellular space. Endophyte infestation of perennial ryegrass, responsible for ryegrass staggers (Chapter 10), confers protection to the ryegrass from harmful insects such as the Argentine stem weevil. Insect resistance of endophyte-infected plants probably explains why old ryegrass pastures in New Zealand contain a very high incidence of infected plants. The endophyte associated with tall fescue, implicated in fescue-induced disorders such as summer fescue toxicosis (Chapter 10), may also have a role in protection of the fescue from insect damage. These observations add a further dimension to the interrelationships of plant toxicants, mycotoxins, pest resistance, and livestock poisoning.

The continuum of evolutionary changes in synthesis of secondary compounds by plants, followed by the evolution of comparable detoxification mechanisms in animals, has been aptly termed the **coevolutionary arms race** (Berenbaum and Feeny, 1981). The most potent toxins tend to be towards the end of a long biosynthetic route, in which substances of lower toxicity are intermediates. This suggests that selection pressure has driven the coevolutionary process. This occurs with mycotoxins (see sterigmatocystin, Chapter 5) and plant toxins. For example, linear **furanocoumarins** in plants of the Umbelliferae family are toxic to insects. The swallow tail butterfly larvae are able to cleave linear furanocoumarins and thus feed successfully on these plants. However, in some Umbelliferae, the furanocoumarins have shifted to an angular configuration (see Furocoumarins, Chapter 16), which are toxic to the butterfly larvae (Berenbaum and Feeny, 1981). This is an example of coevolution at work: insects have evolved enzymes to detoxify plant chemicals (linear furanocoumarins); plants evolve under this selection pressure to form new defensive chemicals (angular furanocoumarins).

Most of the selection pressure on plants to respond to herbivory is provided by plant eating (phytophagous) insects, and toxic effects on large animals may be incidental. Both physical and chemical defenses are effective. Most insects cannot digest cellulose, so cellulose and lignified cell walls act as a broad-spectrum defense against insects (Hochuli, 1996).

Besides the apparent role of secondary substances in protecting plants against herbivory, some plants produce compounds which protect them from the competition of other plants. Compounds which inhibit the growth of other plants are termed **allelopathic substances**. Juglone secreted from the roots of black walnut trees inhibits the growth of many other plants, including common garden vegetables and fruit trees (Coder, 1983). Diffuse knapweed (*Centaurea diffusa*) is an important range weed in western North America. It contains allelopathic substances (Muir and Majak, 1983) which may aid in its successful invasion of grassland. Black knapweed (*Centaurea nigra*) also has allelopathic effects (Vezina and Doyon, 1983). Nicollier *et al.* (1983) suggest that Johnson grass secretes phytotoxins which inhibit the growth of other plants, allowing it to establish pure stands.

Towers (1980) and Arnason *et al.* (1992) reviewed the photodynamic action of photosensitizers in plants. Some plants produce **phototoxic compounds** which, when eaten by insects, result in the insects being killed by subsequent exposure to sunlight. Many of these compounds are polyacetylene compounds. The roots of marigolds (*Tagetes* spp.) produce thiophene derivatives which have nematocidal activity; their toxic effects on nematodes are tremendously enhanced by light. Some insect species overcome phototoxic plant defenses by rolling the leaf around them, protecting themselves from light.

Another aspect of coevolution of plants and herbivores is that in some instances, herbivore predation has beneficial effects on plants. McNaughton (1979) has discussed interrelationships between plants and wild ruminants on the Serengeti Plain of East Africa. Some plant species in this community require animal grazing for their survival. Sod-forming species with a prostrate growth form depend on heavy grazing pressure on more upright species to prevent being crowded out. In pasture management, it is well known that close grazing is required to maintain white clover in swards; with low grazing intensity, grasses become dominant. Herbivores have been suggested to have a direct stimulatory effect on grass productivity arising from plant growth–promoting agents in ruminant saliva (Reardon *et al.*, 1972). It is probable that any effect of saliva on plants is slight. Belsky (1986) in a review article concluded that there is no convincing evidence that herbivory benefits grazed plants. There is considerable controversy on this issue, which has relevance to disputes dealing with the impacts of livestock production on grassland ecosystems (rangelands) and whether or not grasslands "need" to be grazed by large herbivores.

IMPACTS OF TOXICANTS IN FEEDSTUFFS

The impact and significance of toxicants on livestock and humans depend on a variety of factors. Intentional or coincidental changes in these factors can result in a previously minor problem becoming a major one or conversely can eliminate a toxicity problem. A few examples will illustrate these points. In the Middle Ages, **ergotism** was a major public health problem in Europe, caused by ergot infection of the rye grain grown for making bread. Improved agricultural practices led to a marked decline in ergot in grain. The introduction of deep plowing, with the invention of the moldboard plow, resulted in fungal spores being buried too deep for them to germinate. The new cultivation methods, which also improved soil drainage, allowed wheat to grow where previously only rye could be grown as a grain crop. Wheat is much less susceptible than rye to ergot infestation. Thus these agronomic developments had far-reaching impacts on human health in markedly reducing ergotism. In the pioneer days of the U.S., a disease called **milksickness** decimated villages and was a major public health problem. The disease was caused by a toxin secreted in the milk of cows consuming the weed white snakeroot. Most families had their own cow, which frequently grazed in waste areas, roadsides, or stream banks, where the white snakeroot grew. With greater intensification of the dairy industry, the problem was virtually eliminated by changes in grazing management and the use of improved pastures. Interestingly, the problem of white snakeroot toxicity has reappeared because of the current interest in small-scale, part-time farming with its use of less intensive production systems and weed control programs than are common in commercial agriculture (Stotts, 1984).

Geographic differences in toxicity problems may be associated with cropping or pasture management systems. Perennial ryegrass staggers and annual ryegrass toxicity are significant livestock poisoning problems in Australia. In Oregon, ryegrass is extensively grown and grazed by sheep, but poisoning problems are very uncommon. Ryegrass staggers is caused by fungal tremorgens, while annual ryegrass toxic-

FIGURE 4–7 Field burning in western Oregon. Field burning kills the nematodes which in Australia cause annual ryegrass toxicity.

ity is due to a bacterial toxin produced in nematode-infected seed heads. In Oregon, ryegrass is grown as a seed crop, with pasturing by sheep of the vegetative stage during the winter. Therefore, sheep are not exposed to the seed heads except when the grass seed screenings are used. A common practice with Oregon ryegrass fields is burning the stubble to destroy fungi, nematodes, and other pests (Fig. 4–7). The causative agents of the two ryegrass-associated toxicities are kept in check by field burning, while in Australia the fungi, nematodes, and bacteria are not controlled and build up large infestations.

Another example of an agricultural change that has virtually eliminated many toxicity problems in the U.S. is the introduction and almost universal use of **herbicides** to control broadleaved weeds in grain fields (Fig. 4–8). In the past, before herbicides were used, grain fields were frequently heavily contaminated with many weeds, and both human and livestock toxicities occurred because of contamination of the grain with toxic weed seeds. For example, poisoning of livestock due to tarweed, which contains pyrrolizidine alkaloids, and poison hemlock, containing the alkaloid conium, occurred because of grain contamination. In the Pacific Northwest, contamination of grain with vetch seed caused losses of poultry. This type of problem still exists in areas where herbicides are not routinely used. Several outbreaks of pyrrolizidine alkaloid poisoning in Australia occurred in the 1970s due to contamination of grain with crotalaria seeds, which contain pyrrolizidine alkaloids. Hooper (1978) described a situation in which pigs which were fed sorghum grain died from pyrrolizidine alkaloid poisoning. This was shown to be due to *Crotalaria retusa* seeds in the sorghum at a level of about 0.05%, which was equivalent to 1 crotalaria seed per 65,000 sorghum seeds. This batch of sorghum was destined for export for human consumption until the poisoning outbreak in pigs occurred. In modern grain-handling processes, most weed seeds are removed in the cleaning process and constitute the screenings to be fed to livestock. Hence, any toxic weed seeds in the grain become concentrated in the fraction fed to livestock. There is a certain amount of irony associated with efforts in the U.S. to restrict the use of herbicides. There is no known case of injury or death from residues on food, while prior to their use, public health and livestock problems existed because of natural toxins in grain. This situation still exists in many parts of the world. In Afghanistan and India in 1976, epidemics of pyrrolizidine alkaloid poisoning in humans occurred because of *Heliotropium* and *Crotalaria* contamination of wheat used to make bread. Wheat from the affected areas contained about 300 mg of *Heliotropium* seeds per kilogram of wheat. Chronic contamination of foodstuffs has been a problem in South Africa, with *Senecio* and *Crotalaria* the major offending species.

FIGURE 4–8 The use of herbicides in crop production has markedly reduced livestock poisoning problems caused by contamination of grains with seeds of poisonous plants.

The intentional consumption of toxic plants sometimes occurs without the intent of self-poisoning. "Bush tea" is used in a number of tropical areas, such as in the Caribbean region. A variety of wild leaves is picked and brewed into a tea for medicinal purposes. Frequently the leaves contain toxins, such as pyrrolizidine alkaloids that cause irreversible liver damage. Two current trends in the U.S. that increase human exposure to toxicants are the use of herbal teas and "foraging," or collecting edible wild plants.

Herbal teas have been implicated in several human health problems. Huxtable (1989) cited several examples. In Arizona and other parts of the Southwest, an herbal tea, called gordolobo yerba, is widely used for medicinal purposes by Mexican–Americans. Several mortalities in children administered this tea have been reported and are due to the presence in the tea of *Senecio longilobus*, a plant containing a very high concentration of highly toxic pyrrolizidine alkaloids. Comfrey (*Symphytum officinale*) is widely used to make tea; it is of interest that comfrey contains at least eight pyrrolizidine alkaloids (Culvenor *et al.*, 1980) and both the leaves and roots have been shown to be carcinogenic (Hirono *et al.*, 1978). Huxtable (1979) reports a case of two people who thought they were collecting comfrey when in fact they collected *Digitalis* (foxglove) and died from drinking a tea prepared from it. Another case cited by Huxtable was the death of a person drinking *Datura* (jimsonweed) tea on a desert survival exercise. The practice of **foraging** for edible wild plants may increase exposure to toxicants. The fiddleheads of bracken fern are an example of a plant frequently collected. Bracken fern, especially in the young frond stage, is carcinogenic, has caused many tumors in livestock, and is suspected as a cause of human cancer in Japan (Hirono, 1981). Finally, there is the deliberate ingestion of poisonous plants to experience a disturbance of the central nervous system. Many deaths have occurred from the ingestion of poisonous wild mushrooms that were picked for their hoped-for hallucinogenic properties. *Datura* seeds are sometimes chewed for their mind-altering effects.

The recognition of toxicants in some feeds has allowed an easy solution of an existing problem by changes in feed processing procedures. Raw soybeans or soybean meal are unsatisfactory as poultry or swine feeds unless they are heat treated because of the presence of heat-labile toxins such as trypsin inhibitors and lectins. Routine heat treatment of these products prevents any problems due to these factors. The widespread use of soybean meal in the 1950s resulted in a high incidence of parakeratosis in swine. This was found to be due to dietary zinc being rendered unavailable by the high phytic acid content of soybeans. The solution was to increase the level of supplementary zinc used with soybean meal–containing diets. The effects of gossypol in cottonseed meal can be overcome by the addition of iron salts during processing.

Another approach to the problems caused by toxicants in feeds is the modification of crops and forages by plant breeding. Perhaps the best example of this is the work by Canadian scientists to improve rapeseed meal. They have produced a marked drop in the glucosinolate and erucic acid contents of rapeseed to virtually eliminate toxicity problems associated with the feeding of rapeseed meal to livestock. Other plant-breeding developments include low gossypol cotton, low saponin alfalfa, and low alkaloid reed canarygrass. A case where plant breeding has had adverse effects is the development of bird-resistant sorghum which reduces wild bird damage to sorghum. However, the bird resistance is due to an increase in the tannin content of the grain, which markedly lowered its nutritional value. Plant breeders have replaced these high-tannin types with varieties having a lower tannin content but still with some bird resistance.

While problems with many plant toxicants are less now than in the past because of recognition of the cause, changes in agronomic practices, plant breeding, feed processing, and so on, problems caused by mycotoxins are in many situations becoming of increasing importance. Factors responsible for this include continuous cropping with one crop (monoculture) which allows fungal populations to build up, increased corn and soybean production in the U.S. south and southeast where the climate is favorable to fungal growth, and greater recognition of and concern for the effects of mycotoxins. Higher energy costs may prompt farmers to alter crop-drying procedures and store corn at a higher moisture content, allowing fungal growth during storage. Mycotoxin problems also show a marked seasonal variation. Epidemic years, such as 1977 when the Food and Drug Administration (FDA)–approved levels of aflatoxin contamination had to be raised because much of the U.S. corn crop exceeded the tolerated aflatoxin level, are due to climatic factors such as drought, predisposing the stressed plant to fungal infection. The 1993 flooding of the U.S. cornbelt also caused extensive problems with mycotoxins.

In tropical areas of the world, mycotoxins in grains, protein concentrates, and other feedstuffs are a major problem. Warm humid conditions favor fungal growth, and farming practices in many tropical areas are not sophisticated. Crop storage conditions are frequently inadequate. Epidemiological studies in Africa and Asia have shown a strong positive association between liver cancer rates in humans and dietary aflatoxin intake (Eaton and Groopman, 1994), providing circumstantial evidence of a causal relationship between aflatoxin intake and liver cancer incidence. Aflatoxin may potentiate the effects of hepatitis B in causing liver cancer.

IMPACTS OF POISONOUS PLANTS ON LIVESTOCK PRODUCTION

The economic impact of livestock losses from poisonous plants is very difficult to assess. In range areas, with extensive grazing, losses may be unnoticed, or the cause of death of a dead animal may be unknown. If the cause is known by the rancher, the loss will be unreported unless it is of such major proportion that professional help is solicited. Losses are not merely those of dead animals. Losses from poisonous plants include decreased animal productivity, decreased resistance to other stresses, abortions, birth defects and other reproductive problems, and the cost of poisonous plant control, which may include spraying, reseeding, fencing, increased herding, and abandoning of certain ranges. James *et al.* (1992) estimated that the economic loss to livestock producers from poisonous plants in the western U.S. exceeds $300 million annually.

Correct **grazing management** of livestock can control many poisonous plant problems by grazing particular species when their toxicity is lowest, by using certain ranges when the toxic plants they contain are unpalatable and will not likely be grazed, by avoiding the herding of sheep or trailing of cattle through areas infested with poisonous plants, and by avoiding certain plants when livestock are physiologically most susceptible. This latter consideration is of particular relevance with plants that contain teratogens or abortifacients. Keeler (1978) has provided a good discussion of ways to reduce the incidence of congenital deformities by grazing management. The fetus is susceptible to effects of a teratogen during a fairly well-defined period when a particular tissue is undergoing rapid development. Thus a particular type of deformity would be produced by the maternal ingestion of a plant at a specific period, and that period is not necessarily the same for each class of livestock. The approximate period when teratogens produce certain defects is shown in Table 2–1 (Chapter 2).

A few examples will illustrate the relationships between teratogens, fetal susceptibility, and animal management. Various lupin species on western ranges contain quinolizidine alkaloids (e.g., anagyrine) that have teratogenic effects, producing a condition called **crooked calf disease**. The calves are born with twisted or bowed limbs, spinal curvatures, and cleft palates. The susceptible period of gestation is between 40 and 70 days. The alkaloid content of lupins is very high early in the season and decreases markedly with maturity, except for a brief period while mature seeds are in the pods.

Therefore, the likelihood of teratogenicity is highest when young plants or plants in the mature seed stage are grazed by cows whose gestation stage is between 40 and 70 days. By adjusting grazing management accordingly, the teratogenic effects of lupins can be avoided, while the high-quality lupin forage can still be utilized. Poison hemlock (*Conium maculatum*) also causes crooked calf disease when consumed by cows between days 40 and 70 of gestation. However, the piperidine alkaloid content of the plant is highly variable, so there is little likelihood of lowering the dose by selective grazing during a low hazard period. In this case, animal management involves keeping pregnant cows from *Conium* exposure by control of the weed or by fencing off weed patches. Another example of how management and teratogenicity interrelate is the case of *Veratrum californicum* and the cyclops eye condition in lambs. When ewes consume the plant on the fourteenth day of pregnancy, the lambs may be born with gross facial deformities. Veratrum has a restrictive habitat, growing in dense, sharply defined stands in high moist meadows or other moist lightly wooded areas. Sheep ranchers can prevent teratogenicity from veratrum by keeping ewes in areas where the plant does not grow until at least 15 days after the rams have been removed, thus avoiding exposure to the plant during the 1-day period of susceptibility. A final example is the teratogenic effect of locoweed (*Astragalus* spp.). Locoweed produces a variety of deformities in lambs, and the susceptibility periods for each deformity are different. Consumption by ewes of significant amounts of locoweed during almost any stage of gestation may produce birth defects, so there is no safe grazing period for pregnant ewes. The only successful grazing management procedure is to remove sheep from loco-infested ranges, or if this is not feasible, to be prepared to stand a certain amount of loss. Since locoweed is palatable to animals habituated to it, provision of adequate amounts of other pasture or supplementary feed does not prevent losses.

Management to reduce livestock losses from poisonous plants involves both plant management of the range or pasture, and management of the grazing animals. If a pasture contains a high degree of infestation of one or more poisonous plants, thereby posing a significant risk to grazing animals, it is likely that pasture renovation will be economical. On overgrazed ranges where infestation with poisonous plants has occurred, **chemical spraying** and reseeding with desirable vegetation may be warranted. Most poisonous plants are broad-leaved forbs or shrubs and can be killed by phenoxy herbicides without damage to grasses. Caution should be used when pastures and ranges are sprayed for the control of poisonous plants. There is often an increase in palatability of plants following spraying associated with the wilted condition or perhaps an increase in sugars in the sprayed material. Livestock may consume poisonous plants after spraying, whereas previously they did not. It is advisable to remove animals from pastures for 1–2 weeks after spraying for weed control. Williams and James (1983) have reviewed the effects of herbicides on the concentration of toxins in poisonous plants. In some cases, as with *Astragalus* spp. and spring parsley (*Cymopterus watsonii*), spraying with phenoxy herbicides results in a rapid decline in toxicity of the plants, associated with bleaching of the leaves. Within 4–5 weeks of spraying, these plants have lost their toxicity. Conversely, the alkaloid concentration of Barbey larkspur (*Delphinium barbeyi*) is markedly increased for 3 weeks following spraying with Silvex or 2,4,5-T; grazing of treated larkspur should be delayed for at least 4 weeks following spraying. Tansy ragwort (*Senecio jacobaea*) is normally unpalatable to cattle and horses, but, following spraying, may be consumed. Hence, in range and pasture management, consideration should be given to the types of poisonous plants present and the effects of herbicides on their toxicity and on grazing habits of livestock.

In a few cases, biological control of the poisonous plants may be feasible. Tansy ragwort (*S. jacobaea*) has in certain situations been controlled with larvae of the cinnabar moth (*Tyria jacobeae*). St. Johnswort (*Hypericum perforatum*) has been very successfully controlled in California and Oregon by two beetles, *Chrysolina gemellata* and *C. hyperici* (Fig. 4–9). In some cases, it may be feasible to use alternate types of livestock to avoid losses. Sheep can be used to control tansy ragwort in cattle pastures because of their marked resistance to the toxic effects of pyrrolizidine alkaloids. Sheep are less susceptible than cattle to larkspur toxicity, so if other conditions (e.g., type of range and predators) are satisfactory, sheep may be used instead of cattle in areas with larkspur. Sheep and goats can be used to control leafy spurge (*Euphorbia esula*), an important rangeland weed in the western U.S.

FIGURE 4–9 *Chrysolina* beetles are specialist feeders which feed exclusively on St. Johnswort (*Hypericum perforatum*). (Courtesy of S.S. Delfosse)

Poisonous plants are often localized in their distribution, and sometimes a relatively small stand of poisonous plants can have a major effect on livestock because of its location. Chokecherry poisoning often occurs when the plant is located near water sources. Animals may consume chokecherry leaves as they are congregating to drink, and when they drink, the water in the rumen speeds up the hydrolysis of glycosides to yield free cyanide. Larkspur often occurs in patches, and chemical control can be economically advantageous. Control of poisonous plants has numerous advantages besides the obvious one of eliminating losses. There may be greater flexibility in seasonal grazing patterns, and there may be a longer total grazing period. When there are no toxic plants present, it is often feasible to increase the stocking rate, giving a greater economic return from the pasture.

Animal management to avoid losses, as reviewed by Krueger and Sharp (1978), makes use of general characteristics of poisonous plants: they are usually unpalatable, frequently become less toxic as they mature, usually make up a greater proportion of the available forage in the early spring, and become proportionately less abundant as forage species begin growth. Hence, animal management on early spring ranges is especially critical. Schuster (1978) listed the following suggestions for management of livestock to minimize losses from poisonous plants:

1. Learn identification and toxic principles of poisonous plants.
2. Use good grazing management to maintain range in a condition that is not conducive to the development of high densities of poisonous plants.
3. Adjust stocking rates so that animals have ample availability of forage relative to the amount of poisonous plants present.
4. Supplement with salt, minerals, and other nutrients as needed.

5. Avoid grazing livestock in areas where toxic plants are abundant either by herding or fencing off the infested areas.
6. Use a class of livestock not generally poisoned by the plants present.
7. Avoid turning hungry animals on to ranges containing poisonous plants. This is especially pertinent when sheep or cattle are released after being transported.
8. Provide adequate watering facilities to prevent nonselective grazing following water deprivation and subsequent watering.
9. Reduce poisonous plant populations by use of mechanical, chemical, biological, or other control methods.

Hungry animals may eat toxic plants that they would normally avoid. Therefore, when livestock are trailed or shipped during shearing or branding or during other times when they may be hungry or stressed, management is particularly important to avoid exposure to poisonous plants. Sheep are especially likely to consume a toxic dose of the oxalate-containing plant halogeton when they are unloaded hungry and then herded along a road lined with the plant. Sheep can adapt to halogeton; if allowed a few days of exposure to it, a population of oxalate-degrading microorganisms proliferates in the rumen. A management system has been developed in Texas for grazing sheep on ranges heavily infested with sneezeweed (*Helenium hoopesii*). If the range is fenced, a portion of the range can be treated to control sneezeweed and the sheep removed from sneezeweed areas for a few days every 3 weeks. Similarly, if the sheep are under the control of a herder, they can be taken off sneezeweed-infested areas for a few days. Apparently this prevents accumulation of the toxins. In Australia, the darling pea (*Swainsona* spp.) has toxic effects similar to those of locoweed. A high stocking rate of short duration may be used to remove the palatable darling pea before any animals can be permanently affected. If a lower stocking rate is used, certain sheep may tend to concentrate on grazing the *Swainsoma* spp. and develop permanent brain injury.

In some cases, provision of **dietary supplements** to livestock may help to control losses from toxicants. It is often believed, although definitive proof seems to be lacking, that lack of salt and mineral supplements may increase the likelihood of animals consuming poisonous plants. Livestock deficient in phosphorus may develop a depraved appetite (pica) which could result in their consuming toxic plants. Palfrey *et al.* (1967) observed a positive correlation between the incidence of ragwort (*S. jacobaea*) poisoning of cattle and the lack of use of mineral supplements, and suggested that cattle consumed ragwort to obtain the phosphorus lacking in their diet. Another explanation of this finding is that farmers who provide mineral supplements probably also provide superior pasture management. Duby (1975) reported positive field trial results of an "alleviator" of ragwort poisoning, containing a mixture of 19 components, mostly mineral elements. Johnson (1982) examined the effects of a number of these components on ragwort toxicosis in cattle and found no protective activity. Keeler *et al.* (1977) studied the effect of provision of a mineral supplement to cattle on the incidence of lupin-induced fetal malformations and observed no protective effect. Ranchers using the supplement had believed it to be effective, but the apparent protective activity was due to variation in alkaloid content of the lupins and the nature of the management practices used. Protein and mineral supplements in the diet of sheep fed *Astragalus* had no preventative value against the onset and severity of locoweed poisoning (James and Van Kampen, 1974). Similarly, Bachman *et al.* (1992) found that supplementation of animals consuming locoweed with mineral supplements did not reduce toxicity, in contrast to commercial claims made for the products. A similar lack of effect of mineral supplements is noted for larkspur poisoning (Pfister and Manners, 1991). The above examples are indicative of the general failure of dietary supplements to protect against poisonous plant losses. Synthetic antioxidants, which protect against pyrrolizidine alkaloid poisoning in laboratory animals (Garrett and Cheeke, 1984), showed evidence of only slight protective activity when given to horses (Garrett *et al.*, 1984) and cattle (Cheeke *et al.*, 1985) fed tansy ragwort. Positive examples of effectiveness of supplements include the use of supplementary zinc to protect against sporidesmin toxicity (Smith *et al.*, 1978) and lupinosis (Allen and Masters, 1980). Dicalcium phosphate as a supplement may in some cases help to prevent halogeton (oxalate) poisoning in sheep. A dietary supplement, containing synthetic antioxidants and methionine, has been developed to help control sneezeweed poisoning (Kim *et al.*, 1981). An antidote to oak poisoning has also been developed (see Oak Poisoning, Chapter 12).

Adaptation of livestock to poisonous plants prior to exposure is a technique that has sometimes been employed with varying degrees of success. Farmers in South Africa have traditionally predosed cattle orally with blue tulp (*Morea polystachya*) prior to allowing them to graze on tulp-infested pastures. Blue tulp contains cardiac glycosides. Whether or not this predosing is actually useful has not been clearly shown. Strydom and Joubert (1983) compared the toxicity of tulp in cattle that had been predosed with the plant to those that had not and found no influence of the predosing on toxicity. A well-documented example of adaptation is the increased resistance of sheep to halogeton poisoning following prior exposure to oxalate, which increases the population of oxalate-degrading ruminal microbes. In those instances where plants cause irreversible, cumulative damage, as with those containing pyrrolizidine alkaloids, preexposure would, of course, be undesirable.

Toxicity of some plants varies with the site on which they grow. Majak *et al.* (1977) in British Columbia found that timber milk vetch (*Astragalus miser*) in shaded, moist timbered sites was lower in toxin (miserotoxin) than milk vetch growing on open, dry grasslands. Management programs can be devised to account for this type of variation. Soil fertility can influence the concentration of toxicants in plants. In New Zealand, the condensed tannin content of *Lotus pedunculatus* may reach very high levels (8–11% of dry matter), markedly reducing grazing animal productivity. The tannin content is several times higher in lotus grown on acid soils without fertilizer application than when phosphorus and sulfur fertilizers are applied (Barry and Forss, 1983). In contrast, the content of S-methylcysteine sulfoxide (brassica anemia factor) in kale and other forage brassicas is increased following nitrogen and sulfur fertilization (McDonald *et al.*, 1981). The phytoestrogen content of clovers increases with fungal infection of the plants. Because plant toxins are chemical defenses, their concentration is often increased under adverse environmental conditions.

Some plants have properties that make animal management difficult for control of the problem. Water hemlock (*Cicuta* spp.) is of such high toxicity that livestock must be kept from contact with it, either through eradication of the plant or fencing off of infested areas. Larkspurs are quite palatable, while locoweeds may be addictive, making it difficult to control losses when these plants are present.

Control of poisonous plants can sometimes be controversial, depending upon competing interests. In Australia, *Echium plantagineum*, or Paterson's curse, causes pyrrolizidine alkaloid poisoning of cattle and horses. Attempts to introduce biological agents to control echium infestations have been vigorously opposed by beekeepers who fear the loss of a major food source for their bees. Therefore, the desirability of controlling echium is largely a matter of the relative importance of the plant to livestock producers and beekeepers. This particular controversy also illustrates the importance of accurate data on toxicologic properties of poisonous plants. In spite of a long-held belief and clear field evidence that echium is toxic, at the time the controversy on its control erupted, there was little data in the literature showing echium to be toxic to livestock. The situation was further confused because a subsequent extensive feeding trial with sheep showed little evidence of toxicity of echium (Culvenor *et al.* , 1984). Hence, simple feeding trials with poisonous plants, while seemingly mundane to some, can be of critical importance in policy making, and the livestock industry can be negatively affected by the absence of this type of data.

There are possible **interactions** between naturally occurring toxicants. Sheep in Australia may graze on heliotrope for one season and on lupin stubble containing phomopsins the next. Since hepatotoxins are involved in each case, it is likely that the two different toxicants have an additive effect and that sheep with heliotrope-induced liver damage may be more susceptible than others to lupinosis. It is likely that numerous interactions between toxic plants in their effects on livestock may occur that have not yet been identified.

Seawright *et al.* (1972) reported an incident in which sheep were noted to be unusually sensitive to toxicity of carbon tetrachloride, which is used as an anthelmintic. The sheep were on a sparse pasture and had browsed on foliage of *Eucalyptus caleyi. Eucalyptus* spp. have essential oils containing compounds such as cineol, which are cytochrome P450 inducers. Therefore, the increase in toxicity of carbon tetrachloride was probably due to its more rapid metabolism to the toxic metabolites following MFO induction. White *et al.* (1983) found that pretreatment of rats with 10% *Eucalyptus globulus* leaves in the diet prior to oral dosing with *Senecio longilobus* pyrrolizidine alkaloid increased the toxicity of the alkaloid. Hence, exposure of livestock to plants that affect hepatic drug–metabolizing enzyme activity but are not themselves overtly toxic may influence the toxicity of poisonous plants.

EFFECTS OF TOXICANTS ON WILD ANIMALS

Evolutionary adaptation to toxicants appears to have occurred in wild species that have been exposed to particular plants for many generations. Domestic animals have been removed from their native habitat and therefore have had less opportunity to adapt to specific plants. Several examples of adaptation of wild species can be cited. In Western Australia, sheep readily consume plants containing fluoroacetate (*Gastrolobium* and *Oxylobium* spp.); just a few leaves are enough to kill a sheep. Some of the native kangaroos avoid these plants and are not poisoned. Other types of kangaroos, and numerous other marsupials have developed mechanisms to detoxify fluoroacetate (see Chapter 15) and so can tolerate fluoroaceate (1080) levels hundreds of times higher than can be tolerated by the same species native to eastern Australia, where 1080-containing plants do not occur. Native big game animals also appear to be able to safely consume large quantities of poisonous plants native to their habitats, although there are documented cases of poisonings (Fowler, 1983). Locoism has been observed in elk and pronghorn antelope feeding on *Astragalus* spp. and pyrrolizidine alkaloid poisoning has been noted in deer (Fowler, 1983), although Dean and Winward (1974) found that black-tailed deer were resistant to tansy ragwort toxicosis. Wolfe and Lance (1984) described locoweed poisoning in elk in New Mexico. Under normal range conditions, locoweed poisoning was not a significant mortality factor, but could affect population dynamics of elk herds on ranges severely infested by locoweed.

There are a number of behavioral and other adaptations which may account for the general paucity of toxicological problems with poisonous plants in wild animals. Many wild animals have fastidious grazing habits and tend to nibble small quantities of feed from a variety of different plants, minimizing the likelihood of consuming an acutely toxic dose of toxins. Large wild herbivores tend to range over extended areas and are not forced to consume poisonous plants because of lack of other feed, as sometimes occurs with confined domestic animals. Fencing can affect wild species; Fowler (1983) cited an incident in Texas in which the use of woven wire rather than barbed wire fencing resulted in pronghorn antelope being confined to specific ranches rather than being allowed to follow a traditional migratory pattern. During a drought in 1964–1965, a large number of pronghorns on these ranches died from consumption of tarbush (*Flourensia cernua*) because of a lack of other feed. Tarbush contains as-yet unidentified hepatoxic agents (Fredrickson *et al.*, 1994). In South Africa, wild animals confined to game ranches by fencing have suffered mortality from consumption of tannin-containing browse plants which would not be consumed by free-ranging animals (Van Hoven, 1984). Thus, overgrazing, range deterioration, and fencing may enhance the risk of poisoning of both livestock and wild species.

There seems to be little evidence that large wild mammalian herbivores have evolved specific anatomical or metabolic defenses against poisonous plants. Their grazing and browsing behavior probably is the major factor limiting poisoning. The extent of problems associated with natural toxicants in free-ranging wild animals is extremely difficult to assess, but in general it appears that plant poisoning is a minor problem.

REDUCING TOXICANTS BY PLANT BREEDING PROGRAMS

A variety of crops contain toxic factors, and much progress has been made in reducing their significance through plant breeding and selection. A few examples will be given as illustrations. Rapeseed contains glucosinolates and erucic acid which have detrimental effects on livestock. Canadian plant breeders have made great progress in reducing the content of these substances to the extent that "double zero" cultivars are available which are very low in both substances. The meal from these has been called "canola meal" to help overcome resistance in the feed trade to the use of rapeseed meal, which formerly had the reputation of containing toxins. Active plant-breeding programs exist to reduce the level of trypsin inhibitors in triticale, a feed grain with much potential. The trypsin inhibitor content of soybeans has been lowered through plant breeding. Low-gossypol types of cotton have been developed; unfortunately, these are of increased susceptibility to pest damage and so are not grown commercially to a major extent. Plant breeders developed bird-resistant cultivars of sorghum to reduce grain losses to wild birds; however, these varieties are of lower feeding value to livestock because of the deleterious effects of high tannin content. Another example is the potato cultivar "Lenape," which had to be withdrawn from production because of its excessive alkaloid content. The nutritional value of cassava, a tropical feed and forage plant, has been improved by the selection of low cyanogenic glycoside strains.

In the case of forages, potential exists for improvement in their feeding value through selection for reduced toxicant content. Low-saponin alfalfa has been developed which is of improved value for nonruminants. Progress is being made in the development of bloat-free alfalfa by selection for a reduced rate of release of bloat-producing cytoplasmic protein fractions in the rumen. An alternative approach is the incorporation of genes for synthesis of protein precipitants such as phenolics (tannins) into bloat-producing legumes. Since some of the genera that cause bloat do not have tannin-containing species, it is necessary to use mutagenic techniques to attempt to introduce the desired factor. Low-phytoestrogen cultivars of subterranean clover in Australia have reduced the problems of "clover disease" in that country.

Because toxicants in plants may play a role in resistance to pests and diseases, it is not surprising that selection for low toxicant levels may increase susceptibility of plants to insects and diseases. Low-saponin alfalfa is more susceptible to insect damage (Fig. 4–10) than unselected alfalfa. Low-gossypol cotton is of increased susceptibility to pests and diseases. Low-tannin sorghum is susceptible to bird damage. Thus, in some cases, a certain level of toxicants in plants may be necessary so that they can be grown effectively. Plant breeders must balance these requirements with the desirability of reducing toxicant levels.

There is much interest in increased exploitation of tropical forage legumes to improve livestock production in tropical areas. Unfortunately, many of these contain deleterious factors. In some cases, plant breeders have worked on improving agronomic characteristics of new forage legumes, only to find that they are toxic. Australian researchers are now routinely screening new accessions through the use of a rat toxicity test to identify those that are frankly toxic (Aylward *et al.*, 1987). For example, several *Indigofera* species that show promise as tropical legumes are free of indospecine, while others are toxic. With the use of the rat toxicity test, plant breeders can begin their selection of improved agronomic features with nontoxic types rather than using the traditional approach of not discovering if a new strain is toxic until after a great deal of work has been invested in it. *Leucaena* is a tropical forage with a tremendous potential for protein production, but it is toxic because of its mimosine content. Efforts to produce low-mimosine cultivars are in progress. In ruminants, protection against mimosine toxicity has been achieved by introducing mimosine-degrading microbes into the rumen (see Chapter 11).

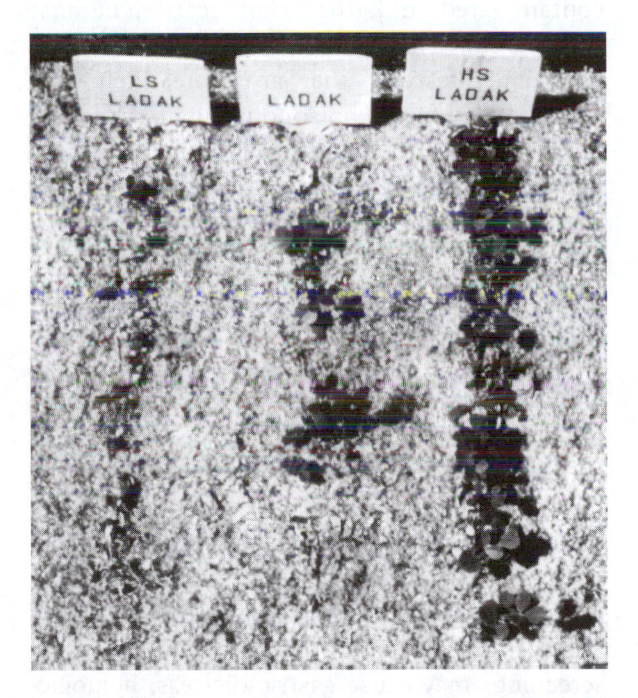

FIGURE 4–10 Alfalfa seedlings of low-saponin (LS ladak), commercial (ladak), and high-saponin (HS ladak) selections of ladak alfalfa. The low-saponin seedlings have been decimated by pea aphids, whereas the high-saponin selection is untouched, suggesting that saponins have a role in protection of plants against insect pests. (Courtesy of E.L. Sorensen)

In the case of range plants, it might appear that there would be little opportunity for genetic modification, but in some cases there may be. Williams and Davis (1982) discussed the use of screening for aliphatic nitro compounds of *Astragalus* spp. before they are introduced into the U.S. from other parts of the world. For example, sicklepod milk vetch (*A. falcatus*) has been introduced to western ranges; it is highly toxic to cattle and sheep. It would be wise in the future to test such plants for toxicants before they are introduced. The U.S. Soil Conservation Service has developed a nontoxic strain of sickle-keeled lupin for seeding on roadsides and rangelands as a replacement for toxic lupin species. In Australia, it has been suggested that it might be feasible to select a low alkaloid type of *Echium* that could be overseeded on areas where the wild plant is widespread.

Thus, there are many opportunities and challenges to reduce the effects of toxicants on livestock through plant modification by plant breeding.

TOXICANT RESIDUES IN ANIMAL PRODUCTS

Toxins consumed by livestock potentially could be transferred to humans by way of consumption of animal products such as meat, milk and eggs. While there are few documented instances where human health has been impacted by toxin residues in animal products, in isolated instances problems have occurred. In most cases, these have involved special circumstances and uncommon or non-conventional agricultural practices.

Milk Transfer of Toxins

Milk contains lipid and aqueous phases, so can be a route of elimination of both water-soluble and fat-soluble toxins and toxin metabolites. The classic example of illness due to a milk-transferred toxin is **milksickness**, the condition in humans resulting from drinking milk from cows that have consumed white snakeroot (see *Eupatorium rugosum*).

The main natural toxin of concern in milk in the U.S. is **aflatoxin**. The hydroxylated metabolite aflatoxin M_1 is the form excreted in milk. It is carcinogenic, and considered an important food chain contaminant. A major source of aflatoxin in milk has been cottonseed and cottonseed meal (Wood, 1992). Milk is monitored by state and/or Food and Drug Administration (FDA) inspectors in the U.S. for aflatoxin contamination.

Other milk-transferred toxins include pyrrolizidine alkaloids and anagyrine, an alkaloid in wild lupins. No human health implications of pyrrolizidine alkaloids in milk have been reported. The proportion of ingested pyrrolizidine alkaloids excreted in milk is low, and is mainly in the form of non-toxic water-soluble metabolites. Thus milk-transferred pyrrolizidine alkaloids are unlikely to be a health hazard.

Meeker and Kilgore (1991) described a case in northern California of skeletal deformity of a baby linked to the consumption of anagyrine-containing goats' milk by the mother during pregnancy. The goats had been grazing the wild lupin *Lupinus latifolius*; the lupins contained high levels of the teratogenic alkaloid anagyrine.

The risk to the population at large from poisonous plant toxins being transferred in milk is very remote. In the modern dairy industry, dairy cattle are managed intensively, and are unlikely to be fed poor quality forages containing poisonous plants. Milk is pooled during processing, so any contaminated milk from one farm would be highly diluted with other milk before it reached consumers. The principal hazard with poisonous plants would be in situations such as with the lupin incident, where a family consumes milk from their own dairy animals. Goats would be a more likely source of contaminated milk than dairy cattle, because they are more likely to be raised in "non-establishment" lifestyles, they are more likely than cows to eat many poisonous plants, and are more resistant to many plant toxins so would likely consume higher quantities of a poisonous plant. For example, they readily consume *Senecio* spp. which contain pyrrolizidine alkaloids.

Glucosinolates from *Brassica* species such as rape and kale can be excreted in the milk of dairy animals. They are unlikely to be a health hazard, but may cause off-flavors in the milk. **Stinkweed** (*Thlaspi arvense*) is a common annual weed of grain crops and waste areas. A member of the Cruciferae family, it contains glucosinolates that are hydrolyzed to yield allylthiocyanate. The seeds of stinkweed (fanweed, field pennycress) in contaminated grain or grain screenings may cause gastric distress, hemoglobinurea (brassica anemia) and tainted milk. The allylthiocyanate in stinkweed is not metabolized by rumen microbes, so may be absorbed and excreted in milk (Majak, 1992). Allylisothiocyanate, or mustard oil, is a more common product of glucosinolate hydrolysis than allylthiocyanate, and is readily metabolized in the rumen (Majak, 1992). Thus many cruciferous weeds and forages do not produce tained milk, even though they contain glucosinolates. The significance of natural toxins in milk has been reviewed by Panter and James (1990).

Natural Toxins in Meat

There is little evidence for and likelihood of toxins from poisonous plants being present in meat in hazardous quantities. One of the rare instances reported of natural toxins in meat is an incident in Australia in which dogs died after being fed horsemeat containing the toxic amino acid indospecine (Hegarty *et al.*, 1988). The horses had been feeding on *Indogofera* spp. and had sufficient indospecine in their muscle tissues to poison dogs, which are highly susceptible to indospecine toxicity.

The growth promotant zearalenol is derived from the *Fusarium* mycotoxin zearalenone. Zearalenone-producing fungi are common in New Zealand pastures (DiMenna *et al.*, 1991). There is concern in New Zealand that detectable residues of zearalenone in meat and milk arising from the natural compound in pastures could affect exports of agricultural products to countries in which use of zearalenol is banned.

POISONOUS PLANTS HAZARDOUS TO HUMANS

Kingsbury (1964, 1980) has summarized much of the scientific knowledge on plants poisonous to humans. Although many common house and ornamental plants are regarded as poisonous, the evidence to support that contention is lacking in many cases. Kingsbury (1980) cites several examples. Holly (*Ilex* spp.) berries are viewed as being toxic to children, although the only recorded reference to *Ilex* toxicity is a second-hand French report from 1889 with no authority stated. This report has been carried forward in books to the present time. Rodrigues *et al.* (1984) presented a case report on two children who consumed a "handful" of holly berries. Moderate to severe gastrointestinal disturbance resulted. Poinsettia (*Euphorbia pulcherrima*) has been reported (Kingsbury, 1964) to have caused the death of a child, whereas feeding experiments (Stone and Collins, 1971) have shown a lack of toxicity. Oehme (1978) dosed rats with several common ornamental plants including chrysanthemum (*Chrysanthemum morifolium*), Jerusalem cherry (*Solanum pseudocapsicum*), dieffenbachia (*Dieffenbachia picta*), geranium (*Pelargonium clonesticum*), and philodendron (*Philodendron cordatum*). The dosage was 5 g of fresh material per kilogram of body weight. Little evidence of toxicity was observed with any of the plants or plant parts. According to Oehme (1978), it is unlikely that adults or children would voluntarily consume these plants at levels equivalent to the dosage used. For example, a 40-lb child would be required to chew and swallow over 100 g of the fresh plant material to take in an equivalent amount.

Some of the plants that have been implicated in human poisoning will be briefly discussed. The majority of the plant intoxications reported at poison control centers in the U.S. result from the ingestion of plants containing gastrointestinal tract irritants. Symptoms range from burning of the mouth and throat because of chewing (usually by toddlers) houseplants such as *Philodendron* spp. and *Diffenbachia* spp. to severe vomiting, intestinal cramping, and diarrhea from consumption of such plant parts as fresh roots and stems of pokeweed (*Phytolacca americana*), *Wisteria* spp. seeds, the berries of *Daphne* spp., or leaves of buttercups (*Ranunculus* spp.).

A number of human fatalities have occurred from the consumption of plants containing cardiac glycosides. These include foxglove (*Digitalis purpurea*), lily of the valley (*Convallaria majalis*), and oleander (*Nerium oleander*). Poisonings have resulted from consumption of the berries, chewing the leaves or flowers, or drinking water from vases containing the flowers of these plants. There is initial irritation to the mouth and vomiting, followed by abdominal pain and diarrhea. The consumption of tea brewed from foxglove leaves, mistaken for comfrey (*Symphytum officinale*), has caused fatalities. The first-year growth of foxglove has a superficial resemblance to a clump of comfrey, a commonly grown herb.

A number of plants contain pyridine and piperidine alkaloids, including wild tobacco (*Nicotiana* spp.) and poison hemlock (*Conium maculatum*). Poisonings result from nibbling on the leaves, using them in "wild salads," or eating the seeds of conium.

Atropine and its related alkaloids have been implicated in a number of poisonings. Jimsonweed (*Datura stramonium*) is the most important source. Contamination of a nationally distributed brand of comfrey tea with an atropine-containing plant(s) caused atropine intoxication in a heavy user of the tea in 1983, resulting in a recall of the tea nationwide. Comfrey contains carcinogenic pyrrolizidine alkaloids; their hazard to users of comfrey products has not been fully assessed.

The seeds of the popular ornamental tree, the horse chestnut (*Aesculus hippocastanum*), are poisonous. They contain a glycoside, aesculin (Williams and Olsen, 1984). The toxic nuts should be collected and removed from areas where children and livestock might gain access to them. Signs of toxicity include depression, incoordination, paralysis, coma, and death. The seeds of Ohio buckeye (*A. glabra*) and yellow buckeye (*A. octandra*) are of much lower toxicity than those of the horse chestnut.

The major source of toxicosis involving convulsions is consumption of the roots of water hemlock (*Cicuta* spp.).

Mushrooms are responsible for a large number of intoxications, particularly following the growth of popularity of hallucinogenic mushroom consumption in the U. S. A discussion of toxic mushrooms is beyond the scope of this book; because of the great similarity in appearance between many edible mushrooms and very deadly ones, the importance of consulting an authoritative source before collecting wild mushrooms cannot be overemphasized.

While this is a brief treatment of common poisonous plants that have been implicated in toxicoses in humans, it is noteworthy that in many instances, particularly for some common ornamental plants widely assumed to be toxic, definitive studies on their toxicity are lacking.

ETHNOBOTANY

Ethnobotany involves the study of the use of plants for food, cultural and medicinal purposes by different ethnic groups, particularly indigenous peoples who have coexisted with plants in a particular environment for many generations. In many cases, these interactions involve toxicants in foodstuffs and knowledge of poisonous and medicinal plants. Many aboriginal peoples have developed knowledge of harmful plants and often have evolved methods of food processing which minimize their toxic effects. Cassava is widely used as a human food in tropical countries. Indigenous peoples developed a variety of processing methods, including pulping or smashing the tubers, soaking the pulped material, and then washing it prior to consumption. This process results in hydrolysis of cyanogenic glycosides and washing out the cyanide. The seeds of *Cycas media*, which contain a carcinogen, have been commonly used as a food by tropical peoples. A variety of processing methods, including fermentation, heating, water extraction, and sun drying, have been used to detoxify the seeds to make a nutritious flour.

Bitter lupin seeds are detoxified by people in the Andes by prolonged extraction in water. The aboriginal people of Australia had an awareness of poisonous plants and developed strategies for avoiding their toxic effects. For example, the seeds of nardoo (*Marsilea drummondii*) were widely used. Nardoo contains potent thiaminase activity, which is destroyed by heat treatment.

Ethnobotany is an active area of scientific investigation, and a journal, *The Journal of Ethnopharmacology* (Elsevier), is devoted to the subject. The history and potentials of ethnobotany have been comprehensively reviewed by Schultes and von Reis (1995). It is a concern that much indigenous knowledge on plant use is being lost as tropical deforestation continues.

REFERENCES

Allen, J.G., and H.G. Masters. 1980. Prevention of lupinosis by the oral administration of zinc sulphate and the effect of such therapy on liver and pancreas zinc and liver copper. Aust. Vet. J. 57:212–215.

Arnason, J.T., B.J.R. Philogene, and G.H.N. Towers. 1992. Phototoxins in plant–insect interactions. pp. 317–341 in: G.A. Rosenthal and M.R. Berenbaum (Eds.). Herbivores. Their Interactions with Secondary Plant Metabolites. Academic Press, San Diego.

Austin, P.J., L.A. Suchar, C.T. Robbins, and A.E. Hagerman. 1989. Tannin-binding proteins in saliva of deer and their absence in saliva of sheep and cattle. J. Chem. Ecol. 15:1335–1347.

Aylward, J.H., R.D. Court, K.P. Haydock, R.W. Strickland, and M.P. Hegarty. 1987. *Indigofera* species with agronomic potential in the tropics. Rat toxicity studies. Aust. J. Agric. Res. 38:177–186.

Bachman, S.E., M.L. Galyean, G.S. Smith, D.M. Hallford, and J.D. Graham. 1992. Early aspects of locoweed toxicosis and evaluation of a mineral supplement or clinoptilolite as dietary treatments. J. Anim. Sci. 70:3125–3132.

Baldwin, I.T., and J.C. Schultz. 1983. Rapid changes in tree leaf chemistry induced by damage: evidence for communication between plants. Science 221:277–279.

Barry, T.N., and D.A. Forss. 1983. The condensed tannin content of vegetative *Lotus pedunculatus*, its regulation by fertilizer application, and effect upon protein solubility. J. Sci. Food Agric. 34:1047–1056.

Basey, J.M., S.H. Jenkins, and P.E. Busher. 1988. Optimal central-place foraging by beavers: tree-size selection in relation to defensive chemicals of quaking aspen. Oecologia 76:278–282.

Belsky, A.J. 1986. Does herbivory benefit plants? A review of the evidence. Am. Nat. 127:870–891.

Berenbaum, M., and P. Feeny. 1981. Toxicity of angular furanocoumarins to swallow tail butterflies: Escalation in a coevolutionary arms race? Science 212:927–929.

Bryant, J.P., P.J. Kuropot, P.B. Reichardt, and T.P. Clausen. 1991. Controls over the allocation of resources by woody plants to chemical antiherbivore defense. pp. 83–102 in: P.T. Palo and C.T. Robbins (Eds.). Plant Defenses Against Mammalian Herbivory. CRC Press, Boca Raton, FL.

Bryant, J.P., P.B. Reichardt, and T.P. Clausen. 1992. Chemically mediated interactions between woody plants and browsing mammals. J. Range Manage. 45:18–24.

Camazine, S. 1985. Olfactory aposematism. Association of food toxicity with naturally occurring odor. J. Chem. Ecol. 11:1289–1295.

Cheeke, P.R. 1994. A review of the functional and evolutionary roles of the liver in the detoxification of poisonous plants, with special reference to pyrrolizidine alkaloids. Vet. Human Toxicol. 36:240–247.

Cheeke, P.R. 1995. Endogenous toxins and mycotoxins in forage grasses and their effects on livestock. J. Anim. Sci. 73:909–918.

Cheeke, P.R., and R.T. Palo. 1995. Plant toxins and mammalian herbivores: co-evolutionary relationships and antinutritional effects. pp. 437–456 in: M. Journet, E. Grenet, M.-H. Farce, M. Theriez, and C. Demarquilly (Eds.). Recent Developments in the Nutrition of Herbivores. Proceedings of the IVth International Symposium on the Nutrition of Herbivores. INRA Editions, Paris.

Cheeke, P.R., M.W. Pederson, and D.C. England. 1978. Responses of rats and swine to alfalfa saponins. Can. J. Anim. Sci. 58:783–789.

Cheeke, P.R., J.A. Schmitz, E.D. Lassen, and E.G. Pearson. 1985. Effects of dietary supplementation with ethoxyquin, magnesium oxide, methionine hydroxy analog, and B vitamins on tansy ragwort (*Senecio jacobaea*) toxicosis in beef cattle. Am. J. Vet. Res. 46:2179–2183.

Coder, K.D. 1983. Seasonal changes of juglone potential in leaves of black walnut (*Juglans nigra* L.). J. Chem. Ecol. 9:1203–1212.

Culvenor, C.C.J., M. Clarke, J.A. Edgar, J.L. Frahn, M.V. Jago, J.E. Peterson, and L.W. Smith. 1980. Structures and toxicity of the alkaloids of Russian comfrey (*Symphytum* x *uplandicum* Nyman), a medicinal herb and item of human diet. Experientia 36:377–379.

Culvenor, C.C.J., M.V. Jago, J.E. Peterson, L.W. Smith, A.L. Payne, D.G. Campbell, J.A. Edgar, and J.L. Frahn. 1984. Toxicity of *Echium plantagineum* (Paterson's Curse). I. Marginal toxic effects in Merino wethers from long-term feeding. Aust. J. Agric. Res. 35:293–304.

Dahl, A.R., and W.M. Hadley. 1991. Nasal cavity enzymes involved in xenobiotic metabolism: effects on the toxicity of inhalants. Crit. Rev. Tox. 21:345–372.

Dean, R.E., and A.H. Winward. 1974. An investigation into the possibility of tansy ragwort poisoning of blacktailed deer. J. Wildl. Dis. 10:166–169.

Dicke, M., P. Van Baarlen, R. Wessels, and H. Dukman. 1993. Herbivory induces systemic production of plant volatiles that attract predators of the herbivore: extraction of endogenous elicitor. J. Chem. Ecol. 19:581–599.

DiMenna, M.E., D.R. Lauren, J.M. Sprosen, and K.S. Maclean. 1991. *Fusarium* and zearalenone on herbage fractions from short and from long pasture. N.Z.J. Agric. Res. 34:445–452.

Duby, G.D. 1975. Tansy ragwort. Mod. Vet. Pract. 56:183–188.

Eaton, D.L., and J.D. Groopman (Eds.). 1994. The Toxicology of Aflatoxins. Human Health, Veterinary and Agricultural Significance. Academic Press, San Diego.

Fowler, M.E. 1983. Plant poisoning in free-living wild animals—A review. J. Wildl. Dis. 19:34–43.

Fowler, S.V., and J.H. Lawton. 1985. Rapidly induced defenses and talking trees; the devil's advocate position. Am. Nat. 126:181–195.

Fredrickson, E., J. Thilsted, R. Estell, and K. Havstad. 1994. Effect of chronic ingestion of tarbush (*Flourensia cernua*) on ewe lambs. Vet. Human Toxicol. 36:409–415.

Furstenburg, D., and W. Van Hoven. 1994. Condensed tannin as anti-defoliate agent against browsing by giraffe (*Giraffa camelopardalis*) in the Kruger National Park. Comp. Biochem. Physiol. 107A:425–431.

Garrett, B.J., and P.R. Cheeke. 1984. Evaluation of amino acids, B vitamins and butylated hydroxyanisole as protective agents against pyrrolizidine alkaloid toxicity in rats. J. Anim. Sci. 58:138–144.

Garrett, B.J., D.W. Holtan, P.R. Cheeke, J.S. Schmitz, and Q.R. Rogers. 1984. Effects of dietary supplementation with butylated hydroxyanisole, cysteine and B vitamins on tansy ragwort (*Senecio jacobaea*) toxicosis in ponies. Am. J. Vet. Res. 45:459–464.

Hagerman, A.E., and C.T. Robbins. 1993. Specificity of tannin-binding salivary proteins relative to diet selection by mammals. Can. J. Zool. 71:628–633.

Harborne, J.B. 1993. Introduction to Ecological Biochemistry. Academic Press, San Diego, CA, USA.

Hegarty, M.P., W.R. Kelly, D. McEwen, O.J. Williams, and R. Cameron. 1988. Hepatotoxicity to dogs of horse meat contaminated with indospecine. Aust. Vet. J. 65:337–340.

Hirono, I. 1981. Natural carcinogenic products of plant origin. CRC Crit. Rev. Toxicol. 8:235–276.

Hirono, I., H. Mori, and M. Haga. 1978. Carcinogenic activity of *Symphytum officinale*. J. Natl. Cancer Inst. 61:865–869.

Hochuli, D.F. 1996. The ecology of plant/insect interactions: implications of digestive strategy for feeding by phytophagous insects. OIKOS 75:133–141.

Hooper, P.T. 1978. Pyrrolizidine alkaloid poisoning—pathology with particular reference to differences in animal and plant species. pp. 161–176 in: R.F. Keeler, K.R. Van Kampen, and L.R. James (Eds.). Effects of Poisonous Plants on Livestock. Academic Press, NY.

Huxtable, R.J. 1989. Human health implications of pyrrolizidine alkaloids and herbs containing them. pp. 41–86 in: P.R. Cheeke (Ed.). Toxicants of Plant Origin. Vol. I. Alkaloids. CRC Press, Boca Raton, FL.

James, L.F., D.B. Nielsen, and K.E. Panter. 1992. Impact of poisonous plants on the livestock industry. J. Range Management 45:3–8.

James, L.F., and K.R. Van Kampen. 1974. Effect of protein and mineral supplementation on potential locoweed (*Astragalus* spp.) poisoning in sheep. J. Am. Vet. Med. Assoc. 164:1042–1043.

Johnson, A.E. 1982. Failure of mineral–vitamin supplements to prevent tansy ragwort (*Senecio jacobaea*) toxicosis in cattle. Am. J. Vet. Res. 43:718–723.

Karban, R., and J.H. Myers. 1989. Induced plant responses to herbivory. Ann. Rev. Ecol. Syst. 20:331–348.

Keeler, R.F. 1978. Reducing incidence of plant-caused congenital deformities in livestock by grazing management. J. Range Manage. 31:355–360.

Keeler, R.F., L.F. James, J.L. Shupe, and K.R. Van Kampen. 1977. Lupine-induced crooked calf disease and a management method to reduce incidence. J. Range Manage. 30:97–102.

Kim, H.L., A.C. Anderson, M.K. Terry, and E.M. Bailey. 1981. Protective effect of butylated hydroxyanisole in acute hymenoxon and bitterweed poisoning. Res. Commun. Chem. Pathol. Pharmacol. 33:365–368.

Kingsbury, J.M. 1964. Poisonous Plants of the United States and Canada. Prentice–Hall, Englewood Cliffs, NJ.

Kingsbury, J.M. 1980. Phytotoxicology. pp. 578–590 in: J. Doull, C.D. Klaassen, and M.O. Amdur (Eds.). Casarret and Doull's Toxicology: The Basic Science of Poisons. Macmillan Publishing Co., NY.

Krebs, C.J., S. Boutin, R. Boonstra, A.R.E. Sinclair, J.N.M. Smith, M.R.T. Dale, K. Martin, and R. Turkington. 1995. Impact of food and predation on the snowshoe hare cycle. Science 269:1112–1115.

Krueger, W.C., and L.A. Sharp. 1978. Management approaches to reduce livestock losses from poisonous plants on rangeland. J. Range Manage. 31:347–350.

Laycock, W.A. 1978. Coevolution of poisonous plants and large herbivores on rangelands. J. Range Manage. 31:335–342.

Longo, V., A. Mazzaccaro, F. Naldi, and P.G. Gerasi. 1991. Drug-metabolizing enzymes in liver, olfactory, and respiratory epithelium of cattle. J. Biochem. Toxicol. 6:123–128.

Lowry, J.B., C.S. McSweeney, and B. Palmer. 1996. Changing perceptions of the effect of plant phenolics on nutrient supply in the ruminant. Aust. J. Agric. Res. 47:829–842.

Majak, W. 1992. Stability of allylthiocyanate and allylisothiocyanate in bovine ruminal fluid. Toxicol. Lett. 63:75–78.

Majak, W., P.D. Parkinson, R.J. Williams, N.E. Looney, and A.L. Van Ryswyk. 1977. The effect of light and moisture on Columbia milkvetch toxicity in lodgepole pine forests. J. Range Manage. 30:423–427.

Marten, G.C. 1978. The animal–plant complex in forage-palatability phenomena. J. Anim. Sci. 46:1470–1477.

McArthur, C., C.T. Robbins, A.E. Hagerman, and T.A. Hanley. 1993. Diet selection by a ruminant generalist browser in relation to plant chemistry. Can. J. Zool. 71:2236–2243.

McDonald, R.C., T.R. Manley, T.N. Barry, D.A. Forss, and A.G. Sinclair. 1981. Nutritional evaluation of kale (*Brassica oleracea*) diets. 3. Changes in plant composition induced by soil fertility practices, with special reference to SMCO and glucosinolate concentration. J. Agric. Sci. 97:13–23.

McNaughton, S.J. 1979. Grazing as an optimization process: Grass–ungulate relationships in the Serengeti. Am. Nat. 113:691–703.

McNaughton, S.J., and J.L. Tarrants. 1983. Grass leaf silicification: Natural selection for an inducible defense against herbivores. Proc. Natl. Acad. Sci. U.S.A. 80:790–791.

Meeker, J.E., and W.W. Kilgore. 1991. Investigations of the teratogenic potential of the lupine alkaloid anagyrine. pp. 41–60 in: R.F. Keeler and A.T. Tu. Toxicology of Plant and Fungal Compounds. Vol. 6, pp. 41–60. Marcel Dekker, Inc., NY.

Mole, S., L.G. Butler, and G. Iason. 1990. Defense against dietary tannin in herbivores: a survey for proline rich salivary proteins in mammals. Bioch. Syst. Ecol. 18:287–293.

Muir, A.D., and W. Majak. 1983. Allelopathic potential of diffuse knapweed (*Centaurea diffusa*) extracts. Can. J. Plant Sci. 63:989–996.

Nastis, A.S., and J.C. Malechek. 1981. Digestion and utilization of nutrients in oak browse by goats. J. Anim. Sci. 53:283–290.

Nebert, D.W., and F.J. Gonzalez. 1987. P450 genes: Structure, evolution, and regulation. Ann. Rev. Biochem. 56:945–993.

Nicollier, G.F., D.F. Pope, and A.C. Thompson. 1983. Biological activity of dhurrin and other compounds from Johnson grass (*Sorghum halepense*). J. Agric. Food Chem. 31:744–748.

Oehme, F.W. 1978. The hazard of plant toxicities to the human population. pp. 67–80 in: R.F. Keeler, K.R. Van Kampen, and L.F. James (Eds.). Effects of Poisonous Plants on Livestock. Academic Press, NY.

Palfrey, G.D., K.S. MacLean, and W.M. Langille. 1967. Correlation between incidence of ragwort (*Senecio jacobaea* L.) poisoning and lack of mineral in cattle. Weed Res. 7:171–175.

Palo, R.T., and C.T. Robbins. 1991. Plant Defenses Against Mammalian Herbivory. CRC Press, Inc., Boca Raton, Florida.

Panter, K.E., and L.F. James. 1990. Natural plant toxicants in milk: A review. J. Anim. Sci. 68:892–904.

Pfister, J.A., and G.D. Manners. 1991. Mineral salt supplementation of cattle grazing tall larkspur-infested rangeland during drought. J. Range Manage. 44:105–111.

Pfister, J.A., F.D. Provenza, and G.D. Manners. 1990. Ingestion of tall larkspur by cattle: Separating effects of flavor from postingestive consequences. J. Chem. Ecol. 16:1697–1705.

Pirozynski, K.A., and D.L. Hawksworth. 1988. Coevolution of Fungi with Plants and Animals. Academic Press, Inc., San Diego, CA, USA.

Provenza, F.D. 1995. Postingestive feedback as an elementary determinant of food preference and intake in ruminants. J. Range Mgmt. 48:2–17.

Quirk, M.F., J.J. Bushell, R.J. Jones, R.G. Megarrity, and K.L. Butler. 1988. Live-weight gains on leucaena and native grass pastures after dosing cattle with rumen bacteria capable of degrading DHP, a ruminal metabolite from leucaena. J. Agric. Sci. 111:165–170.

Radcliffe, J.E. 1986. Gorse—A resource for goats? N. Z. J. Exp. Agric. 14:399–410.

Reardon, P.O., C.L. Leinweber, and L.B. Merrill. 1972. The effect of bovine saliva on grasses. J. Anim. Sci. 34:897–898.

Rhoades, D. F. 1979. Evolution of plant chemical defense against herbivores. pp. 4–54 in: G.A. Rosenthal and D.H. Janzen (Eds.). Herbivores: Their Interaction with Secondary Plant Metabolites. Academic Press, NY.

Robbins, C.T., A.E. Hagerman, P.J. Austin, C. Mcarthur, and T.A. Hanley. 1991. Variation in mammalian physiological responses to a condensed tannin and its ecological implications. J. Mamm. 72:480–486.

Robbins, C.T., S. Mole, A.E. Hagerman, and T.R. Hanley. 1987. Role of tannins in defending plants against ruminants: Reduction in dry matter digestion? Ecology 68:1606–1615.

Rodrigues, T.D., P.N. Johnson, and L.P. Jeffrey. 1984. Hollyberry ingestion. Case report. Vet. Hum. Toxicol. 26:157–158.

Rosenthal, G.A., and M.R. Berenbaum. 1991. Herbivores. Their Interactions with Secondary Plant Metabolites. Vol. I. Academic Press, San Diego.

Rosenthal, G.A., and M.R. Berenbaum. 1992. Herbivores. Their Interactions with Secondary Plant Metabolites. Vol. II. Academic Press, San Diego.

Schultes, R.E., and S. von Reis (Eds.). 1995. Ethnobotany. Evolution of a Discipline. Dioscorides Press (an imprint of Timber Press, Inc.) Portland, OR.

Schuster, J.L. 1978. Poisonous plant management problems and control measures on U.S. rangelands. pp. 23–34 in: R.F. Keeler, K.R. Van Kampen, and L.F. James (Eds.). Effects of Poisonous Plants on Livestock. Academic Press, NY.

Seawright, A.A. 1989. Animal Health in Australia, Vol. 2. Chemical and Plant Poisons. Australian Government Publishing Service, Canberra.

Seawright, A.A., D.P. Steele, and R.E. Menrath. 1972. Seasonal variation in hepatic microsomal oxidative metabolism *in vitro* and susceptibility to carbon tetrachloride in a flock of sheep. Aust. Vet. J. 48:488.

Seldal, T., K.-J. Andersen, and G. Hogstedt. 1994. Grazing-induced proteinase inhibitors: a possible cause for lemming population cycles. Oikos 70:3–11.

Smith, B.L., B.D. Coe, and P.P. Embling. 1978. Protective effect of zinc sulphate in a natural facial eczema outbreak in dairy cows. N. Z. Vet. J. 26:314–315.

Smith, G.S. 1992. Toxification and detoxification of plant compounds by ruminants: an overview. J. Range Manage. 45:25–30.

Spencer, K.C. 1988. Chemical Mediation of Coevolution. Academic Press, Inc., San Diego, CA, USA.

Stone, R.P., and W.J. Collins. 1971. *Euphorbia pulcherrima*: Toxicity to rats. Toxicon 9:301–302.

Stotts, R. 1984. White snakeroot toxicity in dairy cattle. Vet. Med. Small Anim. Clin. 79:118–120.

Strydom, J.A., and J.P.J. Joubert. 1983. The effect of predosing *Homeria pallida* Bak. to cattle to prevent tulp poisoning. J. S. Afr. Vet. Assoc. 54:201–203.

Tallamy, D.W., and M.J. Raupp (Eds.). 1991. Phytochemical Induction by Herbivores. John Wiley and Sons, Inc., New York.

Towers, G.N.N. 1980. Photosensitizers from plants and their photodynamic action. Prog. Phytochem. 6:183–202.

Van Etten, C.H., and H.L. Tookey. 1979. Chemistry and biological effects of glucosinolates. pp. 471–500 in: G.A. Rosenthal and D.H. Janzen (Eds.). Herbivores: Their Interaction with Secondary Plant Metabolites. Academic Press, NY.

Van Hoven, W. 1984. Tannins and digestibility in greater kudu. Can. J. Anim. Sci. 64:(Suppl. 1), 177–178.

Vezina, L., and D. Doyon. 1983. A note on the inhibitor effect of black knapweed (*Centaurea nigra*) residues. Phytoprotection 64:77–81.

White, S.M., B.L. Welch, and J.T. Flinders. 1982. Monoterpenoid content of pygmy rabbit stomach ingesta. J. Range Manage. 35:107–109.

Williams, M.C., and A.M. Davis. 1982. Nitro compounds in introduced *Astragalus* species. J. Range Manage. 35:113–115.

Williams, M.C., and L.F. James. 1983. Effects of herbicides on the concentration of poisonous compounds in plants: A review. Am. J. Vet. Res. 44:2420–2422.

Williams, M.C., and J.D. Olsen. 1984. Toxicity of seeds of three *Aesculus* spp. to chicks and hamsters. Am. J. Vet. Res. 45:539–542.

Wolfe, G.J., and W.R. Lance. 1984. Locoweed poisoning in a northern New Mexico elk herd. J. Range Manage. 37:59–63.

Wood, G.E. 1992. Mycotoxins in foods and feeds in the United States. J. Anim. Sci. 70:3941–3949.

PART II

Toxicants in Animal Feeds

CHAPTER 5

Mycotoxins in Cereal Grains and Protein Supplements

The major cereal grains are, in order of global importance, wheat, rice, corn (maize), barley, oats, sorghum, millet and triticale. **Cereal grains** are the seeds of cultivated grasses. Cereal grains were developed by our ancestors in the Middle East (most grains) and the Americas (maize). Seeds of wild grasses (e.g. wheat grasses) were collected for food, and presumably grasses with particularly large or tasty seeds were selected for planting as agriculture developed. The transition from hunting–gathering to sedentary agriculture in the Middle East about 10,000 years ago occurred when wheat and barley were sufficiently improved to provide enough nutritious seed for storage to last throughout the year.

Because grasses in general are poorly equipped with chemical defenses (see Chapter 4), the grass seeds that our ancestors chose to domesticate have few natural toxins. It is likely that grass seeds were eaten raw during the hunter–gatherer stage, as people roamed the grasslands, and that there would have been a natural selection against seeds that had a bad taste (i.e. contained toxins) or produced obvious toxic effects. Thus, it is apparent that cereal grains would, by the nature of the domestication process, be relatively free of intrinsic toxins. Most other crops were domesticated more recently, when use of fire (cooking) and other detoxification procedures had been developed. Most of the major food crops other than cereal grains (e.g. beans, potatoes, many vegetables) do contain toxins which must be removed or detoxified by cooking, etc., prior to consumption by humans.

The above comments might suggest that toxicants in cereal grains would be of minor importance in human and animal nutrition. In fact, the opposite is true—by far the most important economic and toxicologic effects of natural toxins are associated with grains. The toxins involved are **mycotoxins**—toxins produced by fungi during growth or storage of grains. In the past, ergotism caused by ergot alkaloids in grain has been one of the major pub-

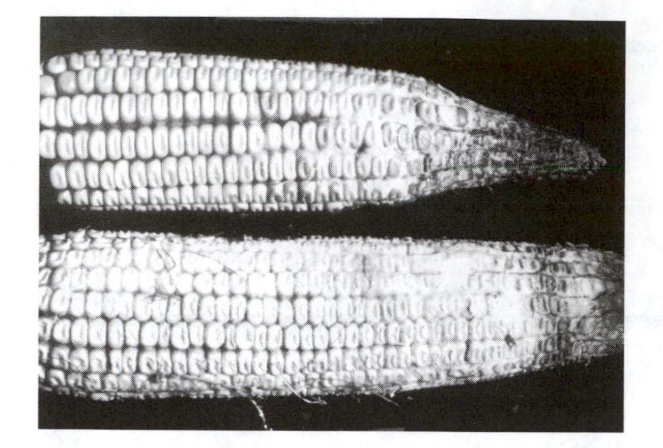

FIGURE 5–1 Mycotoxins produced by molds such as *Fusarium* spp. growing on corn cause major toxicity problems in livestock and poultry production. (Courtesy of E.T. Moran, Jr.)

lic health problems of all time. Currently, contamination of corn and other grains with aflatoxins, fumonisins, zearalenone, trichothecenes, ochratoxin, citrinin and many other mycotoxins causes billions of dollars of losses worldwide. These losses include direct loss of grain, and reduced animal performance and mortalities due to use of mycotoxin-contaminated feed grains. Effects on human health are less easy to document, but several mycotoxins (e.g. aflatoxins, fumonisins) are carcinogenic. Particularly in the tropical developing countries, because of climatic and economic conditions, mycotoxin contamination of grains is likely involved in many human health problems, including liver and other types of cancer. In the U.S. and other developed countries, mycotoxin contamination of feed grains constitutes one of the most significant problems of animal production, particularly of poultry and swine.

Most of the mycotoxins that may occur in cereal grains are also found in other feedstuffs such as soybeans, soybean meal, cottonseed meal, etc. Thus the discussion of mycotoxins in grains will include other feedstuffs as well.

ERGOT ALKALOIDS

Ergot alkaloids are produced by various fungi. There are two major sources of ergot to which livestock may be exposed: the alkaloids produced by fungi which parasitize the seed heads of numerous grasses and grains, and those produced by endophytes which infect various grasses such as tall fescue. The ergotism associated with infected seeds will be discussed in this chapter, while endophyte-produced alkaloids are presented in Chapter 10. Although fungi are the primary source of ergot alkaloids, they are also synthesized by some plants, mainly of the morning glory family (Wilkinson *et al.*, 1987). The seeds of these plants are sometimes consumed for their ergot-induced mind-altering effects.

The term **ergot** is used in general as a common name for species of the *Claviceps* fungi. These fungi parasitize developing grass seeds, producing an enlarged structure called a sclerotium or ergot body, containing ergot alkaloids. The three most significant *Claviceps* are *C. purpurea, C. paspali,* and *C. cinerea. Claviceps purpurea* grows on rye, wheat, barley, and several wild grasses; *C. paspali* grows on grasses of the *Paspalum* species (e.g., dallisgrass and bahiagrass) and *C. cinerea* parasitizes several other grasses (e.g., tobasagrass). Ergot also specifically refers to the sclerotium formed by *C. purpurea* when it grows on rye (*Secale cereale*). Ergot is used for medicinal purposes such as controlling hemorrhaging at childbirth.

Consumption of ergotized grain was responsible for epidemics of human ergotism in Europe in the Middle Ages. **Ergotism in humans** was a major problem in France from the ninth to the fourteenth centuries. Convulsive ergotism was manifested by itching, numbness, severe muscle cramps, sustained spasms and convulsions, and extreme pain. A foot, leg, or less commonly, an arm would be affected, with the victim experiencing feelings of cold alternating with severe burning sensations (St. Anthony's fire). Numbness and dry gangrene followed, with loss of fingers, hands, feet, and even entire limbs. The epidemics decreased due to changes in farming practices. Wheat replaced rye as the major grain crop and was much less susceptible to ergot infection. The advent of deep plowing resulted in the sclerotia being buried so they did not successfully germinate and form spores.

The hallucinogenic drug **lysergic acid diethylamide (LSD)** is an ergot alkaloid. Hallucinogenic effects are also produced by various other ergot alkaloids, including those in ergotized grain. Matossian (1989) has suggested that the frenzied activities of peasants that culminated in the French Revolution may have been due to the hallucinogenic effects and delirium induced by ergot alkaloids in rye bread, the staple food of the French peasants at that time.

Signs of ergotism in animals are similar to those described for humans. Ergot poisoning of livestock from consumption of infected grain and pasture grasses containing grasses ergotized seed heads is relatively common.

FIGURE 5–2 Ergot bodies in triticale seed. (Courtesy of R.J. Metzger)

Life Cycle of Ergot

The *Claviceps* spp. parasitize the ovary of the developing grass flower. The fungus sends filaments throughout the ovary tissue and prevents the development of the seed. At the tips of the filaments spores are formed which are shed in droplets of sticky exudate, or "honeydew." Insects spread the spores to infect additional grass heads. At the same time as spores are produced, the filaments harden into a pink or purple structure which replaces the grain or grass seed. This hard structure is called the **sclerotium** or ergot and is the poisonous stage of the *Claviceps* life cycle (Fig. 5–2). The sclerotia are either harvested with the seed head or are shed to overwinter and produce spores the following spring. These spores begin the infection of the crop again in the second season.

The extent to which grass or grain is infected by ergot depends on numerous factors, so the severity of infection is of high seasonal variability. A cool damp spring favors germination of sclerotia and, by delaying pollination, extends the period of plant susceptibility to infection. Rye and triticale are more susceptible than other grains because they are cross-pollinators, requiring a longer period to become pollinated than wheat, oats, and barley. Shallow cultivation and seeding leave the sclerotia near the soil surface where they germinate readily. "No-till" farming, and lack of crop rotation increase the likelihood of ergot infection of crops. Ergot infection tends to be greatest around the edges of a field because wild grasses near cultivated ground may harbor spores and infect the grain.

Structure and Metabolism of Ergot Alkaloids

The structures of ergot alkaloids are complex. The four main structural groups are clavine alkaloids, lysergic acids, simple lysergic acid amides and peptide alkaloids. The peptide alkaloids are the most physiologically active. The main alkaloids in ergotized grain are ergopeptide alkaloids such as ergotamine (Fig. 5–3) and ergocristine (Scott *et al.*, 1992; Porter *et al.*, 1987).

There is little information available on the metabolism of ergot alkaloids. They disappear rapidly from blood and tissues with a high first-pass clearance by the liver (Moubarak *et al.*, 1996). In contrast, their physiological effects persist for lengthy periods of time. Their mode of action is largely mediated via inhibitory effects on prolactin secretion by the pituitary, by activating the D2-dopamine receptors in pituitary lactotrophs. This is discussed in more detail under Tall Fescue Toxicosis in Chapter 10. Pharmacokinetic studies indicate hepatic metabolism of ergot alkaloids by cytochrome P450 enzymes (Ball *et al.*, 1996)

Effects of Ergot on Livestock

The general effects of ergot on livestock can be categorized as follows:

1. behavioral effects—convulsions, incoordination, lameness, difficulty in breathing, excessive salivation, and diarrhea;

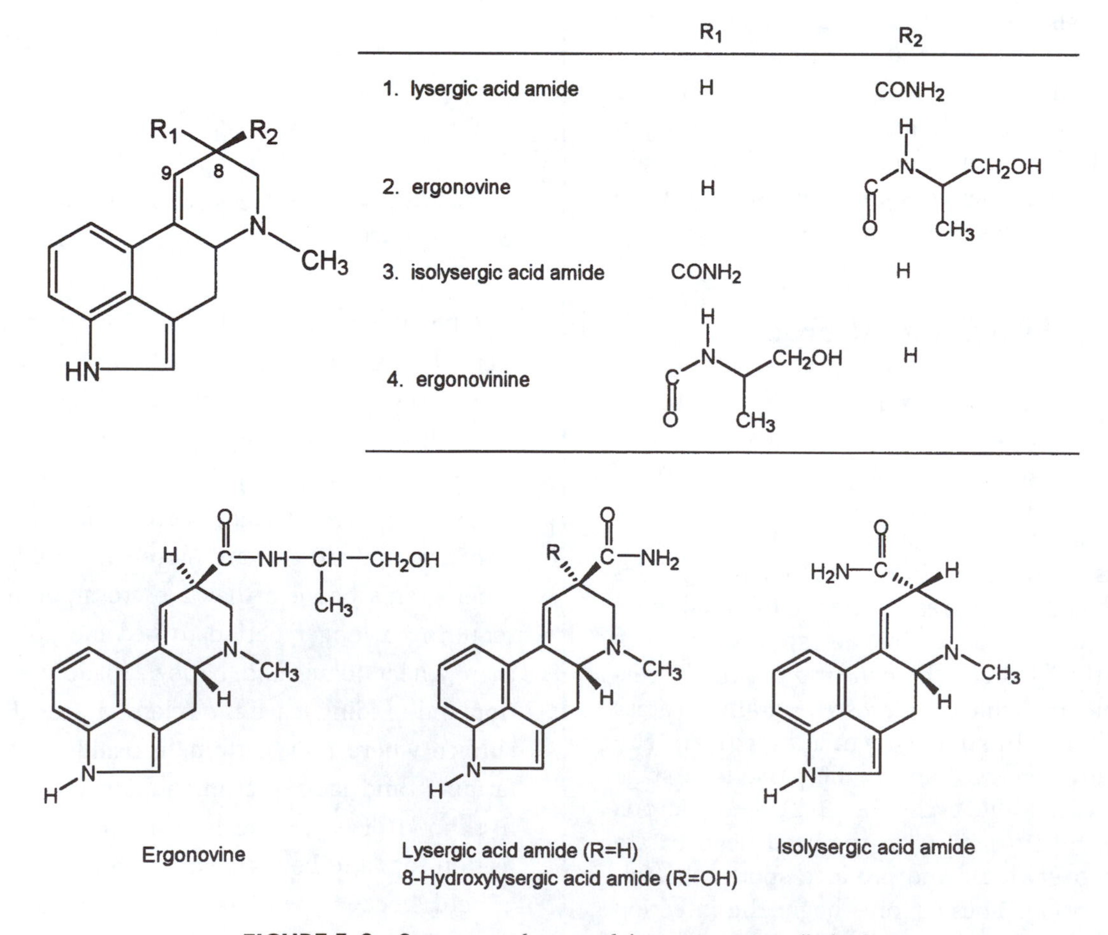

FIGURE 5–3 Structures of some of the major ergot alkaloids.

2. dry gangrene of the extremities;
3. reproductive effects—abortion, high neonatal mortality, reduced lactation; and
4. reduced feed intake and weight gains.

These effects are not all seen in all types of livestock; they are fairly species specific and are modified by the ergot source, amount consumed, period of exposure, and age and stage of production of the animal.

Ergot alkaloids have a direct stimulatory effect on smooth muscle, causing vasoconstriction and elevated blood pressure. During the third trimester of pregnancy, ergot has an oxytocin-like effect; the uterine muscle at this stage is more sensitive to ergot than are other smooth muscles. Ergot alkaloids do not appear to affect intestinal motility (Rotter and Phillips, 1991).

Two general forms of ergotism are the convulsive and gangrenous forms. The **gangrenous form** affects all types of livestock. The extremities (nose, ears, tail, and limbs) are affected due to **vasoconstriction** of arterioles. The early signs are manifested by evidence of pain, such as lameness, and stamping of the feet. The affected areas may feel cool. A sharply defined band encircles the limb, separating the normal tissue from the affected area. Dry gangrene follows, with the affected portion sloughing off, leaving a clean, rapidly healing surface. The digestive tract may also be affected, with inflammation, internal bleeding, vomiting, and constipation or diarrhea. Necrosis of the skin (cutaneous ergotism), resembling photosensitization, occurs with cattle (Coppock *et al.*, 1989).

Cattle may exhibit both convulsive and gangrenous ergotism. The **convulsive form** is primarily with *C. paspali* infection of *Paspalum* spp. grasses (e.g., dallisgrass), and not with *C. purpurea* ergot. The most important ergot problem in livestock in the U.S. is due to cattle consumption of ergotized dallisgrass. The clinical signs are hyperexcitability, belligerency, incoordination, convulsions, and opisthotonus (stargazing posture). Removal of cattle from the affected pasture usually results in recovery in 3–10 days. Gangrenous ergot in cattle is caused by ergot from both *C. paspali* and *C. purpurea*. Gangrene of the ear tips and tail may occur, but generally it is the feet that are affected. Signs include tenderness of the hind feet, with development of gangrene and sloughing of the hooves. There is little effect on reproduction, with abortion and agalactia (which are seen in swine) not observed.

Sheep which consume *C. purpurea* ergot show breathing difficulty, excessive salivation, diarrhea, and internal bleeding within the digestive tract. Sheep tend not to graze on grass flowers and therefore are less affected than cattle due to a difference in grazing behavior. **Horses** grazing grasses infected with *C. paspali* may develop symptoms of convulsive ergotism.

Classical signs of convulsive and gangrenous ergotism are usually not seen in **swine**. Abortions may occur, and newborn pigs have a high rate of mortality due to depressed lactation by affected sows. Swine are less sensitive than other livestock to ergot. Growing pigs fed ergotized grain may have reduced feed intake and lowered rate of gain, with some gut lesions. Mature **poultry** develop comb gangrene as a major symptom. Ergotism in chicks is characterized by depressed growth, poor feathering, nervousness, incoordination and inability to stand, and the beaks, toenails and toes become dark and necrotic (Rotter *et al.*, 1985a,b). Broiler chicks may be slightly more sensitive to ergot than Leghorn chicks (Rotter *et al.*, 1985a).

Hyperthermia (elevated body temperature) and susceptibility to heat stress occur in livestock consuming ergot alkaloids (Ross *et al.*, 1989), similar to signs of tall fescue toxicity (also induced by ergot alkaloids—see Chapter 10). Pronounced depression in serum prolactin levels also occur, accounting for the agalactia syndrome. Ergot alkaloid–induced hyperthermia and hypoprolactemia are discussed in Chapter 10.

Ergot alkaloids are not transferred in the milk of cows consuming ergot.

Treatment and Prevention of Ergotism

Removal of animals from ergot-infected pastures or removal of contaminated grain from feeds is the only effective treatment. Ergot infestation of grain fields can be minimized by using clean seed, crop rotations, and deep cultivation. Growing ergot-resistant grains (wheat, barley, or oats) rather than rye or triticale may be advisable in areas where ergot is a problem. Sclerotia can be removed from grain by standard seed-cleaning techniques. Of course, the screenings from ergotized grain should not be used in feeds. Infected grain can be blended with clean grain to reduce the ergot concentration to a nontoxic level. The **tolerance level for ergot** in grain in the U.S. is 0.3% crude ergot alkaloid. Levels of 0.1% ergot in complete feeds may have adverse effects on livestock performance.

ASPERGILLUS AND *PENICILLIUM* TOXINS

Aflatoxin

Aflatoxins are a family of related bisfuranocoumarin compounds produced primarily by toxigenic strains of *Aspergillus flavus* and *Aspergillus parasiticus*. Only about one-half of the known strains of *A. flavus* and *A. parasiticus* produce aflatoxins. Although other fungi such as *Penicillium* spp., *Rhizopus* spp., *Mucor* spp., and *Streptomyces* spp. are capable of producing aflatoxins, their relevance to livestock production has not been established. The name **aflatoxin** derives from *Aspergillus* (a-), *flavus* (-fla-), and toxin. *Aspergillus flavus* and *A. parasiticus* produce four major toxins: B_1, B_2, G_1, and G_2 (Fig. 5–4). These were named according to their fluorescence properties under shortwave ultraviolet light on thin-layer chromatographic (TLC) plates; B_1 and B_2 fluoresce blue, whereas G_1 and G_2 fluoresce green. Fourteen other aflatoxins are known but most of these are metabolites formed endogenously in animals administered one or more of the four major aflatoxins. Metabolites of toxicological significance include: aflatoxin B_1 8,9-epoxide (AFB_1 8,9-epoxide); aflatoxin M_1 (AFM_1); aflatoxicol; and aflatoxin B_{2a} (AFB_{2a}) (Fig. 5–5). In older literature, the 8,9-epoxide was referred to as 2,3-epoxide.

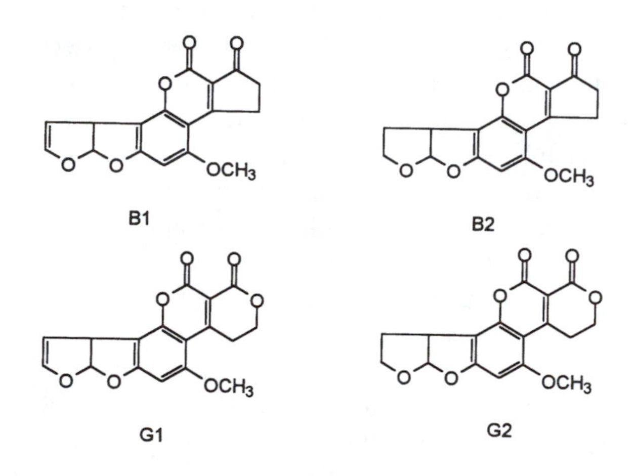

FIGURE 5–4 Structures of the major naturally-occurring aflatoxins.

Aflatoxins were discovered as a result of extensive mortality of turkeys in Great Britain in 1960. Over 100,000 turkeys died in the outbreak, which was called "turkey X disease." Intensive investigation revealed the cause to be mycotoxins in moldy groundnut (peanut) meal that was imported from Brazil for use as a protein supplement in animal diets.[1] A comprehensive review of aflatoxins is provided by Eaton and Groupman (1994).

[1]Retrospective studies suggest that cyclopiazonic acid may also have been involved in turkey X disease—see Cyclopiazonic acid in this chapter.

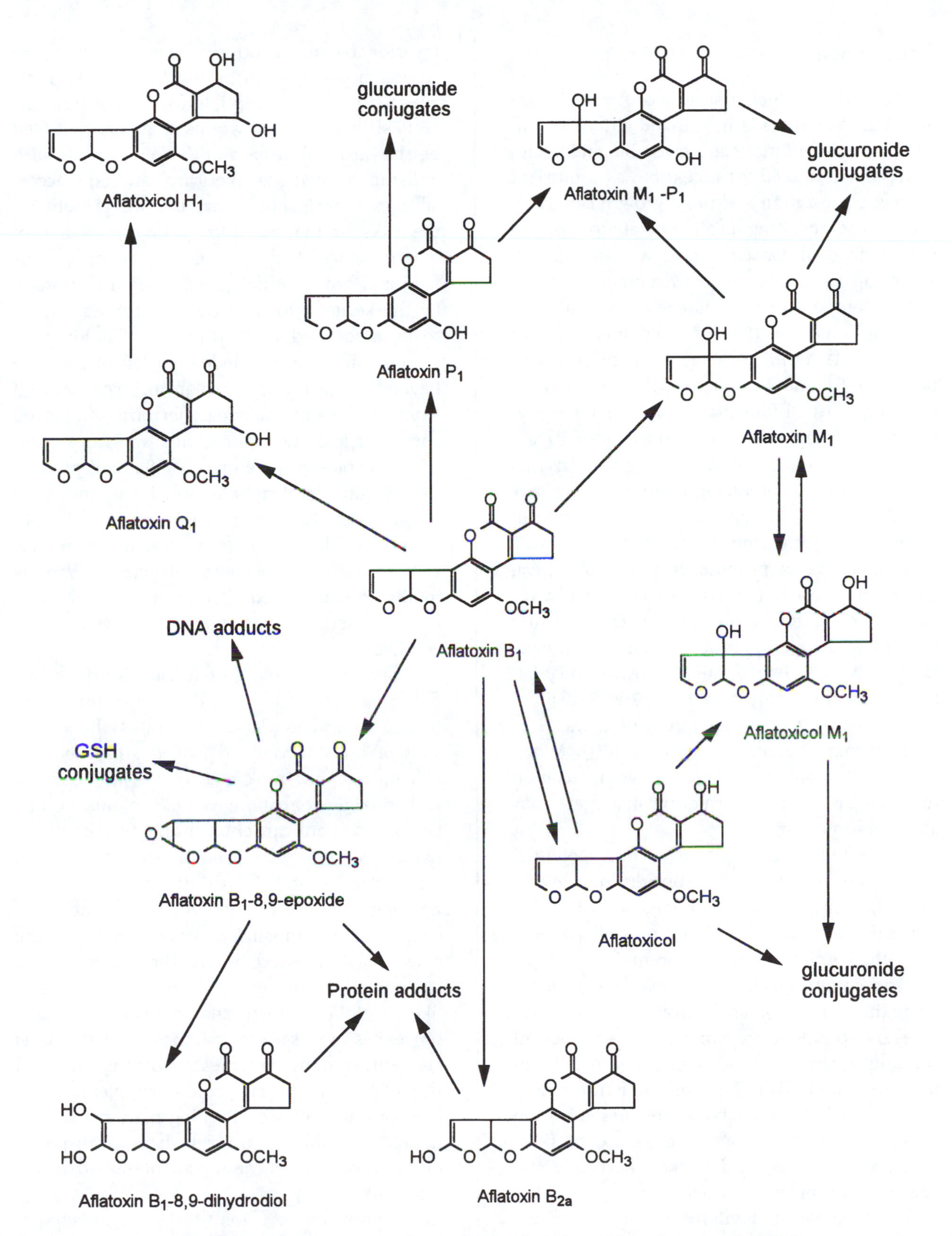

FIGURE 5–5 Metabolic pathways for aflatoxin B_1 (adapted from Eaton *et al.*, 1994).

Occurrence

Aflatoxin-producing strains of *Aspergillus* are distributed worldwide in soil and air. When environmental conditions are favorable and a substrate (feed or seed) is accessible as a nutrient source, colonization and mold growth can easily occur. The resulting profile of aflatoxins and their individual concentrations will vary greatly according to the existing environmental conditions (temperature, moisture, aeration), the substrate, and the type of mold involved. For example, *A. flavus* growing on corn produces primarily B_1 and B_2 whereas *A. parasiticus* on corn produces all four major aflatoxins. On soybeans, only negligible concentrations of B_1 are produced by both species. *Aspergillus flavus* is primarily a seed-colonizing mold and is usually referred to as a storage mold. It is capable of colonizing many important crops including corn, small grains, peanuts, cottonseed, cassava, copra, and most nut crops if moisture and temperature conditions are favorable. The moisture content of the seed is probably the most important factor. In general, mold growth and aflatoxin formation require a moisture content of greater than 14%, a temperature of at least 25° C, and some degree of aeration (O_2). When these requirements are met, mold infestation followed by aflatoxin formation in target crops are likely to occur.

Three major feedstuffs with high potential for invasion by *Aspergillus* spp. during growth, harvest, transportation, or storage are corn, cottonseed, and peanuts. Colonization of soybeans and small grains generally occurs in storage. Storage conditions for soybeans that promote aflatoxin formation, aside from optimal moisture and temperature conditions, are lack of aeration systems or their improper use (temperature differentials can cause moisture migration), kernel damage and spore dissemination caused by storage insects, presence of fines (dust, weed seeds, and broken kernels), and poor sanitary practices in feed areas.

In corn, elevated aflatoxin content in any particular year is usually the result of increased invasion by molds prior to harvest. There are several possible contributing factors. Drought-stressed corn is susceptible to damage from insects such as corn earworms or borers that feed on the husks or kernels of the ear. Kernels with a disrupted seed coat (pericarp) are more accessible to fungal spores that may be present on the silks or that are carried on the bodies of insects. Once in the kernel, the spores germinate and grow, utilizing the nutrients afforded by the kernel. Other factors or stresses known to be associated with increased aflatoxin contamination of corn include corn left in the field beyond maturity, close planting, competition from weeds, and inadequate fertilization. Stored corn, particularly ground, high-moisture corn, has an explosive potential for aflatoxin production. Storage in airtight silos or incorporation of certain preservatives can effectively retard mold growth. Inadequately dried corn not anaerobically stored will probably be invaded by various molds. Prepared feeds left for more than a day or two in feeding bins and troughs are also susceptible.

In cotton, aflatoxin production is primarily a field problem, with insect damage an important factor. *Aspergillus flavus* penetrates the carpel wall of bolls through damaged areas such as exit holes of the pink bollworm. The lygus bug and stinkbug contribute to mold colonization by serving as spore carriers. Chronic field contamination is usually associated with daily mean temperatures of 34°C or greater late in the growing season (July and August in the U.S.) coupled with unusually heavy rainfall. If the cotton is harvested before the moisture has evaporated, aflatoxin production in storage is likely to follow. When aflatoxin-containing cottonseed is processed for oil, most of the toxin is concentrated in the meal. Cottonseed meal (CSM) is a common protein supplement for livestock and poultry. Feeding dairy cattle contaminated CSM is a problem because of possible translocation of the metabolite aflatoxin M_1 into milk.

Peanuts can be colonized by *Aspergillus* spp. in the ground before digging, during curing and drying in windrows or stacks, and in storage. Before digging, fungal invasion has been attributed to drought-induced stress, damaged pods, or overmaturity. After digging, a moisture content of 14–30% is conducive to mold invasion and subsequent aflatoxin formation, but is prevented when the moisture content is very high. The optimum temperature range for aflatoxin production in peanut kernels of undamaged pods is 25°–35°C. When peanuts are being cured, retardation of drying by rainfall or humid weather usually results in some degree of aflatoxin production. A moisture content greater than 30% or less than 10% or a temperature greater than 41°C or less than 12°C restricts growth of mold. Prompt, steady drying to a moisture content of 7–8% within 3–5 days prevents formation of aflatoxins in peanuts. Peanut meal (groundnut meal) used for feed commonly carries large numbers of *Aspergillus* spores. Consequently, under favorable moisture and temperature conditions, spore germination followed by aflatoxin production can easily occur.

Biotransformation of Aflatoxins

The metabolism of AFB_1 has been reviewed by Massey *et al.* (1995) and Eaton *et al.* (1994). Several cytochrome P_{450} subfamilies and specific isoforms metabolize AFB_1. For example, in humans, the subfamily CYP1A, isoform CYP1A2, has a prominent role in detoxification (Massey *et al.*, 1995). This enzyme is induced by polycyclic aromatic hydrocarbons (PAHs). An interesting consequence of PAH induction of CYP1A1 and A2 activity is that in an area of China with major aflatoxin contamination of food, the incidence of hepatoma is lower in smokers than in non-smokers, apparently due to the induction of CYP1A activity by PAH in cigarette smoke (Massey *et al.*, 1995). Some of the major cytochrome P_{450} subfamilies and isoforms involved in AFB_1 metabolism are shown in Table 5–1.

In addition to cytochrome P_{450}, other enzymes are involved in AF metabolism. The epoxide metabolite can be produced by tissue lipoxygenases (lipid hydroperoxide–dependent AFB_1 epoxidation). A major detoxification pathway is conjugation of AFB_1-8,9-epoxide with glutathione, catalyzed by glutathione-S-transferase. Of lesser significance is the activity of epoxide hydrolase, which converts the epoxide to the dihydrodiol (Fig. 5–5). The hydroxylated metabolites such as AFM_1, AFQ_1 and AFP_1 form glucuronides and sulfate conjugates which are excreted.

Species differences in susceptibility to AFB_1 intoxication reflect variations in hepatic enzyme activity, and particularly the cytochrome P_{450} activities.

TABLE 5–1 Cytochrome P_{450} Subfamilies and Isoforms Involved in Metabolism of AFB_1*

P_{450} subfamily	Inducing agent	Isoforms	AFB_1 metabolite
CYP1A	PAH[a]	1A1, 1A2	AFB_1-8,9-epoxide, AFM_1
CYP2B	Phenobarbital	2B1, 2B7	AFB_1-8,9-epoxide
CYP2C	Phenobarbital	2C1, 2C2	AFB_1-8,9-epoxide
CYP3A	Glucocorticoids	3A3, 3A4	AFB_1-8,9-epoxide, AFQ_1

*Adapted from Massey *et al.* (1995).
[a]PAH = polycyclic aromatic hydrocarbons.

Toxicity in Different Species

Acute Intoxication. Acute aflatoxin poisoning of farm animals is less likely to occur than chronic aflatoxicosis. The principal target organ in all species is the liver. Numerous liver functions are affected, and the cumulative impact can be fatal to animals. After administration of a large single dose of AFB_1, hepatocytes undergo progressive changes, which include infiltration with lipids, eventually ending in necrosis (cell death). These toxic effects are believed to be the result of widespread and nonspecific interactions between AFB_1 or its activated metabolites and various cell proteins. Interaction with key enzymes can disrupt basic metabolic processes in the cell, such as carbohydrate or lipid metabolism, and protein synthesis. Modification of permeability characteristics of hepatocytes or subcellular organelles, primarily the mitochondria, contributes to the necrosis. As the liver loses its functionality, other effects appear such as derangement of blood-clotting mechanisms (coagulopathy), icterus (jaundice), and reduction of essential serum proteins which are synthesized in the liver. Impairment of blood clotting and increased capillary fragility result in widespread hemorrhaging, including accumulation of blood in the gastrointestinal tract. In addition to liver damage, higher doses may cause necrosis of kidney tubules in some species. Although the thymus is a target organ, the effects on immunosuppression are more associated with chronic aflatoxicosis than with acute intoxication.

In general, the same biological changes occur in all acutely intoxicated species. However, the susceptibility (toxicity) among different species is highly variable. These differences are believed to be directly related to the ability of animals to metabolize AFB_1 to the reactive metabolite AFB-dihydrodiol via AFB_{2a} or AFB_1 8,9-oxide intermediates (Fig. 5–6). Rabbits and ducks have a high rate of AFB_1 metabolism and are highly sensitive to aflatoxins, whereas sheep and rats have a slower rate of AFB_1 metabolism and are less sensitive (Hsieh *et al.*, 1977). Species with the ability to form greater quantities of aflatoxicol also tend to be more sensitive to acute poisoning. Because the determinants of acute intoxication are specific metabolic pathways and interactions with cellular receptors that are different from chronic aflatoxicosis, the acute LD_{50} is not very useful in predicting actual field contamination problems.

Chronic Intoxication. Chronic poisoning, or **aflatoxicosis**, can result when low levels of toxin are ingested over a prolonged period. The toxic effects are not nearly as specific or clinically evident as in acute intoxication. In general, affected livestock exhibit decreased growth rate, lowered productivity (milk or eggs), and immunosuppression. Also, carcinogenicity has been observed and studied extensively in several nonfarm species. Although both immunosuppression and carcinogenicity are chronic toxicoses, they are discussed separately in following sections.

Reduced growth rate is considered the most common effect associated with chronic aflatoxicosis in farm animals. In young animals fed low levels of toxin, this may be the only detectable abnormality. The lack of other clinical signs frequently causes aflatoxicosis to remain undiagnosed, resulting in serious economic loss. Although the actual mechanism(s) by which aflatoxins decrease growth rate is not known, subtle disturbances of one or more basic metabolic processes (carbohydrate, lipid, or protein metabolism) in the liver are probably involved. Supporting evidence is that aflatoxins increase the dietary requirement for nutrients such as protein. Without the necessary dietary adjustments, suboptimal growth would occur. In addition, a dose-related loss of appetite usually accompanies the decreased growth rate, accounting for at least part of the growth depression.

Liver damage is also prevalent in chronic aflatoxicosis in all species. At necropsy, the liver is usually pale to yellow, and the gall bladder may be enlarged. The occurrence of icterus and hemorrhaging is unpredictable and varies

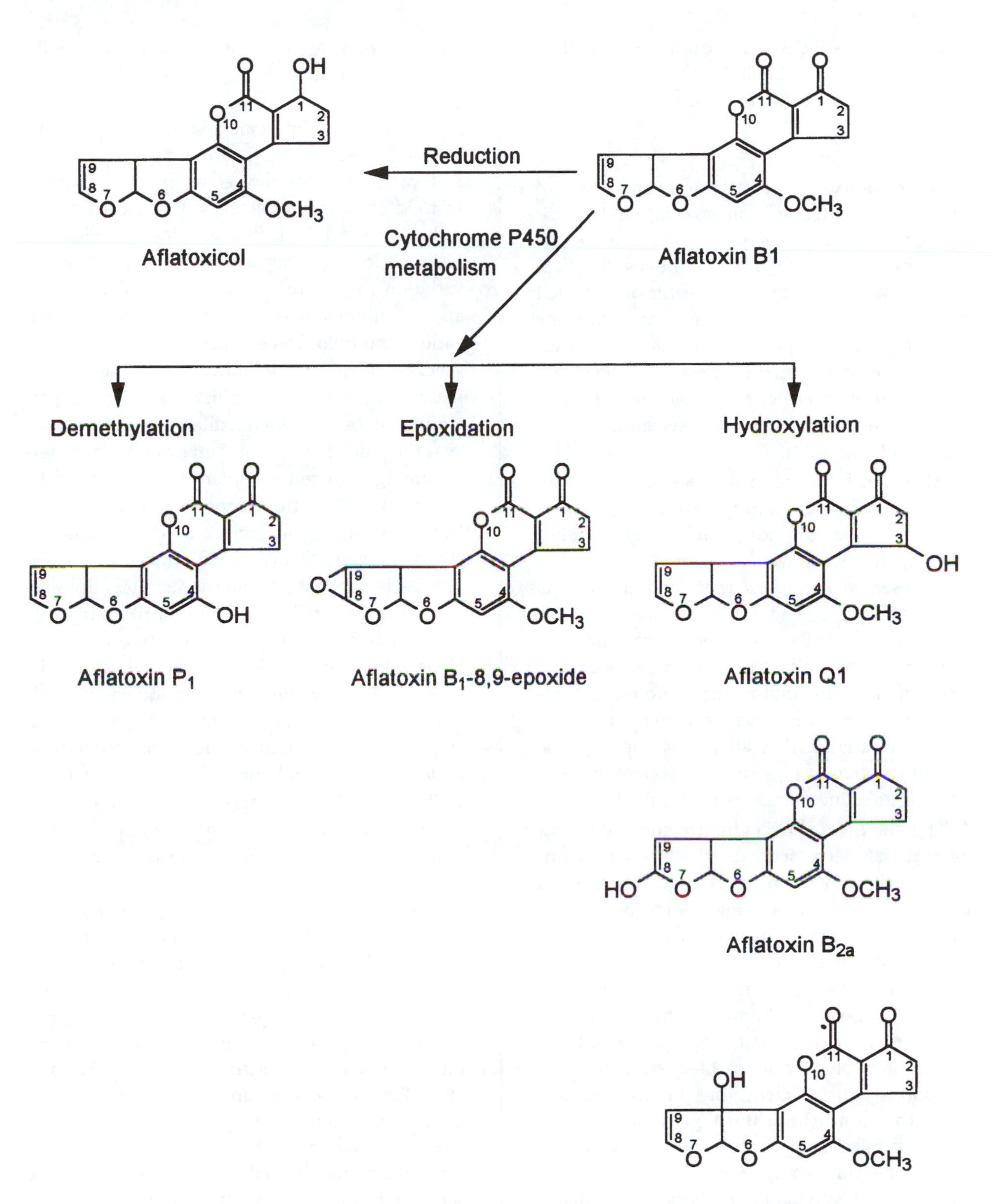

FIGURE 5–6 Summary of major aspects of aflatoxin biotransformation.

greatly with species and dose. Histological changes include subcellular accumulation of lipid, fibrosis, and extensive bile duct proliferation.

Swine. Since corn is a major component of swine diets, particularly in growing and finishing pigs, chronic exposure is a common occurrence in the southern and midwestern U.S. Feed containing 0.4 ppm or greater of AFB_1 fed from weaning to market weight can adversely affect the health of pigs. Among the mildest effects are decreased feed efficiency and poor rate of gain. More severe effects, that may result in death, include acute hepatitis, systemic hemorrhage, and nephrosis.

Although baby pigs are more sensitive than older pigs, exposure during gestation or during suckling is generally not great enough to cause adverse effects. However, stunted growth has been observed in piglets that nursed on sows fed contaminated feed; AFM_1, which is a toxic metabolite of AFB_1 produced in the liver, is transferred into milk. There is no evidence that aflatoxin is detrimental to the reproductive performance of either sows or boars. However, there is indication that aflatoxins suppress the immune system of pigs causing decreased resistance to infectious diseases in the field.

The main toxic effects in growing and finishing pigs are decreased growth rate, liver damage, icterus, and hemorrhaging. Stress from handling can suddenly trigger widespread subcutaneous hemorrhaging; increased pressure in the gluteal muscles of the ham can result in ataxia exemplified by the pig assuming a doglike sitting position. Administration of vitamin K is an effective treatment. Prerequisite to hemorrhage is prolongation of blood-clotting time (coagulopathy). Deficiency of vitamin K exacerbates the condition. Both growth-rate and blood-coagulation processes return to normal when aflatoxin exposure is removed, demonstrating the reversibility of these toxic effects. Moreover, supplementation of the diet with additional vitamin K or menadione has been shown to improve weight gain during this recovery period. Also, supplementation with additional protein affords some protection during exposure.

The lesions of porcine aflatoxicosis generally reflect changes occurring directly in the liver and vasculature or indirectly from emaciation. In mild chronic aflatoxicosis, the liver becomes yellow and firm which coincides with icterus. Histopathological hepatic changes may include variation in hepatocytic size, atypical (enlarged and multiple) nuclei, cytoplasmic lipid vacuolization, and mild fibrosis. Larger doses can cause severe fibrosis, disorganization of the lobular pattern, hyperplasia of biliary ductules, hypertrophy of biliary ducts, dilation of lymphatic vessels, and eventually **hepatocytic necrosis** beginning with centrilobular hepatocytes. In the latter stages of lethal chronic aflatoxicosis, the liver effects generally include severe fibrosis, necrosis, nodules of hyperplastic hepatocytes, extensive hyperplasia of biliary ductules, lipid accumulation, inflammation, and markedly atypical hepatocytes. Extrahepatic effects include edema in the lungs, at the base of the heart, and in the spinal folds of the colon. Also, hemorrhages may be present in the gastrointestinal tract, ventricular endocardium, lymph nodes, kidneys, and the fascia and subcutis of axillary and groin tissue. Compositional changes in blood accompany organ effects. Some of these are useful diagnostic criteria.

Poultry. Avian species are quite variable in sensitivity to chronic aflatoxicosis. Turkey poults and ducklings are the most sensitive; a dietary level of 0.25 ppm impairs their growth, whereas levels of 1.5 ppm in broilers and 4 ppm in Japanese quail are required to reduce growth rate. Age is another factor causing variable toxicity. Poults and ducklings are more sensitive than their adult counterparts, whereas no age effect is apparent in chickens. In general, the toxic effects manifested in avian species are the same as those observed in mammals, including reduced growth rate, impaired blood coagulation, hemorrhage, hepatic necrosis, decreased resistance to infection, and death. While growth retardation is usually the first dose-dependent adverse effect to appear in mammals, this is not the case in birds. **Impaired blood coagulation**, increased susceptibility of carcasses to bruising, and decreased resistance to infection result from dosages below those affecting growth. Dietary concentrations of aflatoxin greater than 2 ppm can significantly diminish egg production in layers, with production decreased to 50% with 10 ppm and to 0% at 20 ppm.

Several interesting **nutritional interactions** involving protein, lipid, and vitamin metabolism have been demonstrated in birds administered aflatoxins chronically. Increasing the dietary protein content counteracts growth impairment in broilers. Increasing the lipid content of the diet reduces mortality in broilers but does not overcome other toxic effects associated with aflatoxicosis. Feeding diets deficient in certain nutrients, such as methionine, lysine, vitamin D_3, and riboflavin, exacerbates chronic aflatoxicosis. In contrast, a deficiency of thiamin or vitamin A affords some protection against aflatoxicosis. Some of the mechanisms underlying these interactions have been elucidated. In the case of vitamin D_3, it is believed that aflatoxins impair the availability of bile salts in the gut, resulting in decreased absorption of fat-soluble vitamins. Decreased liver vitamin A in intoxicated chickens is a direct result of this effect. Because of impaired vitamin D assimilation, secondary effects such as lowered blood calcium have been noted, probably due to lowered calcium absorption from the digestive tract. Rickets were observed in broiler chicks fed aflatoxin-containing diets which were also marginally deficient in vitamin D (Hamilton, 1977). Decreased bone strength in broilers fed aflatoxins was attributed to inadequate mineralization of bone (Huff, 1980). Dietary aflatoxin decreases serum retinol and serum and tissue tocopherol (vitamin E) concentrations, suggesting that aflatoxin exposure could increase the susceptibility of animals to vitamin A and E deficiencies (Harvey *et al.*, 1994).

Ruminants. The metabolic and physiological responses during chronic aflatoxicosis are generally the same as those that occur in nonruminants. Early signs of poisoning include reduction in feed intake followed by weight loss or decreased rate of gain. Other effects associated with chronic intake of aflatoxin-containing feedstuffs in dairy and beef cattle are decreased feed efficiency, immunosuppression, increased susceptibility to stress, and decreased reproductive performance. Calves are more sensitive than adult cattle. A dosage level of about 0.2 mg/kg body weight/day causes reduced rate of gain and impaired blood coagulation in calves. Early metabolic indicators of aflatoxicosis in calves are poor feed utilization and a rapid rise in serum alkaline phosphatase (APT) activity. In young beef cattle (400–500 lb) and in yearling steers, diets containing 7–10 ppm of aflatoxins cause decreased growth and feed efficiency. Recovery of normal rates of feed intake and growth after exposure ceases is very slow. Hepatic changes similar to those in other species are observed.

Although adult cattle are not as sensitive and the signs are not as evident as in growing animals, chronic aflatoxicosis is characterized by unthriftiness, anorexia, a drying and peeling of skin on the muzzle, prolapse of the rectum, liver damage, elevated levels of blood constituents such as cholesterol, bilirubin, and serum indicator enzymes, and edema in the abdominal cavity. Milk production may be dramatically decreased in dairy cows fed aflatoxin-contaminated feed. However, it is not clear to what degree this effect is the direct result of decreased feed intake. No effect on milk fat has been observed.

There is growing evidence of **anti-reproductive effects** of aflatoxins in ruminants. Observed effects include decreased fertility in sheep and abortion and birth of underweight calves in cattle. The mechanisms of these effects have not been delineated, but two unrelated toxic actions have been suggested: (1) an indirect effect on the dam mediated by aflatoxin-induced hypovitaminosis A, and (2) due to structural similarity of aflatoxins and steroid hormones, a direct antagonistic interaction with steroid hormone receptors (Bodine and Mertens, 1983).

There is some evidence that ruminal microorganisms are relatively sensitive to aflatoxins. Results from in vitro systems indicate decreased cellulolysis, volatile fatty acid (VFA) production, and ammonia formation. Both in vitro and in vivo findings are suggestive of a shift in the proportions of VFA produced, resulting in decreased acetic acid and increased butyric acid, and a decreased total VFA production. There are several antimicrobial agents in crude extracts of *A. flavus*, but it is assumed AFB_1 is the principal toxin. Mertens (1979) speculated that the growth and function of ruminal microorganisms could be disrupted by levels of aflatoxins commonly encountered in the field. Moreover, the relative resistance of the ruminant in contrast to monogastrics points to the premise that the rumen microflora may possess some detoxification processes for aflatoxin (Bodine and Mertens, 1983).

Biological Effects of Aflatoxin

Aflatoxins interact with a number of different cell components involved in protein synthesis. These interactions are considered to be the major biochemical lesion in aflatoxin intoxication accounting for the majority of the acute, chronic, or carcinogenic responses in animals. The interactions can be classified into two groups: those inhibiting nucleic acid synthesis (i.e., interference with transcription processes), and those interfering with protein synthesis (i.e., translational processes). Aflatoxins inhibit both processes.

Many of the toxic responses in animals intoxicated with aflatoxins can be attributed to alterations of basic metabolic processes. These include interference with carbohydrate and lipid metabolism and mitochondrial respiration. However, the contribution of alterations in these essential processes to the over-all toxicity has been difficult to assess.

The primary effect on carbohydrate metabolism is decreased hepatic glycogen content simultaneous with increased serum glucose. The decrease in glycogen apparently results from inhibition of key glycogenic enzymes and increased activity of enzymes that deplete glycogen precursors. It is believed that inhibition of enzyme synthesis in the liver accounts for the decreased glycogenesis rather than a direct inhibitory reaction between the enzymes and aflatoxins. In vivo studies on glycogenesis have been complicated somewhat by the normal depletion of glycogen associated with aflatoxin-induced anorexia.

Interference with lipid metabolism occurs in several ways. The most prominent effect is the accumulation of lipid in the liver, resulting in a condition known as **fatty liver**. The effect is believed to be due to impaired transport of lipids (triglycerides, phospholipids, and cholesterol) out of the liver after their synthesis rather than an increase in their synthesis. It is not clear if the effect on lipid transport is secondary (e.g., decreased protein synthesis) or is a primary lesion during aflatoxicosis. In chickens, impaired lipid transport occurs at a dosage considerably lower than that required for reduction of growth rate and RNA synthesis. A second effect of aflatoxins on lipid metabolism is inhibition of fatty acid and cholesterol biosynthesis. Evidence suggests two types of interference can occur: decreased synthesis of enzymes such as fatty acid synthetase, and direct interaction between aflatoxins and enzymes such as thiokinase. A third effect on lipid metabolism is interference with absorption of lipids. In chickens, decreased absorption from the gut is believed to result from decreased bile salts and pancreatic lipase. Increased excretion of lipids in feces supports this conclusion.

The ability of aflatoxins to alter vital biochemical processes lies in their interaction with macromolecules, such as nucleic acids or proteins, and with subcellular organelles, such as mitochondria and ribosomes. These interactions may be either covalent or noncovalent; metabolic activation of aflatoxin is prerequisite to covalent binding to macromolecules.

Aflatoxins have a high affinity for nucleic acids and polynucleotides (RNA, DNA), particularly at nucleophilic sites on the guanine base. Formation of these **aflatoxin-nucleic acid adducts** (covalent bonds) can lead to a number of different biological effects, including cancer. In general, the toxicologic and carcinogenic potency of different aflatoxins appears to correlate well with their affinity for nucleic acids, after metabolic activation.

Carcinogenesis. Aflatoxin is carcinogenic in several species including rats, ducks, mice, trout, and subhuman primates. AFB_1 is considered the most potent carcinogen known, with AFG_1, AFB_2 and AFG_2 in their order of decreasing potency. Dietary levels of AFB_1 as low as 15 ppb fed chronically to rats cause a high rate of hepatic carcinomas (or tumors). Carcinogenesis in livestock and poultry appears to be a rarity, but has been reported in swine fed contaminated feed for long periods (>2 yr). Pigs that survived the acute toxicity phase when fed highly contaminated cottonseed and peanut meal all developed hepatic carcinomas much later (Carnaghan and Crawford, 1964). The lack of observed carcinogenesis in livestock could also be because they are usually marketed long before tumors would become clinically apparent. Rainbow trout are extremely susceptible to AFB_1-induced liver cancer (Fig. 5–7).

FIGURE 5–7 Liver cancer (hepatoma) in a rainbow trout as a result of dietary exposure to 1 ppb aflatoxin. (Courtesy of R.O. Sinnhuber)

Activation of AFB_1 by the formation of the 8,9-oxide and subsequent covalent binding to DNA is considered the first stage in carcinogenesis. Almost all of the AFB_1 metabolite bound to DNA is with the seventh nitrogen of guanine (Fig. 5–8). This interaction provides the basis for altering gene expression which results in the development of carcinomas. Exactly how the structural alteration of DNA changes its physiological function is not known.

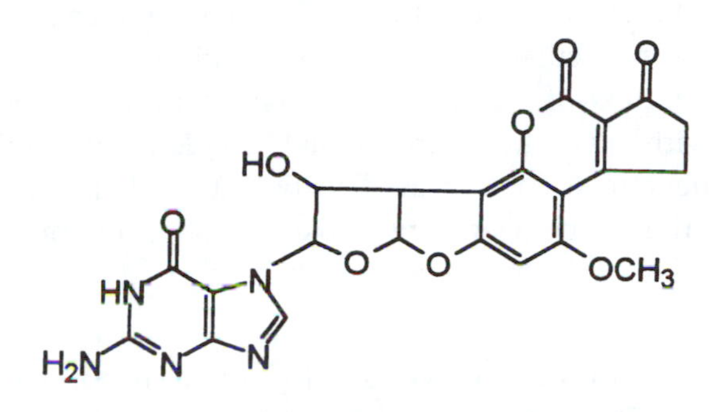

FIGURE 5–8 Covalent binding of aflatoxin B_1 to guanine in DNA, to form a DNA adduct.

As in other aspects of aflatoxicosis, the susceptibility to aflatoxin-induced carcinogenesis is subject to various physiological or environmental factors. Most of these variables are related to alterations in metabolism. Because metabolic activation is essential to produce the carcinogenic response, any change that shifts the balance between activating and detoxifying pathways will affect carcinogenicity. Most of the investigations on this subject have been conducted in rats, and the results are generally inconsistent. For example, feeding diets low in protein has both increased and decreased carcinogenicity in separate studies. Factors found to enhance carcinogenicity include feeding a lipotrope-deficient diet or a diet containing 0.2% methionine. Inhibiting factors include low dietary vitamin A, pretreating with phenobarbital or diethylstilbestrol, and hypophysectomy. In rainbow trout, the most susceptible species to carcinogenesis, feeding diets containing cyclopropene fatty acids increases the carcinogenicity of AFB_1, AFM_1, aflatoxicol, and AFQ because of an effect on liver metabolism (see Chapter 7).

There has been much speculation linking aflatoxin consumption and **liver cancer in humans** based on numerous epidemiological studies which suggest such an association (Hall and Wild, 1994). Liver cancer rates increase with increasing levels of exposure in all populations studied. There may be an interaction between liver cancer, hepatitis B and aflatoxin intake, with increased risk from aflatoxin in individuals who are hepatitis B carriers. These interactions are particularly important in tropical countries, because of widespread contamination of food with mycotoxins and a high incidence of viral hepatitis. There may also be interactions with other carcinogenic mycotoxins such as fumonisins (see Fumonisins, this chapter).

Immunosuppression. Impairment of the immune system has been observed in chronic aflatoxicosis in several species. Two primary effects can occur. First, an interference in the development of acquired immunity (immunogenesis) can develop, and this involves effects on the **cell-mediated immune system**. The thymus and thymus-derived lymphocytes (T cells) are apparently sensitive to AFB_1 and AFM_1. Antigen-cell interactions and phagocytosis appear to be modified as well. Antibody formation is normal unless very high dosages are administered. Both growth rate and cell-mediated immunity as measured by delayed hypersensitivity are decreased by approximately the same dosage of aflatoxins. A serious consequence of impaired immunogenesis is diminished effectiveness of vaccination procedures commonly carried out in swine and poultry.

The second main immunosuppressive effect is impairment of native resistance to infection. Decreased phagocytic activity of macrophages and concentrations of nonspecific humoral substances such as complement and interferon have been implicated. Such deficiencies seriously hinder the animal's ability to defend against invading pathogens. For example, in cows given aflatoxin and then challenged intramammarily with mastitis-causing microorganisms, there was more pronounced teat inflammation and higher bacteria counts in milk than in mastitic cows receiving no aflatoxin (Brown *et al.*, 1981).

Aflatoxins in Milk

Dairy cattle consuming AFB_1-contaminated feed produce milk with detectable levels of aflatoxin, mainly AFM_1. AFM_1 has been demonstrated to cause liver cancer in laboratory animals (van Egmond,1994). Numerous countries, including the United States, have legislation on limits for aflatoxin in feedstuffs for dairy cattle and AFM_1 in milk and milk products. By such means, potential hazards of significant aflatoxin occurrence in the food supply are avoided. In-

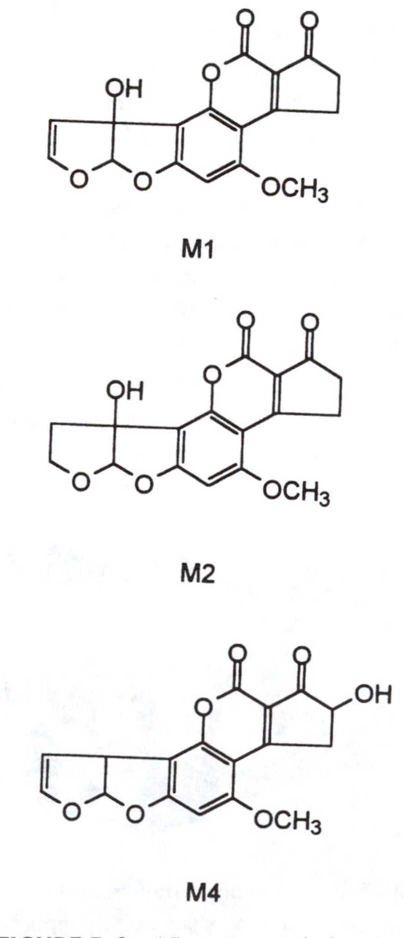

FIGURE 5–9 Aflatoxin metabolites in milk.

gested aflatoxins are rapidly eliminated from the body, so tissue residues in meat are not significant. The US Food and Drug Administration prohibits interstate movement of feed grains containing more than 20 ppb aflatoxin and prohibits the sale of milk with more than 0.5 ppb aflatoxin. AFB_1 metabolites in milk include AFM_1, AFM_2 and AFM_4 (Fig. 5–9).

Feed Processing to Reduce Aflatoxin Exposure

Aflatoxin production in corn and other commodities can occur both pre- and post-harvest. Contamination in the field may sometimes be unavoidable, and is influenced by climatic conditions. Drought and insect damage are the two most frequent causes of fungal infestation of corn. Prevention of mold growth and aflatoxin formation in stored grains is accomplished by proper storage conditions of low humidity and protection against exposure to moisture. Preservatives and anti-fungal agents such as propionic acid may be somewhat effective in reducing aflatoxin formation, but should not be used as substitutes for proper drying and good post-harvest management.

The most effective and practical method of detoxification of aflatoxin-contaminated grain is ammoniation with anhydrous ammonia (Phillips *et al.*, 1994; Park *et al.*, 1988). **Ammoniation** can reduce aflatoxin levels by more than 99% (Phillips *et al.*, 1994). The mechanism of action involves hydrolysis of the lactone ring, with various end-products produced with much lower toxicity than the parent compound. According to Park *et al.* (1988), results of ammoniation studies "demonstrate overwhelming support for the efficacy and safety of ammoniation as a practical solution to aflatoxin detoxification in animal feeds."

Another approach to reducing aflatoxicosis in livestock is the use of **dietary additives** which bind aflatoxins and prevent their absorption from the gut. These include various clay minerals such as bentonites and hydrated sodium calcium aluminosilicates (Lindemann *et al.*, 1993). At a concentration of 0.5% of the diet, the aluminosilicates are very effective at binding aflatoxins and preventing their absorption in both ruminants and nonruminants. They are not effective with other mycotoxins.

Sterigmatocystin (ST) is a biosynthetic precursor of aflatoxin in *A. flavus* and *A. parasiticus*, and is the primary final metabolite in a number of other *Aspergillus* spp. It is considerably less carcinogenic than AFB_1. ST is metabolized by the cytochrome P450 system to an epoxide, which can form DNA adducts and a glutathione conjugate (Olson and Chu, 1993a,b). ST can contribute additively to the toxicity of aflatoxin, so is of potential concern.

Wicklow (1988) reviewed evidence suggesting that the most potent mycotoxins tend to be those towards the end of a long biosynthetic pathway in which substances of lower toxicity are intermediates. This would result from coevolutionary processes, whereby synthesis of increasingly toxic chemical defenses would result from development of detoxifying capabilities in organisms consuming the toxins. Thus the lower toxicity of sterigmatocystin than AFB_1 would be predicted.

Ochratoxin

The ochratoxins are a family of isocoumarin derivatives of the amino acid phenylalanine. Nine ochratoxins have been identified, but only ochratoxin A (OA) and very rarely ochratoxin B

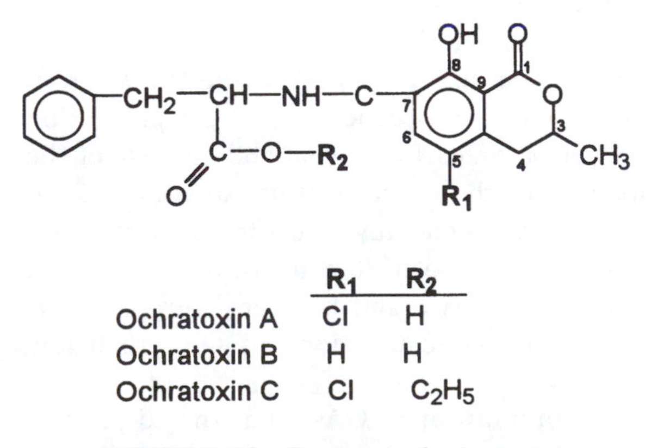

	R_1	R_2
Ochratoxin A	Cl	H
Ochratoxin B	H	H
Ochratoxin C	Cl	C_2H_5

FIGURE 5–10 Structures of ochratoxins.

(Fig. 5–10) are produced under natural conditions. The ochratoxins are produced by several species of *Aspergillus* and *Penicillium*. The toxins were first isolated from *A. ochraceus* (now *A. alutaceus*) from which their name was derived. The most important producers of ochratoxins are *P. viridicatum* and *A. alutaceus*. Ochratoxin A is a potent **nephrotoxin**. It has been confirmed as the cause of kidney disease in livestock (e.g., porcine nephropathy) and suspected as the cause of Balkan nephropathy in humans. Ochratoxicosis has been reviewed in depth by Marquardt and Frolich (1992) and Leeson *et al.* (1995).

Occurrence

The fungi that produce ochratoxins are distributed worldwide and are common in soil, decaying plants, insects, stored seeds, and cereals. Ochratoxicosis, particularly nephropathy (kidney disease) in swine, is quite prevalent in certain regions and is usually associated with the feeding of contaminated barley. Ochratoxin A has also been detected in wheat, oats, barley, corn, beans, peanuts, hay, green coffee beans, and mixed feeds. Up to 27 ppm has been measured in Canadian wheat and Danish barley.

Not all strains of *P. viridicatum* and *A. alutaceus* are toxigenic. Also, some strains produce citrinin as well as OA. Production of OA in the toxigenic strains is primarily governed by temperature and moisture. In general, optimal conditions for OA production are a moisture content of 19–22% and a temperature of about 24°C. Toxin production can still occur at temperatures as low as 4°C. Because the environmental conditions conducive to OA production are not particularly unusual and because of the ubiquity of the fungal strains, surveys of feedstuffs for OA generally reveal frequent contamination in the cooler climates of the world such as North America and northern Europe. However, usual concentrations of OA are below toxic levels.

Ammoniation of OA-contaminated grain is very effective in eliminating its toxicity (Marquardt and Frolich, 1992). Binding agents have not generally been effective or economical. Supplemental ascorbic acid (vitamin C) counteracts the toxicity of OA to some degree (Haazele *et al.*, 1993).

Mode of Action

Ochratoxins interact with a number of macromolecules in the body, but the toxicological significance or relationship to ochratoxicosis of most of these interactions have not been established. The toxin has a high affinity for a number of enzymes, but does not in all cases result in inhibition. It is a competitive inhibitor of protein synthesis, particularly in the spleen, due to its ability to inhibit the enzyme phenylalonyl-tRNA synthetase. This is a consequence of the phenylalanine moiety in OA (Fig. 5–11). OA also binds strongly to serum albumins. The hydrolysis metabolites of OA, Oα and Oβ, also bind to proteins but with much less affinity than OA. Binding to serum proteins may serve to decrease the amount of OA available to other sites of toxic action or affect the binding of other biologically active substrates in serum,

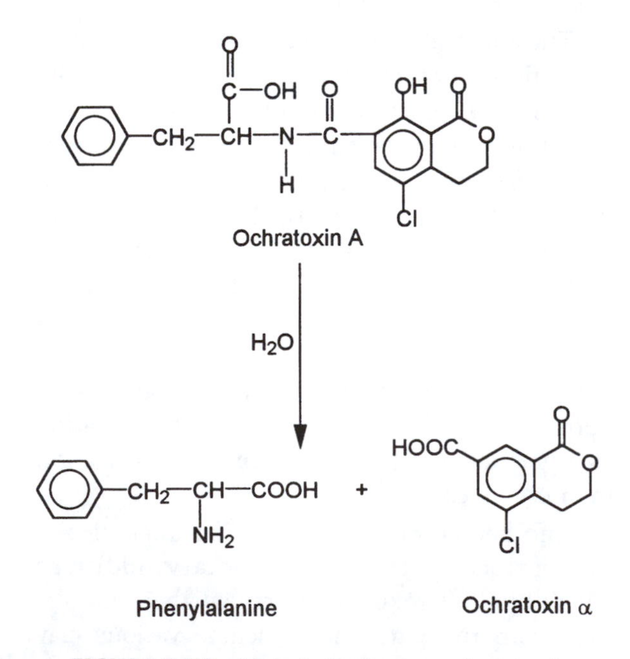

FIGURE 5–11 Hydrolysis of ochratoxin A to phenylalanine and ochratoxin α.

possibly resulting in secondary effects. Binding of OA to the blood albumin contributes to a long retention time of the toxin in animals (Marquardt and Frohlich, 1992). Interestingly, the affinities of different ochratoxins with serum albumins correlate quite well with their acute toxicities.

Interference with **carbohydrate metabolism** is considered to play a key role in ochratoxicosis. Depending on the species, glycogen concentration in organs may either increase or decrease during ochratoxicosis. In rats, liver glycogen is depleted after a single dose of OA, whereas in chickens, glycogen accumulates in liver and muscle during chronic exposure, resulting in glycogen storage disease, or **hyperglycogenation**. The depletion effect in rats has been attributed to decreased glycogenesis (glycogen formation) due to inhibition by OA of a key enzyme system, glycogen synthetase. Impaired transport of glucose into liver and acceleration of glycogenolysis (glucose utilization) have also been suggested as possible causes. In chickens, the accumulated glycogen in liver is not metabolically available due to apparent OA inhibition of the glucagon-mediated mobilization process, whereas the muscle glycogen mobilized by epinephrine is available. Neither the cause of hyperglycogenation nor its role in ochratoxicosis in chickens has been determined. There is some evidence that OA inhibits a cyclic AMP-dependent protein kinase which initiates a series of enzyme reactions leading to glycogenolysis (catabolism to glucose). In the kidney, but not in liver, OA also inhibits gluconeogenesis by inhibiting an enzyme called carboxykinase.

Cell energetics may also be disturbed during ochratoxicosis, as evidenced by in vitro studies indicating that both OA and Oα are respiratory chain inhibitors. This effect is believed to result from competitive interference of the toxin on the uptake of metabolic anions such as ADP and P_i into mitochondria. Additionally, the ADP:O ratio and O_2 consumption are both decreased by OA, indicating impaired ATP production. Succinic dehydrogenase and other Krebs cycle enzymes are also inhibited by OA.

Biological Effects

The kidneys undergo several morphological and functional changes during both acute and chronic intoxication. These changes are quite comparable among all animals tested and also in humans with **Balkan nephropathy**. At necropsy, the kidneys appear gray in color, have a granular surface, and are usually enlarged. Widespread subcutaneous, mesenteric, and perirenal edema occurs in swine. Histological changes include atrophy of the proximal tubules accompanied by disintegration of the brush border, swollen mitochondria, and disorganization of the endoplasmic reticulum. Interstitial cortical fibrosis and glomerular fibrosis are characteristic of chronic ochratoxicosis. These morphological changes are accompanied by various impaired renal functions indicated by elevation of potassium, protein (proteinuria), and glucose (glycosuria) in urine. The earliest indicator is an increased appearance in urine of leucine aminopeptidase (LAP), an enzyme normally located in the brush border of the proximal tubule. Other measurements indicating nephropathy are decreased glomerular filtration rate (GFR), decreased clearance of test substances, increased blood urea nitrogen (BUN), and inability to concentrate urine (decreased osmolality).

Although the liver is not a major target organ, some toxic effects occur there. Fatty degeneration, accumulation of glycogen (glycogen storage disease), especially in chickens, and centrolobular necrosis including lesions in the intrahepatic biliary system have been observed. Decreases in serum protein content and coagulation of blood in chickens are indicative of liver damage. Lipid peroxidation may be involved in the **hepatoxic effects** of OA (Marquardt and Frolich, 1992).

The gastrointestinal tract of several species is affected by ochratoxins and is characterized by enteritis. Also, necrosis of various lymphoid tissues such as the spleen and lymph nodes occurs. The implications of this effect on the immune system are apparent but have not been thoroughly investigated. There is some indication of suppressed antibody production in mice and decreased phagocytic activity in poultry. A case of apparent **immunosuppression** in breeding sows in Germany was linked to consumption of barley and oats containing two *Penicillium* mycotoxins: viomellein and ochratoxin B (Leistner, 1984).

Teratogenic and embryocidal (e.g., fetal resorption) **effects** of OA have been observed experimentally in several laboratory species but not in food-producing animals. The studies in mice revealed deformities of the skull, vertebrae, and ribs after an intraperitoneal injection but not after oral administration. **Carcinogenicity** of ochratoxins has not been demonstrated except in an especially sensitive strain of mouse known as the ddy strain. In these mice, both renal and hepatic tumors were detected with very high levels of OA. Interestingly, simultaneous administration of OA and citrinin enhanced the incidence of renal tumors but not that of liver tumors (Kanisawa, 1984).

Effects in Different Species

Among farm animals, monogastric species (horses, swine, poultry) are much more sensitive to ochratoxins than ruminants. As with other mycotoxins, impaired growth in young animals including calves is the first observable sign of intoxication. Young animals are clearly more sensitive than adults. Ochratoxicosis is generally not diagnosable until postmortem examination of kidneys is conducted.

Adult **ruminants** are afforded protection by rumen microorganisms (Kiessling *et al.*, 1984; Westlake *et al.*, 1989; Xiao *et al.*, 1991a,b), particularly protozoa which hydrolyze the peptide bond of OA, thereby forming less toxic ochratoxin α and phenylalanine (Fig. 5–11). This reaction is catalyzed by the enzymes carboxypeptidase A and chymostrysinogen. The hydrolysis of OA is much greater in the rumen of animals fed hay than in ruminants fed grain (Xiao *et al.*, 1991a,b), suggesting that OA-contaminated grain would be most effectively detoxified if considerable hay were included in the diet. There is some evidence that ochratoxin C (OC), which is equally toxic to OA, may be formed in minor quantities by rumen microorganisms. The beneficial role of **rumen fermentation** in protecting against ochratoxicosis in cattle is evidenced by the wide difference in toxicity of OA when administered orally (13 mg/kg body weight) versus intravenously (1 mg/kg body weight). Even though there is substantial evidence that OA has bacteriostatic activity against gram-positive bacteria, rumen fermentation is not impaired by levels of OA normally found in feed. However, in one experiment, weaned calves administered OA equivalent to 2–40 ppm in feed developed nephritis, enteritis, coagulopathy, and reduced growth, suggesting the rumen may not be as protective in the less-developed ruminant (Pier *et al.*, 1976). Also, there is some indication that OA causes abortion or contributes to other diseases in cattle. However, the general consensus is that natural ochratoxicosis in ruminants is rare.

In swine, the primary syndrome is called **porcine nephropathy** which generally occurs after chronic ingestion of diets containing from 0.2 to 4 ppm OA. The effects on the kidney are comparable to those seen in a variety of other species (e.g., poultry, rats, fish, and monkeys) and are also similar to the human syndrome called **Balkan nephropathy**. Both the swine and human diseases are endemic to certain parts of Europe. Other effects include weight loss or decreased growth and depression. A number of extrarenal effects have been observed experimentally, but these are induced only by high dosages. The kidneys are the sole target organ when naturally occurring levels of exposure are administered. The rat is a useful model for studying chronic but not acute porcine nephropathy. A single, acutely toxic dose administered to rats can be fatal without producing renal effects, but this does not occur in swine.

Avian ochratoxicosis has been documented in broilers, layers, and turkeys. As in other species, the principal effect is nephropathy. The oral LD_{50} varies among different species (e.g., broilers, 3.4 mg/kg; poults, 5.0 mg/kg; quail, 16.5 mg/kg). Interestingly, the sensitivity of poultry (acute and chronic) is greater to ochratoxin A than to either T-2 toxin or aflatoxin B_1; the minimum dietary dose that impairs growth in young broiler chicks is 2 ppm for ochratoxins, 2.5 ppm for AFB_1, and 4 ppm for T-2 toxin. The only other clinical signs that may be observed are neurological effects which include tremors, flailing, and loss of righting reflex. At necropsy, birds may appear dehydrated or emaciated, with some indication of proventricular hemorrhage, which is a result of decreased coagulation. **Coagulopathy** can occur in broilers fed 1 ppm OA for a few weeks. Another unique effect in poultry is **visceral gout** characterized by white urate deposits throughout the body cavity and internal organs. Kidney and liver effects are typically the same as in other species. Hepatic accumulation of glycogen is particularly notable in poultry. Like AFB_1, OA decreases bone strength in young broiler chicks. The bones may become rubbery, apparently from poor mineralization. In layers, levels of OA as low as 0.5 ppm can decrease egg production and feed consumption. Egg fertility is not particularly susceptible; however, the egg embryo appears to be rather sensitive to OA. As little as 0.01 μg/egg is embryocidal.

In turkeys, ochratoxicosis is characterized by mortality, decreased growth rate and feed efficiency, nephropathy, and decreased carcass pigmentation. The latter effect may result in an increase in the number of carcass condemnations.

Metabolic Fate and Residues

Ochratoxin A and aflatoxin B_1 are the two mycotoxins considered to be the most serious food-chain residue problems (Stoloff, 1979). Poultry and swine fed OA-containing feed retain some of the toxin in their tissues. Survey data from slaughterhouses in Europe indicate that 25–35% of pigs suffering from porcine nephropathy contain measurable OA residues. Moreover, these carcasses generally pass the inspection process, thereby gaining access to the human food chain.

Ochratoxin A is absorbed from the stomach and also throughout the remainder of the gastrointestinal tract. In pigs given a single experimental dose of OA, about 66% was absorbed, and the peak concentration in plasma occurred at about 8 hr postadministration (Galtier *et al.*, 1981). In the blood, much of the OA is found bound to serum albumin, especially in cattle, pigs, and man. This sequestration of OA in blood decreases its distribution in the body and also retards urinary excretion. The highest tissue levels of OA are found in blood and kidneys, with lower levels in liver and muscle. In pigs fed 1 ppm dietary OA, the ratio of residue concentrations in kidney versus feed is 1 to 38 at 24 hr after administration (Krogh, 1978). This is a relatively small ratio compared to other mycotoxins. Moreover, the fact OA is more toxic than most other mycotoxins intensifies the hazard from residues in edible tissues, particularly kidneys.

Compared to other species, the pig eliminates OA residues quite slowly; for example, the biological half-life of OA is about 90 hr in pigs compared to about 4 hr in chickens. The rate of elimination is even slower (e.g., 100–110 hr) from kidneys and liver (Galtier *et al.*, 1981). Therefore, the residue problem in swine compared to other species can be attributed to such factors as greater protein binding and greater retention in tissues. Levels of OA as high as 100 ppb have been measured in kidneys of nephropathic swine. Swine known to have OA exposure should be fed OA-free feed for at least 4 weeks before slaughter to allow for OA elimination from tissues. The residue problem in chickens is not considered as great, but levels of up to 30 ppb have been found in chickens with avian nephropathy. In ruminants, OA hydrolysis in the rumen decreases the possibility of residues as well as toxicosis.

Citrinin

Citrinin is a nephrotoxin produced by several species of the genus *Penicillium* and three species of the genus *Aspergillus*. The principal toxigenic fungi in animal feeds is *P. viridicatum*. Citrinin is a quinone methide (Fig. 5–12) that was first identified as a secondary metabolite of *P. citrinum*, from which it derived its name. In crystalline form, citrinin is lemon yellow in appearance and is practically insoluble in water.

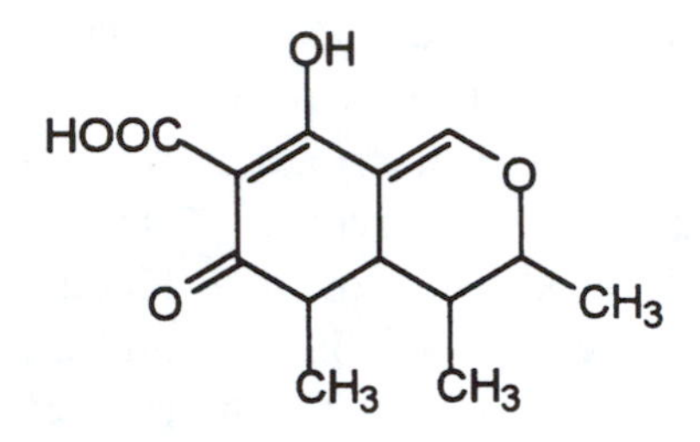

FIGURE 5–12 Structure of citrinin.

Occurrence

The fungi that produce citrinin are found in the temperate regions of the world. Feedstuffs known to be colonized by the toxigenic strains include most grains, such as wheat, oats, barley, rye, and corn. Concentrations as high as 80 ppm have been found in Canadian wheat, but the frequency of contamination at levels high enough to be toxic is rather rare. Contamination of Danish and Canadian grains fed to livestock could be a cause of disease in these regions. Toxicologic assessment of natural contamination of feeds with citrinin has been complicated due to its occurrence with other mycotoxins, including ochratoxin A, patulin, penicillic acid, and aflatoxins.

Fungal infestation and subsequent production of citrinin in feedstuffs are influenced by temperature, substrate, and moisture. In moist grains, citrinin production is maximal at about 25° C but may also occur at lower temperatures (5–12° C). In peanut pods in culture, kernel moisture content and pod damage are major determinants of fungal growth; citrinin production can range up to 1200 ppm. The toxin is not very stable and degrades under certain conditions such as increased heat (60°–70° C) or when the moisture content is increased very much above the optimum for growth. These factors and others undoubtedly impact on the environmental occurrence of citrinin.

Mode of Action

The mode of action of citrinin at the biochemical level has not been completely established, but experimental findings in laboratory rodents have revealed a number of possibilities. The site of these effects is primarily the kidneys and liver. Within a few hours after citrinin administration, DNA, protein, and glutathione (GSH) content in these tissues are decreased. In addition, the respiratory capacity (i.e., uptake of O_2 and the metabolic enzyme, succinic dehydrogenase, are inhibited. Other hepatic changes include decreased glycogen content, increased lipid content (fatty liver), and decreased cholesterol synthesis.

The reduction of GSH concentrations in renal and hepatic tissues was found to occur within 2–4 hr after administration of citrinin to rats (Berndt *et al.*, 1980). Although GSH depletion is clearly related to the hepatoxicity of some chemicals (e.g., acetaminophen, bromobenzene), such a relationship has not been established between citrinin and nephrotoxicity. Interestingly, renal concentrations of GSH returned to normal within 2 to 3 days postadministration and then became elevated by 30–40%. The biological significance of this response is also unclear. Comparatively, ochratoxin A apparently has a less profound effect on tissue GSH content.

Biological Effects

Although citrinin toxicosis in domestic animals is not well characterized, experimental evidence clearly shows that the principal toxic effect in all species tested is nephropathy. In swine, experimental citrinin-induced kidney damage is similar to that observed in field cases of porcine nephropathy, which is attributed to ochratoxin A (OA). This plus the fact that citrinin and OA frequently coexist in moldy feeds has created some serious difficulties in delineating whether citrinin represents a real threat to animal health. Because of its less common occurrence and lower concentrations in feedstuffs, the present consensus of opinion is that citrinin is a contributor rather than a primary cause of porcine and avian nephropathy. There is experimental evidence that citrinin and OA act synergistically, but this phenomenon has not been explored in domestic livestock.

The **nephrotoxic effects** of citrinin are similar in all species. In general, acute tubular necrosis possibly followed by renal failure occur. Grossly, the kidneys are enlarged and appear pale and tan colored. Within the tubular nephron the area most commonly damaged is the proximal tubule. It degenerates, becoming necrotic and even mineralized in some species. These morphological changes are accompanied by alterations in functional indicators, including elevated BUN, decreased GFR, proteinuria, glycosuria, and creatinuria. In addition, daily urinary volume may increase substantially (polyuria) and remain elevated for 2 to 3 days after a single dose of citrinin. Decreased osmolality of urine coincides with the polyuria. Watery excreta occurs in chickens. Depending on the dose, the lesions in the tubules are reversible. However, in rats, severe tubular damage indicated by massive glycosuria is followed by renal failure and death.

The liver is moderately affected by citrinin, particularly in guinea pigs and chickens. Grossly, the livers are enlarged, mottled, and friable. Changes in the gastrointestinal tract include gastric inflammation and cecal ulceration in guinea pigs and hemorrhagic jejunums in chickens. Death of citrinin-intoxicated dogs is due to intestinal intussusception (twisted intestines resulting in obstruction), indicating a neurologic effect.

Citrinin does not appear to be carcinogenic but does promote the renal carcinogenicity of potent carcinogens such as dimethylnitrosamine and perhaps ochratoxin A. It is mutagenic in some microbial systems, but is negative in the *Salmonella typhimurium* Ames assay. Citrinin is not teratogenic in mice, but has potent embryocidal activity in the chicken embryo where the LD_{50} was 80.5 μg/egg.

Although citrinin and ochratoxin A have similar nephrotoxic effects (both pathologically and functionally), there are specific differences distinguishing the two toxins. First, the effects from citrinin are not cumulative. Repeated low dosages of citrinin are not nephrotoxic, where ochratoxin A administered in the same way would produce toxic effects in the kidneys. Second, citrinin does not usually cause enteritis as does ochratoxin A. There are changes in the gastrointestinal tract such as in the guinea pig, but the tract is generally unaffected in most species.

Toxicity in Different Species

Citrinin is not nearly as toxic to farm animals as aflatoxin B_1, ochratoxin A, or T-2 toxin. In **poultry**, dietary levels in excess of 130 ppm are required to produce clinical changes which include growth depression, increased water consumption, and diarrhea as well as nephrotoxic and hepatotoxic effects. The increased water consumption is transitory, occurring after initial administration, and is postulated to be due to an action of citrinin on the sensory thirst mechanism. Layers tolerate dietary levels of citrinin in excess of 250 ppm without reductions in body weight, egg production, or egg quality. In **swine**, levels in excess of 20 mg/kg of body weight can cause reduced growth rate and nephropathy. The toxicity of citrinin has been studied in a variety of laboratory species. In the **rat and mouse**, the intraperitoneal LD_{50}, which is much greater than the oral toxicity, is 60–80 mg/kg body weight. Other clinical signs observed in nonfarm animals include salivation, lacrimation secretions, vomiting (dog), and hyperemia of the ears and mucous membranes. The major effects in **rabbits** given acute doses are diarrhea and kidney lesions (Hanika *et al.*, 1983).

Because citrinin possesses antibiotic activity against Staphylococci and other gram-positive bacteria, it could possibly interfere with rumen function. This subject has not been experimentally explored primarily because there is no field evidence that citrinin toxicosis occurs in ruminants.

Metabolic Fate and Residues

Although its metabolic fate in animals has not been extensively studied, it is clearly evident that citrinin is readily absorbed from the gastrointestinal tract, metabolized, and rapidly eliminated. Metabolism to more water-soluble metabolites has been demonstrated, but their identification has not yet been achieved. In rats, the major part of a dose is excreted in urine within 24 hr as metabolites; these are partially conjugated to GSH which correlates with the depletion of GSH from hepatic and renal tissue (Berndt *et al.*, 1980). Urine is normally the primary route of excretion; however, in citrinin-damaged kidneys the toxin is poorly excreted due to decreased GFR. To compensate, biliary excretion becomes the primary excretory route. The rapid elimination of citrinin and its metabolites prevents accumulation in tissues, thereby decreasing the toxicologic threat to animals. In addition, the possibility of retained residues in edible tissues is reduced.

Cyclopiazonic Acid

Cyclopiazonic acid (CPA) is an indole tetramic acid (Fig. 5–13) produced by many *Aspergillus* and *Penicillium* spp. It is of food safety concern because many of the *A. flavus* strains which produce aflatoxins also produce CPA (Dorner *et al.*,1994). In fact, it appears that the "Turkey X Disease" in England that led to the discovery of aflatoxin was in part due to contamination of the groundnut (peanut) meal with CPA. Bradburn *et al.* (1994) in a retrospective study detected 31 ppm CPA in a sample of groundnut meal implicated in the original outbreak of Turkey X Disease, and speculated that some of the symptoms were caused by CPA rather than by aflatoxin.

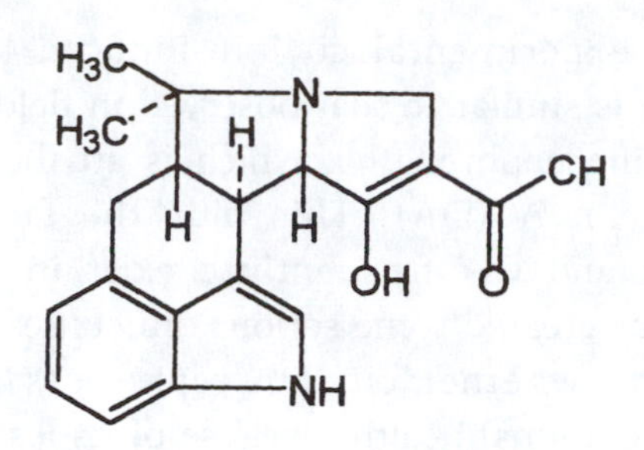

FIGURE 5–13 Structure of cyclopiazonic acid.

CPA has been detected in a large number of foods, including cheese, corn, peanuts, eggs and meat (Dorner *et al.* 1994). Despite its widespread occurrence, no definitive human or animal disease has been unequivocally attributed to CPA.

Experimental administration of CPA to animals has neurologic and muscular effects. Ataxia, spasms, labored breathing and opisthotonus are seen. The mode of action appears to involve an interaction of CPA with skeletal muscle sarcoplasmic reticulum vesicles, with inhibition of Ca^{2+} ATPase, leading to disruption of cellular calcium homeostasis (Goeger and Riley, 1989).

Although CPA does not seem to be significant in terms of livestock and poultry production parameters, there is concern that it could enter the human food supply via milk and eggs produced using CPA-contaminated feed (Dorner *et al.*, 1994).

Kojic Acid

Kojic acid (2-hydroxymethyl-5-hydroxypyrone) is a fungal metabolite produced by many *Aspergillus* and *Penicillium* spp. It is often produced in association with aflatoxin. Giroir *et al.*

(1991) found that the toxicity of kojic acid to broiler chicks was very low, with significant adverse effects observed only at concentrations greater than 2 g per kg feed. It is unlikely to be a significant toxicological hazard to livestock or poultry.

Rubratoxins

Rubratoxins are bisanhydride metabolites (Fig. 5–14) produced by *Penicillium rubrum* and *P. purpurogenum*. The toxins derive their name from *P. rubrum*, from which they were first isolated. There are two known toxins, rubratoxin A (RA) and B (RB). Comparatively, RB is about twice as toxic as RA.

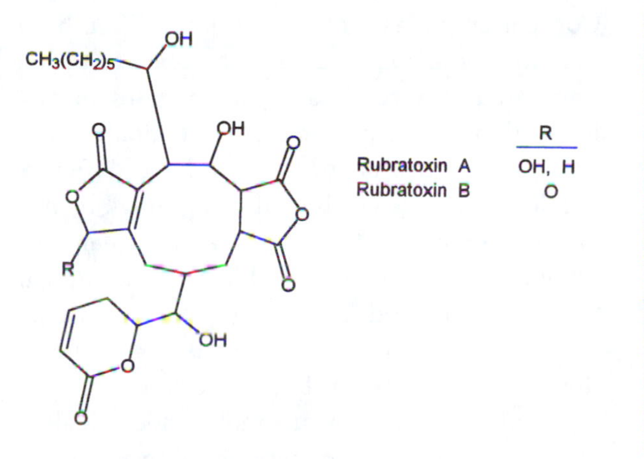

FIGURE 5–14 Structures of rubratoxins.

These fungal toxins were originally found in moldy corn and caused a disease characterized by hepatitis, nephrosis, and hemorrhage when fed to cattle and swine. Because *Aspergillus flavus* was isolated along with *P. rubrum*, the primary attention was directed toward the former and the production of aflatoxins. However, isolates of *P. rubrum* grown on corn and fed to animals are more toxic than *A. flavus* isolates. Even though no field cases of rubratoxicosis have been reported, Pier (1981) speculated that toxic effects may occasionally occur in cattle and swine.

Although the mechanism of action has not been clearly established, biochemical changes at the subcellular level include inhibition of mitochrondrial respiration, ATPase activity, and protein synthesis. The latter involves binding to DNA, decreased RNA polymerase activity, decreased levels of RNA, and disaggregation of polysomes. A structure–activity relationship is apparent from studies with laboratory species. Alteration of functional groups such as hydrogenation of the α,β unsaturated lactone moiety greatly decreases toxicity. Therefore, toxicity is attributed primarily to the parent compound.

The liver is the primary target organ although congestion and hemorrhaging may occur in several other organs including kidneys, spleen, lungs, and gastrointestinal tract. Liver cells show degenerative changes and possibly necrosis. Alterations of hepatic function are indicated by increased prothrombin time (coagulopathy) and bilirubinemia (icterus). Mild degenerative changes in renal tubule epithelium are seen in some species. Clinical signs induced experimentally include depression, anorexia, decreased weight gain, coagulopathy, hemorrhaging, bloody feces, and death.

Experimental rubratoxicosis has been produced in a number of species, including several farm animals. In calves, daily doses of 8 mg/kg of RB decreased liver function, 12 mg/kg caused depression and anorexia, and 16 mg/kg caused acute hepatic failure and death (Pier *et al.*, 1976). Acute toxicity is much less in monogastric species; the LD_{50} is about 400 mg/kg in rats and 83 mg/kg in chicks. The lesser toxicity in these species is thought to be related to less absorption from the gastrointestinal tract, possibly due to greater degradation in the tract. This is also apparent from the greater (10 times) toxicity of RB administered parenterally versus orally in rats. Similarly, chronic feeding studies show that rubratoxins are not very toxic to chickens; 500 ppm in the diet fed for 3 weeks was required to produce weight loss (Wogan *et al.*, 1971). Rubratoxin B is not carcinogenic but is mutagenic, teratogenic, and embryotoxic in mice and egg embryos; the α,β unsaturated lactone ring is necessary for this activity (Mirocha, 1980). Also, there is the potential for immunosuppression due to impairment of complement formation in the liver.

Perhaps the most significant environmental feature of rubratoxins is their ability to act synergistically with other mycotoxins. Experimental evidence in calves, guinea pigs, rats, and dogs indicates that AFB1 and RB administered together produce toxic effects not seen with either toxin alone. Synergism between ochratoxin A and RB in rats has also been demonstrated.

Much of the low oral toxicity of rubratoxins appears to be due to their slow absorption from the gastrointestinal tract, which may include detoxification prior to absorption. Studies in rats show 80% elimination of ^{14}C labeled RB in 7 days in urine and feces equally (Unger and Hayes, 1979). Partial metabolism including conjugation to glucuronides and sulfates precedes elimination. There is some evidence that conjugates excreted in bile are recirculated through the enterohepatic system; this is based on biphasic elimination from liver (half-lives are 14 and 100 hr). Evidence for involvement of hepatic MFO in their metabolism includes localization of ^{14}C labeled RB in the endoplasmic reticulum after administration and decreased toxicity of RB in rodents after stimulation of MFO with barbiturates. The latter finding also demonstrates that RB is more toxic than its metabolites. There is no evidence that rubratoxins should be considered a residue problem in food animals.

Patulin

Patulin is a hemiacetal lactone (Fig. 5–15) produced by several species in the genera *Aspergillus, Penicillium,* and *Byssochlamys*. This toxin is toxic to both plants and animals and has potent antibiotic activity. Most of the known fungal producers of patulin were identified in the 1940s at a time when the search for antibiotics was quite intense. Patulin was originally called

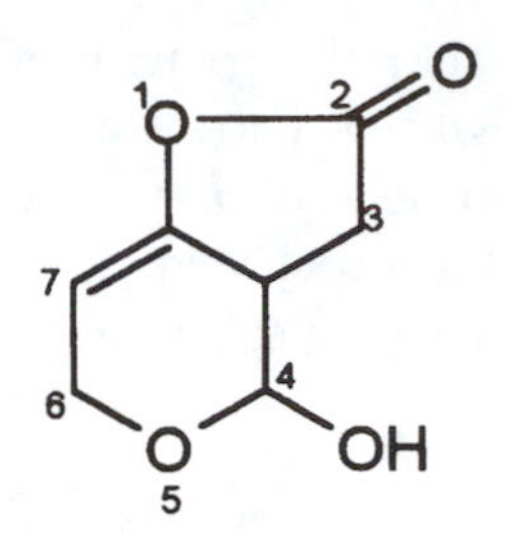

FIGURE 5–15 Structure of patulin.

claviformin, named for *Penicillium claviforme*, from which it was first isolated. The name "patulin" was derived because structural characterization was made in *Penicillium patulum*.

The occurrence of patulin in foods and feeds is not well understood. To date, proven natural contamination in agricultural products has been limited to rotted apples and apple juice or cider. Concentrations as high as 1000 ppm have been found in these commodities. The toxin has been implicated in several cases of toxicoses involving cattle and sheep. From the standpoint of potential contamination of foods and feeds, Ciegler (1977) listed the following fungi: *P. urticae, P. expansum, P. melinii, P. cyclopium, A. clavatus, A. terreus,* and *B. nivea*. Even though these patulin-producing fungi are periodically found in foodstuffs such as cereals and legumes, the toxin itself has not been detected.

Patulin is relatively unstable under alkaline conditions but is quite stable in an acidic environment. This accounts for its long-term stability in some foods and feeds but not in others. During its breakdown, patulin reacts with sulfhydryl-containing amino acids or proteins, forming patulin-cysteine adducts. Although less reactive than patulin, these adducts appear to retain some of the toxic potency of the parent toxin.

No toxicologic studies with patulin have been conducted in domestic animals, but there is some indirect evidence of patulin toxicosis (Ciegler, 1977). Feeding of mold-infested malt was believed to have caused the deaths of over 100 dairy cattle in Japan. The fungi *P. urticae* that produces patulin was identified, and experimental feeding of malt inoculated with this fungi to a bull resulted in neurological signs, brain hemorrhage, and death. Similar effects were observed in rodents injected with patulin. In France, cattle fed wheat contaminated with *A. clavatus* (a patulin producer) developed pulmonary edema and congestion, and several died. Sheep fed extracts of *B. nivea* (also a patulin producer) exhibited anemia, decreased serum protein concentrations, nasal discharge, loss of rumination, pain in the sternal area, anorexia, and weight loss (Lynch, 1979). Postmortem findings included abomasal hemorrhaging and lesions in liver and kidneys.

Patulin is a relatively potent mycotoxin. Experimental findings in laboratory species indicate LD_{50} values ranging from 10 to 35 mg/kg of body weight, depending on animal species and route of administration. Toxicity is lower by the oral route than when injected. In the chick, the LD_{50} is 170 mg/kg. The primary toxic effects are ascites, hydrothorax, and pulmonary edema. It is a dermal irritant when applied topically. Also, lesions develop at the site of subcutaneous injection. At the molecular level, patulin inhibits aerobic respiration, membrane permeability, and ATPase activity. When it interacts with sulfhydryl-containing amino acids such as cysteine during breakdown, the resulting adducts may be toxic. Tissue injury by patulin is a consequence of sulfhydryl depletion (Barhoumi and Burghardt, 1996).

Patulin is also toxic to bacteria, protozoa, and fungi. In fact, it was tested extensively for possible antibiotic use in humans but proved to be too toxic. It inhibits the growth of bacteria and protozoa, but its toxicity on ruminal or intestinal microflora has not been determined. Mertens (1979) speculated that patulin ingestion would be detrimental to gastrointestinal tract microflora.

No metabolism studies in farm animals have been conducted. Studies in rats indicate rapid metabolism and elimination. Therefore, one could speculate patulin has a low potential for residues in food products of animal origin.

FUSARIUM TOXINS

Zearalenone

The zearalenones (Fig. 5–16) are a family of phenolic compounds produced by several species of *Fusarium* that can cause estrogenic effects

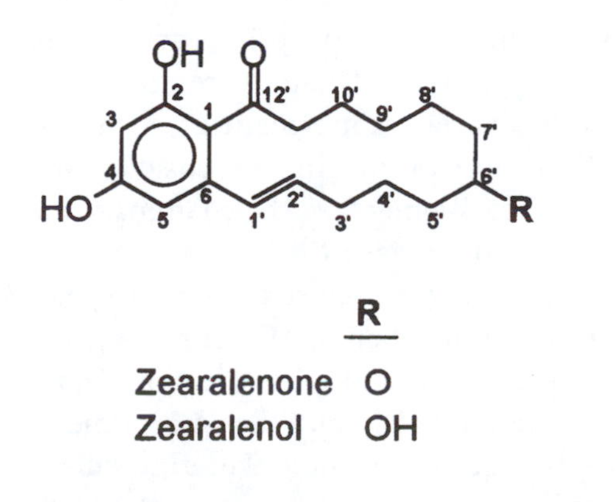

FIGURE 5–16 Structures of zearalenone and zearalenol.

and infertility in animals. These compounds are one of two major categories of *Fusarium* toxins; the other group is the trichothecenes which are discussed in the following section. The chemical name of **zearalenone**, sometimes called **F-2 toxin**, is 6-(10-hydroxy-6-oxo-*trans*-1-undecenyl)-β-resorcyclic acid lactone.

The name zea-ral-en-one is derived as follows: "zea" from the primary host *Zea mays*, "ral" from resorcyclic acid lactone, "-en-" from the double bond at C-1' and C-2', and "-one" from the ketone moiety at C-6.'

Fusarium **spp.** are distributed worldwide and have been shown to infect many important crops and feeds, such as corn, wheat, sorghum, barley, oats, sesame seed, hay, corn silage, and commercial feeds. Corn is clearly the most frequently contaminated crop. Several species of *Fusarium* produce zearalenone, most notably *F.*

roseum (the name of the sexual stage is *Gibberella zeae*). Others include *F. avenaceum, F. nivale, F. sambucinum,* and *F. moniliforme*. Production of zearalenone by *Fusarium* spp. usually occurs in storage when moisture and temperature conditions are optimal. However, in the field, infected corn ears may develop a rot of the crown or cob, aptly named **Gibberella rot**. Toxins other than zearalenone, such as deoxynivalenol, which causes feed refusal and vomiting in swine, are likely to be produced simultaneously but in greater quantities. The ear is most susceptible to Gibberella rot during silking. It is believed that ideal conditions for development of Gibberella rot are chronic rainfall combined with mean temperatures of ≥70°F during silking. Since these climatic conditions are infrequent during the silking stage, Gibberella rot is not an annual problem. However, in the past 25 years there have been three major outbreaks in the midcentral U.S. at about 7-year intervals.

The equivalent of Gibberella rot in wheat, barley, and oats in the field is called **scab**, which is characterized by a dark discoloration of the kernel. Scab is more prevalent on wheat grown in humid and semihumid areas or when sufficient moisture is available during the flowering and early postflowering stages. The amounts of zearalenone produced by either Gibberella rot or scab are usually less than in stored feed. In addition, hyperestrogenic effects are usually not seen in swine fed infected feed because of their sensitivity to the refusal toxins (trichothecenes) which will limit intake.

In stored feeds, copious amounts of zearalenone may be produced by toxigenic molds when conditions are optimal. Whether colonization by *Fusarium* occurs in the field or in storage, growth is optimum at 20°–25°C and a high moisture content (greater than 23%; 45% is optimum). Zearalenone production is stimulated when the temperature drops to about 15°C while the moisture content remains high. These conditions may be encountered in areas such as the midcentral U.S. where cob corn is frequently stored in open cribs. During a wet or humid autumn, if temperatures are warm during the day so that *Fusarium* can become established, the cool night may stimulate the production of zearalenone.

The primary biological effect induced by zearalenone in animals is **hyperestrogenism** which is a chronic toxicosis. Swine, particularly prepubertal females (gilts), are the most commonly affected species, but there is also some evidence for effects in cattle. Estrogenic responses in gilts, which are indistinguishable from those caused by administering excess estradiol, can be caused by feeds containing 1 ppm zearalenone. The syndrome is characterized by swelling of the vulva (vulvovaginitis) and mammary glands, enlargement of the uterus, atrophy of the ovaries, anal prolapse, and vaginal prolapse. Studies in laboratory species indicate that the uterine and mammary effects are induced by an interaction of zearalenone with estrogenic cytosolic receptors in these organs. Moreover, the action of zearalenone on the hypothalamus and pituitary glands appears to be the same as estrogen. In young male swine, zearalenone can cause feminization which includes testicular atrophy, swelling of the prepuce, and enlargement of the mammary glands. In sows, reproductive problems may occur, but greater levels of zearalenone (50–100 ppm) in feed are often required, although Young *et al.* (1990) found reduced litter size when 10 ppm were fed to sows. Disorders include infertility characterized by constant estrus (nymphomania), pseudopregnancy, reduced litter size, smaller and occasionally malformed offspring, and juvenile hyperestrogenism. Other effects in the sow include ovarian abnormalities, death of ova, proliferation of uterine mucosal glands, and ductile development in the mammary gland. In the boar, there is no evidence of reproductive effects. However, reduced spermatogenesis has been noted in other species such as geese.

The sensitivity of gilts in comparison to rats is illustrated by the fact that a 25 times greater dose is required to induce uterine enlargement in young rats. It has been speculated that the lower capacity for metabolism of zearalenone to zearalenol may account for the greater sensitivity of young swine. Although the hydroxylated metabolite zearalenol possesses estrogenic activity, it is excreted more quickly than zearalenone.

Poultry are extremely resistant to zearalenone toxicosis, with virtually no pathology observed with dietary levels as high as 1000 ppm (Leeson *et al.*, 1995).

Cattle and sheep are less sensitive than pigs to the estrogenic effects, with little evidence of practical problems being noted with ruminants. In fact, zearalenone is a widely-used growth promotant for cattle. It is marketed as an implant containing up to 36 mg zearalanol (zeranol), under the trade name Ralgro.

In New Zealand, zearalenone occurs in grasses infected with *Fusarium* spp. (DiMenna *et al.*, 1987, 1991), and has been associated with ewe infertility. Of particular concern to the New Zealand livestock industry is that residues of zearalenone and its metabolites in animal tissues might adversely affect New Zealand exports of animal products to countries which have banned use of synthetic growth promotants such as zeranol (Erasmuson *et al.*, 1994).

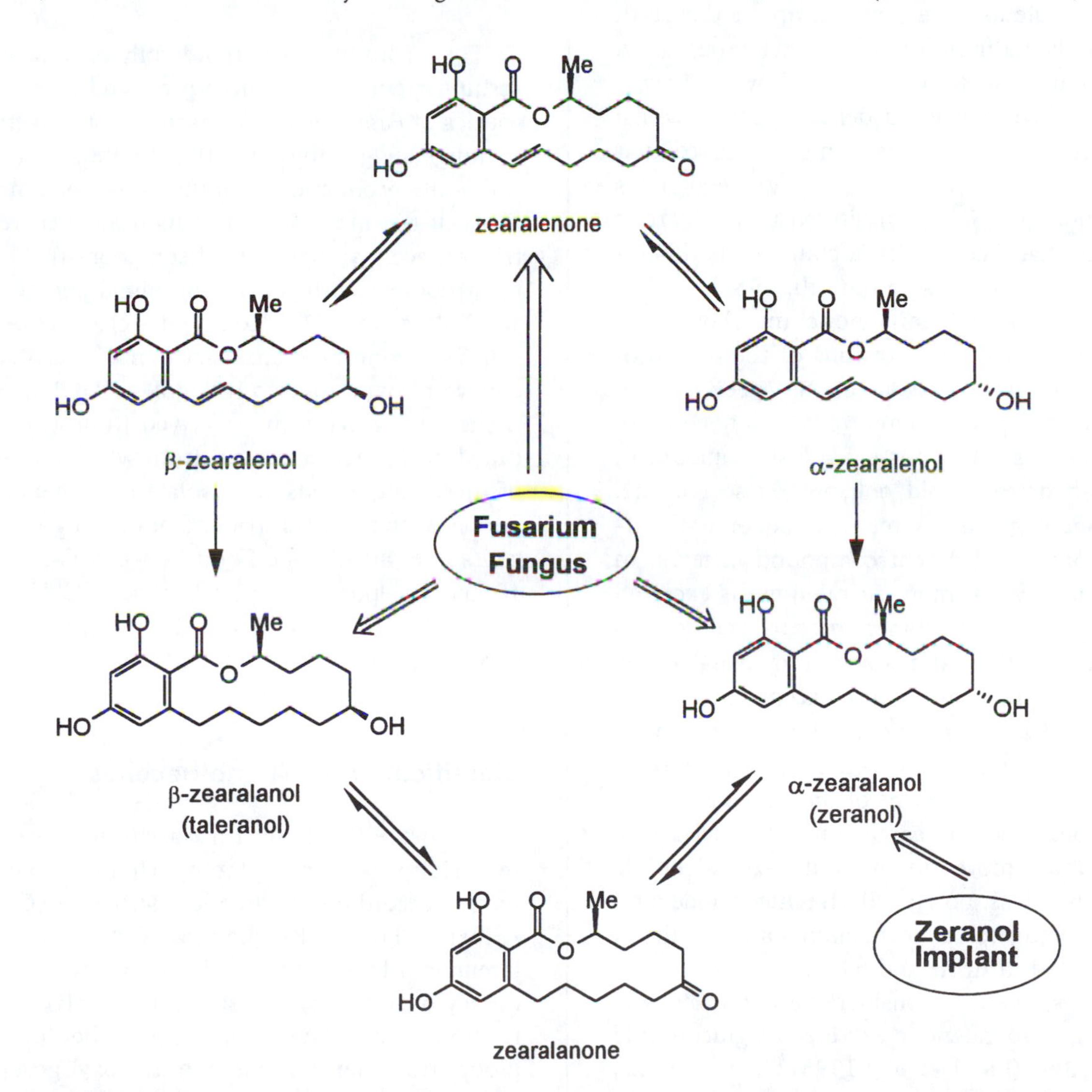

FIGURE 5–17 Metabolism of zearalenone, and relationship to metabolism of the growth promotant zeranol (adapted from Erasmuson *et al.*, 1994).

However, there are techniques capable of separating zearalenone and its metabolites from zeranol and its metabolites (Richard *et al.*, 1993).

Zearalenone is absorbed quite easily from the gastrointestinal tract, as would be expected from its high lipid solubility. Dietary supplementation with binding substances such as anion exchange resins is sometimes used to reduce exposure by decreasing absorption and increasing fecal excretion of zearalenone. Metabolic reduction in the liver gives rise to two stereoisomers of a single metabolite called α- and β-zearalenol; the keto group in the sixth position is reduced to a hydroxyl group (Fig. 5–17). This reduction is catalyzed by an enzyme called 3α-hydroxysteroid dehydrogenase, which exists in several different forms and subcellular locations. This enzyme not only metabolizes zearalenone but is also inhibited by it. There is evidence that feeding alfalfa counteracts this inhibitory effect (James and Smith, 1982).

The extent of zearalenone metabolism to zearalenol and the proportions of the α- and β-isomers are quite variable among species. Compared to rats, pigs metabolize zearalenone more slowly because they possess a lesser amount of the 3α-hydroxysteroid dehydrogenase enzyme. Consequently, pigs excrete zearalenone primarily as conjugated parent compound in feces. In all species tested, more zearalenone is excreted unchanged (as a combination of free and conjugated forms) than as zearalenol. α-Zearalenol is usually the major metabolite except in dairy cattle where β-zearalenol predominates. The physiological importance of α- and β-zearalenol is still unclear, but in terms of their estrogenic potency compared to zearalenone, β-zealalenol is three times more potent and α-zearalenol is considered less potent. Both glucuronide and sulfate conjugates of zearalenone and zearalenol are detected in urine and feces.

In pigs, there is considerable entero-hepatic recycling of zearalenone excreted as glucuronide metabolites (Biehl *et al.*, 1993). This extends the half-life of plasma zearalenone, prolonging the duration of estrogenic effects.

Because zearalenone is metabolized and conjugated, it is eliminated relatively quickly (within a few days) from the body. With high doses, residues of both zearalenone and zearalenol are measurable in liver but do not persist. In lactating cows, less than 1% of a dose is transmitted into milk as free or conjugated forms of zearalenone and zearalenol. In general, neither zearalenone nor its metabolites are significant contaminants of the food chain.

Trichothecenes

The trichothecenes are a family of tetracyclic sesquiterpenoid substances produced by several species of *Fusarium* and at least five other genera of fungi suggesting that the potential for tricothecene production is quite widespread in nature. Of the more than 100 naturally occurring trichothecenes that have been identified, the most notable with regard to animal agriculture are **T-2 toxin**, **HT-2 toxin**, diacetoxyscirpenol (**DAS**), 15-monoacetoxyscirpenol (**15-MAS**), and vomitoxin (deoxynivalenol or **DON**). The trichothecene name was derived from a fungus called *Trichothecium roseum*, from which the first of these compounds was isolated. All members are derivatives of the trichothecane ring system containing an olefinic bond between C-9 and -10 and an epoxy group at C-12 and C-13. The latter is the basis for the name 12,13-epoxytrichothecenes frequently used for these toxins.

Classification of Trichothecenes

Ueno (1977) derived a classification system for trichothecenes, placing them into four groups according to chemical structure. Group A toxins (Fig. 5–18), the largest group, are differentiated by various combinations of hydroxyl or acyloxyl (OAc) substitutions at R_1–R_5 of trichothecene. The simplest member of this group, trichodermol, has one hydroxyl group at R_2. Important members of this class include T-2 toxin, DAS, and monoacetoxyscirpenol.

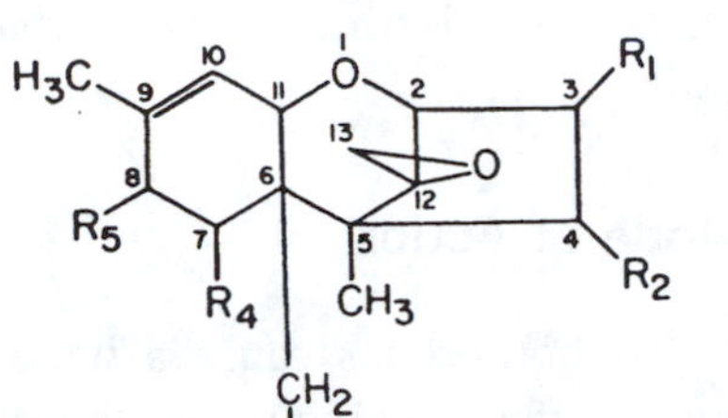

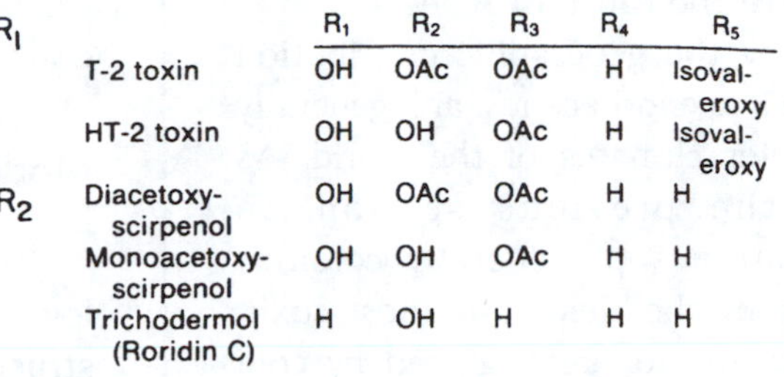

Group A trichothecenes.

	R_1	R_2	R_3	R_4	R_5
T-2 toxin	OH	OAc	OAc	H	Isoval-eroxy
HT-2 toxin	OH	OH	OAc	H	Isoval-eroxy
Diacetoxy-scirpenol	OH	OAc	OAc	H	H
Monoacetoxy-scirpenol	OH	OH	OAc	H	H
Trichodermol (Roridin C)	H	OH	H	H	H

Group B trichothecenes.

	R_1	R_2	R_3	R_4
Nivalenol	OH	OH	OH	OH
Monoacetylnivalenol (fusarenon-X)	OH	OAc	OH	OH
Diacetylnivalenol	OH	OAc	OAc	OH
Deoxynivalenol (vomitoxin)	OH	H	OH	OH

FIGURE 5–18 Structures of Group A and Group B trichothecenes.

Group B trichothecenes (Fig. 5–18) are distinguished by possessing a carbonyl group at R_5 (C-8). Like group A toxins, members of this group possess different combinations of hydroxyl and acyloxyl substitutions at R_1–R_4. The nivalenol analog deoxynivalenol (DON or vomitoxin) is an important member of this group.

Group C is composed of macrocyclic trichothecenes with a smaller number of members owing partially to the lesser number of substitution sites. Genera that produce macrocyclic trichothecenes include *Myrothecium* spp. and *Stachybotrys* spp. but not *Fusarium* spp. A well-known member of this group is satratoxin which is produced by *Stachybotrys atra* and believed to be the cause of the disease in horses called **stachybotryotoxicosis.** Group D is comprised of a single member, crotocin, distinguished from other trichothecenes due to the presence of a second epoxide moiety at C-7 and C-8.

Fusarium spp. that produce group A toxins generally do not produce group B toxins, and vice versa. Although there are exceptions, these fungi tend to be minor contributors to the total production of trichothecenes. Also, *F. roseum*, which is the primary producer of zearalenone, is also known to synthesize both group A and B trichothecenes. However, most species can be subdivided into varieties that produce either group A or group B toxins.

Occurrence

Several diseases of farm animals in different parts of the world are attributed to trichothecenes. The more classical of these are moldy corn toxicosis (T-2 toxin) in North America, red mold disease or Akakabi (nivalenol, vomitoxin, and fusarenon-X) associated with moldy wheat and barley in Japan, bean hull poisoning (neosolaniol and T-2 toxin) in horses in Japan, and stachybotryotoxicosis (satratoxin) in horses, swine, and poultry fed moldy feed and hay in Russia and Hungary. A disease in humans, alimentary toxic aleukia (ATA), was recognized before 1900. Although the cause was not known until recently, several thousand people in the USSR developed ATA and many died after consuming cereals that overwintered in the fields. It is now believed T-2 toxin and/or its derivatives were the cause of ATA.

As indicated by the geographic distribution of these diseases, trichothecenes are generally found in the cooler climates of the world. As described in the previous section on zearalenone, *Fusarium* spp. generally colonize grains or cereals in the field, but most toxin production occurs in storage triggered by cool temperatures (0°–15°C). Head blight disease is the name given to wheat grains infected during the flowering stage. Important environmental factors include high moisture or humidity and both warm and cool temperatures. Greater trichothecene production is usually found in years when the autumn is cool and wet and harvest is delayed. Exposure of cob corn stored in open cribs to moisture coupled with alternating cooling and warming in autumn can result in both fungal colonization (warm temperature) and toxin production (cool temperature). Frost-damaged corn ears tend to be more susceptible to fungal invasion. Trichothecenes are relatively stable, persisting in feeds long after the fungi that produced them have disappeared.

The importance of environmental factors as determinants of the geographical distribution of trichothecenes is exemplified by findings in Japan showing that group A *Fusarium* spp. tend to predominate in the northern districts, whereas group B species tend to frequent the more southern districts. Similarly, laboratory studies indicate the pattern of trichothecenes varies greatly with temperature. For example, *F. tricinctum* produces T-2 toxin and DAS at 8°C and then begins production of HT-2 toxin at 25°C. Comparative growth rates of the various toxigenic species of fungi vary with temperature, so the profile of trichothecenes produced will vary accordingly.

In the U.S., the extent to which farm animals are exposed to trichothecenes, other than DON, is unclear. Consumption of either contaminated corn or wheat is considered the greatest route of potential exposure. In addition to direct feeding of contaminated grains, feed fractions derived from either wet-milling or dry-milling of grains can contain levels of trichothecenes several times greater than the starting grain.

Mode of Action

Trichothecenes cause a wide variety of biological effects owing to the diversity of chemical structures within the group. Whereas these different responses such as neurotoxic, cytotoxic, or immunotoxic effects are related to the type of side-chain substitution, the reactivity of all trichothecenes is due to the presence of both the 12,13-epoxide and the 9-10 double bond in the molecule.

Trichothecenes are considered the most potent inhibitors of **protein synthesis** in eucaryotic cells. The specific site of interference is the translational stage which takes place on the polysomes in the endoplasmic reticulum. Trichothecenes can interfere with all three translational processes, namely, initiation, elongation, and termination. In initiation, ribosomal subunits from a cellular pool join the mRNA at the initiation region to form the first peptide bond. Several initiation factors and an enzyme, peptidyl transferase, are involved. Some trichothecenes inhibit peptidyl transferase and others cause polyribosomes to break down, thereby impairing protein synthesis. After initiation and during elongation, a variable number of amino acids are added to the growing polypeptide chain. This process requires peptidyl transferase and other factors. Termination occurs in response to a termination codon on mRNA followed by release of the polypeptide chain (protein) and mRNA from the ribosome. Trichothecene inhibition of peptidyl transferase and possibly other effects can impair the elongation or termination steps. The site of action of some trichothecenes is preferentially the initiation steps; for others, it is elongation or termination steps. The inhibitory potency varies considerably among the trichothecenes and appears to be closely correlated with the type of R-group substitution on the trichothecene rings.

Although most of the relationships between impaired protein synthesis and clinical effects have not been characterized, one example is the degeneration seen in actively dividing cells such as the thymus, bone marrow, small intestine, testis, and ovary. These effects are in a category of toxic responses sometimes called "radiomimetic" injury, so named because of the similarity to radiation poisoning. Interference with immunologic responses results from damage incurred on the thymus and other lymphoid tissues.

The epoxy group of trichothecenes also can react with sulfhydryl groups on enzymes. This has the potential for numerous biological disturbances. Also, there is some evidence that certain trichothecenes bind to membrane components.

Biological Effects in Different Species

Although trichothecene toxicoses involve a broad spectrum of clinical disorders, those that tend to be common among most species include nausea, vomiting, feed refusal, inflammation, epithelial necrosis, diarrhea, abortion, hemorrhage, hematological changes, and nervous disturbances. The degree to which these effects are manifested varies greatly among different species. Young or immature animals are generally more susceptible than adults, and there does not appear to be a sex difference. Both acute and chronic effects can occur and are generally quite similar in nature.

In swine, DAS and T-2 toxin administered intravenously have LD_{50} values of 0.3 and 1.3 mg/kg, respectively. Acute toxic effects in swine include vomiting, lethargy, frequent defecation, and posterior paresis. Refusal of feed containing greater than 16 ppm T-2 toxin and 10 ppm DAS has been observed. Feed levels from 1 to 10 ppm DAS may cause decreased growth rate and oral necrosis. Levels of T-2 toxin in the range of 8–12 ppm may produce decreased rate of gain and reduced pig and litter size.

In poultry, as in swine, DAS is more toxic than T-2 toxin; these two are the most toxic trichothecenes. Some evidence suggests chickens are less sensitive to oral administration than other animals. Feed containing about 4 ppm of T-2 toxin or DAS can cause oral necrosis, nervous disorders, hepatic hematoma, and reduced weight gain in broilers. A similar dose in young chicks may also cause abnormal feathering. In turkey poults, but not in broilers, reduced feed efficiency accompanies the adverse effect on growth. Poults are not as sensitive as broilers. Greater levels of exposure in broilers can result in severe hematopoietic damage and clinical blood changes such as decreased number of leukocytes and impaired blood coagulation. In laying hens, the toxicosis is similar to that in young chickens except a greater dosage is required. A level of about 20 ppm in feed can cause decreased egg production, egg shell thickness, and shell strength. Return to normal body weight and egg laying occurs gradually over about 2 to 3 weeks, whereas healing of necrotic lesions may take only half as long.

The first natural case of trichothecene toxicosis was discovered in dairy cattle fed moldy corn in 1971 (Hsu *et al.*, 1972). These cattle exhibited a hemorrhagic syndrome after consuming corn containing 1 ppm T-2 toxin. Experimental administration of 0.2 mg T-2 toxin per kilogram body weight to adult cattle caused feed refusal and ruminal ulcers. Hemorrhaging is not produced experimentally by pure T-2, indicating that the 1971 problem may not have been entirely due to T-2 toxin. Other mycotoxins may have been present, but were not analyzed. Calves fed rations containing 10–50 ppm T-2 toxin developed necrotic lesions of the lips, denudation of ruminal papilla, abomasal ulcers, and severe diarrhea. Monoacetoxyscirpenol, a group A toxin (Fig. 5–18), is suspected of causing illness and death in dairy cattle in the north-central U.S. Rumen function does not appear to be adversely affected by trichothecenes. Moreover, there is no evidence that these toxins are metabolized in the rumen.

Feed Refusal and Vomiting

Vomiting (emesis) and feed refusal are associated with most trichothecenes, particularly **DON (vomitoxin)**. Swine are especially sensitive to these toxic effects. Rejection of contaminated feed (feed refusal) causes decreased weight gain and, if enough is consumed, vomiting (emesis) will result. As little as 5% infected kernels or about 10 ppm DON in feed causes refusal in swine. Whether the refusal is due to a localized irritant action in the upper gastrointestinal tract or to palatability is not known. It is not advisable to reverse the refusal by masking moldy feed (e.g., with molasses) because of the possible development of toxic reactions from other trichothecenes that are probably present in the feed.

Vomiting is an acute reaction occurring within a few minutes after either parenteral or oral administration of toxin and may reoccur intermittently over several hours. Swine, dogs, cats, and ducklings appear to be especially sensitive. Ducklings are even sometimes used as a bioassay tool to test for the presence of trichothecenes in feed. Even though DON, a group B toxin, is more associated with vomiting, group A toxins have greater emetic potency. T-2 and HT-2 toxins are about 100 times more potent than DON in ducklings. Whether the vomiting results from gastric irritation or central nervous system stimulation is not known. However, because pretreatment with central depressants such as chlorpromazine suppresses vomiting after fusarenon-X administration, the toxin may act by stimulation of the chemoreceptor trigger zone in the medulla oblongata. After removal of contaminated feed, animals rapidly recover from both feed refusal and vomiting with no apparent permanent damage.

Smith (1992) has suggested that the behavioral changes with trichothecene toxicosis, including loss of appetite, vomiting and lack of muscle coordination, may be mediated through changes in brain neurotransmitters such as serotonin. The hyperaminoacidemia caused by inhibition of hepatic protein synthesis could lead to increased uptake of tryptophan by the brain, due to elevated blood tryptophan levels. This in turn could cause increased serotonin synthesis, and increased serotonergic neuron activity.

Cytoxic Effects

Both T-2 and DAS react upon contact with epithelial tissues such as skin and mucous membranes, causing local cytotoxic effects. Application to the skin causes inflammation, scaling, subepidermal hemorrhaging, and necrosis (**epithelionecrosis**). Guinea pigs are particularly sensitive. Direct contact with contaminated feed has caused local necrotic effects on the snout and legs of pigs. When administered orally, epithelionecrotic effects may occur on the lips or beak, mouth, esophagus, and stomach in all species. The epithelium of the gastrointestinal tract generally undergoes erosion and ulceration, sometimes accompanied by hemorrhaging, resulting in severe gastroenteritis and possibly death. The clinical lesions are exacerbated by the invasion of the necrotic areas by the normal microflora in the area causing infection. Necrotic effects in the intestines are variable among animal species, but diarrhea is common and rectal hemorrhaging sometimes occurs.

The mechanism of epithelionecrosis has not been elucidated. However, there is some evidence that the toxins may increase capillary permeability by direct action on the vessels or by disrupting mast cells, resulting in the release of chemical mediator substances which are biologically reactive and may affect vessel integrity. Group A toxins such as T-2 toxin and DAS have the greatest epitheliotoxicity, with DAS being about five times more toxic than T-2 toxin. Edematous damage of skin tissue is associated with group C macrocyclic trichothecenes, whereas skin reddening is more characteristic of group B toxins.

Blood Effects

Blood effects associated with trichothecene intoxication include hemorrhagic diathesis, decreased hematopoiesis (anemia), and decreased numbers of white blood cells (leukopenia), thrombocytes (thrombocytopenia), and red blood cells (erthrocytopenia). All of these clinical changes are observed in humans suffering from chronic ATA and in cats and chickens administered trichothecenes. Cellular degeneration in bone marrow and spleen, as in other proliferative tissues, is believed to be responsible for the decreased cell numbers in blood. Hemorrhagic diathesis, characterized by blood in the gastrointestinal tract and widespread petechial (focal) hemorrhaging, is thought to be associated with defective blood coagulation. All these effects are probably manifestations of the inhibitory action of trichothecenes on protein synthesis.

In cattle and swine, hemorrhagic diathesis attributed to trichothecene exposure has been observed under field conditions but has been difficult to produce experimentally. However, hemorrhagic bowel lesions usually accompanied by the presence of blood in the feces occurs in cattle, swine, and poultry following acute parenteral administration of T-2 toxin or DAS.

Neurotoxic Effects

Neurological signs resulting from disturbances in the central and autonomic nervous systems include ataxia, impaired righting reflex, hysteroid seizures, and abnormal positioning of wings in poultry treated with T-2 toxin. A decline in body temperature, elevation of blood pressure, and decreased respiratory rate have been measured in laboratory rodents during acute trichothecene intoxication. Posterior paresis (paralysis) occurs in swine acutely intoxicated with DAS. Bean hull poisoning in horses is characterized by staggering and other neurological disturbances. In cats, visual disturbances and meningeal hemorrhage are caused by T-2 toxin. The vomiting reaction in several species may also be a result of a neurological disturbance. A number of nervous disorders associated with ATA in humans are believed to be due to destructive changes in the neurons of the third ventricle and the sympathetic ganglia. These include impaired reflexes and neuropsychiatric symptoms.

Other Biological Effects

Trichothecenes are capable of modifying the immune response. Thymic involution, impaired antibody production, and decreased capacity for clearance of inoculated bacteria are caused by T-2 toxin and DAS. Furthermore, T-2 toxin is able to cross the placental barrier in mice and cause thymic atrophy in the fetus. Immunologic effects have been reviewed by Sharma (1993) and Rotter *et al.* (1996). DON can be immunosuppressive or immunostimulatory, depending upon dose and duration of exposure (Rotter *et al.*, 1996).

There is some evidence in laboratory species that T-2 toxin is both mutagenic and teratogenic. Abortion and retarded growth rate of offspring have been described in cattle, swine, and laboratory species. Long-term feeding studies in rats have not provided consistent results on the question of carcinogenicity. However, benign and malignant tumors of the gastrointestinal tract and brain were found in rats given T-2 toxin in one study, indicating that the possibility does at least exist.

Most (70–80%) of T-2 toxin and its metabolites are excreted via the biliary system into the gastrointestinal tract, and the remainder is excreted in urine. There is some evidence for reabsorption of toxins from the intestines, resulting in enterohepatic recirculation. Moreover, it has been suggested that metabolites of T-2 toxin excreted in bile may be the cause of histopathological lesions in the intestinal tract.

Stachybotryotoxicosis

Stachybotryotoxicosis is a trichothecene-induced disease occurring in animals, particularly horses, after consumption of moldy straw or hay. It has been observed in the USSR, Hungary, Czechoslovakia, Romania, Finland, France, and South Africa. The causative fungi are *Stachybotrys atra* (or *S. alternans*) and possibly *Myrothecium* spp. and *Dendrodochium* spp. Because *Stachybotrys atra* utilizes cellulose, it may be found on any organic matter rich in this nutrient including paper, cotton, plant debris, sugar cane roots, cereal grains, straw and hay. Several toxins collectively called stachybotryotoxins are produced, but not all have been identified. However, the disease is most likely caused by one or more group C macrocyclic trichothecenes including roridin E, verrucarin J, and satratoxin E, G, and H.

Although the disease is best characterized in horses, it is known to occur in cattle, sheep, swine, poultry, and man. In horses, stachybotryotoxicosis progresses in stages after consumption of mold-infested straw or hay over several weeks. Initially, the lips, tongue, and buccal mucosa are irritated due to the epithelionecrotic effects of the toxins. Those areas can become severely necrotic and edematous. The second stage is characterized by coagulopathy, leukopenia, and thrombocytopenia due to toxic effects on hematopoietic processes. Additionally, the necrosis in the mouth area becomes worsened, diarrhea usually develops due to gastrointestinal tract irritation, and animals become quite weak. Death from hemorrhaging and septicemia may occur. A different form of strachybotryotoxicosis develops in horses that ingest a large quantity of toxic fodder. In these cases, clinical signs progress rapidly and include nervous disorders, loss of reflex responses, loss of vision, evidence of circulatory collapse, and finally death.

Other farm animals affected with stachybotryotoxicosis generally exhibit the same toxic signs as horses. In calves, hemorrhaging is widespread throughout the body. Toxic effects in swine include vomiting, tremors, anemia, and abortion. Dermatitis develops around the teat area as well as the mouth area in nursing sows. Skin rash and necrosis have been observed in humans after handling contaminated straw. Farm workers in Hungary were also shown to suffer respiratory distress, nose bleed, and eye irritation after contact with infected litter and fodder. In Hungary, mortalities of sheep occurred when they ingested straw bedding infected with *S. atra* (Harrach *et al.*, 1983).

Studies in laboratory species indicate stachybotryotoxins are also immunosuppressive. Both antibody production and delayed cutaneous hypersensitivity are impaired. These effects have not been studied in domestic animals.

Fumonisins

The fumonisins are mycotoxins produced by the fungus *Fusarium moniliforme* which grows primarily on corn (maize). They were first isolated and identified in South Africa, during studies on the cause of leukoencephalomalacia in horses (Marasas *et al.*, 1988; Kellerman *et al.*, 1990). Fumonisins have since been implicated in numerous human and animal health concerns, and are now recognized as highly significant mycotoxins. They cause liver damage in all species, as well as species-specific target organ toxicity (Smith *et al.*, 1996).

Fumonisins have a simple chemical structure, and are structurally similar to **sphingosine**, a constituent of sphingolipids in nerve tissue. Their biological activity is probably due to disruption of sphingolipid biosynthesis. Six fumonisins have been identified: fumonisins A1, A2, B1, B2, B3 and B4 (Fig. 5–19). The A series are amides while the B series have a free amine group. Various hydroxyl group substitutions account for the difference between individual fumonisins in each series. The major one is fumonisin B1 (FB1).

Fumonisin Toxicosis in Animals

Equine leucoencephalomalacia (ELM) is a lethal disease of horses characterized by neurological effects. It affects horses and other equids, and has been known worldwide since

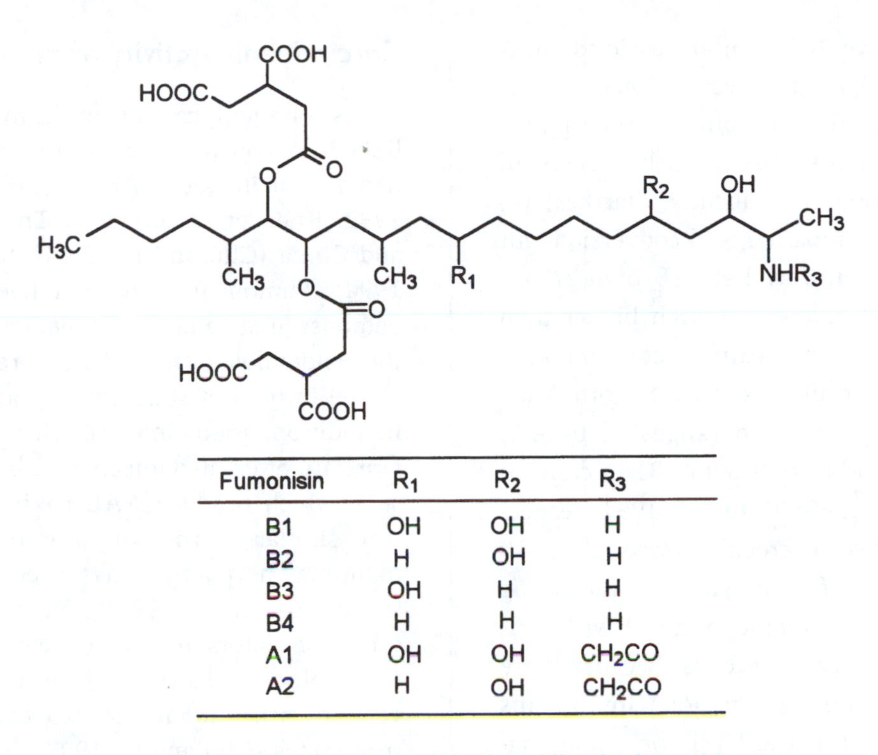

Fumonisin	R_1	R_2	R_3
B1	OH	OH	H
B2	H	OH	H
B3	OH	H	H
B4	H	H	H
A1	OH	OH	CH_2CO
A2	H	OH	CH_2CO

FIGURE 5–19 Structures of the A and B series fumonisins.

the 1800's. The disease results in necrosis of the white matter of the cerebrum. Clinical signs include hypersensitivity, ataxia, posterior weakness, convulsions, and inability to swallow. Centrilobular necrosis of the liver may also occur (Kellerman *et al.*, 1990). The liver lesions are associated with short-term high-dose exposure to fumonisins while brain lesions have a longer-term lower-dose etiology (Ross *et al.*, 1993). The onset of clinical signs is rapid, occurring within a few days of consumption of fumonisin-contaminated grain. The minimum dietary level for induction of ELM is tentatively suggested as 8 ppm FB1 (Ross *et al.*, 1993.) This level has been associated with natural outbreaks of ELM (Thiel *et al.*, 1991).

In pigs, the typical syndrome observed with fumonisin toxicosis is **pulmonary edema**. Characteristic clinical signs are typical of pulmonary insufficiency, such as labored breathing, gasping, cyanosis and death (Osweiler *et al.*, 1992). Consumption of moldy corn or corn screenings is the usual cause. There are year to year variations in fumonisin contamination; in 1989, for example, widespread fumonisin toxicoses in the U.S. occurred. Hepatic lesions are also seen in pigs. It is speculated (Norred, 1993) that disruption of hepatocyte membranes by inhibited sphinogolipid biosynthesis may result in membrane fragments being released into the blood. The lung tissue of pigs is particularly rich in pulmonary intravascular macrophages (PIM) which phagocytize the membrane fragments. This initiates the PIM to release factors that increase capillary permeability, resulting in edema. The absence of lung lesions in other species would thus be explained by their much lower levels of PIM than occur in swine (Norred, 1993). Fumonisins also cause a decrease in heart rate and cardiac output in swine (Smith *et al.*, 1996), which may exacerbate pulmonary hypertension. The pulmonary vasculature is also a target of fumonisin in swine (Casteel *et al.*, 1994).

Poultry are relatively resistant to the toxic effects of FB1 (Bermudez *et al.*, 1995). Fumonisin toxicity in chickens and turkeys has been induced by feeding high levels of FB1 (100–400 ppm). Levels above 200 ppm are toxic to ducklings. Principal signs are liver lesions such as hepatocyte necrosis and biliary hyperplasia (Diaz and Boermans, 1994). Other signs in chicks fed 300 ppm FB1 include diarrhea, reduced growth and impaired feed conversion, gut mucosal lesions and rickets (Brown *et al.*, 1992). These FB1 levels are much higher than might often occur in naturally-contaminated corn. In a survey of midwestern U.S. corn, Murphy *et al.* (1993) reported ranges of 0–14.9, 0–37.9, 0–19.1, and 0–15.8 ppm FB1 in corn for crop years 1988, 1989, 1990 and 1991, respectively. Values for corn screenings were about 10 times higher than for intact corn, indicating that corn screenings should be used with caution in animal and poultry feeds. Because there can be additive effects of various mycotoxins (Kubena *et al.*, 1995), the levels of contamination of concern under field conditions may be lower than toxic levels observed under controlled dosage regimes. Also, *Fusarium moniliforme* produces other mycotoxins, such as **moniliformin**, which may contribute to the toxicity of moldy corn. The interaction of moniliformin and fumonisins may be involved in **spiking mortality syndrome** of broilers (Ledoux *et al.*, 1995), in which young birds exhibit a spike of greater than normal mortality.

Ruminants are less susceptible to fumonisins than are horses, swine and poultry. Osweiler *et al.* (1993) fed diets with 15, 31 and 148 ppm FB1 to feeder cattle, and observed only slight microscopic liver lesions with the highest level. Evidence of minor changes in immune function was noted. Edrington *et al.* (1995) induced acute fumonisin toxicity in lambs by dosing with *Fusarium moniliforme* culture material. Liver and kidney function were impaired. Responses of sheep to levels of fumonisins that might be encountered under normal feeding conditions have not been determined.

Carcinogenic Activity of Fumonisins

Esophageal cancer in humans has been linked to consumption of fumonisin-contaminated corn in South Africa (Gelderblom *et al.*, 1992; Rheeder *et al.*, 1992; Theil *et al.*, 1992) and China (Chu and Li, 1994; Yoshizawa *et al.*, 1994). Fumonisins have not been conclusively demonstrated to be carcinogens in humans, but the epidemiological evidence presented by the above authors is suggestive of an involvement. In addition, grains infected with *Fusarium moniliforme* are often also infected with *Alternaria alternata*, which produce **AAL toxin**, with a structure similar to that of fumonisins. The AAL toxin may have an additive effect with fumonisins (Caldas *et al*, 1994). Fumonisins are liver cancer initiators in rats (Cawood *et al.*, 1994; Gelderblom *et al.*, 1991). *F. moniliforme* can convert nitrate and amines to carcinogenic nitrosamines (Chu and Li, 1994). Thus aflatoxins, fumonisins, AAL toxin nitrosamines may all interact in the carcinogenic action of fungi-infected grains.

Metabolic Mode of Action

Fumonisins are structurally similar to sphingosine (Fig. 5–20). **Sphingosine** is the backbone of the sphingolipids, which have important physiological roles in cell metabolism and membrane structure. Fumonisins inhibit sphingolipid biosynthesis (Wang *et al.*, 1991) by inhibition of sphingamine N-acyltransferase (Ramasamy *et al.*, 1995), an enzyme involved in formation of the precursor (ceramide) of sphingosine (Fig. 5–20). The most sensitive diagnostic indicator of the *in vivo* effects of fumonisins is an elevation of the ratio of sphinganine to sphingosine in urine (Riley *et al.*, 1994) and blood (Wang *et al.*, 1992).

Sphingosine is involved not only in sphingolipid biosynthesis but also functions as a messenger in signal transduction, and inhibits protein kinase C (Norred, 1993). The pathological signs of fumonisin toxicosis can be explained in terms of the effects on the above

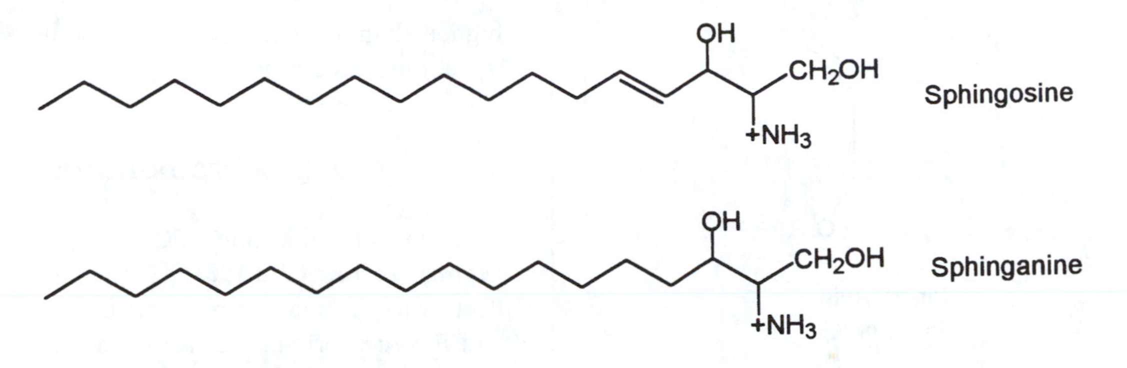

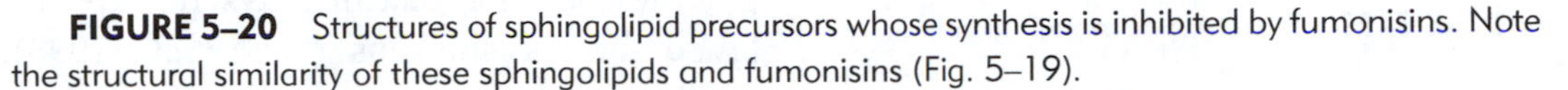

FIGURE 5–20 Structures of sphingolipid precursors whose synthesis is inhibited by fumonisins. Note the structural similarity of these sphingolipids and fumonisins (Fig. 5–19).

functions. The brain has a high content of sphingolipids; thus the degeneration of neuronal cells in ELM is likely due to impaired sphingolipid biosynthesis. Liver damage may be a consequence of derrangements in cell membranes due to depletion of complex sphingolipids, and disruption of normal regulatory mechanisms within cells due to accumulation of sphinganine (Riley *et al.*, 1993).

Fumonisins in Feeds

Fumonisins occur primarily in corn that has been infected by *Fusarium moniliforme* in the field. Other grains do not seem to be affected. Corn screenings contain about 10 times higher fumonisin contents than the intact kernels (Murphy *et al.*, 1993) because the toxin occurs in the outer layers (pericarp) of the seed, which tend to slough off and be segregated as screenings (fines). Physical treatments of corn to remove the pericarps reduces the fumonisin content of the grain (Sydenham *et al.*, 1994). The fumonisins are heat stable (Sydenham *et al.*, 1995). Ammoniation, which is effective in detoxifying aflatoxin, does not reduce the toxicity of fumonisin-contaminated corn (Norred *et al.*, 1991).

Mirocha *et al.* (1992) reported the isolation of FB1 from pasture grass in New Zealand. Domestic deer grazing these pastures suffered from ill thrift and liver dysfunction. Concentrations of 1–9 ppm FB1 were detected. A causal relationship between fumonisins and the deer pathologies was not demonstrated. This report is the first to document the occurrence of fumonisins in grasses.

Horses are more sensitive to fumonisins than other domestic animals. Dietary FB1 levels as low as 8 ppm have been linked to ELM (Ross *et al.*, 1993; Thiel *et al.*, 1991). Toxic thresholds for other species have not been established.

Sydenham *et al.* (1991) in a survey of commercial corn–based human foodstuffs (e.g. cornmeal, corn flakes) from several countries detected widespread contamination with fumonisins, indicating that human exposure occurs. Further data is needed to assess the hazard of such exposure. Fumonisins could be particularly important in Latin American and African cultures where corn is stored under conditions conducive to production of mycotoxins. Interactions among mycotoxins, such as with aflatoxin and fumonisin, could be of particular concern. Both of these toxins cause liver damage and have carcinogenic activity.

Moniliformin

Moniliformin (Fig. 5–21) is a mycotoxin produced by *Fusarium moniliforme*, the same fungus which produces fumonisins. Thus these two mycotoxins frequently occur together in feeds.

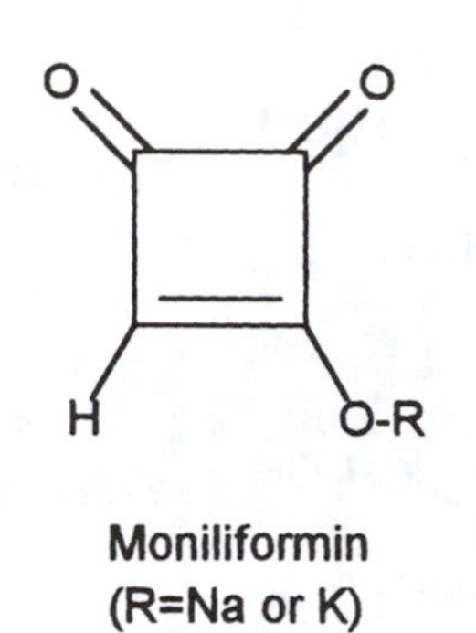

FIGURE 5–21 Structure of moniliformin.

A number of other *Fusarium* spp., such as *F. proliferatum*, also produce moniliforme. It is highly toxic to chickens (Leeson *et al.*, 1995.) The minimum dietary level causing mortality in broilers is 16–27 ppm. Javed *et al.* (1993) investigated the toxic effects of moniliformin and fumonisins. Their toxicities appeared to be additive. Besides reduction in growth, clinical signs of moniliformin toxicosis include depression, weakness, ataxia, dyspnea and gasping terminating in death (Javed *et al.*, 1993). Enlarged heart (cardiomegaly) occurs in broiler chicks fed toxic levels of moniliformin (Ledoux *et al.*, 1995). The cardiotoxic effects may contribute to a **spiking mortality syndrome** of broiler chicks, characterized by a "spike" of high mortality in chicks (Ledoux *et al.*, 1995; Javed *et al.*, 1993). The molecular mode of action of the toxin appears to involve inhibition of oxidative pathways of pyruvate and α-ketoglutarate metabolism in mitochondria (Gathercole *et al.*, 1986), and particularly the inhibition of thiamin-dependent pyruvate dehydrogenase. *F. moniliforme*, in addition to elaborating fumonisins and moniliformin, also produces a thiaminase. Thus there may be an accentuation of moniliformin toxicity by an induced thiamin deficiency (Nagaraj *et al.*, 1994). Supplementation with thiamin partially overcame the toxic effects of *F. proliferatum* culture material to chicks. Nagaraj *et al.* (1994) suggest that the anti-thiamin activity of *Fusarium* spp. is a contributing factor to *Fusarium* mycotoxicoses in poultry, and may explain the commonly observed responses to higher than normal levels of thiamin supplementation in commercial poultry production.

Fusarochromanone

Fusarochromanone (FC) is produced by *Fusarium equiseti* (Wu *et al.*, 1990). The name reflects the chromane ring structure (Fig. 5–22) and *Fusarium* origin. FC has been implicated in the condition of **tibial dyschondroplasia (TD)** of broiler chickens (Wu *et al.*, 1993). TD involves an abnormal cartilage mass in the proximal head of the tibiotarsus, associated with disruption of the blood supply to the growth plate. Normal mineralization of the cartilage does not occur. Affected birds have difficulty in walking. There are numerous factors involved in TD; FC may play a role as a contributory factor (Leeson *et al.*, 1995). FC is also suspected of being involved in Kashin–Beck disease in China, characterized by bone and joint deformation (Xie *et al.*, 1989).

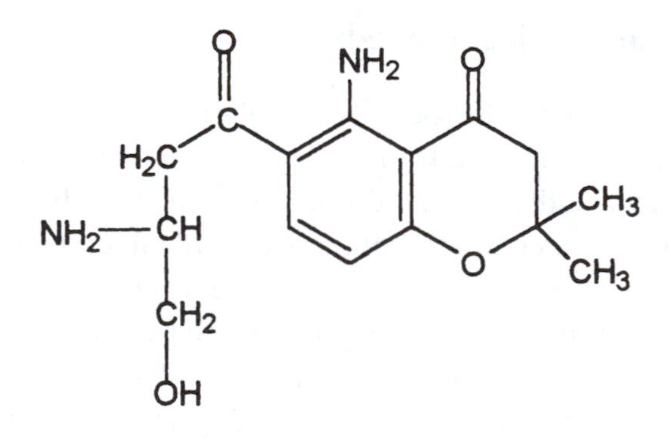

FIGURE 5–22 Structure of fusarochromanone.

Fusaric Acid

Fusaric acid (Fig. 5–23) is produced mainly by *Fusarium moniliforme*. It is a hypotensive agent, and modifies brain neurotransmitter (serotonin, dopamine, norepinephrine) concentrations (Smith and Sousadias, 1993). The toxicity of fusaric acid is very low compared to other *Fusarium* toxins. Chu *et al.* (1993) and Ogunbo

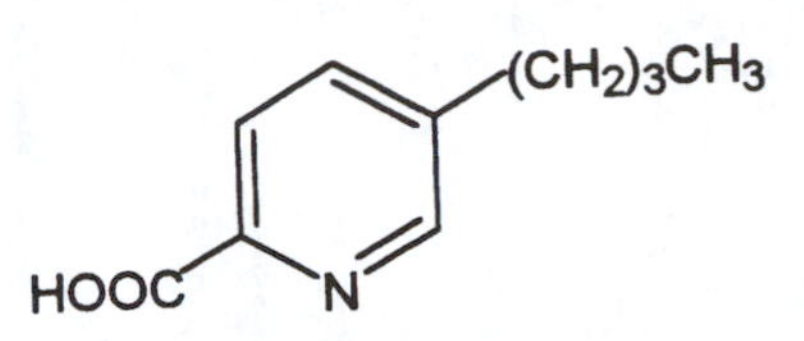

FIGURE 5–23 Structure of fusaric acid.

et al. (1994) fed various levels of fusaric acid to chicks and poults, and observed no effects on performance with levels up to 400 ppm (turkeys) and 150 ppm (chicks). Smith and MacDonald (1991) dosed swine with 200 mg fusaric acid per kg body weight. The pigs exhibited vomiting, lethargy and hypothalamic neurochemical changes. These authors concluded that fusaric acid might act synergistically with trichothecenes to cause feed refusal and vomiting in pigs.

OTHER MYCOTOXINS

Oosporein (Fig. 5–24), a red crystalline substance, was first isolated from the fungus *Oospora colorans,* and is produced by other filamentous fungi including *Chaetomium aureum.* These fungi are common contaminants of cereal grains. Toxicity has been observed only in poultry. The primary lesion involves necrosis of the renal tubules, resulting in gout (deposition of monosodium urate crystals in fluid of soft tissues and joints). Mortality occurs with dietary oosporein levels in excess of about 20 ppm (Leeson *et al.,* 1995). Mortality is a reflection of impaired kidney function (Pegram and Wyatt, 1981).

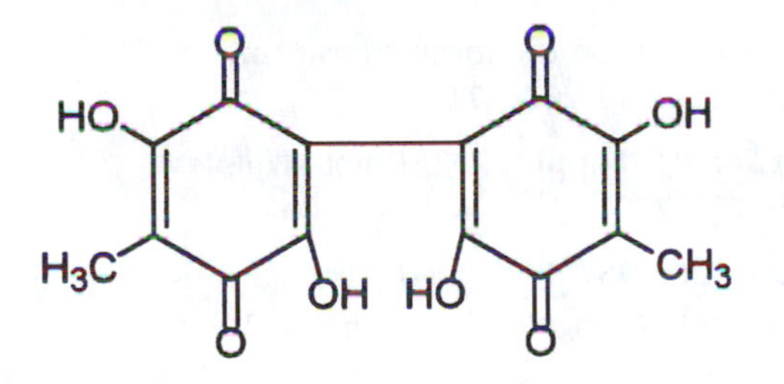

FIGURE 5–24 Structure of oosporein.

Diplodiosis is a disorder of cattle and sheep grazing on harvested corn (maize) fields in winter. It is caused by the consumption of corn infected with the cob rot fungus *Diplodia maydis.* The disease is characterized by ataxia, paresis and paralysis, and stillborn lambs and calves (Kellerman *et al.,* 1988; 1991). Although *Diplodia maydis* occurs world-wide, wherever corn is grown, diplodiosis has occurred only in southern Africa. It is a significant cause of livestock mortality in South Africa.

Another condition reported from South Africa is spinal cord degeneration of dairy cows, associated with the growth of *Aspergillus clavatus* on brewers grains (Van der Lugt, 1994). The presumed mycotoxin has not been identified, although *A. clavatus* is known to produce patulin and a number of tremorgens (Kellerman *et al.,* 1988).

MYCOTOXIN INTERACTIONS

In many instances, feeds may be contaminated with more than one mycotoxin. Mycotoxins may interact with synergistic or additive effects, particularly if they have similar modes of action or target tissues. For example, aflatoxin and fumonisins have additive effects (Harvey *et al.,* 1995). In general, mycotoxins tend to have additive rather than synergistic effects—i.e. the response to a combination of mycotoxins is similar to that which would be predicted from each toxin individually.

REFERENCES

Ergot Alkaloids

Ball, S.E., G. Maurer, M. Zollinger, M. Ladona, and A.E.M. Vickers. 1996. Characterization of the cytochrome P-450 gene family responsible for the *N*-dealkylation of the ergot alkaloid CQA 206-291 in humans. Drug Metab. Disp. 20:56–63.

Coppock, R.W., M.S. Mostrom, J. Simon, D.J. Mckenna, Barry Jacobsen, and H.L. Szlachta. 1989. Cutaneous ergotism in a herd of dairy calves. J. Am. Vet. Med. Assoc. 194:549–551.

Matossian, M.A.K. 1989. Poisons of the Past: Molds, Epidemics and History. Yale University Press, New Haven, CT.

Moubarak, A.S., E.L. Piper, Z.B. Johnson, and M. Flieger. 1996. HPLC method for detection of ergotamine, ergosine, and ergine after intravenous injection of a single dose. J. Agric. Food Chem. 44:146–148.

Porter, J.K., C.W. Bacon, R.D. Plattner, and R.F. Arrendale. 1987. Ergot peptide alkaloid spectra of *Claviceps*-infected tall fescue, wheat, and barley. J. Agric. Food Chem. 35:359–361.

Ross, A.D., W.L. Bryden, W. Bakau, and L.W. Burgess. 1989. Induction of heat stress in beef cattle by feeding the ergots of *Claviceps purpurea*. Aust. Vet. J. 66:247–249.

Rotter, R.G., R.R. Marquardt, and G.H. Crow. 1985a. A comparison of the effect of increasing dietary concentrations of wheat ergot on the performance of Leghorn and broiler chicks. Can. J. Anim. Sci. 65:963–974.

Rotter, R.G., R.R. Marquardt, and J.C. Young. 1985b. Effect of ergot from different sources and of fractionated ergot on the performance of growing chicks. Can. J. Anim. Sci. 65:953–961.

Rotter, R.G., and G.D. Phillips. 1991. Effect of dietary ergot on the mean transit time of digesta in the small intestine of sheep. Can. J. Anim. Sci. 71:767–771.

Scott, P.M., G.A. Lombaert, P. Pellaers, S. Bacler, and J. Lappi. 1992. Ergot alkaloids in grain foods sold in Canada. J. AOAC Intern. 75:773–779.

Wilkinson, R.E., W.S. Hardcastle, and C.S. McCormick. 1987. Seed ergot alkaloid contents of *Ipomoea hederifolia, I. quamoclit, I. cocinea* and *I. wrightii*. J. Sci. Food Agric. 39:335–339.

Aflatoxin

Bodine, A.B., and D.R. Mertens. 1983. Toxicology, metabolism, and physiological effects of aflatoxin in the bovine. pp. 46–50 in: U.L. Diener, R.L. Asquith and J.W. Dickens (Eds.). Aflatoxin and *Aspergillus flavus* in Corn. So. Cooperative Ser. Bull. 279. Craftmaster Printers, Inc., Opelika, AL.

Brown, R.W., A.C. Pier, J.L. Richard, and R.E. Krogstad. 1981. Effects of dietary aflatoxin on existing bacterial intrammary infections in dairy cows. J. Am. Vet. Res. 42:927–933.

Carnaghan, R.B.A., and M. Crawford. 1964. Relationship between ingestion of aflatoxin and primary liver cancer. Br. Vet. J. 120:201–204.

Eaton, D.L., and J.D. Groopman (Eds.). 1994. The Toxicology of Aflatoxins. Academic Press, San Diego.

Hall, A.J., and C.P. Wild. 1994. Epidemiology of aflatoxin-related disease. pp. 233–258 in: D.L. Eaton and J.D. Groopman (Eds.). The Toxicology of Aflatoxins. Academic Press, San Diego.

Hamilton, P.B. 1977. Interrelationship of mycotoxins with nutrition. Fed. Proc., Fed. Am. Soc. Exp. Biol. 36:1899.

Harvey, R.B., L.F. Kubena, and M.H. Elissalde. 1994. Influence of vitamin E on aflatoxicosis in growing swine. Am. J. Vet. Res. 55:572–577.

Hsieh, D.P.H., Z.A. Wong, J.J. Wong, C. Michas, and B. Ruebner. 1977. Comparative metabolism of aflatoxin. pp. 37–50 in: J.V. Rodricks, C.W. Hellsetine and M.A. Mehlman (Eds.). Mycotoxins in Human and Animal Health. Pathotox Publishers, Park Forest South, IL.

Huff, W.E. 1980. Discrepancies between bone ash and toe ash during aflatoxicosis. Poult. Sci. 59:2213–2215.

Lindemann, M.D., D.J. Blodgett, E.T. Kornegay, and G.G. Schurig. 1993. Potential ameliorators of aflatoxicosis in weanling/growing swine. J. Anim. Sci. 71:171–178.

Massey, T.E., R.K. Stewart, J.M. Daniels, and L. Liu. 1995. Biochemical and molecular aspects of mammalian susceptibility to aflatoxin B_1 carcinogenicity. Proc. Soc. Exp. Biol. Med. 208:213–227.

Mertens, D.R. 1979. Biological effects of mycotoxins upon rumen fermentation and lactating dairy cows. pp. 118–136 in: Interactions of Mycotoxins in Animal Production. Natl. Acad. Sci., Washington, D.C.

Olson, J.J., and F.S. Chu. 1993a. Immunochemical studies of urinary metabolites of sterigmatocystin in rats. J. Agric. Food Chem. 41:250–255.

Olson, J.J., and F.S. Chu. 1993b. Urinary excretion of sterigmatocystin and retention of DNA adducts in liver of rats exposed to the mycotoxin: An immunochemical analysis. J. Agric. Food Chem. 41:602–606.

Park, D.L., L.S. Lee, R.L. Price, and A.E. Pohland. 1988. Review of the decontamination of aflatoxins by ammoniation: Current status and regulation. J. Assoc. Off. Anal. Chem. 71:685–703.

Phillips, T.D., B.A. Clement, and D.L. Park. 1994. Approaches to reduction of aflatoxins in foods and feeds. pp. 383–406 in: D.L. Eaton and J.D. Groopman (Eds.). The Toxicology of Aflatoxins. Academic Press, San Diego.

Van Egmond, H.P. 1994. Aflatoxins in milk. pp. 365–381 in: D.L. Eaton and J.D. Groopman (Eds.). The Toxicology of Aflatoxins. Academic Press, San Diego.

Wicklow, D.T. 1988. Metabolites in the coevolution of fungal chemical defence systems. pp. 173–201 in: K.A. Pirozynski and D.L. Hawksworth (Eds.). Coevolution of Fungi with Plants and Animals. Academic Press, San Diego.

Ochratoxin

Galtier, P., M. Alvineric, and J.L. Charpenteau. 1981. The pharmacokinetic profiles of ochratoxin A in pigs, rabbits and chickens. Food Cosmet. Toxicol. 19:735–738.

Haazele, F.M., W. Guenter, R.R. Marquardt, and A.A. Frohlich. 1993. Beneficial effects of dietary ascorbic acid supplement on hens subjected to ochratoxin A toxicosis under normal and high ambient temperatures. Can. J. Anim. Sci. 73:149–157.

Kanisawa, M. 1984. Synergistic effect of citrinin on hepatorenal carcinogenesis of ochratoxin A in mice. pp. 245–254 in: H. Kurata and Y. Ueno (Eds.). Toxigenic Fungi—Their Toxins and Health Hazard. Elsevier, NY.

Kiessling, K-H., H. Pettersson, K. Sandholm, and M. Olsen. 1984. Metabolism of aflatoxin, ochratoxin, zearalenone, and three trichothecenes by intact rumen fluid, rumen protozoa, and rumen bacteria. Appl. Environ. Microbiol. 47:1070–1073.

Krogh, P. 1978. Mycotoxicoses of animals. Mycopathologia 65:43–45.

Leeson, S., G. Diaz, and J.D. Summers. 1995. Poultry Metabolic Disorders and Mycotoxins. University Books, Guelph, Canada.

Leistner, L. 1984. Toxigenic penicillia occurring in feeds and foods. pp. 162–171 in: H.H. Kurata and Y. Ueno (Eds.). Toxigenic Fungi—Their Toxins and Health Hazard. Elsevier, NY.

Marquardt, R.R., and A.A. Frohlich. 1992. A review of recent advances in understanding ochratoxicosis. J. Anim. Sci. 70:3968–3988.

Pier, A.C., S.J. Cysewski, J.L. Richard, A.L. Baetz, and L. Mitchell. 1976. pp. 130–148 in: Experimental Mycotoxicosis in Calves with Aflatoxin, Ochratoxin, Rubratoxin, and T-2 Toxin. Proc. U.S. Anim. Health Assoc.

Stoloff, L. 1979. Mycotoxin residues in edible animal tissues. pp. 157–166 in: Interactions of Mycotoxins in Animal Production. Natl. Acad. Sci., Washington, D.C.

Westlake, K., R.I. Mackie, and M.F. Dutton. 1989. In vitro metabolism of mycotoxins by bacterial, protozoal and ovine ruminal fluid preparations. Anim. Feed Sci. Tech. 25:169–178.

Xiao, H., R.R. Marquardt, A.A. Frohlich, G.D. Phillips and T.G. Vitti. 1991a. Effect of a hay and a grain diet on the rate of hydrolysis of ochratoxin A in the rumen of sheep. J. Anim. Sci. 69:3706–3714.

Xiao, H., R.R. Marquardt, A.A. Frohlich, G.D. Phillips, and T.G. Vitti. 1991b. Effect of a hay and a grain diet on the bioavailability of ochratoxin A in the rumen of sheep. J. Anim. Sci. 69:3715–3723.

Citrinin

Berndt, W.O., A.W. Hayes, and R.D. Phillips. 1980. Effects of mycotoxins on renal function: Mycotoxic nephropathy. Kidney Int. 18:656–664.

Hanika, C., W.W. Carlton, and J. Tuite. 1983. Citrinin mycotoxicosis in the rabbit. Food Chem. Toxicol. 21:487–494.

Leeson, S., G. Diaz, and J.D. Summers. 1995. Poultry Metabolic Disorders and Mycotoxins. University Books, Guelph, Canada.

Nelson, T.S., J.N. Beasley, L.K. Kirby, Z.B. Johnson, G.C. Ballam, and M.M. Campbell. 1981. Citrinin toxicity in growing chicks. Poult. Sci. 60:2165–2166.

Cyclopiazonic Acid

Bradburn, N., R.D. Coker, and G. Blunden. 1994. The aetiology of turkey "X" disease. Phytochem. 35:817.

Dorner, J.W., R.J. Cole, D.J. Erlington, S.Suksupath, G.H. McDowell, and W.L. Bryden. Cyclopiazonic acid residues in milk and eggs. J. Agric. Food Chem. 42:1516–1518.

Goeger, D.E., and R.T. Riley. 1989. Interaction of cyclopiazonic acid with rat skeletal muscle sarcoplasmic reticulum vesicles. Biochem. Pharmacol. 38:3995–4003.

Leeson, S., G. Diaz, and J.D. Summers. 1995. Poultry Metabolic Disorders and Mycotoxins. University Books, Guelph, Canada.

Kojic Acid

Giroir, L.E., W.E. Huff, L.E. Kubena, R.B. Harvey, M.H. Elissalde, D.A. Witzel, A.G. Yersin, and G.W. Ivie. 1991. Toxic effects of kojic acid in the diet of male broilers. Poult. Sci. 70:499–503.

Rubratoxins

Mirocha, C.J. 1980. Rubratoxin, sterigmatocystin and stacchybotrys mycotoxins. pp. 152–176 in: W. Shimoda (Ed.). Conference on Mycotoxins in Animal Feeds and Grains Related to Animal Health. National Technical Information Service, U.S. Dept. of Commerce, Springfield, VA.

Pier, A.C. 1981. Mycotoxins and animal health. Adv. Vet. Sci. Comp. Med. 25:185–243.

Pier, A.C., S.J. Cysewski, J.L. Richard, A.L. Baetz, and L. Mitchell. 1976. Experimental mycotoxicosis in calves with aflatoxin, ochratoxin, rubratoxin, and T-2 toxin. Proc. U.S. Anim. Health Assoc. pp. 130–148.

Unger, P.D., and A.W. Hayes. 1979. Disposition of rubratoxin B in the rat. Toxicol. Appl. Pharmacol. 47:585–591.

Wogan, G.N., G.S. Edwards, and P.M. Newberne. 1971. Acute and chronic toxicity of rubratoxin B. Toxicol. Appl. Pharmacol. 19:712–720.

Patulin

Barhoumi, R., and R.C. Burghardt. 1996. Kinetic analysis of the chronology of patulin- and gossypol-induced cytotoxicity *in vitro*. Fund. Appl. Toxicol. 30:290–297.

Ciegler, A. 1977. Patulin. pp. 609–624 in: J.V. Rodricks, C.W. Hesseltine and M.A. Mehlman (Eds.). Mycotoxins in Human and Animal Health. Pathotox Publishers, Park Forest South, IL.

Lynch, G.P. 1979. Effects of mycotoxins on ruminants. pp. 96–117 in: Interactions of Mycotoxins in Animal Production. Natl. Acad. Sci., Washington, DC.

Mertens, D.R. 1979. Biological effects of mycotoxins upon rumen fermentation and lactating dairy cows. pp. 118–136 in: Interactions of Mycotoxins in Animal Production. Natl. Acad. Sci., Washington, D.C.

Zearalenone

Biehl, M.L., D.B. Prelusky, G.D. Koritz, K.E. Hartin, W.B. Buck, and H.L. Trenholm. 1993. Biliary excretion and enterohepatic cycling of zearalenone in immature pigs. Toxicol. Appl. Pharmacol. 121:152–159.

DiMenna, M.E., D.R. Lauren, P.R. Poole, P.H. Mortimer, R.A. Hill, and M.P. Agnew. 1987. Zearalenone in New Zealand pasture herbage and the mycotoxin-producing potential of *Fusarium* species from pasture. N.Z.J. Agric. Res. 30:499–504.

DiMenna, M.E., D.R. Lauren, J.M. Sprosen, and K.S. MacLean. 1991. *Fusarium* and zearalenone on herbage fractions from short and from long pasture. N.Z.J. Agric. Res. 34:445–452.

Eienne, M., and J.-Y. Dourmad. 1994. Effects of zearalenone or glucosinolates in the diet on reproduction in sows: A review. Livestock Production Science 40:99–113.

Erasmuson, A.F., B.G. Scahill, and D.M. West. 1994. Natural zeranol (a-zearalanol) in the urine of pasture-fed animals. J. Agric. Food Chem. 42:2721–2725.

James, L.J., and T.K. Smith. 1982. Effect of dietary alfalfa on zearalenone toxicity and metabolism in rats and swine. J. Anim. Sci. 55:110–118.

Leeson, S., G. Diaz, and J.D. Summers. 1995. Poultry Metabolic Disorders and Mycotoxins. University Books, Guelph, Canada.

Richard, J.L., G.A. Bennett, P.F. Ross, and P.E. Nelson. 1993. Analysis of naturally occurring mycotoxins in feedstuffs and food. J. Anim. Sci. 71:2563–2574.

Young, L.G., H. Ping, and G.J. King. 1990. Effects of feeding zearalenone to sows on rebreeding and pregnancy. J. Anim. Sci. 68:15–20.

Trichothecenes

Charmley, E., H.L. Trenholm, B.K. Thompson, D. Vudathala, J.W.G. Nicholson, D.B. Prelusky, and L.L. Charmley. 1993. Influence of level of deoxynivalenol in the diet of dairy cows on feed intake, milk production, and its composition. J. Dairy Sci. 76:3580–3587.

Friend, D.W., B.K. Thompson, H.L. Trenholm, H.J. Boermans, K.E. Hartin, and P.L. Panich. 1992. Toxicity of T-2 toxin and its interaction with deoxynivalenol when fed to young pigs. Can. J. Anim. Sci. 72:703–711.

Harrach, B., A. Bata, E. Bajmocy, and M. Benko. 1983. Isolation of satratoxins from the bedding straw of a sheep flock with fatal stachybotryotoxicosis. Appl. Environ. Microbiol. 45:1419–1422.

Hsu, I.C., E.B. Smalley, F.M. Strong, and W.E. Ribellin. 1972. Identification of T-2 toxin in moldy corn associated with a lethal toxicosis in dairy cattle. Appl. Microbiol. 24:682–690.

Rotter, B.A., D.B. Prelusky, and J.J. Pestka. 1996. Toxicology of deoxynivalenol (vomitoxin). J. Toxicol. Environ. Health 48:1–34.

Rotter, B.A., B.K. Thompson, and M. Lessard. 1995. Effects of deoxynivalenol-contaminated diet on performance and blood parameters in growing swine. Can. J. Anim. Sci. 75:297–302.

Sharma, R.P. 1993. Immunotoxicity of mycotoxins. J. Dairy Sci. 76:892–897.

Smith, T.K. 1992. Recent advances in the understanding of *Fusarium* trichothecene mycotoxicoses. J. Anim. Sci. 70:3989–3993.

Trenholm, H.L., B.C. Foster, L.L. Charmley, B.K. Thompson, K.E. Hartin, R.W. Coppock, and M.A. Albassam. 1994. Effects of feeding diets containing *Fusarium* (naturally) contaminated wheat or pure deoxynivalenol (DON) in growing pigs. Can. J. Anim. Sci. 74:361–369.

Ueno, Y. 1977. Trichothecenes: Overview address. pp. 189–207 in: J.V. Rodricks, C.W. Hesseltine and M.A. Mehlman (Eds.). Mycotoxins in Human and Animal Health. Pathotox Publishers, Park Forest South, IL.

Fumonisins

Bermudez, A.J., D.R. Ledoux, and G.E. Rottinghaus. 1995. Effects of fusarium moniliforme culture material containing known levels of fumonisin B_1 in ducklings. Avian Dis. 39:879–886.

Brown, T.P., G.E. Rottinghaus, and M.E. Williams. 1992. Fumonisin mycotoxicosis in broilers: Performance and pathology. Avian Dis. 36:450–454.

Caldas, E.D., A.D. Jones, B.Ward, C.K. Winter, and D.G. Gilchrist. 1994. Structural characterization of three new AAL toxins produced by *Alternaria alternata* f. sp. *lycopersici*. J. Agric. Food Chem. 42:327–333.

Casteel, S.W., J.R. Turk, and G.E. Rottinghaus. 1994. Chronic effects of dietary fumonisin on the heart and pulmonary vasculature of swine. Fund. Appl. Toxicol. 23:518–524.

Cawood, M.E., W.C.A. Gelderblom, J.F. Alberts, and S.D. Snyman. 1994. Interaction of ^{14}C-labelled fumonisin B mycotoxins with primary rat hepatocyte cultures. Fd. Chem. Toxicol. 32:627–632.

Chu, F.S., and G.Y. Li. 1994. Simultaneous occurrence of fumonisin B_1 and other mycotoxins in moldy corn collected from the People's Republic of China in regions with high incidences of esophageal cancer. Appl. Environ. Microbiol. 60:847–852.

Diaz, G.J., and H.J. Boermans. 1994. Fumonisin toxicosis in domestic animals: A review. Vet. Human Toxicol. 36:548–555.

Edrington, T.S., C.A. Kamps-Holtzapple, R.B. Harvey, L.F. Kubena, M.H. Elissalde, and G.E. Rottinghaus. 1995. Acute hepatic and renal toxicity in lambs dosed with fumonisin-containing culture material. J. Anim. Sci. 73:508–515.

Gelderblom, W.C.A., N.P.J. Kriek, W.F.O. Marasas, and P.G. Thiel. 1991. Toxicity and carcinogenicity of the *Fusarium moniliforme* metabolite, fumonisin B_1, in rats. Carcinogenesis 12:1247–1251.

Gelderblom, W.C.A., E. Semple, W.F.O. Marasas, and E. Farber. 1992. The cancer-initiating potential of the fumonisin B mycotoxins. Carcinogenesis 13:433–437.

Kellerman, T.S., W.F.O. Marasas, P.G. Thiel, W.C.A. Gelderblom, M. Cawood, and J.A.W. Coetzer. 1990. Leukoencephalomalacia in two horses induced by oral dosing of fumonisin B_1. Onderstepoort J. Vet. Res. 57:269–275.

Kubena, L.F., T.S. Edrington, C. Kamps-Holtzapple, R.B. Harvey, M.H. Elissalde, and G.E. Rottinghaus. 1995. Influence of fumonisin B_1, present in *Fusarium moniliforme* culture material, and T-2 toxin on turkey poults. Poult. Sci. 74:306–313.

Ledoux, D.R., A.J. Bermudez, G.E. Rottinghaus and, J. Broomhead. 1995. Effects of feeding *Fusarium fujikuroi* culture material, containing known levels of moniliformin, in young broiler chicks. Poult. Sci. 74:297–305.

Marasas, W.F.O., T.S. Kellerman, W.C.A. Gelderblom, J.A.W. Coetzer, P.G. Thiel, and J.J. Van Der Lugt. 1988. Leukoencephalomalacia in a horse induced by fumonisin B_1 isolated from *Fusarium moniliforme*. Onderstepoort J. Vet. Res. 55:197–203.

Mirocha, C.J., C.G. Mackintosh, U.A. Mirza, W. Xie, Y. Xu, and J. Chen. 1992. Occurrence of fumonisin in forage grass in New Zealand. Appl. Environ. Microbiol. 58:3196–3198.

Murphy, P.A., L.G. Rice, and P.F. Ross. 1993. Fumonisin B_1, B_2, and B_3 content of Iowa, Wisconsin, and Illinois corn and corn screenings. J. Agric. Food Chem. 41:263–266.

Norred, W.P. 1993. Fumonisins—Mycotoxins produced by *Fusarium moniliforme*. J. Toxicol. Environ. Health 38:309–328.

Norred, W.P., K.A. Voss, C.W. Bacon, and R.T. Riley. 1991. Effectiveness of ammonia treatment in detoxification of fumonisin-contaminated corn. Fd. Chem. Toxicol. 29:815–819.

Osweiler, G.D., M.E. Kehrli, J.R. Stabel, J.R. Thurston, P.F. Ross, and T.M. Wilson. 1993. Effects of fumonisin-contaminated corn screenings on growth and health of feeder calves. J. Anim. Sci. 71:459–466.

Osweiler, G.D., P.F. Ross, T.M. Wilson, P.E. Nelson, S.T. Witte, T.L. Carson, L.G. Rice, and H.A. Nelson. 1992. Characterization of an epizootic of pulmonary edema in swine associated with fumonisin in corn screenings. J. Vet. Diagn. Invest. 4:53–59.

Ramasamy, S., E. Wang, B. Hennig, and A.H. Merrill, Jr. 1995. Fumonisin B_1 alters sphingolipid metabolism and disrupts the barrier function of endothelial cells in culture. Toxicol. Appl. Pharmacol. 133:343–348.

Rheeder, J.P., W.F.O. Marasas, P.G. Thiel, E.W. Sydenham, G.S. Shephard, and D.J. van Schalkwyk. 1992. *Fusarium moniliforme* and fumonisins in corn in relation to human esophageal cancer in Transkei. Phytopathology 82:353–357.

Riley, R.T., D.M. Hinton, W.J. Chamberlain, C.W. Bacon, E. Wang, A.H. Merrill, Jr., and K.A. Voss. 1994. Dietary fumonisin B_1 induces disruption of sphingolipid metabolism in Sprague–Dawley rats: A new mechanism of nephrotoxicity. J. Nutr. 124:594–603.

Riley, R.T., W.P. Norred, and C.W. Bacon. 1993. Fungal toxins in foods: Recent concerns. Annu. Rev. Nutr. 13:167–189.

Ross, P.F., A.E. Ledet, D.L. Owens, L.G. Rice, H.A. Nelson, G.D. Osweiler, and T.M. Wilson. 1993. Experimental equine leukoencephalomalacia, toxic hepatosis, and encephalopathy caused by corn naturally contaminated with fumonisins. J. Vet. Diagn. Invest. 5:69–74.

Smith, G.W., P.D. Constable, C.W. Bacon, F.I. Meredith, and W.M. Haschek. 1996. Cardiovascular effects of fumononisins in swine. Fund. Appl. Toxicol. 31:169–172.

Sydenham, E.W., G.S. Shephard, P.G. Thiel, W.F.O. Marasas, and S. Stockenstrom. 1991. Fumonisin contamination of commercial corn–based human foodstuffs. J. Agric. Food Chem. 39:2014–2018.

Sydenham, E.W., S. Stockenstrom, P.G. Thiel, G.S. Shephard, K.R. Koch, and W.F.O. Marasas. 1995. Potential of alkaline hydrolysis for the removal of fumonisins from contaminated corn. J. Agric. Food Chem. 43:1198–1201.

Sydenham, E.W., L. Van der Westhuizen, S. Stockenstrom, G.S. Shephard, and P.G. Thiel. 1994. Fumonisin-contaminated maize: physical treatment for the partial decontamination of bulk shipments. Food Addit. Contam. 11:25–32.

Thiel, P.G., W.F.O. Marasas, E.W. Sydenham, G.S. Shephard, and W.C.A. Gelderblom. 1992. The implications of naturally occurring levels of fumonisins in corn for human and animal health. Mycopathologia 117:3–9.

Thiel, P.G., G.S. Shephard, E.W. Sydenham, W.F.O. Marasas, P.E. Nelson, and T.M. Wilson. 1991. Levels of fumonisins B_1 and B_2 in feeds associated with confirmed cases of equine leukoencephalomalacia. J. Agric. Food Chem. 39:109–111.

Wang, E., W.P. Norred, C.W. Bacon, R.T. Riley, and A.H. Merrill, Jr. 1991. Inhibition of sphingolipid biosynthesis by fumonisins. Implications for diseases associated with *Fusarium moniliforme*. J. Biol. Chem. 266:14486–14490.

Wang, E., P.F. Ross, T.M. Wilson, R.T. Riley, and A.H. Merrill, Jr. 1992. Alteration of serum sphingolipids upon dietary exposure of ponies to fumonisins, mycotoxins produced by *Fusarium moniliforme*. J. Nutr. 122:1706–1716.

Yoshizawa, T., A. Yamashita, and Y. Luo. 1994. Fumonisin occurrence in corn from high- and low-risk areas for human esophageal cancer in China. Appl. Environ. Microbiol. 60:1626–1629.

Moniliformin

Gathercole, P.S., P.G. Thiel, and J.H.S. Hofmeyr. 1986. Inhibition of pyruvate dehydrogenase complex by moniliformin. Biochem. J. 233:719–723.

Javed, T., G.A. Bennett, J.L. Richard, M.A. Dombrink-Kurtzman, L.M. Cote, and W.B. Burk. 1993. Mortality in broiler chicks on feed amended with *Fusarium proliferatum* culture material or with purified fumonisin B_1 and moniliformin. Mycopath. 123:171–184.

Ledoux, D.R., A.J. Bermudez, G.E. Rottinghaus, J. Broomhead, and G.A. Bennett. 1995. Effects of feeding *Fusarium fujikuroi* culture material, containing known levels of moniliformin, in young broiler chicks. Poult. Sci. 74:297–305.

Leeson, S., G. Diaz, and J.D. Summers. 1995. Poultry Metabolic Disorders and Mycotoxins. University Books, Guelph, Canada.

Nagaraj, R.Y., W.D. Wu, and R.F. Vesonder. 1994. Toxicity of corn culture material of *Fusarium proliferatum* M-7176 and nutritional intervention in chicks. Poult. Sci. 73:617–626.

Fusarochromanone

Leeson, S., G. Diaz, and J.D. Summers. 1995. Poultry Metabolic Disorders and Mycotoxins. University Books, Guelph, Canada.

Wu, W., M.E. Cook, Q. Chu, and E.B. Smalley. 1993. Tibial dyschondroplasia of chickens induced by fusarochromanone, a mycotoxin. Avian Dis. 37:302–309.

Wu, W., P.E. Nelson, M.E. Cook, and E.B. Smalley. 1990. Fusarochromanone production by *Fusarium* isolates. Appl. Environ. Microbiol. 56:2989–2993.

Xie, W., C.J. Mirocha, R.J. Pawlowsky, Y. Wen, and X. Xu. 1989. Biosynthesis of fusarochromanone and its monoacetyl derivative by *Fusarium equeseti*. Appl. Environ. Microbiol. 55:794–797.

Fusaric Acid

Chu, Q., W. Wu, and E.B. Smalley. 1993. Decreased cell-mediated immunity and lack of skeletal problems in chicks consuming diets amended with fusaric acid. Avian Dis. 37:863–867.

Ogunbo, S., D.R. Ledoux, A.J. Bermundez, and G.E. Rottinghaus. 1994. Effects of fusaric acid on broiler chicks and turkey poults. Poult. Sci. 73(Suppl. 1):154.

Smith, T.K., and E.J. MacDonald. 1991. Effect of fusaric acid on brain regional neurochemistry and vomiting behavior in swine. J. Anim. Sci. 69:2044–2049.

Smith, T.K., and M.G. Sousadias. 1993. Fusaric acid content of swine feedstuffs. J. Agric. Food Chem. 41:2296–2298.

Other Mycotoxins

Kellerman, T.S., J.A.W. Coetzer, and T.W. Naude. 1988. Plant Poisonings and Mycotoxicoses of Livestock in Southern Africa. Oxford University Press, Cape Town.

Kellerman, T.S., L. Prozesky, R. Anitra Schultz, C.J. Rabie, H. Van Ark, B.P. Maartens, and A. Lubben. 1991. Perinatal mortality in lambs of ewes exposed to cultures of *Diplodia maydis* (= *Stenocarpella maydis*) during gestation. Onderstepoort J. Vet. Res. 58:297–308.

Leeson, S., G. Diaz, and J.D. Summers. 1995. Poultry Metabolic Disorders and Mycotoxins. University Books, Guelph, Canada.

Pegram, R.A., and R.D. Wyatt. 1981. Avian gout caused by oosporein, a mycotoxin produced by *Chaetomium trilaterale*. Poult. Sci. 60:2429–2440.

Van der Lugt, J.J., T.S. Kellerman, A. Van Vollenhoven, and P.W. Nel. 1994. Spinal cord degeneration in adult dairy cows associated with the feeding of sorghum beer residues. J. S. Afr. Vet. Med. Assoc. 65:184–188.

Mycotoxin Interactions

Harvey, R.B., T.S. Edrington, F. Kubena, M.H. Elissalde, and G.E. Rottinghaus. 1995. Influence of aflatoxin and fumonisin B_1-containing culture material on growing barrows. Am. J. Vet. Res. 56:1668–1672.

CHAPTER 6

Toxins Intrinsic to Cereal Grains and Other Concentrate Feeds

As previously mentioned in Chapter 5, cereal grains have relatively few intrinsic or inherent toxins. Tannins in grain sorghum and protease inhibitors in triticale and rye are examples of intrinsic toxins. In terms of the broad definition of a toxicant provided in Chapter 1, phytates and non-starch polysaccharides are considered here as toxicants, because they are constituents of grains which have adverse biological and economic effects in animal production.

FIGURE 6–1 Sorghum grain contains condensed tannins, which help protect against wild bird damage. The small seeds are in an open panicle, readily accessible to birds.

PHYTATES

Much of the phosphorus in seeds is in an organic form, as phosphorylated myo-inositol, also known as **phytic acid**. This organic phosphorus represents a storage form of the element, to be used by the germinating seed. Phytic acid can form complexes with divalent mineral ions, called **phytates** (Fig. 6–2). Both phytic acid and phytates are important in nutrition, because they reduce the bioavailability of phosphorus and other minerals. This is particularly important in poultry, swine, humans and other non-ruminants. In cereal grains, 60–80% of the total phosphorus may be bound as phytic acid. In most grains, much of the phytic acid (and phosphorus) occurs in the outer bran layers of the seed, whereas in corn, 90% of the phytate is in the germ portion of the kernel. Thus corn grain and wheat processing by-products (e.g. wheat bran, wheat middlings) are high in phytate. Minerals bound to phytic acid

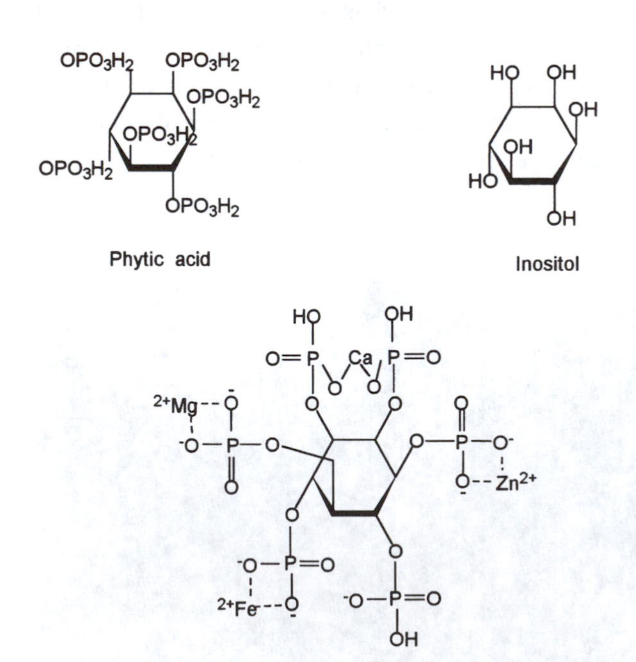

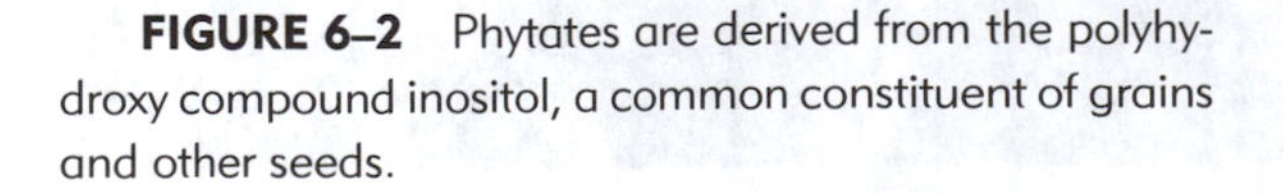

FIGURE 6–2 Phytates are derived from the polyhydroxy compound inositol, a common constituent of grains and other seeds.

and phytate are unavailable for absorption. The enzyme **phytase** cleaves off the phosphates from inositol, making the phosphorus and other minerals available. Phytase is produced by microbes but not by vertebrates. In ruminants, there is adequate ruminal phytase to render all dietary phytate digestible. Some microbial phytase activity occurs in the hindgut of the pig, whereas there is little phytase in poultry, particularly in young chicks. Thus the bioavailability of phytates is lowest in poultry and highest in ruminants. On a practical basis, phosphorus and zinc are the two elements of most significance when "tied-up" as phytates. Other minerals are also complexed with phytic acid; the decreasing order of stability is zinc, copper, cobalt, magnesium and calcium (Erdman, 1979). Phytates are not readily destroyed by heat or soaking, but microbial fermentation does liberate phosphates form phytic acid. Commercial preparations of phytase, most produced from the fungus *Aspergillus niger*, are available for use as feed additives. Commercial phytases are very effective in increasing the availability of plant phosphorus to swine (Lei *et al.*, 1993) and poultry (Simons *et al.*, 1990). Interest in use of phytases has occurred for two main reasons. One, phosphorus is a relatively expensive nutrient and is sometimes in short supply. Thus improving the utilization of phosphorus already present in feedstuffs would enhance efficiency of diet formulation. Secondly, phosphorus pollution of groundwater and streams is a major concern. In many areas, such as the Netherlands and Florida, livestock production is being legislatively curtailed because of nitrogen and phosphorus pollution problems caused by animal wastes. Increasing the bioavailability of plant phosphorus by use of dietary phytase reduces phosphorus excretion and hence reduces phosphorus pollution (Simons *et al.*, 1990).

In human nutrition, phytates in high-fiber diets may have protective effects against **colon cancer** (Graf and Eaton, 1993). The proposed mechanism of action involves the complexing of iron in the colon. By binding iron, phytates reduce the iron-catalyzed production of free radicals in the colon. Lipid peroxidation in the colon epithelium has been linked to neoplasia (Graf and Eaton, 1993). These authors reviewed studies indicating that free iron and lipid peroxidation promote colonic tumorigenesis, and epidemiological evidence associated high phytate diets with reduced colon cancer risk.

NON-STARCH POLYSACCHARIDES (NSP)

The cell walls of cereal grains are composed of a complex of **non-starch polysaccharides (NSP)** such as glucans and pentosans. These complex carbohydrates have greater structural rigidity than starch, so essentially envelop the starchy endosperm by forming the outer layers of the seed. During seed germination, endogenous glucanases and pentosanases are produced, resulting in the release of the free sugars. The composition of the NSP fraction differs among grains. In barley and oats, β-glucans predominate, while in rye, triticale and wheat, pentosans are the major NSP. **Glucans** are polymers of glucose, with linkages other than the α1,4-bonds that link glucose molecules in starch. **Pentosans (arabinoxylans)** are polymers of pentose sugars, such as xylose and arabinose. It has been known for many years that β-glucans in barley cause various problems in poultry, including wet litter, sticky droppings (Fig. 6–3) and pasty vent in chicks. Similarly, it has been appreciated for many years that rye is not well utilized by non-ruminants, although the implication of pentosans in the reduced performance is a recent development. Also, it has not been widely appreciated that NSP could affect the feeding value of wheat. Primarily through the work of Australian scientists (Annison, 1991; Annison and Choct, 1991), it has become more widely known that wheat pentosans may cause significant reductions in apparent metabolizable energy (AME) values, and thus reduce the feeding value of some wheats.

Glucans and pentosans are often referred to as **soluble fiber**. Soluble fiber is of interest in human nutrition because of its apparent cholesterol-lowering properties, as well as anti-constipation activity. In poultry and swine, the main consequences of soluble fiber or NSP are negative. Campbell and Bedford (1992) provide a thorough review of the nutritional effects of soluble fiber. In poultry, pentosans and glucans cause reduced growth, impaired nutrient absorption, and altered gut function, including enhanced microbial growth. The major nutritional effects are a consequence of increased viscosity of gut contents. Glucans and pentosans are

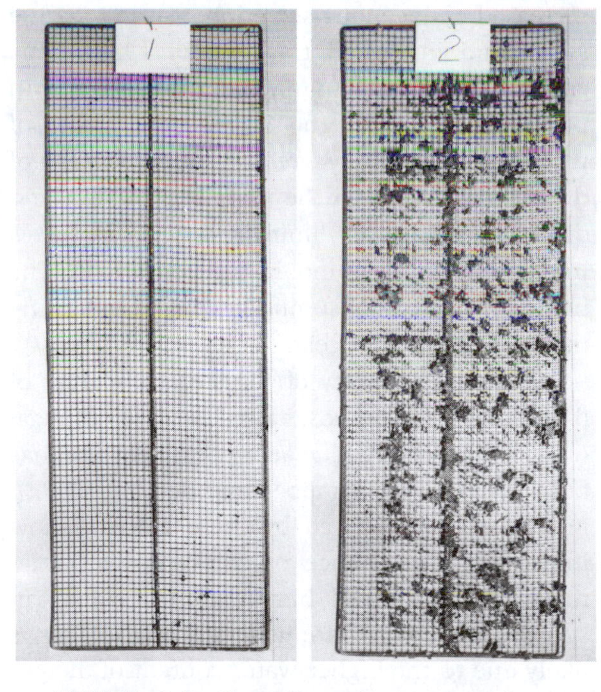

FIGURE 6–3 Cage floors from cages of control chicks fed corn (left) and from birds fed barley (right) showing the sticky droppings of the birds fed barley. (Courtesy of G.H. Arscott)

poorly water soluble, giving rise to viscous, gummy solutions in the gut. This viscosity has important implications at the site of intestinal absorption. The lining of the intestine at the microscopic level consists of projections, the microvilli, lined with a filamentous web, the glycocalyx. This is known as the **unstirred water layer**, analogous to the layer of water next to the interior wall of a water pipe. Viscous glucans and pentosans form a network of gummy material within this layer, essentially gumming up the unstirred water layer where the final processes of digestion and absorption are occurring. The absorption of large molecules, such as fatty acids and monoglycerides, is particularly inhibited by this physical effect. Thus NSP causes a specific **fat malabsorption syndrome**, due to the increased viscosity of the gut contents. The adverse nutritional effects of NSP are primarily explained by their viscosity (Annison and Choct, 1991).

The hydroscopic nature of NSP causes increased water retention in the gut, as well as increased endogenous secretion of water via the saliva and intestinal tract (Low, 1989). This causes significant management problems in poultry because of the increased amount of water excreted. The excreta is soupey instead of dry, so that birds on the floor become wet and dirty. Air quality is diminished, with increases in humidity and ammonia concentrations. The increased water consumption may significantly increase production costs.

There are a variety of commercial sources of β-glucanase and pentosanase. They are of fungal origin, such as *Aspergillus* spp. Their use has been reviewed by Campbell and Bedford (1992). Many studies reviewed by these authors show improved performance of chickens fed glucanase-supplemented barley diets, while responses with swine are generally slight. This is likely due to the higher water content of the gut contents of swine, diluting the viscosity effect which is directly related to the glucan concentration (Campbell and Bedford, 1992).

In Australian research on low ME wheats, Annison (1991) established a correlation between AME and the NSP content. Subsequent work (Choct and Annison, 1992) demonstrated that wheat pentosans were the specific factors involved, and that their effects were due to increased viscosity of gut contents, thus impairing nutrient absorption. Supplementation with pentosanases decreases viscosity and improves growth and feed conversion. The response to enzyme supplementation of diets containing grains with NSP is primarily due to the partial breakdown of glucan and pentosan polymers, reducing their viscosity (Bedford and Classen, 1992). The release of free sugars and their metabolism as sources of energy makes a very minor contribution, if any, to the response (Campbell and Bedford, 1992).

Petterson *et al.* (1991) noted that an enzyme preparation with glucanase and pentosanase activity was quite heat-stable, with a loss of only about 20% of activity when broiler diets containing barley, wheat and rye were pelleted. A significant improvement in production parameters was noted with both mash and pelleted diets with enzyme supplementation. Heat treatment of rye grain before addition of pentosanase greatly increased the growth response of chicks to the enzyme (Teitge *et al.*, 1991).

Besides occurring in grains, β-glucans are the major structural components of the cell walls of fungi. Interestingly, yeast β-glucan preparations increase disease resistance of fish and mammals when used as feed additives (Jorgensen *et al.*, 1993). These authors speculate that animals have developed recognition mechanisms for fungal cell wall glucans as a defense against fungal infections.

PHENOLIC COMPOUNDS IN GRAINS

Sorghum (*Sorghum bicolor*) is a drought-resistant crop adapted to environmental conditions too harsh for the production of corn. Commonly known as **milo** in the United States, grain sorghum is similar to corn in vegetative appearance. The small, round seeds are produced in an open panicle. **Millets** are important food crops in parts of Asia and Africa. Examples are pearl millet (*Pennisetum glaucum*), foxtail millet (*Setaria italica*) and proso millet (*Panicum miliaceum*). Sorghum and millets (Fig. 6–4) are major food crops in the semiarid tropics, an ecological zone encircling the earth and including China, India, most of Africa, Australia, Argentina and parts of the southern U.S. In the developed countries, about 96% of the total sorghum and millet grown is used for animal feed, whereas in the developing countries, only about 8% of these crops is used for livestock, with the rest used directly as human food. About 90% of the rural population of the Sahelian zone in Africa depends on these crops as their source of food energy. Sorghum is grown in these areas because it can tolerate relatively harsh ecological and growing conditions, including drought.

FIGURE 6–4 Seed heads of millet.

More than other grains, sorghum requires chemical defenses. Whereas corn grain is wrapped in a husk, and barley and wheat are protected by fibrous awns, sorghum seed is produced in an open, exposed panicle. Protection against herbivores and pathogens is conferred by **condensed tannins**. Tannins are phenolic substances in plants which react with proteins. The term "tannin" is derived from the use of plant extracts (e.g. oak leaves) to tan leather. In this process, hydrogen bonds are formed between the phenolic hydroxyl groups and the peptide groups of collagen fibrils to form crosslinks between adjacent protein fibers. This crosslinking alters the physical properties of the collagen, producing leather. **Tannins** are defined as phenolic compounds in plants of high molecular weight (500–3000) with a sufficient number of phenolic hydroxyl groups (1–2 per 100 units of molecular weight) to enable formation of crosslinkages with proteins and other macromolecules. **Phenolic compounds** contain one or more aromatic (benzene) rings with one or more hydroxyl groups. **Polyphenolic compounds** have numerous hydroxyl groups. Phenolic compounds are divided into several categories, including simple phenols, phenolic acids, hydrolyzable tannins, condensed tannins, lignins and lignans.

1. **Simple phenols**

These are structurally simple compounds such as catechol and resorcinol (Fig. 6–5), similar in structure to phenol.

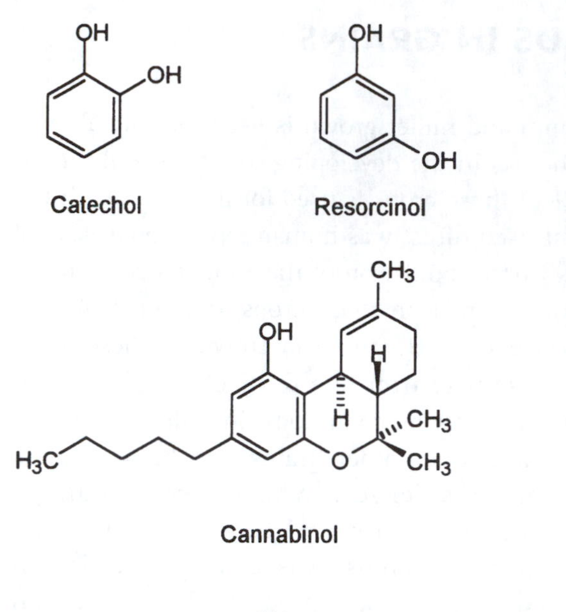

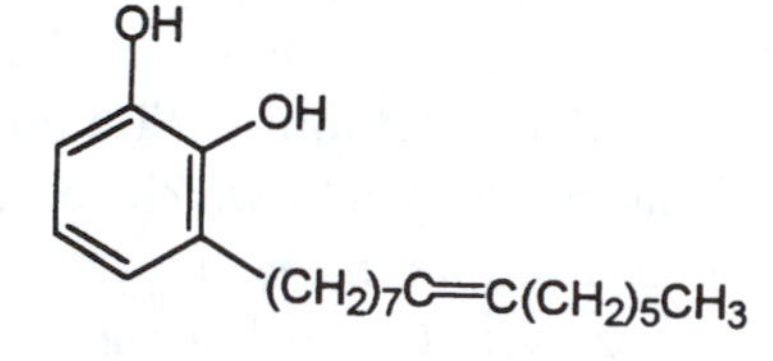

FIGURE 6–5 Examples of simple phenols.

Toxic derivatives of catechol with side-chains attached to the ring include the toxin in poison ivy (urushiol), while the cannabinols in marijuana are derivatives of resorcinol. Simple phenols readily undergo oxidation to produce quinones.

2. **Phenolic acids**

Phenolic acids are simple phenols with one or more carboxyl groups (Fig. 6–6). One of the most abundant is gallic acid. The carboxyl groups may be on the ring (e.g. gallic acid) or in the side chain (e.g. caffeic acid).

3. **Hydrolyzable tannins**

These contain a central core of a polyhydroxyl alcohol, usually glucose, esterified with either gallic acid to form **gallotannins** such as tannic acid (Fig. 6–7) or with ellagic acid to form **ellagitannins**.

The esterified phenolic acids can be hydrolyzed from the central sugar molecule; hence the name **hydrolyzable tannins**.

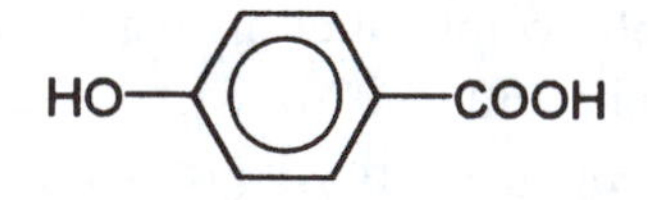

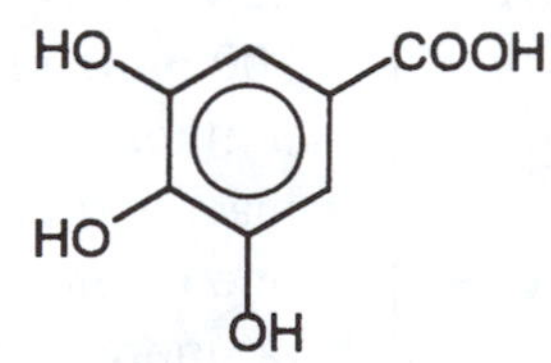

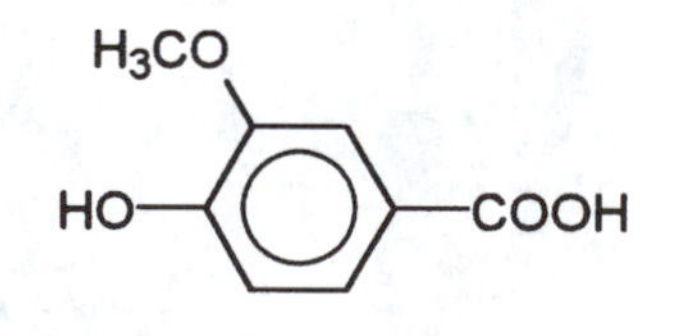

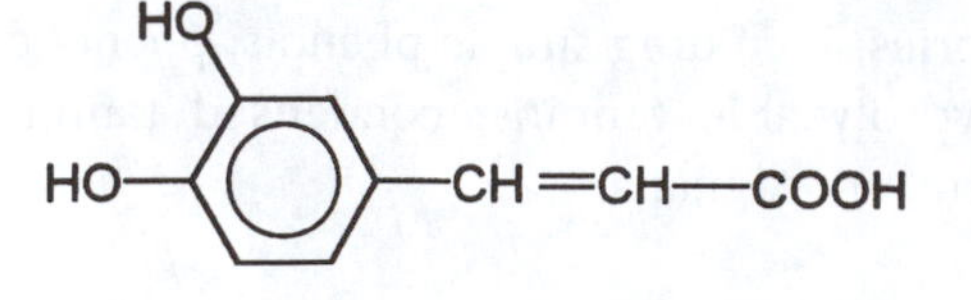

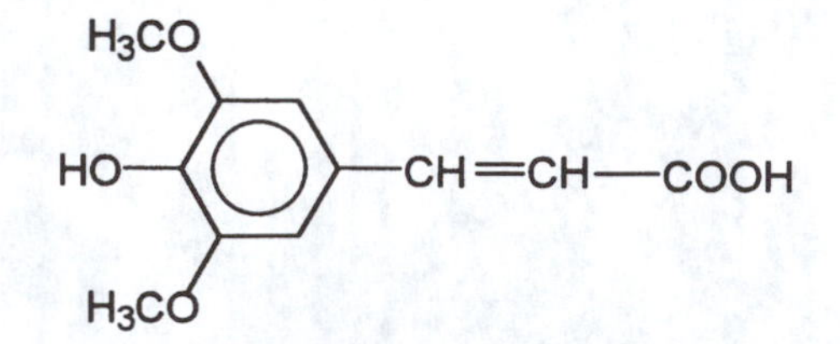

FIGURE 6–6 Examples of phenolic acids.

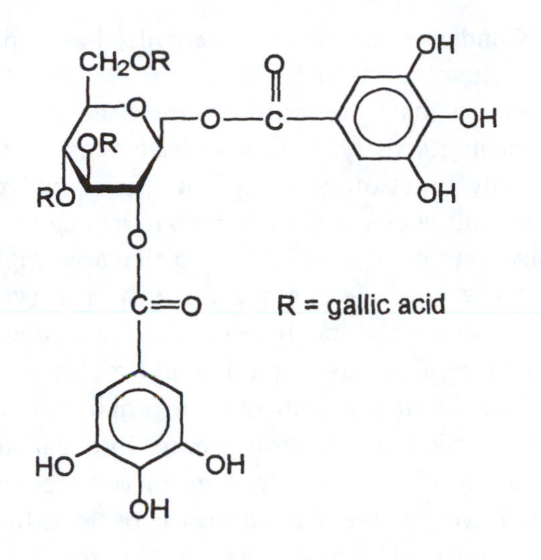

FIGURE 6–7 Tannic acid is composed of a core of glucose, with an esterified phenolic acid (gallic acid) on each hydroxyl group. Note that for simplicity only two of the five gallic acids are drawn in the structure illustrated.

4. **Condensed tannins**

Condensed tannins contain phenols of the flavone type (Fig 6–8). One of the simplest in structure is the dimer **procyanidin.** Additional flavone groups may be added on to procyanidin to produce large condensed tannin molecules. Condensed tannins are also referred to as **proanthocyanidins.** Very large polymers of condensed tannin may occur in plants. Many of them are pigments found in seeds, flower petals, etc. Loss of the astringency of unripe fruit as it ripens is attributed in part to increase polymerization of condensed tannins.

5. **Lignins and lignans**

Lignins are complex phenolic compounds deposited in association with cellulose in plant cell wall material. Although lignin has phenolic groups, its biosynthesis is dissimilar to the phenolic compounds listed above. **Lignans** are dimers of two aromatic rings linked by carbon–carbon bonds between the side chains (Fig. 6–9).

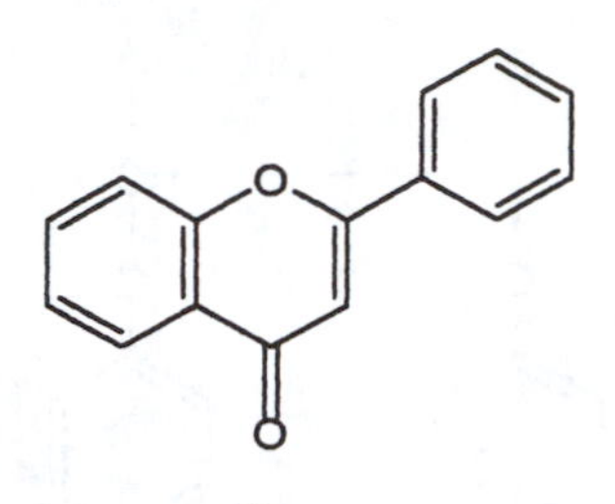

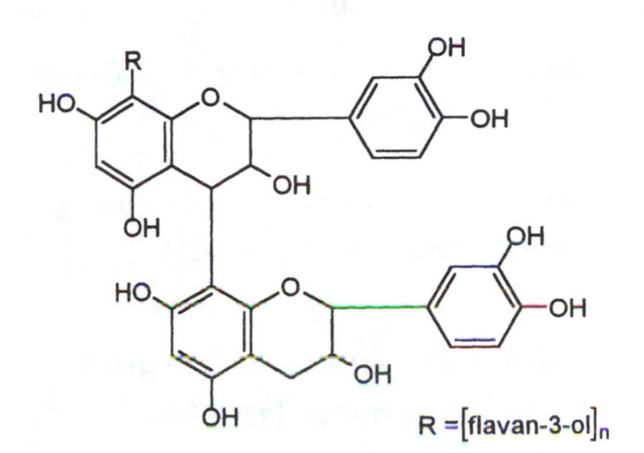

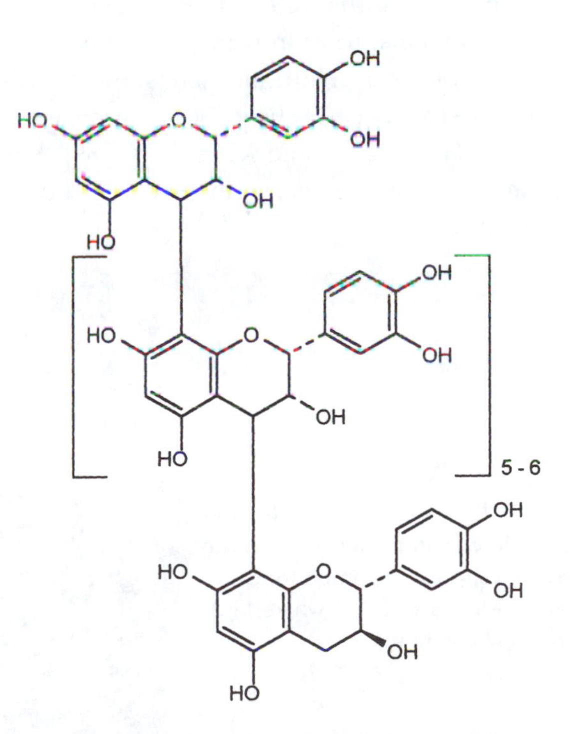

FIGURE 6–8 Condensed tannins are polymers of flavones.

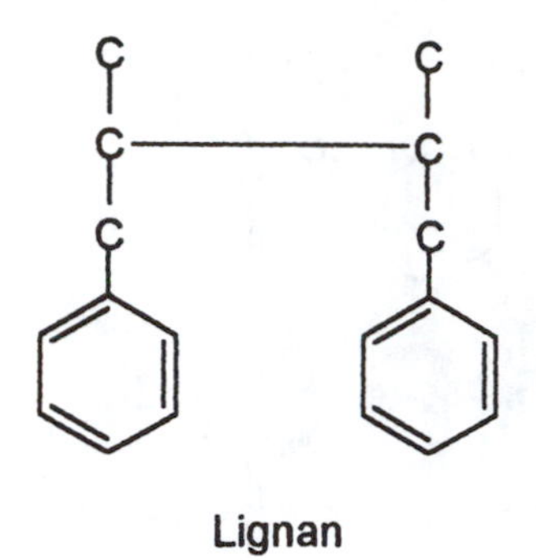

FIGURE 6–9 Lignans are similar in name but not in structure to lignin.

Lignans have been postulated to have phytoestrogen activity (see Chapter 7).

Biological Effects of Sorghum Condensed Tannins

The main polyphenols in grain sorghums are **condensed tannins** (CT). Their biological effects on animals, roles in plants, and nutritional and toxicological significance in sorghum will be discussed. The CT have been described in detail by Hemingway and Karchesy (1989). Sorghum polyphenols were reviewed by Butler (1989a,b).

Condensed tannins are recognized as important chemical defenses of many plants. Most phenolics have antioxidant and free radical-quenching activity. This may have been an important function when plants first evolved, when ultraviolet radiation was much more intense than now. Tannins are particularly important chemical defenses against herbivory by mammals, birds and insects, and they provide resistance to pathogenic fungi and molds.

The CT in sorghum have an important role in protection of the crop against bird damage. Sorghum seed is produced in an open panicle that is very vulnerable to attack by wild birds (Fig. 6–10). The small size of the seeds also causes them to be attractive to birds. In Africa, the red-billed weaver bird causes great destruction. It is considered the most numerous and destructive bird pest in the world, comparable to locusts in the amount of damage it causes. The weaver bird population base in Africa is estimated at 100 billion. In North America, blackbirds, starlings, and sparrows are a problem, while in Asia, sparrows, parakeets, crows, and mynas cause destruction of sorghum. In Latin America, doves, blackbirds, sparrows, parakeets, and parrots are the major problem species. Several bird-resistant lines of sorghums have been developed which have a high tannin content.

FIGURE 6–10 A flock of blackbirds descending upon a sorghum field illustrates the importance of bird-resistance in areas where birds occur in large number.

The tannins are located in the pericarp and testa, near the surface of the seed. The darkness of the seed coat of sorghum is related to the CT content. While it is true that high tannin sorghum tends to have a dark seed color, the seed color is not always a reliable indicator of tannin content (Boren and Waniska, 1992). Yellow seed is usually low in tannin.

Tannins have astringent properties due to their propensity to react with proteins. The interaction of tannins with salivary proteins and glycoproteins in the mouth causes the sensation of **astringency**. Bird-resistant sorghum is grown in areas where low-tannin cultivars would be susceptible to extensive bird damage. Of course, the feeding value of bird-resistant sorghum is lower than for low-tannin types, because of the negative effects of CT. According to Butler (1989b), the feed consumption of livestock fed bird-resistant sorghum is not reduced, but performance is affected because of adverse effects of tannins on protein utilization. Bird damage is most extensive when the grain is in the milk stage. When the grain ripens, the CT content is reduced due to polymerization of the tannins.

Most adverse nutritional effects of CT, including astringency, are due to their tendency to react with proteins. They react with dietary protein, forming indigestible complexes, and complex with digestive enzymes, reducing digestibility of all nutrient components. They cause irritation and erosion of the intestinal mucosa. This results in increased mucus secretion for protection against the mucosal damage. The **hypersecretion of mucus** increases metabolic protein requirements, because of the extra endogenous protein excreted in the feces (Salunkhe *et al.*, 1989).

Protein–tannin interactions have been reviewed by Hagerman *et al.* (1992). Condensed tannins react with proteins and reduce their bioavailability, but hydrolyzable tannins do not. Instead, hydrolyzable tannins are hydrolyzed in the gut, releasing simple phenolics which are absorbed and conjugated for urinary excretion (see Chapter 11). Condensed tannins bind with proteins selectively, and have an especially high affinity for large proteins, conformationally open proteins, and proline-rich proteins (Hagerman *et al.*, 1992). They precipitate proteins most effectively at pH values near the isoelectric point of the protein. Their tendency to interact with proteins increases as the molecular weight of the tannin and the number of phenolic groups increase.

Sorghum tannins seem to be of greatest nutritional significance in monogastric animals, particularly poultry. High-tannin sorghum may cause reduced growth and feed efficiency, and decreased egg production. Extra dietary protein partially overcomes these effects. The feeding of high-tannin sorghums to broiler chicks causes a leg abnormality characterized by an outward bowing of the hock joint (Elkin *et al.*, 1978). It was suggested that absorbed tannins might increase the amount of cross-linking in collagen, thereby altering the organic matrix of the bone. Sell and Rogler (1983) demonstrated that high-tannin sorghums cause increased activities of liver UDPglucuronyltransferase, an enzyme involved in detoxification of phenolics. This enzyme induction indicates that phenolics are absorbed in the chick, and therefore would be available for reacting with protein (e.g., collagen) in vivo. Using radiolabeled sorghum CT, Jimenez-Ramsey *et al.* (1994) found that CT are not absorbed from the chicken digestive tract. They suggested that low molecular weight polyphenols may account for the toxic effects (e.g. leg abnormalities) seen in chickens fed high-tannin sorghum diets.

Components in sorghum other than condensed tannins may affect nutrient bioavailability. Elkin *et al.* (1996) reported wide variations in crude protein digestibilities in chickens fed sorghums with similar tannin contents. They suggested that variations in the content of **kafirins**, or storage proteins in sorghum, might explain the inconsistent relationship between tannin content and digestibility.

In ruminants, there is no clear evidence that sorghum tannins are harmful. The association of protein with tannins reduces ruminal degradation of proteins, increasing the potential for high-quality protein to "bypass" the rumen (Streeter *et al.*, 1993). Tannin–protein complexes are stable in the rumen (pH about 6.5) and dissociate in the small intestine, releasing the protein. This also raises the possibility of the liberated tannin causing damage to the intestinal tract; studies of this aspect are needed. Effects of tannins in forage legumes on rumen fermentation and by-pass protein are discussed in Chapter 11.

The crop residue (stover) of sorghum is often used as a ruminant feedstuff. The digestibility of the residue of bird-resistant sorghum varieties is often lower than for low-tannin varieties, but the effects on digestibility vary with the specific phenolic compounds present in various sorghum cultivars (Mueller-Harvey and Reed, 1992).

Tannins form insoluble complexes with divalent metal ions, and may reduce mineral absorption. For example, the bioavailability of iron in high-tannin sorghum is lower than in low-tannin cultivars (Salunkhe *et al.*, 1989).

Condensed Tannins and Human Health

Morton (1970) has described a number of interesting correlations between the presence of tannins in foods and the incidence of **esophageal cancer** in humans. The consumption of high-tannin sorghum grain in several parts of the world, including the Netherlands Antilles in the West Indies and South Africa where Africans use "kaffir corn", a bird-resistant brown sorghum, has been linked by Morton (1970) to a high incidence of esophageal cancer in these populations. Tannins in tea have also been implicated. The addition of cream or milk to tea eliminates this effect, presumably because of formation of a protein–tannin complex. Morton (1989) speculated that a high incidence of esophageal cancer in Dutch tea drinkers in the 1800's and a low incidence in England during the same period can be explained by the consumption of black tea by the Dutch and tea with milk by the English. The disease became rare in Holland when coffee from the Dutch East Indies (now Indonesia) replaced tea as the beverage of choice.

While Morton (1970) makes an intriguing case for sorghum tannins as the cause of a high incidence of esophageal cancer in various regions of Southern Africa, there are other possible and more plausible explanations. Marasas and coworkers in South Africa suggest that the cause of high rates of esophageal cancer in various regions of Southern Africa such as Transkei is mycotoxin contamination of maize (see Chapter 5). The specific mycotoxin that may be involved is **fumonisin**, produced by *Fusarium moniliforme*, a common mold of corn. Similarly, the high esophageal cancer rate in the Netherlands Antilles has an alternative explanation, related to consumption of **phorbol esters** (see Chapter 15).

Morton (1970, 1989) also has speculated that tannins in wines may cause esophageal cancer. There is a different view that tannins (flavonols) in wine are beneficial to human health, with favorable effects on coronary heart disease and cancer (see Chapter 15).

Phenolic compounds in millet have been linked to antithyroid activity and goiter. This may be associated with the inhibition of thyroid peroxidase by dietary flavanoids (Divi and Doerge, 1996).

Tannins and Salivary Proteins

When growing rats are fed diets based on high CT sorghum grain, they initially respond very poorly, losing weight. However, in a few days they begin to grow at a nearly normal rate (Butler, 1989b). During the adaptation period, the salivary glands enlarge by 3–5 fold, with a large increase in secretion of **proline-rich salivary proteins (PRP)**. The PRP have a very high affinity for tannins and function as tannin-binding proteins (Mehansho *et al.*, 1987). By binding the CT of sorghum, they prevent post-ingestive effects such as inhibition of digestive enzymes. However, because the PRP-CT complex is excreted in the feces, the protein requirement of animals fed high tannin grain is increased. In contrast to the rat, hamsters do not synthesize PRP, and as a consequence are unusually sensitive to dietary tannins (Butler, 1989b). There are numerous species differences in salivary PRP (Mole *et al.*, 1990). The tannin-binding PRP are particularly important in browsing animals such as deer (Austin *et al.*, 1989) and bears (Robbins *et al.*, 1991), which are much more tolerant of tannins in trees and shrubs (browse) than are grazing animals such as cattle. There is considerable animal species diversity in the specificity of PRP. For example, moose and beaver produce PRP that bind only linear condensed tannins common in their preferred feeds such as willow, aspen and birch. Mule deer, with a more generalized diet, have PRP which bind linear and branched chain condensed tannins, gallotannins, but not ellagitannin. The omnivorous black bear has PRP which bind all types of tannins. These relationships are reviewed by Hagerman and Robbins (1993). The evolutionary development of PRP is discussed by McArthur *et al.* (1995). The proline-rich proteins in saliva function not only in binding tannins, but also function in oral homeostasis, including calcium metabolism, lubrication and bacterial aggregation. There are basic and acidic PRP's, with somewhat different activities. Acidic PRP's function in calcium binding and maintaining saturation of tooth surfaces with calcium. Basic PRP form complexes with tannins over a wide pH range and are the major group of PRP interacting with tannins in the digestive tract (McArthur *et al.*, 1995). McArthur *et al.* (1995) speculate that while all animals have acidic PRP, the basic PRP have evolved in animals that on an evolutionary basis have subsisted on nitrogen-poor, tannin-rich diets.

Detoxification of Tannins in Feeds

Heat treatment or cooking does not destroy tannins. Various food processing techniques that reduce the effects of CT in sorghum grain have been developed in many cultures in Africa and Asia. The seeds can be mechanically dehulled; this can be accomplished with a simple mortar and pestle. The hulls, containing most of the CT, are discarded. Various chemical treatments can be used. In the traditional preparation of foods from high-tannin sorghum in Central Africa, the grain is soaked and germinated in an alkaline wet wood ash preparation. The tannins are unstable under alkaline conditions. Soaking in **alkali**, followed by washing the grain, greatly reduces CT levels. Ammoniation of high-tannin sorghum with anhydrous ammonia is an effective detoxification technique (Butler, 1989b; Salunkhe *et al.*, 1989). Alkaline conditions alter tannin structure so they do not react with proteins.

Various tannin-binding feed additives can be used in sorghum diets to reduce the effects of CT. **Polyethylene glycol (PEG)** is an inexpensive tannin-binding substance which might have potential for use under practical conditions (Salunkhe *et al.*, 1989). Supplemental protein is also effective in binding tannin, but often may not be economical. Numerous reports have indicated that supplemental methionine improves the performance of animals fed high-tannin sorghum (Salunkhe *et al.*, 1989). The mode of action may involve provision of methyl groups for detoxification of absorbed phenolic compounds.

Phenolics in Other Grains

Barley contains anthocyanin in the green tissue and proanthocyanidins and catechins in the aleurone cells of the mature seed (Newman *et al.*, 1984). These pigments are phenolic compounds which might reduce protein and amino acid availability in the gut. Proanthocyanidin-free barley cultivars have been developed and give superior chick growth than barley of conventional varieties (Newman *et al.*, 1984). These workers suggest that phenolic compounds in barley are detrimental to nutritional quality and that reduction of these antinutritional constituents by plant breeding could improve the feeding value of this crop.

Resorcinol is *m*-dihydroxybenzene. Derivatives of resorcinol, the **5-alkyl resorcinols**, occur in triticale and may account for some of the growth-depressing effects of this grain (Radcliffe *et al.*, 1981). According to Farrell *et al.* (1983), the resorcinols are of little if any significance in this regard, and adverse effects of triticale on performance of poultry and swine are likely due to trypsin inhibitors. However, Radcliff *et al.* (1983) found that differences in swine growth performance and efficiency of nutrient utilization were not associated with the trypsin inhibitor of triticale. These authors suggest that neither 5-alkyl resorcinols nor trypsin inhibitors are responsible for growth-inhibitory effects of triticale. The general structure of 5-alkyl resorcinols is:

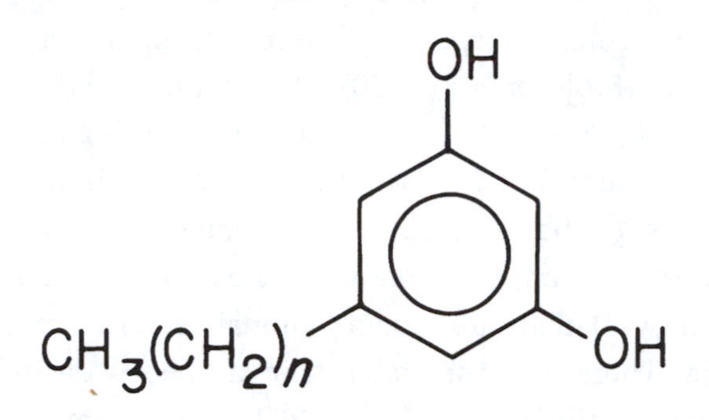

PROTEASE INHIBITORS IN GRAINS

Protease inhibitors (also commonly referred to as **trypsin inhibitors**) are proteins which bind irreversibly to the active sites of proteolytic enzymes (e.g. trypsin, chymotrypsin), thus impairing protein digestion. They are of particular significance in soybeans and other pulses (legume seeds), and will be discussed in more detail in the section on Toxins in Soybeans. All of the cereal grains contain protease inhibitors (Liener, 1980). They are of practical significance only in the case of triticale, a synthetic grain (wheat x rye). Some cultivars or selections of triticale have given poor feeding results, because of high trypsin-inhibitor activity (Erickson *et al.*, 1979). Newer varieties produced since 1980 are largely free of trypsin inhibitor activity. Sosulski *et al.* (1988) reviewed trypsin inhibitors in cereal grains.

TOXINS IN MINOR GRAINS

There are several non-cereal grains that are grown as protein–energy sources, fulfilling a role similar to that provided by the cereal grains. These minor grains include buckwheat, quinoa and amaranth. Quinoa and amaranth were traditional grains of the Incas in Central and South America. Buckwheat has been a minor crop of Europe and North America. There has been some recent interest in these crops as alternatives to conventional, monoculture-grown intensive cereal crops.

Quinoa (*Chenopodium quinoa*) and **grain amaranth** (*Amaranthus* spp.) are basically domesticated strains of the common weeds lamb's-quarters (*Chenopodium album*) and redroot pigweed (*Amaranthus retroflexus*). Lamb's quarters and pigweed are ubiquitous in vegetable gardens, cultivated crops, and in barnyards and fence rows. Both lamb's-quarters and pigweed may contain toxins, including nitrates and oxalates.

Quinoa grain contains **saponins**, which are bitter glycosides with detergent activity. Gee *et al.* (1993) have reviewed the saponins of quinoa, including processing methods to reduce their effects. Saponins can damage intestinal mucosal cells by altering cell membrane permeability, and may increase the uptake of antigens by the small intestine. Processing methods for removal of saponins involve aqueous extraction. "Sweet" varieties with low saponin contents have also been developed (Gee *et al.*, 1993).

There has been considerable interest in **grain amaranth** (*Amaranthus* spp.) as a food crop (Fig. 6–11), largely stimulated by an extensive amaranth breeding project at the Rodale Organic Gardening and Farming Research Center, Kutztown, PA. Interest in amaranth has developed because of the intrinsic appeal to many people of an ancient crop by-passed by agribusiness and conventional agriculture, and the relatively high protein and lysine contents of the grain. Pond *et al.* (1991) summarized research indicating that grain amaranth has a growth-inhibiting factor(s) that are destroyed by moist heat (e.g. autoclaving). The growth inhibitor(s) have not yet been identified, but lectins, trypsin inhibitors and phytates have been suggested as possibly being involved (Pond *et al.*, 1991).

FIGURE 6–11 There is increasing interest in grain amaranth as a "new" food crop.

Some *Amaranthus* spp. are grown as vegetables and may have potential as forages for livestock (Pond and Lehmann, 1989). Toxins of possible concern include oxalates, nitrates, saponins and tannins.

Buckwheat (*Fagopyrum esculentum*) is a fast-growing broadleaved annual grown for its edible seeds and as a forage. It reaches maturity 10–12 weeks from planting, and often yields better on poor soils than cereal grains. The grain is used for pancake flour and as a livestock feed. The production and utilization of buckwheat was reviewed by Pomeranz (1983).

Buckwheat seed and forage may cause photosensitization (fagopyrism) when light-skinned animals which have consumed buckwheat are exposed to sunlight. In mild cases, reddening of unpigmented areas of skin occurs, while in severe cases, nervous symptoms such as excitement, running, squealing or bellowing, convulsions and prostration may occur. The active principle is a dianthroquinone derivative called **fagopyrin**.

Buckwheat contains a significant quantity of **rutin**, a phenolic compound with favorable effects on human health, including antiinflammatory, antimutagenic, anticarcinogenic and hypotensive effects (Oomah and Mazza, 1996).

TOXINS IN ROOTS AND TUBERS

Cassava Cyanogens

Cassava (*Manihot esculenta*), also known as yuca, manioc and tapioca, has enlarged, starchy roots which are grown for food and animal feed (Fig. 6–12). Native to the American tropics, it has been introduced to Africa and Asia, where it is an important food crop.

The major toxicants in cassava are cyanide-yielding glycosides, although the foliage also contains tannins. **Cyanogens** are glycosides of a sugar or sugars (usually glucose) and a cyanide-containing aglycone. All plants produce cyanide during the formation of ethylene, so cyanide is ubiquitous in plants. However, concentrations are very low in most plants. Cyanogenic plants are those in which much higher levels of cyanide occur; a level of 10 mg HCN per kg fresh weight is the minimum for a plant to be considered cyanogenic (Davis, 1991). Cyanogens are not unique to plants; they occur in fungi, algae, bacteria, centipedes, millipedes and certain insects. Besides the glycoside form, cyanide can also occur in cyanolipids in seed oils of certain plants of the Sapindaceae family.

FIGURE 6–12 Cassava roots. Cassava contains cyanogens, which release cyanide when the root is processed.

Cyanogenic glycosides can be hydrolyzed by enzymatic action with the release of HCN (hydrogen cyanide, prussic acid). Hydrogen cyanide is a potent toxin, inhibiting the terminal respiratory enzyme cytochrome oxidase.

The major cyanogens of importance in animal nutrition are the following:

1. **Amygdalin** (laetrile). This glycoside is found in Rosaceae, such as chokecherries, wild cherries, mountain mahogany, saskatoon serviceberries, and the kernels of almonds, apricots, peaches, and apples. **Prunasin** is also found in these plants and in bracken; it has the same structure as

amygdalin except it has one glucose rather than two attached to the aglycone.

2. **Dhurrin**. This occurs in sorghum species such as grain sorghums, forage sorghums (e.g. Sudan grass), and Johnson grass.

3. **Linamarin**. This compound is found in white clover, flax (linseed), cassava, and lima beans.

Hydrogen cyanide is formed when the glycosides are hydrolyzed by plant enzymes in the following steps:

1. cyanogenic glycosides $\xrightarrow{\beta\text{-glucosidase}}$ sugar + aglycone

2. aglycone $\xrightarrow{\text{hydroxynitrile lyase}}$ HCN + aldehyde or ketone

The glycosides occur in vacuoles in plant tissue, while the enzymes are found in the cytosol. Damage to the plant from wilting, trampling, mastication, frost, drought, bruising (cassava), and so on results in the enzymes and glycosides coming together, causing HCN to be formed. These enzymes are also produced by rumen microorganisms. The optimal pH for the enzymes is near neutrality, so release of HCN is more rapid in the rumen than in the highly acid stomach of the nonruminant animal. For this reason ruminants are more sensitive to cyanogens than nonruminants. The specific reactions for each of the major types of cyanogens are illustrated in Fig. 6–13.

Metabolism of Hydrogen Cyanide and Signs of Cyanide Poisoning

Hydrogen cyanide is readily absorbed and enters individual tissue cells. It inhibits **cytochrome oxidase**, the terminal step in electron transport. When cytochrome oxidase is blocked, ATP formation ceases, and the tissues suffer energy deprivation. Death follows rapidly. Signs of cyanide poisoning are dyspnea (difficult or labored breathing), excitement, gasping, staggering, paralysis, convulsions, coma, and death. The pupils are dilated, and mucous membranes are bright. The blood is a bright cherry red. The odor of benzaldehyde or acetone may be detectable in the contents of the rumen if the dead animal is examined immediately.

Cyanide is readily detoxified, so acute toxicity occurs only if detoxification reactions are exceeded. All animal tissues contain an enzyme called **rhodanese**[1] (thiosulfate sulfurtransferase), which catalyzes conversions of cyanide to thiocyanate:

$$\underset{\text{Thiosulfate}}{S_2O_3^{2-}} + CN^- \xrightarrow{\text{rhodanese}} SO_3^{2-} + \underset{\text{Thiocyanate}}{SCN^-}$$

Thiocyanate is excreted in the urine. This reaction is employed in the treatment of cyanide toxicity. Injection (IV) of sodium thiosulfate and sodium nitrate is used. Sodium thiosulfate participates in the above reaction, while nitrate converts hemoglobin to methemoglobin. **Methemoglobin** has a greater affinity for cyanide than does cytochrome oxidase, so it strips the cyanide from the enzyme. The dosage is critical, since there must be a balance between hemoglobin and methemoglobin sufficient to maintain oxygen transport. In most cases, treatment is not practical under farm or ranch conditions because of the rapidity of development of lethal cyanide poisoning.

There are species differences in tissue rhodanese activity, which probably result in differences in susceptibility to cyanide intoxication (Aminlari and Gilanpour, 1991; Aminlari and Shahbazi, 1994). The gut mucosa, liver and kidney generally have the highest rhodanese activities.

Acute cyanide poisoning occurs in both humans and livestock. There are numerous examples of human deaths from consumption of apricot kernels, bitter almonds, and apple seeds. Major livestock poisoning problems involve the

---

[1]The unusual ending "ese" for an enzyme has been retained from the original German name given when the enzyme was first isolated.

1. Amygdalin

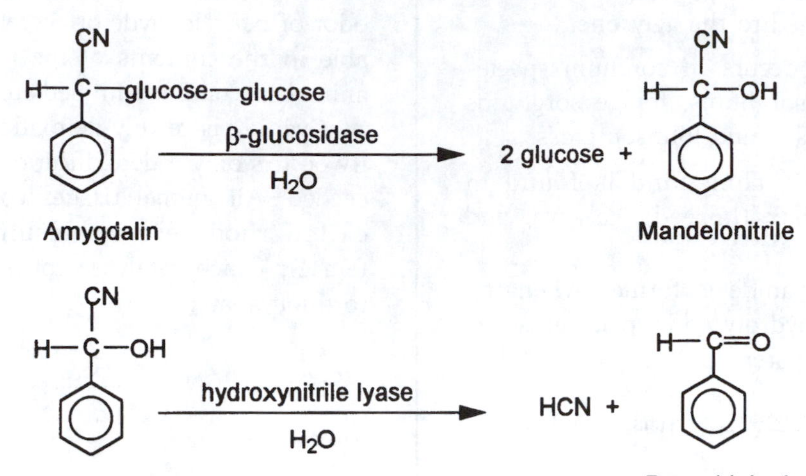

2. Dhurrin

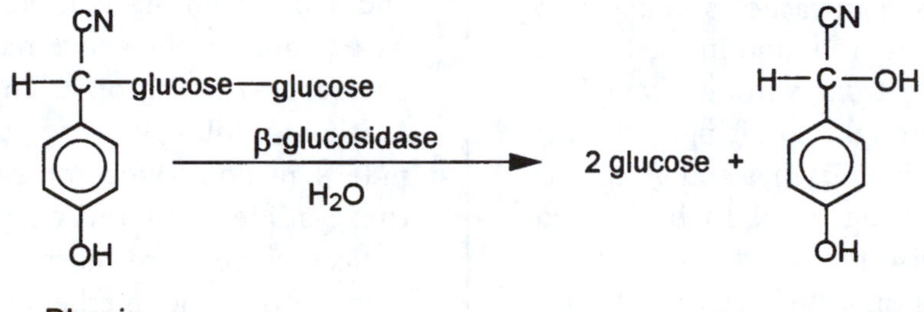

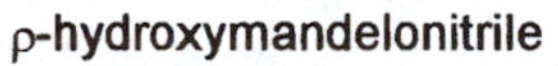

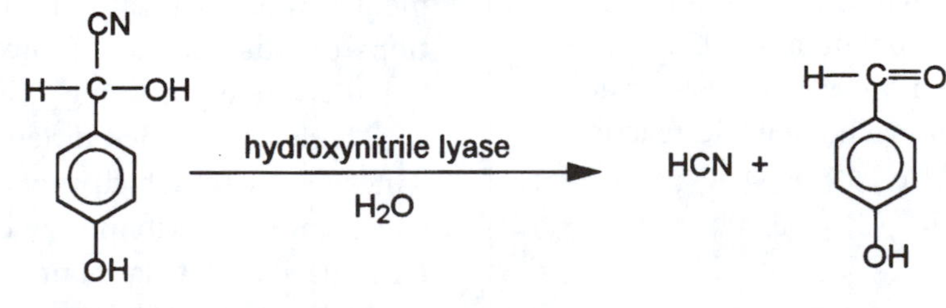

3. Linamarin

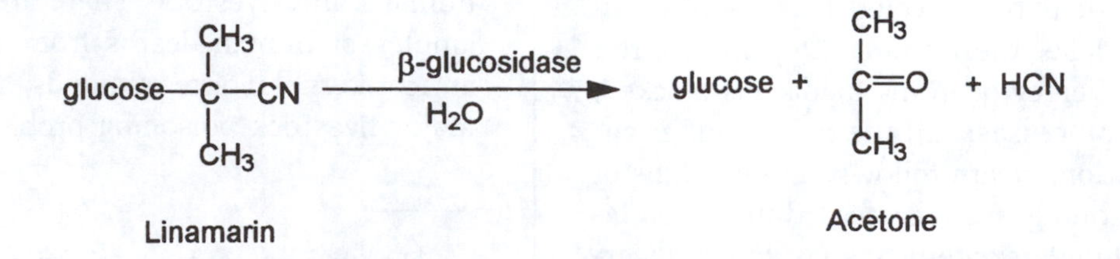

FIGURE 6–13 Major cyanogenic glycosides and their enzymatic conversions resulting in production of free cyanide (HCN).

forage sorghums, chokecherries, and other cyanogen-containing plants on western ranges of the U.S. (see Chapter 15).

Cassava has traditionally been prepared for human consumption in such a way that its toxicity is largely eliminated. It can be grated and soaked in water to activate the conversion of the cyanogenic glycoside to HCN and then washed to leach out the HCN. Alternatively, it can be cooked to destroy the enzymes, so the heat-stable glycoside cannot be converted to HCN. Plant breeders are selecting for low-cyanogen types of cassava. Without proper processing, cassava can contain lethal cyanide concentrations. For example, Espinoza *et al.* (1992) reported an incident in Venezuela in which eight children were acutely poisoned after eating raw cassava roots. Several of the children were critically ill with respiratory failure and cardiovascular collapse; they recovered after receiving sodium nitrate and sodium thiosulfate. An extensive outbreak of acute cyanide toxicosis, with nausea and vomiting as major symptoms, occurred in a drought-sticken part of Tanzania in 1988. Cassava was the only crop to survive the drought, and people consumed it immediately upon harvest, without normal processing to allow cyanide dissipation (Mlingi *et al.*, 1992).

The lethal dose of cyanide is from 0.5 to 3 mg/kg body weight. Examples of the cyanide yield reported from a number of cyanogen-containing plants (from Montgomery, 1980, and Kingsbury, 1964) are shown in Table 6–1.

TABLE 6–1 Representative Cyanide Contents of Various Products

	mg CN/100 g plant tissue
Sorghum forage	250
Wild cherries (e.g., chokecherry)	140–370
Arrow grass	77
Bitter almonds	250
Lima beans	10–300
Bamboo tips	800
Linseed meal	53
Cassava root	53
Cassava leaves	104

Putting these figures in perspective, wild cherry leaves with about 200 mg CN/100 g contain about 10 times the minimum level considered dangerous. Less than 100 g of leaves would be fatal to a 100-lb animal.

The benzaldehyde produced during hydrolysis of cyanogenic glycosides is apparently not of sufficient magnitude to represent a toxicological problem.

Chronic exposure to moderate levels of cyanide occurs in some human populations. In tropical Africa, millions of people consume cassava as a staple dietary item and pulses (seed legumes) that contain cyanogens. A condition called **tropical ataxic neuropathy** occurs in West Africa and is apparently associated with the consumption of cassava. Symptoms are lesions of optic, auditory and peripheral nerves, elevated blood level of thiocyanate, and an increased incidence of goiter. Vitamin B_{12} and methionine supplementation have beneficial effects. The evidence is strong that chronic cyanogen ingestion can lead to neurological problems in humans, and that dietary vitamin B_{12} and sulfur amino acid status can have implications in these disturbances (Montgomery, 1980). Similar neurological disorders have been observed in livestock consuming cyanogenic forage (Wheeler and Mulcahy, 1989). Chronic cassava consumption and protein malnutrition contribute to impaired glucose tolerance and malnutrition-related diabetes (Akanji and Famuyiwa, 1993).

In areas where cassava is a dietary staple, endemic goiter occurs. This is caused by chronically elevated blood levels of thiocyanate, which impairs the uptake of iodine by the thyroid gland.

The foliage of cassava can be utilized as an animal feed. The digestibility can be adversely affected by the tannin content (Reed *et al.*, 1982).

Potato Glycoalkaloids

Potatoes (*Solanum tuberosum*) are grown primarily for human food, but extensive quantities of cull potatoes and potato by-products (peels, French fry trimmings, etc.) are used as livestock feed. *Solanum* spp. such as potatoes, eggplant, nightshades, the Jerusalem cherry (a Christmas ornamental), as well as the closely related tomato, contain steroid alkaloids. Solanine, a glycoalkaloid in potatoes, was discovered in 1826. **Solanine** is a glycoside containing a steroid alkaloid nucleus (the aglycone) with a side chain of sugars. Thus, glycoalkaloids are glycosides of alkaloids. The aglycone is called **solanidine**. In 1954, another glycoalkaloid, **chaconine**, was discovered in potatoes (Jadhav *et al.*, 1981). The structures of these compounds are shown in Fig. 6–14. Solanine and chaconine have the same aglycone (solanidine) but differ with re-

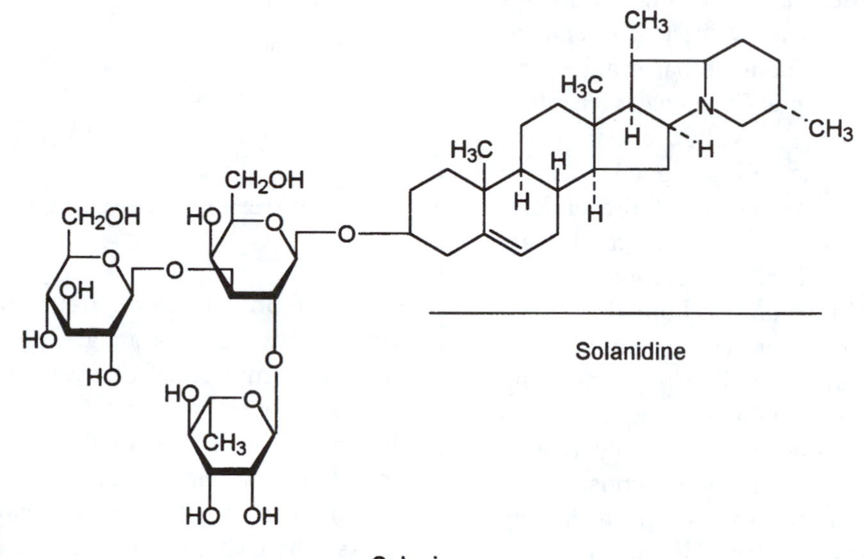

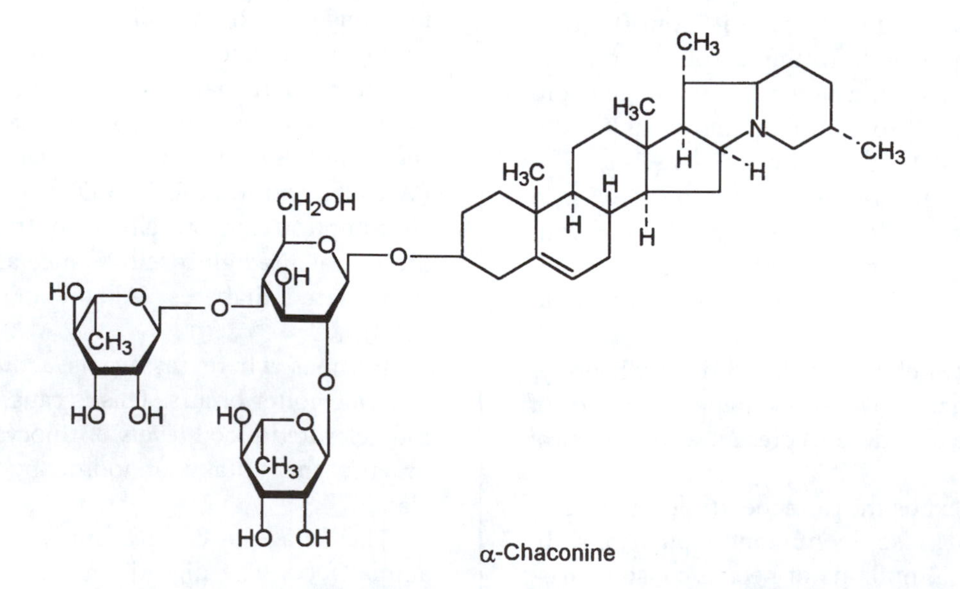

FIGURE 6–14 Representative glycoalkaloids. The glycoside α-solanine contains the aglycone solanidine. The α-chaconine also contains solanidine, but differs from solanine in the carbohydrate side chain.

spect to the carbohydrate side chain. Solanidine is a steroid alkaloid, but also contains the indolizidine nucleus (see swainsonine) so could also be classified as an indolizidine alkaloid. The chemistry and toxicologic properties of steroidal glycoalkaloids have been reviewed by Van Gelder (1991) and Keeler *et al.* (1991).

The two principal effects of solanum alkaloid poisoning are gastrointestinal tract irritation and impairment of the nervous system. The alkaloids are much more toxic when administered parenterally than when given orally because of relatively low gastrointestinal absorption (Nishie *et al.*, 1971)., The glycosides (e.g., solanine) are more toxic than the aglycones (e.g., solanidine). Apathy, drowsiness, salivation, dyspnea (difficult or labored breathing), trembling, weakness, paralysis, and loss of consciousness are manifestations of the effects on the nervous system. Solanum alkaloids are **cholinesterase inhibitors** (Jadhav *et al.*, 1981), which explains their neural effects. Acetylcholine is the neurotransmitter formed at the terminals of all preganglionic nerve fibers and at the endings of the postganglionic fibers of the parasympathetic nerve system and mediates the transfer of impulses from somatic motor nerves to skeletal smooth muscle (neuromuscular junctions). Inhibition of acetylcholinesterase results in the accumulation of acetylcholine in nerve tissue and effector organs, with the result that neural function is impaired. **Neurological signs**, such as ataxia, convulsions, coma, muscle weakness, and involuntary urination are the consequence. Besides solanum alkaloids, common pesticides such as organophosphates are cholinesterase inhibitors.

Gastrointestinal tract effects of solanum alkaloids are manifested by inflammation of intestinal mucosa, hemorrhage or ulceration, abdominal pain, and constipation or diarrhea.

Teratogenic effects of potato alkaloids have been suggested. Sharma and Salunkhe (1989) reviewed reports linking potato consumption to birth defects such as **spina bifida**. An incident of this birth defect had been associated with maternal ingestion of blight-infected potatoes. *Solanum* alkaloids have been extensively investigated for their teratogenic effects (Keeler *et al.*, 1990; Keeler *et al.*, 1991). Numerous *Solanum* alkaloids, including some from potatoes, have teratogenic effects in laboratory animals, especially causing craniofacial malformations. Hamsters are especially susceptible (Gaffield and Keeler, 1996). The alkaloids involved are structurally similar to the steroidal *Veratrum* alkaloids well known to cause facial deformities in livestock (see *Veratrum*, Chapter 14). Are there significant teratogenic risks with potatoes? According to Keeler *et al.* (1991), who have conducted extensive studies with teratogens, "Whether there are real-world teratogenic hazards for humans and animals upon ingestion of these *Solanum* species is a question that cannot yet be answered with complete assurance".

Poisoning of both livestock and humans from potatoes has occurred. The solanum alkaloids are in highest concentration in green sprouts and green potato skins (peels). The **greening of potatoes** occurs when the tubers are exposed to sunlight during growth or after harvest. The green pigment is chlorophyll; the increased concentration of solanum alkaloids in green potatoes is due to the fact that similar environmental conditions promote development of both chlorophyll and glycoalkaloids (Jadhav *et al.*, 1981). The biosyntheses of chlorophyll and glycoalkaloids are unrelated physiological processes (Dao and Friedman, 1994). The level of glycoalkaloids is elevated in potatoes grown without pesticides (Hlywka *et al.*, 1994), reflecting their role as chemical defenses. According to Friedman and Dao (1992), 14–27% of the annual U.S. potato crop is culled because of greening. Jadhav *et al.* (1981) and Kingsbury (1964) cite examples of human deaths from consumption of green potatoes. Over 2000 documented cases of human poisoning from potatoes have been described (Keeler *et al.*, 1991). Livestock have been poisoned after being fed potato sprouts, peelings, and sunburned or spoiled potatoes. Potato vines have also caused toxicity, since the alkaloids are in highest concentration in green tissues. The potato alkaloids are not destroyed as a result of boiling, baking, frying, or drying at high temperatures (Jadhav *et al.*, 1981). The relatively rare occurrence of solanine poisoning has been explained by Nishie *et al.* (1971) as the result of the following factors: (1) solanine is poorly absorbed; (2) solanine is hydrolyzed to a considerable extent in the gut to the less toxic solanidine; and (3) there is a rapid urinary and fecal excretion of metabolites.

In the rumen, potato glycoalkaloids are hydrolyzed to solanidine, which is further metabolized to a 5,6-dihydro analog of solanidine (King and McQueen, 1981).

An interesting consequence of *Solanum* alkaloids was the withdrawal of a new potato variety, Lenape, in the late 1960s. This variety had excellent characteristics for french fry production, but after commercial production began, it was found to have a toxic level of solanine. New potato varieties are now screened before release, and glycoalkaloid levels must be below 20 mg/100 g. Levels above about 14 mg/100 g are bitter, while levels above 20 cause a burning sensation of the mouth and throat. Lenape had a level of about 30 mg/100 g (Jadhav *et al.*, 1981). In a survey of commercial varieties, Friedman and Dao (1992) found α-chaconine contents ranging from 1.17 to 13.5 mg/100 g fresh weight and solanine values ranging from 0.58 to 5.9 mg/100 g. In a cultivar with total glycoalkaloid content of 14.7 mg/100 g in the tubers, the leaves had 145 mg, stems 32–46 mg, and sprouts, 997 mg/100 g. Although potato glycoalkaloids are toxic, low levels are considered necessary for the desired flavor properties of potatoes (Slanina, 1990).

Nicholson *et al.* (1978) reported a study on the potential feeding value of **potato vines**. Potato growers often wish to harvest their crop before the vines have died to minimize the transfer of viruses to the tubers. Conventional potato harvesters cannot handle a large bulk of vines; the usual practice is to remove the vines with mechanical beaters or to kill them with chemicals. Either operation is expensive and returns no revenue. Harvesting the vines for livestock feed would provide a use for them and utilize the nutrients they contain. Nicholson *et al.* (1978) concluded that potato vines harvested prior to senescence are nontoxic and provide a useful level of nutrients to ruminants. Possible pesticide residues are the primary concern, and a modified pest management system would be needed for vines intended for livestock feed.

Potato tubers contain **trypsin and chymotrypsin inhibitors** (Nicholson and Allen, 1989). Potatoes should be cooked for use as animal feed, both to increase digestibility of the starch granules (raw potato starch has a low digestibility) and to inactivate heat-labile protease inhibitors. Potatoes also contain inhibitors of carboxypeptidases (Nicholson and Allen, 1989), a group of pancreatic proteolytic enzymes.

Potatoes contain phenolics, with about 90% being chlorogenic acid (Griffiths *et al.*, 1995). **Chlorogenic acid** is associated with a darkening of the flesh after cooking (referred to as after-cooking blackening) which is due to formation of an iron–chlorogenic acid complex during cooking, which on exposure to air is oxidized from a colorless ferrous to a bluish-grey ferric compound (Griffiths *et al.*, 1995).

Tomatoes contain a toxic solanum alkaloid called tomatine (aglycone is tomatidine). The structure of tomatine, first identified in 1948, is provided by Jadhav *et al.* (1981). The mature fruit is low in alkaloid, but green tomatoes and the vines contain appreciable quantities. Cattle and pigs have been poisoned when fed tomato vines. Tomatine is a cholinesterase inhibitor.

Toxins in Sweet Potatoes and Yams

Sweet potatoes (*Ipomoea batatas*) are grown primarily for human consumption in tropical and subtropical areas. Cull sweet potatoes and the vines are used as animal feed. The tubers contain trypsin inhibitors and must be cooked before use as feed (Bouwkamp, 1985). Sweet potatoes produce stress metabolites (**phytoalexins**) when subjected to stresses such as mechanical damage, insect attack and fungal in-

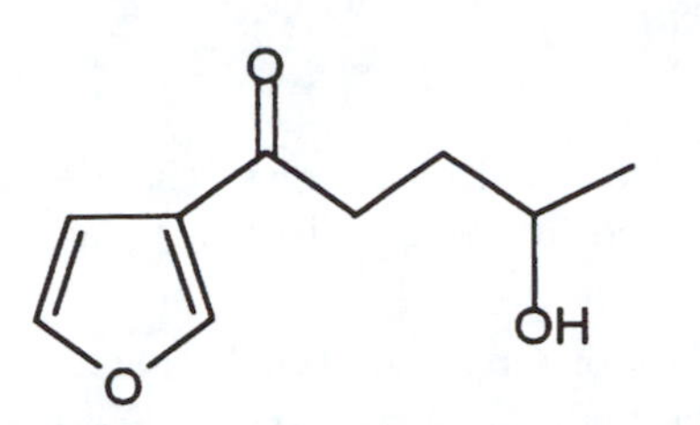

FIGURE 6–15 Ipomeanol, a pneumotoxic 3-substituted furan in moldy sweet potato vines and tubers.

fection (Wilson and Burka, 1983). These are a group of 3-substituted furans such as **ipomeanol** (Fig. 6–15). Consumption of moldy sweet potato tubers or vines by livestock may result in respiratory disease, with signs identical to those of acute bovine pulmonary emphysema (see Chapter 9), which is caused by 3-methyl indole, a metabolite of tryptophan. Ipomeanol is bioactivated by cytochrome P450 in lung tissue, producing metabolites which covalently bind to subcellular constituents (Wilson and Burka, 1983). Low *et al.* (1993) reported mortality from sweet potato poisoning of pigs in Papua New Guinea (PNG) and suggested that the widespread consumption of moldy sweet potatoes by humans in PNG might have an involvement in the existence of chronic respiratory disease in a large proportion of the human population.

Yams (*Dioscorea* spp.) are similar to sweet potatoes in appearance and culture. They contain trypsin inhibitors, so must be cooked before use as a feed for nonruminants (Panigrahi and Francis, 1982). Cocoyam or **taro** (*Colocasia esculenta*) contains oxalates and perhaps other toxins, which are inactivated by cooking (Samarasinghe and Rajaguru, 1992).

Toxins in Onions

Cull onions are sometimes used as livestock feed. Onions contain numerous sulfur-containing compounds that can induce red blood cell hemolysis. Onion toxicity is discussed in the section on forage-induced anemia (Chapter 11).

REFERENCES

Phytates

Erdman, J.W., Jr. 1979. Oilseed phytates: nutritional implications. J. Am. Oil Chem. Soc. 56:736–741.

Graf, E., and J.W. Eaton. 1993. Suppression of colonic cancer by dietary phytic acid. Nutrition and Cancer 19:11–19.

Jongbloed, A.W., Z. Mroz, and P.A. Kemme. 1992. The effect of supplementary *Aspergillus niger* phytase in diets for pigs on concentration and apparent digestibility of dry matter, total phosphorus, and phytic acid in different sections of the alimentary tract. J. Anim. Sci. 70:1159–1168.

Lei, X.G., P.K. Ku, E.R. Miller, M.T. Yokoyama, and D.E. Ullrey. 1993. Supplementing corn–soybean meal diets with microbial phytase maximizes phytate phosphorus utilization in weanling pigs. J. Anim. Sci. 71:3368–3375.

Maga, J.A. 1982. Phytate: Its chemistry, occurrence, food interactions, nutritional significance and methods of analysis. J. Agr. Food Chem. 30:1–9.

Simons, P.C.M., H.A.J. Versteegh, A.W. Jongbloed, P.A. Kemme, P. Slump, K.D. Bos, M.G.E. Wolters, R.F. Beudeker, and G.J. Verschoor. 1990. Improvement of phosphorus availability by microbial phytase in broilers and pigs. Brit. J. Nutr. 64:525–540.

Non-starch Polysaccharides

Annison, G. 1991. Relationship between the levels of soluble nonstarch polysaccharides and the apparent metabolizable energy of wheats assayed in broiler chickens. J. Agr. Food Chem. 39:1252.

Annison, G., and M. Choct. 1991. Anti-nutritive activities of cereal non-starch polysaccharides in broiler diets and strategies minimizing their effects. World's Poult. Sci. J. 47:232–242.

Bedford, M.R., and H.L. Classen. 1992. Reduction in intestinal viscosity through manipulation of dietary rye and pentosanase concentration is effected through changes in the carbohydrate composition of the intestinal aqueous phase and results in improved growth rate and food conversion efficiency of broiler chicks. J. Nutr. 122:560–569.

Campbell, G.L., and M.R. Bedford. 1992. Enzyme applications for monogastric feeds: a review. Can. J. Anim. Sci. 72:449–466.

Choct, M., and G. Annison. 1992. The inhibition of nutrient digestion by wheat pentosans. Brit. J. Nutr. 67:123–132.

Jorgensen, J.B., H. Lunde, and B. Robertsen. 1993. Peritoneal and head kidney cell response to intraperitoneally injected yeast glucan in Atlantic salmon, *Salmo salar* L. J. Fish Diseases 16:313–325.

Low, A.G. 1989. Secretory response of the pig gut to non-starch polysaccharides. Anim. Feed Sci. Tech. 23:55–65.

Petterson, D., H. Graham, and P. Aman. 1991. The nutritive value for broiler chickens of pelleting and enzyme supplementation of a diet containing barley, wheat and rye. Anim. Feed Sci. Tech. 33:1–14.

Teitge, D.A., G.L. Campbell, H.L. Classen, and P.A. Thacker. 1991. Heat pretreatment as a means of improving the response to dietary pentosanase in chicks fed rye. Can. J. Anim. Sci. 71:507–513.

Phenolic Compounds in Grains

Austin, P.J., L.A. Suchar, C.T. Robbins, and A.E. Hagerman. 1989. Tannin-binding proteins in saliva of deer and their absence in saliva of sheep and cattle. J. Chem. Ecol. 15:1335–1347.

Boren, B., and R.D. Waniska. 1992. Sorghum seed color as an indicator of tannin content. J. Appl. Poultry Res. 1:117–121.

Butler, L.G. 1989a. Effects of condensed tannin on animal nutrition. pp. 391–402 in: R.W. Hemingway and J.J. Karchesy (Eds.). Chemistry and Significance of Condensed Tannins. Plenum Press, New York and London.

Butler, L.G. 1989b. Sorghum polyphenols. pp. 95–121 in: P.R. Cheeke (Ed.). Toxicants of Plant Origin. Vol. IV. Phenolics. CRC Press, Boca Raton, FL.

Divi, R.L., and D.R. Doerge. 1996. Inhibition of thyroid peroxidase by dietary flavonoids. Chem. Res. Toxicol. 9:16–23.

Elkin, R.G., W.R. Featherston, and J.C. Rogler. 1978. Investigations of leg abnormalities in chicks consuming high tannin sorghum grains. Poult. Sci. 57:757–762.

Elkin, R.G., M.B. Freed, B.R. Hamaker, Y. Zhang, and C.M. Parsons. 1996. Condensed tannins are only partially responsible for variations in nutrient digestibilities of sorghum grain cultivars. J. Agric. Food Chem. 44:848–853.

Farrell, D.J., C. Chan, and F. McCrae. 1983. A nutritional evaluation of triticale with pigs. Anim. Feed Sci. Tech. 9:49–62.

Hagerman, A.E., and C.T. Robbins. 1993. Specificity of tannin-binding salivary proteins relative to diet selection by mammals. Can. J. Zool. 71:628–633.

Hagerman, A.E., C.T. Robbins, Y. Weerasuriya, T.C. Wilson, and C. McArthur. 1992. Tannin chemistry in relation to digestion. J. Range Manage. 45:57–62.

Hemingway, R.W., and J.J. Karchesy (Eds.). 1989. Chemistry and Significance of Condensed Tannins. Plenum Press, New York and London.

Jimenez-Ramsey, L.M., J.C. Rogler, T.L. Housley, L.G. Butler, and R.G. Elkin. 1994. Absorption and distribution of ^{14}C-labeled condensed tannins and related sorghum phenolics in chickens. J. Agric. Food Chem. 42:963–967.

McArthur, C., C.T. Robbins, A.E. Hagerman, and T.A. Hanley. 1993. Diet selection by a ruminant generalist browser in relation to plant chemistry. Can. J. Zool. 71:2236–2243.

McArthur, C., G.D. Sanson, and A.M. Beal. 1995. Salivary proline–rich proteins in mammals: Roles in oral homeostasis and counteracting dietary tannin. J. Chem. Ecol. 21:663–691.

Mehansho, H., D.K. Ann, L.G. Butler, J. Rogler, and D.M. Carlson. 1987. Induction of proline-rich proteins in hamster salivary glands by isoproterenol treatment and an unusual growth inhibition by tannins. J. Biol. Chem. 262:12344–12350.

Mole, S., L.G. Butler, and G. Iason. 1990. Defense against dietary tannin in herbivores: a survey for proline rich salivary proteins in mammals. Biochemical Systematics and Ecology 18:287–293.

Morton, J.F. 1970. Tentative correlations of plant usage and esophageal cancer zones. Econ. Bot. 24:217–226.

Morton, J.F. 1989. Tannin as a carcinogen in bush-tea: Tea, mate and khat. pp. 403–416 in: R.W. Hemingway and J.J. Karchesy (Eds.). Chemistry and Significance of Condensed Tannins. Plenum Press, New York.

Mueller-Harvey, I., and J.D. Reed. 1992. Identification of phenolic compounds and their relationships to in-vitro digestibility of sorghum leaves from bird-resistant and non-bird–resistant varieties. J. Sci. Food Agric. 60:179–196.

Newman, R.K., C.W. Newman, A.M. El-Negoumy, and S. Aastrup. 1984. Nutritional quality of proanthocyanidin-free barley. Nutr. Rep. Int. 30:809–816.

Radcliffe, B.C., C.F. Driscoll, and A.R. Egan. 1981. Content of 5-alkyl resorcinols in selection lines of triticale grown in South Australia. Aust. J. Exp. Agric. Anim. Husb. 21:71–74.

Radcliffe, B.C., A.R. Egan, and C.J. Driscoll. 1983. Nutritional evaluation of triticale grain as an animal feedstuff. Aust. J. Exp. Agric. Anim. Husb. 23:419–425.

Robbins, C.T., A.E. Hagerman, P.J. Austin, C. McArthur, and T.A. Hanley. 1991. Variation in mammalian physiological responses to a condensed tannin and its ecological implications. J. Mamm. 72:480–486.

Salunkhe, D.K., J.K. Chavan, and S.S. Kadam. 1989. Dietary Tannins: Consequences and Remedies. CRC Press, Boca Raton, FL.

Sell, D.R., and J.C. Rogler. 1983. Effects of sorghum grain tannins and dietary protein on the activity of liver UDP-glucuronyltransferase. Proc. Soc. Exp. Biol. Med. 174:93–101.

Streeter, M.N., G.M. Hill, D.G. Wagner, F.N. Owens, and C.A. Hibberd. 1993. Effect of bird-resistant and non-bird–resistant sorghum grain on amino acid digestion by beef heifers. J. Anim. Sci. 71:1648–1656.

Protease Inhibitors in Grains

Erickson, J.P., E.R. Miller, F.C. Elliott, P.K. Ku, and D.E. Ullrey. 1979. Nutritional evaluation of triticale in swine starter and grower diets. J. Anim. Sci. 48:547–553.

Liener, I.E. 1980. Miscellaneous toxic factors. pp. 429–467 in: I.E. Liener (Ed.)., Toxic Constituents of Plant Foodstuffs, 2nd Edition. Academic Press, NY.

Sosulski, F.W., L.A. Minja, and D.A. Christensen. 1988. Trypsin inhibitors and nutritive value in cereals. Plant Foods for Human Nutr. 38:23–34.

Toxins in Minor Grains

Gee, J.M., K.R. Price, C.L. Ridout, G.M. Wortley, R.F. Hurrell, and I.T. Johnson. 1993. Saponins of quinoa (*Chenopodium quinoa*): Effects of processing on their abundance in quinoa products and their biological effects on intestinal mucosal tissue. J. Sci. Food Agric. 63:201–209.

Oomah, B.D., and G. Mazza. 1996. Flavonoids and antioxidative activities in buckwheat. J. Agric. Food Chem. 44:1746–1750.

Pomeranz, Y. 1983. Buckwheat: structure, composition and utilization. CRC Crit. Rev. Food Sci. Nutr. 19:213–258.

Pond, W.G., and J.W. Lehmann. 1989. Nutritive value of a vegetable amaranth cultivar for growing lambs. J. Anim. Sci. 67:3036–3039.

Pond, W.G., J.W. Lehmann, R. Elmore, F. Husby, C.C. Calvert, C.W. Newman, B. Lewis, R.L. Harrold, and J. Froseth. 1991. Feeding value of raw or heated grain amaranth germplasm. 33:221–236.

Cassava Cyanogens

Akanji, A.O., and O.O. Famuyiwa. 1993. The effects of chronic cassava consumption, cyanide intoxication and protein malnutrition on glucose tolerance in growing rats. Brit. J. Nutr. 69:269–276.

Aminlari, M., and H. Gilanpour. 1991. Comparative studies on the distribution of rhodanese in different tissues of domestic animals. Comp. Biochem. Physiol. 99B:673–677.

Aminlari, M., and M. Shahbazi. 1994. Rhodanese (thiosulfate:cyanide sulfurtransferase) distribution in the digestive tract of chicken. Poult. Sci. 73:1465–1469.

Davis, R.H. 1991. Cyanogens. pp. 202–225 in: J.P.F. DeMello, C.M. Duffus and J.H. Duffus (Eds.). Toxic Substances in Crop Plants. The Royal Society of Chemistry, Cambridge, UK.

Espinoza, O.B., M. Perez, and M.S. Ramirez. 1992. Bitter cassava poisoning in eight children: a case report. Vet. Human Tox. 34:65.

Kingsbury, M. 1964. Poisonous Plants of the United States and Canada. Prentice–Hall, Englewood Cliffs, NJ.

Mlingi, N., N.H. Poulter, and H. Rosling. 1992. An outbreak of acute intoxications from consumption of insufficiently processed cassava in Tanzania. Nutr. Res. 12:677–687.

Montgomery, R.D. 1980. Cyanogens. pp. 143–160 in: I.E. Liener (Ed.)., Toxic Constituents of Plant Foodstuffs. Academic Press, NY.

Reed, J.D., R.E. McDowell, P.J. Van Soest, and P.J. Horvath. 1982. Condensed tannins: a factor limiting use of cassava forage. J. Sci. Food Agric. 33:213–220.

Wheeler, J.L., and C. Mulcahy. 1989. Consequences for animal production of cyanogenesis in sorghum forage and hay—A review. Tropical Grasslands 23:193–202.

Potato Glycoalkaloids

Dao, L., and M. Friedman. 1994. Chlorophyll, chlorogenic acid, glycoalkaloid, and protease inhibitor content of fresh and green potatoes. J. Agric. Food Chem. 42:633–639.

Friedman, M., and L. Dao. 1992. Distribution of glycoalkaloids in potato plants and commercial potato products. J. Agric. Food Chem. 40:419–423.

Gaffield, W., and R.F. Keeler. 1996. Induction of terata in hamsters by solanidane alkaloids derived from *Solanum tuberosum*. Chem. Res. Toxicol. 9:426–433.

Griffiths, D.W., H. Bain, and M.F.B. Dale. 1995. Photo-induced changes in the total chlorogenic acid content of potato (*Solanum tuberosum*) tubers. J. Sci. Food Agric. 68:105–110.

Hellenas, K.-E., C. Branzell, H. Johnsson, and P. Slanina. 1995. Glycoalkaloid content of early potato varieties. J. Sci. Food Agric. 67:125–128.

Hlywka, J.J., G.R. Stephenson, M.K. Sears, and R.Y. Yada. 1994. Effects of insect damage on glycoalkaloid content in potatoes (*Solanum tuberosum*). J. Agric. Food Chem. 42:2545–2550.

Jadhav, S.J., R.P. Sharma, and D.K. Salunkhe. 1981. Naturally occurring toxic alkaloids in foods. CRC Crit. Rev. Toxicol. 9:1–104.

Keeler, R.F., D.C. Baker, and W. Gaffield. 1990. Spirosolane-containing *Solanum* species and induction of congenital craniofacial malformations. Toxicon 28:873–884.

Keeler, R.F., D.C. Baker, and W. Gaffield. 1991. Solanum alkaloids. pp. 607–636. in: R.P. Sharma and D.K. Salunke. Mycotoxins and phytoalexins. CRC Press, Boca Raton, FL.

King, R.R., and R.E. McQueen. 1981. Transformation of potato glycoalkaloids by rumen microorganisms. J. Agric. Food Chem. 29:1101–1103.

Nicholson, J.W.G., and J.G. Allen. 1989. The distribution of trypsin and chymotrypsin inhibitors in potato tubers. Can. J. Anim. Sci. 69:513–515.

Nicholson, J.W.G., D.A. Young, R.E. McQueen, H. Dejong, and F.A. Wood. 1978. The feeding value potential of potato vines. Can. J. Anim. Sci. 58:559–569.

Nishie, K., M.R. Gumbmann and A.C. Keyl. 1971. Pharmacology of solanine. Toxicol. Appl. Pharmacol. 19:81–92.

Sharma, R.P., and D.K. Salunkhe. 1989. Solanum glycoalkaloids. pp. 179–236 in: P.R. Cheeke (Ed.). Toxicants of Plant Origin. Vol. I. Alkaloids. CRC Press, Boca Raton, FL.

Slanina, P. 1990. Solanine (glycoalkaloids) in potatoes: Toxicological evaluation. Fd. Chem. Toxicol. 28:759–761.

Van Gelder, W.M.J. 1991. Steroidal glycoalkaloids in *Solanum*: Consequences for potato breeding and for food safety. pp. 101–134 in: R.F. Keeler and A.T. Tu (Eds.). Handbook of Natural Toxins. Vol. 6. Toxicology of Plant and Fungal Compounds. Marcel Dekker, Inc., New York.

Toxins in Sweet Potatoes and Yams

Bouwkamp, J.C. (Ed.). 1985. Sweet Potato Products: A New Resource for the Tropics. CRC Press, Boca Raton, FL.

Low, S.G., I.McL. Grant, B. Rodoni and W.L. Bryden. 1993. Sweet potato (*Ipomoea zbatatas*) poisoning of pigs in Papua New Guinea. N.Z. Vet. J. 41:218(Abst.).

Panigrahi, S., and B. Francis. 1982. Digestibility and possible toxicity of the yam *Dioscorea alata*. Nutr. Rep. Int. 26:1007–1013.

Samarasinghe, K., and A.S.B. Rajaguru. 1992. Raw and processed wild colocasia corm meal (*Colocasia esculenta* (L.) Schott, var. esculenta) as an energy source for broilers. Anim. Feed Sci. Tech. 36:143–151.

Wilson, B.J., and L.T. Burka. 1983. Sweet potato toxins and related toxic furans. pp. 3–41 in: R.F. Keeler and A.T. Tu (Eds.). Handbook of Natural Toxins. Vol. 1. Plant and Fungal Toxins. Marcel Dekker, Inc., New York.

CHAPTER 7

Toxins in Protein Supplements and Grain Legumes

The major plant protein supplements used in animal feeding are the meals remaining after oil is extracted from oilseeds. In North America, the major oilseeds are soybeans, cottonseed and rapeseed (canola). These oilmeals, while very important as animal feedstuffs, contain deleterious or toxic factors, which are sometimes referred to as ANF's (anti-nutritional factors). Their optimal utilization as feedstuffs requires an understanding of their toxicants and the detoxification effects of feed processing procedures.

SOYBEANS AND SOYBEAN MEAL

Soybeans, in common with most other legume seeds, contain a variety of toxicants. These include protease (trypsin) inhibitors, allergenic proteins, lectins, phytoestrogens, saponins, goitrogens, phytates and oligosaccharides (Liener, 1994).

Trypsin (Protease) Inhibitors

A wide variety of plants contain proteins which inhibit protein digestion in the digestive tract of animals. The trypsin inhibitors of soybeans are the best known and most widely studied. Other plants containing trypsin inhibitors include most types of beans, potatoes, rye, triticale, barley, and alfalfa. Those in soybeans, common beans, and triticale are of particular significance in livestock feeding. Although these factors are commonly referred to as trypsin inhibitors, protease inhibitor is probably a better term, since other enzymes such as chymotrypsin are also affected.

In soybeans, the **protease inhibitors** are of two main categories: those having a molecular weight of 20,000–25,000 with few disulfide bonds and with specific activity toward trypsin, and those having a molecular weight of 6000–10,000 with a high proportion of disulfide bonds and with inhibitory activity to trypsin and chymotrypsin at independent binding sites. These two have been referred to as the **Kunitz inhibitor** and the **Bowman–Birk inhibitor**, respectively. The Kunitz inhibitor has been isolated and its amino acid sequence determined. It contains 181 amino acids, with 2 disulfide bonds and with the active site at amino acid 63. The Kunitz inhibitor combines with trypsin stoichiometrically: 1 mole of inhibitor inactivates 1 mole of trypsin. The reaction is almost instantaneous, and the complex formed is a very tight one. It appears that trypsin reacts with the inhibitor in the same way that it reacts with other proteins it digests, but a number of noncovalent bonds formed at the active site result in an irreversible complex. The Bowman–Birk inhibitor has two active sites, one that binds trypsin and one that complexes chymotrypsin. It is a single polypeptide chain with 71 amino acids, and 7 disulfide bonds.

Trypsin inhibitors occur mainly in seeds, although in some cases they are also found in leaves. Besides soybeans, other legume seeds with trypsin inhibitors include lima, kidney, navy, pinto and common garden beans, peanuts, cowpeas, fava beans, and peas. While all common cereal grains (barley, wheat, corn, rye, rice, and sorghum) contain trypsin inhibitors, they are of practical significance only in the case of triticale. The physiological role of trypsin inhibitors in plants is uncertain, but they may be involved in defense mechanisms. Many plant leaves, after wounding by insects or bacterial infection, show an accumulation of protease inhibitors at the site of damage. They may inhibit insect and bacterial proteases.

Trypsin inhibitors in soybeans are primarily of concern for nonruminant animals. In chicks, rats and pigs, the feeding of raw soybeans as a major part of the diet results in poor growth, poor hair or feather appearance, and disturbed digestive processes (Fig. 7–1). In some species, there may be great enlargement of the pancreas, hypersecretion of pancreatic enzymes and bile, and reduced digestibility of protein, carbohydrate and lipids (Grant, 1989), reducing nutrient retention and efficiency of feed utilization. Although several other toxicants in raw soybeans contribute to the growth depressing effects, the trypsin inhibitors are the major cause. In young pigs and pre-ruminant calves, allergenic proteins account for part of the growth depression, while lectins also contribute.

FIGURE 7–1 Turkey poults fed raw (left) and heat-treated soybeans. The bird fed raw soybeans shows pronounced growth depression and poor feathering. (Courtesy of J.A. Harper)

The responses of nonruminants to trypsin inhibitors in raw soybeans differ somewhat among species. In chicks, rats and mice, **pancreatic enlargement** (hypertrophy) and increased pancreatic secretion occur. In the pig, dog and pre-ruminant calf, pancreatic hypertrophy and hypersecretion do not occur. The normal pancreas size as a % of body weight is considerably larger in species which show pancreatic hypertrophy when fed trypsin inhibitors than in species without pancreatic enlargement (Liener and Kakade, 1980). Species whose pancreas exceeds about 0.3% of body weight show hypertrophy. Liener and Kakade (1980) speculate that because these species have a high trypsin production rate, they are more sensitive to stimuli that increase trypsin secretion.

The growth depression of chicks fed sources of trypsin inhibitors is largely explained by increased endogenous excretion of amino acids. When trypsin becomes irreversibly bound to trypsin inhibitor in the intestine, the pancreas responds by secreting more trypsin (and other enzymes as well). Regulation of pancreatic secretion is mediated by a negative feedback stimulation of the pancreas by the hormone **cholecystokinin (CCK)**. CCK release from cells of the intestinal mucosa is regulated by pancreatic proteases. When the level of free intestinal proteases is reduced by binding to trypsin inhibitor, CCK releasing factors are increased, causing release of CCK into the blood. This in turn increases enzyme synthesis and secretion by the pancreas. Continued stimulation of this pathway by dietary trypsin inhibitor results in pancreatic hypertrophy and hyperplasia (increase in cell number). The hypersecretion of pancreatic enzymes increases the animal's amino acid requirements, especially for the sulfur amino acids. Thus the growth depression in species such as the chick which experience pancreatic hypertrophy is due to an induced amino acid deficiency caused by an excessive synthesis, secretion and fecal excretion of pancreatic enzymes. The digestibility of nutrients is not impaired in these species, because the hypersecretion of enzymes compensates for the inactivation of enzymes by trypsin inhibitor.

In contrast, in animals such as the pig, dog and calf, pancreatic enlargement does not occur. In these species, growth depression occurs because of the reduced digestibility of nutrients, especially proteins. Whereas in the chick there is increased enzyme secretion, in the pig there is hyposecretion of trypsin and other enzymes. Thus the combined effect of reduced enzyme secretion and irreversible binding of enzyme with trypsin inhibitors is to reduce digestibility and hence inhibit growth rate. In species in which pancreatic hypertrophy occurs, the endocrine function of the pancreas is affected, with reduction in pancreatic insulin content and blood insulin levels (Grant, 1989).

Ruminants are less sensitive to the effects of trypsin inhibitors than are nonruminants. In an *in vitro* rumen fermentation study, Holmes *et al.* (1993) found that the trypsin inhibitors in seven grain legumes were not degraded in 18 hours of fermentation, and suggested that this would allow sufficient time for trypsin-inhibitory activity to escape the rumen to the small intestine. Susmel *et al.* (1995) also reported that soybean trypsin inhibitors are degraded in rumen fluid at a lower rate than other soybean proteins. However, Albro *et al.* (1993) observed that substitution of raw soybeans for soybean meal as protein supplements for weaned beef steers did not affect animal performance. At levels which raw soybeans would be used in ruminant diets, any effect of trypsin inhibitor activity is likely to be slight.

Feeding raw soybeans to rats for a prolonged period results in cancerous lesions (adenocarcinoma) and enhanced **carcinogenesis** when fed with or prior to known pancreatic carcinogens (Roebuck, 1987). Low levels of trypsin inhibitor in processed and cooked soybean products used in the human diet might be of concern as potential carcinogens. On the other hand, there is considerable evidence that soybean trypsin inhibitors, particularly the Bowman–Birk inhibitor, have cancer-suppressing activity. These effects are reviewed by Kennedy *et al.* (1993). Isoflavone **phytoestrogens** in soybeans have also been demonstrated to have anti-cancer properties (Messina *et al.*, 1994).

Trypsin inhibitors are readily destroyed by treatment of plant material with moist heat. Over 95% of the activity is destroyed in 15 min of 100° C heat treatment. For on-the-farm processing of soybeans, an extruder in which the beans are forced through a die is effective (Fig. 8–1). The friction in the die builds up sufficient heat to destroy the inhibitors. A simple test (the **urease test**) to determine if trypsin inhibitory activity is present involves incubation of treated soybeans with urea in the presence of an indicator dye. Soybeans contain urease, which is also heat sensitive. If heating has not been adequate to destroy urease, the urea is converted to ammonia, which changes the pH and causes color development. The presence of active urease suggests that active trypsin inhibitors will also be present.

Heat treatment to destroy trypsin inhibitors may have deleterious effects on protein quality. When heated, lysine reacts with sugars to produce brown, indigestible polymers (see **Maillard reaction**, Chapter 8). Thus lysine "tie-up" may occur when raw soybeans or soybean meal are heat-treated to destroy protease inhibitors.

Cultivars of soybeans free of Kunitz trypsin inhibitor have been developed. Although when fed raw they support better growth of chicks (Anderson-Hafermann *et al.*, 1992) and pigs (Herkelman *et al.*, 1992) than raw conventional soybeans, they still require heat treatment to maximize their feeding value (Zhang *et al.*, 1993). Presumably, this is because the Kunitz trypsin inhibitor constitutes 40–50% of the total trypsin inhibitor activity, and other heat-labile toxins such as lectins are also present. Because trypsin inhibitors are readily inactivated by heat treatment, and they contribute to the soybean plant's chemical defense against pests, it may not necessarily be of advantage to develop trypsin inhibitor–free soybeans.

Bovine colostrum contains a large amount of trypsin inhibitor, which assists in the transfer of antibodies to the neonate via the absorption of intact proteins. Presence of trypsin inhibitor reduces the intestinal digestion of the antibodies. Supplementation of colostrum with soybean trypsin inhibitor increases immunoglobulin absorption in calves (Quigley *et al.*, 1995).

Antigenic Proteins in Soybeans

Calves and baby pigs fed milk replacers based on soy products often exhibit symptoms of poor growth and diarrhea. Two storage proteins in soybeans, **glycinin** and **β-conglycinin**, are involved in inducing an antigenic response. These proteins are resistant to digestion, and in the young animal can be absorbed. When soy products are used in milk replacers and creep diets (a creep diet is one fed to suckling animals prior to weaning), glycinin and conglycinin may be absorbed into the intestinal mucosa. The lymphoid tissue in the gut, particularly in areas known as Peyer's patches, respond with the production of two classes of immunoglobulins, IgA and IgM. These antibodies are secreted into the gut and react with antigens to prevent them from being absorbed. Antigens which escape

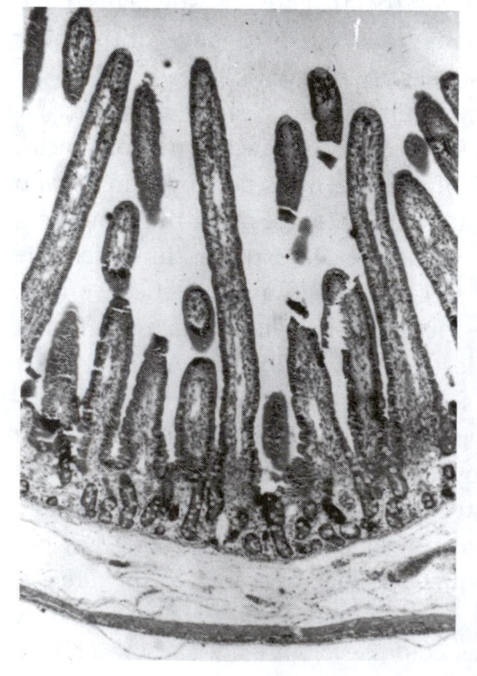

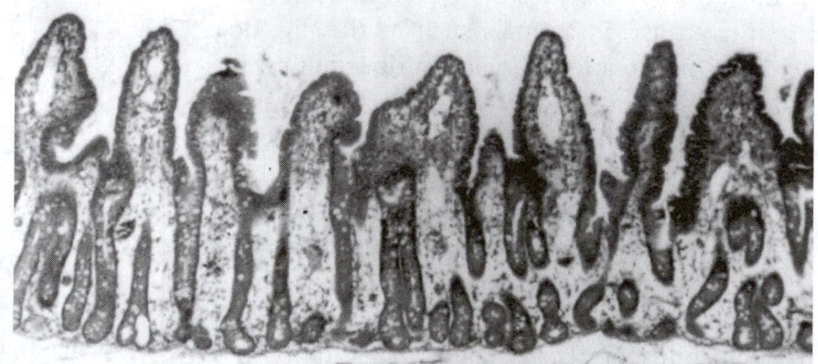

FIGURE 7–2 An electron microscopic view of the normal intestinal mucosa of a preweaning pig (left). Note the elongated, narrow villi. (The midsections of some villi are missing because of the plane of the tissue slicing). The individual cells lining the villi, the enterocytes, are visible. These are formed in the crypts of Lieberkuhn at the base of the villi, and move up the villi as they elongate. In contrast, the mucosa of a newly weaned pig (above) reveals shortened, flattened villi. Nutrient absorption is impaired, and the pig is susceptible to post-weaning diarrhea or enteritis. (Courtesy of D.J. Hampson)

antibody detection initiate an inflammatory response in the mucosa. The villi and microvilli are damaged. The villi become shorter and broader (Fig. 7–2), while the microvilli are stunted and dysfunctional (Dunford *et al.*, 1989; Li *et al.*, 1990). A characteristic change accompanying villi atrophy is an increased depth of the crypts of Lieberkuhn (**crypt hyperplasia**).

When calves or baby pigs are first exposed to soy products, and if absorption of the antigenic proteins occurs, the animals become sensitized to them. Subsequent exposures result in intestinal lesions, causing reduced nutrient absorption, increased gut motility, growth of opportunistic pathogens such as *E. coli*, and diarrhea. The animals grow poorly, are stunted, have diarrhea, and may die. The allergenic responses to soy products in pigs and calves have been reviewed in detail by D'Mello (1991).

Soybean products for use in milk replacers can be specially processed to reduce or eliminate the effects of antigenic proteins. Conventional heat processing, which inactivates trypsin inhibitors, is not effective in denaturing the antigenic proteins. They are inactivated or removed by extraction of soy with hot, aqueous ethanol. Alkali treatment with sodium hydroxide, or treatment with acid, inactivates the antigenic proteins. Several commercial soy products, free of antigens, are available for use in calf and baby pig diets.

The antigenic proteins of soybeans are of increasing significance in animal production. In the swine industry, early weaning of pigs is used to increase reproductive performance. Post-weaning problems such as poor growth and diarrhea are common with early-weaned pigs. A possible strategy to prevent antigenic responses is to obtain a high intake of soybean meal in the creep phase to induce tolerance rather than sensitization. This can be accomplished with concentrated, nutrient-dense creep diets containing 15–22% soybean meal that ensure adequate intake of soybean meal to induce **immune tolerance**, which is achieved when the system no longer has the capacity to express either a cell-mediated or humoral immune response (Friesen *et al.*, 1993). Thus a low intake of soy proteins by the baby animal sensitizes it to later exposure to the antigens, whereas a high intake of soy in the preweaning phase induces tolerance to later exposure. A problem in achieving this is the low consumption of creep diet by the baby pig. In the pig, the induced sensitivity to allergenic proteins tends to be transient, occurring in the immediate post-weaning period. In the calf, the effects are prolonged and permanently damaging. Although glycinin and conglycinin are denatured by rumen fermentation, minute amounts that enter the intestine are sufficient to chronically stimulate the immune system in calves fed soy protein in the post-weaning period (Tukur *et al.*, 1993).

Other Toxicants in Soybeans

In addition to trypsin inhibitors and antigenic proteins, soybeans and soybean meal contain other compounds with potential for deleterious effects. These include phytates, lectins, saponins, goitrogens, phytoestrogens and oligosaccharides. Their significance in animal nutrition will be briefly described. It is increasingly recognized that some of the secondary compounds in soybeans may have beneficial effects on human health. Beneficial effects of trypsin inhibitors, isoflavones, saponins, phytates and other components are discussed in a symposium proceedings (Messina and Erdman, 1995).

Phytates (see also Chapter 6) in soybean meal are important in animal nutrition. Widespread outbreaks of zinc deficiency occurred in poultry and swine in the U.S. in the 1950's, when soybean meal first became an important feed ingredient. Zinc deficiency, characterized by poor growth and extensive skin lesions (parakeratosis), was a consequence of the high phytate activity of soybean meal. The problem was easily overcome by increasing the level of zinc supplementation. Phytates in soybean meal, by reducing phosphorus bioavailability, contribute to pollution problems from animal excreta, as discussed under Phytates in Cereal Grains (Chapter 6).

Saponins are a chemically-diverse group of compounds which form a stable foam in aqueous solution. They occur in plants as glycosides. Their biological properties include irritation of mucus membranes, red blood cell hemolysis, and serum cholesterol–lowering activity. Several saponins have been isolated from soybeans, with soyasapogenin A the major one (Oakenfull and Sidhu, 1989). Soybean saponins do not seem to be of significance to livestock, but may have some value as hypocholesterolemic agents in humans (see Chapter 11). Saponins will be discussed in further detail later as components of legume forages.

Soybeans contain isoflavones with estrogenic activity. The major **phytoestrogens** in soybeans are genistein and daidzein. Phytoestrogens can cause infertility by disturbances in metabolism of endogenous estrogens. The primary plant which has caused phytoestrogen-induced infertility in livestock is subterranean clover (see Chapter 11). Soybean estrogens have not been reported to cause fertility problems in livestock. Setchell *et al.* (1987) suggested that an infertility situation in captive cheetahs was probably due to phytoestrogens in soybean meal in their diet. In humans, soybean phytoestrogens have been suggested to have favorable effects on **estrogen-responsive cancers** of breast, ovarian and prostate tissue (Adlercreutz, 1990). It is hypothesized that the mode of action involves interference with the uptake of estrogens at the tissue level by binding with estrogen receptors. Aldercreutz *et al.* (1991) reported a low incidence of breast and prostate cancer in Japanese people, and linked this observation to the high intake of soybean products in the traditional Japanese diet. They observed very high amounts of isoflavones, mainly genistein, daidzein and equol, in the urine of Japanese subjects consuming a traditional diet. Isoflavones occur in a wide variety of food products made from soybeans (Wang and Murphy, 1994; Dwyer *et al.*, 1994).

Fotsis *et al.* (1993) suggested an alternative anticarcinogenic mode of action of genistein. They demonstrated that genistein is an inhibitor of **angiogenesis** (the generation of new capillaries). Angiogenesis occurs in tumors. Well-vascularized tumors expand both locally and by metastasis, while avascular tumors do not expand beyond 1–2 mm (Fotsis *et al.*, 1993). Thus, the inhibition of angiogenesis by phytoestrogens could contribute to antitumorigenic effects of soybeans and other isoflavone-containing plant products.

Genistein is an inhibitor of tyrosine kinase which phosphorylates tyrosine residues on key proteins involved in signal transduction events in normal and tumor cells. This inhibition is an alternative mode of action for anticarcinogenic effects of soybean isoflavones (Messina *et al.*, 1994).

Genistein and daidzein are the major isoflavones in soybean products. Messina *et al.* (1994) provided a comprehensive review of their anti-cancer activity. Daidzein has a higher bioavailability to humans than genistein (Xu *et al.*, 1994). These authors measured circulating levels of phytoestrogens in humans consuming soy products, and found them within the range that could have potential cancer-inhibiting activity.

Lectins are toxic glycoproteins found in legume seeds, including soybeans. They have a minor role in the toxicity of raw soybeans to poultry and swine. They are of more significance in the common field bean (see Bean Lectins).

Raw soybeans contain **goitrogens** which may cause thyroid enlargement (Grant, 1989). They are not of practical significance in animal feeding.

Oligosaccharides are short-chain polymers of sugars. Raffinose and stachyose contain 3 and 4 sugars, respectively, linked by α-1,6 bonds. They are indigestible by endogenous enzymes, because animals do not produce α1,6-galactosidase. Thus the oligosaccharides pass undigested through the small intestine to the hindgut, where they serve as substrates for fermentation. In humans, this causes digestive problems associated with the consumption of beans, including nausea, diarrhea and gas production (flatulence). In swine and poultry, the presence of oligosaccharides in soybeans and other legume seeds lowers the digestible energy content of the diet from what it would be if a more digestible type of carbohydrate were present. This effect is greater in poultry than in swine, because in pigs there is greater fermentation in the hindgut, and volatile fatty acids produced by fermentation can be absorbed. In poultry, there is little hindgut fermentation, so the undigested oligosaccharides are excreted. Coon *et al.* (1990) found that extraction of soybean meal with 80% ethanol to remove oligosaccharides improved its feeding value for poultry, increasing the metabolizable energy content by 20%, and reducing the rate of transit through the gut by 50%. The longer retention time, due to the lack of the cathartic effect of oligosaccharides, increased dry matter digestibility. The crude protein content of ethanol-extracted meal was 64% vs. 46% for unextracted soybean meal. Ethanol extraction thus may be used to produce a high quality soybean meal for specialty purposes, such as for turkey poults which have high dietary protein requirements. For baby pig diets, ethanol extraction removes both oligosaccharides and antigenic proteins (glycinin and conglycinin).

Raw soybeans contain the enzyme **lipoxidase (lipoxygenase)** which promotes lipid peroxidation (oxidation leading to formation of peroxides). Lipoxidase catalyzes the oxidation of lipids containing *cis-cis*-1,4-pentadiene systems to hydroperoxides (Fig. 7–3). Because it is a specific catalyst for the *cis-cis*-1,4-pentadiene system, lipoxidase will oxidize linoleic, linolenic, and arachidonic acids and their derivatives, but not those of oleic acid. Lipoxidases in grains, and particularly in legumes such as soybeans and alfalfa, readily destroy carotene and xanthophylls, reducing the vitamin A activity of feeds. In the baking industry lipoxidase is used to bleach carotene in dough. Sources of lipoxidase, in decreasing order of concentration, are soybeans, lentils, green peas, peanuts, field beans, wheat, barley, and sunflower seeds. The "beany" flavor of beans is due to the action of lipoxidases on free fatty acids in the seed, giving rise to ketones and aldehydes with undesirable flavors (Marczy *et al.*, 1995). Heat treatment of 80° C will denature the enzyme.

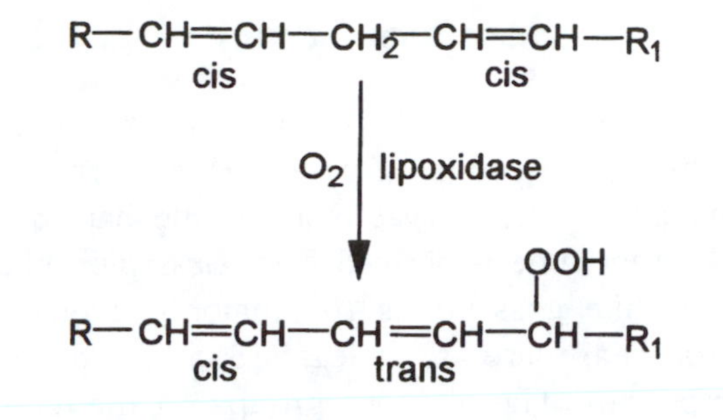

FIGURE 7–3 Lipoxidase catalyzes the oxidation of unsaturated fatty acids, producing toxic and off-flavor peroxides.

COTTONSEED AND COTTONSEED MEAL

Whole cottonseed and cottonseed meal are widely used in animal feeding, particularly for ruminants. The use of cottonseed products is primarily limited by their gossypol content. While cottonseed and cottonseed meal are valuable feedstuffs, they should always be used with an awareness of potential toxicity problems.

Whole cottonseed contains an excellent nutrient balance for dairy cattle, containing about 23% of both protein and fat, and 17% fiber. Hence it is widely used in dairy cow diets. Cottonseed meal has lower protein and lysine contents than soybean meal, so is less suitable than soybean meal for swine and poultry. It is widely utilized as a protein supplement for ruminants.

Cottonseed products contain several deleterious factors, including gossypol, cyclopropene fatty acids, oligosaccharides and tannins.

Gossypol Toxicity

Gossypol (Fig. 7–4) is a phenolic compound found in pigment glands of cottonseed (*Gossypium* spp.). It was isolated and named in 1899; the name is derived from ***Gossypium*** phe**nol**. While gossypol is the major phenolic in cottonseed, there are at least 15 closely related compounds. The phenolic groups on the molecule are chemically reactive, so it reacts readily with various substances, including minerals and amino acids.

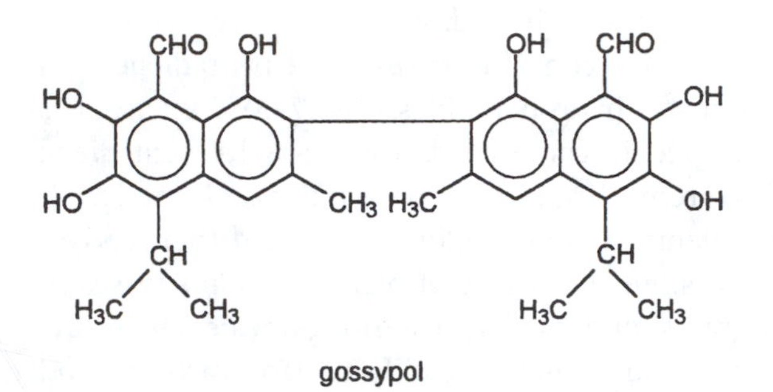

FIGURE 7–4 Structure of gossypol.

The pigment glands (Fig. 7–5) have a cell wall composed of cellulose, pectin, hemicellulose, and uronic acid, and are resistant to rupture. During the extraction of cottonseed oil, processors attempt to leave most of the gossypol in the meal, as it adversely affects the color and quality of the oil. Heat treatment of the seed causes gossypol to bind to protein and remain in the meal.

In animal feeding, the main concern from a toxicological point of view is with free gossypol. The bound gossypol is physiologically inactive, but because it is bound to protein and particularly lysine, it reduces the biological value of the protein. Some of the physiological effects of **free gossypol** are the following:

1. It causes olive green yolks in hen's eggs (Fig. 7–6). This is the most sensitive indicator of physiological activity. The cyclopropenoid fatty acids in cottonseed meal

FIGURE 7–5 (Top): Pigment glands, evident as dark spots in this picture, occur throughout the cotton seed. (Courtesy of R.J. Hron, Sr) (Bottom): Whole cottonseed, with cotton lint attached, is widely used as a protein source for dairy cattle.

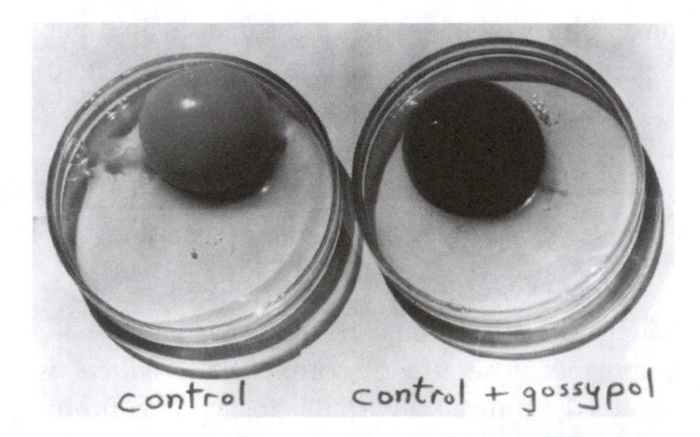

FIGURE 7–6 Egg yolks from a hen fed a control diet (left) and a hen fed cottonseed meal (right). Gossypol causes an olive green pigmentation, by reacting with iron in the yolk, changing the color absorbance of the gossypol pigment.

cause pink albumins. They enhance the yolk discoloration by accelerating the increase in yolk pH during cold storage of eggs. The green pigmentation is due to a reaction between gossypol and yolk iron.

2. High dietary levels of gossypol in layer diets decrease the hatchability of eggs.
3. Gossypol causes a depressed appetite and loss of weight.
4. Liver and lung lesions, including edema and congestion of the lungs, and congestion and hemorrhages of the liver are caused by gossypol.
5. Ascites and tissue edema are observed.
6. The most common effect is cardiac irregularity; death is generally ascribed to cardiac failure.
7. "Thumps," a labored breathing, occurs in swine, with pulmonary edema a common feature in most animals.
8. Anemia may occur, with reduced erythrocytes, hemoglobin, and packed cell volume, due to a complexing of iron by gossypol.
9. Reproduction in both males and females may be impaired.

The biological effects of gossypol are cumulative. For example, pigs fed on a cottonseed meal diet for a year may appear normal, but then abruptly begin to gasp for breath and die with severe anemia. Gossypol toxicity is caused by free gossypol, whereas the bound form (bound to lysine in proteins) is non-toxic. Ruminants are more tolerant than non-ruminants, because of binding of gossypol to proteins in the rumen. There is some evidence that bound gossypol can be converted back to the free form in the intestinal tract (Risco *et al.*, 1992). While ruminants are more resistant, there are numerous reports of gossypol toxicity in calves and lambs (Holmberg *et al.*, 1988; Hudson *et al.*, 1988; Morgan *et al.*, 1988). In an incident in California, over 500 calves in a calf-rearing facility died from gossypol toxicosis when fed a diet with 27% cottonseed meal (Holmberg *et al.*, 1988). Major lesions included ascites, edema, liver damage (acute centrilobular necrosis), cardiovascular lesions and kidney damage. In a feeding trial with Holstein calves, Risco *et al.* (1992) fed 0, 100, 200, 400 and 800 ppm free gossypol in diets containing cottonseed meal to provide the desired gossypol levels. They concluded that 200 ppm was safe, 400 ppm was toxic (cardiovascular lesions) and 800 ppm caused mortality. The 400 ppm group had an increased incidence of pneumonia, suggesting that lung damage increased sensitivity to respiratory infection. For mature dairy cows, free gossypol intake should not exceed about 24 g per day (Barraza *et al.*, 1991). Pelleting cottonseed product–containing diets can reduce the free gossypol level by as much as 70% in whole cottonseed and 48% in cottonseed meal (Barraza *et al.*, 1991). However, Risco *et al.* (1992) found only a 9% reduction of free gossypol in pelleted cottonseed meal–containing diets. Presumably the pelleting conditions (heat, moisture, etc.) would influence the degree of detoxification. For non-ruminants, the level of free gossypol in the diet should not exceed 0.001%, or about 10% dietary cottonseed meal (about 100 ppm free gossypol).

Mature beef cows can be safely fed cottonseed meal as the entire protein supplement, because the level of supplement used is relatively low. High producing dairy cows have a high feed intake, and gossypol toxicosis can occur when whole cottonseed is used, particularly when the diet also contains cottonseed meal. Because of the possibility of gossypol toxicity, whole cottonseed and cottonseed meal should be fed to dairy cows at levels no higher than about 3.6 kg per head per day (Coppock *et al.*, 1987). Jersey cattle are more susceptible to copper toxicity than are other dairy breeds. Jersey cows fed cottonseed meal have increased susceptibility to copper toxicity (Hemken *et al.*, 1996), probably because of gossypol-induced liver damage.

The levels of free and bound gossypol in cottonseed products are variable; an analysis should be obtained if large quantities are to be used in a feeding program. Typical values for free gossypol are 0.5% for whole cottonseed, 0.1–0.5% for solvent extracted cottonseed meal, and 0.05% for expeller (screw press) process meal. The mechanical (screw press) process involves the most heat and pressure, and so has the lowest free gossypol level (because most of it becomes bound with heat treatment). Solvent extraction uses the least heat and pressure, leaving higher free gossypol levels. Whole seed has almost all the gossypol in the free form. Use of extruders or expanders prior to solvent extraction produces a meal with a low free gossypol content.

In ruminants, the free gossypol content of cottonseed products does not correlate well with their toxicity (Calhoun *et al.*, 1995). This is a reflection of differences in rumen by-pass of cottonseed products subjected to different processing methods. Plasma gossypol levels reflect the availability of dietary gossypol and can be used to assess the potential for toxicity of cottonseed products. A level of 5 μg gossypol per ml plasma is the upper safe limit for plasma gossypol in dairy cattle fed cottonseed products for extended periods (Calhoun *et al.*, 1995). In dairy cows fed total mixed rations containing about 15% whole linted cottonseed, the plasma gossypol averages 1.5–3.5 μg/ml.

Gossypol occurs in the plant as a mixture of **(+) and (–) isomers** (Percy *et al.*, 1996). The (–) isomer is the more toxic. There are genetic variations among cotton cultivars in the proportions of the two isomers.

Gossypol reacts with minerals, and especially iron. The olive-green color of egg yolks in stored eggs from hens fed gossypol is due to the color of the **gossypol–iron complex** (egg yolks are rich in iron). Anemia is a sign of gossypol toxicity, due to the "tying-up" of iron in an unavailable gossypol–iron complex. Iron salts (e.g. ferric sulfate) can be added to cottonseed-containing diets to detoxify gossypol by binding the reactive groups with iron (Barraza *et al.*, 1991). A 1:1 ratio of iron to free gossypol is recommended. Even with the addition of iron, the maximum level of free gossypol should not exceed 400 ppm for swine and broilers, and 150 ppm for laying hens. Without iron, these levels should be reduced to 100 ppm and 50 ppm, respectively. High dietary protein also has protective effects, especially in pelleted diets, by binding free gossypol.

Gossypol has some interesting effects on **reproduction**. Much of the interest stems from observations on humans in China and the possible development of a male birth control pill. A Chinese scientist reported that in a 10 year period, not a single child had been born in Wang village in Jiangsu (Hron *et al.*, 1987). It was discovered that a crude, homemade cottonseed oil preparation had been used for cooking in this village. Women lacked menstrual cycles and men were sterile. Subsequent investigation linked these conditions to the consumption of gossypol.

Randel *et al.* (1992) reviewed the effects of cottonseed products and gossypol on reproduction. Gossypol causes males to be infertile because of non-motile sperm and depressed sperm counts. Specific lesions occur in the spermatozoa and in the germinal epithelium. Randel *et al.* (1992) concluded "Extensive damage to the germinal epithelium has been shown in both rams and bulls fed diets containing gossypol and is of major concern". These authors recommend that sources of free gossypol should be avoided in the diets of breeding male ruminants. Testosterone levels in growing bulls are not affected by gossypol consumption, but testicular morphology changes occur (Chase *et al.*, 1994). In nonruminants, the direct toxic effects of gossypol are such that they mask the more chronic anti-fertility effects. Cusack and Perry (1995) found no effect on fertility of bulls fed whole cottonseed to provide calculated daily intakes of 7.6–19.8 g free gossypol. However, they speculated that a high mineral level in the water may have resulted in binding of minerals with free gossypol in the rumen, inactivating it. Thus the mineral intakes via feed and water can be important determinants of the toxic level of gossypol.

In females, gossypol disrupts estrous cycles, pregnancy and early embryo development (Randel *et al.*, 1992), particularly in nonruminants. The ruminant female is somewhat insensitive to the antifertility effects of gossypol (Randel *et al.*, 1992; Gray *et al.*, 1993).

Glandless cotton, lacking the pigment glands that contain gossypol, has been developed, but it has not had a major impact. Gossypol has protective effects against insects and diseases, so the glandless cotton is more susceptible to pests and requires use of more pesticides than regular cotton. Cottonseed meal is a by-product of the third order; cotton is grown primarily for its fiber, with the oil as a valuable by-product. Since the meal is the least valuable component, glandless cotton is not likely to be produced if it increases agronomic problems and expenses in growing the crop. An interesting possibility for producing gossypol-free cottonseed meal arises from work by Dilday (1986) who found a single cotton plant that had gossypol in all parts but the seeds. A plant breeding program has been initiated to transfer the gene(s) for glandless seeds into commercial cotton varieties. This would utilize the insecticidal value of gossypol in protecting the plant, while producing gossypol-free seeds.

The presence of gossypol in cottonseed meal reduces the protein quality of the product. The bound gossypol is that which has reacted with amino acids, mainly lysine. The gossypol–protein interaction occurs through reaction of the formyl groups of gossypol with the epsilon amino groups of lysine and arginine (the **Maillard or browning reaction**). Gossypol may also react with the thiol group of cysteine. The products undergo various reactions and ultimately form insoluble, indigestible polymerization products. Both the phenolic and carbonyl groups of gossypol are chemically very reactive. The phenolic groups can form hydrogen bonds or be oxidized to quinones that react with proteins. Additionally, cottonseed contains **oligosaccharides** such as raffinose which can participate in lysine tie-up by Maillard reactions.

Condensed Tannins in Cottonseed

Cottonseed contains **condensed tannins**, primarily in the hulls. Cottonseed meal may contain about 1.6% condensed tannin (Yu *et al.*, 1993). Terrill *et al.* (1992) suggested that tannins in cottonseed products could have useful effects in ruminant nutrition, by binding proteins in the rumen to reduce their microbial degradation. Condensed tannin–protein complexes formed in the rumen dissociate in the abomasum and small intestine. Yu *et al.* (1993) found an inverse relationship between tannin and gossypol levels in cottonseed, and suggested that they both functioned as chemical defenses against insects. Selection for high tannin–low gossypol cottonseed could improve cottonseed meal for nonruminants while retaining insect resistance, because the hulls, where the condensed tannins are located, can be removed in processing.

Cyclopropene Fatty Acids in Cottonseed

Two **cyclopropene fatty acids**, sterculic and malvalic acids (Fig. 7–7), occur in cottonseed oil at levels of 1–2% of the crude oil. Sterculic and

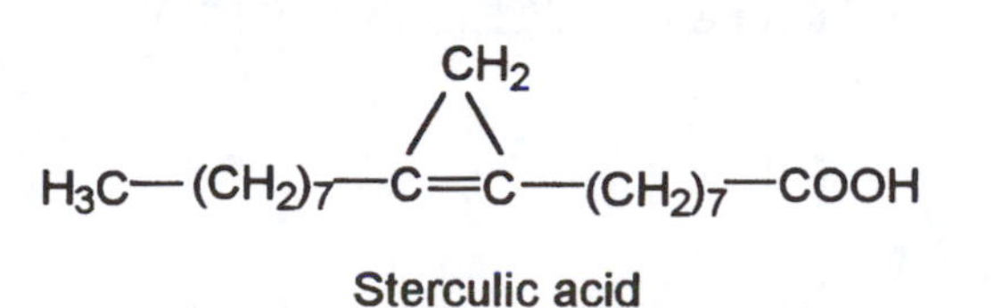

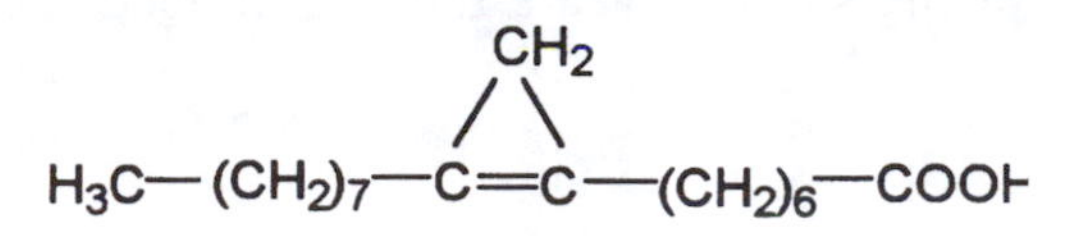

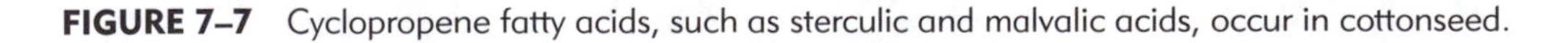
FIGURE 7–7 Cyclopropene fatty acids, such as sterculic and malvalic acids, occur in cottonseed.

malvalic acids occur in seed lipids of plants of the order Malvales, which includes cotton and kapok, a tropical tree which produces a seed fiber (kapok) similar in properties to cotton fiber, and seeds of the baobab tree. Malvalic acid is present in a higher concentration in cottonseed oil than sterculic acid. The oil of the Java olive (*Sterculia foetida*) contains nearly 70% cyclopropene fatty acids and has been used as a source of the cyclopropenoids for experimental purposes. The cyclopropenoid fatty acids in cottonseed oil and meal intensify the effect of gossypol in causing olive green yolks of eggs and also cause the development of **pink albumins** during egg storage. These fatty acids have been shown to be **cocarcinogens (tumor promoters)**, increasing the tumor incidence and decreasing the latent period for tumor formation in trout fed carcinogens (Lee *et al.*, 1971). The cyclopropenes in cottonseed oil have a pronounced effect on the mixed function oxidase system of trout liver (Loveland *et al.*, 1979), resulting in reductions in microsomal protein, cytochrome P450, NADPH-cytochrome c reductase, cytochrome b_5, and aldrin epoxidation, while there was an increase in benzo[α]pyrene hydroxylation activity. The conversion of aflatoxin B_1 to aflatoxicol in trout liver was inhibited, and production of aflatoxin M_1 was completely lacking in fish fed cyclopropenes. Similar but less pronounced changes were seen in rabbits feed cyclopropenoid fatty acids (Eisele *et al.*, 1982). Modification of hepatic drug metabolizing enzymes may be a result of structural changes in microsomal membranes caused by incorporation of cyclopropenoid fatty acids into the membrane structure (Andrianaivo-Rafehivola *et al.*, 1995).

Cyclopropenoid fatty acids inhibit fatty acid **desaturation** in numerous species, causing a rise in the stearate:oleate ratio of the body fat. Pigs fed diets containing cottonseed meal, which often contains residual lipids, produce hard fat with a high melting point. Dairy cattle fed cyclopropene fatty acids protected against biohydrogenation in the rumen have changes in the milk stearate:oleate ratio, probably due to a cyclopropene-mediated inhibition of mammary desaturase enzymes (Cook *et al.*, 1976). Cyclopropenoid fatty acids induce hypercholesterolemia and atherosclerosis in rabbits (Ferguson *et al.*, 1976).

Cyclopropenoid fatty acids may be of possible concern in the human diet. Glandless cottonseeds (lacking gossypol) have been approved by the Food and Drug Administration (FDA) as a nut substitute and snack item for human consumption (Lusas and Jividen, 1987). Since whole cottonseed kernels contain cyclopropenoid fatty acids, which are metabolically active and are cocarcinogens, potential exists for a human health risk. Hendricks *et al.* (1980) fed roasted cottonseed kernels to rainbow trout as part of a complete diet and found reduced growth and induction of liver carcinomas. They concluded that glandless cottonseed products could pose a possible health risk in the human diet.

Sterculic acid is more active than malvalic acid in inducing the biological effects described above. The cyclopropene fatty acids may be incorporated into biomembranes (Pawlowski *et al.*, 1985). They inhibit desaturation of fatty acids in membrane phospholipids (Cao *et al.*, 1993).

Dietary cyclopropenoid fatty acids cause delayed sexual maturity and infertility (Slayden and Stormshak, 1990). Sterculic acid reduces progesterone synthesis in the corpus luteum, by inhibiting the conversion of pregnenolone to progesterone (Tumbelaka *et al.*, 1994).

RAPESEED, CANOLA, AND OTHER BRASSICAS

Rapeseed (Fig. 7–8) is one of the world's major oilseed crops, exceeded in importance only by soybeans and oil palm as sources of edible oil. Rape is a member of the *Brassica* genus of the Cruciferae (Brassicaea) or cabbage family. The terms crucifer and brassica are used interchangeably to refer to these plants. **Crucifer** is usually used with reference to vegetables such as cabbage, cauliflower and Brussel's sprouts, while **brassica** commonly refers to the seed crops such as rape and mustard, and forage brassicas such as kale and forage rape.

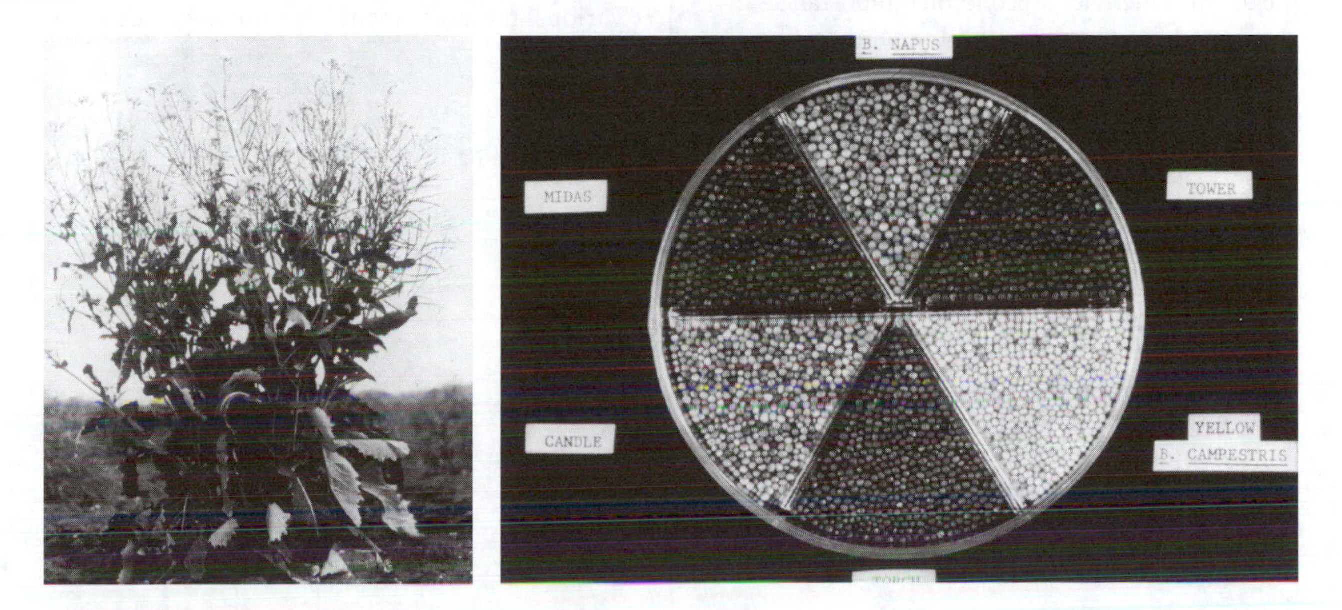

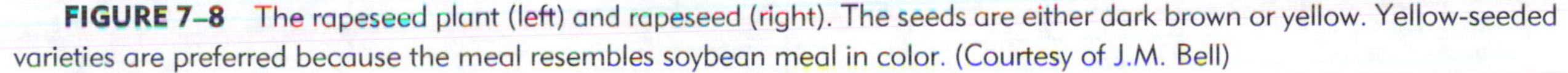
FIGURE 7–8 The rapeseed plant (left) and rapeseed (right). The seeds are either dark brown or yellow. Yellow-seeded varieties are preferred because the meal resembles soybean meal in color. (Courtesy of J.M. Bell)

Brassica spp. are rich in sulfur-containing substances such as glucosinolates and sulfur-containing amino acids (e.g. S-methyl cysteine sulfoxide).

Glucosinolates

Glucosinolates (thioglycosides, glucobrassicans) are glycosides of β-D-thioglucose with an aglycone that yields an isothiocyanate, nitrile, thiocyanate, or similar structure upon hydrolysis (Fig. 7–9). They occur widely in cultivated plants, particularly in the Cruciferae family. Most of the glucosinolate-containing plants that are important in human or animal nutrition are in the genus *Brassica*; examples include cabbage, broccoli, kale, rapeseed, mustard, and turnips.

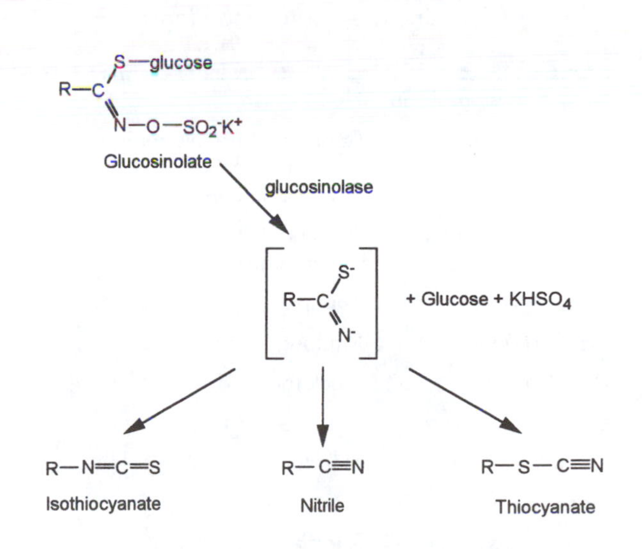

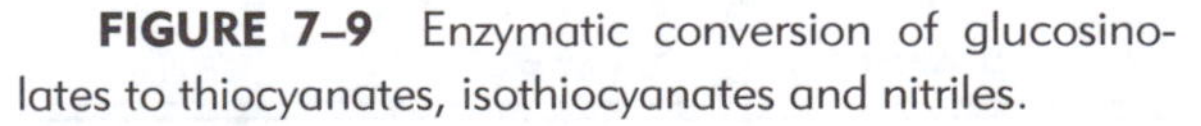
FIGURE 7–9 Enzymatic conversion of glucosinolates to thiocyanates, isothiocyanates and nitriles.

The primary significance of glucosinolates in animal nutrition is that they occur in several oil meals used as protein supplements and have adverse effects on livestock consuming these supplements. Rapeseed is widely grown in Canada, the northern U.S., Europe, and Australia as a source of edible and industrial oils. The residue after oil extraction is used as a protein supplement. Other glucosinolate-containing oil meals of much less economic importance include crambe, mustard, and limnanthes (meadowfoam) meals.

The chemistry of glucosinolates is complex. Reviews by McDanell *et al.* (1988) and Fenwick *et al.* (1989) provide further information on structures and metabolism.

Glucosinolates give rise to "hot" compounds with a biting taste, a property that has long been exploited in the use of condiments such as mustard and horseradish. A list of common vegetable and crop plants containing glucosinolates is given in Table 7–1. A comprehensive review of glucosinolates in food plants (Fenwick *et al.*, 1983) provides detailed information on glucosinolate content of vegetables and the physiological effects of the compounds on humans.

TABLE 7–1 Examples of Glucosinolate-Containing Plants

Amoracia lapathifolia	Horseradish
Brassica oleracea	Cabbage, brussels sprouts, cauliflower, kohlrabi, broccoli, kale
Brassica chinensis	Pak-choi (Chinese white cabbage)
Brassica campestris	Turnips
Brassica napus	Rutabaga, rape
Brassica nigra	Black mustard
Crambe abyssinica	Crambe
Limnanthes alba	Meadowfoam
Nasturtium officinalis	Water cress
Raphanus sativus	Radish
Thlaspi arvense	Stinkweed

The glucosinolates are hydrolyzed by enzymes called glucosinolases, thioglucosidases, or myrosinase (synonymous terms) in a manner similar to the situation with cyanogenic glycosides. The enzyme is found in the plant and is released when the plant material is crushed. It is also produced by rumen microorganisms. The products always include glucose and the acid sulfate ion. The organic aglycone may undergo various rearrangements, producing isothiocyanates, thiocyanates, or nitriles. Other products are sometimes produced. These include oxazolidine-2-thiones such as goitrin (Fig. 7–10). A variety of other compounds with complex structures may be produced. One of the main glucosinolates in rapeseed meal has the trivial name progoitrin. It is converted to goitrin (Fig. 7–10), which has goitrogenic activity. Isothiocyanates, thiocyanates, and nitrites are also produced from rapeseed glucosinolates.

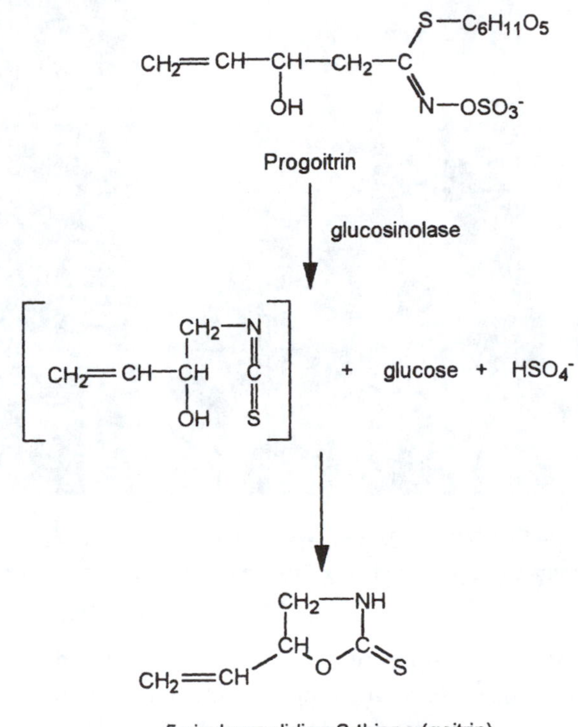

FIGURE 7–10 Conversion of progoitrin to goitrin.

Glucosinolate Metabolites

A major effect of the hydrolysis products of glucosinolates is inhibition of the function of the thyroid gland. The thyroid produces hormones such as thyroxine that are important in regulating the rate of cellular metabolism (metabolic rate). Antithyroid agents have four general effects. They may interfere with iodine uptake by the thyroid gland, interfere with iodination of tyrosine, suppress thyroxine secretion, or act as metabolic antagonists to thyroxine in the tissues. The general reactions involved in thyroid hormone synthesis are as shown in Fig. 7–11. The effects of some of the hydrolysis products of glucosinolates, such as goitrin, thiocyanates, isothiocyanates, and nitriles, are now discussed briefly.

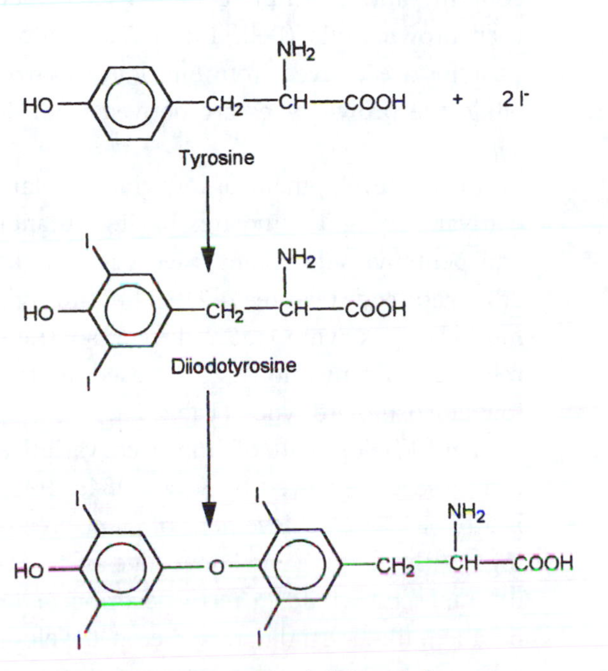

FIGURE 7–11 Reactions in the synthesis of thyroid hormones from tyrosine. Glucosinolates and their derivatives block these reactions, causing thyroid insufficiency and goiter.

1. **Goitrin.** Goitrin from rapeseed meal inhibits thyroid function. It causes reduced growth rate, and hyperplasia and hypertrophy of the thyroid (**goiter**). The oxazolidine-2-thiones (goitrins) act by inhibiting the incorporation of iodine into precursors of thyroxine and by interfering with thyroxine secretion. The antithyroid effect is not overcome by adding increased iodine to the diet.
2. **Thiocyanates and Isothiocyanates.** Thiocyanate inhibits iodine uptake by the thyroid. The effect is most pronounced when dietary iodine is low and can be overcome by increasing the iodine level of the diet. Since isothiocyanates are irritating to mucous membranes, they are probably not consumed as such in toxic amounts. However, if they are consumed as the glucosinolate precursor, with the isothiocyanates released in the gut, they may act as antithyroid agents.
3. **Nitriles.** Nitriles are formed from glucosinolates in crambe meal (*Crambe abyssinica*) and rapeseed meal. Nitriles are toxic, causing poor growth and liver and kidney lesions, including bile duct hyperplasia, liver necrosis, and megalocytosis of the kidney tubular epithelium. The biochemical cause of these effects is not clear. Paik *et al.* (1980) suggested that since nitriles may be metabolized by mixed function oxidases, they may cause enzyme induction and a consequent increase in liver size.

   Rapeseed meal nitriles may be partially converted to thiocyanate (Paik *et al.*, 1980). Dietary thiosulfate overcomes part of the growth-depressing effects of high-nitrile rapeseed meals (Paik *et al.*, 1980).

Effects of Glucosinolates on Humans and Livestock

1. **Humans. Goiter** (enlarged thyroid gland) in humans has been attributed to the consumption of large amounts of cabbage or

other crucifers (Tookey *et al.*, 1980). It is estimated that 96% of all human goiter is caused by uncomplicated iodine deficiency, so in reality the goitrogenic agents play a very minor role in human diseases. While it is plausible that goiter may be accentuated by consumption of brassica vegetables in areas where iodine content of food is low, there is no direct proof that this is the case.

2. **Livestock.** Poultry and swine fed raw rapeseed meal exhibit enlarged thyroids and growth depression. Additional problems noted in poultry include perosis (tendon slipped off hock joint, causing leg to protrude sideways), lowered egg production, off-flavors in eggs, and liver damage. Liver hemorrhage in poultry fed rapeseed meal may be caused by nitriles. Swine may have enlarged livers when fed rapeseed meal. Levels of 5–10% rapeseed meal have been acceptable for nonruminants. Ruminants can tolerate at least 10% dietary rapeseed meal with no ill effect. They are less sensitive to glucosinolates and their derivatives than nonruminants, presumably because of rumen metabolism of these compounds.

Of major significance to the utilization of rapeseed meal by livestock has been the development in Canada of low-glucosinolate cultivars. These include so-called "double zero" varieties, low in both glucosinolates and erucic acid. The meal from these new varieties can be used at a much higher level than previous varieties without reducing animal performance. The difference in feeding value is so apparent that Canadian producers have coined the term **canola meal** for the low-glucosinolate meal. **Canola** is the registered name of the rapeseed containing less than two percent of the total fatty acids in the oil as erucic acid and less than 30 μmoles of alkenyl glucosinolates per gram of oil-free dry matter. Canola meal can be used as a total replacement for soybean meal for some classes and types of livestock, such as finishing pigs. Further improvements are a reduction in the fiber content and the introduction of genes for yellow seed color into commercial varieties. This differentiates canola meal from rapeseed meal, which is brown in color. Also, the brown hulls are virtually indigestible, while the energy and protein in the yellow hulls are of higher digestibility. The yellow hulled strains have a lower hull content, and the hull contains less fiber than brown hulls (Bell, 1993). The yellow varieties are derived from *Brassica campestris* while the brown types are derived from *B. napus*.

In the development of low-glucosinolate cultivars, the glucosinolates having butanol and pentenyl side chains have been markedly reduced (Table 7–2). The indolylmethyl types were not reduced, so their relative concentration is increased in the low-glucosinolate types (Table 7–2).

The development of improved varieties of rapeseed, reviewed by Bell (1984, 1993) is a remarkable achievement of plant breeding. With our present knowledge of toxicity, rapeseed would in retrospect appear to be an unlikely candidate for crop development, as both the oil and the meal contain toxic factors. The rate of return on investment from rapeseed research has exceeded that from hybrid corn and poultry research in the U.S. (Bell, 1984).

The residual glucosinolates in canola are primarily of the indole type (Bell, 1993). The indolyls are not goitrogenic, but do yield nitriles, which cause tissue damage. **Very low glucosinolate (VLG) canola** has been developed, and gives improved poultry and swine performance when compared to conventional canola meal (Bell, 1993). Conversion of commercial canola varieties to the VLG type would be possible and desirable. The VLG canola meal is virtually equal to soybean meal as a protein supplement.

TABLE 7–2 Glucosinolate Content[a] of Selected High- and Low-Glucosinolate Rapeseed Cultivars*

Glucosinolate type	Side chain (R)	Cultivars: Torch	Cultivars: Candle
3-Butenyl	$CH_2{=}CH(CH_2)_2{-}$	31.2	4.5
4-Pentenyl	$CH_2{=}CH(CH_2)_3{-}$	22.9	3.9
2-OH-3-butenyl	$CH_2{=}CH{-}CH(OH){-}CH_2{-}$	22.5	5.2
2-OH-4-pentenyl	$CH_2{=}CH{-}CH_2{-}CH(OH){-}CH_2{-}$	3.8	1.3
3-Indolylmethyl	indol-3-yl $-CH_2-$ (N–H)	0.4	0.3
1-Methyl-3-Indolylmethyl	indol-3-yl $-CH_2-$ (N–OCH_3)	12.3	12.5
Total glucosinolate		93.1	27.7

[a]µmol glucosinolate per gram meal.

*Adapted from Bell (1984).

Glucosinolates in Animal Products

Glucosinolates and their derivatives can be transferred in the milk of dairy animals, causing thyroid enlargement, increased ^{131}I uptake by the thyroid, and decreased blood thyroid hormones in experimental animals fed the milk. Placental transfer can also occur. Throckmorton *et al.* (1981) noted goiter and altered serum thyroid hormones in lambs from ewes fed raw meadowfoam meal (*Limnanthes alba*), while White and Cheeke (1983) noted evidence of thyroid changes in rabbits and goat kids fed milk from goats fed raw meadowfoam meal. Tissue residues of glucosinolates have been detected in cattle fed crambe meal (Van Etten *et al.*, 1977). Embryonic thyroid enlargement in chicks from eggs from hens fed rapeseed meal was attributed to a low iodide content of the eggs due to increased iodide trapping by the enlarged maternal thyroid (March and Leung, 1976).

Beneficial Effects of Glucosinolates and Their Derivatives

Cruciferous vegetables such as cabbage, broccoli and cauliflower have been shown in many studies to have protective effects against cancer, particularly of the colon. Glucosinolates

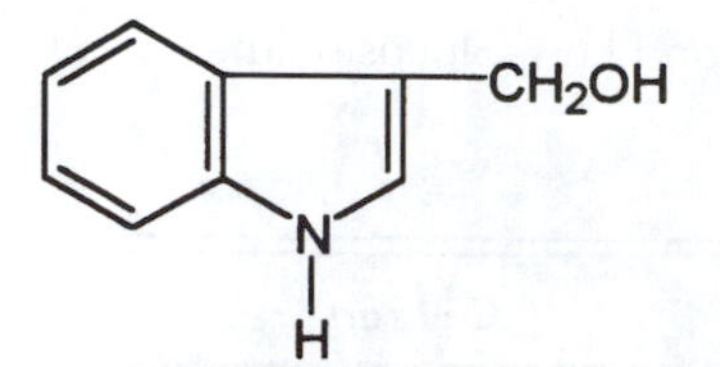

$H_3C—S(=O)—(CH_2)_4—NCS$

Sulforaphane

FIGURE 7–12 Indole-3-carbinol and sulforaphane are glucosinolate derivatives with beneficial effects on human health.

and their degradation products are implicated in this anti-cancer activity. Glucosinolate metabolites are potent inducers of Phase I (cytochrome P450) and Phase II (conjugation) enzyme activities (Bjeldanes *et al.*, 1991). One of the glucosinolate metabolites of greatest interest is **indole-3-carbinol** (Fig. 7–12), an autolysis product of indole glucosinolates. When administered before carcinogen exposure, indole-3-carbinol reduces carcinogenesis (tumor formation), but when administered after tumor formation, it is a promoter of carcinogenesis (Bjeldanes *et al.*, 1991). These contrasting effects are a reflection of induction of drug-metabolizing enzyme activities. Also, glucosinolate indole derivatives bind to the **aromatic hydrocarbon (Ah) receptors**, which could influence intracellular transport of carcinogens (Bjeldanes *et al.*, 1991). Some of the indole derivatives have high affinity for the Ah receptor, and act as Ah receptor agonists (they enhance Ah receptor activity). According to Bjeldanes *et al.* (1991), the beneficial anti-carcinogenic effects of some indoles in crucifers outweigh the potential negative effects of others. Another glucosinolate derivative with anticarcinogenic properties is sulforaphane (Zhang *et al.*, 1992; 1994), a type of isothiocyanate. **Sulforaphane** (Fig. 7–12) is a potent enzyme inducer, accelerating the metabolic disposal of xenobiotics. Sulforaphane is a monofunctional inducer; it induces Phase II enzymes without stimulating Phase I (cytochrome P450). Thus it enhances toxin excretion without increasing bioactivation (Zhang *et al.*, 1994). Consumption of cruciferous vegetables is associated with reductions in tissue cholesterol levels (LeBlanc *et al.*, 1994). At dietary levels lower than those affecting carcinogenesis, indole-3-carbinol affects cholesterol homeostasis, with a reduction in serum cholesterol levels (LeBlanc *et al.*, 1994). Thus it is generally accepted that generous consumption of cruciferous vegetables is advisable for good health, including favorable effects on serum cholesterol levels and cancer prevention (Waldron *et al.*, 1993; Stoewsand, 1995).

Erucic Acid

Erucic acid is one of the major fatty acids found in rapeseed oil. It has been implicated in causing myocardial lesions in rats and is therefore of toxicological interest.

Erucic acid (22:1ω9) has important industrial applications primarily because it is a longer-chain fatty acid than most fatty acids in plant oils. When cleaved at the ethylene bond, it yields brassylic acid, a dicarboxylic acid (Fig. 7–13) that is used in the manufacture of nylon and other polymers. There is no other convenient commercial source for C13 dicarboxylic acid. Rapeseed oil is used in the lubrication of

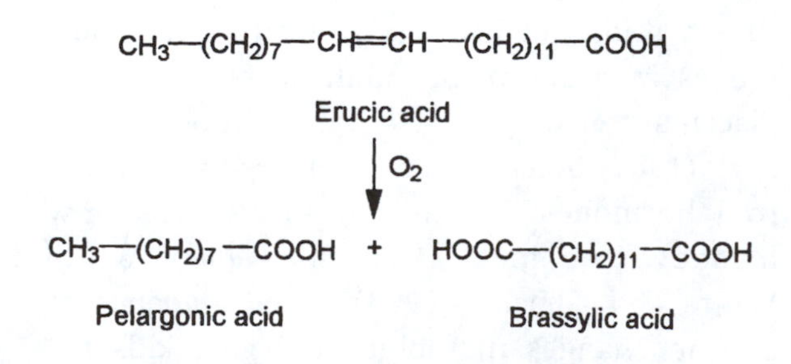

FIGURE 7–13 Conversion of erucic acid in rapeseed oil to derivatives with industrial applications.

high-speed engines, as a flotation agent in mining, and for various other applications for which a long-chain fatty acid is useful. It also has a higher smoke point and vaporization temperature than other common oils. The erucic acid content of rapeseed oil varies from 25 to 35% in Canadian varieties up to 55% in some European varieties. While the high erucic acid content of rapeseed oil is advantageous for industrial applications, it has negative implications in oil used for human consumption.

Feeding high erucic acid rapeseed oil to animals such as mice, rats, swine, guinea pigs, and poultry causes a reduction in growth rate, enlargement of the adrenals, increased mortality of the offspring of rats, moderate hepatic lipidosis, and degenerative changes in the kidney. In addition, myocardial lesions in rats occur. These develop in three stages: severe lipidosis, inflammation, and scar formation and fibrosis. The intracellular lipidosis starts within a few hours after rapeseed oil is fed and reaches a maximum in 3–6 days. Droplets of lipid are deposited in the myocardial fibers. After 4–8 weeks on a rapeseed oil–containing diet, foci of inflammation are found, associated with the presence of lipid-laden fragmented myofibers. At later periods, replacement of muscle fibers by scar tissue occurs, with hemosiderin deposits. The significance of these lesions reported in rats to human utilization of rapeseed oil is not certain. However, the question may become academic, since low erucic acid varieties of rapeseed have been developed. In general, the high erucic acid varieties are grown to produce industrial oil, while low erucic acid types are grown to produce edible oils for margarine, salad oil, and other food uses. The low erucic acid type of oil has been designated canola oil, while the term rapeseed oil refers to the high erucic acid types. The term **canola** was derived from its designation as "**Can**adian **o**il **l**ow **a**cid." Canola oil has received much attention as a "healthy" vegetable oil. It has a higher content of monounsaturated fatty acids (primarily 18:1, oleic acid) than other common oils (corn oil, soybean oil, etc.). Monounsaturated fatty acids have cholesterol-lowering properties, and have minimal adverse effects from peroxide formation. Polyunsaturated fatty acids also have cholesterol-lowering properties, but may increase the formation of free radicals and peroxides. Lipid peroxidation of cell membranes has been linked to pathologies including coronary heart disease and cancer.

Other Deleterious Factors in Grain Brassicas

Rapeseed and canola meals contain other substances with antinutritional properties, including sinapine, tannins, phytic acid and oligosaccharides.

Sinapine is an ester of choline (a B vitamin) and the phenolic acid sinapic acid (Fig. 7–14). Choline has three labile methyl groups. Sinapine is significant as a source of "fishy-flavor" in eggs of layers fed canola meal. Bacteria in the ceca ferment sinapine, producing trimethylamine, which is then absorbed. Most breeds of chickens convert trimethylamine to trimethylamine oxide, which is excreted. Certain strains of hens that lay brown-shelled eggs lack the liver enzyme trimethylamine oxidase, leading to an accumulation of **trimethylamine** and its transfer to the developing egg. This results in a fishy-flavor (amines are the source of the odor

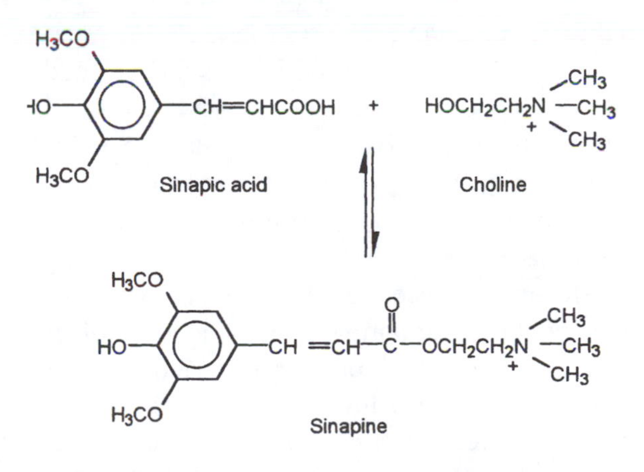

FIGURE 7–14 Formation of sinapine, a constituent of rapeseed, from choline and sinapic acid. Sinapine is responsible for "fishy-flavor" of brown shelled eggs.

of fish tissue). The fishy odor does not develop unless the diet contains more than 0.1% sinapine; canola meal contains 2.5–3% sinapine (Bell, 1993). Reduction in the sinapine content of canola through plant breeding would enhance the value of the meal for the poultry industry.

Canola meal contains about 1.22% total phosphorus, about half of which is bound as **phytate** phosphorus (Bell, 1993), compared to values of about 0.66% and 0.38% for total and phytate phosphorus in soybean meal. Phytic acid occurs primarily in the seed embryo.

Rapeseed and canola contain **polyphenols** (tannins). They are found mainly in the hull and seed coat. Yellow-seeded varieties are lower in polyphenols than brown-seeded types (Slominski *et al.*, 1994a). Digestibilities of non-starch polysaccharides, dry matter and amino acids are higher for yellow seeded types (Slominski *et al.*, 1994a). A so-called **triple low canola** cultivar with low contents of erucic acid, glucosinolates and tannins has been developed. However, there were no apparent benefits of triple low canola in pig diets as compared to double-low varieties (Agunbiade *et al.*, 1991).

Canola meal contains about 2.5% **oligosaccharides** such as raffinose and stachyose. Ethanol extraction of the oligosaccharide fraction did not improve the nutritive value of canola meal for poultry (Slominski *et al.*, 1994b).

Brassica spp. contain an amino acid, **S-methyl cysteine sulfoxide**, also known as the brassica anemia factor. This will be discussed in the section on forage *Brassica* spp. (see Chapter 11).

LINSEED MEAL

Flax (*Linum usitatissimum*) is an oilseed crop, from which linseed oil is obtained. The residue from oil extraction is linseed meal. Linseed oil is unique among the vegetable oils in containing more than 50% linolenic acid (18:3). The double bonds readily oxidize in air, polymerizing into a pliable film. This drying property accounts for its use in paints and numerous industrial uses. Its susceptibility to autooxidation makes it unsuitable as an edible oil. A low-linolenic acid linseed, **linola**, has been developed, with a high content of linoleic acid. Linola oil has much greater oxidative stability than linseed oil, and could become an important source of polyunsaturated oil for human consumption (Batterham *et al.*, 1991).

Although linseed oil is considered unsuitable for direct human consumption, it is of interest because it is one of the richest plant sources of **omega-3 fatty acids**. These fatty acids have beneficial effects on human cardiovascular health. Feeding linseed oil or flax seed to livestock is a means of increasing the omega-3 fatty acid content of the meat (Cunnane *et al.*, 1990). Cunnane *et al.* (1993) reported that a daily intake of 50 g high linolenic acid flaxseed by humans increased the omega-3 fatty acids in blood cells and reduced serum cholesterol.

Flax or linseed meal contains an antagonist of the vitamin pyridoxine (Klosterman, 1974). The pyridoxine antagonist has been identified as the amino acid 1-amino-D-proline. It occurs in linseed meal as a dipeptide of 1-amino-D-proline and glutamic acid (Fig. 7–15), called **linatine**. The 1-amino-D-proline reacts with pyridoxal phosphate, forming a hydrazone, and impairing its function as a cofactor in amino acid metabolism. Pyridoxal phosphate is involved in transamination, decarboxylation, and other reactions of amino acid metabolism. Symptoms of pyridoxine deficiency, including depressed appetite, poor growth, and convulsions, may develop in chickens fed raw linseed meal. Autoclaving and water extraction or pyridoxine supplementation of the meal will overcome the antipyridoxine effects.

Linseed meal contains a moderate level of cyanogenic glycosides, such as **linamarin** (see

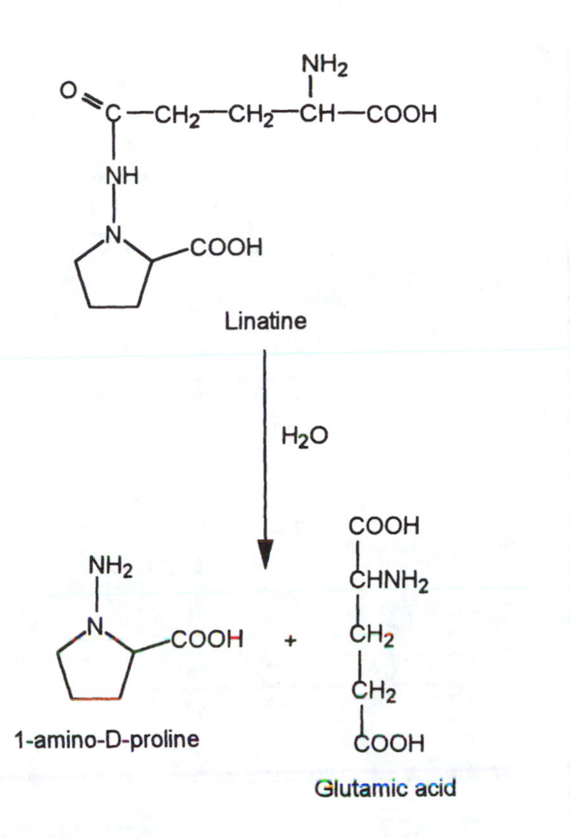

FIGURE 7–15 Linatine, a pyridoxine (vitamin B_6) antagonist in flax seed.

cassava cyanogens, Chapter 6). Drought and other stress conditions increase cyanogen levels in flaxseed (Oomah *et al.*, 1992). The amount present is not usually sufficient to cause cyanide toxicity. However, cyanogens may have a useful role in protecting against selenium toxicity. In areas with chronic selenium toxicity of livestock due to high soil and forage selenium levels, such as the northern Great Plains of the U.S., it has long been known that linseed meal has protective effects against selenium toxicity. Palmer *et al.* (1980) identified the protective factors as the cyanogenic glycosides linustatin and neolinustatin, which are glycosides of linamarin and lotaustralin, respectively. The protective action of the cyanogens is due to the metabolic formation of cyanide. Cyanide reacts with seleno-compounds to produce **selenocyanates**, comparable to the thiocyanates normally produced in cyanide detoxification. Dietary cyanide may also increase severity of selenium deficiency by promoting selenium excretion (Gutzwiller, 1993).

Although linseed meal is unlikely to cause cyanide toxicity, pregnant ewes fed linseed meal may have lambs with **goiter**. This is due to elevated blood levels of thiocyanate, a goitrogenic substance, produced when cyanide is detoxified in the rhodanese reaction with thiosulfate.

GRAIN LEGUMES

Grain legumes, often referred to as **pulse crops** or pulses, are members of the Leguminosae family. Their seeds are harvested for feed and food. The major grain legumes grown in temperate countries are soybeans, field beans, lupins, peanuts, lentils and peas. A variety of others such as the winged bean, rice bean, chickpea, cowpea, etc., are grown in the tropics. With the exception of soybeans, most grain legumes are grown primarily for human food. Cull seeds (e.g. cull pinto beans) are used as livestock feed. Virtually all grain legumes contain trypsin inhibitors and most contain lectins. Toxic amino acids and alkaloids are also encountered in grain legumes. Amylase inhibitors which reduce starch digestibility are common in beans. Anti-nutritional factors in beans and their inactivation by feed processing have been reviewed by van der Poel (1990).

Field Beans (*Phaseolus* spp.)

Many of the common beans used as food, such as the snap, kidney, pinto and navy bean, are varieties of *Phaseolus vulgaris*. Other *Phaseolus* spp. include *P. aureus* (mung bean), *P. lunatus* (lima bean) and *P. calcaratus* (rice bean). Other legume seeds referred to as beans include the garbanzo or chickpea (*Cicer arietinum*) and the winged bean (*Psophocarpus tetragonolobus*). The principal toxins in *Phaseolus* spp. are trypsin inhibitors, lectins and tannins. Refer to "Soybeans and Soybean Meal" for a discussion of trypsin inhibitors.

Hemagglutinins (Lectins)

Hemagglutinins (phytohemagglutinins, lectins) cause the clumping or agglutination of red blood cells in vitro. They were first isolated from castor beans, which contain a potent lectin called ricin. **Lectins** are proteins that have a high affinity for certain sugar molecules. Many contain covalently bound sugars and so can be classed as glycoproteins. Their biological effects are probably due to their affinity for sugars. There are carbohydrate moieties in animal cell membranes which may bind to lectins and be altered in functional properties.

More than 70 lectins have been isolated from legume seeds (Sharon and Lis, 1990). They usually consist of two or four subunits, each with one carbohydrate-binding site. Lectins may have a role as mediators of the symbiosis between N-fixing bacteria and leguminous plants (Sharon and Lis, 1990). They may also function as chemical defenses. Lectins have been extensively used in biomedical research for the separation and characterization of glycoproteins and the study of cell membranes.

Lectins are found in soybeans and other field beans (*Phaseolus*) such as kidney, pinto, and navy beans (Table 7–3). The lectins in raw soybeans are relatively nontoxic, but feeding high levels of raw kidney beans to rats will kill them due to lectin activity. Hemagglutinins cause various adverse effects, including reduced growth, diarrhea, decreased nutrient absorption, and increased incidence of bacterial infection. The major effects seem to be on the intestinal mucosa. The hemagglutinins bind to cells in the intestinal wall and cause a nonspecific interference with nutrient absorption. In addition, there is evidence that lectins impair the immune system, leading to greater sensitivity to bacterial infection. Changes in gut permeability may lead to invasion by normally innocuous intestinal microflora. Pusztai *et al.* (1979) have reported that field bean lectins disrupt the brush borders of the cells lining the duodenum and jejunum (Fig. 7–16) and that an abnormally high rate of tissue–protein catabolism occurs in rats administered lectins. A dramatic proliferation of *Escherichia coli* occurs in the small intestine (Wilson *et al.*, 1980).

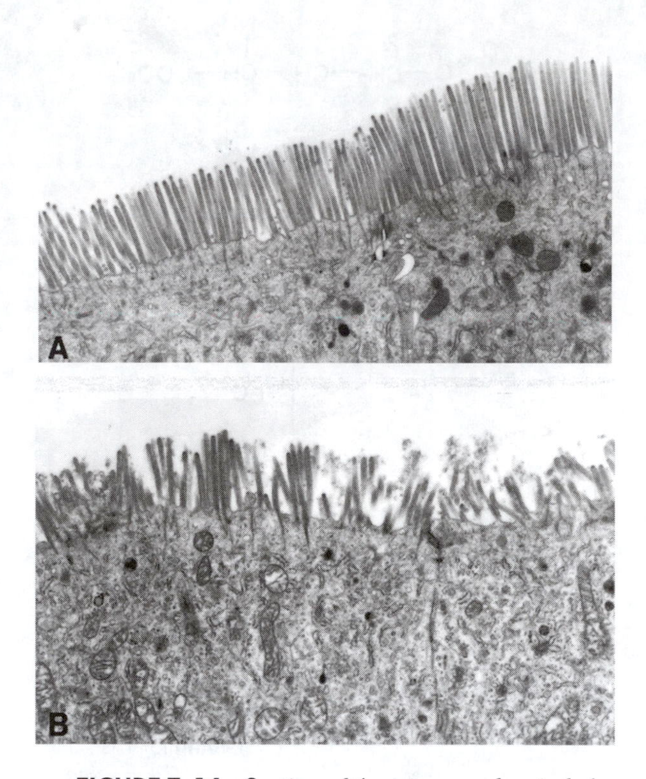

FIGURE 7–16 Section of the jejunum of a pig fed a control diet (top) compared with a similar section from a pig fed raw kidney beans (bottom), showing disruption of the microvilli (x6300). (Courtesy of R. Begbie and T.P. King)

A portion of the growth-inhibiting properties of lectins in chicks can be overcome by dietary supplementation with polyamines such as putrescine (Mogridge *et al.*, 1996). Polyamines promote intestinal cell growth and cell division, reversing some of the adverse effects of lectins on the intestinal mucosa.

TABLE 7–3 Common Legume Seeds Classified as to Toxicity and Lectin Content*

Group A. Highly toxic, high lectin activity		
	Phaseolus cocceneus	Runner bean
	P. vulgaris	Red or brown kidney bean
	P. vulgaris	White or black kidney bean
	P. acutifolius	Tepary bean
Group B. Growth inhibiting, moderate lectin activity		
	Psophocarpus tetragonolobus	Winged bean
	Phaseolus lunatus	Baby lima bean
Group C. Nontoxic, low lectin activity		
	Lens esculentus	Lentils
	Pisum sativum	Green peas
	Cicer arietinum	Chick-peas
	Vigna sinensis	Blackeyed peas
	Cajanus cajan	Pigeon peas
	Phaseolus aureus	Mung beans
	Vicia faba	Broad beans
	Phaseolus angularis	Aduki beans
Group D. Growth depression due to nonlectin factors		
	Glycine max	Soybeans
	Phaseolus vulgaris	Pinto beans

*Adapted from Grant *et al.* (1983).

Lectins are destroyed by moist heat. They are resistant to dry heat, so kidney and other beans should be soaked prior to cooking to ensure moisture penetration of the entire seed.

Cull field beans are sometimes fed to livestock; thus the potential adverse effects of lectins should be considered. Cull beans should be heat-treated for swine feeding; raw beans even at dietary levels as low as 5% have adverse effects in pigs (Myer *et al.*, 1982). The detrimental effects of raw beans on swine and poultry are reduced by the addition of antibiotics, perhaps due to reduced microbe colonization of the gut wall. Myer and Froseth (1983) found that extrusion of cull red beans was as effective as autoclaving in decreasing the growth-depressing effects in swine. Levels as high as 40% extruded beans gave satisfactory growth rates of pigs. Methionine supplementation of heat-treated beans also was effective in overcoming growth depression. Ruminants are less sensitive to dietary lectins (Paduano *et al.*, 1995), but toxicity of steers to kidney beans has been observed (Williams *et al.*, 1984).

Lectins are the most important antinutritional factors in *Phaseolus*, in terms of impairment of animal performance (van der Poel, 1990).

Other Deleterious Factors in Beans

Beans contain tannins, phytates and oligosaccharides. Bean **tannins** have been reviewed by Reddy *et al.* (1985) and van der Poel (1990). The tannins are located mainly in the seed coat, and mainly in colored seeds. They are primarily condensed tannins, and their nutritional effects are similar to those of sorghum tannins.

Phytates in beans are of similar levels and effects as those in soybeans. **Oligosaccharides** (raffinose, stachyose) are present, and account for the flatulence associated with bean consumption.

Cyanogenic glycosides are found in *Phaseolus* spp. (Montgomery, 1980). The highest concentrations occur in the lima bean (*P. lunatus*). Although lima beans have caused cyanide poisoning (Montgomery, 1980), the hazard is slight. Dark-seed varieties have the most cyanide potential; white varieties have very low levels.

Kidney beans contain a factor(s) which impairs the utilization of **vitamin E**. Feeding raw kidney beans to sheep or chickens induces nutritional muscular dystrophy (Hintz and Hogue, 1964). Autoclaving the beans partially overcomes the effect. The activity may be due to an α-tocopherol oxidase present in kidney beans, alfalfa, and soybeans (Liener, 1980).

Fava Beans and Favism

The fava bean (*Vicia faba*), also called faba bean, horse bean, and broad bean, is an important protein source in the human diet (Fig. 7–17). It is grown extensively in Europe, particularly in Italy, Spain, Greece, and other countries of the Mediterranean region. It is also used as a protein supplement for livestock.

FIGURE 7–17 Fava beans, also known as broad beans, are grown extensively in southern Europe and the mideastern region. Favism is a common human disorder in these areas.

Favism

Consumption of fava beans or inhalation of the pollen of the plant sometimes has a deleterious effect, causing a condition called **favism**. It is characterized by acute hemolytic anemia. Symptoms, which may appear within minutes after inhalation of pollen, or 5–24 hr after consumption of beans, include headache, dizziness, nausea, yawning, vomiting, abdominal pain, and elevation of temperature. These symptoms may subside spontaneously or, in severe cases, acute hemolytic anemia with hemoglobinuria and icterus occur. Children are most affected, with a mortality rate of 6–8% reported in the past. Blood transfusion therapy has greatly reduced fatalities. Two peaks of incidence of favism are noted, one when the plant blooms, and the other in summer when the fresh beans appear on the market. The mathematician Pythagoras is said to have met his death at the hands of Greek soldiers rather than cross a field of fava beans. Pythagoras had founded a cult, and forbade his followers to eat fava beans or even walk among them (Crosby, 1969; Marcus and Cohen, 1967).

It is estimated that over 100 million people in the world are susceptible to favism (Mager *et al.*, 1980). This is due to their genetic deficiency of a red blood cell enzyme, **glucose-6-phos-**

phate dehydrogenase (G6PD). The geographic distribution of G6PD deficiency parallels that of malaria, presumably because the enzyme deficiency increases the resistance of the red cells to the causative protozoal organism (Mager *et al.*, 1980). Certain ethnic groups, such as Oriental Jews, Mediterranean Europeans, Arabs, Asians, and Blacks have a high incidence (5–50% of the population) of a low activity of G6PD. Northern Europeans, European Jews, American Indians, and Eskimos have virtually no incidence of the enzyme deficiency. In individuals susceptible to favism, G6PD activity is only 0–6% of normal.

The causative factors are the aglycones of glycosides in the fava beans. Two of the major glycosides are **vicine** and **convicine**; their respective aglycones are divicine and isouramil (Fig. 7–18). The aglycones (divicine, isouramil) may either react with the red cell membrane directly or produce hydrogen peroxide, causing breakdown of the red cell membrane and hemolysis. In normal individuals, this is prevented by reduction of the oxidants by reaction with reduced glutathione (GSH). The supply of GSH is maintained by reactions of the pentose phosphate pathway in which G6PD functions. The major reactions are shown in Fig. 7–19. With a deficiency of G6PD, formation of reduced NADP cannot be increased by increased oxidation of glucose. As a result, reduced GSH cannot be regenerated fast enough, so the oxidants are not destroyed, but are available to attack the red cell membrane.

Of interest is that selenium functions as an essential nutrient as a part of **glutathione peroxidase**, the terminal enzyme in the above scheme. The essential metabolic role of selenium was discovered by Rotruck *et al.* (1973) when they were investigating why dietary selenium would protect against in vitro red cell hemolysis only when the incubation mixture contained glucose.

A consequence of the oxidant action of the fava bean compounds is the formation of **methemoglobin** in the red cell and the appearance of **Heinz bodies** (see Brassica anemia, Chapter 11). These are clumps of denatured he-

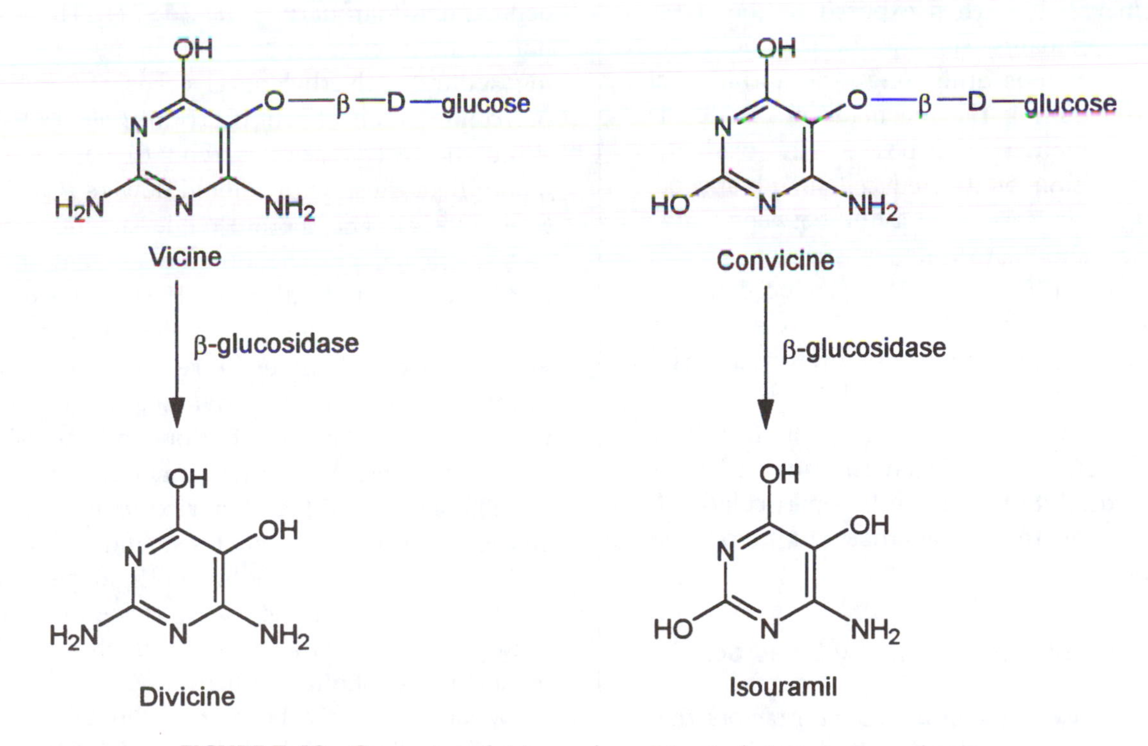

FIGURE 7–18 Conversion of vicine and convicine to their respective aglycones.

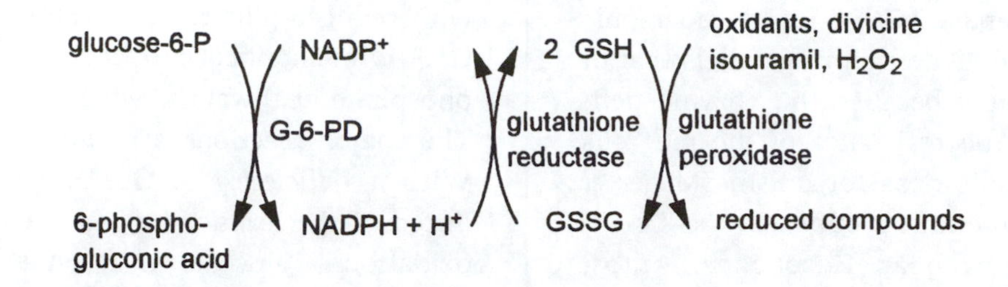

FIGURE 7–19 Enzymatic reactions involved in the hemolysis syndrome in favism. The oxidants from fava beans (divicine, isouramil) attack the red blood cell membrane, causing hemolysis. These oxidants are detoxified by being reduced (addition of hydrogen). The hydrogen is generated from the conversion of glucose-6-phosphate to 6-phosphogluconic acid, catalyzed by glucose-6-phosphate dehydrogenase (G-6-PD). When there is a genetic insufficiency of this enzyme, there is not sufficient hydrogen generated to reduce the oxidants, resulting in hemolysis.

moglobin resulting from the oxidation of its SH groups. The oxidant compounds may react with oxyhemoglobin to produce hydrogen peroxide, which may be the active membrane-rupturing factor. Hydrogen peroxide is converted to water by glutathione peroxidase.

It appears that individuals with a genetic deficiency of red cell G6PD activity can maintain sufficient GSH to cope with normal metabolic requirements, but when exposed to the stress of various oxidants, such as the fava bean glycosides or various drugs (e.g., the antimalarial drug primaquine), the regeneration of GSH is not rapid enough to dispose of the oxidants, and destruction of the red cell membrane occurs. There appears to be a link between G6PD deficiency and historical exposure to malaria (Marquardt, 1989). Oxidants produced in the red blood cells may have protective activity against invasion by the malaria protozoal organism.

Dietary antioxidants such as vitamin E have protective effects against favism-causing factors (Marquardt, 1989). Vitamin E is particularly effective in protecting against red blood cell hemolysis.

Utilization of Fava Beans by Livestock

Unprocessed fava beans contain factors that lower chick growth rate and feed efficiency, alter liver size, and increase pancreas size (Marquardt *et al.*, 1974). In laying hens, dietary fava beans reduce feed efficiency, egg weight, and egg production rate (Campbell *et al.*, 1980). Both thermostable and thermolabile factors are involved. The thermolabile factors include tannins, protease inhibitors, and lectins. Tannins are the major thermolabile antinutritional factor and account for over 50% of the chick growth depression (Marquardt *et al.*, 1977). They are not totally inactivated by heat; the tannin-binding additive polyethylene glycol is effective in overcoming their effects (Garrido *et al.*, 1991b). White flowered varieties tend to have lower tannin than those with colored flowers (Garrido *et al.*, 1991a). The thermostabile factors, vicine and convicine, are the other major antinutritional factors. Muduuli *et al.* (1981) studied the effects of dietary vicine on laying chickens. Feeding vicine caused a reduction in the number of developing ova, egg, and yolk weights, and reduced the fertility and hatchability of the eggs. Vicine consumption also elevated plasma lipid and peroxide levels, increased erythrocyte hemolysis, and the birds had heavier livers with higher lipid peroxide and lowered glutathione levels. Vicine from fava beans therefore has a marked influence on the metabolism of the laying hen.

Whole plant fava beans have potential as a silage crop. Thorlacius and Beacom (1981) compared silage from fava beans, oats, corn, and field peas, in trials with lambs. The dry matter intake and rate of gain were greater for lambs fed fava bean silage than for those fed corn or oat silage, while the digestibility of dry matter and protein was generally higher for fava bean silage than for oats and corn silage. The field pea and fava bean silages were similar in most respects. Promising results with fava bean silage were also obtained by McKnight and MacLeod (1977) and Ingalls *et al.* (1979).

Peas and Lentils

Peas (*Pisum sativum*) and lentils (*Lens culinaris*) are grain legumes grown primarily for human consumption. Cull peas and lentils, when unsuitable for direct human consumption, are often fed to livestock. Although they contain a variety of toxicants, the levels of most are sufficiently low to not be of concern. **Trypsin inhibitors** in peas have slight growth-depressing effects in pigs and poultry (Leterme *et al.*, 1990). They are highest in winter varieties. White-flowered peas have higher feeding value than dark-flowered varieties, because of higher tannin levels in dark-flowered peas.

Grain Lupins

Lupins (*Lupinus* spp.) are very widely distributed leguminous plants. In North America, there are many species of wild lupins, many of which are poisonous to livestock. These will be discussed in Chapter 11. Lupins are commonly grown garden ornamentals. Several species are grown as grain crops, particularly in Europe and Australia, mainly as protein supplements for livestock. The main grain lupins are *L. albus* (white lupin), *L. angustifolius* (blue lupin), and *L. luteus* (yellow lupin). *L. mutabilis* is widely grown for human food in the Andes of South America (Ecuador, Bolivia, Peru).

Lupin Alkaloids

Lupins contain **quinolizidine alkaloids** (Fig. 7–20), based on the bicyclic quinolizidine

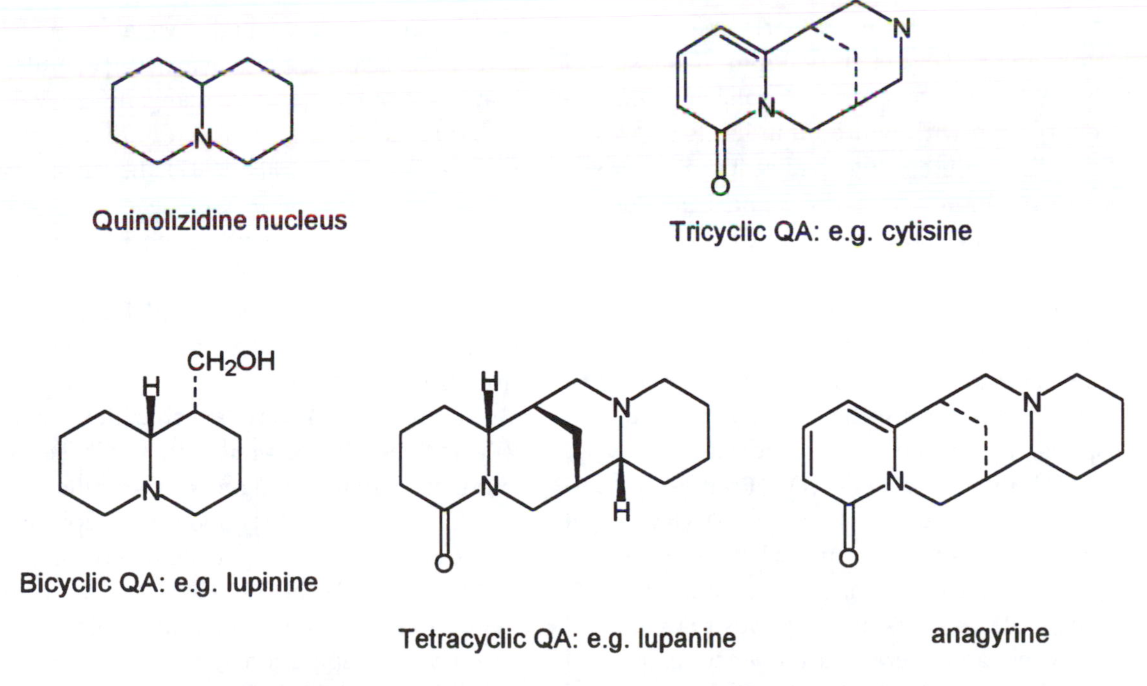

FIGURE 7–20 Examples of quinolizidine alkaloids (QA) found in lupins.

ring. The alkaloids include bicyclic, tricyclic and tetracyclic structures. The main alkaloids in grain lupins are: bicyclic: lupinine; tricyclic: angustifoline, albine, cytisine; and tetracyclic: sparteine, lupanine. The major alkaloid in *L. albus, L. mutabilis,* and *L. angustifolius* is lupanine (Petterson *et al.*, 1987). The grain lupins do not contain anagyrine, an alkaloid in wild lupins which has a teratogenic effect (crooked calf disease) in cattle (see Chapter 11). Thus grain lupins have no teratogenic properties.

The primary significance of quinolizidine alkaloids in grain lupins is that they adversely affect the performance of livestock and poultry. Swine are more sensitive than other livestock to lupin alkaloids (Cheeke and Kelly, 1989). Pigs exhibit feed refusal, vomiting, and reduced growth rates when fed lupins with significant alkaloid levels. Godfrey *et al.* (1985) showed that pigs could tolerate a level of 0.20 g/kg (0.02%) of dietary lupin alkaloids; feed intake and growth rate progressively declined as alkaloid level increased above this threshold.

Poultry are much more tolerant of lupin alkaloids than are swine. Levels of white lupin (*L. albus*) seed up to 30% of the diet have little or no growth depressing effects in poultry, whereas with swine reduced performance is noted with about 10% dietary lupin (Cheeke and Kelly, 1989). Rabbits showed no significant growth depression with white lupin levels up to 62% of the diet, whereas pigs had reduced gains at 10% dietary lupin, with the same batch of lupin seed used for both species (Cheeke and Kelly, 1989).

The mode of action of lupin alkaloids in reducing animal performance may involve adverse effects on feed intake, or post-absorptive metabolic effects. Alkaloids are the primary constituents responsible for the sensory perception by humans of lupin bitterness (Du Pont *et al.*, 1994). Lupins selected for low alkaloid levels are referred to as sweet lupins, whereas unselected types are bitter lupins. Sweet lupins are more susceptible to pests and diseases than are the bitter types, and there is a tendency for reversion to higher alkaloid levels. In the Andean regions of South America, the *L. mutabilis* grown for human consumption are bitter lupins. The seeds are debittered by being boiled in water for 30 minutes followed by steeping in running water for 3 days. While this process removes the alkaloids, it also reduces the dry matter by about 30% because of the leaching out of water-soluble nutrients.

The mode of action of quinolizidine alkaloids such as sparteine and lupanine is that they inhibit ganglionic transmission by blocking nicotinic cholinergic receptors. They decrease cardiac contractility and block uterine smooth muscle contraction (Petterson *et al.*, 1994). The alkaloids are excreted largely unchanged in the urine, although there is some cytochrome P450–mediated oxidation to form N-oxides. There is genetic variation among humans in their ability to metabolize and detoxify quinolizidine alkaloids, related to point mutations in the 2D6 isozyme of cytochrome P450 (Petterson *et al.*, 1994).

Other Deleterious Factors in Lupins

Lupins may accumulate high levels of **manganese**, particularly when grown on acid soils, which tend to have high available manganese levels. *L. albus* is a manganese accumulator; it may have 20 times the manganese level of *L. angustifolius* when both species are grown in the same soil (Hung *et al.*, 1987). Levels as high as 6900 ppm manganese have been reported in *L. albus* (Oram *et al.*, 1979). Such high levels may cause manganese toxicity in animals. Signs include reduced feed intake and lowered growth rate, and anemia caused by manganese inhibition of iron absorption.

Lupins contain **oligosaccharides** such as α-galactosides (Brenes *et al.*, 1993). These may accumulate in the hindgut and stimulate microbial fermentation. The hindgut of lupin-fed pigs can be grossly enlarged (Cheeke and Kelly, 1989). Brenes *et al.* (1993) found that supplementation of a lupin-containing chick diet with an enzyme preparation with α-galactosidase activity increased feed intake and weight gain.

Lupins contain **saponins**, but the levels are low and do not contribute measurable antinutritive effects (Muzquiz *et al.*, 1993; Ruiz *et al.*, 1995).

Heat treatment of lupin seed does not improve performance of swine (Batterham *et al.*, 1986) or poultry (Brenes *et al.*, 1993), suggesting the absence of heat-labile toxicants.

Lupinosis

It is common to graze ruminant animals on lupin fields after the seed has been harvested. A disease, lupinosis, sometimes occurs in animals grazing lupin stubble. **Lupinosis** is caused by the ingestion of toxins produced by the fungus *Phomopsis leptostromiformis*, which grows on lupin. It infects the green lupin plant initially, but persists on the stubble. Lupinosis is characterized by severe liver damage. It has been known for over a century and has been observed in Germany, Poland, South Africa, New Zealand, and Australia. It is of most importance in Australia, where lupins are widely grown as a grain crop, and sheep are grazed on the stubble.

The first signs of lupinosis are a lack of appetite and a loss of weight and condition (Fig. 7–21). With acute toxicity, following a high intake of toxic lupin, the liver increases in weight due to a massive accumulation of fat in the liver cells. The liver is greatly enlarged, bright yellow or orange in color, and very greasy when cut. The gall bladder is increased in size several fold over normal. The subcutaneous tissue and body fat are tinged with orange or yellow.

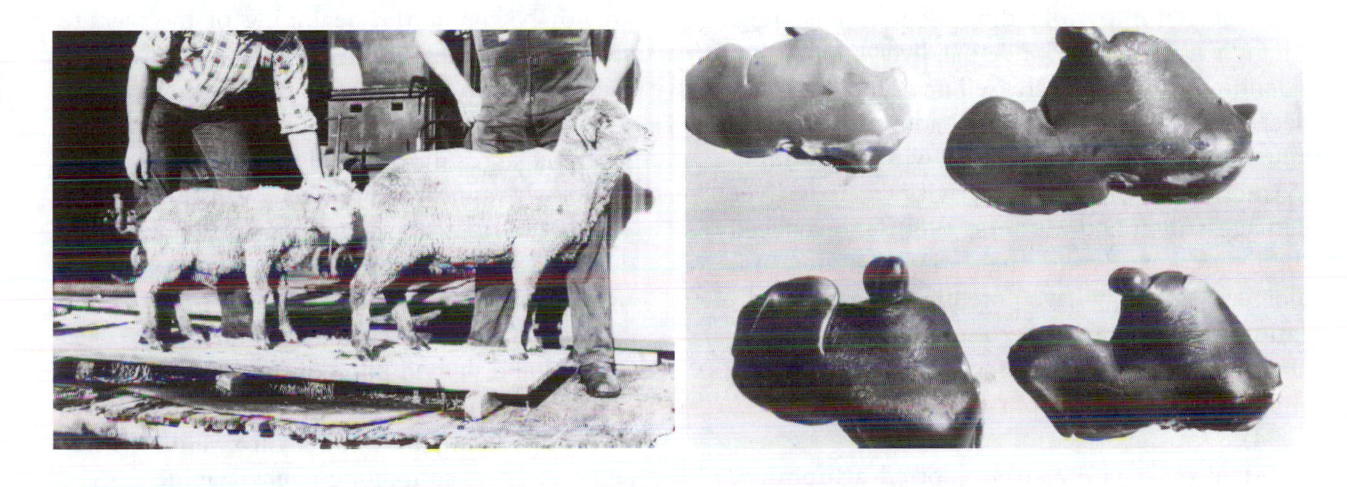

FIGURE 7–21 (Left): The effects of chronic lupinosis. These sheep were the same age (12 months) on the same property. The sheep on the left had grazed lupin stubble for 3 months, while the sheep on the right had been on clover pasture. (Right): Livers from sheep with chronic lupinosis. Note the extreme paleness of the top left liver. All showed significant microscopic damage. (Courtesy of J.G. Allen)

Chronic lupinosis is characterized by liver necrosis and the appearance of signs of liver dysfunction such as jaundice. The membranes of the eye and mouth may be intensely yellow. Affected animals appear dejected and depressed and lag behind the rest of the flock. The detection of these stragglers, by regularly moving the flock, provides an early indication of a lupinosis problem and is a management technique recommended whenever lupin stubble is grazed. In cases of chronic lupinosis, the liver is coppery or tan colored and is smaller than normal. It feels hard and fibrotic and has a granular appearance. In Australia, sheep exposed to lupinosis toxins may also consume **pyrrolizidine alkaloids** in such plants as *Heliotropium* *europaeum*. Because both have hepatoxic effects, it is likely that they could have additive or synergistic activity. Besides liver damage, there is evidence of lesions of muscle, kidney and adrenal cortex (Allen and Randall, 1993).

The growth of the *Phomopsis* fungus on lupin may be weather related, with warm, wet, humid conditions favoring its growth. In Australia it was formerly believed that lupinosis was associated with summer rains, but this was because the varieties then grown had coarse, unpalatable stems which were consumed only when they were moistened and softened by rain. All lupin stubble is potentially toxic. High sheep stocking rates seem to increase the lupinosis incidence.

Sheep that develop lupinosis develop elevated **liver copper** levels. This is a reflection of liver damage; similar elevations in liver copper are observed in pyrrolizidine alkaloid poisoning. Allen *et al.* (1979) found that liver copper and selenium levels are increased and zinc levels are decreased in lupinosis. Oral administration of zinc sulfate to sheep treated with toxic lupin extracts had protective activity against liver damage (Allen and Masters, 1980) and reduced the liver copper levels, suggesting that zinc supplementation may have potential practical application. A mineral mix with low copper and high zinc has been developed for supplementation of sheep grazing toxic lupin stubble (White *et al.*, 1994).

Allen *et al.* (1992) has reported a **lupin-associated myopathy** in sheep that is identical in appearance to white muscle disease (a selenium-vitamin E–responsive myopathy), but which does not respond to and is not prevented by selenium or vitamin E supplementation. It is not known if the lupin-associated myopathy is caused by a direct action of the lupinosis toxins on the muscles or by these toxins interfering with selenium availability to the muscles. Van Rensburg *et al.* (1975) have observed cardiac lesions in lupinosis that they attributed to direct action of the toxin.

Lupinosis also affects cattle (Mackie *et al.*, 1992). Allen (1981) described two conditions in cattle affected with lupinosis. First, there is a fatty liver syndrome that affects only cows in late pregnancy or those that have recently calved and is the more commonly observed condition. The animals show fatty livers and signs of ketosis. Allen (1981) suggests that the lupinosis toxins induce a mild depression of appetite, which is sufficient to produce a glucose deficit. Thus, this syndrome appears to be a secondary nutritional ketosis.

The second syndrome in cattle is a cirrhotic liver condition similar to that seen in chronic lupinosis in sheep. In sheep, the lupinosis is generally obvious while the animals are still grazing the lupin stubble, and photosensitization is rarely seen. Conversely, in cattle the animals may not exhibit signs of lupinosis for some time after grazing on lupin ceases. Generally, the condition occurs when new grass growth begins at the beginning of the pasture season. **Photosensitization** is commonly seen. The udder is often affected, and the cow refuses to let the calf nurse. The photosensitization is probably a result of the increased chlorophyll intake when grass growth begins. The increased phylloerythrin load cannot be excreted by the damaged liver, and secondary photosensitization occurs.

The toxic agents causing lupinosis have been isolated and named phomopsin A and phomopsin B. The **phomopsins** are cyclic hexapeptides involving didehydro- and hydroxyamino acids and a chlorine-containing phenylalanine-derived subunit (Allen and Hancock, 1989). They have a colchicine-like action in arresting liver cell mitosis in metaphase, by inhibiting microtubule polymerization. The aromatic ring of phomopsin is the active site of the toxicologic effects. Phomopsin is extremely toxic; Peterson *et al.* (1987) found the minimum lethal dose in sheep to be 10 μg/kg body weight.

Lupinosis can be prevented by growing varieties such as Ultra, that are resistant to infection by *Phomopsis*. Unfortunately, *Phomopsis*-resistant cultivars do not thrive in all areas where lupins are grown. Animal management techniques are required to prevent stock losses.

Sheep introduced on to lupin stubbles should be herded daily and closely observed for signs of incipient lupinosis. Late-pregnant or recently calved cows should not be grazed on lupin stubbles. Reproductive performance of sheep is impaired when ewes are exposed to phomopsin at the time of breeding (Barnes *et al.*, 1996). Reproductive impairment may be a direct effect from disruption of microtubules by phomopsins, with interference in cell division, or other aspects of embryonic development, or could be an indirect metabolic effect associated with phomopsin-induced liver damage.

Lathyrus spp. and Lathyrism

Lathyrism is a crippling disease in humans caused by the consumption of seeds of *Lathyrus* spp., principally *L. sativus* (chickling pea, grasspea). It has been a major public health problem in India and continues to afflict people in poorer sections of that country. *Lathyrus* is also grown in Bangladesh, Nepal, Pakistan, China, and Ethiopia, and at one time was a food crop in France, Spain and other western European countries. *Lathyrus* spp. are hardy, vigorous legumes that grow in poor soil under drought conditions, and land that floods. The crop is grown in India, in spite of knowledge of its toxicity, as insurance against the failure of other crops. In 1956–1957, 4 million acres of *L. sativus* were under cultivation in India (Padmanaban, 1980). The seeds are used in making flour for bread and as a vegetable. At one time, landlords distributed wages to tenants in the form of *L. sativus* seeds. Several Indian states have now banned the sale of the seeds, but the crop is still widely grown. Deliberate adulteration of other grains with *Lathyrus* seeds is common, as a means of selling the forbidden crop. Epidemics of lathyrism in India occurred in 1958 and 1974. Outbreaks of lathyrism have occurred in the 1980's after severe droughts in Bangladesh and Ethiopia. Lathyrism is of two types: osteolathyrism and neurolathyrism. **Osteolathyrism**, in which skeletal deformities and aortic rupture occur, is seen mainly in livestock, particularly horses, consuming seeds of *L. odoratus*, the annual sweet pea. The perennial sweet pea, *L. latifolius* (Fig. 7–22), is also toxic and is widely naturalized in North America, as is the flat pea, *L. sylvestris*. The rough pea, *L. hirsutus*, grows in the southern U.S. and contains lathyrogens. It was formerly grown as an annual winter forage or cover crop in southern areas of the U.S. These plants will be discussed under Forage Legumes (Chapter 11). **Neurolathyrism**, in which the nervous system is affected, is caused by neurotoxins in *L. sativus* and primarily affects humans.

FIGURE 7–22 Seed pods of *Lathyrus latifolius*, the perennial sweet pea.

Osteolathyrism

This is the principal form of lathyrism that affects livestock. Consumption of seeds of *L. odoratus*, *L. sylvestris*, *L. hirsutus*, or related species of sweet pea causes skeletal deformity and aortic rupture due to defective synthesis of cartilage and connective tissue. Malformations of long bones are caused by irregular hyperplastic cartilage formed in the epiphysis (the area of the proliferative zone of cartilage at the end of a bone). Aortic rupture due to formation of aortic aneurysms (weakness of the artery wall) is due to defective collagen and elastin synthesis.

The lathyrogen in *L. odoratus* and related species is **β-amino propionitrile (BAPN)**, an amino acid derivative:

$$H_2N—CH_2—CH_2—C \equiv N$$

It exists in plants in the form of β-(γ-L-glutamyl) amino propionitrile, but the glutamyl group is not necessary for lathyrogenic activity. BAPN interferes with the cross-linking of collagen and elastin molecules; consequently, the connective tissue loses its normal structure. Cross-linking of connective tissue fibers involves oxidation of the epsilon amino groups of lysine residues to hydroxylysine in the collagen or elastin proteins. The enzyme lysyl oxidase acts on the specific lysine residues involved in cross-linking; this enzyme also requires cupric ions for its activity. Lysyl oxidase is irreversibly inhibited by BAPN, thus accounting for the defective collagen and elastin formed in osteolathyrism. Copper deficiency produces very similar symptoms because of insufficient lysyl oxidase activity. BAPN toxicity is not counteracted by supplementary dietary copper. The reaction involving lysyl oxidase is as follows:

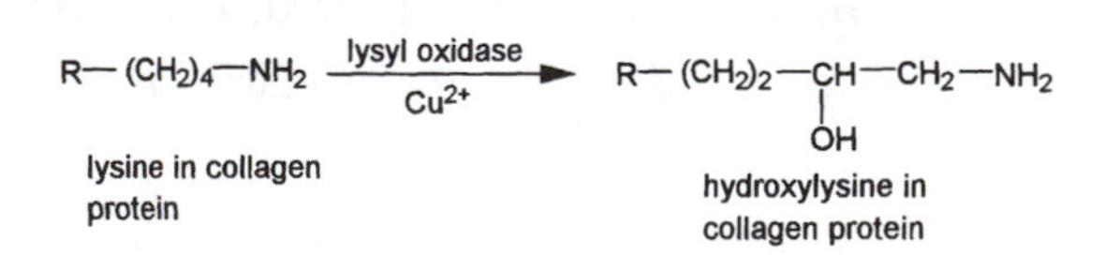

Autoclaving or other heat treatment reduces the toxicity of *Lathyrus* seeds to poultry and livestock (Rotter *et al.*, 1990). Dietary additives such as cysteine and synthetic antioxidants did not reduce the toxicity of *Lathyrus odoratus* to chicks and quail (Raharjo *et al.*, 1988).

Neurolathyrism

Neurolathyrism is a paralysis of the legs due to nerve damage in the spinal cord caused by neurotoxins in *L. sativus*. The principal neurolathyrogen is **β-N-oxalyl-L-α-β-diaminopropionic acid (ODAP)**:

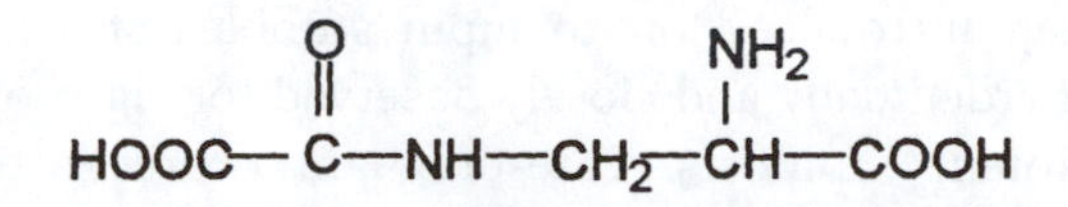

Another common name for ODAP is β-oxalylaminoalanine (BOAA).

Neurolathyrism is a major public health problem in India and other parts of Asia. The disease is associated with both the consumption of seeds of *L. sativus* and environmental factors such as exposure to a wet environment and overwork. It generally appears whenever a diet containing *L. sativus* seeds is consumed for 3–6 months. The symptoms are muscular rigidity, weakness, and paralysis of the leg muscles, leading to death in extreme cases. The onset is unusually sudden, with development of stiffness or paralysis of the lower limbs. In mild cases, there is difficulty in walking. People with more advanced cases require a stick for support. This may advance to the two-stick stage, and finally the victims can only crawl (Fig. 7–23). Lathyrism affects primarily young men between 20 and 30 years of age. This may be interrelated with the environmental factors involved in the disease.

One of the major difficulties in studying neurolathyrism is that experimental animals do not generally respond to the neurotoxin. Day-old chicks have been used and exhibit head retraction and convulsions. Older chickens are not affected. The squirrel monkey also responds to the neurotoxins with muscle tremors and convulsions and has been used as an experimental model. From studies with experimental animals, ODAP appears to be of low toxicity, which emphasizes the probable importance of other factors (preexisting disease, malnutrition, and stress of physical overexertion) in development of neurolathyrism in humans (Parker *et al.*, 1979).

Ross *et al.* (1989) suggested that ODAP (BOAA) is an **excitotoxin**, a compound which causes death of neurons by causing the release of excess glutamate. Glutamate and aspartate are the major excitatory amino acids in the

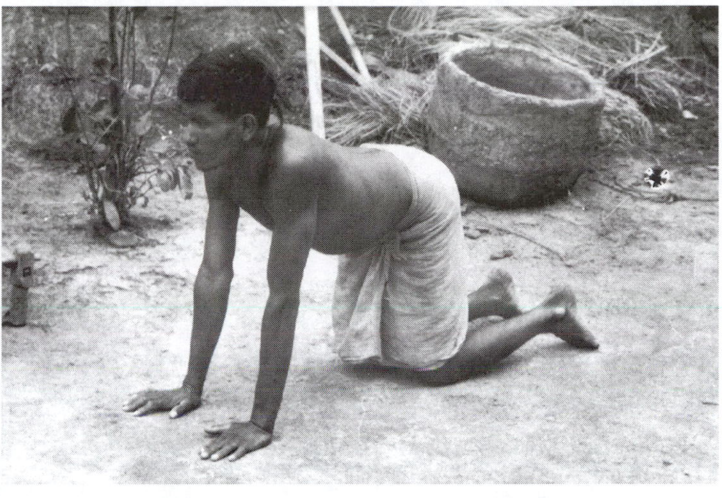

FIGURE 7–23 The two-stick stage of lathyrism (left) progresses to severe lathyrism in which the victim can no longer walk. (Courtesy of M. Mohan Ram)

brain, responsible for about 75% of the brain's excitatory neurotransmission. Brain tissue has numerous types of **glutamate receptors**. Chronic stimulation of glutamate receptors causes an influx of calcium ions leading to cell death (see General Biological Effects of Natural Toxins in Chapter 3). Thus the signs of neurolathyrism may result from selective brain cell death caused by ODAP-stimulated disruption of glutamate receptors. An effect of ODAP on glutamate receptors has been demonstrated by Ross *et al.* (1989).

Another mode of action of neurolathyrogens may involve free radical damage in the hippocampal region of the brain (Willis *et al.*, 1994). Administration of free radical scavengers prevented neurotoxicity and hippocanpal lesions in rats dosed with BOAA (Willis *et al.* , 1994). The antioxidant ascorbic acid protects guinea pigs from neurolathyrism (Dunham *et al.*, 1995), providing further evidence that free radical formation may have a role in neurolathyrism.

Lathyrus sativus seeds can be treated to reduce the toxicity. Steeping and boiling in hot water removes most of the neurotoxin. They are often used in the raw form, e.g., as paste balls, so that the toxin is retained. Fermentation of lathyrus seeds markedly reduces toxin levels (Kuo *et al.*, 1995). Thus, by seemingly simple changes in preparation of the seeds for consumption, toxicity could be avoided. The ideal solution would be development of cultivars of *L. sativus* free of the toxin because the crop has agronomic features (drought resistance and hardiness) that would make it an excellent food plant if it were nontoxic. There has been progress in selection for low lathyrogen levels (Aletor *et al.*, 1994; Deshpande and Campbell, 1992). Low lathyrogenic *Lathyrus* seeds have been evaluated in broiler (Rotter *et al.*, 1991) and swine (Castell *et al.*, 1994) diets. Growth depression associated with other anti-nutritive constituents such as trypsin inhibitors limited animal performance. These effects could be overcome through heat processing.

Other Lathyrogens

Vicia sativa (common vetch) (Fig. 7–24) contains a neurolathyrogen, **β-cyano-L-alanine (BCA)**:

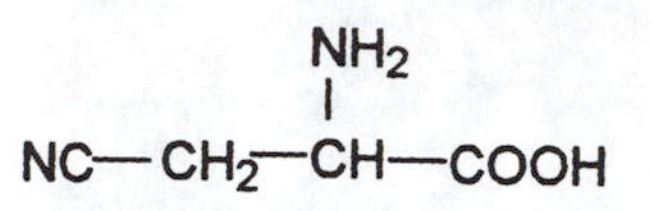

A large portion of the BCA is present as a dipeptide with glutamic acid, α-glutamyl-β-cyano-L-alanine.

FIGURE 7–24 Vetch is an annual viney legume, used as a forage. The seeds contain a neurolathyrogen, β-cyano-L-alanine.

Vetch causes neurolathrysim when fed to poultry (Arscott and Harper, 1963; Harper and Arscott, 1962; Farran *et al.*, 1995). Contamination of grain with vetch seed has resulted in a number of poultry losses. Chicks fed vetch develop convulsions, blindness, and a pronounced, plaintive chirping, resembling a pyridoxine deficiency. Pyridoxine supplementation delays the onset of vetch toxicity symptoms. Vetch was once a widely grown crop, as a component of an oats–peas–vetch forage mixture grown for hay and silage.

β-Cyano-L-alanine may be a contributing factor to neurolathyrism in India, as vetch is sometimes a contaminant of lathyrus seeds. Vetch seed is sometimes deliberately added as an adulterant to other pulses. Tate and Enneking (1992) reported the case of vetch seed being substituted for red lentils, to be used for human consumption. The impetus is a monetary one; vetch seed is often cheaper than lentils and other pulses. Selection of vetch and lathyrus free of neurotoxins would eliminate the toxicity hazards as well as provide new food sources.

Another *Vicia* pulse crop, the narbon bean (*V. narbonensis*) has potential as a human and animal foodstuff. It contains antinutritional factors which reduce growth performance of pigs and poultry (Eason *et al.*, 1990). It contains a dipeptide, α-glutamyl-S-ethenyl cysteine (En-

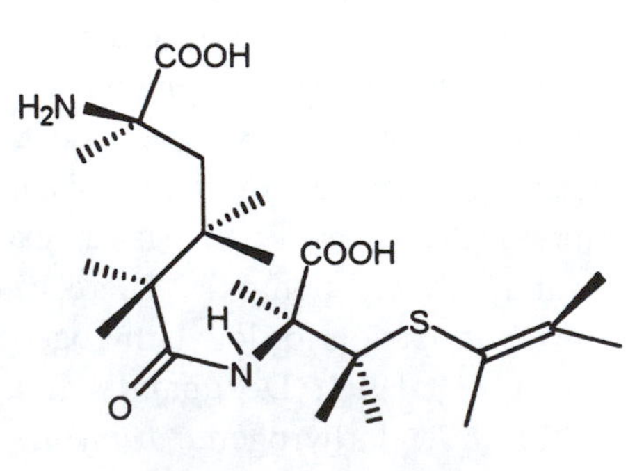

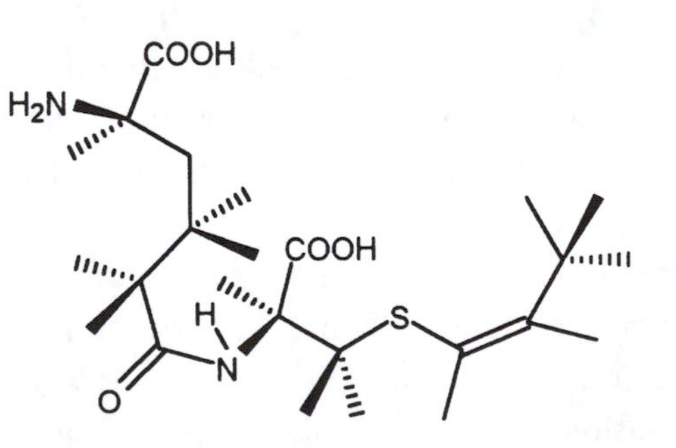

FIGURE 7–25 A toxic S-containing dipeptide, γ-glutamyl-S-ethenyl-cysteine, occurs in *Vicia narbonensis*. It has a structural resemblance to a dipeptide in chives (*Allium schoenoprasum*), which causes erythrocyte hemolysis.

neking, 1995), which is responsible for its adverse effects. This amino acid derivative has some similarity in structure to sulfur compounds in *Brassica* and *Allium* spp. (Fig. 7–25) which cause hemolytic anemia (see Chapter 11). Red blood cell hemolysis has been observed in pigs fed narbon beans (Enneking, 1995).

Lathyrogens and Guam Disease

Guam disease is a fatal neurological disorder of residents of Guam and nearby islands. The disease presents a syndrome of amyotrophic lateral sclerosis (ALS), Parkinson's disease and Alzheimers disease dementia. Spencer *et al.* (1987) suggested that the traditional consumption of the plant *Cycas circinalis* may play a role in the etiology of Guam disease. *Cycas* seeds contain a neurotoxic lathyrogen-like amino acid, **β-N-methylamino-L-alanine (BMAA)**:

CH_2—CH—COO^-
NH_2^+ NH_2
CH_3

β-methylamino-L-alanine (BMAA)

Subsequent work (Kisby *et al.*, 1992; Spencer *et al.*, 1991) has supported a potential role of BMAA and **cycasin** (methylazoxymethanol β-D-glycoside) in Guam disease. Guam disease is believed to have a long latency period of several decades between exposure to neurotoxins and manifestation of the disease syndrome. However, there are other possible causes, including a transmissible agent such as a virus (Sacks, 1993). The disorder is gradually disappearing, as residents of the affected areas adopt Western customs and diet (Stone, 1993). Thus the decline in Guam disease parallels the decline in use of *Cycas* as food, but does not prove conclusively that the neurotoxins in the plant are causative agents. However, the study of this condition has been important in assessing the role of dietary neurotoxins in degenerative neurological diseases such as ALS (Lou Gerig's disease), Parkinson's disease and Alzheimer's disease. Monmaney (1990) has provided an extensive popular press article on the history of Guam disease and the evidence for an involvement of cycad consumption.

Other Toxins in Legume Seeds

Indigofera spicata, or creeping indigo, is a tropical legume with potential as a forage crop. It contains a toxic amino acid, **indospicine**, which is structurally similar to arginine (Fig. 7–26). It inhibits arginine incorporation into tissue proteins and causes liver damage in cattle and sheep consuming the plant, with necrosis and nodular cirrhosis. In Australia, a disorder known as Birdsville horse disease affects horses grazing *Indigofera* spp. It is a neurological condition without liver damage. The meat from horses with *Indigofera* poisoning is toxic to dogs, which are apparently very susceptible to indos-

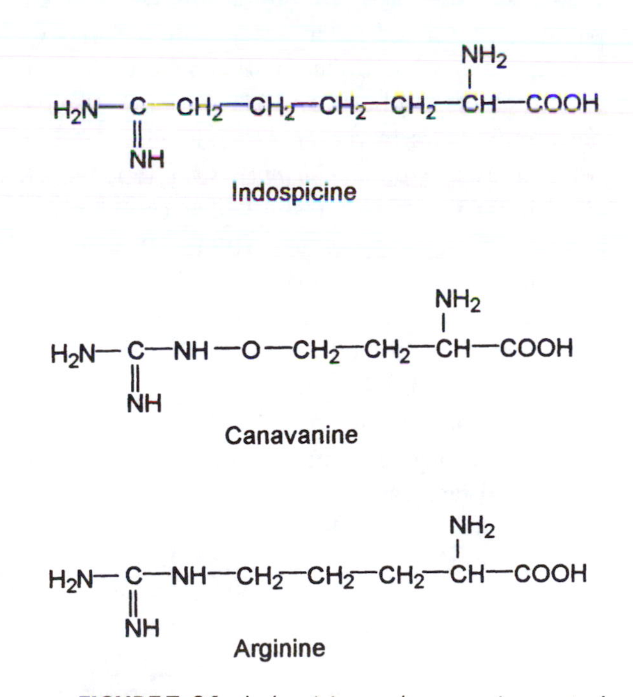

FIGURE 7–26 Indospicine and canavanine are toxic amino acids structurally similar to arginine.

picine toxicity (Hegarty *et al.*, 1988). Numerous *Indigofera* spp. with agricultural potential are non-toxic (Aylward *et al.*, 1987).

Seeds of the jack bean (*Canavalia ensiformis*) contain several toxicants, including a heat-labile lectin, **concanavalin A**, and a toxic amino acid, **canavanine** (Fig. 7–26). Canavanine is an arginine antagonist (D'Mello and Walker, 1991). Birds are more sensitive than mammals, because they lack urea cycle enzymes, and thus cannot synthesize arginine to overcome the inhibitory effect. This is also true with indospicine, which is an arginine antagonist. D'Mello (1991) has discussed the canavanine–arginine interaction in detail. In insects, canavanine may be incorporated into structural proteins, causing disrupted function (Rosenthal, 1991). Thus canavanine may be an important insecticidal component of some plants' chemical defenses.

Canavanine is found in a number of other legume seeds. Alfalfa sprouts contain about 1.5% of their dry weight as canavanine. A severe lupus erythematosus-like syndrome occurs in monkeys fed alfalfa sprouts (Malinow *et al.*, 1982). An increase in severity of systemic lupus erythematosus (skin inflammation with raised patches, prolonged fever, soft tissue lesions) in humans has been linked with consumption of alfalfa tablets (Roberts and Hayashi, 1983). Canavanine occurs in the seeds and foliage of the tropical tree legume *Sesbania sesban* (Shqueir *et al.*, 1989). Vetch (*Vicia* spp.) also contains canavanine (Enneking *et al.*, 1993). Dietary canavanine has a pronounced effect on reducing voluntary feed intake (Enneking *et al.*, 1993). A rapid reduction in feed intake leading to growth depression occurs in chicks and pigs fed jack beans (Belmar and Morris, 1994a,b), attributed in part to canavanine and in part to other antinutritional factors.

Jackbeans have a high content of urease, the enzyme that converts urea to ammonia. **Jackbean urease** is the most common source of the enzyme for use in biomedical research. In the tropics, jackbean meal is sometimes added to urea-containing ruminant diets as a source of urease (probably unnecessary, because rumen microbes produce abundant urease). Pimentel and Cook (1988) immunized chicks with jackbean urease, with the hypothesis that antibodies produced in the gut to jackbean urease would suppress microbial urease in the gut, thus reducing ammonia absorption. Absorbed ammonia may depress animal performance, because energy is required for urea cycle activity to detoxify absorbed ammonia. One amino group from an amino acid is necessary for synthesis of urea from every molecule of absorbed ammonia. In the work of Pimentel and Cook (1988), growth rate was increased in chicks immunized with jackbean urease.

Legume seeds contain proteins which can induce allergenic reactions such as **anaphylaxis**. Anaphylactic shock is a severe reaction to foreign antigens which trigger a massive release of histamine from basophils and mast cells, producing vasodilation, edema and inflammatory response. The main targets are the gastrointestinal tract (food allergies), the skin (urticaria), the respiratory system and the circulatory system (anaphylactic shock). These responses can occur very rapidly after exposure of a sensitized individual to the antigens. Peanuts (*Arachis hypogaea*) are well-known for their allergenic effects in humans (Hopkins, 1995), with skin (rash), digestive tract (nausea, vomiting), and respiratory (wheezing) signs, with anaphylactic shock and collapse in severe cases.

REFERENCES

Soybeans and Soybean Meal

Adlercreutz, H. 1990. Diet, breast cancer, and sex hormone metabolism. Annal. N.Y. Acad. Sci. 595:281–290.

Adlercreutz, H., H. Honjo, A. Higashi, T. Fotsis, E. Hamalainen, T. Hasegawa, and H. Okada. 1991. Urinary excretion of lignans and isoflavonoid phytoestrogens in Japanese men and women consuming a traditional Japanese diet. Am. J. Clin. Nutr. 54:1093–1100.

Albro, J.D., D.W. Weber, and T. Del Curto. 1993. Comparison of whole, raw soybeans, extruded soybeans, or soybean meal and barley on digestive characteristics and performance of weaned beef steers consuming mature grass hay. J. Anim. Sci. 71:26–32.

Anderson-Hafermann, J.C., Y. Zhang, and C.M. Parsons. 1992. Effect of heating on nutritional quality of conventional and Kunitz trypsin inhibitor–free soybeans. Poult. Sci. 71:1700–1709.

Coon, C.N., K.L. Leske, O. Akavanichan, and T.K. Cheng. 1990. Effect of oligosaccharide-free soybean meal on true metabolizable energy and fiber digestion in adult roosters. Poult. Sci. 69:787–793.

D'Mello, J.P.F. 1991. Antigenic proteins. pp. 107–125 in: J.P.F. D'Mello, C.M. Duffus, and J.H. Duffus (Eds.). Toxic Substances in Crop Plants. The Royal Society of Chemistry, Cambridge.

Dunford, B.R., D.A. Knabe, and W.E. Haensly. 1989. Effect of dietary soybean meal on the microscopic anatomy of the small intestine in the early-weaned pig. J. Anim. Sci. 67:1855–1863.

Dwyer, J.T., B.R. Goldin, N. Saul, L. Gualtieri, S. Barakat, and H. Adlercreutz. 1994. Tofu and soy drinks contain phytoestrogens. J. Am. Diet. Assoc. 94:739–743.

Fotsis, T., M. Pepper, H. Adlercreutz, G. Fleischmann, T. Hase, R. Montesano, and L. Schweigerer. 1993. Genistein, a dietary-derived inhibitor of in vitro angiogenesis. Proc. Natl. Acad. Sci. USA 90:2690–2694.

Friedman, M. (Ed.). 1986. Nutritional and Toxicological Significance of Enzyme Inhibitors in Foods. Plenum Press, NY.

Friesen, K.G., R.D. Goodband, J.L. Nelssen, F. Blecha, D.N. Reddy, P.G. Reddy, and L.J. Kats. 1993. The effect of pre- and postweaning exposure to soybean meal on growth performance and on the immune response in the early-weaned pig. J. Anim. Sci. 71:2089–2098.

Grant, G. 1989. Anti-nutritional effects of soyabean: a review. Prog. Food Nutr. Sci. 13:317–348.

Herkelman, K.L., G.L. Cromwell, T.S. Stahly, T.W. Pfeiffer, and D.A. Knabe. 1992. Apparent digestibility of amino acids in raw and heated conventional and low trypsin–inhibitor soybeans for pigs. J. Anim. Sci. 70:818–826.

Holmes, J.H.G., R.M. Dixon, C. Smith, S. Entonu, H. Chau, S. Sanpote, and B. Acharyya. 1993. Resistance of trypsin inhibitors to fermentation by rumen microflora. pp. 183–186 in: A.F.B. van der Poel, J. Huisman, and H.S. Saini (Eds.). Recent Advances of Research in Antinutritional Factors in Legume Seeds. Wageningen Pers, Wageningen, The Netherlands.

Kennedy, A.R., B.F. Szuhaj, P.M. Newberne, and P.C. Billings. 1993. Preparation and production of a cancer chemopreventive agent, Bowman–Birk inhibitor concentrate. Nutrition and Cancer 19:281–302.

Lalles, J.P., H. Salmon, N.P.M. Bakker, and G.H. Tolman. 1993. Effects of dietary antigens on health, performance and immune system of calves and piglets. pp. 253–270 in: A.F.B. van der Poel, J. Huisman, and H.S. Saini (Eds.). Recent Advances of Research in Antinutritional Factors in Legume Seeds. Wageningen Pers, Wageningen, The Netherlands.

Li, D.F., J.L. Nelssen, P.G. Reddy, F. Blecha, J.D. Hancock, G.L. Allee, R.D. Goodband, and R.D. Klemm. 1990. Transient hypersensitivity to soybean meal in the early-weaned pig. J. Anim. Sci. 68:1790–1799.

Liener, I.E. 1994. Implications of antinutritional components in soybean foods. Critical Rev. Food Sci. Nutr. 34:31–67.

Liener, I.E., and M.L. Kakade. 1980. Protease inhibitors. pp. 7–71 in: I.E. Liener (Ed.). Toxic Constituents of Plant Foodstuffs, 2nd Edition. Academic Press, NY.

Marczy, J.S., M.L. Simon, L. Mozsik, and B. Szajani. 1995. Comparative study on the lipoxygenase activities of some soybean cultivars. J. Agric. Food Chem. 43:313–315.

Messina, M., and J.W. Erdman, Jr. (Eds.). 1995. First International Symposium on the Role of Soy in Preventing and Treating Chronic Disease. J. Nutr. 125:567S–808S.

Messina, M.J., V. Persky, K.D.R. Setchell, and S. Barnes. 1994. Soy intake and cancer risk: A review of the in vitro and in vivo data. Nutrition and Cancer 21:113–131.

Oakenfull, D., and G.S. Sidhu. 1989. Saponins. pp. 97–141 in: P.R. Cheeke (Ed.). Toxicants of Plant Origin. Vol. II. CRC Press, Boca Raton, FL.

Quigley, J.D. III, K.R. Martin, H.H. Dowlen, and K.C. Lamar. 1995. Addition of soybean trypsin inhibitor to bovine colostrum: Effects on serum immunoglobulin concentrations in Jersey calves. J. Dairy Sci. 78:886–892.

Roebuck, B.D. 1987. Trypsin inhibitors: potential concern for humans? J. Nutr. 117:398–400.

Setchell, K.D.R., S.J. Gosselin, M.B. Welsh, J.O. Johnston, W.F. Balistreri, L.W. Kramer, B.L. Dresser, and M.J. Tarr. 1987. Dietary estrogens: A probable cause of infertility and liver disease in captive cheetahs. Gastroenterol. 93:225–233.

Susmel, P., M. Spanghero, S. Marchetti, and S. Moscardini. 1995. Trypsin inhibitor activity of raw soya bean after incubation with rumen fluid. J. Sci. Food Agric. 67:441–445.

Tukur, H.M., J.P. Lalles, C. Mathis, I. Caugant, and R. Toullec. 1993. Digestion of soybean globulins, glycinin, α-conglycinin and β-conglycinin in the preruminant and ruminant calf. Can. J. Anim. Sci. 73:891–905.

van der Poel, A.F.B., J. Huisman, and H.S. Saini. (Eds.). 1993. Recent advances of research in antinutritional factors in legume seeds. Wageningen Pers, Wageningen, The Netherlands.

Wang, H., and P.A. Murphy. 1994. Isoflavone content in commercial soybean foods. J. Agric. Food Chem. 42:1666–1673.

Xu, X., H.-J. Wang, P.A. Murphy, L. Cook, and S. Hendrich. 1994. Daidzein is a more bioavailable soymilk isoflavone than is genistein in adult women. J. Nutr. 124:825–832.

Zhang, Y., C.M. Parsons, K.E. Weingartner, and W.B. Wijeratne. 1993. Effect of extrusion and expelling on the nutritional quality of conventional and Kunitz trypsin inhibitor–free soybeans. Poult. Sci. 72:2299–2308.

Cottonseed and Cottonseed Meal

Andrianaivo-Rafehivola, A.A., M.-H. Siess, and E.M. Gaydou. 1995. Modifications of hepatic drug metabolizing enzyme activities in rats fed baobab seed oil containing cyclopropenoid fatty acids. Fd. Chem. Toxic. 33:377–382.

Barraza, M.L., C.E. Coppock, K.N. Brooks, D.L. Wilks, R.G. Saunders, and G.W. Latimer, Jr. 1991. Iron sulfate and feed pelleting to detoxify free gossypol in cottonseed diets for dairy cattle. J. Dairy Sci. 74:3457–3467.

Calhoun, M.C., S.W. Kuhlmann, and B.C. Baldwin, Jr. 1995. Assessing the gossypol status of cattle fed cotton feed products. pp. 147A–158A in: Proc. Pac. Northwest Anim. Nutr. Conf., Portland, OR.

Cao, J., J.-P. Blond, and J. Bezard. 1993. Inhibition of fatty acid Δ^6- and Δ^5-desaturation by cyclopropene fatty acids in rat liver microsomes. Biochimica Biophysica Acta 1210:27–34.

Chase, C.C., Jr., P. Bastidas, J.L. Ruttle, C.R. Long, and R.D. Randel. 1994. Growth and reproductive development in Brahman bulls fed diets containing gossypol. J. Anim. Sci. 72:445–452.

Cook, L.J., T.W. Scott, S.C. Mills, and A.R. Johnson. 1976. Effects of protected cyclopropene fatty acids on the composition of ruminant milk fat. Lipids 11:705–711.

Coppock, C.E., J.K. Lanham, and J.L. Horner. 1987. A review of the nutritive value and utilization of whole cottonseed, cottonseed meal and associated by-products for dairy cattle. Anim. Feed Sci. Tech. 18:89–129.

Cusack, P.M.V., and V. Perry. 1995. The effect of feeding whole cottonseed on the fertility of bulls. Aust. Vet. J. 72:463–466.

Dilday, R.H. 1986. Development of a cotton plant with glandless seeds, and glanded foliage and fruiting forms. Crop Sci. 26:639–641.

Eisele, T.A., P.M. Loveland, D.L. Kruk, T.R. Meyers, J.E. Nixon, and R.O. Sinnhuber. 1982. Effect of cyclopropenoid fatty acids on the hepatic microsomal mixed-function-oxidase system and aflatoxin metabolism in rabbits. Food Chem. Toxicol. 20:407–412.

Ferguson, T.L., J.H. Wales, R.O. Sinnhuber, and D.J. Lee. 1976. Cholesterol levels, atherosclerosis and liver morphology in rabbits fed cyclopropenoid fatty acids. Food Cosmet. Toxicol. 14:15–18.

Gray, M.L., L.W. Greene, and G.L. Williams. 1993. Effects of dietary gossypol consumption on metabolic homeostasis and reproductive endocrine function in beef heifers and cows. J. Anim. Sci. 71:3052–3059.

Hemken, R.W., Z. Du, and W. Shi. 1996. Use of proteinates to reduce competition from other trace minerals. pp. 91–94 in: T.P. Lyons and K.A. Jacques (Eds.). Proc. Alltech 12th Annual Symposium. Nottingham University Press, Nottingham, UK.

Hendricks, J.D., R.O. Sinnhuber, P.M. Loveland, N.E. Pawlowski, and J.E. Nixon. 1980. Hepatocarcinogenicity of glandless cottonseeds and cottonseed oil to rainbow trout (*Salmo gairdnerii*). Science 208:309–311.

Holmberg, C.A., L.D. Weaver, W.M. Guterbock, J. Genes, and P. Montgomery. 1988. Pathological and toxicological studies of calves fed a high concentration cotton seed meal diet. Vet. Pathol. 25:147–153.

Hron, R.J., S.P. Koltun, J. Pominski, and G. Abraham. 1987. The potential commercial aspects of gossypol. J. Am. Oil Chem. Soc. 64:1315–1319.

Hudson, L.M., L.A. Kerr and W.R. Maslin. 1988. Gossypol toxicosis in a herd of beef calves. J. Am. Vet. Med. Assoc. 192:1303–1305.

Lee, D.J., J.H. Wales, and R.O. Sinnhuber. 1971. Promotion of aflatoxin-induced hepatoma growth in trout by methyl malvalate and sterculate. Cancer Res. 31:960–963.

Loveland, P.M., J.E. Nixon, N.E. Pawlowski, T.A. Eisele, L.M. Libbey, and R.O. Sinnhuber. 1979. Aflatoxin B_1 and aflatoxicol metabolism in rainbow trout and the effects of dietary cyclopropene. J. Environ. Pathol. Toxicol. 2:707–718.

Lusas, E.W., and G.M. Jividen. 1987. Glandless cottonseed: A review of the first 25 years of processing and utilization research. J. Am. Oil Chem. Soc. 64:839–854.

Morgan, S., E.L. Stair, T. Martin, W.C. Edwards, and G.L. Morgan. 1988. Clinical, clinicopathologic, pathologic, and toxicologic alterations associated with gossypol toxicosis in feeder lambs. Am. J. Vet. Res. 49:493–499.

Pawlowski, N.E., J.D. Hendricks, M.L. Bailey, J.E. Nixon, and G.S. Bailey. 1985. Structural-bioactivity relationship for tumor promotion by cyclopropenes. J. Agric. Food Chem. 33:767–770.

Percy, R.G., M.C. Calhoun, and H.L. Kim. 1996. Seed gossypol variation within *Gossypium barbadense* L. cotton. Crop Sci. 36:193–197.

Randel, R.D., C.C. Chase, Jr., and S.J. Wyse. 1992. Effects of gossypol and cottonseed products on reproduction of mammals. J. Anim. Sci. 70:1628–1638.

Risco, C.A., C.A. Holmberg, and A. Kutches. 1992. Effect of graded concentrations of gossypol on calf performance: Toxicological and pathological considerations. J. Dairy Sci. 75:2787–2798.

Slayden, O., and Fredrick Stormshak. 1990. *In vivo* and *in vitro* effects of a cyclopropenoid fatty acid on ovine corpus luteum function. Endocrinol. 127:3166–3171.

Terrill, T.H., A.M. Rowan, G.B. Douglas, and T.N. Barry. 1992. Determination of extractable and bound condensed tannin concentrations in forage plants. Protein concentrate meals and cereal grains. J. Sci. Food Agric. 58:321–329.

Tumbelaka, L.I., O. Slayden, and F. Stormshak. 1994. Action of a cyclopropenoid fatty acid on the corpus luteum of pregnant and nonpregnant ewes. Biol. Reprod. 50:253–257.

Yu, F., T.N. Barry, P.J. Moughan, and G.F. Wilson. 1993. Condensed tannin and gossypol concentrations in cottonseed and in processed cottonseed meal. J. Sci. Food Agric. 63:7–15.

Rapeseed and Canola

Agunbiade, J.-A., J. Wiseman, and D.J.A. Cole. 1991. Nutritional evaluation of triple low rapeseed products for growing pigs. Anim. Prod. 52:509–520.

Bell, J.M. 1984. Nutrients and toxicants in rapeseed meal: A review. J. Anim. Sci. 58:996–1010.

Bell, J.M. 1993. Factors affecting the nutritional value of canola meal: A review. 1993. Can. J. Anim. Sci. 73:679–697.

Bjeldanes, L.F., J.-Y. Kim, K.R. Grose, J.C. Bartholomew, and C.A. Bradfield. 1991. Aromatic hydrocarbon responsiveness–receptor agonists generated from indole-3-carbinol *in vitro* and *in vivo*: Comparisons with 2,3,7,8-tetrachlorodibenzo-*p*-dioxin. Proc. Natl. Acad. Sci. 88:9543–9547.

Fenwick, G.R., R.K. Heaney, and R. Mawson. 1989. Glucosinolates. pp. 1–41 in: P.R. Cheeke (Ed.). Toxicants of Plant Origin. Vol. II. CRC Press, Boca Raton, FL.

Fenwick, G.R., R.K. Heaney, and W.J. Mullin. 1983. Glucosinolates and their breakdown products in food and food plants. CRC Crit. Rev. Food Sci. Nutr. 18:123–201.

LeBlanc, G.A., J.D. Stuart, S.E. Dunn, and W.S. Baldwin. 1994. Effect of the plant compound indole-3-carbinol on hepatic cholesterol homoeostasis. Fd. Chem. Toxic. 32:633–639.

March, B.E., and Leung, P. 1976. Effects of alterations in maternal thyroid metabolism on embryonic thyroid development in the chick. Can. J. Physiol. Pharmacol. 54:249–253.

McDanell, R., A.E.M. McLean, A.B. Hanley, R.K. Heaney, and G.R. Fenwick. 1988. Chemical and biological properties of indole glucosinolates (glucobrassicins): A review. Fd. Chem. Toxic. 26:59–70.

Paik, I.K., A.R. Robblee, and D.R. Clandinin. 1980. Products of the hydrolysis of rapeseed glucosinolates. Can. J. Anim. Sci. 60:481–493.

Slominski, B.A., L.D. Campbell, and W. Guenter. 1994a. Carbohydrates and dietary fiber components of yellow- and brown-seeded canola. J. Agric. Food Chem. 42:704–707.

Slominski, B.A., L.D. Campbell, and W. Guenter. 1994b. Oligosaccharides in canola meal and their effect on nonstarch polysaccharide digestibility and true metabolizable energy in poultry. Poult. Sci. 73:156–162.

Stoewsand, G.S. 1995. Bioactive organosulfur phytochemicals in *Brassica oleracea* vegetables—A review. Fd. Chem. Toxic. 33:537–543.

Throckmorton, J.C., P.R. Cheeke, N.M. Patton, G.H. Arscott, and G.D. Jolliff. 1981. Evaluation of meadowfoam (*Limnanthes alba*) meal as a feedstuff for broiler chicks and weanling rabbits. Can. J. Anim. Sci. 61:735–742.

Tookey, H.L., C.H. Van Etten, and M.E. Daxenbichler. 1980. Glucosinolates. pp. 103–142 in: L.E. Liener (Ed.). Toxic Constituents of Plant Foodstuffs, 2nd Edition. Academic Press, NY.

Van Etten, C.H., M.E. Daxenbichler, W. Schroeder, L.H. Princen, and T.W. Perry. 1977. Tests for epiprogoitrin, derived nitriles and goitrin in body tissues from cattle fed crambe meal. Can. J. Anim. Sci. 57:75–80.

Waldron, K.W., I.T. Johnson, and G.R. Fenwick (Eds.). 1993. Food and Cancer Prevention: Chemical and Biological Aspects. Royal Soc. Chem., Cambridge, UK.

White, R.D., and P.R. Cheeke. 1983. Meadowfoam (*Limnanthes alba*) meal as a feedstuff for dairy goats and toxicologic activity of the milk. Can. J. Anim. Sci. 63:391–398.

Zhang, Y., T.W. Kensler, C.-G. Cho, G.H. Posner, and P. Talalay. 1994. Anticarcinogenic activities of sulforaphane and structurally related synthetic norbornyl isothiocyanates. Proc. Natl. Acad. Sci. 91:3147–3150.

Zhang, Y., P. Talalay, C.-G. Cho, and G.H. Posner. 1992. A major inducer of anticarcinogenic protective enzymes from broccoli: Isolation and elucidation of structure. Proc. Natl. Acad. Sci. 89:2399–2403.

Linseed

Batterham, E.S., L.M. Andersen, D.R. Baigent, and A.G. Green. 1991. Evaluation of meals from Linola™ low-linolenic acid linseed and conventional linseed as protein sources for growing pigs. Anim. Feed Sci. Tech. 35:181–190.

Batterham, E.S., L.M. Andersen, and A.G. Green. 1994. Pyridoxine supplementation of Linola^{TM} meal for growing pigs. Anim. Feed Sci. Tech. 50:167–174.

Cunnane, S.C., S. Ganguli, C. Menard, A.C. Liede, M.J. Hamadeh, Z.-Y. Chen, T.M.S. Wolever, and D.J.A. Jenkins. 1993. High α-linolenic acid flaxseed (*Linum usitatissimum*): some nutritional properties in humans. Brit. J. Nutr. 69:443–453.

Cunnane, S.C., P.A. Stitt, S. Ganguli, and J.K. Armstrong. 1990. Raised omega-3 fatty acid levels in pigs fed flax. Can. J. Anim. Sci. 70:251–254.

Gutzwiller, A. 1993. The effect of a diet containing cyanogenic glycosides on the selenium status and the thyroid function of sheep. Anim. Prod. 57:415–419.

Klosterman, H.J. 1974. Vitamin B6 antagonists of natural origin. J. Agric. Food Chem. 22:13–19.

Oomah, B.D., G. Mazza, and E.O. Kenaschuk. 1992. Cyanogenic compounds in flaxseed. J. Agric. Food Chem. 40:1346–1348.

Palmer, I.S., O.E. Olson, A.W. Halverson, R. Miller, and C. Smith. 1980. Isolation of factors in linseed oil meal protective against chronic selenosis in rats. J. Nutr. 110:145–150.

Field Beans

Grant, G., L.J. More, N.H. McKenzie, J.C. Stewart, and A. Pusztai. 1983. A survey of the nutritional and haemagglutination properties of legume seeds generally available in the UK. Brit. J. Nutr. 50:207–214.

Hintz, H.F., and D.E. Hogue. 1964. Kidney beans (*Phaseolus vulgaris*) and the effectiveness of vitamin E for the prevention of nutritional muscular dystrophy in the chick. J. Nutr. 84:283–287.

Liener, I.E. 1980. Miscellaneous toxic factors. pp. 429–467 in: I.E. Liener (Ed.). Toxic Constituents of Plant Foodstuffs, 2nd Edition. Academic Press, NY.

Mogridge, J.L., T.K. Smith, and M.G. Sousadias. 1996. Effect of feeding raw soybeans on polyamine metabolism in chicks and the therapeutic effect of exogenous putrescine. J. Anim. Sci. 74:1897–1904.

Montgomery, R.D. 1980. Cyanogens. pp. 143–160 in: I.E. Liener (Ed.). Toxic Constituents of Plant Foodstuffs, 2nd Edition. Academic Press, NY.

Myer, R.O., and J.A. Froseth. 1983. Heat-processed small red beans (*Phaseolus vulgaris*) in diets for young pigs. J. Anim. Sci. 56:1088–1096.

Myer, R.O., J.A. Froseth, and C.N. Coon. 1982. Protein utilization and toxic effects of raw beans (*Phaseolus vulgaris*) for pigs. J. Anim. Sci. 55:1087–1098.

Paduano, D.C., R.M. Dixon, J.A. Domingo, and J.H.G. Holmes. 1995. Lupin (*Lupinus angustifolius*), cowpea (*Vigna unguiculata*) and navy bean (*Phaseolus vulgaris*) seeds as supplements for sheep fed low quality roughage. Anim. Feed Sci. Tech. 53:55–69.

Pusztai, A., E.M.W. Clarke, and T.P. King. 1979. The nutritional toxicity of *Phaseolus vulgaris* lectins. Proc. Nutr. Soc. 38:115–120.

Reddy, N.R., M.D. Pierson, S.K. Sathe, and D.K. Salunkhe. 1985. Dry bean tannins: A review of nutritional implications. J. Am. Oil Chem. Soc. 62:541–549.

Sharon, N., and H. Lis. 1990. Legume lectins—A large family of homologous proteins. FASEB J. 4:3198–3208.

Van der Poel, A.F.B. 1990. Effect of processing on antinutritional factors and protein nutritional value of dry beans (*Phaseolus vulgaris* L.). A review. Anim. Feed Sci. Tech. 29:179–208.

Williams, P.E.V., A.J. Pusztai, A. MacDearmid, and G.M. Innes. 1984. The use of kidney beans (*Phaseolus vulgaris*) as protein supplements in diets for young rapidly growing beef steers. Anim. Feed Sci. Tech. 12:1–10.

Wilson, A.B., T.P. King, E.M.W. Clarke, and A. Pusztai. 1980. Kidney bean (*Phaseolus vulgaris*) lectin induced lesions in the rat small intestine. 2. Microbiological studies. J. Comp. Pathol. 90:597–602.

Fava Beans and Favism

Campbell, L.D., G. Olaboro, R.R. Marquardt, and D. Waddell. 1980. Use of fababeans in diets for laying hens. Can. J. Anim. Sci. 60:395–405.

Crosby, D.G. 1969. Natural toxic background in the food of man and his animals. J. Agric. Food Chem. 17:532–538.

Garrido, A., A. Gomez-Cabrera, J.E. Guerrero, and R.R. Marquardt. 1991a. Chemical composition and digestibility in vitro of *Vicia faba* L. cultivars varying in tannin content. Anim. Feed Sci. Tech. 35:205–211.

Garrido, A., A. Gomez-Cabrera, J.E. Guerrero, and J.M. van der Meer. 1991b. Effects of treatment with polyvinylpyrrolidone and polyethylene glycol on Faba bean tannins. Anim. Feed Sci. Technol. 35:199–203.

Ingalls, J.R., H.R. Sharma, T. Devlin, F.B. Bareeba, and K.W. Clark. 1979. Evaluation of whole plant fababean forage in ruminant rations. Can. J. Anim. Sci. 59:291–301.

Mager, J., M. Chevion, and G. Glaser. 1980. Favism. pp. 266–294 in: I.E. Liener (Ed.). Toxic Constituents of Plant Foodstuffs, 2nd Edition. Academic Press, NY.

Marcus, J.R., and G. Cohen. 1967. The riddle of the dangerous bean. Harper's Monthly Mag. 234(1405):98–102.

Marquardt, R.R. 1989. Vicine, convicine and their aglycones divicine and isouramil. pp. 161–200 in: P.R. Cheeke (Ed.). Toxicants of Plant Origin. Vol. II. CRC Press, Boca Raton, FL.

Marquardt, R.R., L.D. Campbell, S.C. Strothers, and J.A. McKirdy. 1974. Growth response of chicks and rats fed diets containing four cultivars of raw or autoclaved fababeans. Can. J. Anim. Sci. 54:177–182.

Marquardt, R.R., A.T. Ward, L.D. Campbell, and P.E. Cansfield. 1977. Purification, identification and characterization of a growth inhibitor in fababeans (*Vicia faba* L. var. *minor*). J. Nutr. 107:1313–1324.

McKnight, D.R., and G.K. MacLeod. 1977. Value of whole plant fababean silage as the sole forage for lactating cows. Can. J. Anim. Sci. 57:601–603.

Muduuli, D.S., R.R. Marquardt, and W. Guenter. 1981. Effect of dietary vicine on the productive performance of laying chickens. Can. J. Anim. Sci. 61:757–764.

Rotruck, J.T., A.L. Pope, M. E. Ganther, A.B. Swanson, D.G. Hafeman, and W. G. Hoekstra. 1973. Selenium: Biochemical role as a component of glutathione peroxidase. Science 179:588–590.

Thorlacius, S.O., and S.E. Beacom. 1981. Feeding value for lambs of fababean, field pea, corn and oat silages. Can. J. Anim. Sci. 61:663–668.

Peas and Lentils

Leterme, P., Y. Beckers, and A. Thewis. 1990. Trypsin inhibitors in peas: Varietal effect and influence on digestibility of crude protein by growing pigs. Anim. Feed Sci. Tech. 29:45–55.

Johns, D.C. 1987. Influence of trypsin inhibitors in four varieties of peas (*Pisum sativum*) on the growth of chickens. N. Z. J. Agric. Res. 30:169–175.

Grain Lupins

Allen, J.G. 1981. An evaluation of lupinosis in cattle in Western Australia. Aust. Vet. J. 57:212–215.

Allen, J.G., and G.R. Hancock. 1989. Evidence that phomopsins A and B are not the only toxic metabolites produced by *Phomopsis leptostromiformis*. J. Appl. Toxicol. 9:83–89.

Allen, J.G., and H.G. Masters. 1980. Prevention of ovine lupinosis by the oral administration of zinc sulfate and the effect of such therapy on liver and pancreas zinc and liver copper. Aus. Vet. J. 56:168–171.

Allen, J.G., H.G. Masters, and S.R. Wallace. 1979. The effect of lupinosis on liver copper, selenium and zinc concentrations in Merino sheep. Vet. Rec. 105:434–436.

Allen, J.G., and A.G. Randall. 1993. The clinical biochemistry of experimentally produced lupinosis in the sheep. Aust. Vet. J. 70:283–288.

Allen, J.G., P. Steele, H.G. Masters, and W.J. Lambe. 1992. A lupinosis-associated myopathy in sheep and the effectiveness of treatments to prevent it. Aust. Vet. J. 69:75–81.

Barnes, A.L., K.P. Croker, J.G. Allen, and N.D. Costa. 1996. Lupinosis of ewes around the time of mating reduces reproductive performance. Aust. J. Agric. Res. 47:1305–1314.

Batterham, E.S., L.M. Andersen, B.V. Burnham, and G.A. Taylor. 1986. Effect of heat on the nutritional value of lupin (*Lupinus angustifolius*) seed. Brit. J. Nutr. 55:169–177.

Brenes, A., R.R. Marquardt, W. Guenter, and B.A. Rotter. 1993. Effect of enzyme supplementation on the nutritional value of raw, autoclaved, and dehulled lupins (*Lupinus albus*) in chicken diets. Poult. Sci. 72:2281–2293.

Cheeke, P.R., and J.D. Kelly. 1989. Metabolism, toxicity and nutritional implications of quinolizidine (lupin) alkaloids. pp. 189–201 in: J. Huisman, T.F.B. van der Poel, and I.E. Liener (Eds.). Recent Advances of Research in Antinutritional Factors in Legume Seeds. Pudoc Wageningen.

DuPont, M.S., M. Muzquiz, I. Estrella, G.R. Fenwick, and K.R. Price. 1994. Relationship between the sensory properties of lupin seed with alkaloid and tannin content. J. Sci. Food Agric. 65:95–100.

Godfrey, N.W., A.R. Mercy, Y. Emms and H.G. Payne. 1985. Tolerance of growing pigs to lupin alkaloids. Aust. J. Exp. Agric. 25:791–795.

Hung, T.V., P.D. Handson, V.C. Amenta, W.S.A. Kyle, and R.S.-T. Yu. 1987. Content and distribution of manganese in lupin seed grown in Victoria and in lupin flour, spray-dried powder and protein isolate prepared from the seeds. J. Sci. Food Agri. 41:131–139.

Mackie, J.T., R.S. Rahaley and R. Bennett. 1992. Lupinosis in yearling cattle. Aust. Vet. J. 69:172–174.

Muzquiz, M., C.L. Ridout, K.R. Price, and G.R. Fenwick. 1993. The saponin content and composition of sweet and bitter lupin seed. J. Sci. Food Agric. 63:47–52.

Oram, R.N., D.J. David, A.G. Green, and B.J. Read. 1979. Selection in *Lupinus albus* L. for lower seed manganese concentration. Aust. J. Agric. Res. 30:467–476.

Peterson, J.E., M.V. Jago, A.L. Payne, and P.L. Stewart. 1987. The toxicity of phomopsin for sheep. Aust. Vet. J. 64:293–298.

Petterson, D.S., Z.L. Ellis, D.J. Harris, and Z.E. Spadek. 1987. Acute toxicity of the major alkaloids of cultivated *Lupinus angustifolius* seed to rats. J. App. Toxicol. 7:51–53.

Petterson, D.S., B.N. Greirson, D.G. Allen, D.J. Harris, B.M. Power, L.J. Dusci, and K.F. Ilett. 1994. Disposition of lupanine and 13-hydroxylupanine in man. Xenobiotica 24:933–941.

Ruiz, R.G., K.R. Price, M.E. Rose, A.E. Arthur, D.S. Petterson, and G.R. Fenwick. 1995. The effect of cultivar and environment on saponin content of Australian sweet lupin seed. J. Sci. Food Agric. 69:347–351.

Van Rensburg, I.B.J., W.F.O. Marasas, and T.S. Kellerman. 1975. Experimental *Phomopsis leptostromiformis* mycotoxicosis of pigs. J.S. Afr. Vet. Med. Assoc. 46:197–204.

White, C.L., D.G. Masters, D.I. Paynter, J. McC. Howell, S.P. Roe, M.J. Barnes, and J.G. Allen. 1994. The effects of supplementary copper and a mineral mix on the development of lupinosis in sheep. Aust. J. Agric. Res. 45:279–291.

Lathyrus spp. and Lathyrism

Aletor, V.A., A.A. El-Moneim, and A.V. Goodchild. 1994. Evaluation of the seeds of selected lines of three *Lathyrus* spp. for β-*N*-oxalylamino-L-alanine (BOAA), tannins, trypsin inhibitor activity and certain in-vitro characteristics. J. Sci. Food Agric. 65:143–151.

Arscott, G.H., and J.A. Harper. 1963. Relationship of 2,5-diamino-4,6-diketopyrimidine, 2,4-diaminobutyric acid and a crude preparation of β-cyano-L-alanine to the toxicity of common and hairy vetch seed fed to chicks. J. Nutr. 80:251–254.

Castell, A.G., R.L. Cliplef, C.J. Briggs, C.G. Campbell, and J.E. Bruni. 1994. Evaluation of lathyrus (*Lathyrus sativus* L.) as an ingredient in pig starter and grower diets. Can. J. Anim. Sci. 74:529–539.

Deshpande, S.S., and C.G. Campbell. 1992. Genotype variation in BOAA, condensed tannins, phenolics and enzyme inhibitors of grass pea (*Lathyrus sativus*). Can. J. Plant Sci. 72:1037–1047.

Dunham, W.B., C.S. Tsao, R. Barth, and Z.S. Herman. 1995. Protection by dietary ascorbate of guinea pigs from neurolathyrism. Nutrition Res. 15:993–1004.

Eason, P.J., R.J. Johnson, and G.H. Castleman. 1990. The effects of dietary inclusion of narbon beans (*Vicia narbonensis*) on the growth of broiler chickens. Aust. J. Agric. Res. 41:565–571.

Enneking, D. 1995. The toxicity of *Vicia* species and their utilisation as grain legumes. Centre for Legumes in Mediterranean Agriculture (CLIMA) Occasional Publication No. 6, University of Western Australia, Nedlands W.A.

Farran, M.T., M.G. Uwayjan, A.M.A. Miski, F.T. Sleiman, F.A. Adada, V.M. Ashkarian and O.P. Thomas. 1995. Effect of feeding raw and treated common vetch seed (*Vicia sativa*) on the performance and egg quality parameters of laying hens. Poult. Sci. 74:1630–1635.

Harper, J.A., and G.H. Arscott. 1962. Toxicity of common and hairy vetch seed for poults and chicks. Poult. Sci. 41:1968–1974.

Kisby, G.E., S.M. Ross, P.S. Spencer, B.G. Gold, P.B. Nunn, and D.N. Roy. 1992. Cycasin and BMAA: Candidate neurotoxins for western Pacific amyotrophic lateral sclerosis/parkinsonism–dementia complex. Neurodegeneration 1:73–82.

Kuo, Y.-H., H.-M. Bau, B. Quemener, J.K. Khan, and F. Lambein. 1995. Solid-state fermentation of *Lathyrus sativus* seeds using *Aspergillus oryzae* and *Rhizopus oligosporus* sp T-3 to eliminate the neurotoxin β-ODAP without loss of nutritional value. J. Sci. Food Agric. 69:81–89.

Monmaney, T. 1990. This obscure malady. The New Yorker. Oct. 29, 1990. pp. 85–113.

Padmanaban, G. 1980. Lathyrogens. pp. 239–263 in: I.E. Liener (Ed.). Toxic Constituents of Plant Foodstuffs, 2nd Edition. Academic Press, NY.

Parker, A.J., T. Mehta, N.S. Zarghami, P.K. Cusick, and B.E. Haskell. 1979. Acute toxicity of the *Lathyrus sativus* neurotoxin, L-3-oxalylamino-2-aminopropionic acid, in the squirrel monkey. Toxicol. Appl. Pharmacol. 47:135–143.

Raharjo, Y.C., P.R. Cheeke, and G.H. Arscott. 1988. Effects of dietary butylated hydroxyanisole and cysteine on toxicity of *Lathyrus odoratus* to broiler and Japanese quail chicks. Poult. Sci. 67:153–155.

Ross, S.M., D.N. Roy, and P.S. Spencer. 1989. β-*N*-Oxalylamino-L-alanine action on glutamate receptors. J. Neurochem. 53:710–715.

Rotter, R.G., R.R. Marquardt, and C.G. Campbell. 1991. The nutritional value of low lathyrogenic lathyrus (*Lathyrus sativus*) for growing chicks. Brit. Poult. Sci. 32:1055–1067.

Rotter, R.G., R.R. Marquardt, R.K.-C. Low, and C.J. Briggs. 1990. Influence of autoclaving on the effects of *Lathyrus sativus* fed to chicks. Can. J. Anim. Sci. 70:739–741.

Sacks, O.W. 1993. Guam ALS–PDC: Possible causes. Science 262:826.

Spencer, P.S., C.N. Allen, G.E. Kisby, A.C. Ludolph, S.M. Ross, and D.N. Roy. 1991. Lathyrism and western Pacific amyotrophic lateral sclerosis: Etiology of short and long latency motor system disorders. Advances Neurology 56:287–299.

Spencer, P.S., G.E. Kisby, S.M. Ross, D.N. Roy, J. Hugon, A.C. Ludolph, and P.B. Nunn. 1993. Guam ALS-PDC: Possible causes. Science 262:825–826.

Spencer, P.S., P.B. Nunn, J.Hugon, A.C. Ludolph, S.M. Ross, D.N. Roy, and R.C. Robertson. 1987. Guam amyotrophic lateral sclerosis-parkinsonism–dementia linked to a plant excitant neurotoxin. Science 237:517–522.

Stone, R. 1993. Guam: Deadly disease dying out. Science 261:424–426.

Tate, M.E., and D. Enneking. 1992. A mess of red pottage. Nature 359:357–358.

Willis, C.L., B.S. Meldrum, P.B. Nunn, B.H. Anderton, and P.N. Leigh. 1994. Neuroprotective effect of free radical scavengers on β-N-oxalylamino-L-alanine (BOAA)-induced neuronal damage in rat hippocampus. Neuroscience Letters 182:159–162.

Other Toxins in Legume Seeds

Aylward, J.H., R.D. Court, K.P. Haydock, R.W. Strickland, and M.P. Hegarty. 1987. *Indigofera* species with agronomic potential in the tropics. Rat toxicity studies. Aust. J. Agric. Res. 38:177–186.

Belmar, R., and T.R. Morris. 1994a. Effects of the inclusion of treated jack beans (*Canavalia ensiformis*) and the amino acid canavanine in chick diets. J. Agric. Sci. 123:393–405.

Belmar, R., and T.R. Morris. 1994b. Effects of raw and treated jack beans (*Canavalia ensiformis*) and of canavanine on the short-term feed intake of chicks and pigs. J. Agric. Sci. 123:407–414.

D'Mello, J.P.F. 1991. Toxic amino acids. pp. 21–48 in: J.P.F. D'Mello, C.M. Duffus, and J.H. Duffus (Eds.). Toxic Substances in Crop Plants. The Royal Society of Chemistry, Cambridge.

D'Mello, J.P.F., and A.G. Walker. 1991. Detoxification of jack beans (*Canavalia ensiformis*): studies with young chicks. Anim. Feed Sci. Tech. 33:117–127.

Enneking, D., L.C. Giles, and M.E. Tate. 1993. L-Canavanine; A natural feed-intake inhibitor for pigs (isolation, identification and significance). J. Sci. Food Agric. 61:315–325.

Hegarty, M.P., W.R. Kelly, D. McEwan, O.J. Williams, and R. Cameron. 1988. Hepatotoxicity to dogs of horse meat contaminated with indospicine. Aust. Vet. J. 65:337–340.

Hopkins, J. 1995. The very intolerant peanut. Fd. Chem. Toxic. 33:81–86.

Pimentel, J.L., and M.E. Cook. 1988. Improved growth in the progeny of hens immunized with jackbean urease. Poult. Sci. 67:434–439.

Roberts, J.L., and J.A. Hayashi. 1983. Exacerbation of SLE associated with alfalfa ingestion. New Engl. J. Med. 308:1361.

Rosenthal, G.A. 1991. The biochemical basis for the deleterious effects of L-canavanine. Phytochem. 30:1055–1058.

Shqueir, A.A., D.L. Brown, and K.C. Klasing. 1989. Canavanine content and toxicity of Sesbania leaf meal for growing chicks. Anim. Feed Sci. Tech. 25:137–147.

CHAPTER 8

Toxins Produced by Feed Processing

One of the main objectives of feed processing is to destroy or inactivate toxins. **Heat treatment**, by extrusion, pelleting, popping, steam-rolling and roasting inactivates many toxins including the trypsin inhibitors in raw soybeans and lectins in beans. **Ammoniation** or

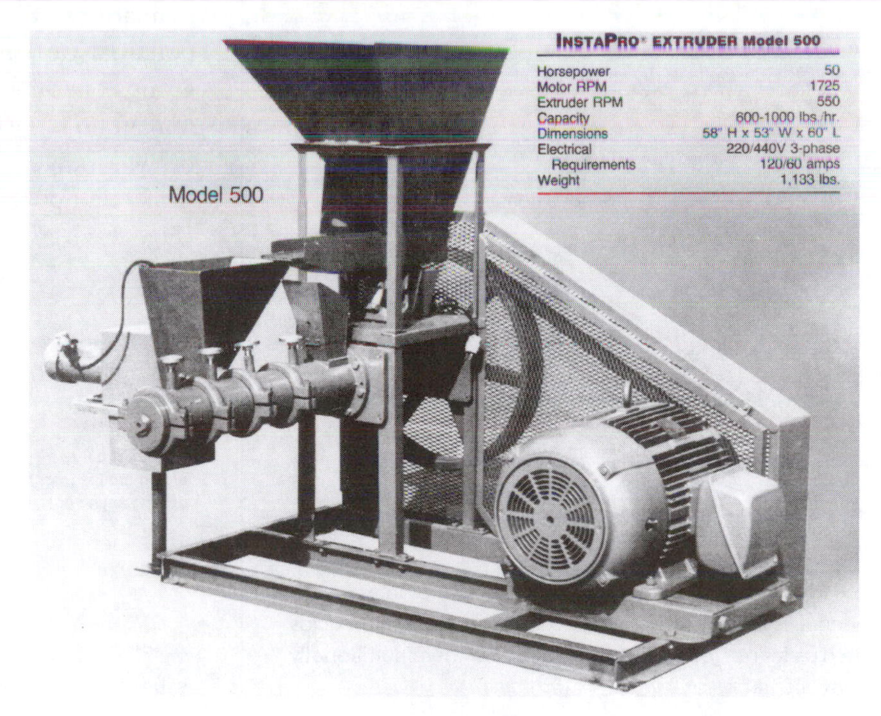

FIGURE 8–1 An extruder, used to process soybeans and other feedstuffs which require heat treatment to destroy inhibitors. The feed is extruded through a die, with heat resulting from the friction involved. (Courtesy of Triple "F," Inc.)

other alkali treatment is used to improve digestibility of low quality roughages, and to detoxify aflatoxins in feeds. However, under certain circumstances, these procedures can cause the formation of toxins.

BROWNING REACTIONS

Proteins may undergo chemical reactions in which brown, indigestible polymers are produced. There are two general types: enzymatic and non-enzymatic browning. **Enzymatic browning** is promoted by the enzyme polyphenol oxidase in plant tissue. **Polyphenol oxidase** catalyzes the conversion of phenolic acids such as caffeic acid and chlorogenic acid to quinones. The quinones react non-enzymatically to polymerize or form covalent bonds with amino, thiol and methylene groups. The epsilon amino of lysine and the thioether group of methionine can covalently bond with quinones to render these amino acids nutritionally unavailable, as shown in Fig. 8–2. Enzymatic browning occurs in fruit, causing it to turn brown when mechanically damaged and exposed to oxygen (e.g. browning of apples when peeled or cut). Various food processing methods are used to prevent this type of browning, including heating to inactivate polyphenol oxidase, and addition of sulfites to foods (now being discontinued because of potential for asthmatic crisis in susceptible individuals).

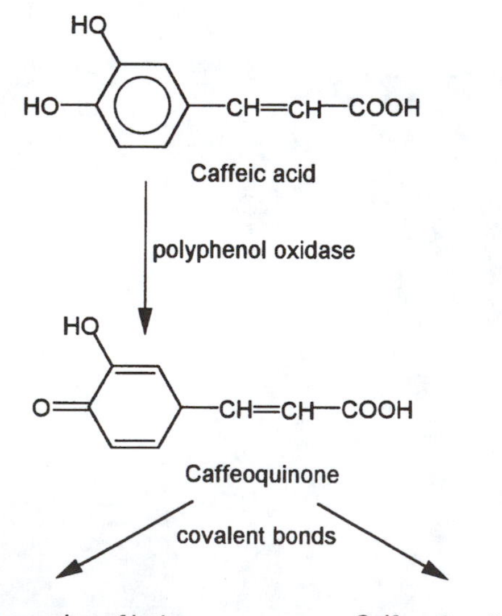

FIGURE 8–2 Polyphenol oxidases in plants convert phenolics to quinones, which can react with amino acids such as lysine and methionine to render them nutritionally unavailable.

Red clover (*Trifolium pratense*), a common and productive forage crop, quickly turns brown when harvested. Jones *et al.* (1995) demonstrated that red clover has polyphenol oxidase and soluble phenols. No other common legume forage has polyphenol oxidase activity. The browning of red clover is accompanied by a loss of proteolysis activity, probably because of the binding of oxidized phenols (quinones) to proteolytic enzymes.

Non-enzymatic browning (Maillard reaction) is more important in animal feeds. In the Maillard reaction, reducing sugars (those with a potential carbonyl group) react with functional groups of amino acids, and particularly the epsilon amino of lysine. Further reactions producing brown, indigestible polymers occur. This results in a "tie-up" of lysine in an unavailable form (Fig. 8–3), thus reducing the protein quality.

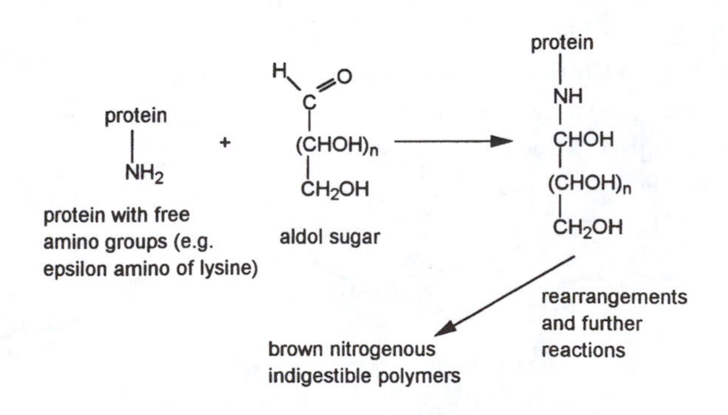

FIGURE 8–3 The browning or Maillard reaction results in the "tie-up" of lysine in an unavailable form.

The Maillard reaction is stimulated by heat and moisture. Whenever proteins are heated, such as when soybean meal is heated to inactivate trypsin inhibitors, browning and **lysine tie-up** will occur. Thus it is important that heat processing be carefully controlled to ensure detoxification but minimize amino acid tie-up. Methionine and tryptophan can also be rendered unavailable by Maillard reactions. Plant protein sources such as soybean meal and cottonseed meal that contain oligosaccharides, such as stachyose and raffinose, or with free glucose, are most susceptible to browning. Besides the reduction in protein quality, some of the products may be toxic, causing growth inhibition and possible mutagenic effects (Friedman, 1994, 1996).

Non-enzymatic browning may also have useful implications in animal production. Cleale *et al.* (1987) used controlled browning of soybean meal with xylose (as a reducing sugar) to increase the by-pass protein value of soybean meal in ruminants. Nakamura *et al.* (1992) evaluated xylose-browned soybean meal as a protein source for lactating dairy cattle, and found that the protein from the non-enzymatically browned soybean meal supported the same level of milk production at half the amount of supplemental soybean meal.

Foods and feeds are sometimes treated with alkali. Potatoes and fruits may be industrially peeled using strong alkali solutions, and the by-products containing the alkali are often used in animal feeds. **Alkali treatment** of proteins can result in the formation of crosslinked amino acids. For example, **lysinoalanine (LAL)** is formed by the reaction of the epsilon amino of lysine with cysteine or serine. LAL formation results in a reduction in available lysine, and a decrease in digestibility of the modified protein in the intestine (Friedman, 1992) and in the rumen (Nishino *et al.*, 1995). There is evidence that LAL is toxic, causing kidney damage in rats. Friedman (1992) suggests that this effect involves the binding of copper by LAL, within the epithelial cells of the proximal tubules of the kidney. According to Friedman (1994): "These observations raise concern about the nutritional quality and safety of alkali-treated proteins".

AMMONIATED FEED TOXICOSIS

Ammoniation of low quality roughages such as cereal straw and corn stover is an economical way of improving digestibility and providing rumen-fermentable nitrogen. Straw stacks are covered with black plastic, and anhydrous ammonia introduced into the straw at about 3% of the straw dry weight. After several weeks of exposure to ammonia, lignin in the cell wall material is degraded, increasing fiber digestibility. Ammonia is a mold inhibitor and is sometimes used as a preservative for damp hay. While ammoniation is useful in increasing the nutritive value of low quality roughages such as straw, it can produce toxins under certain circumstances.

Ammoniated molasses was introduced commercially in the U.S. in the 1950's as a nitrogen–energy supplement. It was withdrawn from the market because it produced a hyperexcitability problem in cattle. In the 1980's, ammoniation of straw became common, and bovine hyperexcitability was again observed. It has been variously called bovine hysteria, bovine bonkers, bovine hyperexcitability, crazy cow syndrome and **ammoniated hay toxicosis**. Affected animals develop neurological signs, such as hyperactivity, incoordination, tremors and convulsions. They may suddenly gallop in circles and run into fences, gates and other objects, often injuring themselves (Kerr *et al.*, 1987; Perdok and Leng, 1987; Weiss *et al.*, 1986). The causative agent, formed by the reaction between ammonia and reducing sugars (e.g. glucose) in the presence of heat, is **4-methyl imidazole**:

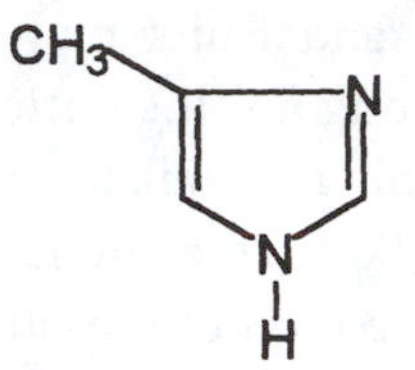

Metabolically, 4-methyl imidazole inhibits cytochrome P450 by binding with the heme iron (Karangwa *et al.*, 1990). It has a relatively long half-life in the tissues (9 hours) for a small molecule, which probably reflects binding to macromolecules (Karangwa *et al.*, 1990). Imidazole formation does not occur if the temperature of ammoniated roughage remains below 70° C during ammoniation (Perdok and Leng, 1987). Because of the involvement of reducing sugars, ammoniation of good quality roughages with abundant reserves of fermentable carbohydrate (hence sugars) should be avoided. There is evidence that the toxin is transferred in the milk, so caution should be used when feeding ammoniated roughages to dairy cattle (Perdok and Leng, 1987). This would most likely occur in tropical developing countries, where low quality roughages such as rice straw are fed to dairy animals. Brazil *et al.* (1994) reported an outbreak of ammoniated forage toxicosis in beef calves nursing cows fed on ammoniated green (immature) barley. Calves are apparently more susceptible than older animals because the cows showed no clinical signs. The barley had been baled moist, and had started to spoil. The large round bales were covered with plastic and anhydrous ammonia introduced as a mold inhibitor. The barley was drought-stressed, so probably had high levels of soluble sugars.

Ammoniated forage toxicosis can be avoided by proper selection of forages to be ammoniated. They should be low quality (e.g. straw), with a low content of reducing sugars. Failed cereal crops should not be ammoniated because of the high sugar levels in the seed heads. Ammoniation should be performed when environmental temperatures are low, to minimize imidazole formation.

BOVINE SPONGIFORM ENCEPHALOPATHY

In the 1980's, a new cattle disease (mad cow disease, raging cow disease) appeared in Great Britain, with potentially drastic implications for the beef industry and consumers. Affected animals develop various neurological symptoms, including exaggerated limb movements, muscular jerking, anxiety and frenzied movements, culminating in death. The brain shows extensive damage, with degeneration of neurons, hypertrophy of connective tissue and a sponge-like appearance (spongiosus). The condition is called **bovine spongiform encephalopathy (BSE)**, and is apparently one of a series of diseases in humans and animals known as transmissible spongiform encephalopathy (TSE). Although the etiology is not certain, it is generally believed by scientists investigating it that BSE was introduced into British cattle by the feeding of meat meal prepared from sheep infected with scrapie, a TSE. **Scrapie** causes sheep to act in a frenzied manner, with signs similar to those of BSE. Scrapie has been recognized in sheep for many years. It has a long incubation period of several years, and appears to be transmitted to the offspring of affected animals. When scrapie is identified in a flock of sheep, the entire flock is slaughtered. The disease also resembles some human conditions such as **Creutzfeldt–Jakob disease (CJD)** and **Kuru**. The CJD disease occurs in the general human population at a low incidence, and causes fatal brain degeneration. Kuru is a disease of a stone-age tribe in the remote highlands of New Guinea. It is believed that the mode of transmission was cannibalism, including eating the brain in funeral rituals. The disease is now rare, because of educational programs to eliminate cannibalism. An outbreak of CJD occurred in France in the early 1990's. It occurred in patients suffering from congenital dwarfism, who were given human growth hormone extracted from the pituitary glands of cadavers. Apparently one or more of the pituitary glands from which growth hormone was extracted was infected with CJD (Aldhous, 1992).

The BSE disease has been thoroughly reviewed by Dealler and Lacey (1990). Among the major concerns are a long incubation period of several years so it is difficult to identify affected animals, and the extreme resistance of the causative agent to destruction. Disinfectants, ionizing radiation and cooking do not destroy infectivity. The causative agent is not conclusively known. Two candidates are that it is a virino or a prion. A **virino** is a small DNA fragment associated with a protein coat, while a **prion** is a protein containing no nucleic acid. The existence of prions is controversial and not accepted by many authorities. The term prion was coined by Prusiner (1982). Wills (1991) has provided an extensive review of the prion theory explanation of BSE.

The origin of the BSE outbreak in Great Britain is believed to have been a change in 1981–82 in the processing of offal. The change involved substitution of a lower temperature continuous flow process without solvent extraction of fats for the previously used high temperature batch process with solvent extraction of fats (Wilesmith, 1994). Apparently the older process effectively destroyed the causative agent.

The disease has occurred in various ruminant species (cattle, African antelope sp. in zoos) fed meat and bone meal. It has also occurred in domestic cats and big cats (mountain lion, cheetah) in zoos. Pigs, dogs and poultry have not contracted the condition, even when fed infective meat and bone meal (Wilesmith, 1994).

A ban on the use of ruminant-derived meat and bone meal in Britain, and vigilant monitoring of exposed herds, have brought the BSE condition under control (Wilesmith, 1994).[1]

[1]Apparent transmission of mad cow disease to humans may have occurred in Britain in 1996.

NITROSAMINES AND NITROSAMIDES

Nitrosamines and related compounds have been of concern as potential carcinogens in the human diet and have caused problems in livestock production as well. They are formed by the reaction of amines with nitrite and thus are likely to be formed when foods or feeds are preserved with nitrates or nitrites. Nitrates are readily reduced to nitrites. The major livestock poisoning problems have involved mink fed fish meal preserved with high levels of sodium nitrite. Massive liver cancer (hepatoma) occurred in mink fed the treated fish meal. Methylamines are abundant in fish and react with nitrate to form dimethylnitrosamine:

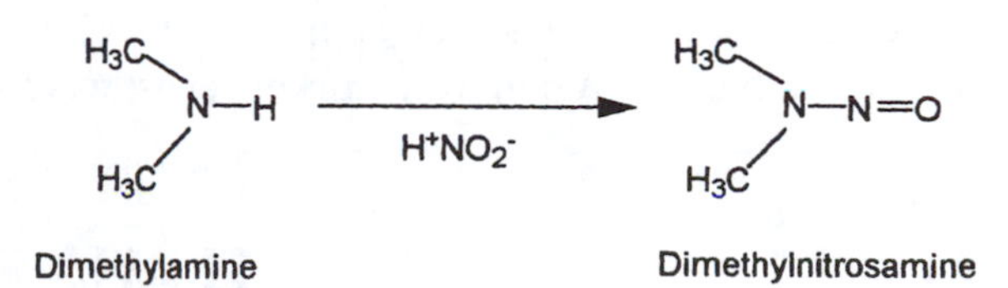

Nitrosamides, which are less stable than nitrosamines, are also powerful carcinogens and may be formed in foods. An example of nitrosamide formation from methylurea is the following:

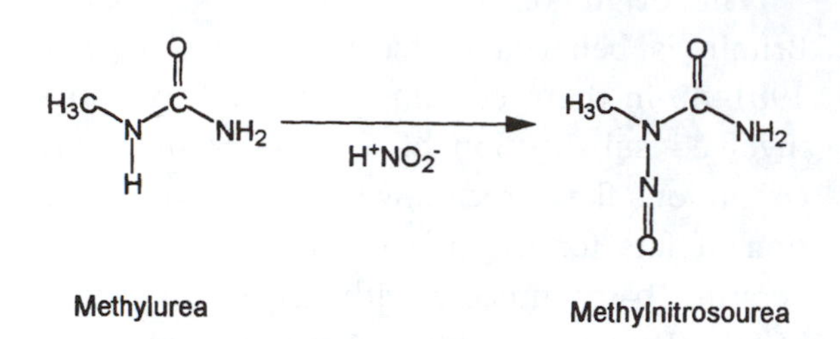

Nitrosamines are formed by nitrosation of secondary, tertiary, and some primary amines, and quaternary ammonium compounds. As these are widely distributed in foods, widespread occurrence of nitrosamines is possible. Nitrites and nitrates are widely used as preserving agents in the production of cured meats and fish. Nitrite is added to cured meats as a preservative, to form a pink color, and for beneficial effects on flavor. Considerable controversy surrounds its use, and efforts to reduce the employment of nitrates and nitrites in the food industry have been implemented. Nitrosamines have also been found in beer and cheese. In livestock feeds, fish meal preserved with nitrate or nitrite is the major source of nitrosamines. Losses from liver cancer of mink and other fur animals fed fish meal have occurred in Canada, the U.S, and Norway.

RAW EGG–INDUCED BIOTIN DEFICIENCY

Avidin is a glycoprotein with a molecular weight of about 43,500. It is secreted by the oviduct of birds into the egg white, and binds with the B vitamin biotin in a tight complex, resisting digestion and absorption. Biotin deficiency in many animals can be induced by feeding raw eggs, egg white, or avidin. Avidin is denatured by moist heat and is inactivated when eggs are cooked. Avidin-induced biotin deficiency has occurred in fur animals. Mink fed the viscera from turkey-processing plants developed an achromatrichial condition in which the pelts were light in color. This was called "turkey waste graying." Eventually the condition was traced to the presence of eggs from cull hen turkeys in the turkey viscera (Stout and Adair, 1969). The avidin in the eggs created a biotin deficiency that could be overcome by either cooking the turkey waste or supplementing the diet with biotin. When raw turkey eggs were fed, severe biotin deficiency characterized by fur graying, "spectacle eye", loss of fur, exudates from the eyes, nose, and mouth, and encrustation of paws was observed. Spray-dried eggs retained sufficient avidin to cause biotin deficiency in mink (Wehr *et al.*, 1980). Turkey eggs contain about four times as much avidin as hen eggs. Since avidin occurs in the egg white and biotin is in the yolk, a biotin deficiency is most readily induced when egg whites rather than whole eggs are fed.

An interesting case (Anon., 1963) of avidin-induced biotin deficiency in humans was that of an Italian laborer who had eaten about 6 dozen raw eggs weekly since childhood, with little other food but 1 to 4 quarts of wine daily! He experienced exfoliative dermatitis and conjunctivitis for many years; the symptoms disappeared after 2 weeks of hospitalization on a liberal diet and injections of biotin.

TRANS FATTY ACIDS

Unsaturated fatty acids of naturally occurring fats generally have a cis configuration of their double bonds. During partial hydrogenation of vegetable oil to produce margarine and partial hydrogenation in the rumen, trans isomers of unsaturated fatty acids may be formed. **Elaidic acid**, a trans monoene of oleic acid, has been extensively used in studies of the biological effects of trans fatty acids. The potential adverse effects of dietary **trans fatty acids** have been of some concern in human nutrition (Mensink *et al.*, 1992; Mensink and Katan, 1990). There is some evidence that they may behave more like saturated than unsaturated fatty acids in metabolism. They have been suggested as possible risk factors in atherosclerosis and cancer. Trans isomers may have hypercholesterolemic effects, and epidemiological studies have suggested a positive association between intake of trans fatty acids and cancer. The public health implications of these findings, if any, have not been resolved. There would appear to be little significance of these compounds to livestock, except that they are produced in the rumen, are found in ruminant body fat, and thus are found in food products of animal origin.

GIZZEROSINE

Gizzard erosion is a disorder that has been observed in poultry for many years. It is characterized by erosion or necrosis of the gizzard lining, often with ulceration into the muscular gizzard wall. The entire lining of the gizzard may have a dry, scaly, brownish color rather than the normal greenish yellow color. The upper part of the small intestine may be ulcerated as well. The condition is related to dietary ingredients, particularly fish meal or grains contaminated with mycotoxins.

The gizzard erosion factor in fish meal has been identified by Japanese scientists as a dipeptidelike compound composed of histamine and lysine (Masumuru *et al.*, 1985). This compound, called **gizzerosine**, is formed by a Maillard (browning) reaction when fish meal is heated. Gizzerosine induces an abnormally high secretion of hydrochloric acid in the proventriculus, apparently acting as an analog of histamine, which is the physiological regulator of gastric acid secretion. It is approximately ten times as potent as histamine in stimulating gastric acid secretion. Gizzerosine is metabolized much more slowly than histamine and has a higher affinity than histamine for gastric histamine receptors (Ito *et al.*, 1988). Sugahara *et al.* (1988) suggest a maximum concentration of 0.4 ppm gizzerosine in practical poultry diets. Higher levels cause retarded growth and mortality.

Mycotoxins also cause gizzard erosion. Some mycotoxins cause damage to the liver and bile duct, decreasing the normal secretion of bile. Bile is periodically regurgitated into the gizzard, neutralizing gastric acid. Absence of normal bile levels thus results in hyperacidity of the gizzard and mucosal damage. In addition, fish oils not adequately protected with antioxidants may cause gizzard erosion, probably due to peroxidation of membrane tissue. Vitamin E and synthetic antioxidants protect against this damage. Aflatoxin B_1 can potentiate the effects of gizzerosine (Diaz and Sugahara, 1995).

REFERENCES

Browning Reactions

Cleale, R.M., R.A. Britton, T.J. Klopfenstein, M.L. Bauer, D.L. Harmon, and L.D. Satterlee. 1987. Induced non-enzymatic browning of soybean meal. II. Ruminal escape and net portal absorption of soybean protein treated with xylose. J. Anim. Sci. 65:1319–1326.

Friedman, M. (Ed.). 1991. Nutritional and Toxicological Consequences of Food Processing. Plenum Press, New York.

Friedman, M. 1992. Dietary impact of food processing. Annu. Rev. Nutr. 12:119–137.

Friedman, M. 1994. Improvement in the safety of foods by SH-containing amino acids and peptides. A review. J. Agric. Food Chem. 42:3–20.

Friedman, M. 1996. Food browning and its prevention: An overview. J. Agric. Food Chem. 44:631–653.

Jones, B.A., R.D. Hatfield, and R.E. Muck. 1995. Screening legume forages for soluble phenols, polyphenol oxidase and extract browning. J. Sci. Food Agric. 67:109–112.

Nakamura, T., T.J. Klopfenstein, F.G. Owen, R.A. Britton, R.J. Grant, and T.S. Winowiski. 1992. Nonenzymatically browned soybean meal for lactating dairy cows. J. Dairy Sci. 75:3519–3523.

Nishino, N., S. Uchida, and M. Ohshima. 1995. Formation of lysinoalanine following alkaline processing of soya bean meal in relation to the degradability of protein in the rumen. J. Sci. Food Agric. 68:59–64.

Ammoniated Feed Toxicosis

Brazil, T.J., J.M. Naylor, and E.D. Janzen. 1994. Ammoniated forage toxicosis in nursing calves: A herd outbreak. Can. Vet. J. 35:45–47.

Karangwa, E., G.E. Mitchell, Jr., and R.E. Tucker. 1990. Pharmacokinetics of 4-methylimidazole in sheep. J. Anim. Sci. 68:3277–3284.

Kerr, L.A., A.W. Groce, and K.W. Kersting. 1987. Ammoniated forage toxicosis in calves. J. Am. Vet. Med. Assoc. 191:551–552.

Perdok, H.B., and R.A. Leng. 1987. Hyperexcitability in cattle fed ammoniated roughages. Anim. Feed Sci. Tech. 17:121–143.

Weiss, W.P., H.R. Conrad, C.M. Martin, R.F. Cross, and W.L. Shockey. 1986. Etiology of ammoniated hay toxicosis. J. Anim. Sci. 63:525–532.

Bovine Spongiform Encephalopathy

Aldhous, P. 1992. French officials panic over rare brain disease outbreak. Science 258:1571–1572.

Dealler, S.F., and R.W. Lacey. 1990. Transmissible spongiform encephalopathies: The threat of BSE to man. Food Microbiol. 7:253–279.

Prusiner, S.B. 1982. Novel proteinaceous infectious particles cause scrapie. Science 216:136–144.

Wilesmith, J.W. 1994. Bovine spongiform encephalopathy and related diseases: An epidemiological overview. N. Z. Vet. J. 42:1–8.

Wills, P.R. 1991. Prion diseases and the frame-shifting hypothesis. N. Z. Vet. J. 39:41–45.

Raw Egg–Induced Biotin Deficiency

Anon. 1963. Vitamin Manual. Upjohn Company, Kalamazoo, MI.

Stout, F.M., and J. Adair. 1969. Biotin defiency in mink fed poultry by-products. Am. Fur Breeder 42:10.

Wehr, N.B., J. Adair, and J.E. Oldfield. 1980. Biotin deficiency in mink fed spray-dried eggs. J. Anim. Sci. 50:877–885.

Trans Fatty Acids

Mensink, R.P., and M.B. Katan. 1990. Effect of dietary trans fatty acids on high-density and low-density lipoprotein cholesterol levels in healthy subjects. New Eng. J. Med. 323:439–445.

Mensink, R.P., P.L. Zock, M.B. Katan, and G. Hornstra. 1992. Effect of dietary cis and trans fatty acids on serum lipoprotein(a) levels in humans. J. Lipid Res. 33:1493–1501.

Gizzerosine

Diaz, G.J., and M. Sugahara. 1995. Individual and combined effects of aflatoxin and gizzerosine in broiler chickens. Brit. Poult. Sci. 36:729–736.

Ito, Y., H. Terao, T. Noguchi, and H. Naito. 1988. Gizzerosine raises the intracellular cyclic adenosine-3', 5'-monophosphate level in isolated chicken proventriculus. Poult. Sci. 67:1290–1294.

Masumura, T., M. Sugahara, T. Noguchi, K. Mori, and H. Naito. 1985. The effect of gizzerosine, a recently discovered compound in overheated fish meal, on the gastric acid secretion in the chicken. Poult. Sci. 64:356–361.

Sugahara, M., T. Hattori, and T. Nakajima. 1988. Effect of synthetic gizzerosine on growth, mortality and gizzard erosion in broiler chicks. Poult. Sci. 67:1580–1584.

PART III

Forage-Induced Toxicoses

CHAPTER 9

Toxicoses Associated with Forages in General

PASTURE (FROTHY) BLOAT

Bloat is a distension of the rumen as a result of the inability of the animal to eructate gases produced in the normal processes of rumen fermentation (Fig. 9–1). The quantity of gas produced varies according to the amount of fermentable substrate and types of microorganisms in the rumen, but in general 30–50 liters of rumen gas per hour are produced in cattle in the period of 3–4 hr after feeding, with a daily production of about 400 liters in cattle and 50 liters in sheep. The principal gases are carbon dioxide and methane. In pasture bloat, these gases are trapped in the form of a stable foam. The eructation mechanism is inhibited by the presence of foam at the base of the esophagus. This is a protective reflex, because eructation of foam would result in death from inhalation of foam into the lungs. Bloat-producing plants, primarily legumes, contain substances which cause the production of a stable foam in the

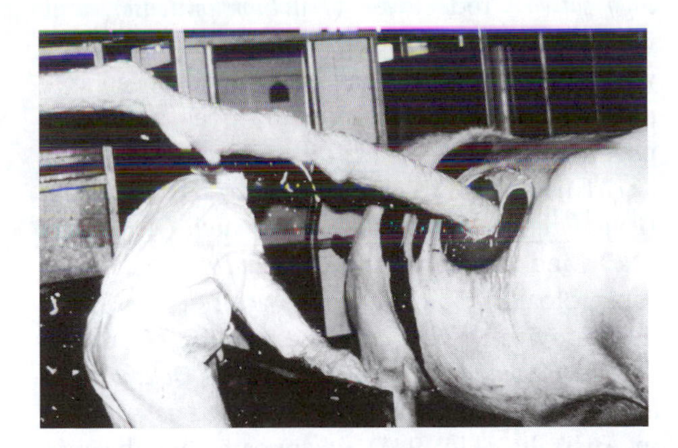

FIGURE 9–1 An experimentally bloated animal, showing the high intraruminal pressures that can develop with frothy (pasture) bloat. This fistulated animal was fed a bloat-inciting diet; when the cap of the rumen fistula was removed, the rumen contents were forced out by the high gas pressure in the bloated animal. (Courtesy of K.A. Beauchemin and K.-J. Cheng)

rumen. In addition, there are animal factors involved, as not all animals on a bloat-inciting pasture develop the condition. Salivary secretions may have an effect. The rumen microorganisms have been postulated to be involved in the bloat complex. Sheep are less susceptible than cattle to bloat, in part because of differences in rumen motility, rumen passage rate and clearance of rumen gases (Colvin and Backus, 1988). High pasture intakes following a period off-feed, such as cows following estrus, may result in increased bloat incidence (Phillips *et al.*, 1996).

FIGURE 9–2 Tropical legumes generally do not cause bloat in part because they often have tough fibrous leaves which reduce the rate of cell rupture. Here, cattle in Australia are grazing *Lablab purpureus*, which has occasionally caused bloat. (Courtesy of M.P. Hegarty)

Plant Factors Involved in Bloat

Bloat primarily occurs on legume pastures, but there are documented cases of the condition on lush grass pasture. In addition, it is a problem on wheat pastures (Branine and Galyean, 1990). Some legumes are well known for their bloat-producing potential, while others are not bloat-producers. The most important bloating species in temperate regions are alfalfa (*Medicago sativa*), red clover (*Trifolium pratense*), and white clover (*Trifolium repens*). Subterranean clover (*T. subterraneum*) has occasionally been implicated in bloat, but generally does not cause problems. Birdsfoot trefoil (*Lotus corniculatus*), sainfoin (*Onobrychis viciifolia*), lespedeza (*Lespedeza stipulacea*), cicer milk vetch (*Astragalus cicer*), and crown vetch (*Coronilla varia*) do not cause bloat. Tropical legumes are not bloat-producers, with the occasional exception of lab lab (*Lablab purpureus*, also referred to as *Dolichos lablab*), a tropical leguminous vine used as a forage in Australia (Fig. 9–2). It appears that the primary plant factors involved in bloat production are **cytoplasmic proteins** that occur in the bloat-producing forages. These have been referred to as fraction I, fraction II, and 18 S proteins. Howath *et al.* (1977) found a good relationship between bloat incidence and total soluble protein level. Other factors associated with the total soluble protein content in causing bloat include the rate at which cell rupture occurs in the rumen, the soluble carbohydrate level in the forage, the rumen pH, and the presence or absence of protein precipitants. Rapid release of cell contents, allowing a high fermentation rate (and thus a high rate of rumen gas production), favors development of bloat. The soluble proteins form the membrane of the gas bubbles, producing a stable foam. Chloroplast fragments may serve as nucleation sites for bubble formation. The formation of a stable foam is influenced by ruminal pH. According to Hall and Majak (1989), the optimal pH for maximum foam stability is between 5 and 6. Vigorous rumen fermentation enhances bloat, not only because of high gas production but also because high VFA production lowers rumen pH, increasing foam stability. The soluble carbohydrate content of the forage is important in determining the rate of fermentation and may also exert an influence on the **rumen protozoa**. In the presence of high forage soluble carbohydrate, protozoa numbers may increase markedly and then suddenly die. They may store excessive starch and then burst, or they may die because of changes in rumen environment such as pH. The protozoa cell contents may contribute to the formation of a stable foam. A major reason for the lack of bloat with some legumes such as birdsfoot trefoil and sainfoin appears to be their content of protein precipitants, such as tannins. By denaturing soluble proteins, tannins prevent foam formation. Plant-breeding efforts to incorporate tannins into bloat-producing forages such as alfalfa are in progress (Hall and Majak, 1989). Howarth *et al.* (1977) also suggest that selection for a reduced ratio of soluble to total protein might be effective.

Lush legume pastures, especially immature plant tops, are more likely to cause bloat than more mature forage or hay. This is because the immature foliage is high in soluble protein and carbohydrate and has a rapid rate of cell rupture in the rumen. Drying causes protein denaturation, thus reducing the level of soluble protein. Thus legume hay rarely causes bloat. Climatic factors can influence bloat. Warm days and cold nights lead to a high starch content of legume tissue, which may trigger bloat outbreaks. Alfalfa that has received a "killing frost" as low as -9.6^{o} C can still cause bloat (Hall and Majak, 1991). The lack of bloat with most tropical legumes may be due to the presence of tannins in some cases, and in others the tough fibrous nature of the leaves may retard the rate of cell rupture. Alfalfa has a greater bloat potential early in the morning compared to later in the day (Hall and Majak, 1995), presumably because of diurnal variations in plant protein, soluble carbohydrate and chlorophyll concentrations. The risk of bloat is lower when cattle are left on an alfalfa pasture continuously than when grazing is interrupted (Majak *et al.*, 1995).

Animals likely to bloat have a vigorous microbial population with a high fermentation potential so that when an abundance of forage with high fermentable carbohydrate is consumed, rapid release of cell contents results in rapid fermentation (thus high rumen gas production) and viscous rumen contents, aiding in the trapping of gases in stable foam (Majak *et al.*, 1983). The rumen bacteria colonize particulate matter such as chloroplast fragments. Majak *et al.* (1983) summarized the rumen conditions that lead to onset of bloat as the following: (1) colonization of chloroplast particles and other particulates by rumen bacteria; (2) accumulation of these particles in fluid of the dorsal sac of the rumen; (3) predisposition to frothiness by entrapment of gas bubbles among the suspended particles; (4) provision of an active inoculum for the rapid disintegration and release of mesophyll cell contents from ingested forage; (5) flotation of digesta because of microbial gas production; and (6) inability of the animal to clear fermentation gases because of entrapment by buoyant, frothy digesta.

Majak and Hall (1990) noted that occurrence of bloat in cattle fed bloat-producing alfalfa was associated with low sodium and high potassium, magnesium and calcium concentrations in the rumen fluid. They suggested that the latter three minerals cause colloidal aggregation whereas sodium promotes colloid dispersion. The increased colloidal aggregation of chloroplast particles induced by these ion changes could increase bloat. In New Zealand, Turner (1981) observed that a high ratio of potassium to sodium in forage was associated with increased bloat incidence. However, Hall and Majak (1992) found that neither sodium nor potassium salts given as supplements influenced bloat in alfalfa-fed cattle.

Saponins are glycosides in alfalfa and other bloat-producing legumes (see Chapter 11). Although they produce a stable rumen foam, they are not believed to have a significant role in pasture bloat (Majak *et al.*, 1980). A possible effect of saponins on bloat could be mediated via effects on the rumen protozoa. Saponins are very effective in killing rumen protozoa (see Chapter 11). Variations in forage saponin concentrations could possibly lead to situations in which rumen protozoa are killed, tipping the balance from a non-bloating to a bloating situation. A role of protozoa in bloat is indicated by the observation that defaunation (removal of protozoa) of cows reduces the incidence and severity of bloat (Clarke *et al.*, 1969). The lack of increased incidence of bloat when cows were fed high-saponin alfalfa (Majak *et al.*, 1980) may be a reflection of saponin-induced defaunation of the rumen.

Prevention of Bloat

The use of **antifoaming agents** such as poloxalene (Bloat Guard) in legume pastures has greatly reduced the bloat problem (Fig. 9–3). These blocks have a high molasses content and serve as the only source of salt, so cattle lick them frequently during the day, ensuring the continual presence of the antifoaming agent in the rumen. One block per five head of cattle should be used, and the blocks should be well distributed throughout the pasture. When used in this manner, "bloat blocks" are highly effective in preventing bloat (Hall *et al.*, 1994b). In areas of extensive grazing with scattered patches of bloat-producing legumes, the blocks

FIGURE 9–3 A steer licking a Bloat Guard block (top). When a water extract of alfalfa is shaken, a stable foam forms. When Bloat Guard is added, the alfalfa extract does not foam (bottom).

may be less effective, or it may not be feasible to use them because of the nature of the terrain and plant distribution.

Other methods of minimizing bloat incidence include using a mixture of grasses and legumes rather than a pure stand of legume. The bloat potential is highest for lush, immature forage which has a rapid rate of cell rupture in the rumen, so it may be advisable to graze legumes only when they are approaching maturity. Strip-grazing (provision of a narrow strip of fresh pasture daily or several times daily) to force consumption of the entire plant rather than selective grazing of tops is another useful management practice. In New Zealand, daily allotments of pastures strip-grazed by dairy cattle are sometimes sprayed with vegetable oil or fats to prevent bloat; the lipids serve as antifoaming agents. Although providing hay or other fibrous roughage as a supplement to cattle on lush pasture is a common practice, often the intake is not sufficient to decrease bloat incidence. Coarse roughage would presumably stimulate eructation. The ultimate solution will be the development, through plant breeding or biotechnology, of legume cultivars that have a minimal bloat potential. This might be accomplished through selection for reduced proportions or rate of release of soluble protein, or incorporation of tannins into the genetic structure of the plant. Since no species with tannins have been found in some of the plant genera that cause bloat, such as *Medicago* (alfalfa and medics), novel techniques such as mutagenesis or "gene-splicing" procedures are being explored as methods of introducing protein precipitants. Li *et al.* (1996) estimate that a phenolic (proanthocyanidin) content of 1–5 mg per g dry matter in forage legumes is a threshhold for bloat safety. In some cases, it may be possible to plant bloat-producing legumes with other forages that do contain tannins, such as sainfoin; the ingested tannins complex with cytoplasmic proteins in the rumen and prevent foaming. In parts of Australia, bloat in cattle grazing pastures dominated by white clover has been controlled by the concurrent ingestion of carpet grass, which is rich in tannins.

Goplen *et al.* (1993) in Canada have a breeding program to develop a **bloat-safe alfalfa** cultivar. Their program is based on the observation that the major foaming agents involved are proteins, and the major cellular site of these proteins is in the mesophyll cells of leaves. The mesophyll cells of bloat-safe legumes are more resistant to rupture from chewing or rumen microbial digestion than those from bloat-causing legumes. For example, the release of proteins from alfalfa leaves through cell rupture is more rapid than from leaves of birdsfoot trefoil, cicer milk vetch, and sainfoin, which are nonbloating legumes. Howarth *et al.* (1982) found differences in rate of cell rupture among alfalfa cultivars, indicating that selection for a bloat-free alfalfa may be possible. They measured the rate of cell rupture using nylon bag digestion of alfalfa in rumen contents. Rapid dry matter disappearance is associated with a rapid rate of cell rupture. Results from the Canadian plant-breeding program to develop bloat-safe alfalfa based on rate of cell rupture indicate that a reduction in cell-rupture rate has been achieved, and the bloat-safe alfalfa has a low bloat potential (Hall and Majak, 1989; Goplen *et al.*, 1993; Hall *et al.*, 1994a).

Supplementation of cattle with the ionophore monensin reduces bloat incidence (Katz *et al.*, 1986; Lowe *et al.*, 1991). Katz *et al.* (1986) attribute the effect to reduced protozoa numbers; hence less gas production. Monensin also reduces the foam stability (Tanner *et al.*, 1995). Lowe *et al.* (1991) used a controlled-release capsule given intra-ruminally, delivering 300 mg monensin per day over a 100 day period. These capsules are widely used in Australia and New Zealand to control bloat and improve milk production of dairy cows on pasture.

PULMONARY EMPHYSEMA

Acute bovine pulmonary emphysema (ABPE) is a respiratory disease of cattle caused primarily by absorbed metabolites of the amino acid tryptophan. It occurs after a sudden dietary change, generally from somewhat sparse feed to a lush, succulent pasture. In the western U.S., it occurs most frequently in range cattle that are transferred from dry summer range to improved pastures or hay meadows in the late summer and autumn. In Britain, where the disease is called fog fever, it occurs after stock are turned into lush fields of kale, rape, or other succulent feeds. The term "fog fever" arises from the occurrence of the condition after cattle are given access to lush regrowth (foggage) pasture. In the U.S., there is a pronounced seasonal incidence of ABPE, with most cases in the autumn following a dry summer. The condition is rare in cattle less than 1-year old and is most prevalent in cattle over 2 to 3 years of age. This is apparently the cumulative result of repeated insults by the causative agent. The largest, heavily lactating cows are most likely to develop ABPE, perhaps because of their high intake of lush forage. The clinical signs are observed 2–10 days after an abrupt feed change. They include an increased rate of respiration, with severe dyspnea (labored breathing). Cattle may stand with their head extended and lowered, tongue protruded, frothing at the mouth, and mouth breathing (Fig. 9–4). An audible expiratory grunt may be heard, giving rise to the common name "grunts" for the condition. Other local terms are cow asthma, green grass sickness, and summer pneumonia. The lungs are grossly enlarged, with considerable quantities of edematous fluid and interstitial emphysema (swelling due to air).

The causative agent is **3-methylindole (3-MI)**, a metabolite produced by rumen fermentation of tryptophan. This was a serendipitous discovery at Washington State University;

FIGURE 9–4 A cow showing typical signs of acute bovine pulmonary emphysema. (Courtesy of J.R. Carlson)

calves were being dosed with different amino acids, and those given tryptophan developed emphysema. The 3-MI is absorbed and is either responsible for the lung damage itself or is metabolized by the pulmonary **mixed function oxidase (MFO)** system (cytochrome P450) to other active metabolites. Postulated routes of metabolism are shown in Fig. 9–5. The toxic dose of tryptophan is about 0.25–0.35 g/kg body weight; 0.1 g 3-MI/kg has the same effect. It is not known for certain why a sudden diet change provokes increased production of 3-MI.

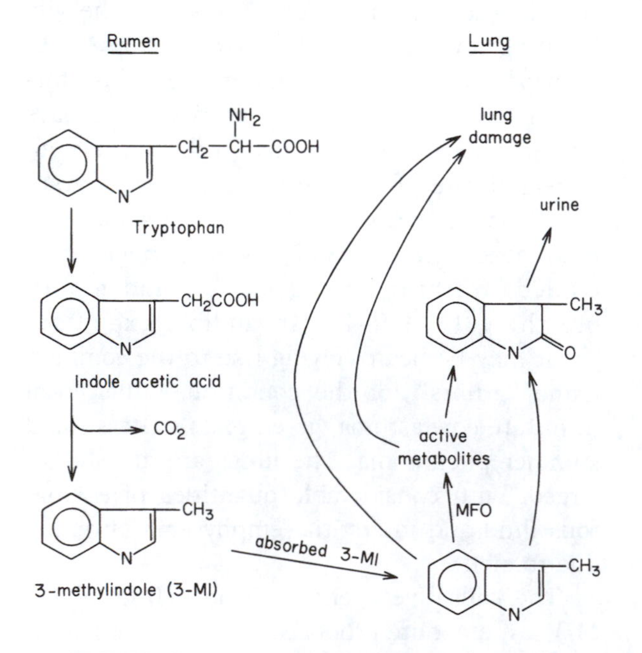

FIGURE 9–5 Metabolism of tryptophan in acute bovine pulmonary emphysema.

It could be the combined result of an increase in available tryptophan and a diet-induced change in rumen microflora. Abrupt change from dry to lush forage is accompanied by pronounced changes in the composition and digestibility of the forage consumed. The fiber content may be halved and the crude protein content doubled in lush grass compared to mature forage of the same species. These changes, along with a large increase in feed intake, could markedly alter rumen fermentation.

The clinical signs of ABPE generally appear from 1 to 14 days after the pasture change, and death may occur 2–4 days later. Mortality is about 30% of affected animals. Animals surviving the episode of clinical signs may recover without serious permanent lung damage.

Evidence that dietary tryptophan is the initiating factor includes the observations that oral administration of tryptophan produces clinical signs and pulmonary lesions indistinguishable from those seen in field cases of ABPE, and that lush grass can contain sufficient tryptophan to produce toxic amounts of 3-MI. Possibly other indoles in forage could also contribute to 3-MI production. Rumen fermentation of tryptophan to 3-MI is a 2-step reaction, with deamination and decarboxylation producing indole acetic acid (IAA) and a subsequent decarboxylation producing 3-MI (Fig. 9–5). Carlson and coworkers at Washington State University have isolated a *Lactobacillus* sp. from rumen fluid which converts IAA to 3-MI (Carlson and Breeze, 1984). There are a number of organisms which metabolize tryptophan to IAA. Dietary factors influence the rate of conversion of tryptophan to 3-MI, with the rate being more rapid on forage-based than on concentrate-based diets.

While 3-MI does have some direct effect on lung tissue, it appears that the mechanism of action is metabolism of 3-MI by pulmonary MFO to produce active metabolites (Skiles and Yost, 1996). Pretreatment of animals with MFO inducers such as phenobarbital increases severity of lung injury, while MFO suppressors such as piperonyl butoxide inhibit 3-MI metabolism and prevent lung injury (Carlson and Breeze, 1984). Pretreatment with cysteine to enhance lung glutathione levels reduces severity of lung damage, while administration of the glutathione depleter diethyl maleate increases lung injury, suggesting a role of glutathione in detoxification of 3-MI metabolites. The bioactivation of 3-MI, formation of metabolites and mechanisms of action have been reviewed by Carlson and Yost (1989). A possible role of free radical intermediates, lipid peroxidation and altered phospholipid synthesis in lung tissue has been proposed by Bray and Kirkland (1990) and Acton *et al.* (1992). Skiles *et al.* (1991) reported that the active metabolite may be a methylene imine, which is detoxified by glutathione conjugation.

There is no effective treatment of animals with ABPE. Prevention involves avoiding a sudden dietary change from a relatively poor pasture to a lush one. In the western range areas of the U.S., supplementation of cows on dry summer range may be advisable. Providing supplementary hay or other feed for the first 2 weeks after cattle are moved to lush grass and interrupting the grazing pattern by gathering and corralling the animals at intervals may be helpful management practices. Carlson and co-workers at Washington State have shown that modification of rumen metabolism may have potential as a means of preventing ABPE. Administration of ionophore antibiotics (e.g. monensin, lasalocid) in a feed supplement, along with feeding dry hay or wilted forage when cows are first turned out on ABPE-inciting pasture, is an effective means of preventing the disorder (Potchnoba *et al.*, 1992). Ionophores inhibit the growth of the gram-positive ruminal bacteria that metabolize tryptophan to 3-MI. Instead, when ionophores are used, tryptophan is converted to indole. While indole can cause hemolysis, it is unlikely that acute indole toxicosis would result from the use of ionophores as ABPE prophylactic agents (Potchoiba *et al.*, 1992).

3-MI, also known as **skatole**, is produced by microbial fermentation in the hindgut of many animals, and contributes to the odor of feces. It can be absorbed from the hindgut, but in most cases does not seem to cause toxicoses. Administration of 3-MI to horses causes severe obstructive respiratory disease (Turk *et al.*, 1983), but the disorder does not occur under practical circumstances. 3-MI is produced in the large intestine of humans, and it occurs in cigarette smoke (Carlson and Yost, 1989).

Bovine pulmonary emphysema has also been encountered with feedstuffs containing 3-substituted furans, such as moldy sweet potatoes and purple mint. **Moldy sweet potato poisoning** of cattle (Hill and Wright, 1992) results in lung disease that is caused by **3-substituted furans** such as **ipomeanol** (see Chapter 6). These compounds are stress metabolites of the sweet potato produced in response to fungal infections. Pulmonary emphysema in pigs consuming moldy sweet potatoes has been observed in Papua New Guinea (Low *et al.*, 1993), with some indication that humans may be affected as well. The purple mint (*Perilla frutescens*), an herb introduced into the U.S. from the Orient, has caused outbreaks of pulmonary emphysema in Oklahoma and Arkansas (Wilson *et al.*, 1978, 1990). It contains 3-substituted furans very similar in structure to those of moldy sweet potatoes.

SILICA UROLITHIASIS (URINARY CALCULI)

Cattle and sheep on the semiarid rangelands of the northern Great Plains of North America are susceptible to the development of **urinary calculi** composed mainly of silica (Fig. 9–6). This condition is also a problem in parts of Australia and the Russian steppes. Calculi are present in the urinary tracts of at least 50% of the cattle in North American range herds

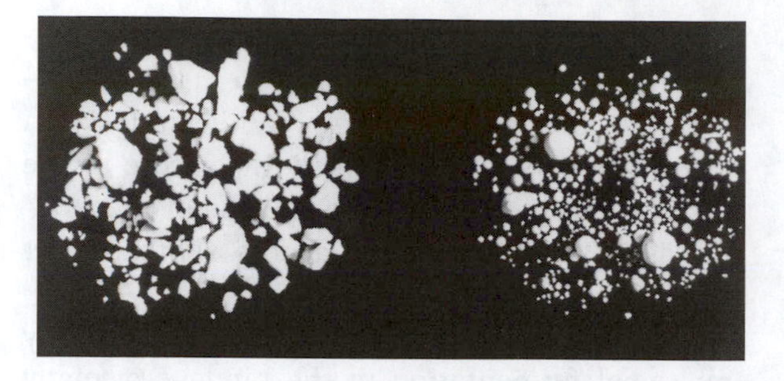

FIGURE 9–6 Mixed selection of calculi removed from the kidneys (left) and bladders (right) of steers with subclinical urolithiasis. The largest bladder calculus is 3 mm in diameter. (From Bailey [1981]; courtesy of C.B.M. Bailey)

(Bailey, 1981). Calculi form in both sexes but are a problem only in males. Displacement of large calculi from the bladder to the urethra causes an obstruction to the normal flow of urine. Within a few days of obstruction, urine pressure causes the bladder or urethra to rupture, and urine enters the abdominal cavity, causing distension of the abdomen, known as water belly. Obstruction and rupture of the bladder is fatal unless an opening is surgically constructed into the urethra to permit draining of the urine. This opening will usually stay open long enough to restore the animal to a condition fit for slaughter for meat. However, the obstruction generally occurs among range cattle when they are not under close observation, so mortality of affected animals is high. Obstruction does not occur in females because the short, wide urethra and rapid urine flow cause passage of calculi before they reach obstructive size. Obstruction is more common in steers than in bulls, since castration retards urethral development. In North America, obstructive urolithiasis is most common in range calves during the winter, while in Australia, it occurs most frequently in the hottest months. This is probably due to climatic effects on water intake. In North America, cold temperatures reduce water intake, while in Australia, water is likely to be in short supply in the summer. A low output of water in the urine relative to the excretion of silicic acid, giving rise to a high concentration of silicic acid in the urine, is the primary cause of calculi formation (Bailey, 1981).The main component of the calculi is silica (silicon dioxide), comprising about 75% of the total weight. The organic component contains protein and some carbohydrate.

The silica content of range grasses is low at the start of the growing season and progressively increases, even after plant growth ceases (Bailey, 1981). Plants exposed to weathering, which removes soluble components, have the highest silica content. The material may contain over 6.5% total silica. This is found as silica associated with plant cell walls, or as free silica, consisting of unpolymerized silicic acid. Calves on native range hay can ingest as much as 500 g of silica per day, while sheep in Australia can consume as much as 40 g/day. A portion of the ingested silica is solubilized in the rumen fluid to keep the rumen saturated with silicic acid. This acid is absorbed and is the form of silicon found in the blood. In range cattle, because of the low level of protein and minerals other than silica in the forage, water resorption in the kidney is high and urine output is low. This concentrates the silicic acid in the urine.

Bailey (1981) has shown that when water intake is below 3.5 kg per kilogram of grass intake, the urine is supersaturated with silicic acid and formation of calculi will occur. Water intakes below this level are frequent in preweaning calves, and calculi are often present by weaning. Mortality is most common from 6 to 8 months postweaning. An effective way of reducing the urinary calculi problem is to induce animals to drink more water to reduce the concentration of silicic acid in the urine. This can be accomplished by increasing the salt intake. However, when provided loose or as a lick, salt will not be consumed in sufficient amounts to influence water intake. Ingestion of salt at a rate of about 1 g/kg body weight increases water intake sufficiently to eliminate calculus formation. This can be accomplished by providing a palatable grain supplement, with 15% salt, in a creep feeder for calves beginning at about 4 months of age. This should be fed continuously through the first winter until the calves are about 1 year of age.

Diets with high calcium to phosphorus ratios that result in alkaline urine promote silica urolithiasis (Stewart *et al.*, 1990). Further work by these researchers demonstrated that reducing the calcium to phosphorus level and feeding 1% dietary ammonium chloride as a supplement to acidify the urine reduced urinary calculi formation (Stewart *et al.*, 1991).

NITRATE–NITRITE TOXICITY

Common crop and pasture plants and numerous weeds may accumulate toxic levels of nitrates. In ruminants, nitrate is readily reduced to nitrite, which is absorbed and causes toxicosis. Nitrite ions oxidize ferrous iron in hemoglobin to produce **methemoglobin** (ferric iron). Methemoglobin cannot react with oxygen, so anoxia occurs. Clinical signs of toxicity may be seen when methemoglobin levels reach 30–40% of total hemoglobin, while death occurs at levels of 80–90% methemoglobin. Clinical signs of **nitrate–nitrite toxicity** include dyspnea (labored breathing), cyanotic mucous membranes, and evidence of abdominal pain. The outstanding feature is dark brown or chocolate-colored blood due to the methemoglobin. Common sources of toxic levels of nitrate are forage plants and water. Abnormal accumulation of nitrate in plants is provoked by high levels of nitrogen fertilization, drought conditions, and herbicide treatment with phenoxyacetic herbicides such as 2,4-D. Nitrate accumulates in the vegetative tissue, particularly in stalks with less in the leaves. Seeds (grains) do not generally contain toxic nitrate levels. Some plants are more likely than others to accumulate toxic levels of nitrate. Among weeds, pigweed (*Amaranthus* spp.) nightshades, and Johnson grass are known as nitrate accumulators. Sudan grass, oats, rape, wheat, barley and corn accumulate nitrate. Heavy fertilization of grass pastures, especially during cool, cloudy weather, may result in toxic nitrate levels. Water sources may be contaminated from barn and feedlot runoff, silage juice, or nitrogen fertilizers. **Forage nitrate levels** of 0.5% and above are potentially dangerous, with **acute poisoning** likely to occur if the nitrate level exceeds 1%. Levels of 200 ppm nitrate in water are potentially hazardous, with 1500 ppm causing acute toxicity.

When silage is made from high-nitrate forage, toxic silo gas may be produced. Anaerobic silage fermentation results in reduction of nitrate to oxides of nitrogen such as nitrogen dioxide (NO_2) and nitrogen tetraoxide (N_2O_4). These gases are yellowish-brown and may collect in barns in sufficient concentration to kill animals.

Chronic nitrate toxicity has been suggested to cause reduced growth, vitamin A deficiency, abortion, infertility, goiter, and other nonspecific problems. In general, these effects are not documented with experimental evidence. The intake of nitrate required to produce abortion in cattle is very similar to the lethal dose for the dam; thus nitrate-induced abortions should be accompanied by signs of nitrate toxicosis in the cows (Norton and Hogan, 1993). The principal concern with nitrates is acute toxicity.

A commercial product containing selected strains of *Propionibacteria* which reduce nitrate to N_2O or N_2 (denitrification) is marketed in the U.S.[1] The microbial preparation can be fed to ruminants which might be exposed to high-nitrate forages, to reduce their susceptibility to nitrate poisoning.

---

[1]Laporte Biochem International, 6120 W. Douglas Ave., Milwaukee, WI 53218; FAX 414-464-6430.

POLIOENCEPHALOMALACIA

Polioencephalomalacia (PEM), also known as cerebrocortical necrosis, is a non-infectious disease of ruminants (polio: Gr. gray; encephalo: Gr. brain; malacia: Gr. softness; thus, softening of the gray matter of the brain). Cattle and sheep are affected, with young animals between 2 and 7 months of age being the most susceptible (Edwin and Jackman, 1982). Clinical signs of the disorder include aimless wandering, disorientation, blindness, recumbency, and opisthotonus (star gazing posture). The brain is edematous, with yellowish discoloration of the cerebral cortex. Affected animals may respond dramatically to administration of thiamin, with doses of 200–500 mg for calves, 1000–2000 mg for adult cattle, and 100–500 mg for sheep, depending on body weight. Although PEM occurs under a variety of circumstances, metabolic depletion of thiamin (vitamin B_1) seems to be involved in each case. Some of the factors implicated in PEM include grain overload of the rumen, high sulfate in drinking water, and consumption of certain plants such as *Kochia scoparia*. PEM occurs with both forage and concentrate diets, so the discussion of it in the Forages section is arbitrary.

Polioencephalomalacia is a serious disorder of feedlot cattle fed diets high in grain. It is associated with high levels of thiaminase activity in the rumen of affected animals (Edwin *et al.*, 1982), leading to an **induced thiamin deficiency**. Several thiaminase-producing bacterial species, such as *Clostridium sporogenes* and *Bacillus* spp., have been implicated (Edwin *et al.*, 1982; Haven *et al.*, 1983). Haven *et al.* (1983) reported that the predominant gram-negative nonspore-forming microflora normally resident in the rumen may be the source of thiaminase, and that these organisms may for some reason develop metabolic alterations which lead to increased thiamin destruction. The mechanism by which thiaminase degrades thiamin is discussed in greater detail in Chapter 15 (Bracken Poisoning).

Kochia scoparia (**kochia**, summer cypress, burning bush, fireweed, Mexican fireweed) grows in arid lands of the U.S. West and Southwest and has been responsible for a number of livestock poisonings. It can be used successfully as a forage plant. It is most toxic during periods of drought and at seed maturation. Supplemental feed should be provided at these times, especially if rain stimulates seed production (Dickie and James, 1983). Signs of toxicosis include icterus and photosensitization, progressive central nervous system dysfunction, hepatic cirrhosis, gastrointestinal tract inflammation, and PEM. The occurrence of PEM suggests involvement of a thiaminase or a hepatotoxin which may lead to impaired thiamin utilization. Dickie and James (1983) found evidence of a thiamin-destroying principle but suggested that it is not the major factor in kochia toxicity. Hepatic necrosis and fibrosis, with photosensitization, suggests the presence of a hepatotoxin. Kochia thrives on arid, infertile, high sodium soils where few other plants will grow. The high sulfate content of kochia generally results in it having a slight laxative effect and it may have a role in the PEM syndrome, as discussed below. Other toxicants which might be associated with kochia are oxalates, saponins and nitrates.

Rankins and Smith (1991) and Rankins *et al.* (1991a,b,c) conducted extensive studies of **kochia toxicity** in cattle and sheep. Lambs fed kochia hay showed changes in liver enzymes reflective of mild hepatotoxicosis without cholestasis; histopathology after 80 days of kochia consumption showed liver lesions (diffuse hepatocyte swelling) and kidney damage (Rankins *et al.*, 1991a). Serum insulin and prolactin were reduced in lambs (Rankins *et al.*, 1991a) and steers (Rankins *et al.*, 1991b,c). Rankins *et al.* (1991a,b,c) indicate that the hepatotoxic agents are alkaloid in nature, but specific alkaloids in kochia have not been identified. The kidney damage is probably due to oxalates, which can be present in kochia hay at levels as high as 6% (Rankins *et al.*, 1991a).

In the Great Plains areas of Canada and the U.S., outbreaks of PEM have been associated with high intakes of **sulfate** from feed or water (Gould *et al.*, 1991; Hamlen *et al.*, 1993). Ground water in the Great Plains often has high sulfate levels, and areas of saline soils containing sulfates are common. Several possible modes of action of sulfate in PEM have been proposed. Gould *et al.* (1991) indicate that sulfate is metabolized to sulfide in the rumen. Sulfide as hydrogen sulfide can be absorbed, and is highly toxic to the nervous system. Hamlen *et al.* (1993) suggest that hydrogen sulfide may be absorbed by the lungs from eructated rumen gases. The PEM-inducing effect of kochia might be due to its high sulfate content (Gould *et al.*, 1991). Gooteratne *et al.* (1989) demonstrated that high sulfate water depleted tissue copper and thiamin in cattle. Sulfides are known to reduce copper absorption through the formation of thiomolybdate-copper complexes. The mode of action of dietary sulfate in reducing tissue thiamin has not been elucidated. Olkowski *et al.* (1992) proposed that absorbed sulfur compounds cause lipid peroxidation in brain tissue, increasing the thiamin requirement. When cattle are exposed to high sulfate levels in feed or water, there is a 10–12 day adaptation period before the rumen microorganisms are capable of generating potentially toxic levels of sulfide (Cummings *et al.*, 1995a,b). Total dietary sulfur concentrations of 0.2–0.3% are recommended in ruminant feeds; 0.4% sulfur is considered the maximal tolerable limit (Cummings *et al.*, 1995a).

Bulgin *et al.* (1996) reported an episode of acute sulfur poisoning of sheep grazing an alfalfa field that had been sprayed with 60 kg sulfur/ha. Signs of severe respiratory distress were attributed to inhalation of hydrogen sulfide from eructated rumen gases. Polioencephalomalacia developed in animals which survived acute poisoning.

HYPOMAGNESEMIA (GRASS TETANY)

Hypomagnesemia (low blood magnesium) is a metabolic disorder of ruminants caused by a deficiency of absorbable magnesium. Magnesium reserves in the animal body are not extensive, so an adequate dietary supply is necessary. Magnesium is a cofactor for numerous enzymes involved in energy metabolism, and is involved in nerve impulse transmission at neuromuscular junctions. Signs of hypomagnesemia (grass tetany) include nervousness and muscle twitching in mild cases, and incoordination, collapse, convulsions, coma and death in severe cases. Cows in early lactation are quite susceptible to grass tetany, because of the high magnesium requirements for lactation.

Grass tetany is most commonly encountered with lush, rapidly-growing temperate grass pastures. Often the pastures have been heavily fertilized, and cold, wet weather may play a role. Numerous factors are involved, all having the effect of reducing absorbable magnesium. These include high forage potassium and ammonia, and low sodium and magnesium. In areas of intensive dairy production, the heavy application of dairy manure to crop land can lead to elevated forage potassium levels (Fisher *et al.*, 1994), which could result in impaired ruminal absorption of magnesium and a greater incidence of hypomagnesemia (Dua and Care, 1995).

In the context of toxins, there is evidence that spring grass contains organic acids that complex magnesium and reduce its availability. Stout *et al.* (1967) suggested that ***trans*-aconitic acid**, which is found in high concentration in forage in the spring, may chelate magnesium and reduce its availability. Therefore, it may be a causative agent in inducing grass tetany, or hypomagnesemia. Camp *et al.* (1968) found that administration of *trans*-aconitic acid to sheep reduced serum magnesium levels but did not cause grass tetany. Bohman *et al.* (1969) reported that *trans*-aconitic acid plus potassium chloride when administered orally to cattle would induce grass tetany. Bohman *et al.* (1983), in a study of grass tetany in cattle on wheat pasture, found a sharp and sudden rise in the *trans*-aconitic acid content of the wheat at the time when the cows developed tetany, providing further evidence of a role of aconitate in the etiology of the disease. *Trans*-aconitic acid is rapidly hydrogenated (Fig. 9–7) in the rumen to **tricarballylic acid** (Russell and van Soest, 1984; Russell and Forsberg, 1986). Tricarballylic acid is the form absorbed and may sequester magnesium for urinary excretion or inhibit aconitase (Russell and Mayland, 1987). Tricarballylate inhibits acetate oxidation in vitro, suggesting an aconitase-inhibitory effect. Administration of tricarballylic acid to rats increased the urinary excretion of magnesium. Organic

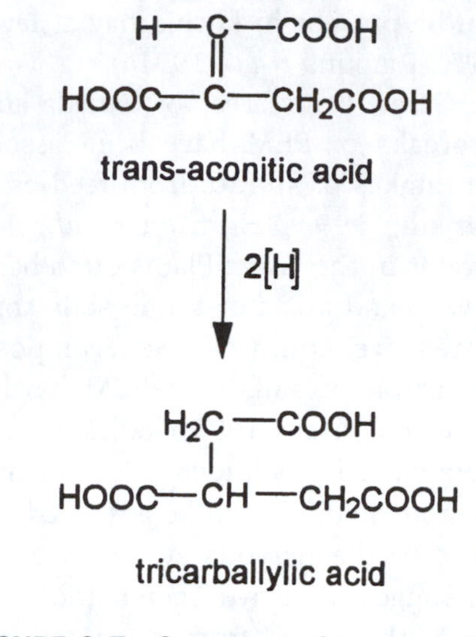

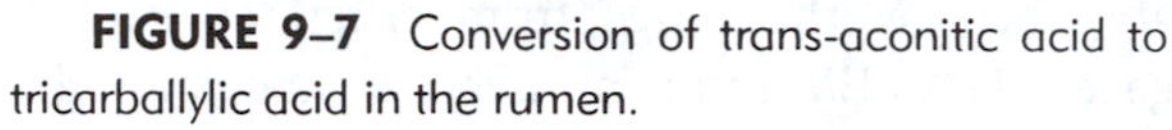

FIGURE 9–7 Conversion of trans-aconitic acid to tricarballylic acid in the rumen.

acids are not the primary herbage factors involved in grass tetany, but could contribute to low magnesium availability or increased urinary excretion of magnesium.

LAMINITIS

Laminitis, or founder, is inflammation of the laminae in the hoof, and general hoof inflammation. The laminae are leaflike structures that support the coffin bone (Fig. 9–8). When they become inflamed and swollen, blood flow in the hoof is impaired. Elevated blood pressure (hypertension) precedes development of the symptoms. Blood is shunted through a circulatory by-pass away from the arterioles in the hoof, through dilated (enlarged) **arteriovenous anastomoses (AVAs)** (a vessel that directly interconnects an artery and a vein, acting as a shunt to bypass a capillary bed). Absorbed toxins, including histamine, lactic acid and bacterial endotoxins produced in the gut, can cause AVA dilation, reduced blood supply to the hoof and inflammation of laminae.

Laminitis is particularly important in horses and cattle. The etiology in each case will be considered.

Equine Laminitis

Laminitis in horses is a serious problem, because virtually all uses of the horse require that it have sound hooves and legs. Laminitis is often a sequel to other disorders, such as grain or lush pasture overload, other gastrointestinal disturbances, infections, retained placenta and plant-induced toxicosis (e.g. black walnut poisoning). Garner *et al.* (1975) developed a technique for inducing laminitis for research purposes, by dosing with starch to cause carbohydrate overload of the hindgut. Excessive carbohydrate in the hindgut results in rapid

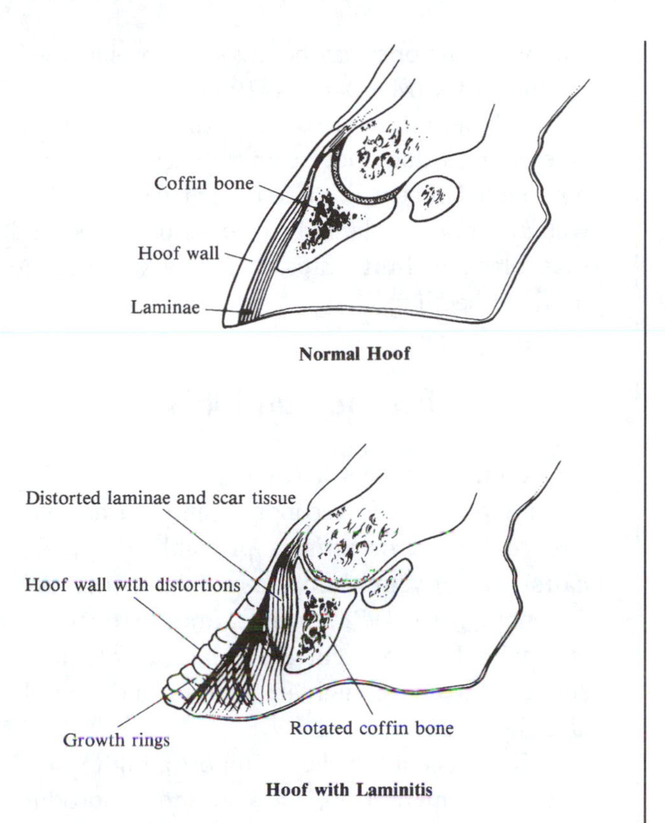

FIGURE 9–8 In the normal hoof, the hoof wall and the laminae are parallel (top). In laminitis, the laminae are distorted, causing rotation of the coffin bone (bottom). Pain, distorted hoof structure and lameness result.

proliferation of microbes, and production of large quantities of lactic acid and bacterial endotoxins. Absorbed lactic acid and endotoxin cause venoconstriction in the hoof, resulting in the expansion of AVA shunts at the level of the coronary band (Moore *et al.* , 1989). Rowe *et al.* (1994) found that oral administration of the antibiotic virginiamycin to horses prevented laminitis induced by grain feeding, by inhibiting the Gram-positive bacteria which produce lactic acid.

The initial event in laminitis is venoconstriction, impeding blood flow in the laminar capillaries (Moore and Allen, 1996), with the increase in capillary pressure causing fluid to move into the interstitial space. This is accompanied by development of thrombosis (clotting) in the affected blood vessels in the hoof, probably due to activation of platelets by endotoxins or other products of hindgut fermentation (Weiss *et al.,* 1994, 1995, 1996). The reduced laminar blood flow is accompanied by shunting of blood away from the affected area by the AVA by-pass.

Many horses and especially ponies (Fig. 9–9) are susceptible to development of laminitis when grazing lush, succulent pasture with high protein and soluble carbohydrate contents. Overloading the hindgut with readily fermentable nutrients in low fiber forages results in excessive lactic acid and bacterial endotoxin production.

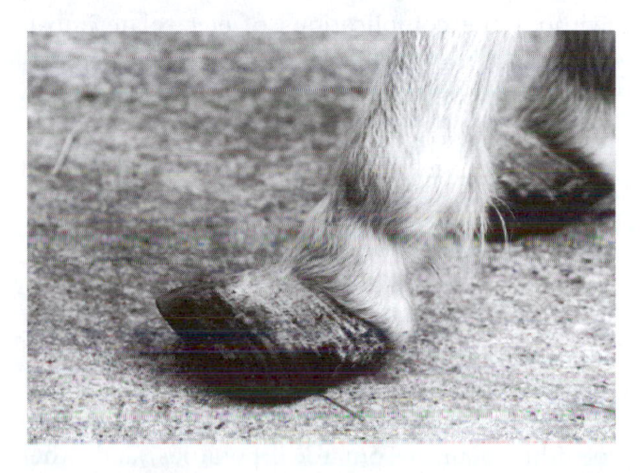

FIGURE 9–9 Hoof of a foundered pony, showing abnormal growth rings and placement of the weight on the heel, to minimize pain.

Consumption of **black walnut** (*Juglans nigra*) wood shavings by horses causes laminitis. True *et al.* (1978) reported that racehorses bedded down with chips or sawdust developed laminitis. MacDaniels (1983) discussed the toxic effect on herds of racehorses in Wisconsin and Michigan where severe allergenic reactions to the shedding of walnut pollen occurred as well as to the shedding of leaves in the autumn. In these instances, the problems were sufficiently serious that all the walnut trees were cut down, the stumps bulldozed, and the soil removed from the paddocks and stables. Ralston and Rich (1983) reported an incident in Colorado involving horse stalls bedded with pine shavings containing 20% black walnut shavings as a contaminant. Within 12 hr of exposure, the horses exhibited signs of toxicosis, including reluctance to move, early acute laminitis, slight edema of the limbs, and depression. Uhlinger (1989) reported an outbreak of black walnut toxicosis in horses bedded with black walnut shavings, with signs of severe abdominal pain and laminitis.

Water extracts of black walnut wood cause laminitis when administered to horses (Galey *et al.*, 1990, 1991). Galey *et al.* (1991) suggested that this procedure provides a good model for induction of laminitis for research purposes, without the complications of electrolyte imbalance, shock and colic that may occur when carbohydrate overload is used to induce the condition.

The active laminitis-inducing principle of black walnut has not been identified. Black walnuts and other members of the walnut family (Persian or English walnuts, butternuts, hickories, and pecans) contain **juglone** (5-hydroxy-1,4-naphthoquinone), a phenolic derivative of naphthoquinone (Fig. 9–10). Juglone is an **allelopathic** substance which inhibits the growth of other plants. Tomatoes, potatoes, and other vegetables, as well as many other trees and shrubs, are inhibited in their growth by juglone released into the soil from black walnut roots. As a result, competition to walnuts from other plants is reduced.

True and Lowe (1980) found that large oral doses of juglone given to horses induced mild clinical signs of laminitis, but the naturally occurring condition could not be duplicated. Precursors of juglone may be involved in black walnut toxicosis (MacDaniels, 1983).

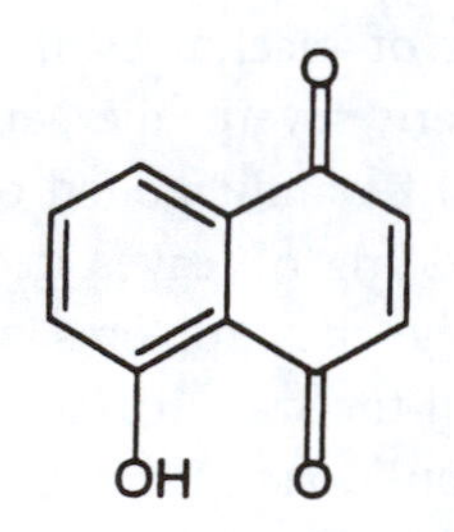

FIGURE 9–10 Juglone, a toxic naphthoquinone in black walnut.

Black walnut toxicosis in other animals besides horses does not seem to occur. It is recommended that walnut trees, particularly black walnuts, not be planted in horse pastures, and that black walnut chips and sawdust not be used for bedding.

Bovine Laminitis

Laminitis in beef and dairy cattle is an important problem for economic and animal welfare reasons. Laminitis is probably the major cause of bovine lameness (Vermunt and Greenough, 1994). Typically, laminitis is associated with high concentrate diets, or with lush, succulent pasture high in protein and soluble carbohydrates.

High concentrate diets support a high rate of ruminal fermentation, with abundant production of VFA, including lactic acid. **Lactic acid** is a stronger acid than the other VFA, and contributes to a drop in rumen pH. The lowered rumen pH is accompanied by changes in the rumen microbes from gram-negative to gram-positive bacteria such as *Streptococcus bovis* and *Lactobacillus* spp. **Endotoxins** are released from the cell walls of lysed gram-negative bacteria. Absorbed endotoxin may cause allergenic reactions in the arterioles of the hoof, impairing blood supply to the tissue. Lactic acid and histamine absorbed from the rumen contribute to the circulatory dysfunction (Vermunt and Greenough, 1994). Dilated AVAs in the corium of the hoof shunt blood away from the hoof tissue. Histamine stimulates AVA dilation. Dietary administration of the antibiotic virginiamycin, which inhibits gram-positive bacteria, is effective in reducing lactic acid production in the rumen (Godfrey *et al.*, 1992). Ruminal administration of a lactate-utilizing microbe, *Megasphaera elsdenii*, also has shown potential in reducing rumen lactate concentrations (Kung and Hession, 1995).

Nutrition is a major factor in the development of laminitis. As in the equine founder syndrome, carbohydrate overload of the gut resulting in excessive microbial growth is the principal contributing factor. In ruminants, carbohydrate overload of the rumen produces changes in the microflora, high rates of production of VFA and lactic acid, lowered rumen pH, and increased microbial toxin production (endotoxins, histamine). High producing dairy cows, with high feed intakes of high energy, low roughage diets, are particularly likely to develop laminitis.

Cattle on heavily fertilized pastures are also susceptible to laminitis. Vermunt (1992) speculated that elevated pasture nitrate levels could be involved. Nitrate is converted to nitrite in the rumen. Absorbed nitrite can produce pronounced vasodilation (Vermunt, 1992), which by promoting the AVA shunt in the hoof could contribute to laminitis. High protein intakes have been implicated in laminitis (Vermunt, 1992). The mechanism of action of protein in this effect has not been elucidated. The laminitis-inducing effects of lush, heavily fertilized pastures may be a consequence of both the high protein and the high soluble carbohydrate contents.

Block (1994) suggested that because laminitis in dairy cattle is related to blood acid–base balance, the dietary cation–anion balance could play a role in the disorder. Further research is needed to validate this suggestion.

GREEN FEED–INDUCED DIARRHEA

Ruminants on lush spring pasture often have profuse watery diarrhea. Plants with high contents of water and fermentable sugars, and low cell wall material, have a rapid rate of flow through the rumen,with a shift in fermentation to the hindgut. Rapid fermentation in the hindgut with high VFA production may result in osmotic diarrhea (Pethick and Chapman, 1991). Thus water is drawn into the colon, producing profuse watery diarrhea.

REFERENCES

Bloat

Branine, M.E., and M.L. Galyean. 1990. Influence of grain and monensin supplementation on ruminal fermentation, intake, digesta kinetics and incidence and severity of frothy bloat in steers grazing winter wheat pasture. J. Anim. Sci. 68:1139–1159.

Clarke, R.T.J., C.S.W. Reid, and P.W. Young. 1969. Bloat in cattle. XXXII. Attempts to prevent legume bloat in dry and lactating cows by partial or complete elimination of the rumen holotrich protozoa with dimetridazole. N.Z.J. Agric. Res. 12:446–466.

Colvin, H.W., Jr., and R.C. Backus. 1988. Bloat in sheep (*Ovis aries*). Comp. Biochem. Physiol. 91A:635–644.

Goplen, B.P., R.E. Howarth, and G.L. Lees. 1993. Selection of alfalfa for a lower initial rate of digestion and corresponding changes in epidermal and mesophyll cell wall thickness. Can. J. Plant Sci. 73:111–122.

Hall, J.W., and W. Majak. 1989. Plant and animal factors in legume bloat. pp. 93–106 in: P.R. Cheeke (Ed.). Toxicants of Plant Origin. Vol. III. CRC Press, Boca Raton, FL.

Hall, J.W., and W. Majak. 1991. Relationship of weather and plant factors to alfalfa bloat in autumn. Can. J. Anim. Sci. 71:861–866.

Hall, J.W., and W. Majak. 1992. Rapid screening of feed supplements for the prevention of legume bloat. Can. J. Anim. Sci. 72:613–617.

Hall, J.W., and W. Majak. 1995. Effect of time of grazing or cutting and feeding on the incidence of alfalfa bloat in cattle. Can. J. Anim. Sci. 75:271–273.

Hall, J.W., W. Majak, D.G. Stout, K.-J. Cheng, B.P. Goplen, and R.E. Howarth. 1994a. Bloat in cattle fed alfalfa selected for a low initial rate of digestion. Can. J. Anim. Sci. 74:451–456.

Hall, J.W., I. Walker, and W. Majak. 1994b. Evaluation of two supplements for the prevention of alfalfa bloat. Can. Vet. J. 35:702–705.

Howarth, R.E., B.P. Goplen, S.A. Brandt, and K.J. Cheng. 1982. Disruption of leaf tissues by rumen microorganisms: An approach to breeding bloat-safe forage legumes. Crop Sci. 22:564–568.

Howarth, R.E., W. Majak, D.E. Waldern, S.A. Brandt, A.C. Fesser, B.P. Goplen, and D.T. Spurr. 1977. Relationships between ruminant bloat and the chemical composition of alfalfa herbage. I. Nitrogen and protein fractions. Can. J. Anim. Sci. 57:345–357.

Katz, M.P., T.G. Nagaraja, and L.R. Fina. 1986. Ruminal changes in monensin- and lasalocid-fed cattle grazing bloat-provocative alfalfa pasture. J. Anim. Sci. 63:1246–1257.

Li, Y.-G., G. Tanner, and P. Larkin. 1996. The DMACA-HCl protocol and the threshold proanthocyanidin content for bloat safety in forage legumes. J. Sci. Food Agric. 70:89–101.

Lowe, L.B., G.J. Ball, V.R. Carruthers, R.C. Dobos, G.A. Lynch, P.J. Moate, P.R. Poole, and S.C. Valentine. 1991. Monensin controlled-release intraruminal capsule for control of bloat in pastured dairy cows. Aust. Vet. J. 68:17–20.

Majak, W., and J.W. Hall. 1990. Sodium and potassium concentrations in ruminal contents after feeding bloat-inducing alfalfa to cattle. Can. J. Anim. Sci. 70:235–241.

Majak, W., J.W. Hall, and W.P. McCaughey. 1995. Pasture management strategies for reducing the risk of legume bloat in cattle. J. Anim. Sci. 73:1493–1498.

Majak, W., R.E. Howarth, K.J. Cheng, and J.W. Hall. 1983. Rumen conditions that predispose cattle to pasture bloat. J. Dairy Sci. 66:1683–1688.

Majak, W., R.E. Howarth, A.C. Fesser, B.P. Goplen, and M.W. Pedersen. 1980. Relationships between ruminant bloat and the composition of alfalfa herbage. II. Saponins. Can. J. Anim. Sci. 60:699–708.

Phillips, C.J.C., N.L. James, and J.P. Murray-Evans. 1996. Effect of forage supplements on the incidence of bloat in dairy cows grazing high clover pastures. Vet. Record 139:162–165.

Tanner, G.J., P.J. Moate, L.H. Davis, R.H. Laby, L. Yuguang, and P.J. Larkin. 1995. Proanthocyanidins (condensed tannin) destabilise plant protein foams in a dose dependent manner. Aust. J. Agric. Res. 46:1101–1109.

Turner, M.A. 1981. Dietary potassium–sodium imbalance as a factor in the aetiology of primary ruminal tympany in dairy cows. Vet. Res. Comm. 5:159–164.

Pulmonary Emphysema

Acton, K.S., H.J. Boermans, and T.M. Bray. 1992. The role of prostaglandin H synthase in 3-methylindole-induced pneumotoxicity in goat. Comp. Biochem. Physiol. 101C:101–108.

Bray, T.M., and J.B. Kirkland. 1990. The metabolic basis of 3-methylindole-induced pneumotoxicity. Pharmac. Ther. 46:105–118.

Carlson, J.R., and R.G. Breeze. 1984. Ruminal metabolism of plant toxins with emphasis on indolic compounds. J. Anim. Sci. 58:1040–1049.

Carlson, J.R., and G.S. Yost. 1989. 3-methylindole-induced acute lung injury resulting from ruminal fermentation of tryptophan. pp. 107–123 in: P.R. Cheeke (Ed.). Toxicants of Plant Origin. Vol. III. CRC Press, Boca Raton, FL.

Hill, B.D., and H.F. Wright. 1992. Acute interstitial pneumonia in cattle associated with consumption of mould-damaged sweet potatoes (*Ipomoea batatas*). Aust. Vet. J. 69:36–37.

Low, S.G., I. McL. Grant, B. Rodini, and W.L. Bryden. 1993. Sweet potato (*Ipomea batatas*) poisoning of pigs in Papua New Guinea. N. Z. Vet. J. 41:218 (abst.).

Potchoiba, M.J., J.R. Carlson, M.R. Nocerini, and R.G. Breeze. 1992. Effect of monensin and supplemental hay on ruminal 3-methylindole formation in adult cows after abrupt change to lush pasture. Am. J. Vet. Res. 53:129–133.

Skiles, G.L., D.J. Smith, M.L. Appleton, J.R. Carlson, and G.W. Yost. 1991. Isolation of a mercapturate adduct produced subsequent to glutathione conjugation of bioactivated 3-methylindole. Toxicol. Appl. Pharmacol. 108:531–537.

Skiles, G.L., and G.S. Yost. 1996. Mechanistic studies on the cytochrome P450-catalyzed dehydrogenation of 3-methylindole. Chem. Res. Toxicol. 9:291–297.

Turk, M.A., R.G. Breeze, and A.M. Gallina. 1983. Pathologic changes in 3-methylindole-induced equine bronchiolitis. Am. J. Path. 110:209–218.

Wilson, B.J., J.E. Garst, R.D. Linnabary, and A.R. Doster. 1978. Pulmonary toxicity of naturally occurring 3-substituted furans. pp. 311–323 in: R.F. Keeler, K.R. Van Kampen, and L.F. James (Eds.). Effects of Poisonous Plants on Livestock. Academic Press, N.Y.

Wilson, W.C., J. Simon, and J.E. Garst. 1990. The effects of selected bulky substituents on the pulmonary toxicity of 3-furyl ketones in mice. J. Anim. Sci. 68:1072–1076.

Silica Urolithiasis (Urinary Calculi)

Bailey, C.B. 1981. Silica metabolism and silica urolithiasis in ruminants: A review. Can. J. Anim. Sci. 61:219–235.

Stewart, S.R., R.J. Emerick, and R.H. Pritchard. 1990. High dietary calcium to phosphorus ratio and alkali-forming potential as factors promoting silica urolithiasis in sheep. J. Anim. Sci. 68:498–503.

Stewart, S.R., R.J. Emerick, and R.H. Pritchard. 1991. Effects of dietary ammonium chloride and variations in calcium to phosphorus ratio on silica urolithiasis in sheep. J. Anim. Sci. 69:2225–2229.

Nitrate–Nitrite Toxicity

Norton, J.H., and J.P. Hogan. 1993. Lack of association between abortion and blood ammonia and methaemoglobin concentrations in dairy cows grazing improved pastures on the Atherton Tableland. Aust. Vet. J. 70:194–195.

Polioencephalomalacia

Bulgin, M.S., S.D. Lincoln, and G. Mather. 1996. Elemental sulfur toxicosis in a flock of sheep. J. Am. Vet. Med. Assoc. 208:1063–1065.

Cummings, B.A., D.R. Caldwell, D.H. Gould, and D.W. Hamar. 1995a. Identity and interactions of rumen microbes associated with dietary sulfate-induced polioencephalomalacia in cattle. Am. J. Vet. Res. 56:1384–1389.

Cummings, B.A., D.H. Gould, D.R. Caldwell, and D.W. Hamar. 1995b. Ruminal microbial alterations associated with sulfide generation in steers with dietary sulfate-induced polioencephalomalacia. Am. J. Vet. Res. 56:1390–1395.

Dickie, C.W., and L.F. James. 1983. *Kochia scoparia* poisoning in cattle. J. Am. Vet. Med. Assoc. 183:765–768.

Edwin, E.E., and R. Jackman. 1982. Ruminant thiamine requirement in perspective. Vet. Res. Commun. 5:237–250.

Edwin, E.E., R. Jackman, and P. Jones. 1982. Some properties of thiaminases associated with cerebrocortical necrosis. J. Agric. Sci. 99:271–275.

Gooneratne, S.R., A.A. Olkowski, R.G. Klemmer, G.A. Kessler, and D.A. Christensen. 1989. High sulfur related thiamine deficiency in cattle: A field study. Can. Vet. J. 30:139–146.

Gould, D.H., M.M. McAllister, J.C. Savage, and D.W. Hamar. 1991. High sulfide concentrations in rumen fluid associated with nutritionally induced polioencephalomalacia in calves. Am. J. Vet. Res. 52:1164–1169.

Hamlen, H., E. Clark, and E. Janzen. 1993. Polioencephalomalacia in cattle consuming water with elevated sodium sulfate levels: A herd investigation. Can. Vet. J. 34:153–158.

Haven, T.R., D.R. Caldwell, and R. Jensen. 1983. Role of predominant rumen bacteria in the cause of polioencephalomalacia (cerebrocortical necrosis) in cattle. Am. J. Vet. Res. 44:1451–1455.

Olkowski, A.A., S.R. Gooneratne, C.G. Rousseaux, and D.A. Christensen. 1992. Role of thiamine status in sulphur induced polioencephalomalacia in sheep. Res. Vet. Sci. 52:78–85.

Rankins, D.L., Jr., and G.S. Smith. 1991. Nutritional and toxicological evaluations of kochia hay (*Kochia scoparia*) fed to lambs. J. Anim. Sci. 69:2925–2931.

Rankins, D.L., Jr., G.S. Smith, and D.M. Hallford. 1991a. Altered metabolic hormones, impaired nitrogen retention, and hepatotoxicosis in lambs fed *Kochia scoparia* hay. J. Anim. Sci. 69:2932–2940.

Rankins, D.L., Jr., G.S. Smith, and D.M. Hallford. 1991b. Serum constituents and metabolic hormones in sheep and cattle fed *Kochia scoparia* hay. J. Anim. Sci. 69:2941–2946.

Rankins, D.L., Jr., G.S. Smith, and D.M. Hallford. 1991c. Effects of metoclopramide on steers fed *Kochia scoparia* hay. J. Anim. Sci. 69:3699–3705.

Hypomagnesemia

Bohman, V.R., F.P. Horn, B.A. Stewart, A.C. Mathers, and D.L. Grunes. 1983. Wheat pasture poisoning. 1. An evaluation of cereal pastures as related to tetany in beef cows. J. Anim. Sci. 57:1352–1363.

Bohman, V.R., A.L. Lesperance, G.D. Harding, and D.L. Grunes. 1969. Induction of experimental tetany in cattle. J. Anim. Sci. 29:99–102.

Camp, B.J., J.W. Dollahite, and W.L. Schwartz. 1968. Biochemical changes in sheep given *trans*-aconitic acid. Am. J. Vet. Res. 29:2009–2013.

Dua, K., and A.D. Care. 1995. Impaired absorption of magnesium in the aetiology of grass tetany. Brit. Vet. J. 151:413–426.

Fisher, L.J., N. Dinn, R.M. Tait, and J.A. Shelford. 1994. Effect of level of dietary potassium on the absorption and excretion of calcium and magnesium by lactating cows. Can. J. Anim. Sci. 74:503–509.

Russell, J.B., and N. Forsberg. 1986. Production of tricarballylic acid by rumen microorganisms and its potential toxicity in ruminant tissue metabolism. Brit. J. Nutr. 56:153–162.

Russell, J.B., and H.F. Mayland. 1987. Absorption of tricarballylic acid from the rumen of sheep and cattle fed forages containing *trans*-Aconitic acid. J. Sci. Food Agric. 40:205–212.

Russell, J.B., and P.J. Van Soest. 1984. In vitro ruminal fermentation of organic acids common in forage. Appl. Environ. Microbiol. 47:155–159.

Schwartz, R., M. Topley, and J.B. Russell. 1988. Effect of tricarballylic acid, a nonmetabolizable rumen fermentation product of *trans*-aconitic acid, on Mg, Ca and Zn utilization of rats. J. Nutr. 118:183–188.

Stout, P.R., J. Brownell, and R.G. Burau. 1967. Occurrences of *trans*-aconitate in range forage species. Agron. J. 59:21–24.

Laminitis

Block, E. 1994. Manipulation of dietary cation–anion difference on nutritionally related production diseases, productivity, and metabolic responses of dairy cows. J. Dairy Sci. 77:1437–1450.

Galey, F.D., V.R. Beasley, D. Schaeffer, and L.E. Davis. 1990. Effect of an aqueous extract of black walnut (*Juglans nigra*) on isolated equine digital vessels. Am. J. Vet. Res. 51:83–88.

Galey, F.D., H.E. Whiteley, T.E. Goetz, A.R. Kuenstler, C.A. Davis, and V.R. Beasley. 1991. Black walnut (*Juglans nigra*) toxicosis: A model for equine laminitis. J. Comp. Path. 104:313–326.

Garner, H.E., J.R. Coffman, A.W. Hahn, D.P. Hutcheson, and M.E. Tumbleson. 1975. Equine laminitis of alimentary origin: an experimental model. Amer. J. Vet. Res. 36:441–444.

Godfrey, S.I., M.D. Boyce, and J.B. Rowe. 1992. Changes within the digestive tract of sheep following engorgement with barley. Aust. J. Agric. Res. 44:1093–1101.

Kung, L., Jr., and A.O. Hession. 1995. Preventing in vitro lactate accumulation in ruminal fermentations by inoculation with *Megasphaera elsdenii*. J. Anim. Sci. 73:250–256.

MacDaniels, L.H. 1983. Perspective on the black walnut toxicity problem—Apparent allergies to man and horse. Cornell Vet. 73:204–207.

Moore, J.N., and D. Allen, Jr. 1996. The pathophysiology of acute laminitis. Vet. Med. 91:936–393.

Moore, J.N., D. Allen, Jr., and E.S. Clark. 1989. Pathophysiology of acute laminitis. Vet. Clinics N. America: Equine Practice 5:67–72.

Ralston, S.L., and V.A. Rich. 1983. Black walnut toxicosis in horses. J. Am. Vet. Med. Assoc. 183:1095.

Rowe, J.B., M.J. Lees, and D.W. Pethick. 1994. Prevention of acidosis and laminitis associated with grain feeding in horses. J. Nutr. 124:2742S–2744S.

True, R.G., and J.E. Lowe. 1980. Induced jugalone toxicosis in ponies and horses. Am. J. Vet. Res. 41:944–945.

True, R.G., J.E. Lowe, J.E. Heissen, and W. Bradley. 1978. Black walnut shavings as a cause of acute laminitis. Proc. Am. Assoc. Equine Pract. 24:511–515.

Uhlinger, C. 1989. Black walnut toxicosis in ten horses. J. Am. Vet. Med. Assoc. 195:343–344.

Vermunt, J.J. 1992. "Subclinical" laminitis in dairy cattle. N.Z. Vet. J. 40:133–138.

Vermunt, J.J., and P.R. Greenough. 1994. Predisposing factors of laminitis in cattle. Brit. Vet. J. 150:151–164.

Weiss, D.J., R.J. Geor, G. Johnston, and A.M. Trent. 1994. Microvascular thrombosis associated with the onset of acute laminitis in ponies. Amer. J. Vet. Res. 55:606–612.

Weiss, D.J., L. Monreal, A.M. Angles, and J. Monasterio. 1996. Evaluation of thrombin–antithrombin complexes and fibrin fragment D in carbohydrate-induced acute laminitis. Res. Vet. Sci. 61:157–159.

Weiss, D.J., A.M. Trent, and G. Johnston. 1995. Prothrombotic events associated with the prodromal stages of acute laminitis. Amer. J. Vet. Res. 56:986–991.

Green Feed–Induced Diarrhea

Pethwick, D.W., and H.M. Chapman. 1991. The effects of *Arctotheca calendula* (capeweed) on digestive function of sheep. Aust. Vet. J. 68:361–363.

CHAPTER 10

Mycotoxins Associated with Forages

TALL FESCUE TOXICOSES

Tall fescue (*Festuca arundinacea*) is a vigorous, coarse perennial grass which grows in pronounced clumps. While somewhat unpalatable, it is readily consumed by livestock in the absence of more palatable forage and in many areas is an excellent pasture species. It forms a sod that is particularly resistant to trampling, and it is quite drought resistant. In the U.S. Pacific Northwest, for instance, it is one of the few pasture grasses that forms a sufficiently dense sod to permit grazing of cattle during the wet winters, while it will continue to grow during the dry summers when other species become dormant without irrigation. It is grown extensively as a pasture and hay grass in the Pacific Northwest, Missouri, Kentucky, and throughout the southeastern U.S.

Two major problems in cattle have been associated with use of tall fescue pastures. These are summer fescue toxicosis, and fescue foot. An additional problem in cattle, fat necrosis, has been linked to the grazing of fescue pastures. In horses, the main signs of fescue toxicosis involve reproductive dysfunctions, such as delayed parturition and agalactia. Fescue poisoning is widespread in the U.S. and has been reported in Argentina, Australia, and New Zealand. In Australia, fescue foot is the main problem observed, and sheep as well as cattle are affected (Seawright, 1989).

Since about 1985, it has become clear that most or all of the toxicological problems associated with tall fescue are caused by endophyte-produced alkaloids. Although a number of types of alkaloids have been determined in tall fescue, the ergopeptine alkaloids (e.g. ergovaline) are primarily responsible for toxicity syndromes.

The history of **tall fescue toxicosis** in the U.S. began in 1931, when an agronomist from the University of Kentucky observed a tall, vig-

FIGURE 10–1 While tall fescue is one of the most important temperate forage grasses, it is also well known as a source of toxicity syndromes, when infected with endophytic fungi. (Courtesy of R.W. Hemken)

orous ecotype of tall fescue growing on a mountainside pasture. He collected seeds and propagated them, releasing the ecotype as a new variety, Kentucky 31 tall fescue (Bacon, 1995). This new productive variety was vigorously promoted by extension agents, and in a short time became the dominant temperate grass in the lower midwest and southeastern U.S. Unfortunately, animal performance did not match the vigor of the grass. Cattle grazing on Kentucky 31 tall fescue pastures often grew poorly, and developed a chronic unthrifty condition, especially in the summer. This was referred to as summer slump, summer syndrome, and summer fescue toxicosis. In addition, cattle on fescue pastures during the winter sometimes developed lameness and gangrene of the extremities (hooves, tail, ear-tips). This condition was called fescue foot. It appeared identical to classic ergotism, but there was no evidence of *Claviceps* infection of the seed heads, and the condition occurred during the winter when no seed heads were present.

In the 1970's, USDA and university researchers identified that in certain fields no fescue toxicosis occurred, whereas in adjacent fields the livestock showed toxicity signs. A comparison of the toxic and non-toxic grass revealed that in the toxic pastures the grass plants were infected with **endophytes**, or fungi which live entirely within the tissue spaces of plants. Their presence could be detected only microscopically (Fig. 10–2).

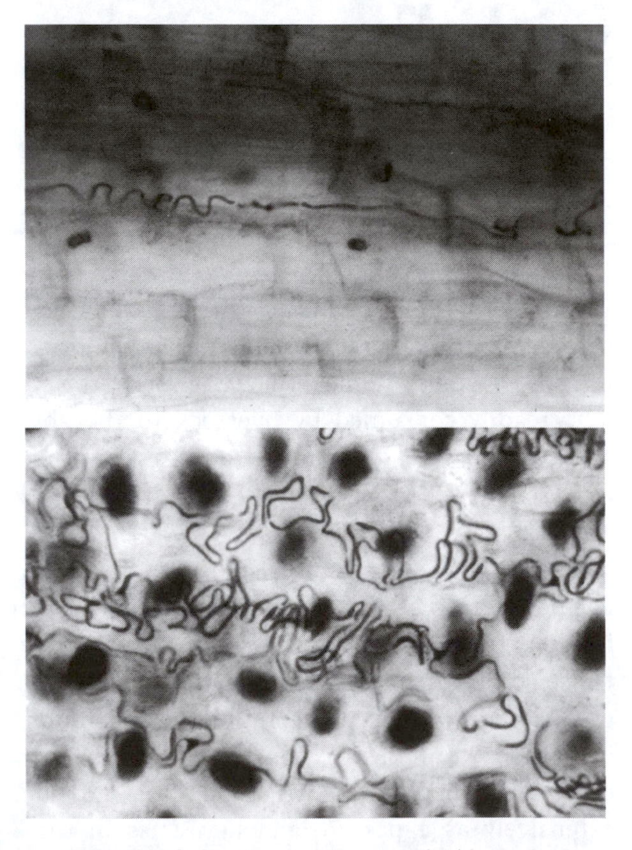

FIGURE 10–2 The dark lines are the *Neotyphodium coenophialum* endophytic fungi growing in the tissue spaces of tall fescue leaves (top) and seeds (bottom). The dark round objects in the seeds are cell nuclei. The endophyte lives in the spaces between cells. (Courtesy of R.E. Welty)

The endophyte that infects tall fescue is a member of the Balansiae tribe of *Clavicipitaceae*, which live systemically in plant tissue, in contrast to the Claviceps tribe which infect the flower. The tall fescue endophyte was originally called *Epichloe typhina*, but then was reclassified as ***Acremonium coenophialum***, and most recently is classified as ***Neotyphodium coenophialum***. Lyons *et al.* (1986) reported that this endophyte produces ergot alkaloids of the ergopeptide class, with **ergovaline** accounting for about 90% of the ergopeptide alkaloids. It is now generally accepted that fescue foot and summer fescue toxicosis are caused by ergot alkaloids (mainly ergovaline) in endophyte-infected tall fescue. Endophyte-free fescue is non-toxic.

The *Neotyphodium* endophyte enjoys a mutualistic relationship with the grass. Living within the plant tissues, it has free access to nutrients. The plant provides for reproduction of the fungus via the infected seeds. The endophyte has lost the ability to reproduce; it has no reproductive phases. Endophytes can be propagated only via infected seeds, and cannot move from an infected plant to a non-infected one. Thus the plant provides the fungus with a nutrient-rich protected environment, and provides a means for its reproduction, while the fungus produces chemicals (ergot alkaloids) which function as chemical defenses of the grass. They may also have some metabolic effects such as an influence on plant hormones and growth regulators (Joost, 1995). The result is that compared to endophyte-free fescue, endophyte-infected grass is more vigorous, pest-resistant, drought-resistant and tolerant of adverse soil and environmental conditions. It is even greener in color. For these reasons, endophyte-infected fescue is widely grown for turf purposes—i.e. lawns, playing fields, playgrounds, parks, highway banks, etc. Thus both endophyte-free (forage type) and endophyte-infected (turf type) fescues are grown. While ideally only endophyte-free fescue should be grown for forage, there are complicating factors. There are over 15 million hectares of endophyte-infected tall fescue pasture in the U.S. (Strickland *et al.*, 1993). Much of the fescue pasture land is rough, hilly country which would be difficult to reseed. Furthermore, in many areas, the endophyte-free fescue does not persist because of its lower vigor. For these reasons, livestock will continue to be exposed to toxic tall fescue for many years to come.

Further detail on the history of tall fescue and *Neotyphodium* are provided by Bacon (1995), Ball *et al.* (1993), Joost and Quisenberry (1993), Strickland *et al.* (1993) and Stuedemann and Hoveland (1988).

Fescue toxicoses can be categorized as thermoregulatory disturbances, altered lipid metabolism, and reproductive effects. The **thermoregulatory disturbances** can be subclassified as hypothermia (fescue foot) and hyperthermia (summer fescue toxicosis).

Hypothermia (Fescue Foot)

Cattle grazing tall fescue pastures are subject to gangrene of the extremities, referred to as **fescue foot** (Fig. 10–3). Lameness, arched back, and diarrhea may occur from a few days to several months after cattle are put into tall fescue pastures. More severe signs include loss of body weight, emaciation, rough hair coat, and gangrene of the tail tip, rear hooves, and ears. The hooves and portions of the tail may slough off. Blood vessels in affected areas are congested, and some tissues contain perivascular hemorrhages.

The cause of fescue foot is peripheral vasoconstriction due to the effects of ergot alkaloids on smooth muscle contraction. Reduction in blood flow to peripheral tissues leads to anoxia and tissue death, and subsequent gangrene. Fescue foot occurs primarily during the winter, when blood supply to the extremities tends to be reduced. Presence of ergot alkaloids would intensify this effect.

There is evidence of increased incidence of laminitis in horses exposed to endophyte-infected tall fescue (Rohrbach *et al.*, 1995).

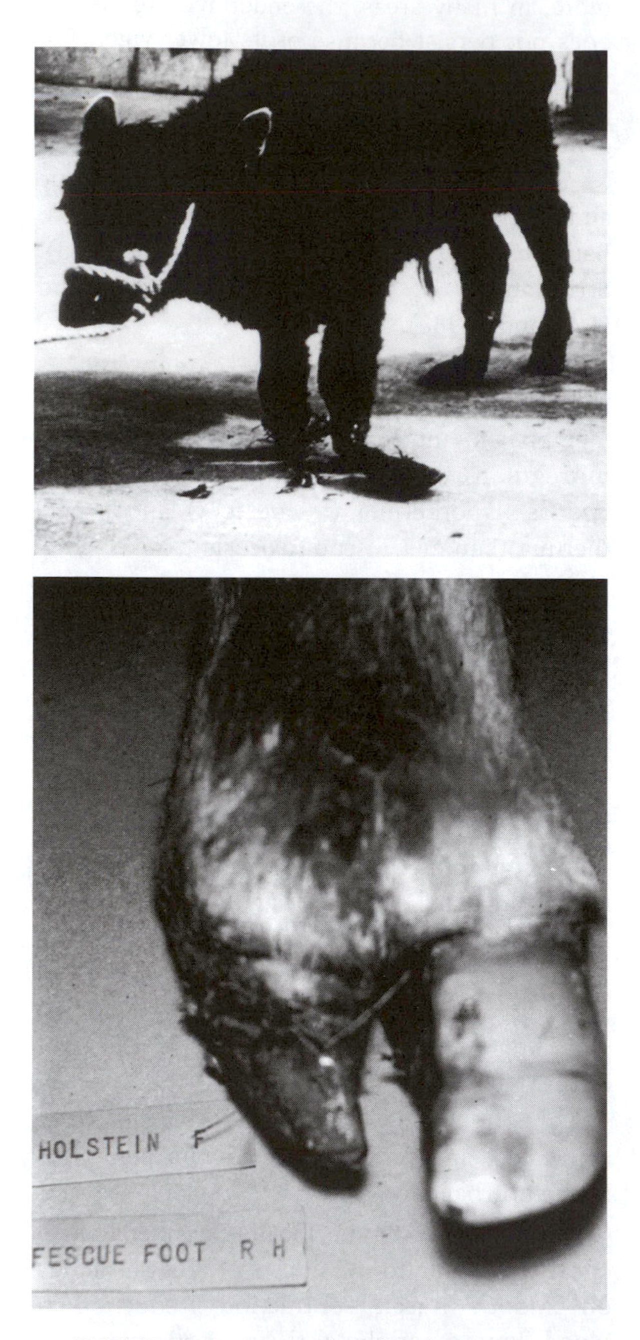

FIGURE 10–3 Fescue foot. (Top): A severe case, with gangrene of the lower limbs. (Courtesy of D.R. Jacobson) (Bottom): A more typical case, with gangrene of the hoof claw. (Courtesy of G.B. Garner)

Hyperthermia (Summer Fescue Toxicosis)

Hyperthermia (elevated body temperature) is a characteristic toxicological effect of ergot alkaloids (Peet *et al.*, 1991). **Summer fescue toxicosis** is manifested by poor animal growth or actual weight loss, a dull, rough hair coat caused by failure to shed the winter coat, excessive salivation, a high respiration rate, elevated body temperature and a susceptibility to heat stress. Affected cattle avoid grazing during the day, spend more time grazing at night, and seek shade or mud wallows to escape heat stress. Part of the reduced growth rate may be a consequence of less time spent grazing and thus lower feed intake (Howard *et al.*, 1992).

The normal physiological response to heat stress is an increase in blood flow to peripheral tissues (skin and extremities) to increase dissipation of heat from the body surface. Surface blood vessels are dilated. In fescue toxicosis, blood flow to peripheral tissues is reduced (Rhodes *et al.*, 1991). The animal is unable to dissipate heat normally, causing body temperature to rise. The vasoconstrictive effects of ergot alkaloids (Oliver *et al.*, 1993) contribute to this effect. Ergot alkaloids also reduce water uptake from the gut (Faichney and Barry, 1986), thus reducing evaporative cooling. Reduced voluntary water intake occurs in fescue toxicosis (Fiorito *et al.*, 1991). Rhodes *et al.* (1991) observed a reduction in blood flow to the digestive tract as well as to peripheral tissues; reduced visceral blood flow might account for impaired water uptake.

The most dramatic clinical sign of fescue toxicosis is severe **hypoprolactemia** (low serum prolactin). Blood prolactin levels may be as low as 1–2% of normal. It is likely that the molecular mode of action of ergot alkaloids is mediated via effects on prolactin. Ergot alkaloids inhibit prolactin secretion by the pituitary (Strickland *et al.*, 1992). Strickland *et al.* (1993) speculate that inhibited prolactin secretion is caused by agonistic (stimulatory) interaction of tall fescue toxins with the **D2-dopamine receptor** on the lactotroph (there are 5 subclasses of dopamine receptors, D1-D5). **Dopamine** regulates prolactin secretion from lactotrophs of the anterior pituitary by activating the D-2 dopamine receptor, causing inhibition of prolactin secretion. By acting as a dopamine agonist (dopaminergic effect), ergot alkaloids suppress prolactin secretion. **Prolactin** has a number of physiological roles, including effects on milk secretion, reproduction, gut motility, and the appetite center. Thus many of the signs of fescue toxicosis may be mediated by prolactin suppression. Prolactin affects lipogenesis (Strickland *et al.*, 1993); altered fat metabolism (fat necrosis) occurs in fescue toxicosis. Seasonal changes in hair coat are influenced by prolactin and melatonin; the lack of shedding of the winter hair coat is explainable on this basis.

Fescue toxicosis affects various neuroendocrine compounds besides prolactin, including melatonin, serotonin and dopamine (Porter *et al.*, 1990, 1994). Ergot alkaloids are structurally similar to biogenic amines such as serotonin, dopamine, epinephrine and norepinephrine, and may interact with biogenic amine receptors, altering neurotransmitter homeostasis. Cattle grazing toxic tall fescue have neurotransmitter imbalances in the pituitary and pineal glands. The pineal gland secretes melatonin which functions in modulating seasonal changes in hair growth, reproduction, temperature acclimation, etc. Thus the involvement of tall fescue toxins in **neurotransmitter imbalance** has profound effects on animal well-being, affecting growth, reproduction, and the ability to respond to seasonal changes in photoperiod and environmental temperature.

Fescue toxicosis affects hair pigmentation of cattle. For example, the black hair of Angus cattle may turn brown or bronze (Lipham *et al.*, 1989). This is probably due to effects of melatonin via reduced prolactin levels (Porter *et al.*, 1993). Seasonal changes in hair coat are related to photoperiod-induced changes in prolactin and melatonin. By its dopaminergic action, ergovaline in toxic tall fescue could have similar effects.

There are cattle **breed differences** in susceptibility to fescue toxicosis. Brahman-type animals are more resistant than European breeds to both heat stress and fescue toxicosis. With-in breed variations in susceptibility have not been found (Hohenboken *et al.*, 1991; Gould and Hohenboken, 1993). Nutting *et al.* (1992) noted that while Brahman cattle are less sensitive than Angus animals to the heat stress syndrome, they are more affected than Angus cows to fescue foot syndrome and poor reproduction.

Fat Necrosis

Rumsey *et al.* (1979) described **fat necrosis** in beef cattle grazing tall fescue pastures in Georgia. The necrotic fat lesions in these cattle were generally in the abdominal cavity and ranged in size from small nodules embedded in normal depot fat to large, irregularly shaped masses that surrounded and constricted the intestines and reproductive organs (Fig. 10–4). The fat lumps were hard and, when cut in a cross section, had a dry, hard, cheesy opaque appearance in contrast to the normal surrounding fat. Calcification was sometimes noted. The necrotic fat was higher in ash, cholesterol, calcium, and magnesium than normal fat and much lower in ether extract (91.9 vs 47.7%). It is possible that fescue-induced fat necrosis is related to the vasoconstriction observed in fescue foot. Rumsey *et al.* (1979) suggest that vasoconstriction may be a direct causative factor or may lead to a general febrile (feverish) condition, which then may trigger a lipolytic process. The febrile condition associated with summer fescue toxicosis would also be consistent with this theory. Fat necrosis generally is seen only in animals which have been exposed to toxic tall fescue over several grazing seasons. It seems to be associated with fescue that has been heavily fertilized with nitrogen or poultry litter (Rumsey *et al.*, 1979; Stuedemann *et al.*, 1985). The con-

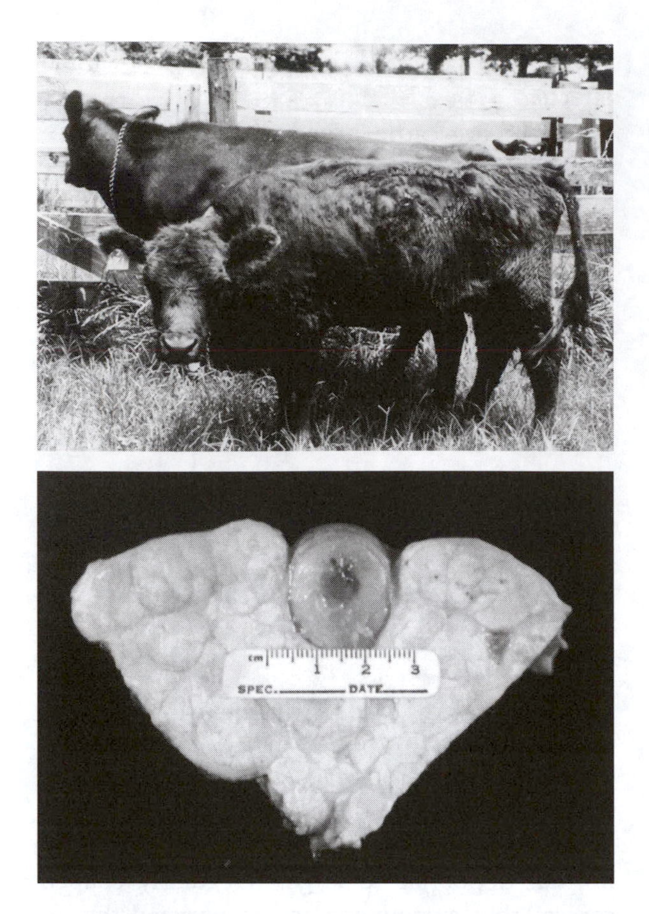

FIGURE 10–4 (Top): Cattle with fat necrosis. They had been grazing tall fescue pastures for five years. Note the lumps of hard subcutaneous fat. (Bottom): A deposit of hard fat constricting the intestine. (Courtesy of J.A. Stuedemann)

dition is of quite limited distribution and economic importance. Mortalities may occur because of constriction of the colon; the condition may also cause dystocia because of constriction of the reproductive tract.

Several **metabolic changes** have been noted with fat necrosis. Affected animals have reduced serum cholesterol (Stuedemann *et al.*, 1985) and elevated serum amylase (Nutting *et al.*, 1992). Cattle on toxic tall fescue pasture have higher saturated fatty acid contents of body fat, which Townsend *et al.* (1991) speculate could play a role in development of fat necrosis, because saturated fatty acids produce a hard body fat. Nutting *et al.* (1992) reported pancreatic lesions in cattle grazing toxic tall fescue, and suggested that pancreatitis could contribute to fat necrosis, in part by lipid hydrolytic enzymes released from the pancreas into the blood.

Reproductive Effects

Pronounced impairment of reproduction occurs in livestock consuming endophyte-infected tall fescue (Porter and Thompson, 1992). Horses are the only livestock whose reactions to the toxic tall fescue are almost exclusively related to poor reproduction. Mares on toxic tall fescue may experience prolonged gestation, dystocea, agalactia (lack of lactation), thickened, edematous placentas and have large, weak foals with elongated hooves (McCann *et al.*, 1992). Foal survival is very low (Putnam *et al.*, 1991). Abortion may also occur. Clinically, affected mares have low serum prolactin and progesterone (McCann *et al.*, 1992) and decreased triiodothyronine concentrations (Boosinger *et al.*, 1995). Low prolactin causes agalactia while suppression of progesterone may be a contributing factor to prolonged gestation. The fescue effects are manifested when the mares are consuming toxic fescue from day 300 of pregnancy. Removal of mares from toxic fescue by day 300 results in a normal parturition (Putnam *et al.*, 1990). Pregnant mares on toxic fescue exhibit no direct signs of toxicity except for intermittent diarrhea and excessive sweating, presumably in response to hyperthermia (Putnam *et al.*, 1991).

The reproductive effects in horses are caused by the ergopeptine alkaloids in endophyte-infected tall fescue. Similar signs occur in horses consuming the ergot bodies of *Claviceps purpurea* (Riet-Correa *et al.*, 1988). Mechanisms involved include reduced serum prolactin, reduced placental blood flow because of vasoconstrictive effects, and ergot-stimulated contraction of uterine muscle causing abortion. Administration of **bromocriptine**, a synthetic ergot alkaloid, to pregnant mares produces the same adverse reproductive effects as endophyte-infected tall fescue (Ireland *et al.*, 1991). Domperidone, a D2 dopamine receptor antagonist, prevents the inhibitory effects of ergovaline on prolactin release, and may offer potential as a treatment for fescue toxicosis in horses (Redmond *et al.*, 1994; Cross *et al.*, 1995).

Impaired reproduction of cattle on endophyte-infected tall fescue occurs (Paterson *et al.*, 1995). Reduced calving rates may be due to altered luteal function and reduced levels of circulating progesterone (Porter and Thompson, 1992). Lowered milk production may occur. General reproductive efficiency is impaired, without the pronounced effects on the fetus and parturition events that occur in horses. Impaired reproduction may be in part a consequence of general unthriftiness and weight loss of cattle on toxic tall fescue (Paterson *et al.*, 1995).

Relatively little information is available on the effects of tall fescue and ergot alkaloids on male reproduction. Bromocriptine (a synthetic ergot alkaloid)-induced hypoprolactemia in rams reduces expression of sexual behavior such as libido (Regisford and Katz, 1994). The absence of much data suggests that effects of toxic tall fescue on male reproduction are probably minor.

Alkaloids in Tall Fescue

A number of alkaloids have been isolated from tall fescue. Two alkaloids that are apparently synthesized by the grass rather than by endophytes are the diazaphenanthrene alkaloids perloline and perlolidine (Fig. 10–5). **Perloline** is a yellowish green fluorescent alkaloid that was first isolated from perennial ryegrass in 1943 by New Zealand researchers.

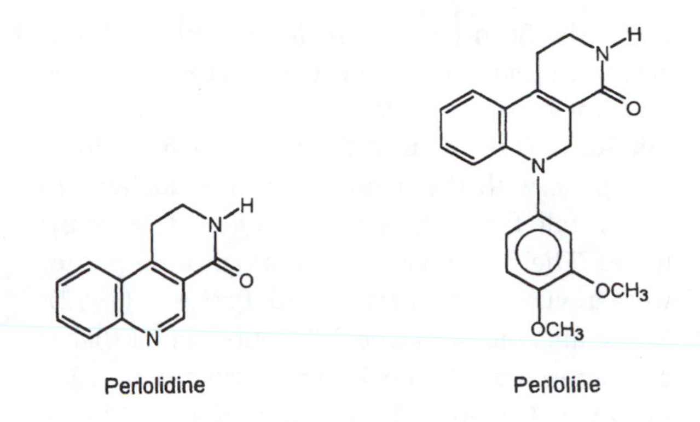

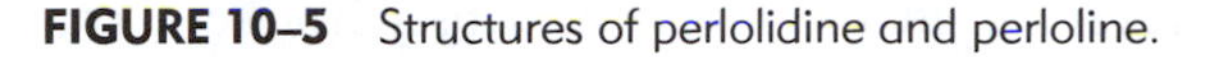
FIGURE 10–5 Structures of perlolidine and perloline.

Perloline has been demonstrated to have physiological effects on animals. After parenteral administration of the alkaloid, symptoms of convulsions, muscular incoordination, increased pulse and respiration rates, mild photosensitization, and coma have been seen in sheep. Boling *et al.* (1975) demonstrated that administration of 0.5% perloline to lambs reduced the digestibility of protein and cellulose and reduced nitrogen retention. Production of volatile fatty acids in the rumen was reduced. The body temperature of the perloline-fed lambs tended to be higher than for the control group. Inhibition of rumen cellulolytic organisms by perloline has also been demonstrated. These alkaloids do not have a major role in tall fescue toxicoses.

The two classes of alkaloids which have received the most attention as possible causes of fescue toxicosis are the loline and ergopeptine

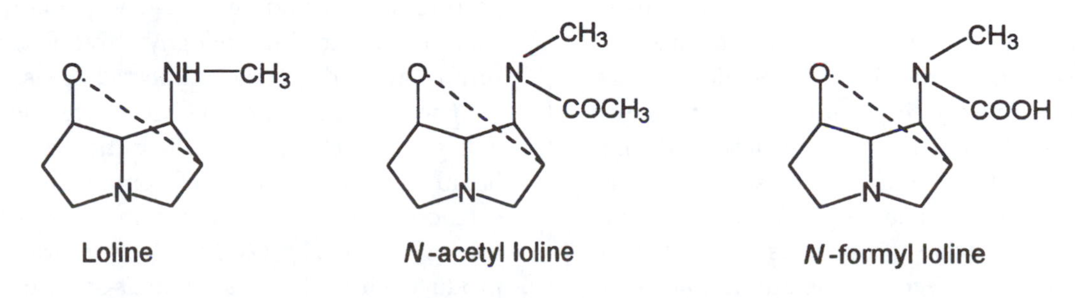

FIGURE 10–6 The lolines in tall fescue are saturated, non-hepatoxic pyrrolizidine alkaloids.

alkaloids. Both classes are found only in *Neotyphodium coenophialum*-infected fescue, indicating that they are of endophyte origin. The **loline alkaloids** (Fig. 10–6) are saturated pyrrolizidine alkaloids, with the major ones in endophyte-infected tall fescue being N-acetyl and N-formyl loline. Their chemistry and biological properties were reviewed by Powell and Petroski (1992). These alkaloids lack the 1,2 double bond that is characteristic of hepatoxic pyrrolizidine alkaloids (see Chapter 12). The role, if any, of loline alkaloids in the fescue toxicosis has been controversial. They appear to have a minor role, with the ergopeptine alkaloids being the major causative agents of fescue toxicoses. Loline alkaloids may have mild vasoconstrictive properties (Strickland *et al.*, 1993), so could contribute to the fescue foot and hyperthermia syndromes. Loline alkaloids may have a slight effect in inhibiting prolactin release from the pituitary, but their activity as D2 dopamine receptor agonists is very minor compared to that of ergovaline (Strickland *et al.*, 1994).

N-formyl loline is the major loline alkaloid (70–80% of total) in endophyte-infected tall fescue, with 15–20% N-acetyl loline, and minor quantities of other lolines (Yates *et al.*, 1990). N-acetyl and N-formyl loline are converted to loline in the rumen by microbial action (Westerndorf *et al.*, 1993), and loline is absorbed from the rumen. This is a bioactivation, because loline is physiologically more active than the N-acetyl and N-formyl derivatives (Westendorf *et al.*, 1993). Loline has been measured in blood and urine (TePaske *et al.*, 1993).

The **ergopeptine alkaloids** (Fig. 10–7) in endophyte-infected tall fescue appear to be the primary causative agents of fescue toxicosis. Their vasoconstrictive and dopamine agonist properties account for the physiological effects and signs of toxicity. Ergovaline is the principal ergopeptide alkaloid in toxic tall fescue (Lyons *et al.*, 1986). Others include ergosine, ergonine and lysergic acid amide (ergine). Lysergic acid amide has **sedative properties** (see sleepygrass) as well as effects on the autonomic nervous system including hypersalivation, emesis, dizziness and diarrhea (Oliver *et al.*, 1993). These effects, including a lethargic, partially-sedated state, are seen in animals consuming a high intake of toxic fescue forage or seed. Lysergic acid diethylamide (LSD), a common substance of abuse, is structurally similar to the alkaloids in tall fescue.

The endophyte *Neotyphodium coenophialum* occurs in the stems, leaf sheaths and seeds but not in the leaf blades. Rottinghaus *et al.* (1991) determined the ergovaline content of different plant parts and stages of growth. **Ergovaline** levels averaged 258 μg/kg (ppb) in leaf blades, 494 μg/kg in stems with leaf sheaths, and 895 μg/kg in seed heads. Occurrence of ergovaline in the leaf blades in the absence of endophyte indicates translocation of the alkaloids within the plant. Nitrogen fertilization increased ergovaline levels. Ergovaline content of the fescue increases with plant maturity, when the seed heads develop. Harvesting hay or silage before seed head development may be an effective means of avoiding toxicoses. Straw remaining after harvesting tall fescue seed may contain significant amounts of ergot alkaloids (Stamm *et al.*, 1994).

Ergovaline is partially degraded in the rumen to unidentified degradation products (Westendorf *et al.*, 1993; Moyer *et al.*, 1993). Moyer *et al.* (1993) suggested that because the endophyte resides within the middle lamella of plant cells, ergovaline may be protected against rumen degradation by being sequestered within the poorly digested plant cell wall. Endophyte-infected seed survives passage through the digestive tract of horses and cattle (Shelby and Schmidt, 1991). Animals grazing on infected pasture should undergo a 3-day quarantine period before entering endophyte-free fescue pasture to avoid dispersal of infected seeds.

The level of ergovaline in tall fescue which will produce clinical signs is difficult to predict because of the effect of environmental temperature on hypo- and hyperthermic responses. A level of 200 μg/kg (200 ppb) of ergovaline may produce clinical signs in heat-stressed cattle (Rottinghaus *et al.*, 1991).

Lysergic acid

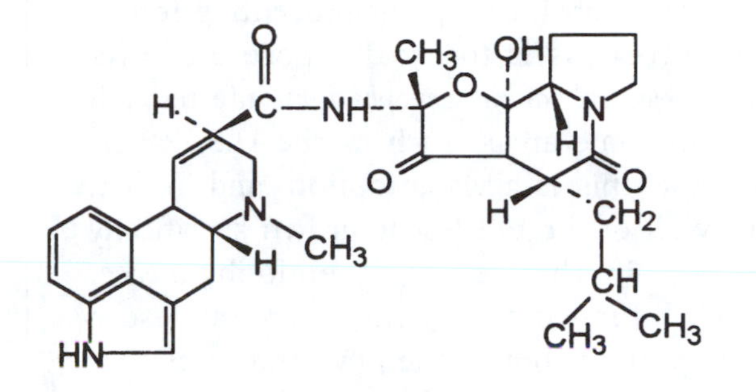

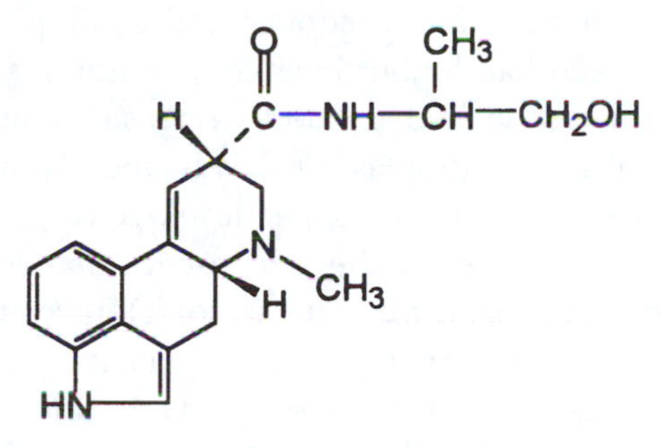

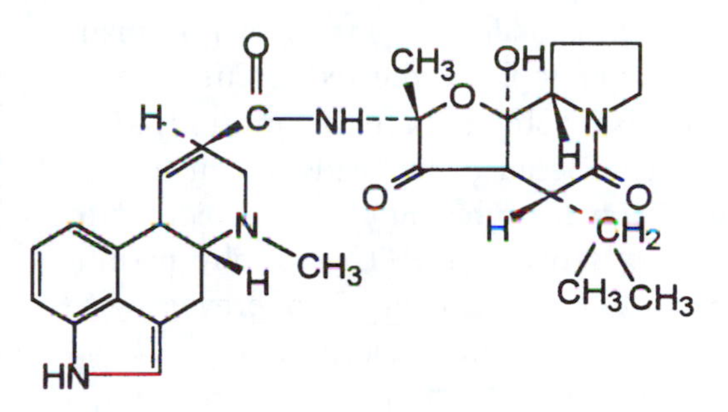

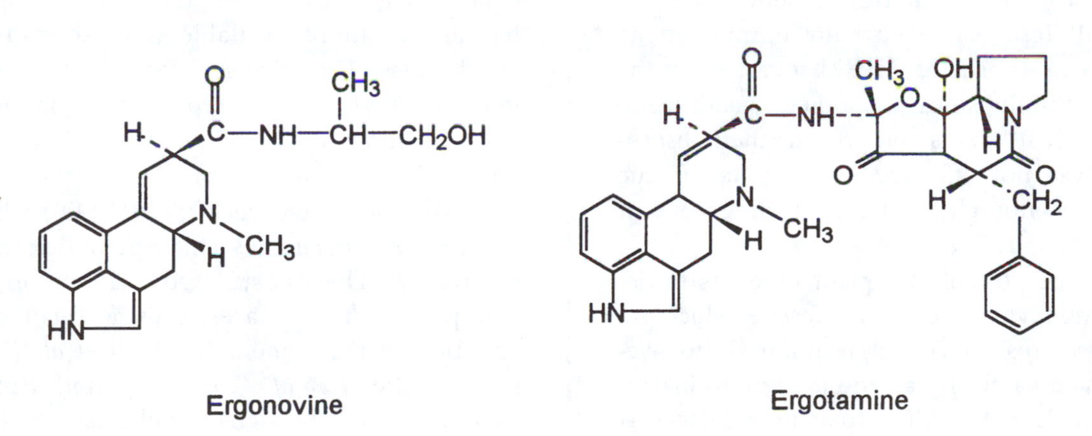

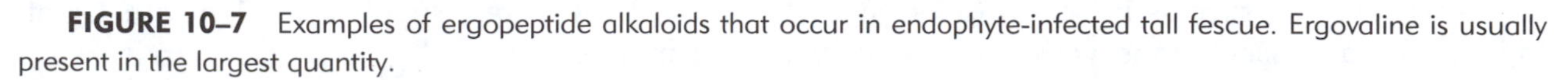

FIGURE 10–7 Examples of ergopeptide alkaloids that occur in endophyte-infected tall fescue. Ergovaline is usually present in the largest quantity.

Prevention of Fescue Toxicoses

The most obvious way of preventing fescue toxicity problems is to avoid exposure of livestock to *Neotyphodium coenophialum*–infected tall fescue. In some areas, such as the U.S. Pacific Northwest, this is a viable option, and in fact, virtually all of the tall fescue grown specifically for forage in this region is endophyte-free. However, in much of the U.S. where tall fescue is an important forage, the environmental conditions are too rigorous for endophyte-free fescue to be grown. In most of the southeastern U.S., endophyte-free fescue is too susceptible to pathogens, insects and environmental stresses to persist in a pasture. Practical conditions dictate that the major forage grass grown in many areas will be endophyte-infected fescue.

Toxicosis problems can be minimized in various ways. Because the seeds are highest in ergovaline content, grazing can be avoided during this vegetative stage. Topography permitting, pastures can be clipped to prevent seed formation. Clover and other forage can be interseeded in fescue, to dilute the intake of ergot alkaloids. Hay and silage can be prepared from the grass before seed development occurs. Hay can be ammoniated with anhydrous ammonia. Kerr *et al.* (1990) and Chestnut *et al.* (1991) reported that **ammoniation** of endophyte-infected tall fescue hay virtually eliminated its toxicity. Chestnut *et al.* (1992) investigated the use of an aluminosilicate as a dietary additive to bind tall fescue toxins and prevent their absorption; it was not effective. Ensiling tall fescue forage does not affect the ergovaline content (Turner *et al.*, 1993).

It may be possible for plant breeders to develop endophyte-infected tall fescue which retains insect resistance but is nontoxic to livestock. The alkaloid **peramine** is toxic to insects but has little mammalian toxicity. Roylance *et al.* (1994) reported success in selecting for endophyte-fescue associations which had high peramine and low ergovaline concentrations. These alkaloids are independently regulated and their production is controlled by both plant and endophyte genotype.

Nutrient Interactions with Tall Fescue

There is some evidence that tall fescue may affect mineral metabolism in cattle. Stoszek *et al.* (1979) reported that cattle on tall fescue pasture containing 6.6 ppm **copper** developed pronounced copper deficiency, while those on an adjacent quack grass (*Agropyron repens*) pasture containing a lower copper level (4.6 ppm) maintained normal blood copper and ceruloplasmin levels, and had higher liver copper stores. Cattle on tall fescue had a rapid decrease in plasma copper and ceruloplasmin levels and a progressive decline in liver copper level. Average daily gains were less on the tall fescue than on the quack grass pasture. The factor(s) involved in the apparent impairment of copper metabolism in animals grazing tall fescue was not identified. Coffey *et al.* (1992) reported a decrease in serum copper in cattle grazing endophyte-infected tall fescue, but animal performance was not improved with copper supplementation. Hathaway *et al.* (1981) showed that the **selenium** in tall fescue was more available than selenium in quack grass. The adverse effects of endophyte-infected tall fescue on reproduction in horses were not modified by supplementing with selenium (Monroe *et al.*, 1988).

According to Lauriault *et al.* (1990), tall fescue toxicosis resembles **thiamin deficiency** in ruminants. They postulated that endophytes may produce thiaminases, causing thiamin destruction in the rumen. Lauriault *et al.* (1990) and Dougherty *et al.* (1991) reported a reduction in some of the signs of tall fescue toxicosis in cattle by dietary supplementation with 1 g of thiamin per day.

PERENNIAL RYEGRASS STAGGERS

Perennial ryegrass staggers is a disorder of animals grazing perennial ryegrass pastures. It is characterized by neurological signs such as incoordination, staggering, head shaking, and collapse (Fig. 10–8). Animals generally appear normal until disturbed. The neurological effects are temporary, and although affected animals usually regain normal composure within a short time, the condition complicates animal management practices such as moving sheep from one pasture to another. Mortality may occur from misadventure if affected animals fall over cliffs, into ditches or ponds, and so on. The disease is particularly prevalent in New Zealand but occurs in most areas where ryegrass is an important pasture species. The causative agents are compounds called **tremorgens**. Although a number of tremorgens have been identified, the most important is **lolitrem B**, an alkaloid produced by the endophytic fungus *Neotyphodium lolii* (DiMenna *et al.*, 1992; Miles *et al.*, 1992). Staggers may also be caused by other tremorgens such as janthitrem B, produced by *Penicillium* spp. growing on dead plant litter in ryegrass pastures (Gallagher *et al.*, 1977; Mantle *et al.*, 1977; Wilkins *et al.*, 1992). The tremorgens (Fig. 10–9) are indole-diterpene neurotoxins (Miles *et al.*, 1992). Other tremorgens with possible involvement in ryegrass staggers include lolitrems A, C, D, and E (Munday-Finch *et al.*, 1995) and other precursors of the lolitrems such as paxitriols, lolitriol and paxilline (DiMenna *et al.*, 1992; Miles *et al.*, 1992; Miles *et al.*, 1994).

FIGURE 10–8 A sheep exhibiting neurological effects of ryegrass staggers. (Courtesy of R.E.G. Keogh and M.E. Soulsby)

As with *Neotyphodium* infection of tall fescue, endophytes in ryegrass improve the vigor of the grass. In New Zealand, persistence of ryegrass in pastures is greatly facilitated by endophyte infection which protects the young plant shoots (tillers) from the Argentine stem weevil. The major alkaloid with insect-deterring properties

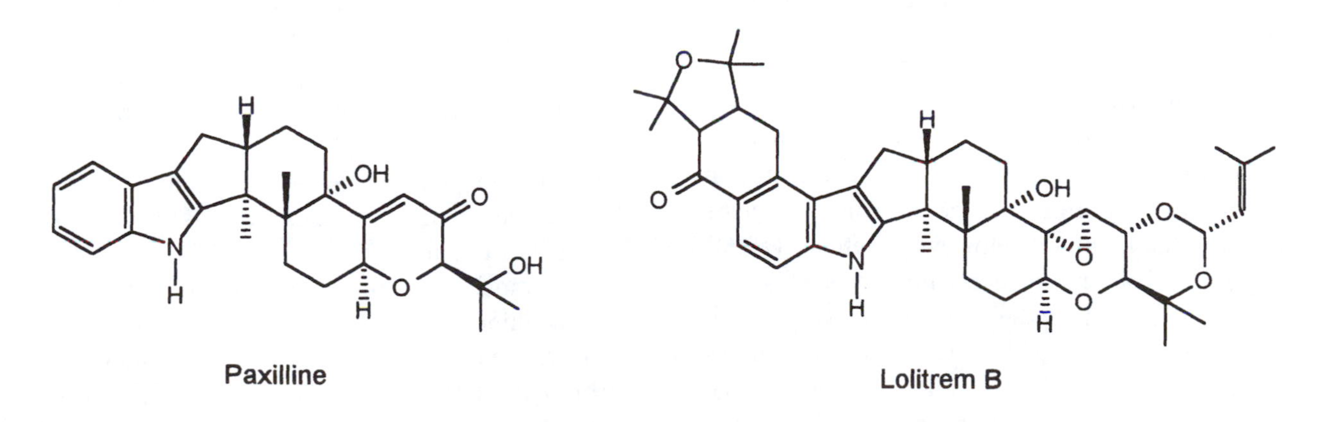

FIGURE 10–9 Lolitrem B is the major tremorgen in endophyte-infected perennial ryegrass. Paxilline is a component in the biosynthesis of lolitrem B, as well as having tremorgenic activity.

is **peramine**, which has low toxicity to mammals (Rowan and Gaynor, 1986). Plant breeding efforts to develop endophyte-infected ryegrass with high insect resistance but low mammalian toxicity show promise (Fletcher, 1993).

A selected *Neotyphodium* strain, called Endosafe, was developed in New Zealand and incorporated into commercially-available ryegrass varieties. It was selected to not produce lolitrem B, but to produce peramine, which protects the grass against Argentine stem weevil. Unfortunately, after seed had been released for commercial use, it was discovered that the Endosafe strain produced toxic levels of ergovaline. Cultivars containing Endosafe were withdrawn from the market.

Lolitrem B concentrations in *Neotyphodium lolii*–infected ryegrass plants are highest in the leaf sheaths and lowest in leaf blades (DiMenna *et al.*, 1992). Thus the staggers syndrome is most often seen in closely grazed pastures. Signs of ryegrass staggers develop when the lolitrem B concentrations exceed 2 to 2.5 $\mu g/g$ of dry matter (DiMenna *et al.*, 1992). Animal management to avoid close grazing will minimize staggers incidence (Fletcher, 1993).

Decreased serum prolactin has been noted in lambs grazing endophyte-infected ryegrass (Fletcher and Barrell, 1984). This effect is likely caused by ergot alkaloids rather than tremorgens. The ergot alkaloids of tall fescue, such as ergovaline, have dramatic prolactin-lowering effects. Ergovaline is also produced by endophytes in ryegrass. However, it would be desirable to determine whether tremorgens have a specific effect on prolactin. Piper (1989) noted that serum aspartate amino transferase was elevated in sheep with ryegrass staggers, which suggests hepatobiliary damage. However, Fletcher (1993) suggested that because serum levels of aspartate amino transferase are also elevated with muscle damage, the likely cause of the changes in serum enzymes is secondary muscle damage from bruising and other physical injuries during episodes of staggers. Fletcher (1993) also reported elevated serum creatine kinase activity along with elevated aspartate amino transferase, suggesting enzyme leakage from muscle cells.

Perennial ryegrass staggers is a serious problem in New Zealand. It has been known for many years in Oregon (Shaw and Muth, 1949), where the condition has also been noted in horses fed ryegrass straw (Hunt *et al.*, 1983). Oregon is a major grass seed–producing area, with tall fescue and perennial ryegrass the two major species grown. Endophyte-infected cultivars of both species make up an increasing share of the total seed production. These endophyte-infected varieties are used for turf (lawns, playgrounds, highway banks, etc.) because of the enhanced vigor of the grass. However, because the seed fields are grazed by sheep during the winter, problems with ryegrass staggers in Oregon are increasing (Pearson *et al.*, 1996). Endophyte-infected forage cultivars have also been introduced into the United States from New Zealand, leading to increased incidents of staggers. These cultivars are appropriate for New Zealand, because of their resistance to the Argentine stem weevil but may be inappropriate for the United States where the stem weevil does not occur. Other recent reports of ryegrass staggers involve cattle and sheep in northern California (Galey *et al.*, 1991), fallow deer in Australia (Mitchell and McCaughan, 1992), and cattle in Argentina (Odriozola *et al.*, 1993).

There is individual animal variability in susceptibility to ryegrass staggers (Foot *et al.*, 1994), offering the possibility of selection for more resistant animals (Morris *et al.*, 1995).

Consumption of endophyte-infected ryegrass causes **diarrhea** in sheep in New Zealand (Fletcher, 1993) and Australia (Foot *et al.*, 1994). This has major economic implications, particularly as it causes wool-staining and flystrike. The toxic agent responsible appears to be a translocated metabolite such as peramine or paxilline, rather than lolitrem B which occurs in the leaf sheaths and is not translocated throughout the plant (Fletcher, 1993). **Hyperthermia** (elevated body temperature and susceptibility to heat stress) is also observed in cattle and sheep grazing endophyte-infected ryegrass. Hyperthermia is most likely caused by ergot alkaloids (e.g. ergovaline) produced by *Neotyphodium* spp. The dual content of lolitrem B and ergovaline of *Neotyphodium lolii*–infected ryegrass can have interactive effects. In Australia, large numbers of sheep have drowned when grazing endophyte-infected ryegrass pastures (Foot *et al.*, 1994). In these cases, sheep experiencing ergovaline-induced hyperthermia during periods of high environmental temperature wade into watering ponds to alleviate heat stress, and the lack of coordination induced by lolitrem B results in their drowning. Other effects noted in Australia with endophyte-infected ryegrass include high lamb mortality and an allelopathic effect of the ryegrass in inhibiting growth of pasture legumes, thereby reducing total pasture productivity (Cunningham *et al.*, 1993). High lamb mortality probably is caused by reduced prolactin secretion in ewes, and reduced milk production, associated with the ergopeptine alkaloids produced by *Neotyphodium* spp. Watson *et al.* (1993) suggest that the apparent allelopathic effect of endophyte-infected ryegrass on clover is actually a change in competitive balance between the grass and clover influenced by an interaction with insect pests of both species.

Neotyphodium lolii infection of perennial ryegrass results in a complex of animal effects, with staggers caused by lolitrems, hyperthermia and lowered serum prolactin caused by ergovaline and other ergopeptine alkaloids, and diarrhea, possibly caused by other mycotoxins such as peramine and paxilline. In Oregon, respiratory disease often accompanies ryegrass staggers in cattle (Pearson *et al.*, 1996), but a pneumotoxic agent in the grass has not been identified.

OTHER ENDOPHYTE-RELATED SYNDROMES

Sleepygrass (*Stipa robusta*) is a perennial bunch grass found on rangelands of the southwestern United States (Colorado, Arizona, Texas, and New Mexico). Consumption of the grass by cattle and horses causes a profound stuporous condition which may last several days. Animals may enter a state of deep somnolence from which they cannot be roused. When the horse was the principal means of travel, considerable inconvenience and sometimes danger occurred when animals consumed the plant. Reportedly, the U.S. Cavalry was sometimes trapped in Indian attacks when their horses had eaten sleepygrass (USDA, 1976). Horse traders sometimes fed sleepygrass to wild ponies and then sold them as halterbroken (Smalley and Crookshank, 1976). Kingsbury (1964) summarized early research on the toxic signs. Mildly poisoned animals are dejected, inactive, and withdrawn. With a large dose they become somnolent, with a drooping head, closed eyes, and irregularity of gait if forced to move. They slobber copiously and may urinate frequently, even while lying down. Severely poisoned animals lie flat on their side with their head on the ground. They are in profound slumber and can be woken only momentarily. A type of catatonia as well as sleepiness is induced; animals may freeze in one position, with a hoof raised, for instance, and remain absolutely motionless, oblivious to pestering flies, for as long as 45 minutes (USDA, 1976). In the past, horses were the main animals affected; sleepygrass is now mainly a problem with cattle (USDA, 1976). Sheep are not affected as severely as cattle and horses. Epstein *et al.* (1964) reported that diacetone alcohol present in the plant is the soporific agent. However, more recent work, reviewed by Petroski *et al.* (1992), suggests that ergot alkaloids (Fig. 10–7) produced by *Acremonium* endophytes in *Stipa robusta* are the causative agents. The predominant ergot alkaloid identified in sleepygrass is **lysergic acid amide**, which is similar in structure to the well-known hallucinogen LSD (lysergic acid diethylamide). Petroski *et al.* (1992) collected sleepygrass forage near Cloudcroft, NM, a mountainous area with many reports of sleepygrass toxicosis. All samples contained *Acremonium* endophyte. The plant material was extracted and a number of ergot alkaloids identified. The concentrations of lysergic acid amide and isolysergic acid amide, both of which produce pronounced sedative effects in humans, were sufficiently high to account for the effects of sleepygrass on cattle and horses (assuming that dose responses are similar to those of humans). For example, they calculated that at an expected intake of 1% of body weight of sleepygrass, horses would consume approximately six times the dose expected to produce a sedative effect in humans. Lysergic acid amide also causes hypersalivation (Oliver *et al.*, 1993), a prominent sign of sleepygrass toxicosis. Lysergic acid and isolysergic acid amides have also been isolated from endophyte-infected tall fescue at levels approximately 10% of those in sleepygrass (Petroski *et al.*, 1992). This would likely explain the depression and dullness sometimes noted in animals fed endophyte-infected tall fescue. Powell and Petroski (1992) have reviewed alkaloids found in endophyte-infected grasses, including sleepygrass.

Various other endophyte-infected grasses have caused syndromes similar to those of ryegrass staggers, fescue toxicosis and sleepygrass toxicosis. In South Africa, *Melica decumbens* (staggers grass, dronk grass) causes a staggers condition (Kellerman *et al.*, 1988.) It was shown to contain paxilline analogs but no lolitrems (Miles *et al.*, 1993) and was endophyte infected. In China, the syndrome known as drunken horse disease is caused by endophyte-infected *Achnatherum (Stipa) inebrians* (drunken horse grass). It contains high levels of ergonovine, which causes sloughing of extremities of sheep during intoxication (similar to fescue foot), and significant levels of lysergic acid amide (Miles *et al.*, 1996). The latter compound likely is responsible for the symptoms in horses. Mortality of severely affected animals may occur (Miles *et al.*, 1996). A neurotoxic disorder (mal seco) of horses in Argentina and Chile may be associated with infection of Argentine fescue (*Festuca argentina*) with an *Acremonium* endophyte (Uzal *et al.*, 1996).

A condition called **paspalum staggers** occurs in cattle, and occasionally in sheep and horses, in several parts of the world where *Paspalum* spp. such as dallisgrass (*P. dilatatum*) and bahiagrass (*P. notatum*) are grown as pasture grasses, including New Zealand, Australia, South Africa, the U.S., Portugal, and Italy (Mantle *et al.*, 1977). The grass is frequently infected by ergot (*Claviceps paspali*), and for many years ergot alkaloids were believed to be the causative agent of paspalum staggers. Mantle *et al.* (1977) dosed cattle and sheep with *Claviceps paspali* sclerotia and produced the condition, characterized by head tremors, incoordination, and collapse when disturbed. The sclerotia were subjected to various extraction procedures to isolate the tremorgenic fraction. The ergot component was without activity. A mixture of tremorgens was isolated that appeared to be the active fraction. Thus, it appears that paspalum staggers is caused by tremorgens produced by *Claviceps paspali* infection of *Paspalum* seed heads (Tyler *et al.*, 1992; Botha *et al.*, 1996).

Bermuda grass (*Cynodon dactylon*) is an important forage in the southern U.S. and other tropical and subtropical areas. Periodic episodes of **bermuda grass tremors** in cattle occur in the southern U.S. Clinical signs include muscle tremors and twitching in shoulder and flank areas, incoordination when disturbed, general weakness of the hindlimbs, and inability to stand. Mortality is generally from drowning or other misadventure (Strain *et al.*, 1982). The causative factors have not been identified, but the tremorgenic effects suggest endophyte involvement. Transient staggers in horses ingesting bermuda grass occurs in California (Galey *et al.*, 1993).

FACIAL ECZEMA AND SPORIDESMIN

Facial eczema is secondary photosensitization caused by the hepatoxic mycotoxin **sporidesmin**, contained in spores (Fig. 10–10) produced by the fungus *Pithomyces chartarum* that grows in the dead litter of ryegrass pastures (Smith and Embling, 1991). It is mainly of importance in New Zealand, where it is of major concern to the sheep and dairy industries. Affected animals develop severe dermatitis of light-skinned areas such as the face and udder. Productivity of affected animals is markedly impaired, including reduced fertility of sheep (Morris *et al.*, 1991).

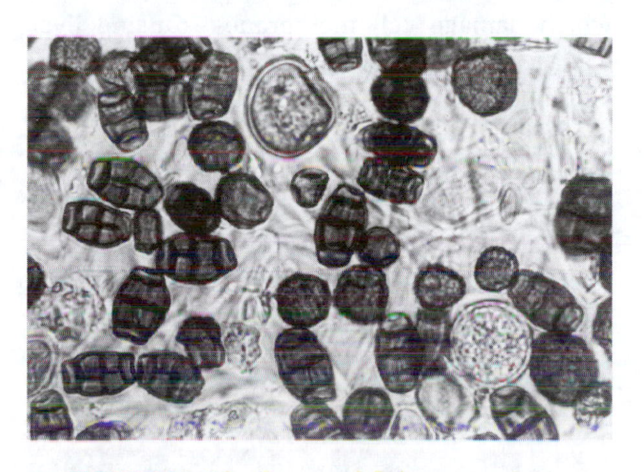

FIGURE 10–10 Spores of *Pithomyces chartarum*, the fungus which produces sporidesmin. The spores are often referred to as being shaped like hand grenades. (Courtesy of H.H. Meyer)

The danger periods for facial eczema follow warm wet weather that favors fungal growth. The spore numbers in the pasture litter rise very rapidly under favorable fungal growth conditions and can be seen as a cloud of black dust when the pasture is disturbed (e.g., mowing). The major toxin produced in the fungal spores is sporidesmin A (Fig. 10–11). There are several others (sporidesmin B, C, D, E, F, G, and H), but they are of low biological activity.

The toxicity of sporidesmin requires the disulfide bridge, which in reaction with cellular

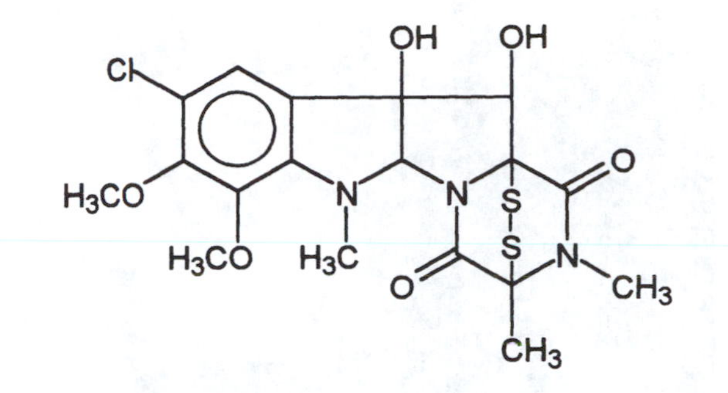

FIGURE 10–11 Structure of sporidesmin.

glutathione, can generate tissue oxidants. In the non-toxic sporidesmin D, the disulfide bridge is broken and the two sulfur groups are methylated (Briggs *et al.*, 1994). Sporidesmin A and D are the major sporidesmins in toxic pasture.

In sheep, the first observable signs are photodynamic lesions on the ears and face. The animals become restless, shake their heads, and may rub their eyes and ears against solid objects and on the ground. The ears become swollen, red, and droopy, and the lips and eyelids swell. Scabs may form over these areas. The sheep stop grazing, particularly if the lips are badly irritated. All parts of the skin exposed to sunlight, including the line where the wool parts down the back, may be affected. The animals show photophobia and seek out any shade available. Newly shorn sheep are especially vulnerable.

In dairy cattle, the lesions are most frequent on the udder and teats and down the inside of the hind legs, especially in Jerseys, and on the white areas of Holsteins. Sometimes the skin of the face peels off over a large area (Fig. 10–12). Prior to observable signs of facial eczema, milk production falls off markedly.

Most affected animals recover, especially if treatment programs are initiated. Adequate shade should be provided. Even a few minutes of exposure to strong sunlight can cause intense skin damage. Affected dairy cows should be

FIGURE 10–12 A severe case of facial eczema showing the skin peeling off the face. (Courtesy of H.H. Meyer)

dried off (cessation of lactation) immediately. This reduces appetite, and hence the intake of toxin, and allows maximum diversion of nutrients to tissue repair. It also lessens the intake of chlorophyll, thus reducing the phylloerythrin load on the liver. Affected animals should be treated to prevent fly strike.

Most reports of facial eczema are from New Zealand. Occasional problems in North America have occurred, such as an outbreak of facial eczema in Oregon sheep (Hansen *et al.*, 1994). Although *Pithomyces chartarum* occurs worldwide, in many areas the local strains do not produce sporidesmin (Collin and Towers, 1995). Collin *et al.* (1996) found that strains of *P. chartarum* which do not produce sporidesmin are not toxic to sheep even when dosed at very high levels, indicating that sporidesmin is the only toxin produced by this species.

New Zealand researchers have shown that zinc supplementation has protective effects against facial eczema. Dosing with zinc salts at the time animals are exposed to toxic pastures will reduce both the number of animals that develop facial eczema and the amount of liver damage. To achieve protection, dose rates with zinc are high, equivalent to 20–25 mg/kg body weight/day. Zinc oxide or chelated (e.g., with EDTA) zinc are recommended; zinc sulfate has corrosive effects on the gastrointestinal tract and should not be used. There is only a three- or fourfold margin of safety between protective and toxic dose rates. In ruminants, fibrosis of the pancreas occurs with high zinc intakes. Therefore, zinc dosing is not recommended as a routine treatment, but can be used for short-term protection on farms with occasional facial eczema problems.

The metabolic defects induced by sporidesmin may be initiated by the intrahepatic generation of oxygen free radicals (superoxide anions) which damage cell membranes, causing liver damage (Munday and Manns, 1989). The secondary photosensitization is a result of liver damage (Fig. 10–13) caused by sporidesmin. Of a variety of metals, zinc was among the most effective inhibitors of active oxygen generation from sporidesmin, suggesting that the protective effect of zinc against sporidesmin toxicity is likely a result of inhibition of superoxide radical generation. Copper strongly catalyzes the oxidation of sporidesmin (Munday and Manns,

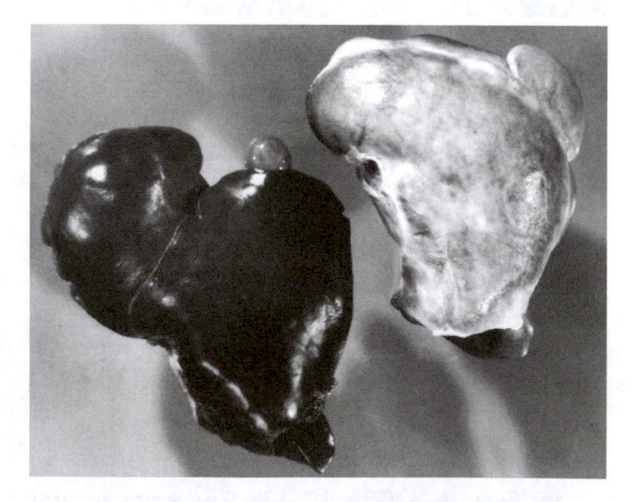

FIGURE 10–13 Liver and gall bladder from an animal with facial eczema (right) contrasted with those from a nonaffected animal (left). The liver may be almost white in a severe case of facial eczema. (Courtesy of H.H. Meyer)

1989). Besides zinc, administration of iron salts to animals also protects against sporidesmin toxicity. These elements reduce copper absorption, thus reducing the amount of copper available in the liver to serve in a catalytic role in sporidesmin oxidation. Sporidesmin contains a disulfide group, which can be reduced (addition of hydrogen) to form a dithiol. Zinc binds with the dithiol group of reduced sporidesmin, preventing its autooxidation and thus reducing cell damage (Munday, 1989). The report of Munday and Manns (1989) should be consulted for further information on the effects of copper, zinc, and iron on sporidesmin metabolism.

There are genetic and species differences in susceptibility to sporidesmin toxicosis. For example, goats are considerably more resistant to sporidesmin than sheep (Smith and Embling, 1991). Saanen goats were more susceptible than feral and Angora goats. Selection for resistance to facial eczema has been effective in sheep (Morris *et al.*, 1989) and dairy cattle (Morris *et al.*, 1991) in New Zealand. In sheep, breed differences have been noted, with Merinos more resistant to sporidesmin intoxication than the British breeds (Smith *et al.*, 1980). Metabolism of sporidesmin involves destruction of the disulfide bridge. Genetic differences in susceptibility may reflect differences in the rate of degradation of the disulfide bridge (Briggs *et al.*, 1994).

Poor productive performance of sheep (ovine ill-thrift) in Nova Scotia (Brewer *et al.*, 1971) has been associated with mycotoxins similar to sporidesmin. Brewer *et al.* (1972) reported that *Chaetomium* spp. of fungi on pasture grasses produce an antibiotic, **chetomin**, that appears to be involved in ovine ill-thrift. They proposed that chetomin is poorly absorbed from the rumen and thus accumulates, where it exerts an antibiotic effect, impairing rumen fermentation.

FUSARIUM TOXICOSIS

Fusarium toxins such as zearalenone and trichothecenes are common mycotoxins found in cereal grains and are important causes of mycotoxicosis in North America (see Chapter 5). They are not commonly associated with forage-based feeding systems. However, *Fusarium* toxins have been found to be of significance in grazing animals in New Zealand (DiMenna *et al.*, 1987; 1991). They have been linked to impaired sheep fertility and poor growth (ill-thrift). **Zearalenone** at levels found in New Zealand forage impairs ewe fertility (Smith *et al.*, 1990; Miles *et al.*, 1996), whereas the **trichothecenes** are probably the cause of ill-thrift. Of particular concern to the New Zealand livestock industry is that zearalenone is chemically similar to the growth promotant α-zearalanol (zeranol), which is banned in some countries. The occurrence of tissue residues of natural origin could adversely affect New Zealand exports of animal products. However, techniques are available which are capable of separating zearalenone and its metabolites from zeranol and its metabolites (Richard *et al.*, 1993).

KIKUYUGRASS POISONING

Kikuyugrass (*Pennisetum clandestinum*) is a widely grown tropical forage grass. It may contain a number of deleterious factors, including saponins, nitrates, and oxalates. It may cause profuse ruminal foam and bloat, presumably because of the saponins. A condition of unknown etiology called **kikuyu poisoning** has been observed in cattle, goats, sheep, and horses, mainly in New Zealand, Australia, and South Africa. Signs of toxicosis generally occur 24 to 48 h after animals consume the toxic pasture. Clinical signs include anorexia, depression, piloerection, drooling, colic, grinding of teeth, cessation of ruminal and intestinal movement, and lack of fecal excretion. Sham-drinking is a distinctive feature; animals congregate at water and attempt but fail to drink (Newsholme *et al.*, 1983). Muscle twitching, a high-stepping gait, and occasionally convulsions are seen. The most striking lesion is an intense necrosis of the ruminal and omasal mucosa. Mortality of affected animals is about 80%. Necrosis of the ruminal epithelium and omasal mucosa is extensive. South African researchers (Newsholme *et al.*, 1983) noted that kikuyu poisoning of cattle occurred in association with previous army worm invasion of pastures. This association was also noted in New Zealand (Smith and Martinovich, 1973). However, in Australia, kikuyu poisoning occurs in the absence of the army worm (Wong *et al.*, 1987; Peet *et al.*, 1990). Newsholme *et al.* (1983) suggested that mycotoxins might be involved. Wong *et al.* (1987) found no evidence of fungal involvement, but noted that the lush grass growth following drought-breaking rains was most toxic, suggesting to them that a plant toxin is involved. In South Africa, kikuyu poisoning and poor animal performance are associated with high nitrogen fertilization, irrigation or rain, hot weather, and rapidly growing grass (Pienaar *et al.*, 1993a,b.) Sheep are apparently somewhat more resistant than other livestock (Peet *et al.*, 1990). The etiology of kikuyu poisoning is unresolved and may involve several factors. Poor animal performance on kikuyu pastures may involve high soluble nitrogen and nitrate, high oxalate, saponins, and high fiber (Pienaar, 1993a). Mineral imbalance caused by severe deficiency of calcium and sodium, and very high potassium levels which induce magnesium deficiency, are also factors in poor animal performance (Miles *et al.*, 1995).

ANNUAL RYEGRASS TOXICOSIS

Annual ryegrass toxicosis is a disease of livestock caused by a group of highly toxic glycolipids called **corynetoxins**. These toxins are very similar in chemical structure to the tunicamycin antibiotics (Jago and Culvenor, 1987); the signs and lesions of annual ryegrass toxicosis can be produced by administration of tunicamycin (Finnie and Jago, 1985). The signs of toxicosis include neurological disturbances, a high-stepping gait, incoordination, and convulsions. Although superficially similar to perennial ryegrass staggers, annual ryegrass toxicosis is a lethal condition with extensive brain damage, particularly to the cerebellum. The corynetoxins affect the nervous system, and their effects become most obvious when animals are stressed or excited. Signs of toxicity may appear as soon as 2 days or as late as 12 weeks after stock are introduced into toxic annual ryegrass pastures. If the animals are not inspected regularly, the first symptoms seen may be a number of mortalities. Close inspection of the herd or flock daily may reveal some animals with neurological signs following disturbance. The animal typically has a high stepping gait (the stair-climbing gait) with the head held high. There is a loss of coordination of the hind legs, and it may collapse, with convulsions and spasms (Fig. 10–14). It may appear to recover after a time, regain its feet, and return to the herd or flock. In more severe cases, it may regain its feet but remain standing only by propping itself on spread legs. In the terminal stages, animals remain lying on the ground, with spasms and convulsions, and with their feet in a paddling motion. Death usually occurs within about 24 hr. Pathological changes include diffuse fat deposits in the liver, hemorrhages in various or-

FIGURE 10–14 Sheep suffering from annual ryegrass toxicity, showing the mounds of earth caused by the "paddling" action. (Courtesy of P. Vogel)

gans, and vascular damage in the brain, particularly in the cerebellum. Sublethal intakes of cornytoxins cause reduced animal performance and wool growth (Davies *et al.*, 1996), probably by impairing liver function.

The biological activities of corynetoxins are virtually identical to those of the closely related tunicamycin antibiotics (Jago and Culvenor, 1987). Both groups of compounds strongly inhibit UDP-*N*-acetylglucosamine:dilicholphosphate *N*-acetylglucosamine phosphate transferase, an enzyme essential for lipid-linked N-glycosylation of glycoproteins. Hence, annual ryegrass toxicity may be the result of the depletion or reduced activity of essential N-glycosylated glycoproteins, which include enzymes, hormones, structural components of the cell membrane, and extracellular matrices and membrane receptors (Jago *et al.*, 1983).

Corynetoxins are not detoxified by rumen fermentation (Vogel and McGrath, 1986). Slight protective effects by dosing sheep with cobalt sulfate have been achieved (Davies *et al.*, 1993, 1995). Administration of cobalt was tested because of earlier studies indicating protective effects of cobalt against *Phalaris* staggers.

The production of corynetoxins involves a complex of factors, including the grass, a nematode, a bacterium, and probably a virus (McKay and Ophel, 1993). Seedlings of annual ryegrass may become infected with a nematode, *Anguina agrostis,* which can infect the plant shortly after germination. The nematode larvae crawl up the plant to the growing tip and are passively carried as the plant grows. When the grass begins to flower, the larvae burrow into the developing flower where they become mature nematodes. The infected flower does not produce seed; the seed is replaced by a gall in which the adult nematodes lay eggs that hatch into larvae in the gall (Fig. 10–15). The larvae remain dormant until the following season, when they become active in the soil and infect ryegrass seedlings. The nematode itself is nontoxic. However, if the nematodes are infected with *Clavibacter toxicus* (formerly *Corynebacterium rathayi*), corynetoxins are produced and the seed galls are toxic (Riley and Ophel, 1992). There is evidence that the toxin is produced only if the bacteria are infected with a bacteriophage (Ophel *et al.*, 1993; Riley and Gooden, 1991). The bacteria produce a yellow slime on the seed heads. The slime can be seen as a yellowness in ryegrass fields; on close inspection it can be seen as a glistening sticky yellow mass oozing from the seed heads. The most reliable means of ascertaining the presence of toxic nematode galls is microscopic evaluation of grass seed heads (McKay and Riley, 1993). The toxic nematode galls are orange in color, while the non-toxic galls are dark

FIGURE 10–15 A parasitized seed head of annual ryegrass (left) that can produce annual ryegrass toxicity, contrasted with a normal seed head (right). (Courtesy of D.J. Schneider)

brown or black. The nematodes remain viable in the galls for several years. When dry seed containing galls which has been stored for several years is placed on a microscope slide and water added, emerging nematodes can be seen within a few minutes.

The toxins (Fig. 10–16) have been identified as glycolipids containing an amino sugar(s) with 3-hydroxy C-17 fatty acid residues (Vogel *et al.*, 1981).

Control of annual ryegrass toxicosis involves preventing nematode infection of the grass. Crop rotation, field burning, clipping immature seed heads, and fallowing are methods of eliminating nematodes. Transport of infected plant material to non-infected areas should be

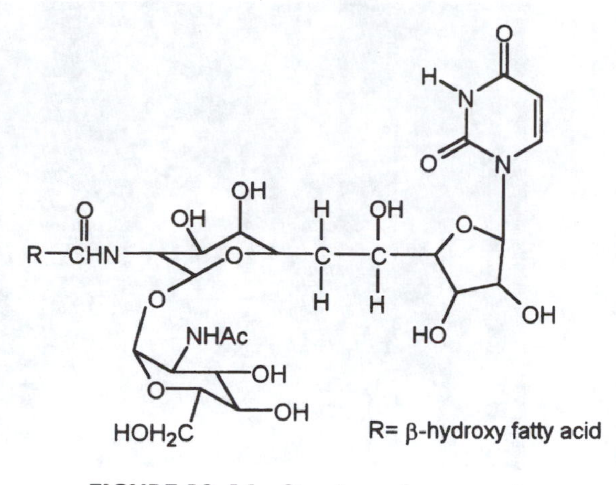

FIGURE 10–16 Structure of corynetoxin.

avoided. Whirlwinds may transfer the nematode galls for short distances.

An interesting finding has been the spontaneous decline in annual ryegrass toxicity in parts of Australia. This apparently is due to infection of the nematodes with a pathogenic fungus (Riley, 1994). The fungus, *Dilophospora alopecuri*, colonizes the ryegrass seed head and infects nematode galls. It also produces spores which infect nematodes on nearby plants. The bacterium *C. toxicus* also provides biological control of the nematode, which seldom survives in galls colonized by *C. toxicus* (Riley, 1994), but the fungus seems to be more effective. Riley (1994) indicates that *D. alopecuri* shows good potential as a biological control agent for the nematodes responsible for annual ryegrass toxicity.

Annual ryegrass toxicosis has been a problem mainly in Australia and South Africa. Although the disease was reported many years ago in Oregon (Shaw and Muth, 1949; Galloway, 1961), it has not been observed recently because of the widespread use of open field burning to dispose of grass straw and stubble. Although annual ryegrass is the main grass involved, recent reports from Australia (Finnie, 1991; Bourke *et al.*, 1992; McKay *et al.*, 1993; Davis *et al.*, 1995; Bertozzi and McKay, 1995) implicate several other grasses, including blowngrass (*Agrostis avenacea*) and annual beardgrass (*Polypogon monspeliensis*). The disorders associated with these grasses are locally known as floodplain staggers (blowngrass) and Stewarts Range Syndrome (beardgrass).

Moldy wheat fed to pigs caused toxicosis with signs resembling annual ryegrass toxicity (Bourke, 1987). Tunicamycin antibiotics produced by unidentified fungi were isolated as the causative agents (Cockrum *et al.*, 1987).

SLAFRAMINE

Fungi and molds play a role in several toxicoses involving forage legumes. In some cases (e.g. slaframine), the fungi infecting the forage produce the toxin *de novo*. In other cases (e.g. dicumarol), fungal infection is involved in the conversion of a plant chemical to a toxin. A further involvement is that fungal infection of some forages increases their concentration of toxins (e.g. subterranean clover and phytoestrogens).

For many years farmers in the U.S. midwest were aware that clover pasture and hay often induced excessive salivation or "slobbers" in livestock. Red clover hay was particularly implicated; white clover pastures also were a problem. When draft horses were important as work animals, slobbering was an unpleasant part of farming when the horses were on white clover pasture. When they were bridled for work, the farmer was drenched with a cup or more of saliva! The factor in clover that causes slobbering is a metabolite of a red clover fungal pathogen, *Rhizoctonia leguminicola*. This compound, called slaframine, is an indolizidine alkaloid (Fig. 10–17). This fungus also produces swainsonine (Hagler and Croom, 1989), the toxic factor in locoweeds (see *Astragalus*, Chapter 13).

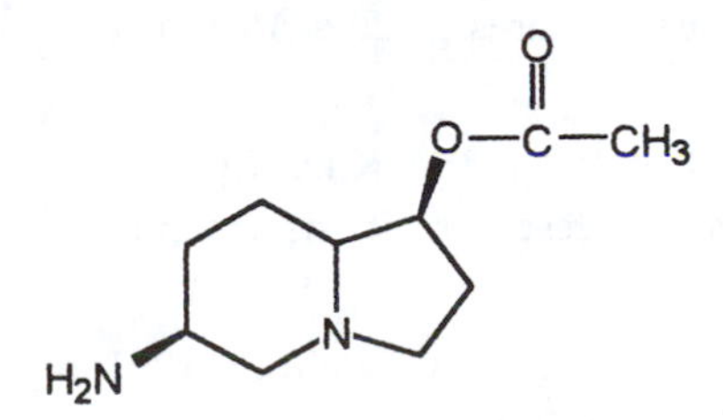

FIGURE 10–17 Structure of slaframine.

The problems associated with slaframine have been noted in Missouri, Illinois, Indiana, Ohio, Virginia, Pennsylvania, and Wisconsin. The offending red clover hay, while not visibly moldy, contains a dark brown fungal mycelium. The fungal infestation is known as black patch disease of red clover. It develops most rapidly in rainy weather and in periods of high humidity. The fungus contaminates clover seeds, is spread via the seeds, and may overwinter on clover stubble. In areas of endemic black patch infection, careful monitoring of livestock fed red clover is advisable. Prompt removal of the toxic forage from livestock generally alleviates all signs of intoxication. Degradation of the toxin in hay may occur; Hagler and Behlow (1981) noted that levels of 50–100 ppm slaframine in fresh red clover hay were reduced to about 7 ppm after 10 months of storage.

Biological Effects of Slaframine

Slobber or salivary syndrome which occurs primarily in cattle and horses is a cholinergic action induced by slaframine. The clinical signs in intoxicated cattle, in addition to the excessive salivation, are lacrimation (eye discharge), bloat, frequent urination, and watery diarrhea. Excessive amounts of saliva may appear soon after exposure to slaframine and possibly continue for several days after consumption of the toxin has ceased. Decreased milk production occurs in affected dairy cattle. Weight loss and abortion may occur. Guinea pigs have been used as the primary assay animals. In 10–20 min following ingestion of the compound, guinea pigs experience a salivation episode of 6–8 hr. In cattle, it may last up to 3 days.

Slaframine is extremely toxic; the acute oral LD_{50} in guinea pigs is less than 1 mg/kg of body weight. Toxic effects other than salivation observed in guinea pigs and other species include increased pancreatic flow, bile flow, and gastric acidity; decreased heart rate, cardiac output, respiration rate, body temperature, and metabolic rate; and uterine hemorrhage and fetal abortion. There are no distinguishable lesions or long-term effects resultant from slaframine intoxication. Spontaneous recovery is complete in 2–3 days. Atropine and certain antihistamines are quite effective in alleviating some of the clinical signs of the toxicosis.

Metabolism of the alkaloid is an important factor in slaframine toxicosis. The toxin possesses no biological activity itself; it is activated by enzymes in the liver. The active metabolite is thought to be a quaternary amine that resembles acetylcholine, thereby accounting for its cholinergic activity (Hagler and Croom, 1989).

Advantageous effects of slaframine have been suggested by Croom *et al.* (1990). They proposed that slaframine might have potential therapeutic effects in treating digestive disorders associated with the feeding of high-concentrate, low-roughage diets to ruminants, as a means of increasing salivary flow rate. Slaframine administration did not improve the utilization of low quality roughage by sheep (Bird *et al.*, 1993). Froetschel *et al.* (1995) found that slaframine administration to steers altered digesta passage rate and shifted some digestive activities from the rumen to the lower tract.

Physiological effects of slaframine and related indolizidines have been reviewed by Hagler and Croom (1989) and Croom *et al.* (1995). Croom *et al.* (1995) suggest that swainsonine and perhaps other *R. leguminicola* metabolites are also involved in the slobbers syndrome.

REFERENCES

Tall Fescue Toxicoses

Bacon, C.W. 1995. Toxic endophyte–infected tall fescue and range grasses: Historic perspectives. J. Anim. Sci. 73:861–870.

Ball, D.M., J.F. Pedersen, and G.D. Lacefield. 1993. The tall-fescue endophyte. American Scientist 81:370–379.

Boling, J.A., L.P. Bush, R.C. Buckner, L.C. Pendlum, P.B. Burrus, S.G. Yates, S.P. Rogovin, and H.L. Tookey. 1975. Nutrient digestibility and metabolism in lambs fed added perloline. J. Anim. Sci. 40:972–976.

Boosinger, T.R., J.P. Brendemuehl, D.L. Bransby, J.C. Wright, R.J. Kemppainen, and D.D. Kee. 1995. Prolonged gestation, decreased triiodothyronine concentration, and thyroid gland histomorphologic features in newborn foals of mares grazing *Acremonion coenophialum*–infected fescue. Am. J. Vet. Res. 56:66–69.

Chestnut, A.B., P.D. Anderson, M.A. Cochran, H.A. Fribourg, and K.D. Gwinn. 1992. Effects of hydrated sodium calcium aluminosilicate on fescue toxicosis and mineral absorption. J. Anim. Sci. 70:2838–2846.

Chestnut, A.B., H.A. Fribourg, K.D. Gwinn, P.D. Anderson, and M.A. Cochran. 1991. Effect of ammoniation on toxicity of *Acremonium coenophialum* infested tall fescue. Anim. Feed Sci. Tech. 35:227–236.

Coffey, K.P., J.L. Moyer, L.W. Lomas, J.E. Smith, D.C. La Rue, and F.K. Brazle. 1992. Implant and copper oxide needles for steers grazing *Acremonium coenophialum*–infected tall fescue pastures: Effects on grazing and subsequent feedlot performance and serum constituents. J. Anim. Sci. 70:3203–3214.

Cross, D.L., L.M. Redmond, and J.R. Strickland. 1995. Equine fescue toxicosis: Signs and solutions. J. Anim. Sci. 73:899–908.

Dougherty, C.T., L.M. Lauriault, N.W. Bradley, N. Gay, and P.L. Cornelius. 1991. Induction of tall fescue toxicosis in heat-stressed cattle and its alleviation with thiamin. J. Anim. Sci. 69:1008–1018.

Faichney, G.J., and T.N. Barry. 1986. Effects of mild heat exposure and suppression of prolactin secretion on gastro-intestinal tract function and temperature regulation in sheep. Aust. J. Biol. Sci. 39:85–97.

Fiorito, I.M., L.D. Bunting, G.M. Davenport, and J.A. Boling. 1991. Metabolic and endocrine responses of lambs fed *Acremonium coenophialum*–infected or noninfected tall fescue hay at equivalent nutrient intake. J. Anim. Sci. 69:2108–2114.

Gould, L.S., and W.D. Hohenboken. 1993. Differences between progeny of beef sires in susceptibility to fescue toxicosis. J. Anim. Sci. 71:3025–3032.

Hathaway, R.L., J.E. Oldfield, and M. Buettner. 1981. Effect of selenium in a mineral-salt mixture on heifers grazing tall fescue and quackgrass pastures. Proc. West. Sect. Am. Soc. Anim. Sci. 32:32–33.

Hohenboken, W.D., P.L. Berggren-Thomas, W.E. Beal, and W.H. McClure. 1991. Variation among Angus cows in response to endophyte-infected fescue seed in the diet, as related to their past calf production. J. Anim. Sci. 69:85–90.

Howard, M.D., R.B. Muntifering, N.W. Bradley, G.E. Mitchell, Jr., and S.R. Lowry. 1992. Voluntary intake and ingestive behavior of steers grazing Johnstone or endophyte-infected Kentucky-31 tall fescue. J. Anim. Sci. 70:1227–1237.

Ireland, F.A., W.E. Loch, K.Worthy, and R.V. Anthony. 1991. Effects of bromocriptine and perphenazine on prolactin and progesterone concentrations in pregnant pony mares during late gestation. J. Reprod. Fert. 92:179–186.

Joost, R.E. 1995. *Acremonium* in fescue and ryegrass: Boon or bane? A review. J. Anim. Sci. 73:881–888.

Joost, R., and S. Quisenberry (Eds.). 1993. *Acremonium*/grass interactions. Proc. of the Int. Symp. on *Acremonium*/grass Interactions. Agric., Ecosyst. and Environ. 44:1–321.

Kerr, L.A., C.P. McCoy, C.R. Boyle, and H.W. Essig. 1990. Effects of ammoniation of endophyte fungus–infested fescue hay on serum prolactin concentration and rectal temperature in beef cattle. Am. J. Vet. Res. 51:76–78.

Lauriault, L.M., C.T. Dougherty, N.W. Bradley, and P.L. Cornelius. 1990. Thiamin supplementation and the ingestive behavior of beef cattle grazing endophyte-infected tall fescue. J. Anim. Sci. 68:1245–1253.

Lipham, L.B., F.N. Thompson, J.A. Stuedemann, and J.L. Sartin. 1989. Effects of metoclopramide on steers grazing endophyte-infected fescue. J. Anim. Sci. 67:1090–1097.

Lyons, P.C., R.D. Plattner, and C.W. Bacon. 1986. Occurrence of peptide and clavine ergot alkaloids in tall fescue grass. Science 232:487–489.

McCann, J.S., A.B. Caudle, F.N. Thompson, J.A. Stuedemann, G.L. Heusner, and D.L. Thompson, Jr. 1992. Influence of endophyte-infected tall fescue on serum prolactin and progesterone in gravid mares. J. Anim. Sci. 70:217–223.

Monroe, J.L., D.L. Cross, L.W. Hudson, D.M. Hendricks, S.W. Kennedy, and W.C. Bridges, Jr. 1988. Effect of selenium and endophyte-contaminated fescue on performance and reproduction in mares. J. Equine Vet. Sci. 8:148–153.

Moyer, J.L., N.S. Hill, S.A. Martin, and C.S. Agee. 1993. Crop quality and utilization. Degradation of ergoline alkaloids during in vitro ruminal digestion of tall fescue forage. Crop Sci. 33:264–266.

Nutting, D.F., E.A. Tolley, L.A. Toth, S.D. Ballard, and M.A. Brown. 1992. Serum amylase activity and calcium and magnesium concentrations in young cattle grazing fescue and Bermuda grass pastures. Am. J. Vet. Res. 53:834–839.

Oliver, J.W., L.K. Abney, J.R. Strickland, and R.D. Linnabary. 1993. Vasoconstriction in bovine vasculature induced by the tall fescue alkaloid lysergamide. J. Anim. Sci. 71:2708–2713.

Paterson, J., C. Forcherio, B. Larson, M. Samford, and M. Kerley. 1995. The effects of fescue toxicosis on beef cattle productivity. J. Anim. Sci. 73:889–898.

Peet, R.L, M.R. McCarthy, and M.J. Barbetti. 1991. Hyperthermia and death in feedlot cattle associated with the ingestion of *Claviceps purpurea*. Aust. Vet. J. 68:121.

Porter, J.K., J.A. Stuedemann, F.N. Thompson, B.A. Buchanan, and H.A. Tucker. 1994. Melatonin and pineal neurochemicals in steers grazed on endophyte-infected tall fescue: Effects of metoclopramide. J. Anim. Sci. 71:1526–1531.

Porter, J.K., J.A. Stuedemann, F.N. Thompson, Jr., and L.B. Lipham. 1990. Neuroendocrine measurements in steers grazed on endophyte-infected fescue. J. Anim. Sci. 68:3285–3292.

Porter, J.K., and F.N. Thompson, Jr. 1992. Effects of fescue toxicosis on reproduction in livestock. J. Anim. Sci. 70:1594–1603.

Powell, R.G., and R.J. Petroski. 1992. The loline group of pyrrolizidine alkaloids. pp. 320–338 in: S.W. Pelletier (Ed.). The Alkaloids: Chemical and Biological Perspectives. John Wiley and Sons, New York.

Putnam, M.R., D.I. Bransby, J. Schumacher, T.R. Boosinger, L. Bush, R.A. Shelby, J.T. Vaughan, D. Ball, and J.P. Brendemuehl. 1991. Effects of the fungal endophyte *Acremonium coenophialum* in fescue on pregnant mares and foal viability. Am. J. Vet. Res. 52:2071–2074.

Putnam, M.R., J.P. Brendemuehl, T.R. Boosinger, D.I. Bransby, D.D. Kee, J. Schumacher, and R.A. Shelby. 1990. The effect of short term exposure to and removal from the fescue endophyte *Acremonium coenophialum* on pregnant mares and foal viability. pp. 255–258 in: S.S. Quisenberry and R.E. Joost (Eds.). Proceedings of the Int. Symposium on Acremonium/grass Interactions. LA Ag. Exp. Sta., Baton Rouge, LA.

Redmond, L.M., D.L. Cross, J.R. Strickland, and S.W. Kennedy. 1994. Efficacy of domperidone and sulpiride as treatments for fescue toxicosis in horses. Am. J. Vet. Res. 55:722–729.

Regisford, E.G.C., and L.S. Katz. 1994. Effects of bromocriptine treatment on the expression of sexual behavior in male sheep (*Ovis aries*). J. Anim. Sci. 72:591–597.

Rhodes, M.T., J.A. Paterson, M.S. Kerley, H.E. Garner, and M.H. Laughlin. 1991. Reduced blood flow to peripheral and core body tissues in sheep and cattle induced by endophyte-infected tall fescue. J. Anim. Sci. 69:2033–2043.

Riet-Correa, F., M.C. Mendez, A.L. Schild, P.N. Bergamo, and W.N. Flores. 1988. Agalactica, reproductive problems and neonatal mortality in horses associated with the ingestion of *Claviceps purpurea*. Aust. Vet. J. 65:192–193.

Rohrbach, B.W., E.M. Green, J.W. Oliver, and J.F. Schneider. 1995. Aggregate risk study of exposure to endophyte-infected (*Acremonium coenophialum*) tall fescue as a risk factor for laminitis in horses. Am. J. Vet. Res. 56:22–26.

Rottinghaus, G.E., G.B. Garner, C.N. Cornell, and J.L. Ellis. 1991. HPLC method for quantitating ergovaline in endophyte-infected tall fescue: Seasonal variation of ergovaline levels in stems with leaf sheaths, leaf blades, and seed heads. J. Agric. Food Chem. 39:112–115.

Roylance, J.T., N.S. Hill, and C.S. Agee. 1994. Ergovaline and peramine production in endophyte-infected tall fescue: Independent regulation and effects of plant and endophyte genotype. J. Chem. Ecol. 20:2171–2183.

Rumsey, T.S., J.A. Stuedemann, S.R. Wilkinson, and D.J. Williams. 1979. Chemical composition of necrotic fat lesions in beef cows grazing fertilized "Kentucky 31" tall fescue. J. Anim. Sci. 48:673–682.

Seawright, A.A. 1989. Animal Health in Australia, Vol. 2. Chemical and Plant Poisons. Australian Government Publishing Service, Canberra.

Shelby, R.A., and S.P. Schmidt. 1991. Survival of the tall fescue endophyte in the digestive tract of cattle and horses. Plant Disease 75:776–778.

Stamm, M.M., T. DelCurto, M.R. Horney, S.D. Brandyberry, and R.K. Barton. 1994. Influence of alkaloid concentration of tall fescue straw on the nutrition, physiology and subsequent performance of beef steers. J. Anim. Sci. 72:1068–1075.

Stoszek, M.J., J.E. Oldfield, G.E. Carter, and P.H. Weswig. 1979. Effect of tall fescue and quackgrass on copper metabolism and weight gains of beef cattle. J. Anim. Sci. 48:893–899.

Strickland, J.R., D.L. Cross, G.P. Birrenkott, and L.W. Grimes. 1994. Effect of ergovaline, loline, and dopamine antagonists on rat pituitary cell prolactin release in vitro. Am. J. Vet. Res. 55:716–721.

Strickland, J.R., D.L. Cross, T.C. Jenkins, R.J. Petroski, and R.G. Powell. 1992. The effect of alkaloids and seed extracts of endophyte-infected tall fescue on prolactin secretion in an in vitro rat pituitary perfusion system. J. Anim. Sci. 70:2779–2786.

Strickland, J.R., J.W. Oliver, and D.L. Cross. 1993. Fescue toxicosis and its impact on animal agriculture. Vet. Human Toxicol. 35:454–464.

Stuedemann, J.A., and C.S. Hoveland. 1988. Fescue endophyte: History and impact on animal agriculture. J. Prod. Agric. 1:39–44.

Stuedemann, J.A., T.S. Rumsey, J. Bond, S.R. Wilkinson, L.P. Bush, D.J. Williams, and A.B. Caudle. 1985. Association of blood cholesterol with occurrence of fat necrosis in cows and tall fescue summer toxicosis in steers. Am. J. Vet. Res. 46:1990–1995.

TePaske, M.R., R.G. Powell, R.J. Petroski, M.D. Samford, and J.A. Paterson. 1993. Quantitative analyses of bovine urine and blood plasma for loline alkaloids. J. Agric. Food Chem. 41:231–234.

Townsend, W.E., M.E. Snook, J.A. Stuedemann, and R.L. Wilson. 1991. Effect of level of endophyte infection, nitrogen fertilization rate, grazing period, and paddock exchange on some chemical properties of four bovine tissues. J. Anim. Sci. 69:2871–2882.

Turner, K.E., C.P. West, E.L. Piper, S.A. Mashburn, and A.S. Moubarak. 1993. Quality and ergovaline content of tall fescue silage as affected by harvest stage and addition of poultry litter and inoculum. J. Prod. Agric. 6:423–427.

Westendorf, M.L., G.E. Mitchell, Jr., R.E. Tucker, L.P. Bush, R.J. Petroski, and R.G. Powell. 1993. In vitro and in vivo ruminal and physiological responses to endophyte-infected tall fescue. J. Dairy Sci. 76:555–563.

Yates, S.G., R.J. Petroski, and R.G. Powell. 1990. Analysis of loline alkaloids in endophyte-infected tall fescue by capillary gas chromatography. J. Agric. Food Chem. 38:182–185.

Perennial Ryegrass Staggers

Cunningham, P.J., J.Z. Foot, and K.F.M. Reed. 1993. Perennial ryegrass (*Lolium perenne*) endophyte (*Acremonium lolii*) relationships: the Australian experience. Agric. Ecosys. Environ. 44:157–168.

DiMenna, M.E., P.H. Mortimer, R.A. Prestidge, A.D. Hawkes, and J.M. Sprosen. 1992. Lolitrem B concentrations, counts of *Acremonium lolii* hyphae, and the incidence of ryegrass staggers in lambs on plots of *A. lolii*–infected perennial ryegrass. N. Z. J. Agric. Res. 35:211–217.

Fletcher, L.R. 1993. Grazing ryegrass/endophyte associations and their effect on animal health and performance. p. 115–120 in: D.E. Hume, G.C.M. Latch, and H.S. Easton (Eds.). Proc. 2nd Int. Symp. on *Acremonium*/Grass Interactions: Plenary Papers. AgResearch, Grasslands Research Centre, Palmerston North, New Zealand.

Fletcher, L.R., and G.K. Barrell. 1984. Reduced liveweight gains and serum prolactin levels in hoggets grazing ryegrasses containing *Lolium* endophyte. N. Z. Vet. J. 32:139–140.

Foot, J.Z., O.J. Woodburn, J.R. Walsh, and P.G. Heazlewood. 1994. Responses in grazing sheep to toxins from perennial ryegrass/endophyte associations. pp. 375–380 in: S.M. Colegate and P.R. Dorling (Eds.). Plant-associated Toxins. Agricultural, Phytochemical and Ecological Aspects. CAB International, Wallingford, UK.

Galey, F.D., M.L. Tracy, A.L. Craigmill, B.C. Barr, G. Markegard, R. Peterson, and M. O'Connor. 1991. Staggers induced by consumption of perennial ryegrass in cattle and sheep from northern California. J. Am. Vet. Med. Assoc. 199:466–470.

Gallagher, R.T., R.G. Keogh, G.C.M. Latch, and C.S.W. Reid. 1977. The role of fungal tremorgens in ryegrass staggers. N. Z. J. Agric. Res. 20:431–440.

Hunt L.D., L. Blythe, and D.W. Holtan. 1983. Ryegrass staggers in ponies fed processed ryegrass straw. J. Am. Vet. Med. Assoc. 182:285–286.

Mantle, P.G., P.H. Mortimer, and E.P. White. 1977. Mycotoxic tremorgens of *Claviceps paspali* and *Penicillium cyclopium*: a comparative study of effects on sheep and cattle in relation to natural staggers syndromes. Res. Vet. Sci. 24:49–56.

Miles, C.O., S.C. Munday, A.L. Wilkins, R.M. Ede, and N.R. Towers. 1994. Large-scale isolation of lolitrem B and structure determination of lolitrem E. J. Agric. Food Chem. 42:1488–1492.

Miles, C.O., A.L. Wilkins, R.T. Gallagher, A.D. Hawkes, S.C. Munday, and N.R. Towers. 1992. Synthesis and tremorgenicity of paxitriols and lolitriol: possible biosynthetic precursors of lolitrem B. J. Agric. Food Chem. 40:234–238.

Mitchell, P.J., and C.J. McCaughan. 1992. Perennial ryegrass staggers in fallow deer (*Dama dama*). Aust. Vet. J. 69:258–259.

Morris, C.A., N.R. Towers, M. Wheeler, and N.C. Amyes. 1995. A note on the genetics of resistance or susceptibility to ryegrass staggers in sheep. N.Z. J. Agric. Res. 38:359–363.

Munday-Finch, S.C., C.O. Miles, A.L. Wilkins, and A.D. Hawkes. 1995. Isolation and structure elucidation of lolitrem A, a tremorgenic mycotoxin from perennial ryegrass infected with *Acremonium lolii*. J. Agric. Food Chem. 43:1283–1288.

Odriozola, E., T. Lopez, C. Campero, and C. Gimenez Placeres. 1993. Ryegrass staggers in heifers: a new mycotoxicosis in Argentina. Vet. Hum. Toxicol. 35:144–146.

Pearson, E.G., C.B. Andreasen, L.L. Blythe, and A.M. Craig. 1996. Atypical pneumonia associated with ryegrass staggers in calves. J. Am. Vet. Med. Assoc. 209:1137–1142.

Piper, E.L. 1989. Liver function and ryegrass staggers. N. Z. Vet. J. 37:173–174.

Rowan, D.D., and D.L. Gaynor. 1986. Isolation of feeding deterrents against Argentine stem weevil from ryegrass infected with *Acremonium lolii*. J. Chem. Ecol. 12:647–658.

Shaw, J.N., and O.H. Muth. 1949. Some types of forage poisoning in cattle and sheep. J. Am. Vet. Med. Assoc. 114:315–317.

Watson, R.N., R.A. Prestidge, and O.J.-P. Ball. 1993. Suppression of white clover by ryegrass infected with *Acremonium* endophyte. pp. 218–221 in: D.E. Hume, G.C.M. Latch, and H.S. Easton (Eds.). Proc. 2nd. Int. Symp. on *Acremonium*/grass Interactions. Ag Research, Grasslands Research Center, Palmerston North, New Zealand.

Wilkins, A.L., C.O. Miles, R.M. Ede, R.T. Gallagher, and S.C. Munday. 1992. Structure elucidation of janthitrem B, a tremorgenic metabolite of *Penicillium janthinellum*, and relative configuration of the A and B rings of janthitrems B, E, and F. J. Agric. Food Chem. 40:1307–1309.

Other Endophyte-Related Syndromes

Botha, C.J., T.S. Kellerman, and N. Fourie. 1996. A tremorgenic mycotoxicosis in cattle caused by *Paspalum distichum* (L.) infected by *Claviceps paspali*. J. S. Afr. Vet. Assoc. 67:36–37.

Epstein, W., K. Gerber, and R. Karler. 1964. The hypnotic constituent of *Stipa vasey*, sleepy grass. Experientia 20:390.

Galey, F.D., B. Barr, and M. Anderson. 1993. Grassland staggers of California. N.Z. Vet. J. 41:217 (Abst.).

Kellerman, T.S., J.A.W. Coetzer, and T.W. Naude. 1988. Plant Poisonings and Mycotoxicoses of Livestock in Southern Africa. Oxford University Press, Cape Town.

Kingsbury, J.M. 1964. Poisonous Plants of the United States and Canada. Prentice–Hall, Englewood Cliffs, NJ.

Mantle, P.G., P.H. Mortimer, and E.P. White. 1977. Mycotoxic tremorgens of *Claviceps paspali* and *Penicillium cyclopium*: A comparative study of effects on sheep and cattle in relation to natural staggers syndromes. Res. Vet. Sci. 24:49–56.

Miles, C.O., G.A. Lane, M.E. DiMenna, I. Garthwaite, E.L. Piper, O.J.-P. Ball, G.C.M. Latch, L.P. Bush, F.K. Min, I. Fletcher, and P.S. Harris. 1996. High levels of ergonovine and lysergic acid amide in toxic *Achnatherum inebrians* accompany infection by an *Acremonium*-like endophytic fungus. J. Agric. Food Chem. 44:1285–1290.

Miles, C.O., S.C. Munday, A.L. Wilkins, R.M. Ede, L.P. Meagher, and I. Garthwaite. 1993. Chemical aspects of ryegrass staggers. N.Z. Vet. J. 211:216–217 (Abst.).

Oliver, J.W., L.K. Abney, J.R. Strickland, and R.D. Linnabary. 1993. Vasoconstriction in bovine vasculature induced by the tall fescue alkaloid lysergamide. J. Anim. Sci. 71:2708–2713.

Petroski, R.J., R.G. Powell, and K. Clay. 1992. Alkaloids of *Stipa robusta* (sleepygrass) infected with an *Acremonium* endophyte. Nat. Toxins 1:84–88.

Powell, R.G., and R.J. Petroski. 1992. Alkaloid toxins in endophyte-infected grasses. Nat. Toxins 1:163–170.

Smalley, H.E., and H.R. Crookshank. 1976. Toxicity studies on sleepy grass, *Stipa robusta* (Vasey) Scribn. Southwest. Vet. 29:35–39.

Strain, G.M., C.L. Seger, and W. Flory. 1982. Toxic bermuda grass tremor in the goat: An electroencephalographic study. Am. J. Vet. Res. 43:158–162.

Tyler, J.W., R.A. Shelby, E.A. Sartink, D.F. Wolfe, J.E. Steiss, D.C. Sorjonen, T.A. Powe, and J.A. Spano. 1992. Naturally occurring neurologic disease in calves fed *Claviceps sp.* infected dallis grass hay and pasture. Prog. Vet. Neurol. 3:99–105.

USDA. 1976. Spaced-out cows swear off grass. Agric. Res. January, pp. 1–16.

Uzal, F.A., M.P. Woodman, C.G. Giraudo, C.A. Robles, and D.L. Doxey. 1996. An attempt to reproduce 'mal seco' in horses by feeding them *Festuca argentina*. Vet. Rec. 139:68–70.

Wolfe, M.H. 1979. More notes on sleepygrass. Rangelands 1:144.

Facial Eczema and Sporidesmin

Brewer, D., F.W. Calder, T.M. MacIntyre, and A. Taylor. 1971. Ovine ill-thrift in Nova Scotia. 1. The possible regulation of the rumen flora in sheep by the fungal flora of permanent pasture. J. Agric. Sci. 76:465–477.

Brewer, D., J.M. Duncan, W.A. Jerram, C.K. Leach, S. Safe, A. Taylor, Z.C. Vining, R. McG. Archibald, R.G. Stevenson, C.J. Mirocha, and C.M. Christensen. 1972. Ovine ill-thrift in Nova Scotia. 5. The production and toxicology of chetomin, a metabolite of *Chaetomium* spp. Can. J. Microbiol. 18:1129–1137.

Briggs, L.R., N.R. Towers, and P.C. Molan. 1994. Development of an enzyme-linked immunosorbent assay for analysis of sporidesmin A and its metabolites in ovine urine and bile. J. Agric. Food Chem. 42:2769–2777.

Collin, R.G., B.L. Smith, and N.R. Towers. 1996. Lack of toxicity of a nonsporidesmin-producing strain of *Pithomyces chartarum* in cell culture and when dosed to lambs. N.Z. Vet. J. 44:131–134.

Collin, R.G., and N.R. Towers. 1995. Competition of a sporidesmin-producing *Pithomyces* strain with a non-toxigenic *Pithomyces* strain. N.Z. Vet. J. 43:149–152.

Hansen, D.E., R.D. McCoy, O.R. Hedstrom, S.P. Snyder, and P.B. Ballerstedt. 1994. Photosensitization associated with exposure to *Pithomyces chartarum* in lambs. J. Am. Vet. Med. Assoc. 204:1668–1671.

Morris, C.A., N.R. Towers, A.G. Campbell, H.H. Meyer, C. Wesselink, and M. Wheeler. 1989. Responses achieved in Romney flocks selected for or against suceptibility to facial eczema, 1975–87. N. Z. J. Agric. Res. 32:379–388.

Morris, C.A., N.R. Towers, B.L. Smith, and B.R. Southey. 1991. Progeny testing bulls for susceptibility to facial eczema. N.Z.J. Agr. Res. 34:413–417.

Munday, R. 1989. Toxicity of thiols and disulphides: Involvement of free-radical species. Free Radical Biol. Med. 7:659–673.

Munday, R.V., and E. Manns. 1989. Protection by iron salts against sporidesmin intoxication in sheep. N.Z. Vet. J. 37:65–68.

Smith, B.L., and P.P. Embling. 1991. Facial eczema in goats: The toxicity of sporidesmin in goats and its pathology. N. Z. Vet. J. 39:18–22.

Smith, B.L., T.A. Stanbridge, and P.P. Embling. 1980. Sheep breed differences in pentobarbitone sleeping-time and response to experimental sporidesmin intoxication. N.Z. Vet. J. 28:35–36.

Fusarium Toxicosis

DiMenna, M.E., D.R. Lauren, P.R. Poole, P.H. Mortimer, R.A. Hill, and M.P. Agnew. 1987. Zearalenone in New Zealand pasture herbage and the mycotoxin-producing potential of *Fusarium* species from pasture. N. Z. J. Agric. Res. 30:499–504.

DiMenna, M.E., D.R. Lauren, J.M. Sprosen, and K.S. MacLean. 1991. *Fusarium* and zearalenone on herbage fractions from short and from long pasture. N. Z. J. Agric. Res. 34:445–452.

Erasmuson, A.F., B.G. Scahill, and D.M. West. 1994. Natural zeranol (α-zearalanol) in the urine of pasture-fed animals. J. Agric. Food Chem. 42:2721–2725.

Miles, C.O., A.F. Erasmuson, A.L. Wilkins, N.R. Towers, B.L. Smith, I. Garthwaite, B.G. Scahill, and R.P. Hansen. 1996. Ovine metabolism of zearalenone to α-zearalanol (zeranol). J. Agric. Food Chem. 44:3244–3250.

Richard, J.L., G.A. Bennett, P.F. Ross, and P.E. Nelson. 1993. Analysis of naturally occurring mycotoxins in feedstuffs and food. J. Anim. Sci. 71:2563–2574.

Smith, J.F., M.E. DiMenna, and L.T. McGowan. 1990. Reproductive performance of Coopworth ewes following oral doses of zearalenone before and after mating. J. Reprod. Fert. 89:99–106.

Kikuygrass Poisoning

Martinovich, D., and B. Smith. 1973. Kikuyu poisoning of cattle. 1. Clinical and pathological findings. N.Z. Vet. J. 21:55–63.

Miles, N., J.F. de Villiers, and T.J. Dugmore. 1995. Macromineral composition of kikuyu herbage relative to the requirements of ruminants. S. Afr. Vet. J. 66:206–212.

Newsholme, S.J., T.S. Kellerman, G.C.A. Van Der Westhuizen, and J. T. Soley. 1983. Intoxication of cattle on kikuyu grass following army worm (*Spodoptera exempta*) invasion. Onderstepoort J. Vet. Res. 50:157–167.

Peet, R.L., J. Dickson, and M. Hare. 1990. Kikuyu poisoning in goats and sheep. Aust. Vet. J. 67:229–230.

Pienaar, J.P., N.M. Tainton, and J.B.J. van Ryssen. 1993a. Factors affecting the voluntary feed intake of sheep grazing *Pennisetum clandestinum* (kikuyu) pastures: observations from forage analysis. Afr. J. Range For. Sci. 10:140–144.

Pienaar, J.P., N.M. Tainton, J.B.J. van Ryssen, and J.P. Swiegers. 1993b. Factors affecting the voluntary feed intake of sheep grazing *Pennisetum clandestinum* (kikuyu) pastures: observations in the animal. Afr. J. Range For. Sci. 10:145–150.

Smith, B., and D. Martinovich. 1973. Kikuyu poisoning of cattle. 2. Epizootiological aspects. N. Z. Vet. J. 21:85–89.

Wong, P.T.W., I.J. Roth, and A.R.B. Jackson. 1987. Kikuyu poisoning of cattle in New South Wales and its relationship to pasture fungi on kikuyu grass (*Pennisetum clandestinum*). Aust. Vet. J. 64:229–232.

Annual Ryegrass Toxicosis

Bertozzi, T., and A.C. McKay. 1995. Incidence on *Polypogon monspeliensis* of *Clavibacter toxicus* and *Anguina* sp., the organisms associated with "flood plain staggers" in South Australia. Aust. J. Exp. Agric. 35:567–569.

Bourke, C.A. 1987. A naturally occurring tunicamycin-like intoxication in pigs eating water damaged wheat. Aust. Vet. J. 64:127–128.

Bourke, C.A., M.J. Carrigan, and S.C.J. Love. 1992. Flood plain staggers, a tunicaminyluracil toxicosis of cattle in northern New South Wales. Aust. Vet. J. 69:228–229.

Cockrum, P.A., C.C.J. Culvenor, J.A. Edgar, M.V. Jago, A.L. Payne, and C.A. Bourke. 1987. Toxic tunicaminyluracil antibiotics identified in water-damaged wheat responsible for the death of pigs. Aust. J. Agric. Res. 39:245–253.

Davis, E.O., G.E. Curran, W.T. Hetherington, D.A. Norris, G.A. Wise, I.J. Roth, A.A. Seawright, and W.L. Bryden. 1995. Clinical, pathological and epidemiological aspects of flood plain staggers, a corynetoxicosis of livestock grazing *Agrostis avenacea*. Aust. Vet. J. 72:187–190.

Davies, S.C., C.L. White, and I.H. Williams. 1995. Increased tolerance to annual ryegrass toxicity in sheep given a supplement of cobalt. Aust. Vet. J. 72:221–224.

Davies, S.C., C.L. White, I.H. Williams, J.G. Allen, and K.P. Croker. 1996. Sublethal exposure to corynetoxins affects production of grazing sheep. Aust. J. Exp. Agric. 36:649–655.

Davies, S.C., I.H. Williams, C.L. White, J.G. Allen, J.G. Edgar, P.A. Cockrum, and P. Stewart. 1993. Protective effect of cobalt against apparent liver damage in sheep exposed to toxic annual ryegrass. Aust. Vet. J. 70:186–187.

Finnie, J.W. 1991. Corynetoxin poisoning in sheep in the south-east of South Australia associated with annual beard grass (*Polypogon monspeliensis*). Aust. Vet. J. 68:370.

Finnie, J.W., and M.V. Jago. 1985. Experimental production of annual ryegrass toxicity with tunicamycin. Aust. Vet. J. 62:248–249.

Galloway, J.H. 1961. Grass seed nematode poisoning in livestock. J. Am. Vet. Med. Assoc. 139:1212–1214.

Jago, M.V., and C.C.J. Culvenor. 1987. Tunicamycin and corynetoxin poisoning in sheep. Aust. Vet. J. 64:232–235.

Jago, M.V., A.L. Payne, J.E. Peterson, and T.J. Bagust. 1983. Inhibition of glycosylation by corynetoxin, the causative agent of annual ryegrass toxicity: A comparison with tunicamycin. Chem.-Biol. Interact. 45:223–234.

McKay, A.C., and K.M. Ophel. 1993. Toxigenic *Clavibacter/Anguina* associations infecting grass seedheads. Annu. Rev. Phytopathol. 31:153–169.

McKay, A.C., K.M. Ophel, T.B. Reardon, and J.M. Gooden. 1993. Livestock deaths associated with *Clavibacter toxicus/Auguina sp.* infection in seedheads of *Agrostis avenacea* and *Polypogon monspeliensis*. Plant Dis. 77:635–641.

McKay, A.C., and I.T. Riley. 1993. Sampling ryegrass to assess the risk of annual ryegrass toxicity. Aust. Vet. J. 70:241–243.

Ophel, K.M., A.F. Bird, and A. Kerr. 1993. Association of bacteriophage particles with toxin production by *Clavibacter toxicus,* the causal agent of annual ryegrass toxicity. Phytopathology 83:676–681.

Riley, I.T. 1994. *Dilophospora alopecuri* and decline in annual ryegrass toxicity in Western Australia. Aust. J. Agric. Res. 45:841–850.

Riley, I.T., and J.M. Gooden. 1991. Bacteriophage specific for the *Clavibacter* sp. associated with annual ryegrass toxicity. Lett. Appl. Microbiol. 12:158–160.

Riley, I.T., and K.M. Ophel. 1992. *Clavibacter toxicus* sp. nov., the bacterium responsible for annual ryegrass toxicity in Australia. Int. J. Syst. Bacteriol. 42:64–68.

Shaw, J.N., and O.H. Muth. 1949. Some types of forage poisoning in Oregon cattle and sheep. J. Am. Vet. Med. Assoc. 114:315–317.

Vogel, P., and M.G. McGrath. 1986. Corynetoxins are not detoxicated by *in vitro* fermentation in ovine rumen fluid. Aust. J. Agric. Res. 37:523–526.

Vogel, P., D.S. Petterson, P.H. Berry, J.L. Frahn, N. Anderton, P.A. Cockrum, J.A. Edgar, M.V. Jago, G.W. Lanigan, A.L. Payne, and C.C.J. Culvenor. 1981. Isolation of a group of glycolipid toxins from seed heads of annual ryegrass (*Lolium rigidum* Gaud.) infected by *Corynebacterium rathayi.* Aust. J. Exp. Biol. Med. Sci. 59:455–467.

Slaframine

Bird, A.R., W.J. Croom, Jr., J.V. Bailey, B.M. O'Sullivan, W.M. Hagler, Jr., G.L.R. Gordon and, P.R. Martin. 1993. Tropical pasture hay utilization with slaframine and cottonseed meal: Ruminal characteristics and digesta passage in wethers. J. Anim. Sci. 71:1634–1640.

Croom, W.J., Jr., M.A. Froetschel, and W.M. Hagler, Jr. 1990. Cholinergic manipulation of digestive function in ruminants and other domestic livestock: A review. J. Anim. Sci. 68:3023–3032.

Croom, W.J., Jr., W.M. Hagler, Jr., M.A. Froetschel, and A.D. Johnson. 1995. The involvement of slaframine and swainsonine in slobbers syndrome: A review. J. Anim. Sci. 73:1499–1508.

Froetschel, M.A., M.N. Streeter, H.E. Amos, W.J. Croom, Jr., and W.M. Hagler, Jr. 1995. Effects of abomasal slaframine infusion on ruminal digesta passage and digestion in steers. Can. J. Anim. Sci. 75:157–163.

Hagler, W.M., and R.F. Behlow. 1981. Salivary syndrome in horses: Identification of slaframine in red clover hay. Appl. Environ. Microbiol. 42:1067–1073.

Hagler, W.M., Jr., and W.J. Croom, Jr. 1989. Slaframine: occurrence, chemistry and physiological activity. pp. 257–279 in: P.R. Cheeke (Ed.). Toxicants of Plant Origin. Vol. I. CRC Press, Boca Raton, FL.

CHAPTER 11

Toxins Intrinsic to Forages

TOXINS INTRINSIC TO GRASSES

Examples of toxins intrinsic to various forage grasses include tryptamine alkaloids in *Phalaris* spp., oxalates in numerous tropical grasses, cyanogenic glycosides in forage sorghums such as sudangrass, and photosensitizing agents such as saponins in various tropical grasses. The occurrence, chemical nature, and toxicologic effects of these substances in livestock have been reviewed by Cheeke (1994). Grasses in general are not well endowed with defensive chemicals. Forage legumes, on the other hand, contain a wide diversity of toxins.

FIGURE 11–1 Sheep grazing forage sorghum. Sorghums contain cyanogenic glycosides, which under certain environmental conditions (e.g. frost) can cause cyanide poisoning. (Courtesy of C. Mulcahy)

Phalaris Poisoning

In North America, the principal *Phalaris* spp. used as livestock forage is **reed canarygrass** (*P. arundinacea*). It is an important forage on wet, poorly drained soils, as the plant survives prolonged flooding. It is often unpalatable and supports lower animal growth performance than would be predicted from its nutrient content, as well as causing diarrhea. It contains at least eight different alkaloids including five **indole alkaloids** (gramine and four tryptamines) and three **β-carbolines** (Fig. 11–2). These alkaloids all contain an indole nucleus. Reed canarygrass also contains hordenine, a N-containing phenol. Technically, hordenine is not an alkaloid be-

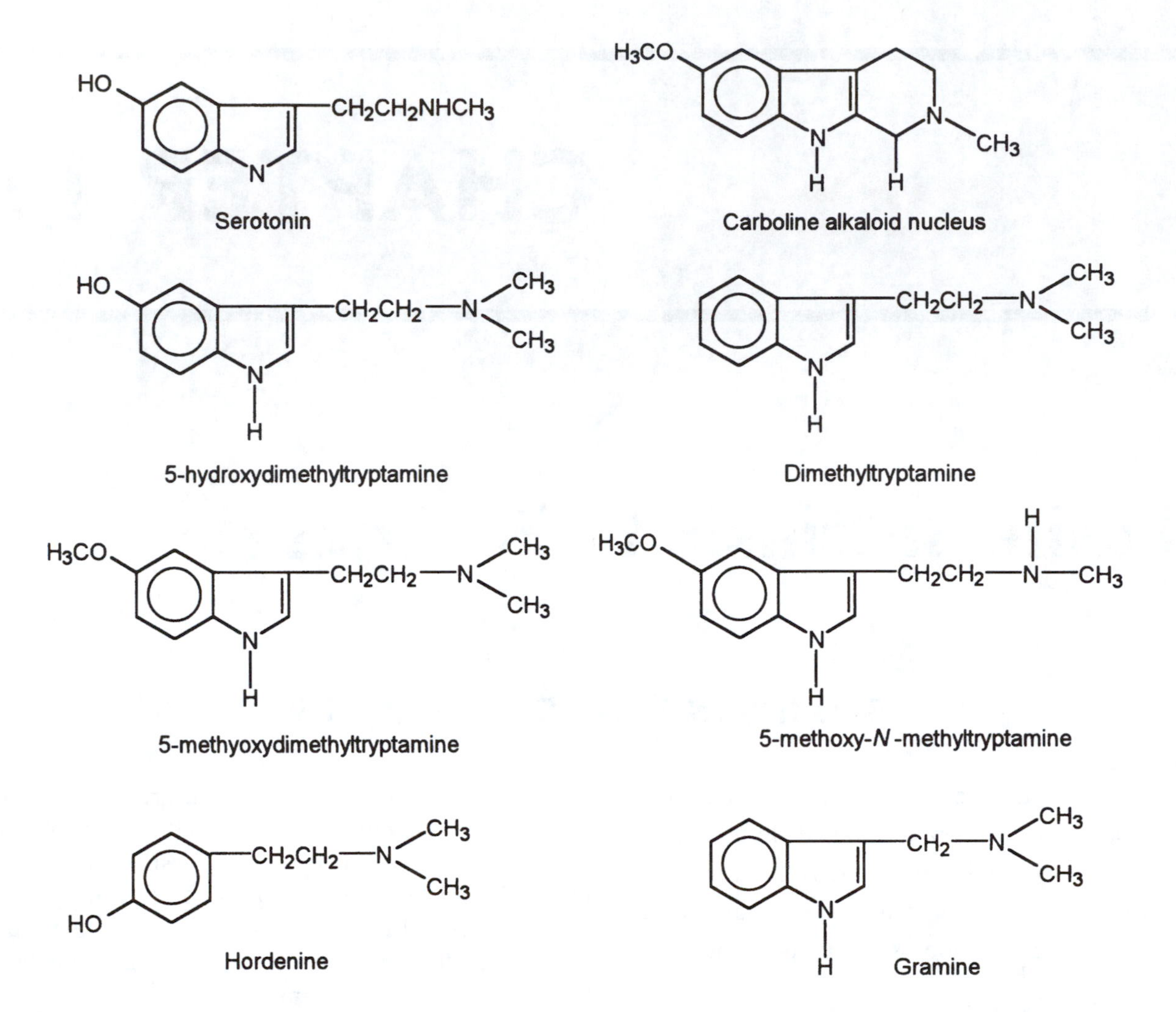

FIGURE 11–2 Toxic alkaloids in *Phalaris* spp., showing their structural similarity to serotonin, a neurotransmitter.

cause it does not contain N in a heterocyclic ring, but it is often referred to as an alkaloid in the literature. The tryptamine and β-carboline alkaloid concentrations can be readily modified by plant selection. Marten *et al.* (1981) demonstrated improved animal performance with low-alkaloid cultivars. Lambs grazing the low-alkaloid cultivars did not develop the diarrhea typically seen with reed canarygrass. Marten *et al.* (1981) concluded that the threshold level for total indole alkaloid concentration in reed canarygrass at or above which diarrhea and reduced growth rate occurs is about .2% of the dry weight. Reduced animal performance and increased diarrhea incidence become progressively greater with alkaloid levels higher than 0.2%. The low palatability (Fig. 11–3) of reed canarygrass is associated with the hordenine content (Marten *et al.*, 1976).

Canadian researchers (Wittenberg *et al.*, 1992; Duynisveld and Wittenberg, 1993) have evaluated low-alkaloid reed canarygrass cultivars as forage. In general, animal performance was improved with the low-alkaloid cultivars. Environmental factors such as water stress and continuous grazing increase the gramine content of reed canarygrass (Duynisveld and Wittenberg, 1993). The alkaloid content is lower in hay than in the fresh plant, but is not reduced

FIGURE 11–3 Plots of reed canarygrass cultivars at the University of Minnesota showing palatability differences. Sheep are walking through a plot of an unpalatable cultivar to graze a plot (foreground) that is palatable. The differences in acceptability are related to tryptamine alkaloid content. (Courtesy of G.C. Marten)

by the ensiling process (Tosi and Wittenberg, 1993).

Reed canarygrass has not commonly been implicated in lethal toxicity in North America but has been a cause of incoordination and death in sheep in New Zealand (Simpson *et al.*, 1969). Other *Phalaris* spp. in the United States have caused animal mortalities. East and Higgins (1988) reported an outbreak of *Phalaris* poisoning in sheep grazing *P. aquatica* (formerly *tuberosa*) in California. Ewes developed chronic neurological signs such as head nodding and a stiff gait with muscle tremors, with varying degrees of incoordination and falling. Numerous animals died with characteristic lesions of the brain and nervous system. In some cases, development of clinical signs occurred months after exposure to *Phalaris* ceased. A similar outbreak occurred in Louisiana in sheep and cattle grazing *P. caroliniana* (Nicholson *et al.*, 1989). Delayed development of clinical signs is characteristic of *Phalaris* staggers (Bourke *et al.*, 1987). An incidence of *Phalaris* poisoning in feedlot lambs in California was reported by Lean *et al.* (1989). The corn and alfalfa pellets were found to contain tryptamine alkaloids, but the original source of them was not identified. Alfalfa fields presumably could contain *Phalaris* spp. as weeds.

It has been suggested (Majak *et al.*, 1979; Parmar and Brink, 1976) that the tryptamines in reed canarygrass might be implicated in a pasture-induced type of acute bovine pulmonary emphysema prevalent in British Columbia and the intermountain states of the U.S. Whether the canarygrass alkaloids are metabolized to 3-methylindole, the causative agent in bovine pulmonary emphysema, has not been demonstrated.

***Phalaris* poisoning** has been extensively studied in Australia. Two major syndromes are encountered; a sudden death syndrome and a nervous syndrome (Bourke *et al.*, 1990). The sudden death syndrome is characterized by sudden collapse and cardiac arrest (Bourke *et al.*, 1988). The cause of the cardiac arrest is not known but does not seem to be associated with the tryptamine alkaloids (Kennedy *et al.*, 1986; Bourke *et al.*, 1988). Bourke and Carrigan (1992) suggested as many as four possible factors involved, including a cardio-pulmonary toxin, a thiaminase and amine cosubstrate, cyanogenic compounds, and nitrate compounds. Anderton *et al.* (1994) have suggested that **N-methyltyramine** is the causative factor of cardiac arrest. N-methyltyramine is similar in structure to hordenine (N-dimethyl tyramine) except it has one methyl instead of two. However, the N-methyl tyramine has about 10 times higher hypertensive action than hordenine (Anderton *et al.*, 1994). The tryptamine and tyramine alkaloids are monoamine oxidase (MAO) inhibitors; this activity may contribute to the cardiotoxic effects.

The neurologic syndrome, commonly referred to as ***Phalaris* staggers**, seems to be caused by dimethylated indole amines (tryptamines) in *P. aquatica* (Bourke *et al.*, 1990). Bourke *et al.* (1990) suggest that the nervous syndrome induced by *P. aquatica* results from a direct action of *Phalaris* alkaloids on serotonergic receptors in specific brain and spinal cord nuclei. Characteristic indole-like pigments are found in the brain and spinal cord in animals with *Phalaris* staggers. These are probably post-effect metabolites of the tryptamine alkaloids (Bourke *et al.*, 1990). Signs of toxicosis often show a delayed onset after *Phalaris* consumption has ceased. Tryptamine alkaloids are metabolized by MAO, with removal of the amine group. Because β-carbolines are strong MAO inhibitors (McKenna and Towers, 1984), they may enhance the toxicity of other *Phalaris* alkaloids by reducing their rate of detoxification.

Tryptamine alkaloids, including those from *Phalaris* spp., are used as psychoactive and recreational drugs, because of their effects on the brain. Various plants containing tryptamine alkaloids are used by shamans (witch doctors) in numerous cultures. Venoms from some toads contain these alkaloids and are used as recreational drugs for their psychoactive effects (McKenna and Towers, 1984).

Oxalates

Various **tropical grasses** contain soluble oxalates (Fig. 11–4) in sufficient concentration to induce calcium deficiency in grazing animals. These include buffelgrass (*Cenchrus ciliaris*), pangolagrass (*Digitaria decumbens*), setaria (*Setaria sphacelata*), and kikuyugrass (*Pennisetum clandestinum*). Numerous problems have been noted with these grasses in Australia, primarily with horses. Oxalates react with calcium to produce insoluble calcium oxalate, reducing calcium absorption. This leads to a disturbance in the absorbed calcium:phosphorus ratio, resulting in mobilization of bone mineral to alleviate the hypocalcemia. Prolonged mobilization of bone mineral results in **nutritional secondary hyperparathyroidism** (NSHP), or osteodystrophy fibrosa (Fig. 11–5), in horses consuming these tropical grasses (Blaney *et al.*, 1981a,b, 1982; McKenzie *et al.*, 1981). Cattle and sheep are less affected because of degradation of oxalate in the rumen (Allison *et al.*, 1981). However, cattle mortalities from oxalate poisoning due to acute

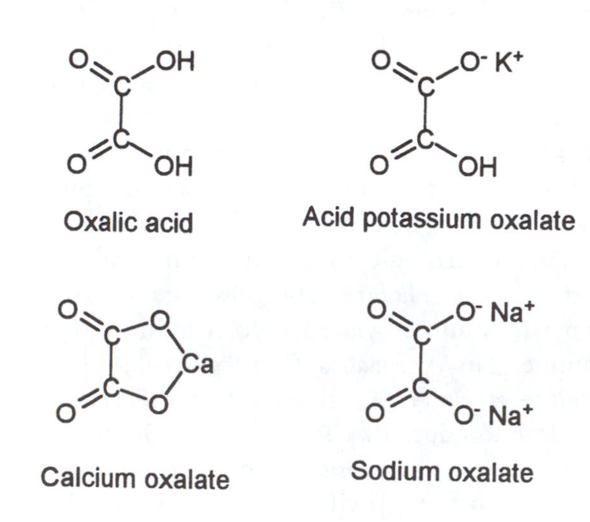

FIGURE 11–4 Structures of oxalates occurring in plants.

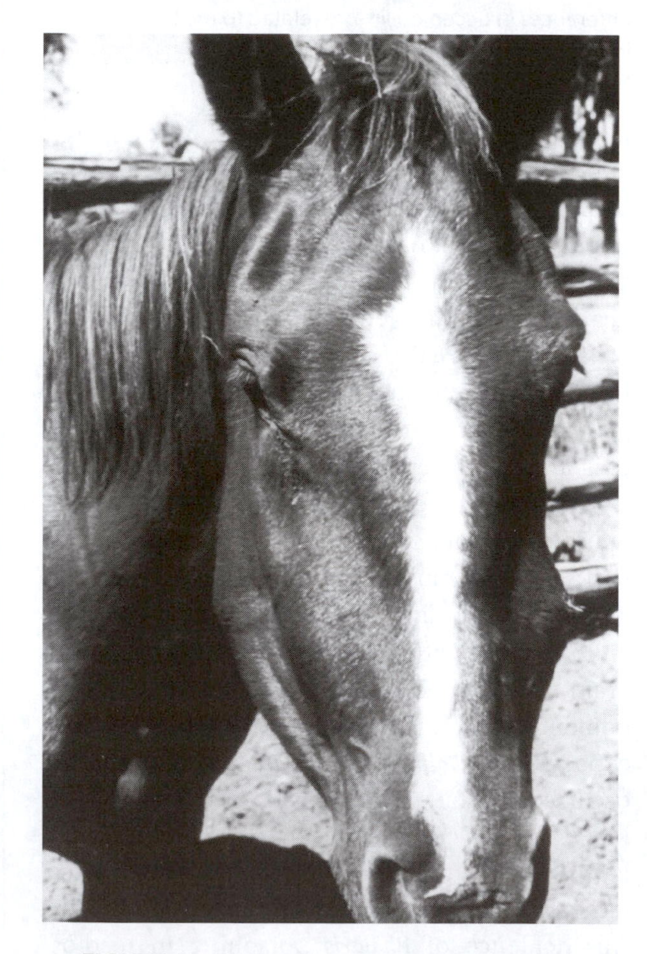

FIGURE 11–5 Nutritional secondary hyperparathyroidism or osteodystrophy fibrosa ("bighead") in a horse that had been grazing on tropical grass pasture species which are high in oxalate and low in calcium. (Courtesy of R.A. McKenzie)

hypocalcemia have occurred on setaria pastures (Jones *et al.*, 1970) and sheep have been poisoned while grazing buffelgrass (McKenzie *et al.*, 1988).

Levels of .5% or more soluble oxalate in forage grasses may induce NSHP in horses. Levels of 2% or more soluble oxalate can lead to acute toxicosis in ruminants (McKenzie *et al.*, 1988). The oxalate content of grasses is highest under conditions of rapid growth with concentrations as high as 6% or more of dry weight.

Oxalate poisoning caused by poisonous plants such as *Halogeton* is discussed in Chapter 12.

Cyanogens

Cyanide poisoning associated with the **forage sorghums** such as sudangrass is well known. Sorghums contain a cyanogenic glycoside, dhurrin, from which free cyanide can be released by enzymatic action (Fig. 11–6). The glycoside occurs in epithelial cells of the plant, whereas the glycoside-hydrolyzing enzymes are in the mesophyll cells (Wheeler and Mulcahy, 1989). Damage to the plant from wilting, trampling, frost, drought, and so on results in the breakdown of cell structure, causing exposure of the glycoside to the inherent hydrolytic enzymes and formation of free cyanide. The cyanide potential and risk of poisoning decreases as forage sorghums (Fig. 11–1) mature (Wheeler *et al.*, 1990). Nutritional value of forage sorghums also declines markedly with maturity, so a management strategy of delayed grazing to avoid toxicosis may result in poor animal performance (Wheeler and Mulcahy, 1989).

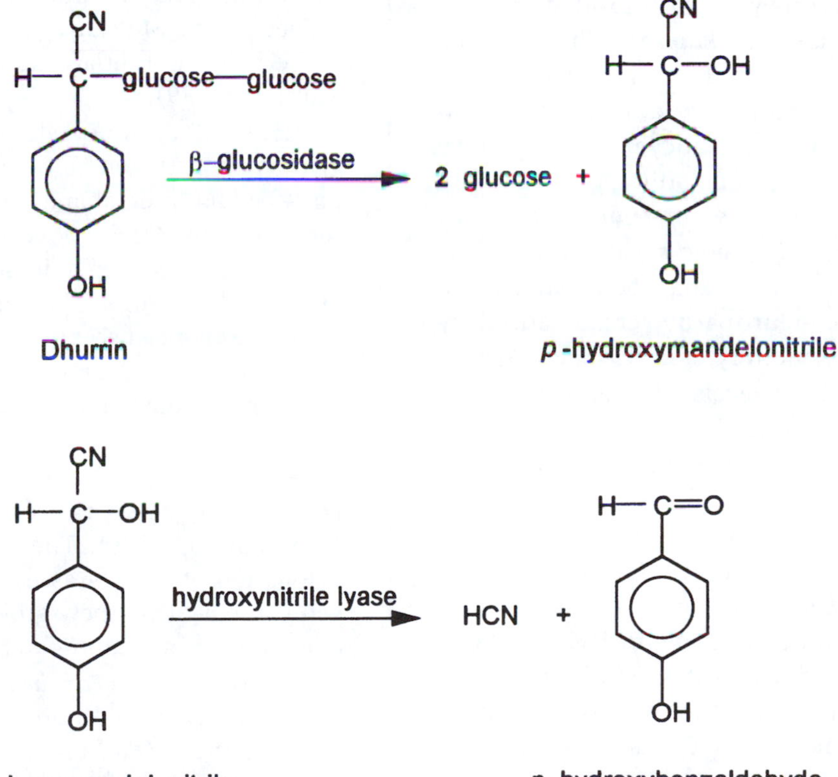

FIGURE 11–6 Enzymatic production of hydrogen cyanide (HCN) from cyanogenic glycoside (dhurrin) in forage sorghums.

Acute cyanide toxicosis is caused by the inhibition of cytochrome oxidase, a terminal respiratory enzyme in all cells. When cytochrome oxidase is inhibited, the cells suffer from rapid ATP deprivation. Signs include labored breathing, excitement, gasping, staggering, convulsions, paralysis, and death. The blood is bright red due to its high content of oxyhemoglobin. The myocardium is the most severely affected tissue; death is caused by acute cerebral anoxia resulting from cardiac failure.

Although cyanide is extremely toxic, acute toxicosis occurs only after detoxification processes are overwhelmed. Cyanide is readily detoxified in animal tissues by reacting with thiosulfate to form thiocyanate:

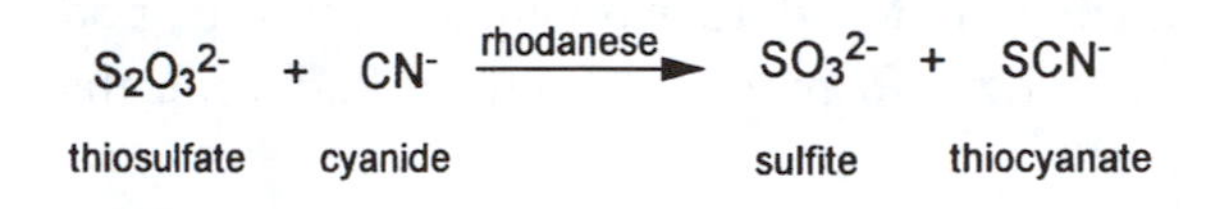

This reaction is catalyzed by thiosulfate:cyanide sulfur transferase (rhodanese). The resulting thiocyanate is excreted in the urine. With chronic exposure to dietary cyanogens, chronically elevated blood thiocyanate may have physiological effects, including goitrogenicity and possible involvement in neurologic disturbances. In humans, chronic consumption of cyanogenic cassava as a staple food may lead to **tropical ataxic neuropathy**, characterized by lesions of optic, auditory, and peripheral nerves and elevated blood levels of thiocyanate (see Chapter 6). A similar condition occurs with prolonged cyanide exposure in cattle and horses, with ataxia and degenerative lesions of the central nervous system (Adams *et al.*, 1969; Wheeler and Mulcahy, 1989). **Cystitis and urinary incontinence** also occur in chronic cyanide exposure in horses (Knight, 1968) and cattle (McKenzie and McMicking, 1977). Bradley *et al.* (1995) reported on extensive losses of sheep grazing sorghum pasture. The animals exhibited weakness, ataxia, head shaking, knuckling of fetlocks, inability to rise and opisthotonos. Lambs born to affected ewes had contracted limbs (arthrogryposis). Neuroaxon degeneration in the brain and spinal cord was noted. Although elevated blood thiocyanate occurs in these chronic conditions, a role of thiocyanate in their etiology has not been demonstrated. Sulfur deficiency induced by the increased excretion of thiocyanate in animals on sorghum pasture is common in Australia (Wheeler and Mulcahy, 1989), causing reduced feed intake and a decline in animal performance. Provision of salt containing 5% sulfur overcomes the deficiency. Caution is needed with sulfur supplementation because excessive sulfur can lead to copper deficiency, induced by ruminal formation of thiomolybdates that complex with copper and reduce its absorption.

The likelihood of acute cyanide toxicosis may be greater with sorghum hay than with grazing of the fresh plant. Dry matter may be consumed much more rapidly in the form of hay than as fresh pasture. Ground and pelleted sorghum hay may be especially toxic because of the rapid rate of intake and cyanide release (Wheeler and Mulcahy, 1989). Ensiling reduces the cyanide potential markedly (Wheeler and Mulcahy, 1989).

Rates of cyanide release from cyanogenic glycosides in the rumen are influenced by diet, post-feeding interval and the chemical nature of the glycosides (Majak *et al.*, 1990). Cyanide production is most rapid in fasted animals, suggesting that animals on limited feed intake might be more susceptible to cyanide intoxication when turned onto sorghum pasture than animals previously on full feed.

Cyanogens occur in other grasses such as *Cynodon* spp. (e.g. Bermuda grass), which are tropical forage grasses (Georgiadis and McNaughton, 1988). These authors noted that defoliation of these grasses increases the cyanogen content of the new growth, suggesting that cyanide plays a role in deterring herbivory.

White clover (*Trifolium repens*) is mainly a temperate pasture legume, having a prostrate growth pattern (Fig. 11–7). It spreads by stolons on the ground surface, and is resistant to heavy grazing pressure. It is one of the major pasture legumes in North America, Europe, New Zealand and parts of Australia and South America.

FIGURE 11–7 White clover, a common pasture legume, contains cyanogenic glycosides.

White clover contains moderate amounts of the cyanogenic glycosides linamarin and lotaustralin (see Chapter 6). It does not generally cause livestock problems due to its cyanogenic activity. In midsummer, it often becomes unpalatable and animals will refuse to graze it. This may be due to bitterness associated with the cyanogens; the cyanide potential increases with drought stress (Vickery *et al.*, 1987). Plant-breeding studies have shown that the cyanogen content of white clover can be easily modified by selection to produce low-cyanogen types. However, the cyanogens are not a problem of sufficient magnitude to warrant a serious effort to reduce them. The major livestock problem associated with white clover is bloat.

Saponins

Saponins are glycosides widely distributed among plants of economic importance. They have either a steroidal or triterpenoid structure (Fig. 11–8). Many forage legumes grown in temperate areas contain saponins. Saponins have a wide variety of biological effects, with both positive and negative implications. They show potential as dietary additives in lowering serum cholesterol levels in humans, thus possibly reducing the risk of atherosclerosis. They have a positive effect on rumen fermentation by inhibiting the protozoa. On the negative side, they inhibit the productive performance of non-

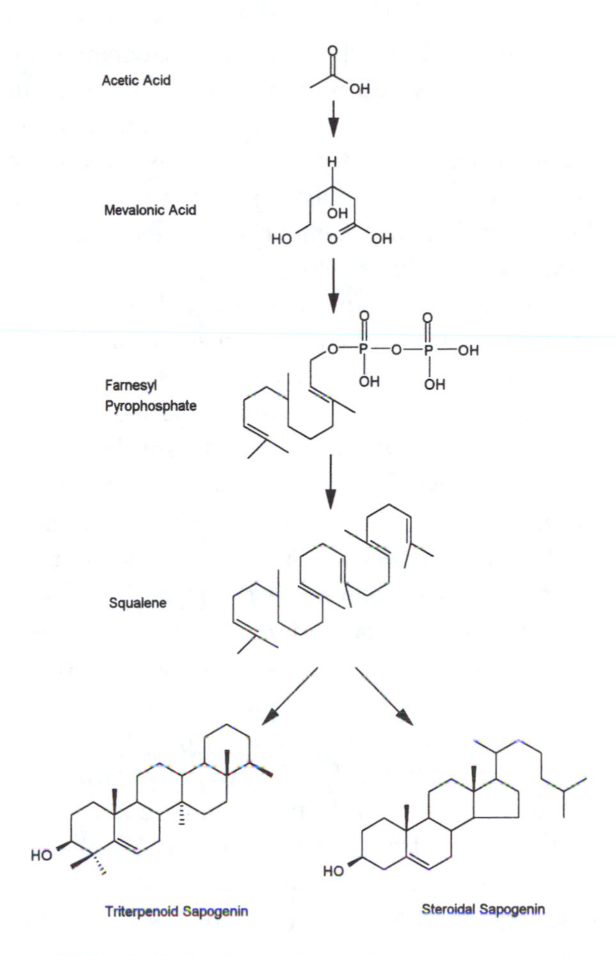

FIGURE 11–8 Biosynthesis of triterpenoid and steroid saponins.

ruminant animals (swine, poultry) fed diets containing alfalfa.

Saponins are characterized by a bitter taste and foaming properties (honeycomb foam in water). They are well known for their in vitro hemolytic effects, a property that has been used in analysis of plant material for saponin content. Saponins have had industrial and commercial applications, including use in soft drinks, shampoo, fire extinguishers, soap, and the synthesis of steroid hormones (e.g., birth control pills). Because the aglycone is nonpolar, while the carbohydrate side chain is water soluble, they have strong **detergent properties**. Because of their foaming properties, they are extremely toxic to fish, and have been used in traditional tropical rain forest cultures as fish poisons.

Many foods and herbs contain saponins (Table 11–1). A main commercial source of saponin is the bark of a South American tree, ***Quillaja saponaria*** (soap tree). The bark contains about 10% quillaja saponin, and has been used as a foaming agent in beverages and confectionary products (Price *et al.*, 1987). In several South American countries, quillaja bark is sold by street vendors for home preparation of shampoo. The other major commercial source of saponins is yucca (***Yucca schidigera***), a desert plant (Fig. 11–9) native to the southwestern U.S. and Mexico. It contains steroidal saponins (Kaneda *et al.*, 1987) such as smilagenin and sarsasapogenin. **Sarsaponin** (a commonly-used term for yucca saponins) is the glycoside of sarsasapogenin. **Yucca extract** is a common feed additive which has ammonia-binding properties. When added to poultry and swine diets, yucca extract binds ammonia released by microbial action in the excreta, thus helping to prevent elevated ammonia levels in confinement poultry and swine facilities. Sarsaponins may be the active ammonia-binding constituents of yucca extract; glycoprotein fractions may also be involved.

TABLE 11–1 Foods and Herbs Containing Saponins

Foods	Herbs and flavorings
Soybeans	Fenugreek
Other beans and peas	Liquorice
Oats	Nutmeg
Potatoes	*Quillaja*
Tomatoes	*Yucca*
Onions	*Saponaria*
Asparagus	*Gypsophila*
Peanuts	Ginseng
Spinach	Many other herbs[a]
Cucumbers	
Yams	

[a]For a comprehensive list of saponin-containing herbs, see Price *et al.* (1987).

FIGURE 11–9 *Yucca schidigera* is harvested commercially in the deserts of Baja California, in Mexico, for the production of yucca extract used in the feed and beverage industries.

The triterpenoid saponins in **alfalfa** (Fig. 11–10) are probably of the most interest in animal production. The principal saponins in common alfalfa cultivars are medicagenic acid, soyasapogenol A and B, and lucernic acid. They differ in the hydroxyl and carboxyl groups attached to the sapogenin (aglycone). Medicagenic acid appears to be the major saponin in alfalfa with antinutritional effects.

The biological activity of saponins is influenced by the carbohydrate side chain. In general, saponins with a glycoside link to one sugar have greater activity than those linked to two or more sugars (Oleszek, 1990).

The saponin content of alfalfa forage is influenced by environmental factors and follows a seasonal cycle, being high in midsummer and low in spring and fall. Consequently, the

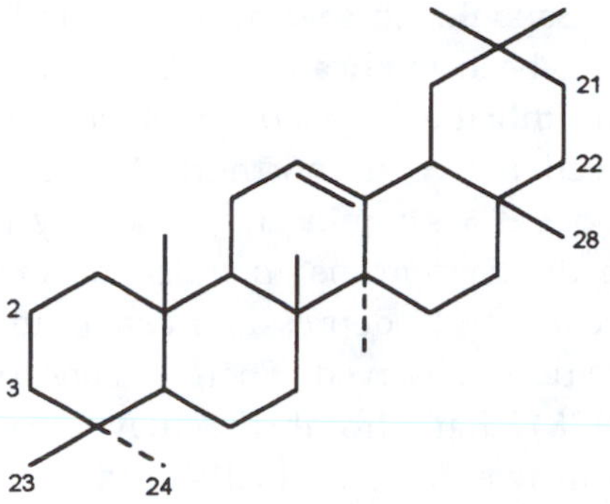

Saponin	C position of groups OH	COOH
Medicagenic acid	2	2
Soyasapogenol A	4	--
Soyasapogenol B	3	--
Lucernic acid	3	1

Structures of alfalfa saponins

FIGURE 11–10 Structures of some of the triterpenoid saponins in alfalfa.

saponin content of second-cutting alfalfa is usually higher than that of first or third cuttings, which may contribute to the generally poorer animal performance observed with second-cutting alfalfa. Saponin in the roots is transferred to the foliage in response to flowering and various environmental factors. Plant breeders have developed strains or cultivars of alfalfa with both high and low saponin content. The low-saponin selections show potential as improved types of alfalfa for animal feeding. Among the effects of saponins on animals are growth inhibition of swine and poultry, reduced palatability, and increased excretion of cholesterol.

Low levels of dietary alfalfa meal reduce the growth rate of poultry and swine, primarily due to effects on palatability and feed intake (Cheeke *et al.*, 1983), rather than to metabolic effects. Use of low-saponin alfalfa strains or cultivars increases the level of alfalfa that can be fed to nonruminants without reducing growth performance. Saponins are bitter and irritate mucous membranes of the mouth and digestive tract. These effects depress voluntary feed intake. The saponin zahnic acid in alfalfa is especially bitter and unpalatable (Oleszek *et al.*, 1992).

Saponins may have effects on rumen fermentation. Lu and Jorgensen (1987) reported a reduction in **ruminal protozoa** numbers when alfalfa saponins were administered to sheep. Klita *et al.* (1996) also observed reduction of rumen protozoa following administration of alfalfa saponins to sheep. They attributed the antiprotozoal effect to the reaction of saponins with cholesterol in the protozoal cell membranes, causing lysis of the protozoa. Bacterial membranes do not contain cholesterol. Reduction in protozoal numbers generally increases rumen bacterial numbers, because protozoa are predators of bacteria. Wallace *et al.* (1994) observed similar anti-protozoal activity with yucca saponins. Possibly saponin-containing plants could have a role as natural defaunating agents in ruminants. **Defaunation** (the elimination of rumen protozoa) can improve efficiency of rumen fermentation.

Saponins may influence nutrient digestion and absorption by interacting with mucosal cell membranes, causing permeability changes or the loss of activity of membrane-bound enzymes (Oleszek *et al.*, 1994). Saponins influence mineral absorption. Southon *et al.* (1988) found that iron absorption is inhibited by dietary saponin. Similar results were reported by West *et al.* (1978).

Saponins are found in a variety of human feedstuffs (Oakenfull and Sidhu, 1989; Price *et al.*, 1987). The food plants highest in saponin content include chick-peas (*Cicer arietinum*), soybeans, alfalfa sprouts, and common beans. The active components of ginseng, a widely used plant in herbal medicine, are saponins called ginsenosides (Li *et al.*, 1996). Oats contain small amounts of saponins called avenacosides (Onning *et al.*, 1996). They are of minor significance in human and animal nutrition.

Saponins form insoluble complexes with **cholesterol**. Dietary alfalfa or alfalfa saponins cause a reduction in tissue and serum cholesterol, reduced cholesterol absorption, and increased cholesterol fecal excretion. Saponins may bind with bile salts that are needed for cholesterol absorption or, because of their surface-active properties, may cause bile salts to bind to polysaccharides in fiber. Dietary saponins increase the fecal excretion of bile acids and neutral sterols. The effect of saponins in lowering serum cholesterol is a result of preventing its reabsorption after it has been excreted in the bile (reduction in entero-hepatic recycling of cholesterol). By binding bile acids, saponins may reduce micelle formation, thus impairing the absorption of fat and fat-soluble vitamins such as vitamins A and E (Jenkins and Atwal, 1994).

The feeding of alfalfa meal to laying hens as a method of reducing egg cholesterol levels has been examined. Any effect of dietary alfalfa or alfalfa saponins on egg cholesterol levels is slight. Nakaue *et al.* (1980) found no difference in egg cholesterol from hens fed diets containing high- or low- saponin alfalfa meal.

For many years, alfalfa saponins were thought to be involved in **bloat** in ruminants. Evidence for such an involvement was that saponins are surface-active agents producing stable foams, they are found in legumes that cause bloat and are not found in nonbloating legumes, and experimental administration of alfalfa saponins has caused bloat. However, Majak *et al.* (1980) found no difference in the incidence of bloat in cattle receiving high- or low-saponin alfalfa. Other factors, such as cytoplasmic protein fractions, appear to be those responsible for bloat. Saponins may actually help to prevent bloat by suppressing rumen protozoa. Defaunation of cows reduces the incidence and severity of bloat (Clarke *et al.*, 1969).

Saponins in several pasture and range weeds have been implicated in toxicological problems. **Alfombrilla** (*Drymaria arenaroides*) is a perennial weed native to northern Mexico. It contains about 3% saponin and is acutely poisonous to cattle. Symptoms include anorexia, diarrhea, arched back, depression, coma, and death. Enteritis is observed. There is concern (Williams, 1978) that alfombrilla may invade the U.S. Southwest. Alfombrilla is a member of the Caryophyllaceae family. Saponins are the principal toxins in other members of this family, including corn cockle (*Agrostemma githago*), bouncing bet or soapwort (*Saponaria officinalis*), and cow cockle (*Saponaria vaccaria*). Broomweed or **snakeweed** (*Gutierrezia sarothrae*) is a perennial resinous shrub found in desert ranges of the southwestern U.S, particularly New Mexico and west Texas. Its toxicity is believed to be due to its saponin content. Toxicity signs include listlessness, anorexia and weight loss, a rough hair coat, gastroenteritis, and liver and kidney lesions. In addition, it causes abortions or the birth of small, weak offspring in cattle. Saponins in broomweed may be the abortifacient fraction. Both male (Edrington *et al.*, 1993a) and female (Edrington *et al.*, 1993b) fertility are adversely affected by ingestion of snakeweed. Other toxins in snakeweed include flavonoids, terpenes and resins (Smith *et al.*, 1994).

Saponin-Induced Photosensitization

Numerous tropical and warm-season grasses cause hepatogenous or **secondary photosensitization** in grazing animals. These include various *Panicum* and *Brachiaria* spp. (Graydon *et al.*, 1991; Miles *et al.*, 1992). Liver damage characterized by deposition of crystalline substances in and around the bile ducts is observed. Affected animals develop photophobia and severe dermatitis. The plant factors responsible for the crystal formation are steroidal saponins (Bridges *et al.*, 1987). New Zealand researchers (Miles *et al.*, 1991, 1992; Munday *et al.*, 1993) have demonstrated that the crystalline material consists of calcium salts of **steroidal saponin**

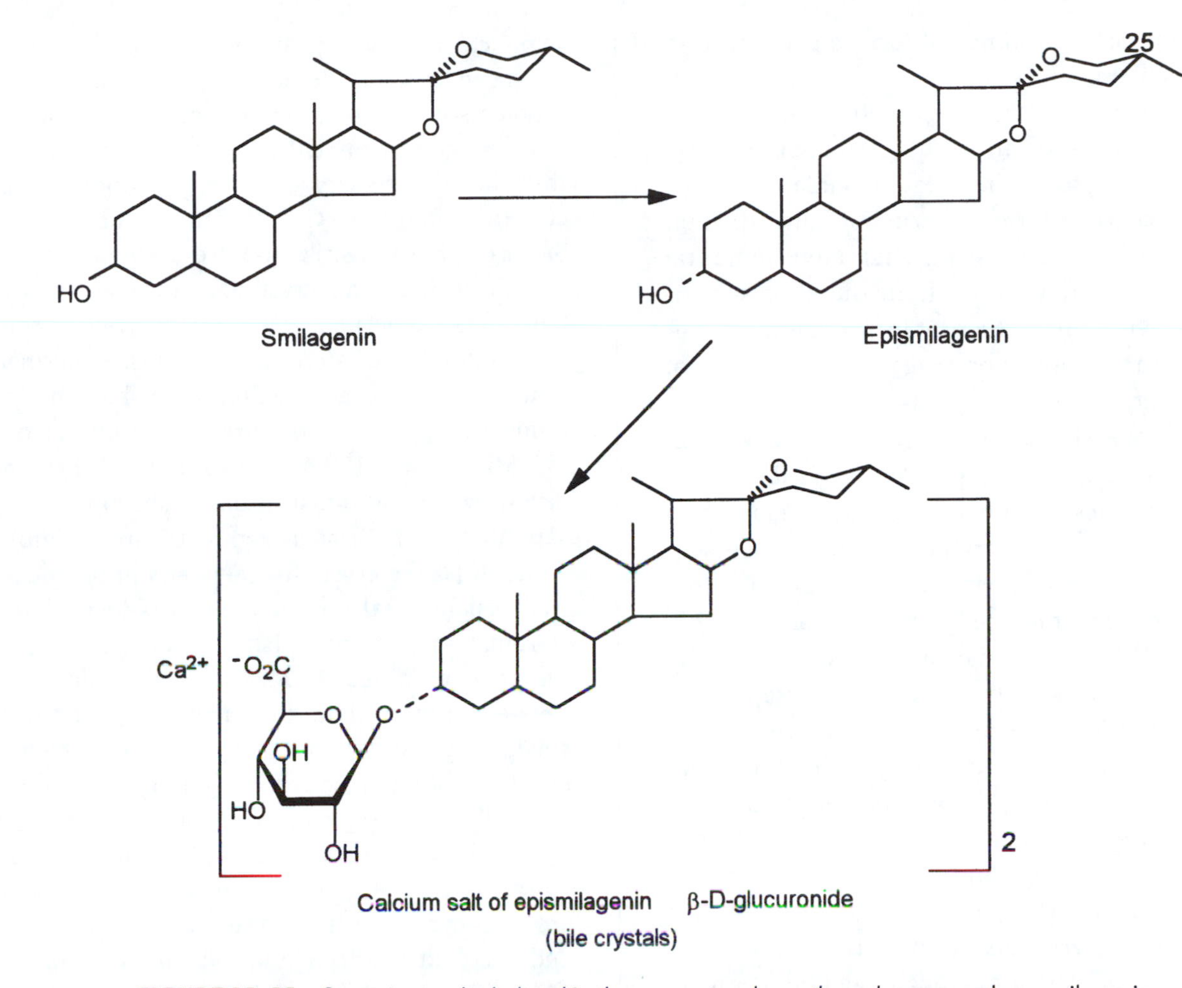

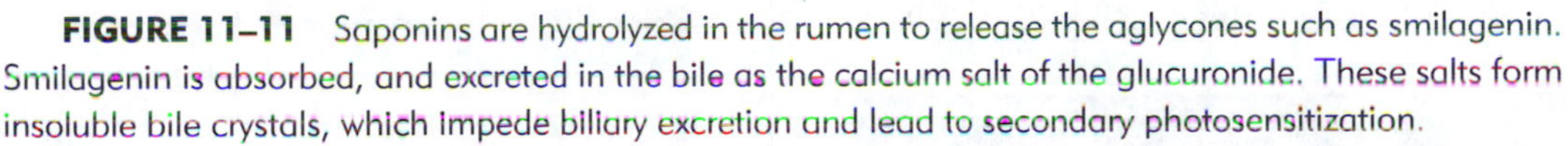

FIGURE 11–11 Saponins are hydrolyzed in the rumen to release the aglycones such as smilagenin. Smilagenin is absorbed, and excreted in the bile as the calcium salt of the glucuronide. These salts form insoluble bile crystals, which impede biliary excretion and lead to secondary photosensitization.

glucuronides (Fig. 11–11). These authors caution that although the bile crystals are saponin derivatives, plant saponins are not necessarily the cause of the liver damage and photosensitization. Experimental production of photosensitization by administration of isolated saponins takes much greater quantities than would be obtained from consumption of the plants (Kellerman *et al.*, 1991). Miles *et al.* (1992) suggest that other hepatotoxins such as mycotoxins (e.g., sporidesmin) may have synergistic effects with saponins. Interactions between saponins and other hepatotoxic agents could explain the sporadic incidence of photosensitization. Mullenax (1991) suggested that the photosensitization outbreaks seen with cattle on *Brachiaria decumbens* pastures may be caused by the mycotoxin sporidesmin rather than intrinsic factors in the grass. However, Smith and Miles (1993) provide convincing evidence that steroidal saponins are the plant factor primarily involved. Sporidesmin and other hepatotoxic mycotoxins could certainly have a synergistic effect. Kleingrass (*Panicum coloratum*) is widely grown in Texas, where it has caused outbreaks of hepatogenous photosensitization in sheep (Bridges *et al.*, 1987) and horses (Cornick *et al.*, 1988). Saponins were identified as causative agents, but the sporadic incidence of the condition suggests that other interacting factors such as mycotoxins or environmental factors contribute to the problem.

Puoli *et al.* (1992) reported an incident of photosensitization in lambs grazing switchgrass (*Panicum virgatum*) in West Virginia. The outbreak occurred during a period of unusually hot, dry weather, suggesting that environmental conditions may play a role in the photosensitization syndrome, either by increasing the sensitivity of the animals or increasing the content of hepatoxic factors in the grass.

Photosensitization and liver damage occur in cattle grazing signalgrass (*Brachiaria decumbens*) in the tropics. Cattle on lush signalgrass pastures often exhibit very poor growth, photosensitization, liver damage, and a marked reluctance to consume the grass (Low *et al.*, 1993, 1994). This lack of palatability, as well as the other signs, could be caused by saponins. Saponins in forages are bitter, and are mucous membrane irritants. Irritation of the oral cavity would likely have a depressing effect on feed intake. Ruminal metabolism of *Brachiaria* saponins releases sapogenins which cause liver and kidney damage (Lajis *et al.*, 1993).

Tribulus terrestris or puncture vine (Fig. 11–12) is a prostrate annual weed common on semi-arid rangelands throughout the world. It has been responsible for major animal health problems in South Africa, causing a photosensitization condition called geeldikkop (yellow thick head [Afrikaans]). A similar condition occurs in Australia (Glastonbury *et al.*, 1984). Signs of photosensitization are accompanied by hepatic lesions and crystalline deposits in the bile ducts (Kellerman *et al.*, 1991). *Tribulus* contains a number of steroidal saponins. Kellerman *et al.* (1991) induced geeldikkop in sheep by administering a crude mixture of *Tribulus* saponins. Miles *et al.* (1994a,b) also induced the condition by oral administration to sheep of crude saponins from *Tribulus terrestris*, and demonstrated that the crystalline deposits in bile ducts were calcium salts of the glucuronides of the steroidal sapogenins epismilagenin and episarsasapogenin. Miles (1994a) studied ruminal and hepatic metabolism of *Tribulus* saponins in sheep. Ingested saponins were rapidly hydrolyzed in the rumen, releasing free sapogenins. Most of the sapogenins were rapidly converted by ruminal microflora to epismilagenin and episarsasapogenin, which are absorbed. These are conjugated with gluronic acid in the liver, and excreted in the bile as calcium salts of the sapogenin glucuronides. These salts are insoluble and precipitate in the bile ducts as crystalline deposits. The disruption of bile excretion results in retention of phylloerythrin and classic hepatic (secondary) photosensitization.

FIGURE 11–12 Puncture vine (*Tribulus terrestris*) contains steroidal saponins, responsible for biliary crystals and secondary photosensitization. Note the seed pods, which contain extremely sharp spines arranged in a pyramidal form. These can puncture tires, feet, etc.; hence, the name puncture vine.

A condition called tribulus staggers occurs in Australia (Bourke, 1984; Bourke *et al.*, 1992). Bourke *et al.* (1992) attribute this neurological condition to tryptamine alkaloids in *Tribulus terrestris*.

Several other plants cause photosensitization accompanied by biliary crystals, with steroid saponins implicated in each case as the causative agents. In Norway, a photosensitization disease of sheep called **alveld** (elf fire, from the folk explanation of elves shooting sheep with arrows) is caused by consumption of an herb, ***Narthecium ossifragum***. Calcium salts of glucuronides of steroid sapogenins in bile crystals and saponins in the plant have been identified

(Miles *et al.*, 1993). The succulent desert plant ***Agave lecheguilla*** also contains steroid saponins and causes bile crystals and photosensitization (Camp *et al.*, 1988). ***Nolina texana***, commonly known as beargrass or **sacahuiste**, is common in the U.S. southwest. It causes liver damage and photosensitization accompanied by biliary crystals. The blooms are the most toxic part of the plant (Rankins *et al.*, 1993). Administration of sacahuiste blossoms to sheep resulted in severe liver damage and mortality. Saponins appear to be the active component, although there is evidence of a toxic volatile constituent as well (Rankins *et al.*, 1993). The sporadic incidence of saponin-induced photosensitization may perhaps be influenced by dietary calcium level. With excess calcium, there may be a greater tendency for calcium salts to be formed in the bile.

ALKALOIDS IN FORAGE LUPINS

Lupins (Fig. 11–13) native to North American rangelands are important forages for range livestock. However, they do contain toxic **quinolizidine alkaloids** which can cause acute toxicity, mainly in sheep, and teratogenic effects in cattle (crooked calf disease). In the 1940–1960 period, wild lupins were responsible for extensive poisonings of range sheep in the western U.S. Because of the marked decline of the range sheep industry since 1960, outbreaks of lupin poisoning are less common than formerly.

FIGURE 11–13 An example of a lupin, showing the characteristic appearance of the leaves and flowers.

At least 100 spp. of lupins grow in western North America (Kingsbury, 1964). The principal toxic species are *L. leucophyllus* (woolly leaved lupin), *L. leucopsis* (big bend lupin), *L. argenteus* *(silvery lupin), and* L. sericeus (silky lupin). The greatest concentration of alkaloids is in the seeds. The preflowering plants are generally low in toxicity and provide good forage. Sheep are the main livestock affected; **lupin poisoning** causes greater losses of sheep than any other poisonous plant in Montana, Idaho, and Utah (Kingsbury, 1964). Consumption of a large quantity of lupin in a short period of time disposes sheep to poisoning. A lethal dose is 0.25–0.5% of body weight for seeds and about 1.5% of body weight for pods and seeds. Symptoms appear within a few hours. The breathing is heavy and labored, often with snoring, and the animal becomes comatose and dies. Sometimes there may be trembling and convulsions. Death is from respiratory paralysis. Lupin poisoning can be controlled by good animal management, avoiding conditions under which large amounts of lupin would be consumed in a short period of time. These would include avoiding moving hungry sheep through heavy stands of lupin and avoiding unloading or bedding down sheep in areas dominated by lupin. Cattle and horses are less commonly poisoned, probably because they find lupin pods less palatable and because they are not herded and so are less likely than sheep to be exposed to and consume a large quantity at one time. The specific alkaloids in wild lupin responsible for acute toxicosis of range sheep have not been well identified. They may include lupanine and sparteine.

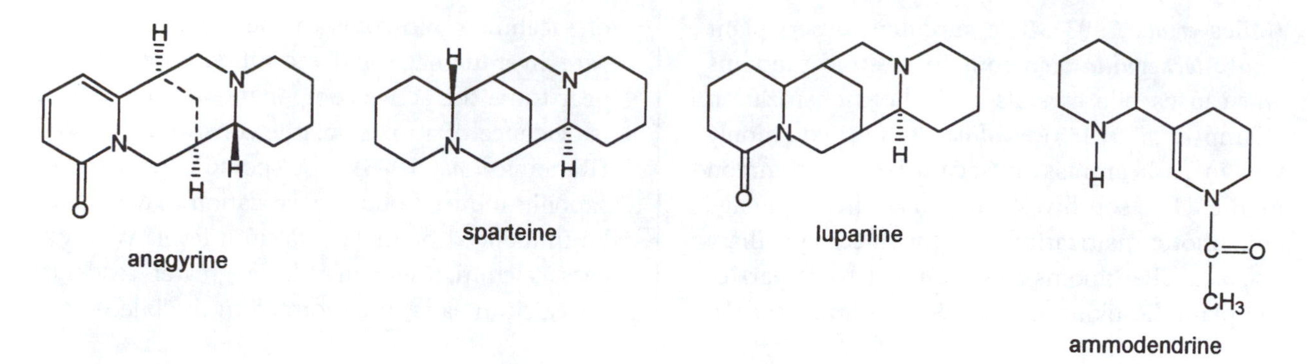

FIGURE 11–14 Examples of alkaloids found in lupins. Anagyrine and ammodendrine are teratogens, causing crooked calf disease.

Goats are much more resistant than sheep to lupin poisoning (Gardner and Panter, 1994). They showed no signs of toxicity after receiving an oral dose of *L. formosus* that was toxic to sheep. The blood levels of piperidine alkaloids (ammodendrine, N-methylammodendrine and N-acetylhystrine) were much lower in goats than in cattle and sheep receiving similar dosages of lupin. *L. formosus* contains **piperidine alkaloids** (e.g. ammodendrine) rather than quinolizidine alkaloids (Fig. 11–14). Goats are also resistant to the quinolizidines found in *L. caudatus* (Gardner and Panter, 1993).

The U.S. Soil Conservation Service has developed a variety of non-toxic sickle-keeled lupin (Hederma) for use in reseeding roadsides, eroded areas, and rangelands to avoid toxicities in wildlife and grazing livestock.

Crooked Calf Disease

In the western range areas of the U.S., a high incidence of skeletal deformities in calves has been observed in some years. This has been referred to as crooked calf disease (Fig. 11–15). The condition is characterized by twisted or bowed limbs (arthrogryposis), twisted neck, spinal curvature, cleft palate, or a combination of these conditions. Researchers at the U.S. Department of Agriculture (USDA) Poisonous Plants Laboratory in Logan, Utah, identified consumption of lupin by pregnant cows as the causative factor in most cases. *Lupinus sericeus, L. caudatus*, and *L. laxiflorus* are high in the causative alkaloids. The teratogenic condition may develop if cows consume these lupins between days 40 and 70 of pregnancy. The principal alkaloid responsible for crooked calf disease is **anagyrine** (Fig. 11–14).

FIGURE 11–15 Crooked calf disease, caused by maternal consumption of teratogenic lupins during days 40–70 of pregnancy.

The concentration of anagyrine is high in the early stages of growth of the plant and is also high in mature seeds. A pregnant cow is at greatest risk when grazing teratogenic lupin early in the plant growth phase or when seed pods have formed. Crooked calf disease can be avoided by cattle and range management through altering breeding schedules or grazing rotations so that cows are not exposed to lupins high in anagyrine when they are from 40- to 70-days pregnant. Sheep are not affected by the teratogenic properties of anagyrine, even when high levels of lupin are fed to pregnant ewes.

Panter and Keeler (1993) have proposed that cattle are uniquely sensitive to anagyrine, perhaps due to ruminal metabolism of anagyrine to a teratogenic metabolite. The mode of action may involve an immobilizing effect on the fetus (Panter and Keeler, 1993). The crooked calf syndrome is characterized by a lack of fetal mobility, resulting in skeletal malformations caused by the fetus remaining in one position for an extended period (Panter *et al.*, 1990; Bunch *et al.*, 1992).

L. formosus is teratogenic to cattle, but contains only traces of quinolizidine alkaloids (Keeler and Panter, 1989). It has a high concentration of the piperidine alkaloid **ammodendrine**, which is teratogenic. Ammodendrine is structurally similar to coniine, the teratogenic piperidine alkaloid in *Conium maculatum* (poison hemlock). Sheep and goats are affected by the teratogenic effects of the piperidine alkaloids in *L. formosus*, but not by quinolizidine alkaloids (Panter *et al.*, 1994).

Kilgore *et al.* (1981) reported an apparent case of human teratogenicity associated with lupin alkaloids in northern California. A baby boy with severe bone deformities in his arms and hands was born to a woman who had regularly consumed goat's milk during her pregnancy. The goats were grazing in an area where *Lupinus latifolius* formed the principal forage available. Goat kids from these goats were deformed, and puppies from dogs fed the goat's milk also showed deformities. Analysis of lupin forage from the area where the incident occurred showed it to contain a very high concentration of anagyrine (Meeker and Kilgor, 1987). When the lupins were fed to a lactating goat, anagyrine was detected in the milk. While it is impossible to prove that the deformities in the baby were due to lupin alkaloids transferred to the mother through consumption of goat's milk, the circumstantial evidence strongly implicated that etiology. The parents initially attributed the problem to the spraying of local forests with herbicides. In the affected infant, red cell aplasia (lack of formation of red blood cells) also occurred (Ortega and Lazerson, 1987).

Because goats are quite resistant to the acute toxic effects of quinolizidine alkaloids (Gardner and Panter, 1993), high concentrations of alkaloids could occur in milk without any signs of toxicity in goats consuming large quantities of lupins (Meeker and Kilgore, 1987, 1991).

TOXIC AMINO ACIDS

Mimosine and *Leucaena* Toxicity

Leucaena leucocephala, commonly referred to as leucaena, ipil-ipil (Southeast Asia), and kao haole (Hawaii), is a tropical legume with great potential as a protein source for livestock (Fig. 11–16). It is vigorous, rapidly growing, drought-tolerant, palatable, high yielding, and its leaves contain from 25 to 35% crude protein. It can be grown in tropical pastures or can be harvested and dried, with the leaf meal used in place of alfalfa meal in poultry and swine diets. It can be harvested in "cut and carry" feeding systems of small farmers and used for feeding cattle, goats, and rabbits. These potential attributes are presently limited by the occurrence of the toxic amino acid mimosine in leucaena.

Mimosine is structurally very similar to tyrosine, but does not seem to act as a tyrosine antagonist. In nonruminant animals, mimosine causes poor growth, alopecia (loss of hair), eye cataracts, and reproductive problems. Levels of leucaena meal about 5–10% of the diet for swine, poultry, and rabbits generally result in

FIGURE 11–16 *Leucaena leucocephala* is a tropical forage legume containing the toxic amino acid mimosine.

poor animal performance. The biochemical mode of action of mimosine in producing its toxic effects is not clear. Hegarty (1978), who has worked extensively on mimosine and leucaena in Australia, suggests that it may act as an amino acid antagonist, it may complex with pyridoxal phosphate, or it may complex metals such as zinc. Stunzi *et al.* (1979) have shown that under physiological conditions, mimosine binds copper and zinc ions more strongly than do most amino acids. Jones *et al.* (1978) have reported that mineral supplements containing zinc reduce the toxicity of leucaena in cattle and suggested that some of the toxic signs (e.g., skin lesions) resemble zinc deficiency. Puchala *et al.* (1996) observed that mimosine causes a marked depletion in blood zinc and magnesium, with the chelate excreted in the urine. They suggested that toxicity signs such as hair loss and dermatitis may be caused by induced zinc deficiency. Mimosine is an inhibitor of cystathionine synthetase and cystathionase, enzymes involved in the conversion of methionine to cystine. This inhibition could be a factor in mimosine-induced alopecia.

Ruminant animals grazing leucaena may show various symptoms such as poor growth, loss of hair, swollen and raw coronets above the hooves, lameness, mouth and esophageal lesions, depressed serum thyroxine levels, and goiter (Fig. 11–17). Some of these symptoms

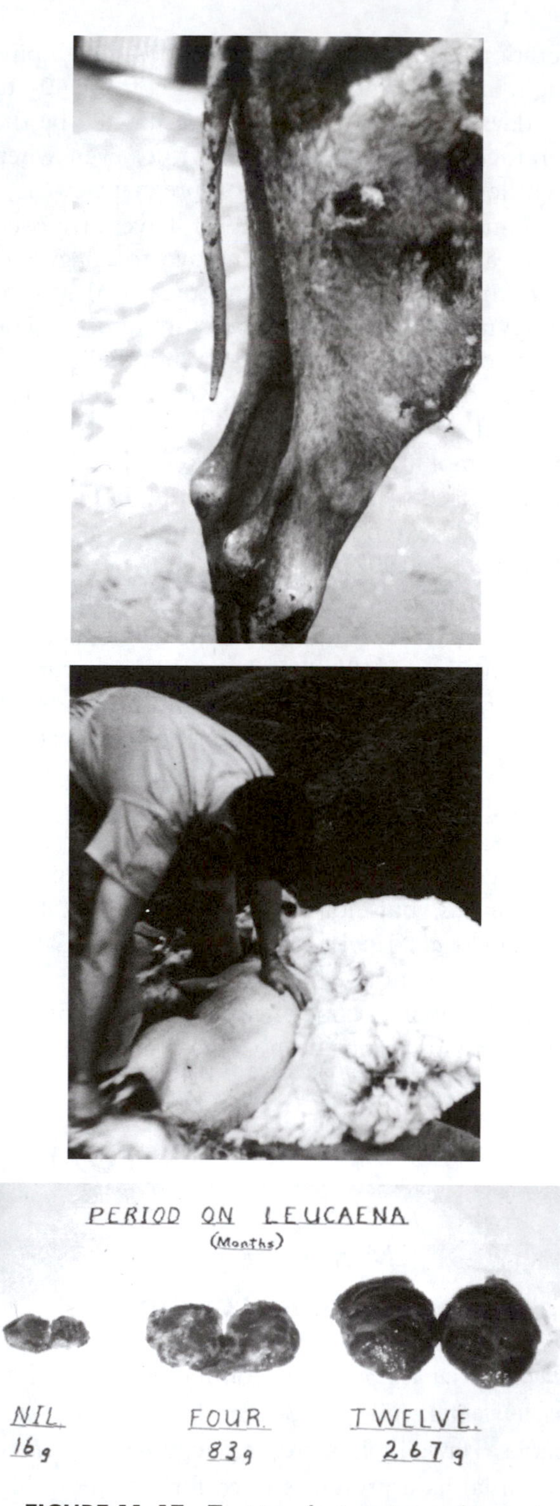

FIGURE 11–17 Toxicity of mimosine and its metabolites. (Top): Alopecia (loss of hair) and dermatitis are characteristic signs of mimosine toxicity. (Middle): Mimosine has been investigated as a chemical defleecing agent. (Bottom): Enlarged thyroid glands in cattle fed leucaena for various periods of time. (Courtesy of R.J. Jones)

sions, depressed serum thyroxine levels, and goiter (Fig. 11–17). Some of these symptoms may be due to mimosine itself and others to metabolites (Fig. 11–18) produced in the rumen. Hegarty *et al.* (1976) reported that in the rumen, mimosine is converted to 3-hydroxy-4(IH)-pyridone (3,4-DHP). **3,4-DHP** is a goitrogen, impairing the incorporation of iodine into iodinated compounds in the thyroid gland.

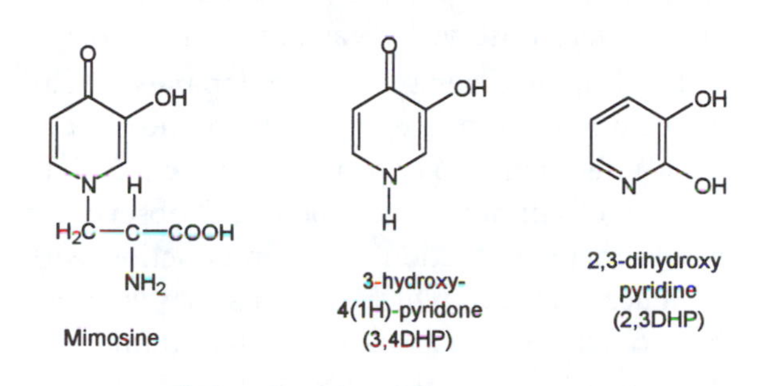

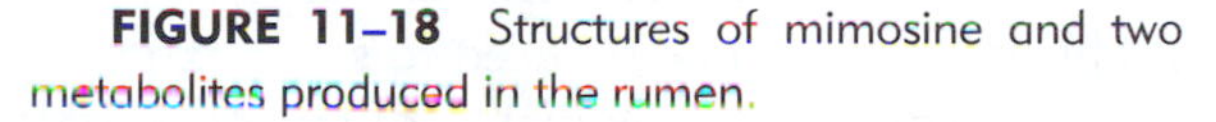
FIGURE 11–18 Structures of mimosine and two metabolites produced in the rumen.

Another ruminal metabolite is 2,3-dihydroxy pyridine (**2,3-DHP**).

In some parts of the world, such as Australia, consumption of leucaena by ruminants causes the problems previously discussed. In other places, such as Hawaii and Indonesia, cattle and goats can consume leucaena with impunity. Jones (1981) found that 3,4-DHP is further metabolized in animals in Hawaii to nontoxic compounds. Mimosine and 3,4-DHP rapidly disappear when incubated in vitro with rumen fluid from Hawaiian ruminants, but are much more stable when inoculum from Australian animals is used. In subsequent work, it was demonstrated that transfer of rumen fluid from animals in Hawaii and Indonesia to Australian ruminants resulted in a complete elimination of the toxic effects of leucaena (Jones and Lowry, 1984; Megarrity and Jones, 1986) (Fig. 11–19). The ruminal bacteria that metabolized mimosine and 3,4-DHP were strict anaerobe gram-negative short rods. These organisms were isolated by Allison *et al.* (1990), and named *Synergistes jonesii* 78-1 (Allison *et al.*, 1992; McSweeney *et al.*, 1993) in honor of Raymond

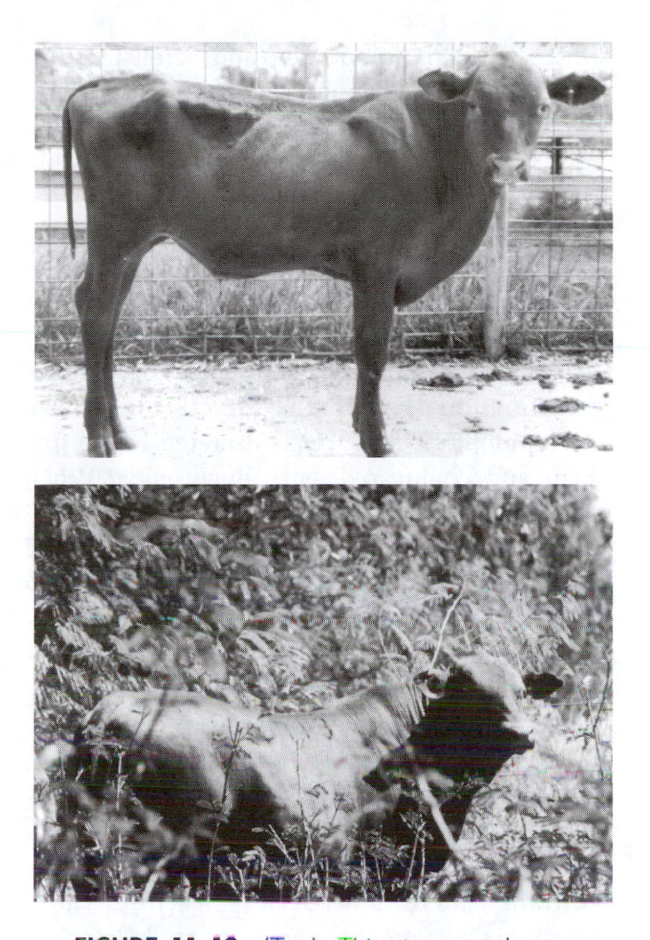

FIGURE 11–19 (Top): This steer was kept on a leucaena pasture, and shows typical signs of mimosine toxicity, including alopecia, dermatitis and poor growth. (Bottom): The same steer on the same pasture a few months after receiving an intraruminal infusion of rumen fluid from a Hawaiian goat. The goat rumen contained mimosine-degrading bacteria. (Courtesy of R.J. Jones)

Jones, CSIRO, Australia, whose pioneering studies demonstrated the geographical variation in capacity for ruminal metabolism of mimosine and 3,4-DHP. *S. jonesii* is an unusual rumen microbe, in that it is the only rumen bacterium so far identified which utilizes arginine and histidine as major energy-yielding substrates (McSweeney *et al.*, 1993). Although it is a strict anaerobe, *S. jonesii* is readily transferred among animals in a herd, so once it is introduced into a few animals, the entire herd can be inoculated. The microbe is excreted in the feces, and is spread among animals via dust in cattle yards where leucaena is dormant and defoliated during the winter, the mimosine-metabolizing microbes survive in the rumen from one grazing season to the next (Hammond *et al.*, 1989; Hammond, 1995).

In ruminants lacking mimosine-degrading ruminal microbes, mimosine, 3,4-DHP and 2,3-DHP are excreted in the urine. Detection of these metabolites in the urine using colorimetric tests is a simple means of detecting whether or not animals have the effective microbes (Hammond, 1995). DHP is excreted in the free form and conjugated as a glucuronide. When animals are dosed with *S. jonesii*, urinary DHP output declines dramatically (Jones, 1994) and feed intake rises correspondingly. Circulating DHP has a marked negative effect on feed intake, accounting for much of the growth depression in non-adapted animals (Jones, 1994).

Dominguez-Bello and Stewart (1990) reported that a wide range of ruminal microbes in Venezualan sheep had DHP-metabolizing capability, including a *Clostridium* sp. (Dominguez-Bello and Stewart, 1991). Presumably, these organisms have coevolved in the rumen microflora of animals in areas where leucaena is native (Central and South America). The successful introduction of these microbes into non-adapted animals has had a major impact on the utilization of leucaena as a protein-rich tropical forage.

While the introduction of rumen microorganisms capable of mimosine-DHP degradation into susceptible ruminants appears to be a solution to the toxicity of leucaena to ruminants, the problem of mimosine toxicity to nonruminants remains. A solution to the mimosine problem could be the development of low-mimosine cultivars. This has not yet been successfully achieved; while there are some low-mimosine types, they are unproductive and of low vigor. Indonesian workers (Lowry *et al.*, 1983) demonstrated enzymatic degradation of mimosine in leaf tissue. Wilting to allow enzymatic destruction of mimosine might be a means of reducing its toxicity to nonruminant herbivorous animals such as rabbits. However, plant hydrolase activity may convert up to 30% of mimosine to DHP during the process of mastication (Lowry *et al.*, 1983), so that reduction of mimosine levels might result in DHP toxicity.

Leucaena is a teratogenic plant, with mimosine the primary teratogenic agent. Deweede and Wayman (1970) fed mimosine to pregnant rats and noted a variety of fetal deformities. In swine fed leucaena during gestation, fetal resorptions and polypodia of the forelimbs of the fetuses were observed (Wayman *et al.*, 1970).

Feeding leucaena meal to Japanese quail caused a reduction in egg production. Ross *et al.* (1980) fed leucaena meal and an equivalent amount of mimosine to quail and observed a much greater reduction in egg production with the leucaena meal, which suggests that leucaena may contain toxic factors other than mimosine.

The effects of leucaena and mimosine on nonruminants can be reduced to some extent by supplementation of the diet with ferrous sulfate (D'Mello and Acamovic, 1989). Mimosine forms a complex with iron which is excreted in the feces.

Other Toxic Amino Acids

Canavanine and indospicine are structural analogs of arginine (Fig. 7–26). **Canavanine** is widely distributed in legumes, but is of toxicological importance mainly in jack bean seeds (*Canavalia ensiformis*). Although the seeds of jackbeans are toxic to nonruminants, ruminants are not affected by canavanine because it is metabolized by ruminal microbes (Dominguez-Bello and Stewart, 1990). *Indigofera spicata*, or creeping indigo, is a tropical legume with potential as a forage and soil improvement crop (Aylward *et al.*, 1987). It contains **indospicine**, which inhibits arginine incorporation into tissue proteins and causes hepatic necrosis in cattle and sheep consuming the plant. In Australia, Birdsville horse disease is a neurological disorder of horses grazing *Indigofera*. This condition has also been observed in Florida (Morton, 1989), in horses on grass pastures containing patches of creeping indigo. Neurological signs include ataxia, incoordination, seizures and collapse, often accompanied by opaque cornea, ulceration and redness of gums, and abortion in mares.

The **flatpea** (*Lathyrus sylvestris*) is a perennial legume forage grown on infertile hill lands and reclaimed strip mines of the Appalachian region of the U.S. (Foster, 1990). Flatpea contains several potentially toxic amino acids, including diaminobutyric acid and other lathyrogens such as oxalyl diaminopropionic acid (ODAP). Flatpea hay is toxic to sheep (Rowe *et al.*, 1993), with signs of nervous and behavioral abnormalities, including "saw-horse like" stance, depression, reluctance to move, trembling and incoordination. Seizures occurred in some animals. These signs were noted with a diet containing 35% flatpea hay. Rasmussen *et al.* (1993) observed similar signs in sheep fed diets with 50% and 100% flatpea hay. These workers reported evidence of detoxification of the toxic amino acids by rumen microbes, and indications of ruminal microbe adaptation to flatpea in the diet. In view of these reports of toxicity, flatpea should be used cautiously as a forage. There appears to be potential for reducing toxicity hazard by plant selection for lower lathyrogen concentrations (Foster, 1990), and by adaptation of animals to allow ruminal detoxification mechanisms to develop (Rasmussen *et al.*, 1993). Lathyrogens in *Lathyrus* seeds are discussed in Chapter 7.

ALSIKE CLOVER POISONING

Alsike clover (*Trifolium hybridum*) is a short-lived perennial that has been widely grown in the eastern and northern midwestern states of the U.S. and is widely grown in the northern farming areas of Canada. It is especially well adapted to cool climates and heavy, poorly drained clay soils. In the early part of this century, alsike clover was responsible for widespread toxicity problems in livestock, particularly in horses. Two syndromes have been recognized: photosensitization and a fatal syndrome known as alsike clover poisoning (Nation, 1989). The **photosensitization** condition is one observed with a number of forage legumes, and is sometimes known as **trifoliosis**. Typical signs of secondary photosensitization occur; a causative hepatoxic agent in alsike clover has not been identified.

Horses are the only species known to be affected by **alsike clover poisoning**. The condition usually occurs after at least one year of exposure to alsike clover pasture or hay, although with pastures consisting mainly of alsike clover, poisoning can occur within a few weeks of initial exposure. Alsike clover is somewhat unpalatable to horses, so as other vegetation is overgrazed, the clover can come to predominate. Signs include jaundice, neurological disturbances such as head-pressing and aimless walking, anorexia and loss of body condition. Animals exhibiting these signs usually die. Pathology includes a greatly enlarged liver, grey-brown or green-yellow in color. There is pronounced fibrosis.

Nation (1989, 1991) has reviewed the literature on alsike clover poisoning, and has concluded that while a specific disease entity occurs, the evidence is not sufficiently strong to incriminate alsike clover as the causative agent. It is recommended that alsike clover not be included in seed mixes for horse pastures. In 1996, alsike clover poisoning occurred on Prince Edward Island, Canada. There was evidence that endophytic bacteria of the genus *Capnocytophaga* may have had a role in the etiology of the poisoning (B. Christie and T. Sturz, Ag Canada, Personal Communication.)

VETCH TOXICITY

Hairy vetch (*Vicia villosa*) toxicosis in cattle has been observed in various parts of the world, including Oklahoma (Panciera, 1978), South Africa (Burroughs *et al.*, 1983), Argentina (Odriozola *et al.*, 1991) and Australia (Harper *et al.*, 1993). Clinical signs include dermatitis, alopecia, conjunctivitis with edema of the eyelids, diarrhea and ill-thrift. Pathology observed includes extensive infiltration of many organs by lymphoreticular cells, plasma cells, multinucleated giant cells and eosinophils (Panciera, 1978). The disorder occurs sporadically, with a mortality rate of about 50% of affected animals. The morbidity in affected herds was 6–8%. Hairy vetch is also toxic to horses (Anderson and Divers, 1983), leading to systemic granulomatous inflammation. Edema, particularly around the lips and eyes, is noted. Conjunctivitis and corneal ulceration occur. Most cases of poisoning have occurred in mid- to late spring when the vetch is approaching maturity.

The causative agent(s) and mode of action of tissue injury have not been identified. An immune system–mediated pathogenesis has been suggested (Panciera *et al.*, 1992), with a constituent of vetch postulated to act as an antigen or haptan to evoke a systemic immunologic reaction.

SWEET CLOVER POISONING

Sweet clover (*Melilotus alba* and *Melilotus officinalis*) contains a glycoside called melilotoside, an ether of glucose and coumarin (Fig. 11–20). **Coumarin** is metabolized by various molds, such as *Penicillium nigricans* and *Pennicillium jensi*, producing **dicumarol**. Dicumarol is an inhibitor of vitamin K and induces a vitamin K deficiency. Because vitamin K functions in blood clotting, a

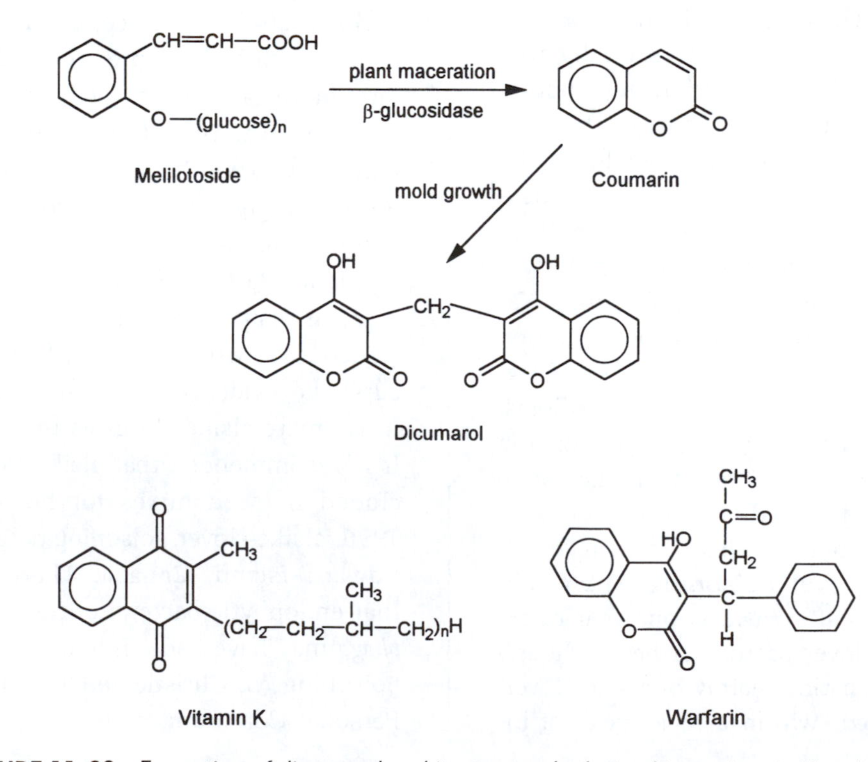

FIGURE 11–20 Formation of dicumarol and its structural relationship to vitamin K and warfarin.

deficiency is characterized by susceptibility to hemorrhaging.

Sweet clover has been widely grown in much of the northern U.S. and the prairie provinces of Canada as a forage and a soil-building (green manure) crop. In the early 1920s, widespread incidence of "bleeding disease" occurred in Ontario and the midwest. It was found to be associated with the consumption of moldy sweet clover hay. The condition was subsequently referred to as **sweet clover poisoning**. Investigators at The University of Wisconsin eventually elucidated the situation by demonstrating that a species of sweet clover which was not bitter did not cause sweet clover poisoning when moldy. It was shown that this species lacked coumarin. Addition of coumarin to alfalfa would result in poisoning only if the mixture was allowed to mold. The toxin was extracted, identified, and given the name dicumarol. In association with these investigations, another vitamin K antagonist was produced. It was given the name *warfarin* (for Wisconsin Alumni Research Foundation), and has been extensively used as a rat poison.

The widespread occurrence of sweet clover poisoning in the early 1920s was due to a combination of circumstances. The acreage of sweet clover for hay was rapidly expanding, and the summers were particularly wet, leading to a lot of moldy hay. Because of its succulent stem, sweet clover is difficult to cure without some molding.

Sweet clover poisoning occurs almost exclusively in cattle. The predominant sign is hemorrhaging, either external or internal. Internal hemorrhaging results in obvious subcutaneous swellings caused by pooling of blood. The mucous membranes are pale, and the animal becomes progressively weaker and dies without a struggle. Before internal hemorrhaging occurs, cattle fed moldy sweet clover hay have a prolonged blood-clotting time. Minor surgery such as dehorning or castration may lead to profuse hemorrhaging and death. Kingsbury (1964) cites an instance in which 21 of 22 cattle that were dehorned died from hemorrhage.

Vitamin K is involved in the activation of prothrombin. Dicumarol inhibits this effect, so there is a deficiency of prothrombin in the blood of affected animals. The basic reactions involved in blood clotting are shown in Fig. 11–21. Thrombin acts as an enzyme to split one or more peptides from fibrinogen, altering its solubility and causing it to precipitate as a clot.

Prothrombin is activated by the addition of a carboxyl group to glutamic acid residues in the molecule. The added carboxyl group provides a calcium-binding site for chelation of a calcium ion between two carboxyl groups (Fig. 11–22).

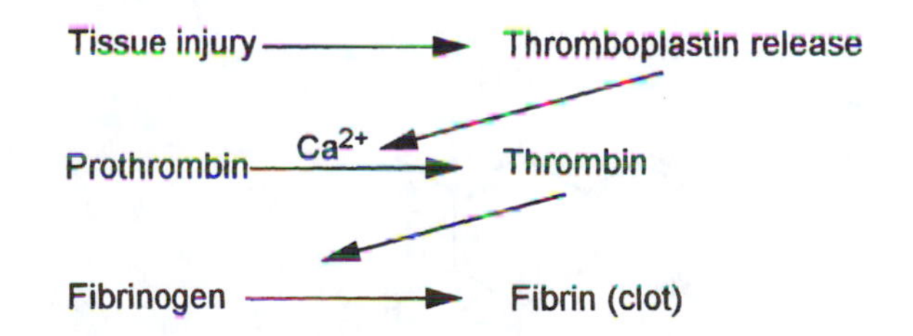

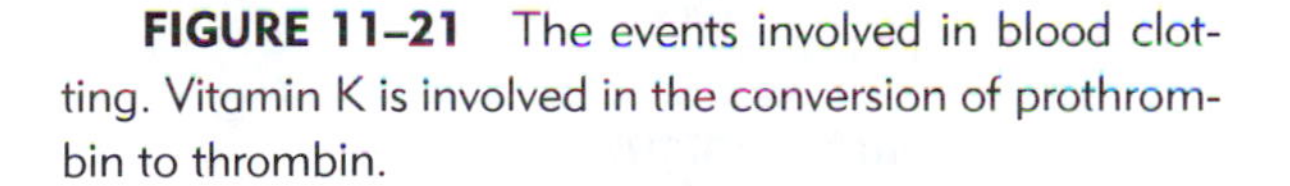
FIGURE 11–21 The events involved in blood clotting. Vitamin K is involved in the conversion of prothrombin to thrombin.

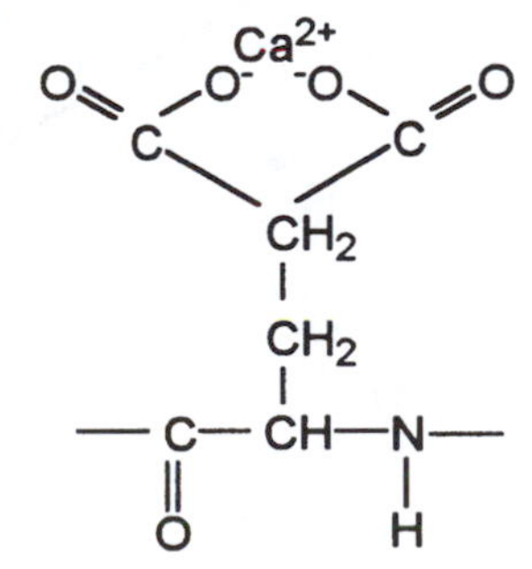

FIGURE 11–22 Vitamin K is involved in the addition of carboxyl groups to prothrombin to create calcium binding sites.

The addition of calcium activates prothrombin to the active enzyme thrombin. The carboxylase reaction for the addition of the carboxyl group to glutamic acid residues in prothrombin is vitamin K–dependent. The hydroquinone (reduced) form of vitamin K is required, and an epoxide-containing quinone is produced in the reaction. This is regenerated to the hydroquinone form by the vitamin K cycle in the liver. Dicumarol and warfarin inhibit 2,3-epoxide reductase which converts the epoxide to the quinone (Fig. 11–23). Thus dicumarol and warfarin cause vitamin K deficiency by inhibiting the regeneration of the active hydroquinone form of vitamin K.

Levels of above 10 ppm dicumarol in sweet clover hay are suggestive of possible toxicity problems. Dicumarol levels tend to be highest in small and in round bales, where opportunity for mold growth is greater than in stacks (Benson *et al.*, 1981). Crimping or conditioning of sweet clover hay should speed up drying and might reduce the amount of mold growth, although Benson *et al.* (1981) found no difference in dicumarol level of crimped vs noncrimped hay. Ammoniation of stored sweet clover hay using anhydrous ammonia reduces dicumarol concentrations by inhibiting mold growth (Sanderson *et al.*, 1985).

Sweet clover poisoning can be treated with injections of vitamin K and also by whole blood transfusion (Radostits *et al.*, 1980). It is a relatively minor problem now as the causative factors are known, and the plant is grown to a

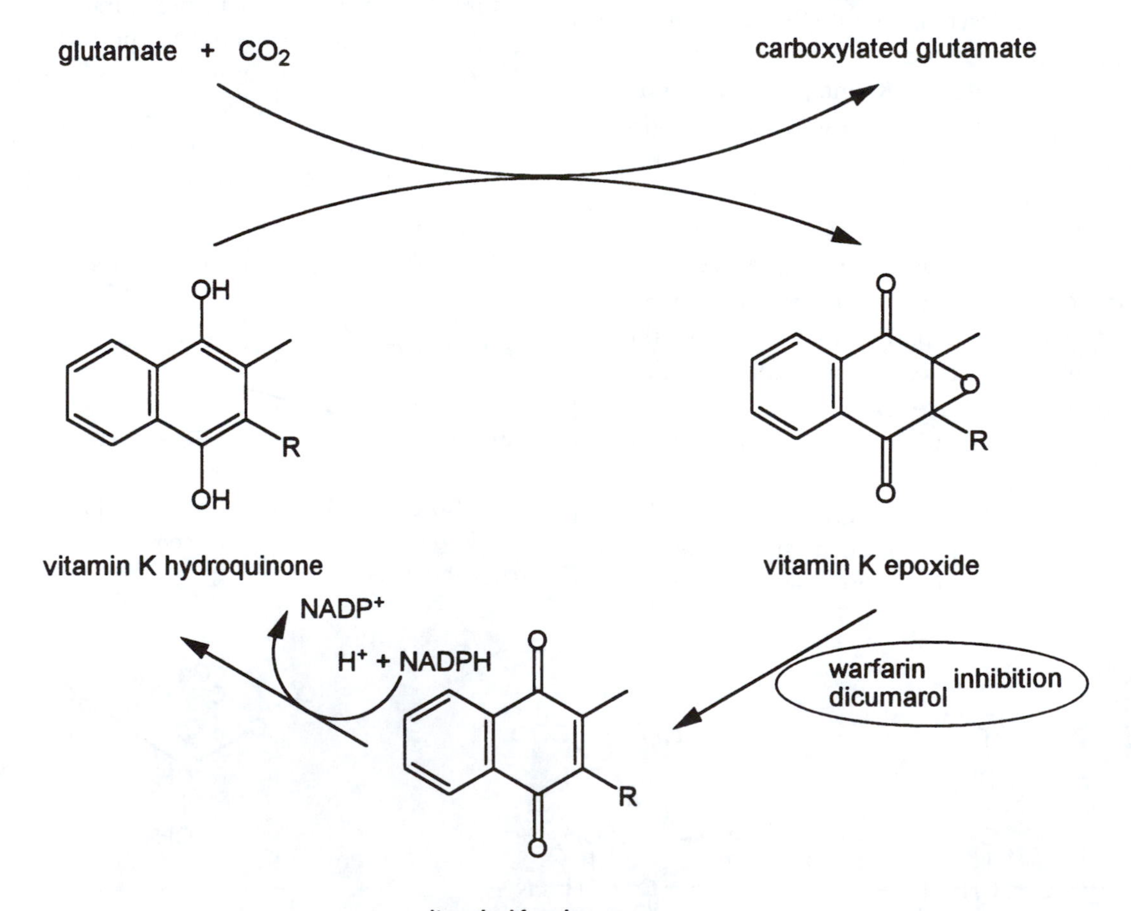

FIGURE 11–23 Dicumarol and warfarin create a vitamin K deficiency by inhibiting the vitamin K cycle for the regeneration of the metabolically active form of vitamin K.

lesser extent than in years past. Low-coumarin varieties of sweet clover, such as Polara, have been developed (Goplen, 1971).

Dicumarol problems in livestock are not exclusively caused by moldy sweet clover. Pritchard *et al.* (1983) in England reported a hemorrhagic syndrome in cattle that was caused by dicumarol in sweet vernal grass (*Anthoxanthum odoratum*) hay. Sweet vernal grass contains coumarin and *o*-coumaric acid, both of which are converted to dicumarol by *Aspergillus* spp. fungi. Both sweet clover and sweet vernal grass have a pleasant or sweet odor caused by the presence of coumarins. *Ferula communis*, a robust, tall, perennial umbelliferous plant that grows in the Mediterranean region, causes a hemorrhagic condition in livestock. It contains ferulenol, a 4-hydroxy-coumarin (Tligui and Ruth, 1994).

Coumarin compounds have commercial applications. Because of their pleasant and persistent vanilla-like odor, they have been used in perfumes and cosmetics (Scheel, 1978), and in condiments. Coumarin is used as a fixative and enhancing agent for the odor of essential oils in perfumes, soap, toothpaste, and hair preparations, and is used in tobacco products to enhance the natural taste (Cohen, 1979). Coumarin additions to food have not been permitted in the U.S. since 1954; it is permitted at 5 ppm in liquors (Cohen, 1979). Hepatoxic effects have been reported in laboratory animals fed high levels of coumarin.

ISOFLAVONES AND COUMESTANS (PHYTOESTROGENS)

Subterranean clover (*Trifolium subterran*) has been widely sown for sheep pasture in many parts of Australia. It is also extensively grown on hill pasture lands in the U.S. Pacific Northwest. It is a winter annual, sprouting with the autumn rains and providing forage during the winter and spring. In late spring, it produces seeds in burs that are pushed into the ground at maturity, so the plant reseeds itself. This characteristic provides the origin of its name. Sub clover, as it is commonly known, has greatly increased pasture productivity in regions where it is well adapted.

In the early 1940s, as sub clover became an important pasture species in Western Australia, a dramatic decrease in the fertility of sheep to a level of about 30% fertility was noted. The infertility was expressed as a failure to conceive and was accompanied by a cystic glandular hyperplasia of the cervix and uterus. Lactation in nonpregnant ewes and wethers suggested that a plant estrogen was involved in the so-called "clover disease." Australian researchers in the early 1950s extracted nearly 5 tons of fresh clover, from which they were able to isolate and identify two isoflavones, genistein and formononetin, that had estrogenic activity. These and other plant estrogens are referred to as **phytoestrogens**. Since that time, a great deal of Australian research has helped to identify the modes of action of phytoestrogens (Adams, 1989, 1995).

Pasture species that cause livestock problems because of their phytoestrogen content include sub clover, red clover (*Trifolium pratense*), and alfalfa (*Medicago sativa*). The estrogens in clovers are usually **isoflavones**, while alfalfa contains **coumestans**. Structures of some common phytoestrogens are shown in Fig. 11–24. Their resemblance to estradiol is apparent. The phytoestrogens occur in plant tissue as water-soluble glycosides. The isoflavones are synthesized by plants from phenylalanine, while the coumestans are synthesized from cinnamic acid.

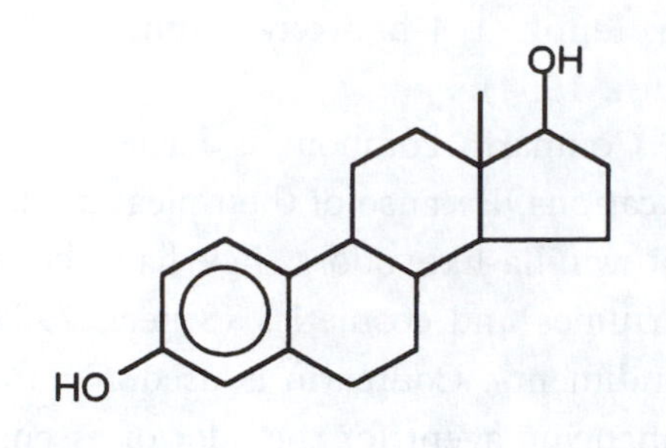

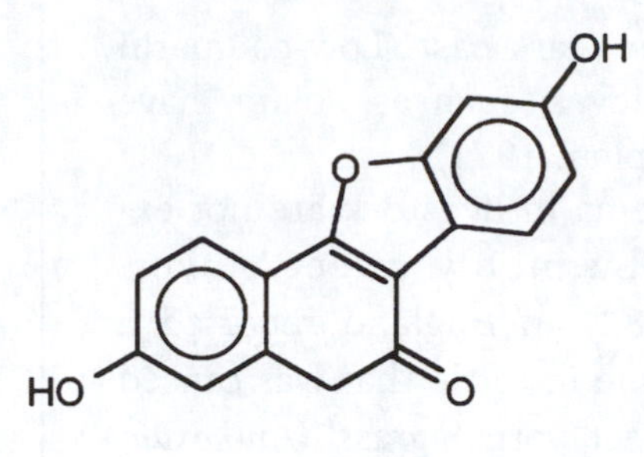

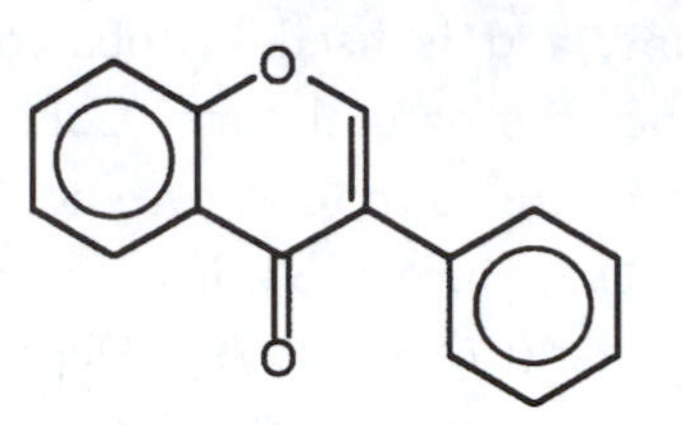

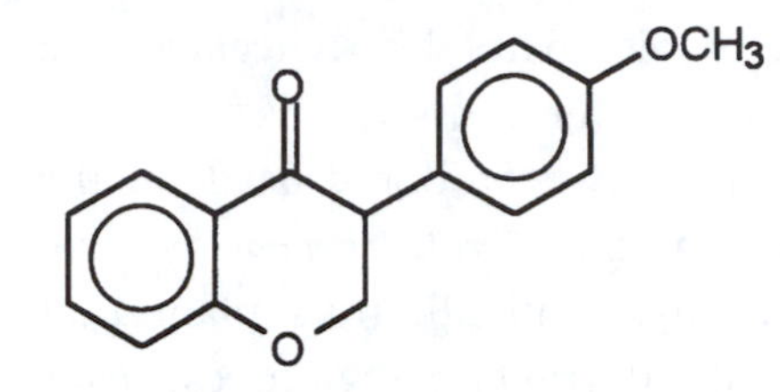

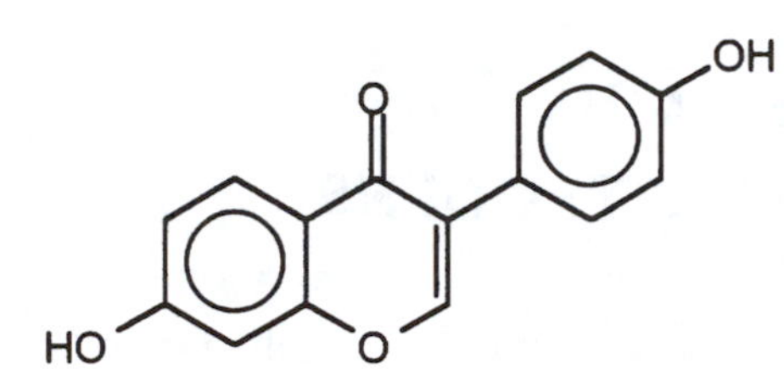

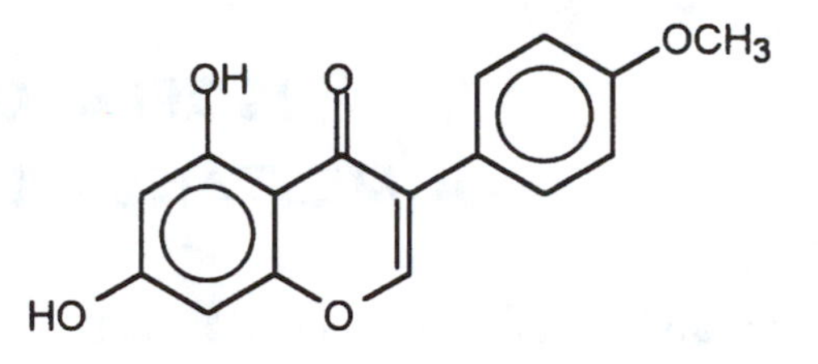

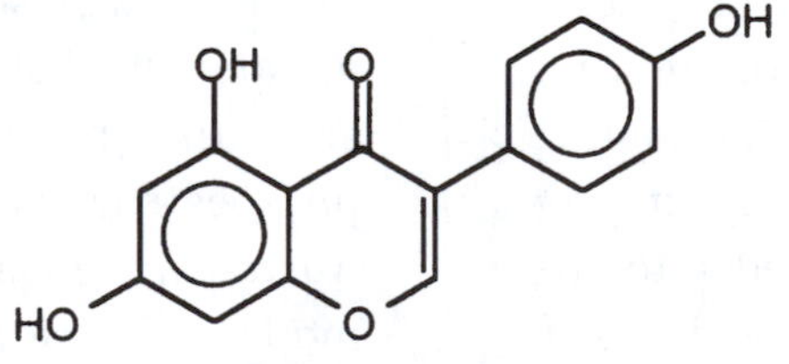

FIGURE 11–24 Structures of some isoflavones, and their structural resemblance to estradiol.

The mouse uterine weight bioassay has been extensively used in studies of phytoestrogens. Plant extracts or the isolated estrogens are injected into immature female mice, and 24 hr later the uterine weight is measured. Estrogens cause an increase in uterine weight (Fig. 11–25). Examples of some typical dose responses are shown in Table 11–2.

The equivalent potencies at a dosage required to produce a 25-mg uterus were estrone, 6900; coumestrol, 35; genistein, 1; daidzein, 0.75; biochanin A, 0.46; and formononetin, 0.26.

These results show that the phytoestrogens have an exceedingly low potency as compared to estrone. However, they can produce signifi-

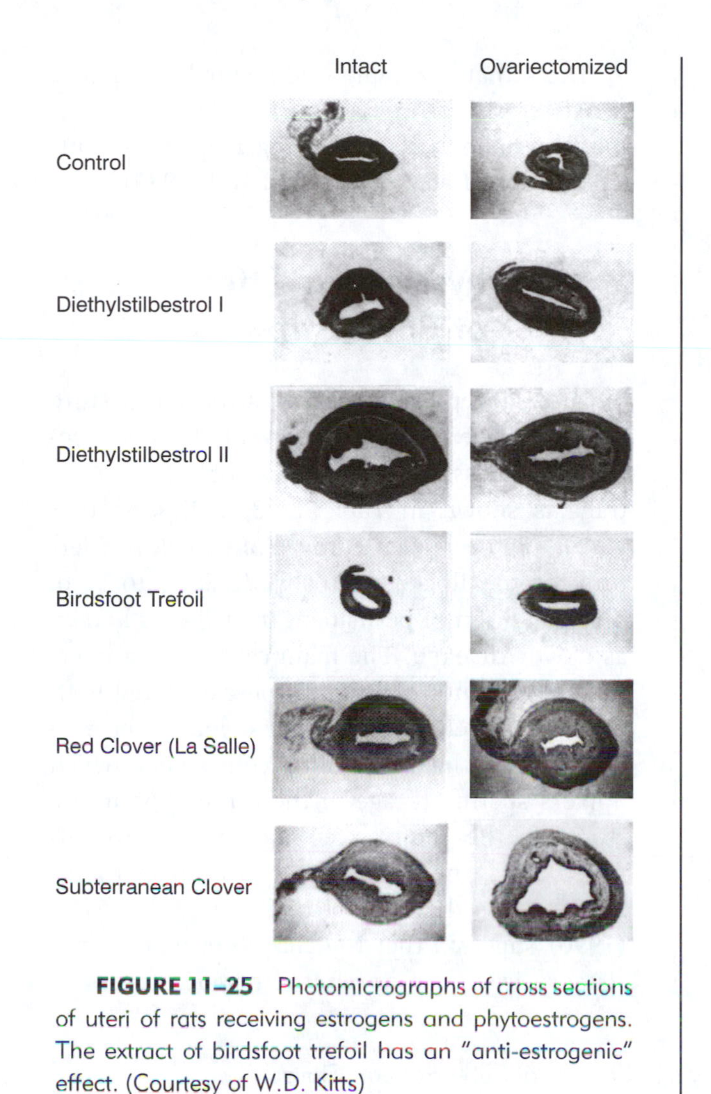

FIGURE 11–25 Photomicrographs of cross sections of uteri of rats receiving estrogens and phytoestrogens. The extract of birdsfoot trefoil has an "anti-estrogenic" effect. (Courtesy of W.D. Kitts)

cant biological effects through an additive action with endogenous estrogen, and they may occur in plants at very high levels. The isoflavone content of sub clover may reach 5% of the dry weight.

A puzzling observation was that the estrogenic activity of sub clover pastures, as assessed by teat length of wethers, was correlated with the **formononetin** content of the pasture. Formononetin has a very low estrogenic activity (Table 11–2). The explanation resides in rumen metabolism. In the sheep rumen, biochanin A and genistein are degraded to p-ethylphenol and a phenolic acid, whereas formononetin is demethylated to daidzein and then metabolized to equol (Fig. 11–26). Equol is estrogenic. Hence, formononetin is bioactivated by sheep rumen microorganisms to a more potent estrogen (Davies and Hill, 1989). There is also evidence that formononetin is directly estrogenic in sheep (Wang *et al.*, 1994). The same metabolism of formononetin to equol occurs in the rumen of cows (Dickinson *et al.*, 1988); the absorbed equol is excreted more rapidly in cattle, so they are less susceptible than sheep to the estrogenic effects of clover isoflavones. Lundh (1990, 1995) and Lundh *et al.* (1990) demonstrated differences between sheep and cattle in glucuronide conjugation of phytoestrogens, but concluded that the most likely explanation for the greater sensitivity of sheep to phytoestrogens is species differences in estrogen receptor activity.

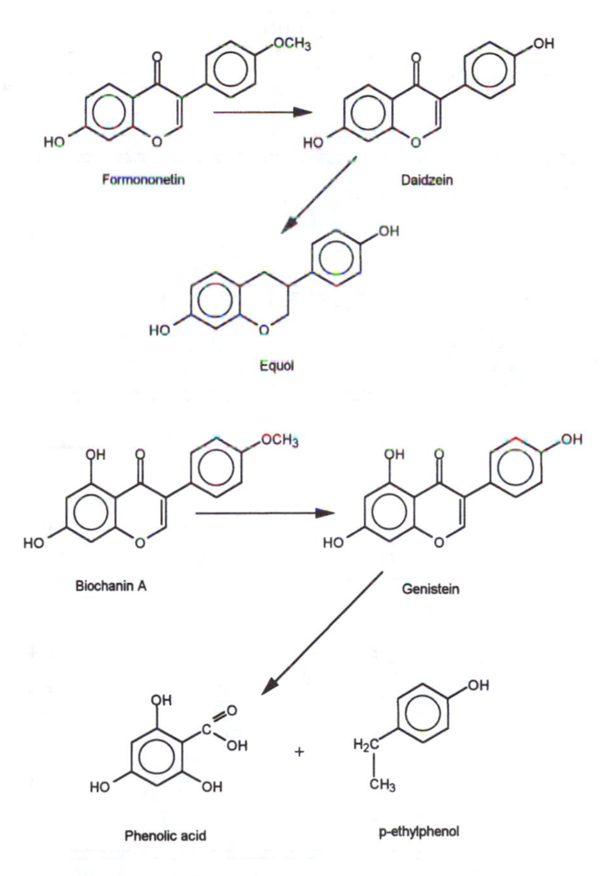

FIGURE 11–26 Metabolism of subterranean clover isoflavones in the sheep rumen.

TABLE 11–2 Dose Response with Phytoestrogens Using Mouse Uterine Weight Bioassay*

Compound	μg/mouse	Uterine weight (mg)
Control	0	9.6
Estrone	0.5	14.7
	1	23.8
	2	45.3
Coumestrol	100	13.8
	200	24.2
	400	40.7
Genistein	5,000	19.4
	8,000	27.0
	12,000	32.4
Daidzein	5,000	17.3
	10,000	24.8
	15,000	31.2
Biochanin A	10,000	20.3
	20,000	27.9
	40,000	45.5
Formononetin	15,000	16.8
	25,000	23.2
	40,000	26.1

*Adapted from Livingston (1978).

Diets that are rapidly fermented promote a more rapid ruminal metabolism of formononetin than those lacking readily fermentable carbohydrate (Davies and Hill, 1989).

Physiological Effects of Phytoestrogens

After sheep have grazed estrogenic pasture for several years, the fertility of the flock becomes depressed. A typical example from Australia is shown in Table 11–3, indicating that with a high-estrogen cultivar of sub clover, fertility of the flock eventually declined to zero. This condition of permanent infertility is known as **clover disease.** The main cause of the infertility is a failure of fertilization associated with poor sperm penetration to the oviduct. The **cervical mucus** has an altered consistency which impairs sperm storage in the cervix. Sperm are stored in the cervix after mating; in clover-affected ewes, the number of sperm present after 24 hr is less than 5% of that expected. Adams (1990) suggests that the change in mucus consistency is due to an altered responsiveness to

TABLE 11–3 Effect of Formononetin Content of Subterranean Clover on Percentage of Ewes Lambing*

	Pasture type				
		Subterranean clover			
Year	Non-estrogenic control	Woogenellup (0.15)[a]	Geraldton (0.79)[a]	Dinninup (1.19)[a]	Dwalaganup (1.30)[a]
1	91	76	87	78	89
2	73	84	78	72	56
3	86	69	56	53	30
4	59	41	42	35	6
5	84	63	53	41	8
6	85	67	52	38	0

*From Neil *et al.* (1974).

[a]Percentage formononetin.

stimulation with estrogen. Therefore, the cervix and vagina of ewes with clover disease fail to respond normally to endogenous estrogen stimulation to "prime" the cervix during the breeding season.

In ewes affected by clover disease, the cervix shows structural and functional changes. The cervical tissue changes in morphology to look more like uterine tissue than a cervix. The normal cervix has folds; in clover disease, the folds of the cervix fuse together, trapping epithelial tissue to make it look like glands (Fig. 11–27). Permanent infertility develops with prolonged exposure to the phytoestrogens.

If sheep are bred while grazing estrogenic pasture, fertility depression can occur. The infertility does not persist if they are subsequently maintained on nonestrogenic pastures. This temporary infertility is especially pertinent to the coumestans. Problems in breeding have occurred with dairy cattle fed alfalfa. These included decreased fertility because of cystic ovaries and irregular estrous cycles, as well as precocious mammary and genital development in heifers. Coumestans suppress estrus and inhibit ovulation, probably by lowering ovarian estrogen secretion. Sheep grazing estrogenic clover do not show the normal seasonal changes in serum LH (Chamley *et al.*, 1981).

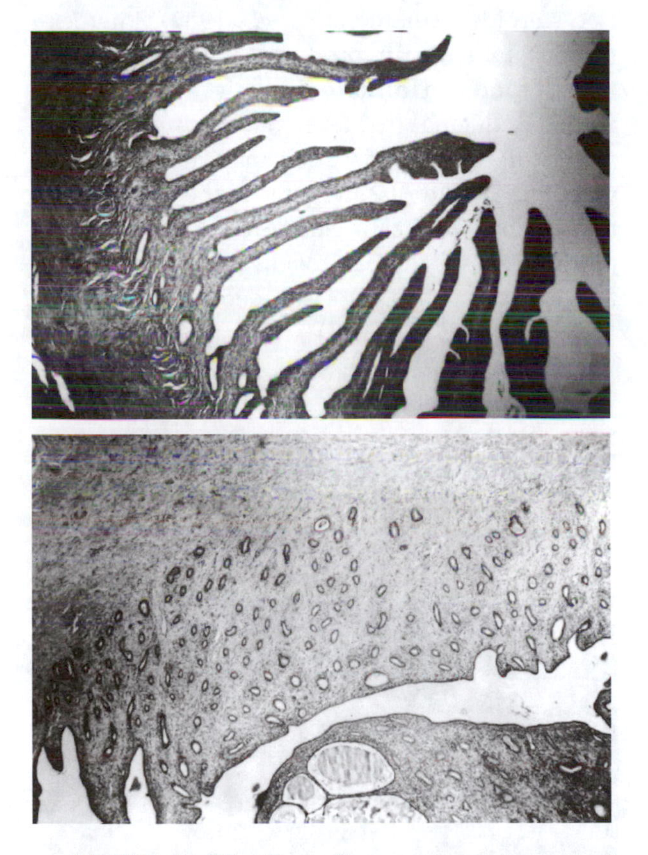

FIGURE 11–27 (Top): Cross section of a normal cervix of a ewe. (Bottom): Cross section of a ewe with clover disease. The animal had grazed estrogenic subterranean clover for 3 years. The folds of the cervix tissue have folded together, trapping epithelial tissue to give the appearance of glands. (Courtesy of N.R. Adams)

The fertility and sperm production of rams does not seem to be affected by their grazing on estrogenic pastures. Wethers may develop enlarged teats and begin lactating. Teat enlargement of wethers has been used as a sensitivity index of the potency of pastures.

An interesting situation is that phytoestrogens may be involved in the regulation of reproduction of Calilfornia quail. Leopold *et al.* (1976) reported that during dry years, stunted desert plants produced high levels of estrogenic isoflavones that inhibited quail reproduction. In normal or wet years, the plants grow abundantly and the levels of phytoestrogens are low, resulting in higher reproduction rates in the quail. Hughes (1988) has proposed an evolutionary role of phytoestrogens as defensive chemicals to modulate the fertility of vertebrate herbivores to limit herbivory.

Phytoestrogens may act as "antiestrogens." A high blood level of phytoestrogens may inhibit the release of gonadotropic hormones from the pituitary and may compete with endogenous estrogens for receptor sites in target tissues such as the uterus and cervix.

Problems associated with phytoestrogens and ewe fertility in Australia are much less than they once were. A major reason has been the development of low-formononetin cultivars of sub clover. Also, animal management to limit estrogenic exposure is practiced. However, moderate depression of fertility is still observed, and it has been estimated that about 4 million ewes fail to lamb each year in Australia because of phytoestrogen exposure (Adams, 1995). The problem will persist for many years because of the difficulty of eliminating the high-estrogen cultivars from existing pastures. Sub clover has a high percentage of hard seeds which resist germination, and therefore a reservoir of seeds of the original cultivars persists in pastures that have been newly seeded to the low-formononetin types. There is genetic variability among sheep in their susceptibility to clover-induced infertility, and selection of ewes for resistance to clover disease shows promise (Croker *et al.*, 1989). Little (1996) developed a strategic grazing plan in which phytoestrogen levels in sub clover are monitored, with the least estrogenic pasture reserved for young ewes, to reduce likelihood of permanent infertility.

Alfalfa tablets have become a common dietary supplement for humans, available in health food stores. Elakovich and Hampton (1983) analyzed three brands of commercial alfalfa tablets for their phytoestrogen content, which ranged from 20 to 190 ppm. They warned that this level of intake, in conjunction with other extraneous estrogen sources such as birth control pills and estrogen replacement therapy, could be potentially harmful. The benefits, if any, of consuming alfalfa tablets are unclear.

Phytoestrogens and Human Health

Estrogens have two opposing effects on **cancer**, depending on dosage. Large doses inhibit breast cancer tumor development and suppress growth of tumors already present, but small doses promote tumor development (Kaldas and Hughes, 1989).

Adlercreutz (1990a,b) has proposed that the high fat, low fiber Western diet promotes a high incidence of hormone-dependent cancers, including breast, prostate and colon cancer. A western-type diet results in elevated plasma levels of sex hormones and decreases the **steroid hormone–binding globulin (SHBG)** fraction of the blood, thus increasing the plasma level of free steroids (Adlercreutz *et al.*, 1987). Phytoestrogens in soybean products and other plant-based foods stimulate hepatic synthesis of SHBG, reducing the level of free steroid hormones in the blood. They may also interfere with the uptake of estrogens at the tissue level by binding with estrogen receptors. Thus dietary sources of phytoestrogens may have protective effects against estrogen-dependent cancers (Adlercreutz *et al.*, 1991).

FORAGE-INDUCED ANEMIA

Brassica Anemia

Plants in the *Brassica* genus, such as kale, rape, cabbage, cauliflower, and turnips, are important livestock feeds. Some, such as kale, forage rape, and turnips, are grown specifically for grazing by animals (Fig. 11–28), while with crops like cauliflower and brussels sprouts food-processing wastes and postharvest stubble may be used as feed for livestock. Kale and rape have been grown extensively in countries such as Britain and New Zealand, particularly as winter feeds for sheep. Turnip production has recently developed in the irrigated areas of Washington

FIGURE 11–28 Cattle grazing on forage brassica.

and Oregon to prevent soil erosion and provide winter grazing for cattle. Turnip seed is applied by plane before wheat is harvested; after the wheat is combined, the stubble is irrigated to germinate the turnips. The turnips are grazed throughout the winter. Two types of sulfur-containing compounds limit the feeding value of these brassica crops. These are the **glucosinolates** and an amino acid, **S-methylcysteine sulfoxide (SMCO)**. The growth performance of ruminants grazed on kale and other brassicas is lower than would be predicted from their nutrient content. It is generally believed that SMCO is the main factor responsible for the poor performance.

When ruminants are fed mainly on a diet of kale, cabbage, or other brassicas, they may develop a severe hemolytic anemia after 3–4 weeks. The first clinical sign of the disease is the appearance of stainable granules within the red blood cells. These are termed **Heinz-Ehrlich bodies**, also called *Heinz bodies* (Fig. 11–29). The Heinz bodies are composed of hemichrome, an oxidized form of hemoglobin which is unstable and rapidly precipitates (Andersen *et al.*, 1994). The formation of hemichrome is the first sign of oxidative damage to hemoglobin. In brassica anemia, the hemoglobin level falls from a normal level of about 11 g/100 ml to 8 g or lower. If the animals are removed from brassica pasture, the hemoglobin levels return to normal in 3–4 weeks and the Heinz bodies disappear. If the animals are left on the brassica pasture, surviving animals make a spontaneous but incomplete recovery, followed by cycles of anemia and partial recovery. Other clinical signs are hemoglobinuria (red urine) when the hemoglobin level falls below 6 g/100 ml, loss of appetite, diarrhea, and jaundice. The liver becomes swollen and pale, with extensive hemosiderin deposits and necrosis. Kidneys and spleen also show massive hemosiderin deposits.

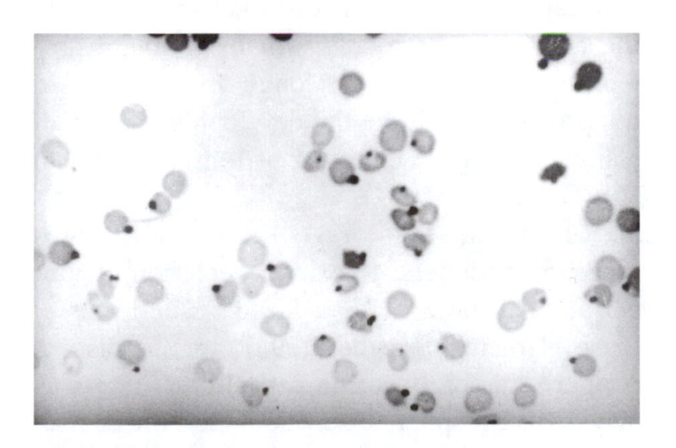

FIGURE 11–29 Brassica anemia, showing Heinz bodies in red blood cells. The Heinz bodies are clumps of denatured hemoglobin. (Courtesy of R.H. Smith)

The SMCO is a fairly rare amino acid, found mainly in brassicas, garlic, and onion (Benevenga *et al.*, 1989). In the brassicas it may occur at levels as high as 4–6% of the dry matter. SMCO is not the primary hemolytic agent; it is metabolized in the rumen, producing dimethyl disulfide (Fig. 11–30).

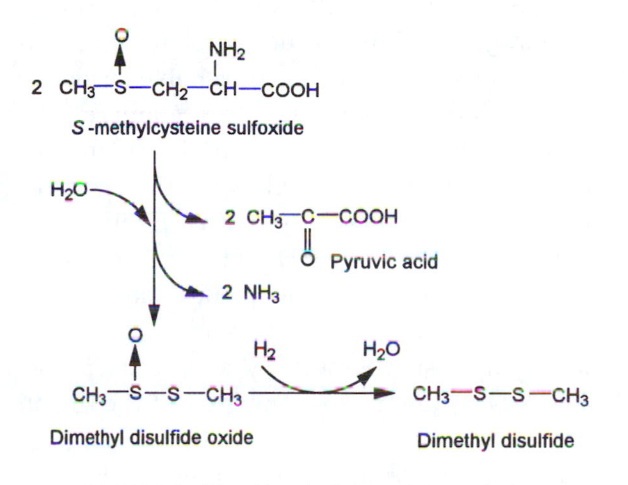

FIGURE 11–30 Metabolism of S-methylcysteine sulfoxide in the rumen. The arrow bond (S → O) indicates a coordinate covalent bond in which one atom (sulfur) provids both electrons needed to form a covalent bond.

Dimethyl disulfide is an oxidant that attacks the red cell membrane. It is inactivated by reacting with reduced glutathione (GSH), producing CH3SH (methylmercaptan) (Fig. 11–31).

Glutathione peroxidase is a selenium-containing enzyme that transfers hydrogen from reduced glutathione to oxidants, thus eliminating their capacity to oxidize lipids in cell mem-

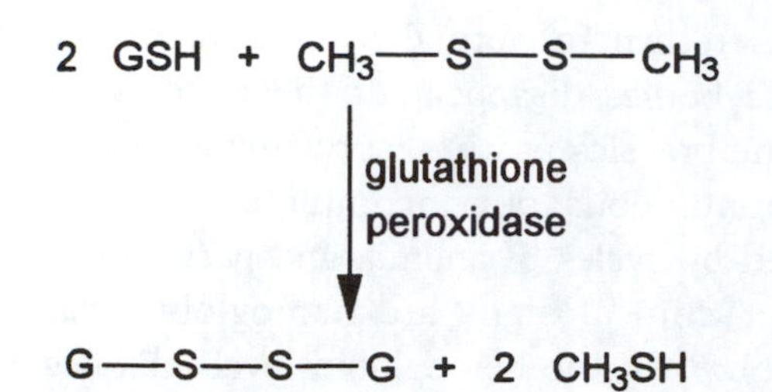

FIGURE 11–31 Enzymatic detoxification of dimethyl disulfide in the red blood cell. This oxidizing agent is reduced (gain of hydrogen) by the addition of hydrogen from reduced glutathione (GSH), catalyzed by the selenium-containing enzyme glutathione peroxidase.

branes. This system is also involved in detoxifying the glycoside derivatives in favism, as previously discussed in Chapter 7.

In the normal red blood cell, hemoglobin is maintained in the reduced state by GSH. With low GSH or low activity of glucose-6-phosphate dehydrogenase (G6PD), hemoglobin under oxidative stress may suffer irreversible oxidative changes and be precipitated as granules—the Heinz bodies. There are genetic differences among sheep in their erythrocyte GSH content and in G6PD activity (Goto *et al.*, 1993). Animals with low blood GSH and G6PD levels are more susceptible to development of anemia.

Typically in brassica poisoning there is a hemolytic crisis followed by a recovery phase. This may be due to an adaptation of the rumen microorganisms so less dimethyl disulfide is produced. Cattle are much more sensitive to brassica-induced anemia than are sheep and goats (Greenhalgh *et al.*, 1969), while non-ruminants, including hindgut fermentors such as rabbits, guinea pigs and horses, are not affected because the SMCO is absorbed in the intestine, anterior to the site of microbial fermentation, the hindgut.

Rapeseed and canola grown for their seeds have similar SMCO levels as forage rape varieties (Griffiths *et al.* , 1994). In Europe, mortality of deer grazing winter rape, with signs of brassica anemia, have been noted (Griffiths *et al.*, 1994).

Part of the growth depression of ruminants on brassica pastures may be caused by glucosinolates (Duncan and Milne, 1992a,b). **Nitriles** are the major ruminal degradation products of glucosinolates (Duncan and Milne, 1992b). Absorbed nitriles cause liver and kidney damage. Rumen microbes that metabolize nitriles are present in sheep adapted to brassica pasture (Duncan and Milne, 1992b). The adaptation process may account for the lag period of poor performance when ruminants are abruptly admitted to brassica pasture.

The SMCO content of brassicas tends to increase with increasing plant maturity. Through the winter, the toxicity of kale, cabbage, turnips, and other brassicas increases. It is advisable not to use these crops for pasture in late winter. Nitrogen fertilization tends to increase the SMCO content. McDonald *et al.* (1981) developed a fertilization program to produce kale with a low SMCO content by growing it on soils with low sulfate sulfur concentrations (5–10 mg sulfate sulfur/kg) and nitrogen fertilization to stimulate incorporation of the sulfur into plant protein. Barry *et al.* (1984) demonstrated that low SMCO kale produced in this manner was of lower toxicity to sheep than normal SMCO kale, but still produced signs of toxicity. Lambs fed kale showed large increases in serum thyroxin and growth hormone concentrations, suggesting that ruminants counteract the protein inactivation caused by dimethyl disulfide by increasing the activity of hormones which stimulate protein synthesis. The best long-term solution to the problem would be the development through plant breeding of cultivars of brassicas low in SMCO (Prache, 1994).

Ruminants fed forage brassicas for prolonged periods tend to exhibit depletion of copper and selenium, with reduced tissue levels of these elements (Prache, 1994). In the case of **copper**, the high sulfur content of brassicas may reduce copper absorption. **Selenium** may be depleted because of increased turnover of glutathione peroxidase. However, lowered glutathione peroxidase status does not seem to increase the susceptibility of ruminants to brassica-induced anemia (Duncan *et al.*, 1995).

SMCO has been demonstrated to have an antihypercholesterolemic effect (Kendler, 1987). The precursor of SMCO, S-methyl-L-cysteine, is found in appreciable quantities in legume seeds (Eyre *et al.*, 1983). Eyre and co-workers demonstrated that this derivative of cysteine had no sparing effect on the methionine requirement of rats, but had no deleterious effects either.

Onion Poisoning of Livestock

Feeding onions to livestock may cause anemia of the same type as the brassica-induced anemia. The causative agent is not the same and is not an amino acid, but is discussed here as it is very similar to the brassica-poisoning situation.

The toxic factors in onions are a number of sulfur-containing compounds, mainly methyl- and propylcysteine sulfoxides. When the onion tissue is macerated, these substances are hydrolyzed to **thiosulfinates**, which are the main flavor components of onions. These compounds may decompose to a number of disulfides, including dipropyl disulfide (also known as n-propyl disulfide) and dipropenyl disulfide (Fig. 11–32). **Dipropenyl disulfide**, an unsaturated compound, appears to be the most toxic (Munday and Manns, 1994). These disulfides, like the dimethyl disulfides arising from brassicas, are oxidizing agents that cause hemolysis of the erythrocyte membrane. They are detoxified by the GSH-G6PD system described for brassica anemia. The concentration of these compounds is increased when the soil is high in sulfur (Randle *et al.*, 1994). Signs of onion poisoning, such as anemia and Heinz bodies in erythrocytes, are due to the oxidant activity, which is described in detail by Munday and Manns (1994).

$CH_3CH_2CH_2SSCH_2CH_2CH_3$

dipropyl disulfide

$CH_2{=}CHCH_2SSCH_2CH{=}CH_2$

di(2-propenyl)disulfide

$CH_3CH{=}CHSSCH{=}CHCH_3$

di(1-propenyl)disulfide

FIGURE 11–32 Some of the major hemolytic thiosulfinates in onions.

Onions are commonly fed to sheep and cattle in onion-producing areas of the world. These may be cull onions or, at times, the main crop when overproduction prevents their marketing in normal channels. The onions are scattered on the ground in piles and are avidly consumed by livestock. Anemia may develop within a week of the onset of onion feeding. Clinical signs are hemoglobinuria, diarrhea, staggering, and collapse. Cattle are most susceptible, with horses intermediate, and sheep and goats somewhat resistant. Sheep can normally be fed cull onions free choice with no toxicity, whereas for cattle, levels above 25% of the diet may be hazardous. Onions fed to cattle should be chopped, to prevent choking on onion bulbs.

Onion poisoning of dogs is reported frequently in the veterinary literature (e.g. Yamoto and Maede, 1992). Dogs and cats are very sensitive to onion toxicity because of a lack of GSH-G6PD protection against oxidative damage in their erythrocytes (Munday and Manns, 1994). Kaplan (1995) reported a case of onion-induced anemia in cats, resulting from the feeding of baby food containing onion powder.

Onions also contain SMCO, which would presumably contribute to the anemia problem. **Garlic** contains S-allylcysteine sulfoxide, which is metabolized to allyl disulfide oxide[1]. The reputed cholesterol-lowering properties of garlic may be due to a reaction of the disulfide group with the sulfhydryl group of CoA, leading to inhibition of lipid synthesis. A review of the effect

[1]The term *allyl* is synonymous with 2-propenyl.

of garlic consumption on blood lipids and atherosclerosis concludes that there are favorable effects (Lau *et al.*, 1983).

Onion poisoning of livestock has been described by Kirk and Bulgin (1979) and Hutchinson (1977). Wild onions also cause hemolytic anemia in livestock (Van Kampen *et al.*, 1970).

Maple Poisoning of Horses

The consumption of wilted or dried leaves and the bark of the red maple (*Acer rubrum*) can cause methemoglobinemia, hemolytic anemia and Heinz body formation in horses (Tennant *et al.*, 1981; George *et al.*, 1982). The signs of toxicity are very similar to those seen with the brassica anemia factor (Fig. 11–33). The condition occurs sporadically in the northeastern U.S. and is associated with the browsing of red maple leaves. The causative agent has not been identified.

Two syndromes of **red maple poisoning** occur: a peracute form characterized by marked methemoglobinemia resulting in sudden death, and the hemolytic condition with Heinz body formation and hemolysis (McConnico and Brownie, 1992). Methemoglobin is formed by oxidant activity; it is detoxified by methemoglobin reductase, which requires NADH as a cofactor. Horses have a low activity of this enzyme, whereas cattle, sheep and goats rapidly reduce methemoglobin (McConnico and Brownie, 1992). Ascorbic acid (vitamin C) nonenzymatically reduces methemoglobin to hemoglobin. Administration of large doses (30–50 mg/kg BW) of ascorbic acid intravenously to horses poisoned by red maple leaves appeared to be effective in treating the toxicity (McConnico and Brownie, 1992).

FIGURE 11–33 Red maple poisoning. Serum from a normal horse (right), and from a horse with severe red maple–induced hemolysis (left). (Courtesy of Lisle

In all cases of red maple poisoning reported, only wilted or dried leaves have been toxic (Tennant *et al.*, 1981; George *et al.*, 1982; Plumlee, 1991; McConnico and Brownie, 1992; Stair *et al.*, 1993), suggesting that a plant constituent is converted to an oxidant during drying.

PHENOLIC COMPOUNDS AND TANNINS IN FORAGE AND BROWSE SPP.

Several temperate forage legumes contain significant concentrations of condensed tannins, also known as **proanthocyanidins**. These legumes include the trefoils (*Lotus* spp.) such as birdsfoot trefoil (*L. corniculatus*), sainfoin (*Onobrychis viciifolia*), crown vetch (*Coronilla coronata*,

Lespedeza spp., and sulla (*Hedysarum coronarium*). The condensed tannins in these forages have both positive and negative effects on grazing animals. At high concentrations, their effects are negative, while at low to moderate concentrations, they may have beneficial effects on livestock. Their effects are due primarily to their ability to form complexes with proteins and other macromolecules. Phenolic compounds have large numbers of free hydroxyl groups which can form hydrogen bonds with proteins and carbohydrates. The tendency of tannins to react with proteins increases as the molecular weight of the tannin increases.

At high concentrations, condensed tannins in forage legumes reduce animal performance by adversely affecting feed intake and nutrient digestibility. Tannins are **astringent**, reacting with taste receptors in the oral cavity to cause an astringent or "puckery" feeling in the mouth. This is the characteristic taste of unripe fruit (e.g. green apples). Browsing animals have enlarged salivary glands and secrete proline-rich salivary proteins which form complexes with dietary tannins (Austin *et al.*, 1989; Mole *et al.*, 1990; Mehansho *et al.*, 1987).

Proteins containing proline have a strong affinity for tannins. Proline is an imino acid with a secondary amine group. The carbonyl oxygen adjacent to a secondary amine nitrogen is a very good hydrogen bond acceptor, so proline-rich proteins form especially strong hydrogen bonds with tannins (Van Hoven and Furstenburg, 1992). Thus the secretion of **tannin-binding proline-rich salivary proteins** reduces the astringent effects of tannins, allowing browsers to consume high-tannin forage. Grazers such as cattle are more sensitive to the anti-palatability effects of dietary tannins than browsers such as goats and deer because they lack the salivary tannin-binding proteins.

Condensed tannins reduce **digestibility** of protein and carbohydrates in the rumen. They form tannin-protein complexes that are stable over the pH range 3.5 to 7.0 (Mangan, 1988), thus reducing digestibility of forage proteins in the rumen by protecting them from microbial enzymes. The complexes dissociate at pH <3.0 and at pH >8.5, releasing the protein in the abomasum and small intestine. Because forage protein is of higher biological value than rumen microbe protein, the net effect on protein utilization may be positive. The tannin-protein complexes reduce the conversion of forage protein to ammonia in the rumen, preventing nitrogen losses associated with absorbed ammonia (Tanner *et al.*, 1994). Condensed tannins in *Lotus* spp. increase the ruminal by-pass of ryegrass protein when the two forages are fed together (Waghorn and Shelton, 1995).

Condensed tannins react with microbial enzymes in the rumen (Jones *et al.*, 1994), forming enzyme-tannin stable complexes. This "tie-up" of enzymes reduces digestibility of forage cell wall components. Tannins may bind directly to bacterial enzymes, and may also impair the growth of rumen microbes. Chiquette *et al.* (1988) observed that with exposure to high tannin trefoil, some of the rumen microbes became encased in glycocalyx-enclosed microcolonies. This defensive mode of growth limits their secretion of exoenzymes, impairs their attachment to and penetration of plant tissues, and reduces their growth rates. Binding of tannins to cell coat polymers of rumen microbes has been reported by Jones *et al.* 1994). Condensed tannins may also form indigestible complexes with cell wall carbohydrates, including cellulose and hemicellulose (Reed, 1995).

As discussed earlier (Chapter 9), condensed tannins in legume forages prevent pasture **bloat** in ruminants, by forming tannin-cytoplasmic complexes during mastication and in the rumen (Mangan, 1988), eliminating the foaming properties of the forage proteins in the rumen. Another favorable effect of tannins is to inhibit gastrointestinal parasites (Niezen *et al.*, 1995).

Tannin-degrading microbes have been isolated. Oswa (1992) found that koalas, which feed exclusively on tannin-rich *Eucalyptus* spp., have tannin–protein complex-degrading enterobacteria in the cecum and feces. Such microbial action could assist koalas and other Eucalyptus-consuming herbivores in utilizing dietary protein from tannin-rich browse (Oswa and Sly, 1992). Brooker *et al.* (1994) isolated from feral goats browsing tannin-rich *Acacia* spp. a tannin-degrading microbe which they named *Streptococcus caprinus*. The organism could not be isolated from domestic sheep or goats, but was readily introduced into domestic sheep by dosing them with feral goat rumen fluid (Miller *et al.*, 1995). However, introduction of a pure culture of *S. caprinus* into the rumen of *Acacia*-consuming sheep was ineffective in improving nitrogen utilization, suggesting that a microbial consortium of several species may be required (Miller *et al.*, 1996). There are numerous rumen bacteria which can metabolize hydrolyzable tannins and phenolic monomers (Nelson *et al.*, 1995).

The formation of complexes of condensed tannins with protein and carbohydrates in the gut interferes with the detergent system (acid and neutral detergent fiber) of forage analysis, and may result in artifactual negative digestibility values for tannin-rich forages (Degen *et al.*, 1995; Reed, 1995). An in vitro gas production procedure used in conjunction with a tannin-binding polymer (PVP) can be used to assess phenolic effects on rumen fermentation (Khazaal and Orskov, 1994).

The two major forages in the U.S. containing condensed tannins are birdsfoot trefoil and lespedeza. **Trefoil** is more tolerant than alfalfa of poorly drained, acid and infertile soils, and is grown in the northeast, midwest and Pacific northwest states. The *Lespedeza* spp. are important legumes for pasture, hay and soil improvement in the southeast states. In general, the condensed tannins in birdsfoot trefoil are beneficial to livestock production. Trefoil is a non-bloating legume, and the tannins reduce the degradation of protein in the rumen, increasing the rumen by-pass protein and the quantity of absorbed amino acids in the intestine (Waghorn *et al.*, 1987). Intake is not adversely affected. In contrast, the tannin concentration in **lespedeza** is sufficient to impair animal performance through anti-palatability effects (Terrill *et al.* , 1989). *L. pedunculatus* (big trefoil) is grown to a very limited extent in the U.S., but is a relatively important forage on acid, poorly drained soils in New Zealand. Barry and colleagues have published extensively on the effects of condensed tannins in *L. pedunculatus* on livestock (Barry and Duncan, 1984; Barry and Manley, 1984; Barry, 1985; Barry *et al.*, 1986a,b; Barry and Manley, 1984; McNabb *et al.*, 1993). The tannin content of *L. pedunculatus* in New Zealand is influenced by soil fertility and environmental conditions. On acid soils of low fertility, condensed tannin concentrations may be 8–10% of dry weight, whereas on more fertile soils the level is 2–4% (Barry and Forss, 1983). At a level of 5.5% condensed tannin, the voluntary trefoil intake of sheep was significantly reduced (Waghorn *et al.*, 1994a), while higher concentrations reduce both feed intake and nutrient digestibility (Barry and Duncan, 1984). Improvement in protein utilization, with an increased flow of amino acids to the abomasum, was noted with 5.5% tannin-trefoil (Waghorn *et al.*, 1994b). A level of 3–4% condensed tannin is probably in the optimal range for best animal performance (Terrill *et al.*, 1992).

Supplementation of tannin-containing forage with **polyethylene glycol (PEG)**, a polymer which binds tannins, improves animal performance by binding tannins and preventing their adverse effects on protein utilization and digestive enzyme activity (see Chapter 6). Use of PEG is probably not economic or practical under field conditions, but is a useful experimental tool to study the effect of tannins on forage utilization.

Hydrolyzable tannins occur in oak leaves (*Quercus* spp.) and several tropical legumes used as forage, including *Acacia* spp. which are important browse spp. in Africa and Australia. Hydrolyzable tannins undergo microbial and acid hydrolyses in the gut with the release of phenolic acids such as gallic acid. Gallic acid can be further metabolized to pyrogallol (see Oak Poisoning, Chapter 12) which is absorbed in sufficient quantities to cause liver damage (Reed, 1995). The major urinary metabolites of gallic acid and its derivatives are glucuronides (Murdiati *et al.*, 1992). Because hydrolyzable tannins are readily hydrolyzed in the rumen to simple phenolic acids, they do not affect nutrient digestibility in ruminants (Hagerman *et al.*, 1992).

Various **phenolic acids** such as p-coumaric acid, ferulic acid, sinapic acid, cinnamic acid, caffeic acid and coniferyl alcohol occur in the cell walls of forages (Jung and Fahey, 1983). They are involved in crosslinking of fiber constituents, and by their antimicrobial activity may inhibit the digestibility of cell wall material. Phenolic acids are metabolized by rumen microbes, with chemical reduction of the hydroxy groups. Some of the products of this reductive process include benzoic acid, 3-phenylpropionic acid (PPA), and cinnamic acid. These reduced phenolics are absorbed and detoxified by conjugation with glycine, producing hippuric acid (Fig. 11–34). Absorbed PPA appears to inhibit hepatic gluconeogenesis from propionate (Cremin *et al.*, 1995), and thus may reduce metabolic efficiency. On the other hand, there is evidence that PPA is a growth factor for some rumen bacteria (Lowry *et al.*, 1993). The nitrogen cost for the excretion of hippuric acid, via metabolic provision of glycine, may reduce animal performance on low protein forages such as tropical grasses (Lowry *et al.*, 1993). Thus the adverse effects of phenolics on animal performance is associated not only with their protein-complex activity, but also with the metabolic costs of detoxification and excretion.

Dietary tannins can impair the absorption of mineral elements, but the effects on mineral bioavailability with forage legumes appears to be minor (Waghorn *et al.*, 1994a).

In the tropics, numerous legume and non-legume **fodder trees and shrubs** are used as fodder for livestock. They may be harvested by hand and fed to livestock fed in confinement (the cut and carry system) or they may by grown in pastures. The major constraint to the use of tree and shrub legumes for fodder is their phenolic content (Kumar and Vaithiyanathan, 1990; Topps, 1992). Values as high as 52% of total dry matter as phenolics have been

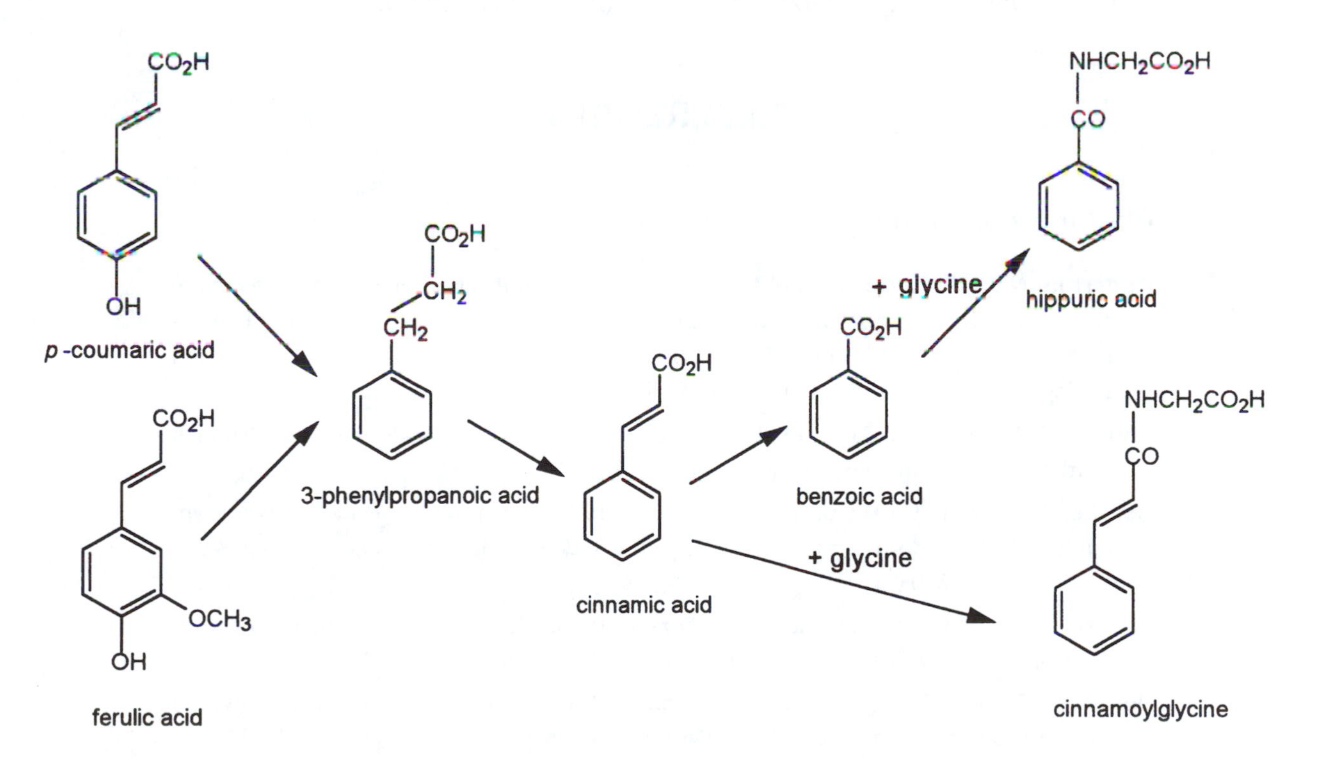

FIGURE 11–34 Metabolism of phenolic acids in ruminants. Absorbed phenolic acids are conjugated with glycine and excreted in the urine.

reported (Reed, 1986; Rittner and Reed, 1992). Because they contain a wide array of specific phenolic compounds, and varying proportions of condensed and hydrolyzable tannins, the nutritional effects are difficult to predict.

Phenolic-protein complexes formed in the gut can interfere with the detergent system of forage analysis involving neutral detergent fiber (NDF) and acid detergent fiber (ADF) determination, especially in tree fodder with high phenolic content. Insoluble tannin-protein complexes are measured as NDF, frequently resulting in negative NDF digestibility values (Reed, 1995). Another effect of dietary phenolics is to cause the excretion of a red or dark urine with flaky sediments (Kumar and Vaithiyanathan, 1990). This may lead to the erroneous conclusion that a pathological condition is involved. Presumably, the red pigments are metabolites of dietary phenolics. The proanthocyanidins, for example, are the precursors of red anthocyanidin pigments of flowers. **Red urine pigments** have also been observed in animals grazing red clover (Niezen *et al.*, 1992). The color forms rapidly with exposure to air, suggesting that an oxidative reaction is involved.

Foliage of the temperate tree legume black locust (*Robinia pseudoacacia*) is used in China and India as a fodder. Black locust leaves have a high condensed tannin content, resulting in a very low digestibility of black locust leaf meal (Ayres *et al.*, 1996; Horton and Christensen, 1981). Black locust bark contains a toxic lectin, called robin, which is poisonous to livestock. Poisoning occurs when horses eat the bark of black locust trees to which they have been tied, and cattle and horses have been poisoned from eating black locust foliage from sprouting stumps. The lectin content of the bark is highest in autumn, winter and early spring (Nsimba-Lubaki and Peumans, 1986).

Hybrid poplars (*Populus* spp.) are grown as a plantation crop for wood fiber production. The abundant foliage is readily consumed by livestock. Poplar foliage contains phenolics which reduce protein digestibility in animals (Ayres *et al.*, 1996).

REFERENCES

Phalaris Poisoning

Anderton, N., P.A. Cockrum, and J.A. Edgar. 1994. Identification of a toxin suspected of causing sudden death in livestock grazing *Phalaris* pastures. pp. 269–274 in: P.R. Dorling and S.M. Colegate (Eds.). Plant-associated Toxins. Agricultural, Phytochemical and Ecological Aspects. CAB International, Wallingford, UK.

Bourke, C.A., and M.J. Carrigan. 1992. Mechanisms underlying *Phalaris aquatica* "sudden death" syndrome in sheep. Aust. Vet. J. 69:165–167.

Bourke, C.A., M.J. Carrigan, and R.J. Dixon. 1988. Experimental evidence that tryptamine alkaloids do not cause *Phalaris aquatica* sudden death syndrome in sheep. Aust. Vet. J. 65:218–220.

Bourke, C.A., M.J. Carrigan, and R.J. Dixon. 1990. The pathogenesis of the nervous syndrome of *Phalaris aquatica* toxicity in sheep. Aust. Vet. J. 67:356–358.

Bourke, C.A., M.J. Carrigan, J.T. Seaman, and J.V. Evers. 1987. Delayed development of clinical signs in sheep affected by *Phalaris aquatica* staggers. Aust. Vet. J. 64:31–32.

Cheeke, P.R. 1995. Endogenous toxins and mycotoxins in forage grasses and their effects on livestock. J. Anim. Sci. 73:909–918.

Duynisveld, G.W., and K.M. Wittenberg. 1993. Evaluation of Rival, Venture and Frontier reed canarygrass as pasture forage. Can. J. Anim. Sci. 73:89–100.

East, N.E., and R.J. Higgins. 1988. Canary grass (*Phalaris* sp.) toxicosis in sheep in California. J. Am. Vet. Med. Assoc. 192:667–669.

Kennedy, D.J., P.D. Cregan, J.R.W. Glastonbury, D.T. Golland, and D.G. Day. 1986. Poisoning of cattle grazing a low-alkaloid cultivar of *Phalaris aquatica,* Sirolan. Aust. Vet. J. 63:88–89.

Lean, I.J., M. Anderson, M.G. Kerfoot, and G.C. Marten. 1989. Tryptamine alkaloid toxicosis in feedlot sheep. J. Am. Vet. Med. Assoc. 195:768–771.

Majak, W., R.E. McDiarmid, A.L. Van Ryswyk, K. Boersma, and S.G. Bonin. 1979. Alkaloid levels in reed canarygrass grown on wet meadows in British Columbia. J. Range Manage. 32:322–326.

Marten, G.C., R.M. Jordan, and A.W. Hovin. 1976. Biological significance of reed canarygrass alkaloids and associated palatability variation to grazing sheep and cattle. Agron. J. 68:909–914.

Marten, G.C., R.M. Jordan, and A.W. Hovin. 1981. Improved lamb performance associated with breeding for alkaloid reduction in reed canarygrass. Crop Sci. 21:295–298.

McKenna, D.J., and G.H.N. Towers. 1984. Biochemistry and pharmacology of tryptamines and β-carbolines. A mini review. J. Psychoactive Drugs 16:347–358.

Nicholson, S.S., B.M. Olcott, E.A. Usenik, H.W. Casey, C.C. Brown, L.E. Urbatsch, S. E. Turnquist, and S.C. Moore. 1989. Delayed phalaris grass toxicosis in sheep and cattle. J. Am. Vet. Med. Assoc. 195:345–346.

Parmar, S.S., and V.C. Brink. 1976. Tryptamine levels in pasturage implicated in bovine pulmonary emphysema. Can. J. Plant Sci. 56:175–184.

Simpson, B.H., R.D. Jolly, and S.H.M. Thomas. 1969. *Phalaris arundinacea* as a cause of deaths and inco-ordination in sheep. N.Z. Vet. J. 17:240–244.

Tosi, H.R., and K.M. Wittenberg. 1993. Harvest alternatives to reduce the alkaloid content of reed canarygrass forage. Can. J. Anim. Sci. 73:373–380.

Wittenberg, K.M., G.W. Duynisveld, and H.R. Tosi. 1992. Comparison of alkaloid content and nutritive value for tryptamine- and β-carboline-free cultivars of reed canarygrass (*Phalaris arundinacea L.*). Can. J. Anim. Sci. 72:903–909.

Oxalates

Allison, M.J., H.M. Cook, and K.A. Dawson. 1981. Selection of oxalate-degrading rumen bacteria in continuous cultures. J. Anim. Sci. 53:810–816.

Allison, M.J., K.A. Dawson, W.R. Mayberry, and J.G. Foss. 1985. *Oxalobacter formigenes* gen. nov., sp. nov.: oxalate-degrading anaerobes that inhabit the gastrointestinal tract. Arch. Microbiol. 141:1–7.

Blaney, B.J., R.J.W. Gartner, and T.A. Head. 1982. The effects of oxalate in tropical grasses on calcium, phosphorus and magnesium availability to cattle. J. Agric. Sci. 99:533–546.

Blaney, B.J., R.J.W. Gartner, and R.A. McKenzie. 1981a. The effects of oxalate in some tropical grasses on the availability to horses of calcium, phosphorus and magnesium. J. Agric. Sci. 97:507–514.

Blaney, B.J., R.J.W. Gartner, and R.A. McKenzie. 1981b. The inability of horses to absorb calcium from calcium oxalate. J. Agric. Sci. 97:639–641.

Jones, R.J., A.A. Seawright, and D.A. Little. 1970. Oxalate poisoning in animals grazing the tropical grass *Setaria sphacelata*. J. Aust. Inst. Agri. Sci. 36:41–43.

McKenzie, R.A., A.M. Bell, G.J. Storie, F.J. Keenan, K.M. Cornack, and S.G. Grant. 1988. Acute oxalate poisoning of sheep by buffel grass (*Cenchrus ciliaris*). Aust. Vet. J. 65:26.

McKenzie, R.A., R.J.W. Gartner, B.J. Blaney, and R.J. Glanville. 1981. Control of nutritional secondary hyperparathyroidism in grazing horses with calcium and phosphorus supplementation. Aust. Vet. J. 57:554–557.

Cyanogens

Adams, L.G., J.W. Dollahite, W.M. Romane, T.L. Bullard, and C.H. Bridges. 1969. Cystitis and ataxia associated with *Sorghum* ingestion by horses. J. Am. Vet. Med. Assoc. 155:518–524.

Bradley, G.A., H.C. Metcalf, C. Reggiardo, T.H. Noon, E.J. Bicknell, F. Lozano-Alarcon, R.E. Reed, and M.W. Riggs. 1995. Neuroaxonal degeneration in sheep grazing *Sorghum* pastures. J. Vet. Diagn. Invest. 7:229–236.

Georgiadis, N.J., and S.J. McNaughton. 1988. Interactions between grazers and a cyanogenic grass, *Cynodon plectostachyus*. OIKOS 51:343–350.

Knight, P.R. 1968. Equine cystitis and ataxia associated with grazing of pastures dominated by sorghum species. Aust. Vet. J. 44:257.

Majak, W., R.E. McDiarmid, J.W. Hall, and K.-J. Cheng. 1990. Factors that determine rates of cyanogenesis in bovine ruminal fluid in vitro. J. Anim. Sci. 68:1648–1655.

McKenzie, R.A., and L.I. McMicking. 1977. Ataxia and urinary incontinence in cattle grazing sorghum. Aust. Vet. J. 53:496–497.

Vickery, P.J., J.L. Wheeler, and C. Mulcahy. 1987. Factors affecting the hydrogen cyanide potential of white clover (*Trifolium repens* L.) Aust. J. Agric. Res. 38:1053–1059.

Wheeler, J.L., and C. Mulcahy. 1989. Consequences for animal production of cyanogenesis in sorghum forage and hay—a review. Tropical Grasslands 23:193–202.

Wheeler, J.L., C. Mulcahy, J.J. Walcott, and G.G. Rapp. 1990. Factors affecting the hydrogen cyanide potential of forage sorghum. Aust. J. Agric. Res. 41:1093–1100.

Saponins

Bourke, C.A. 1984. Staggers in sheep associated with the ingestion of *Tribulus terrestris*. Aust. Vet. J. 61:360–363.

Bourke, C.A., G.R. Stevens, and M.J. Carrigan. 1992. Locomotor effects in sheep of alkaloids identified in Australian *Tribulus terrestris*. Aust. Vet. J. 69:163–165.

Bridges, C.H., B.J. Camp, C.W. Livingston, and E.M. Bailey. 1987. Kleingrass (*Panicum coloratum* L.) poisoning in sheep. Vet. Pathol. 24:525–531.

Camp, B.J., C.H. Bridges, D.W. Hill, B. Patamalai, and S. Wilson. 1988. Isolation of a steroidal sapogenin from a sheep fed *Agave lecheguilla*. Vet. Human Toxicol. 30:533–535.

Cheeke, P.R., J.S. Powley, H.S. Nakaue, and G.H. Arscott. 1983. Feed preference responses of several avian species fed alfalfa meal, high and low saponin alfalfa, and quinine sulfate. Can. J. Anim. Sci. 63:707–710.

Clarke, R.T.J., C.S.W. Reid, and P.W. Young. 1969. Bloat in cattle. XXXII. Attempts to prevent legume bloat in dry and lactating cows by partial or complete elimination of the rumen holotrich protozoa with dimetridazole. N.Z.J. Agric. Res. 12:446–466.

Cornick, J.L., G.K. Carter, and C.H. Bridges. 1988. Kleingrass-associated hepatotoxicosis in horses. J. Am. Vet. Med. Assoc. 193:932–935.

Edrington, T.S., G.I. Flores-Rodriguez, G.S. Smith, and D.M. Hallford. 1993a. Effect of ingested snakeweed (*Gutierrezia microcephala*) foliage on reproduction, semen quality, and serum clinical profiles of male rats. J. Anim. Sci. 71:1520–1525.

Edrington, T.S., G.S. Smith, T.T. Ross, D.M. Hallford, M.D. Samford, and J.P. Thilsted. 1993b. Embryonic mortality in Sprague–Dawley rats induced by snakeweed (*Gutierrezia microcephala*). J. Anim. Sci. 71:2193–2198.

Glastonbury, J.R.W., F.R. Doughty, S.J. Whitaker, and E. Sergeant. 1984. A syndrome of hepatogenous photosensitisation, resembling geeldikkop, in sheep grazing *Tribulus terrestris*. Aust. Vet. J. 61:314–316.

Graydon, R.J., H. Hamid, P. Zahari, and C. Gardiner. 1991. Photosensitisation and crystal-associated cholangiohepatopathy in sheep grazing *Brachiaria decumbens*. Aust. Vet. J. 68:234–236.

Hostettmann, K., and A. Marston. 1995. Saponins. Cambridge University Press, Cambridge, UK.

Jenkins, K.J., and A.S. Atwal. 1994. Effects of dietary saponins on fecal bile acids and neutral sterols, and availability of vitamins A and E in the chick. J. Nutr. Biochem. 5:134–137.

Kaneda, N., H. Nakanishi, and E.J. Staba. 1987. Steroidal constituents of *Yucca shidigera* plants and tissue cultures. Phytochemistry 26:1425–1429.

Kellerman, T.S., G.L. Erasmus, J.A.W. Coetzer, J.M.M. Brown, and B.P. Maartens. 1991. Photosensitivity in South Africa. VI. The experimental induction of geeldikkop in sheep with crude steroidal saponins from *Tribulus terrestris*. Onderstepoort J. Vet. Res. 58:47–53.

Klita, P.T., G.W. Mathison, T.W. Fenton, and R.T. Hardin. 1996. Effects of alfalfa root saponins on digestive function in sheep. J. Anim. Sci. 74:1144–1156.

Lajis, N.H., A.S.H. Abdullah, S.J.S. Salim, J.B. Bremner, and M.N. Khan. 1993. *Epi*-sarsasapogenin and *epi*-smilagenin: two sapogenins isolated from the rumen content of sheep intoxicated by *Brachiaria decumbens*. Steroids 58:387–389.

Li, T.S.C., G. Mazza, A.C. Cottrell, and L. Gao. 1996. Ginsenosides in roots and leaves of American ginseng. J. Agric. Food Chem. 44:717–720.

Low, S.G., W.L. Bryden, S.B. Jephcott, and I. McL. Grant. 1993. Photosensitization of cattle grazing signal grass (*Brachiaria decumbens*) in Papua New Guinea. N.Z. Vet. J. 41:220–221.

Low, S.G., S.B. Jephcott, and W.L. Bryden. 1994. Weaner illthrift of cattle grazing signal grass (*Brachiaria decumbens*) in Papua New Guinea. pp. 567–571 in: S.M. Colegate and P.R. Dorling (Eds.). Plant-associated Toxins. Agricultural, Phytochemical and Ecological Aspects. CAB International, Wallingford, UK.

Lu, C.D., and N.A. Jorgensen. 1987. Alfalfa saponins affect site and extent of nutrient digestion in ruminants. J. Nutr. 117:919–927.

Majak, W., R.E. Howarth, A.C. Fesser, B.P. Goplen, and M.W. Pedersen. 1980. Relationships between ruminant bloat and the composition of alfalfa herbage. II. Saponins. Can. J. Anim. Sci. 60:699–708.

McDonough, S.P., A.H. Woodbury, F.D. Galey, D.W. Wilson, N. East, and E. Bracken. 1994. Hepatogenous photosensitization of sheep in California associated with ingestion of *Tribulus terrestris* (puncture vine). J. Vet. Diagn. Invest. 6:392–395.

Miles, C.O., S.C. Munday, P.T. Holland, B.L. Smith, P.P. Embling, and A.L. Wilkins. 1991. Identification of a sapogenin glucuronide in the bile of sheep affected by *Panicum dichotomiflorum* toxicosis. N.Z. Vet. J. 39:150–152.

Miles, C.O., A.L. Wilkins, G.L. Erasmus, and T.S. Kellerman. 1994a. Photosensitivity in South Africa. VIII. Ovine metabolism of *Tribulus terrestris* saponins during experimentally induced geeldikkop. Onderstepoort J. Vet. Res. 61:351–359.

Miles, C.O., A.L. Wilkins, G.L. Erasmus, T.S. Kellerman, and J.A.W. Coetzer. 1994b. Photosensitivity in South Africa. VII. Chemical composition of biliary crystals from a sheep with experimentally induced geeldikkop. Onderstepoort J. Vet. Res. 61:215–222.

Miles, C.O., A.L. Wilkins, S.C. Munday, A. Flaoyen, P.T. Holland, and B.L. Smith. 1993. Identification of insoluble salts of the β-D-glucuronides of episarsasapogenin and epismilagenin in the bile of lambs with alveld, and examination of *Narthecium ossifragum, Tribulus terrestris* and *Panicum miliaceum* for sapogenins. J. Agr. Food Chem. 41:914–917.

Miles, C.O., A.L. Wilkins, S.C. Munday, P.T. Holland, B.L. Smith, M.J. Lancaster, and P. P. Embling. 1992. Identification of the calcium salt of epismilagenin β-D-glucuronide in the bile crystals of sheep affected by *Panicum dichotomiflorum* and *Panicum schinzii* toxicoses. J. Agric. Food Chem. 40:1606–1609.

Mullenax, C. H. 1991. *Brachiaria decumbens* toxicity—grass or fungus? Vet. Hum. Toxicol. 33:464–465.

Munday, S.C., A.L. Wilkins, C.O. Miles, and P.T. Holland. 1993. Isolation and structure elucidation of dichotomin, a furostanol saponin implicated in hepatogenous photosensitization of sheep grazing *Panicum dichotomiflorum*. J. Agric. Food Chem. 41:267–271.

Nakaue, H.S., R.R. Lowry, P.R. Cheeke, and G.H. Arscott. 1980. The effect of dietary alfalfa of varying saponin content on yolk cholesterol level and layer performance. Poult. Sci. 59:2744–2748.

Oakenfull, D., and G.S. Sidhu. 1989. Saponins. pp. 97–141 in: P.R. Cheeke (Ed.). Toxicants of Plant Origin. Vol. II. Glycosides. CRC Press, Boca Raton, FL.

Oleszek, W. 1990. Structural specificity of alfalfa (*Medicago sativa*) saponin haemolysis and its impacts on two haemolysis-based quantification methods. J. Sci. Food Agric. 53:477–485.

Oleszek, W., M. Jurzysta, M. Ploszynski, I.J. Colquhoun, K.R. Price, and G.R. Fenwick. 1992. Zahnic acid tridesmoside and other dominant saponins from alfalfa (*Medicago sativa* L.) aerial parts. J. Agric. Food Chem. 40:191–196.

Oleszek, W., J. Nowacka, J.M. Gee, G.M. Wortley, and I.T. Johnson. 1994. Effects of some purified alfalfa (*Medicago sativa*) saponins on transmural potential difference in mammalian small intestine. J. Sci. Food Agric. 65:35–39.

Onning, G., Q. Wang, B.R. Westrom, N.-G. Asp, and B.W. Karlsson. 1996. Influence of oat saponins on intestinal permeability *in vitro* and *in vivo* in the rat. Brit. J. Nutr. 76:141–151.

Price, K.R., I.T. Johnson, and G.R. Fenwick. 1987. The chemistry and biological significance of saponins in foods and feedingstuffs. CRC Critical Reviews Food Sci. Nutr. 26:27–135.

Puoli, J.R., R.L. Reid, and D.P. Belesky. 1992. Photosensitization in lambs grazing switchgrass. Agron. J. 84:1077–1080.

Rankins, Jr., D.L., G.S. Smith, T.T. Ross, J.S. Caton, and P. Kloppenburg. 1993. Characterization of toxicosis in sheep dosed with blossoms of sacahuiste (*Nolina microcarpa*). J. Anim. Sci. 71:2489–2498.

Smith, B.L., and C.O. Miles. 1993. A role for *Brachiaria decumbens* in hepatogenous photosensitization of ruminants? Vet. Hum. Toxicol. 35:256–257.

Smith, G.S., T.T. Ross, D.M. Hallford, J.P. Thilsted, E.C. Staley, J.A. Greenberg, and R.J. Miller. 1994. Toxicology of snakeweeds (*Gutierrezia sarothrae* and *G. microcephala*. Proc. West. Sec. Am. Soc. Anim. Sci. 45:98–102.

Southon, S., A.J.A. Wright, K.R. Price, S.J. Fairweather-Tait, and G.R. Fenwick. 1988. The effect of three types of saponin on iron and zinc absorption from a single meal in the rat. Brit. J. Nutr. 59:389–396.

Wallace, R.J., L. Arthaud, and C.J. Newbold. 1994. Influence of *Yucca shidigera* extract on ruminal ammonia concentrations and ruminal microorganisms. Appl. Environ. Microbiol. 60:1762–1767.

Waller, G.R., and K. Yamasaki (Eds.). 1996. Saponins Used in Food and Agriculture. Plenum, New York.

West, L.G., J.L. Greger, A. White, and B.J. Nonnamaker. 1978. In vitro studies on saponin-mineral complexation. J. Food Sci. 43:1342–1343.

Williams, M.C. 1978. Toxicity of saponins in alfombrilla (*Drymaria arenariodes*). J. Range Manage. 31:182–184.

Williams, M.C., L.F. James, and C. Luis. 1980. Seasonal concentration and toxicity of saponins in alfombrilla. J. Range Manage. 32:157–158.

Alkaloids in Forage Lupins

Bunch, T.D., K.E. Panter, and L.F. James. 1992. Ultrasound studies of the effects of certain poisonous plants on uterine function and fetal development in livestock. J. Anim. Sci. 70:1639–1643.

Gardner, D.R., and K.E. Panter. 1993. Comparison of blood plasma alkaloid levels in cattle, sheep, and goats fed *Lupinus caudatus*. J. Natural Toxins 2:1–11.

Gardner, D.R., and K.E. Panter. 1994. Ammodendrine and related piperidine alkaloid levels in the blood plasma of cattle, sheep and goats fed *Lupinus formosus*. J. Natural Toxins 3:107–116.

Keeler, R.F., and K.E. Panter. 1989. Piperidine alkaloid composition and relation to crooked calf disease–inducing potential of *Lupinus formosus*. Teratology 40:423–432.

Kilgore, W.W., D.G. Crosby, A.L. Craigmill, and N.K. Poppen. 1981. Toxic plants as possible human teratogens. Calif. Agric. (Nov.–Dec.), 6.

Kingsbury, J.M. 1964. Poisonous Plants of the United States and Canada. Prentice–Hall, Englewood Cliffs, NJ.

Meeker, J.E., and W.W. Kilgore. 1987. Identification and quantitation of the alkaloids of *Lupinus latifolius*. J. Agric. Food Chem. 35:431–433.

Meeker, J.E., and W.W. Kilgore. 1991. Investigations of the teratogenic potential of the lupine alkaloid anagyrine. pp. 41–60 in: R.F. Keeler and A.T. Tu (Eds.). Handbook of Natural Toxins. Vol. 6. Toxicology of Plant and Fungal Compounds. Marcel Dekker, Inc., New York.

Ortega, J.A., and J. Lazerson. 1987. Anagyrine-induced red cell aplasia, vascular anomaly, and skeletal dysplasia. J. Pediatrics 111:87–89.

Panter, K.E., T.D. Bunch, R.F. Keeler, D.V. Sisson, and R.J. Callan. 1990. Multiple congenital contractures (MCC) and cleft palate induced in goats by ingestion of piperidine alkaloid–containing plants; reduction in fetal movement as the probable cause. Clin. Toxicol. 28:69–83.

Panter, K.E., D.R. Gardner, and R.J. Molyneux. 1994. Comparison of toxic and teratogenic effects of *Lupinus formosus, L. arbustus* and *L. caudatus* in goats. J. Natural Toxins 3:83–93.

Panter, K.E., and R.F. Keeler. 1993. Quinolizidine and piperidine alkaloid teratogens from poisonous plants and their mechanism of action in animals. Vet. Clinics of N. Amer.: Food Anim. Prac. 9:33–40.

Panter, K.E., R.F. Keeler, L.F. James, and T.D. Bunch. 1992. Impact of plant toxins on fetal and neonatal development: A review. J. Range Manage. 45:52–57.

Mimosine and *Leucaena* Toxicity

Allison, M.J., A.C. Hammond, and R.J. Jones. 1990. Detection of ruminal bacteria that degrade toxic dihydroxypyridine compounds produced from mimosine. Appl. Environm. Microbiol. 56:590–594.

Allison, M.J., W.R. Mayberry, C.S. McSweeney, and D.A. Stahl. 1992. *Synergistes jonesii,* gen. nov., sp. nov.: A rumen bacterium that degrades toxic pyridinediols. System. Appl. Microbiol. 15:522–529.

Deweede, S., and O. Wayman. 1970. Effect of mimosine on the rat fetus. Teratology 3:21–28.

D'Mello, J.P.F., and T. Acamovic. 1989. *Leucaena leucocephala* in poultry nutrition—A review. Anim. Feed Sci. Tech. 26:1–28.

Dominguez-Bello, M.G., and C.S. Stewart. 1990. Degradation of mimosine, 2,3-dihydroxypyridine and 3-hydroxy-4(1H)-pyridone by bacteria from the rumen of sheep in Venezuela. FEMS Microbiol. Ecol. 73:283–290.

Dominguez-Bello, M.G., and C.S. Stewart. 1991. Characteristics of a rumen *Clostridium* capable of degrading mimosine, 3(OH)-4-(1H)-pyridone and 2,3 dihydroxypyridine. System. Appl. Microbiol. 14:67–71.

Hammond, A.C. 1995. Leucaena toxicosis and its control in ruminants. J. Anim. Sci. 73:1487–1492.

Hammond, A.C., M.J. Allison, M.J. Williams, G.M. Prine, and D.B. Bates. 1989. Prevention of leucaena toxicosis of cattle in Florida by ruminal inoculation with 3-hydroxy-4-(1H)-pyridone-degrading bacteria. Amer. J. Vet. Res. 50:2176–2180.

Hegarty, M.P. 1978. Toxic amino acids of plant origin. pp. 575–585 in: R.F. Keeler, K.R. Kampen, and L.R. James (Eds.). Effects of Poisonous Plants on Livestock. Academic Press, N.Y.

Hegarty, M.P., R.D. Court, G.S. Christie, and C.P. Lee. 1976. Mimosine in *Leucaena leucocephala* is metabolised to a goitrogen in ruminants. Aust. Vet. J. 52:490–492.

Jones, R.J. 1981. Does ruminal metabolism of mimosine explain the absence of *Leucaena* toxicity in Hawaii? Aust. Vet. J. 57:55.

Jones, R.J. 1994. Management of anti-nutritive factors—with special reference to leucaena. pp. 216–231 in: R.C. Gutteridge, and H.M. Shelton (Eds.). Forage Tree Legumes in Tropical Agriculture. CAB International, Wallingford, Oxon OX10 8DE, UK.

Jones, R.J., C.G. Blunt, and B.I. Nurnberg. 1978. Toxicity of *Leucaena leucocephala*: The effect of iodine and mineral supplements on penned steers fed a sole diet of *Leucaena*. Aust. Vet. J. 54:387–392.

Jones, R.J., and J.B. Lowry. 1984. Australian goats detoxify the goitrogen 3-hydroxy-4(1H) pyridone (DHP) after rumen infusion from an Indonesian goat. Experientia 40:1435–1436.

Lowry, J.B., M. Tangendjaja, and B. Tangendjaja. 1983. Optimising autolysis of mimosine to 3-hydroxy-4(1H)-pyridone in green tissues of *Leucaena leucocephala*. J. Sci. Food Agric. 34:529–533.

McSweeny, C.S., M.J. Allison, and R.I. Mackie. 1993. Amino acid utilization by the ruminal bacterium *Synergistes jonesii* strain 78-1. Arch. Microbiol. 159:131–135.

Megarrity, R.G., and R.J. Jones. 1986. Successful transfer of DHP-degrading bacteria from Hawaiian goats to Australian ruminants to overcome the toxicity of Leucaena. Aust. Vet. J. 63:259–262.

Puchala, R., S.G. Pierzynowski, T. Sahlu, and S.P. Hart. 1996. Effects of mimosine administered to a perfused area of skin in Angora goats. Brit. J. Nutr. 75:69–79.

Ross, E., O. Wayman, and F.G. Oishi. 1980. The use of Japanese quail to study the effect of *Leucaena leucocephala* and mimosine on egg production. Proc. West. Sect. Am. Soc. Anim. Sci. 31:129–132.

Stunzi, H., D.A. Perrin, T. Teitei, and R.L.N. Harris. 1979. Stability constants of some metal complexes formed by mimosine and related compounds. Aust. J. Chem. 32:21–30.

Wayman, O., I.I. Iwanaga, and W.I. Hugh. 1970. Fetal resorption in swine caused by *Leucaena leucocephala* (Lam.) De Wit in the diet. J. Anim. Sci. 30:583–588.

Other Toxic Amino Acids

Aylward, J.H., R.D. Court, K.P. Haydock, R.W. Strickland, and M.P. Hegarty. 1987. Indigofera species with agronomic potential in the tropics. Rat toxicity studies. Aust. J. Agric. Res. 38:177–186.

Dominguez-Bello, M.G., and C.S. Stewart. 1990. Effects of feeding *Canavalia ensiformis* on the rumen flora of sheep, and of the toxic amino acid canavanine on rumen bacteria. System. Appl. Microbiol. 13:388–393.

Foster, J.G. 1990. Flatpea (*Lathyrus sylvestris* L.): A new forage species? A comprehensive review. Advances in Agronomy 43:241–313.

Morton, J.F. 1989. Creeping indigo (*Indigo spicata* Forsk.) (Fabaceae)—A hazard to herbivores in Florida. Econ. Bot. 43:314–327.

Rasmussen, M.A., M.J. Allison, and J.G. Foster. 1993. Flatpea intoxication in sheep and indications of ruminal adaptation. Vet. Hum. Toxicol. 35:123–127.

Rowe, L.D., G.W. Ivie, J.R. DeLoach, and J.G. Foster. 1993. The toxic effects of mature flatpea (*Lathyrus sylvestris* L. CV. Lathco) on sheep. Vet. Hum. Toxicol. 35:127–133.

Alsike Clover Poisoning

Nation, P.N. 1989. Alsike clover poisoning: A review. Can. Vet. J. 30:410–415.

Nation, P.N. 1991. Hepatic disease in Alberta horses: A retrospective study of "alsike clover poisoning" (1973–1988). Can. Vet. J. 32:602–607.

Vetch Toxicity

Anderson, C.A., and T.J. Divers. 1983. Systemic granulomatous inflammation in a horse grazing hairy vetch. J. Am. Vet. Med. Assoc. 183:569–570.

Burroughs, G.W., J.A. Neser, T.S. Kellerman, and F.A. Vanniekerk. 1983. Suspected hybrid vetch (*Vicia villosa* crossed with *Vicia dasycarpa*) poisoning of cattle in the Republic of South Africa. J. S. Afr. Vet. Assoc. 54:75–80.

Harper, P.A.W., R.W. Cook, P.A. Gill, G.C. Fraser, L.M. Badcoe, and J.M. Power. 1993. Vetch toxicosis in cattle grazing *Vicia villosa* spp. *dasycarpa* and *V. benghalensis*. Aust. Vet. J. 70:140–144.

Odriozola, E., E. Paloma, T. Lopez, and C. Campero. 1991. An outbreak of Vicia villosa (hairy vetch) poisoning in grazing Aberdeen Angus bulls in Argentina. Vet. Hum. Toxicol. 33:278–280.

Panciera, R.J. 1978. Hairy vetch (*Vicia villosa* Roth) poisoning in cattle. pp. 555–563 in: R.F. Keeler, K.R. Van Kampen, and L.F. James (Eds.). Effects of Poisonous Plants on Livestock. Academic Press, NY.

Panciera, R.J., L. Johnson, and B.I. Osbourn. 1966. A disease of cattle grazing hairy vetch pasture. J. Am. Vet. Med. Assoc. 148:804–808.

Panciera, R.J., D.A. Mosier, and J.W. Ritchey. 1992. Hairy vetch (*Vicia villosa* Roth) poisoning in cattle: update and experimental induction of disease. J. Vet. Diagn. Invest. 4:318–325.

Sweet Clover Poisoning

Benson, M.E., H.H. Casper, and L.J. Johnson. 1981. Occurrence and range of dicoumerol concentration in sweet clover. Am. J. Vet. Res. 42:2014–2015.

Cohen, A.J. 1979. Critical review of the toxicology of coumarin with special reference to interspecies differences in metabolism and hepatoxic response and their significance to man. Food Cosmet. Toxicol. 17:277–289.

Goplen, B.P. 1971. Polara, a low coumarin cultivar of sweetclover. Can. J. Plant Sci. 51:249–251.

Kingsbury, J.M. 1964. Poisonous Plants of the United States and Canada. Prentice–Hall, Englewood Cliffs, NJ.

Pritchard, D.G., L.M. Markson, P.J. Brush, J.A.A. Sawtell, and P.A. Bloxham. 1983. Haemorrhagic syndrome of cattle associated with the feeding of sweet vernal (*Anthoxanthum odoratum*) hay containing dicoumarol. Vet. Rec. 113:78–84.

Radostits, O.M., G.P. Searcy, and K. Mitchell. 1980. Moldy sweet-clover poisoning in cattle. Can. Vet. J. 21:155–158.

Sanderson, M.A., D.W. Meyer, and H. Casper. 1985. Dicoumarol concentrations and forage quality of sweet clover forage treated with propionic acid or anhydrous ammonia. J. Anim. Sci. 61:1243–1252.

Scheel, L.D. 1978. The toxicology of sweet clover and coumarin anticoagulants. pp. 121–142 in: T.D. Wyllie and L.G. Morehouse (Eds.). Mycotoxic Fungi, Mycotoxins and Mycotoxicosis: An Encyclopedia Handbook. Marcel Dekker, NY.

Tligui, N., and G.R. Ruth. 1994. *Ferula communis* variety *brevifolia* intoxication of sheep. Am. J. Vet. Res. 55:1558–1563.

Phytoestrogens

Adams, N.R. 1989. Phytoestrogens. pp. 23–51 in: P.R. Cheeke (Ed.). Toxicants of Plant Origin, Vol. IV. CRC Press, Boca Raton, FL.

Adams, N.R. 1990. Permanent infertility in ewes exposed to plant oestrogens. Aust. Vet. J. 67:197–201.

Adams, N.R. 1995. Detection of the effects of phytoestrogens on sheep and cattle. J. Anim. Sci. 73:1509–1515.

Adlercreutz, H. 1990a. Diet, breast cancer, and sex hormone metabolism. Annal N.Y. Acad. Sci. 595:281–290.

Adlercreutz, H. 1990b. Western diet and Western diseases: some hormonal and biochemical mechanisms and associations. Scand. J. Clin. Lab Invest. 50:3–23.

Adlercreutz, H., K. Hockerstedt, C. Bannwart, S. Bloigu, E. Hamalainen, T. Fotsis, and A. Ollus. 1987. Effect of dietary components, including lignans and phytoestrogens,on enterohepatic circulation and liver metabolism of estrogens and on sex hormone binding globulin (SHBG). J. Steroid Biochem. 27:1135–1144.

Adlercreutz, H., H. Honjo, A. Higashi, T. Fotsis, E. Hamalainen, T. Hasegawa, and H. Okada. 1991. Urinary excretion of lignans and isoflavonoid phytoestrogens in Japanese men and women consuming a traditional Japanese diet. Am. J. Clin. Nutr. 54:1093–1100.

Chamley, W.A., N.R. Adams, R.D. Hooley, and R. Carson. 1981. Hypothalamic-pituitary function in normal ewes and ewes which grazed oestrogenic subterranean clover for several years. Aust. J. Biol. Sci. 34:239–244.

Cox, R.I. 1978. Plant estrogens affecting livestock in Australia. pp. 451–464 in: R.F. Keeler, K.R. Van Kampen, and L.F. James (Eds.). Effects of Poisonous Plants on Livestock. Academic Press, NY.

Croker, K.P., R.J. Lightfoot, T.J. Johnson, N.R. Adams, and M.J. Carrick. 1989. The effects of selection for resistance to clover infertility on the reproductive performances of Merino ewes grazed on oestrogenic pastures. Aust J. Agric. Res. 40:165–176.

Davies, H.L., and J.L. Hill. 1989. The effect of diet on the metabolism in sheep of the tritiated isoflavones formononetin and biochanin A. Aust. J. Agric. Res. 40:157–163.

Dickinson, J.M., G.R. Smith, R.D. Randel, and I.J. Pemberton. 1988. In vitro metabolism of formononetin and biochanin A in bovine rumen fluid. J. Anim. Sci. 66:1969–1973.

Elakovich, S.D., and J.M. Hampton. 1983. Analysis of coumestrol, a phytoestrogen, in alfalfa tablets sold for human consumption. J. Agric. Food Chem. 32:173–175.

Hughes, Jr., C.L. 1988. Phytochemical mimicry of reproductive hormones and modulation of herbivore fertility by phytoestrogens. Environmental Health Perspectives 78:171–175.

Kaldas, R.S., and C.L. Hughes, Jr. 1989. Reproductive and general metabolic effects of phytoestrogens in mammals. Reprod. Toxicol. 3:81–89.

Leopold, A.S., M. Erwin, J. Oh, and B. Browning. 1976. Phytoestrogens. Adverse effects on reproduction in California quail. Science 191:98–99.

Little, D.L. 1996. Reducing the effects of clover disease by strategic grazing of pastures. Aust. Vet. J. 73:192–193.

Livingston, L. 1978. Forage plant estrogens. J. Toxicol. Environ. Health 4:301–324.

Lundh, T. 1995. Metabolism of estrogenic isoflavones in domestic animals. Proc. Soc. Exp. Biol. Med. 208:33–39.

Lundh, T.J.-O. 1990. Conjugation of the plant estrogens formononetin and daidzein and their metabolite equol by gastrointestinal epithelium from cattle and sheep. J. Agric. Food Chem. 38:1012–1016.

Lundh, T.J.-O., H.I. Pettersson, and K.A. Martinsson. 1990. Comparative levels of free and conjugated plant estrogens in blood plasma of sheep and cattle fed estrogenic silage. J. Agric. Food Chem. 38:1530–1534.

Neil, H.G., R.J. Lightfoot, and H.E. Fels. 1974. Effect of legume species on ewe fertility in South Western Australia. Proc. Aust. Soc. Anim. Prod. 10:136.

Wang, W., Y. Tanaka, Z. Han, and J. Cheng. 1994. Radioimmunoassay for quantitative analysis of formononetin in blood plasma and rumen fluid of wethers fed red clover. J. Agric. Food Chem. 42:1584–1587.

Forage-Induced Anemia

Andersen, H.J., L. Pellett, and A.L. Tappel. 1994. Hemichrome formation, lipid peroxidation, enzyme inactivation and protein degradation as indexes of oxidative damage in homogenates of chicken kidney and liver. Chemico-Biological Interactions 93:155–169.

Barry, T.N., T.R. Manley, K.R. Millar, and R.H. Smith. 1984. The relative feeding value of kale (*Brassica oleracea*) containing normal and low concentrations of S-methyl-L-cysteine sulphoxide (SMCO). J. Agric. Sci. 102:635–643.

Benevenga, N.J., G.L. Case, and R.D. Steele. 1989. Occurrence and metabolism of S-methyl-L-cysteine and S-methyl-L-cysteine sulfoxide in plants and their toxicity and metabolism in animals. pp. 203–228 in: P.R. Cheeke (Ed.). Toxicants of Plant Origin. Vol. III. CRC Press, Boca Raton, FL.

Duncan, A.J., and J.A. Milne. 1992a. Effect of long-term intra-ruminal infusion of the glucosinolate metabolite allyl cyanide on the voluntary food intake and metabolism of lambs. J. Sci. Food Agric. 58:9–14.

Duncan, A.J., and J.A. Milne. 1992b. Rumen microbial degradation of allyl cyanide as a possible explanation for the tolerance of sheep to brassica-derived glucosinolates. J. Sci. Food Agric. 58:15–19.

Duncan, A.J., B. Roncin, and D.A. Elston. 1995. Effect of blood glutathione status on the susceptibility of sheep to haemolytic anaemia induced by the brassica anti-metabolite, dimethyl disulphide. Anim. Sci. 60:93–98.

Eyre, M.D., D.E. Phillips, I.M. Evans, and A. Thompson. 1983. The nutritional role of S-methyl-L-cysteine. J. Sci. Food Agric. 34:696–700.

Fredrickson, E.L., R.E. Estell, K.M. Havstad, W.L. Shupe, and L.W. Murray. 1995. Potential toxicity and feed value of onions for sheep. Livestock Production Sci. 42:45–54.

George, L.W., T.J. Divers, E.A. Mahaffey, and M.J.H. Suarez. 1982. Heinz body anemia and methemoglobinemia in ponies given red maple (*Acer rubrum L.*) leaves. Vet. Pathol. 19:521–533.

Goto, I., N.S. Agar, and Y. Maede. 1993. Relation between reduced glutathione content and Heinz body formation in sheep erythrocytes. Am. J. Vet. Res. 54:622–626.

Greenhalgh, J.F.D., G.A.M. Sharman, and J.N. Aitken. 1969. Kale anaemia. I. The toxicity to various species of animal of three types of kale. Res. Vet. Sci. 10:64–72.

Griffiths, D.W., W.H. Macfarlane-Smith, and B. Boag. 1994. The effect of cultivar, sample date and grazing on the concentration of S-methylcysteine sulphoxide in oilseed and forage rapes (*Brassica napus*). J. Sci. Food Agric. 64:283–288.

Hutchinson, T.W.S. 1977. Onions as a cause of Heinz body anemia and death in cattle. Can. Vet. J. 18:358–360.

Kaplan, A.J. 1995. Onion powder in baby food may induce anemia in cats. J. Am. Vet. Med. Assoc. 207:1405.

Kendler, B.S. 1987. Garlic (*Allium sativum*) and onion (*Allium cepa*): A review of their relationship to cardiovascular disease. Preventive Med. 16:670–685.

Kirk, J.H., and M.S. Bulgin. 1979. Effects of feeding cull domestic onions (*Allium cepa*) to sheep. Am. J. Vet. Res. 40:397–399.

Lau, B.H.S., M.A. Adetumbi, and A. Sanchez. 1983. *Allium sativum* (garlic) and atherosclerosis: A review. Nutr. Res. 3:119–128.

Lincoln, S.D., M.E. Howell, J.J. Combs, and D.D. Hinman. 1992. Hematologic effects and feeding performance in cattle fed cull domestic onions (*Allium cepa*). J. Am. Vet. Med. Assoc. 200:1090–1094.

McConnico, R.S., and C.F. Brownie. 1992. The use of ascorbic acid in the treatment of 2 cases of red maple (*Acer rubrum*)–poisoned horses. Cornell Vet. 82:293–300.

McDonald, R.C., T.R. Manley, T.N. Barry, D.A. Forss, and A.G. Sinclair. 1981. Nutritional evaluation of kale (*Brassica oleracea*) diets. 3. Changes in composition induced by soil fertility practices, with special reference to SMCO and glucosinolate concentrations. J. Agric. Sci. 97:13–23.

Munday, R., and E. Manns. 1994. Comparative toxicity of prop(en)yl disulfides derived from *Alliceae*: Possible involvement of 1-propenyl disulfides in onion-induced hemolytic anemia. J. Agric. Food Chem. 42:959–962.

Plumlee, K.H. 1991. Red maple toxicity in a horse. Vet. Hum. Toxicol. 33:66–67.

Prache, S. 1994. Haemolytic anaemia in ruminants fed forage brassicas: A review. Vet. Res. 25:497–520.

Randle, W.M., E. Block, M.H. Littlejohn, D. Putman, and M.L. Bussard. 1994. Onion (*Allium cepa* L.) thiosulfinates respond to increasing sulfur fertility. J. Agric. Food Chem. 42:2085–2088.

Stair, E.L., W.C. Edwards, G.E. Burrows, and K. Torbeck. 1993. Suspected red maple (*Acer rubrum*) toxicosis with abortion in two Percheron mares. Vet. Hum. Toxicol. 35:229–230.

Tennant, B., S.G. Dill, L.T. Glickman, E.J. Mirro, J.M.King, D.M. Polak, M.C. Smith, and D.C. Kradel. 1981. Acute hemolytic anemia, methemoglobinemia, and Heinz body formation associated with ingestion of red maple leaves by horses. J. Am. Vet. Med. Assoc. 179:143–150.

Van Kampen, K.R., L.F. James, and A.E. Johnson. 1970. Hemolytic anemia in sheep fed wild onions (*Allium validum*). J. Am. Vet. Med. Assoc. 156:328–332.

Yamoto, O., and Y. Maede. 1992. Susceptibility to onion-induced hemolysis in dogs with hereditary high erythrocyte reduced glutathione and potassium concentrations. Am. J. Vet. Res. 53:134–137.

Phenolics and Tannins in Forages

Austin, P.J., L.A. Suchar, C.T. Robbins, and A.E. Hagerman. 1989. Tannin-binding proteins in saliva of deer and their absence in saliva of sheep and cattle. J. Chem. Ecol. 15:1335–1347.

Ayres, A.C., R.P. Barrett, and P.R. Cheeke. 1996. Feeding value of tree leaves (hybrid poplar and black locust) evaluated with sheep, goats and rabbits. Anim. Feed Sci. Tech. 57:51–62.

Barry, T.N. 1985. The role of condensed tannins in the nutritional value of *Lotus pedunculatus* for sheep. 3. Rates of body and wool growth. Brit. J. Nutr. 54:211–217.

Barry, T.N., T.F. Allsop, and C. Redekopp. 1986a. The role of condensed tannins in the nutritional value of *Lotus pedunculatus* for sheep. 5. Effects on the endocrine system and on adipose tissue metabolism. Brit. J. Nutr. 56:607–614.

Barry, T.N., and S.J. Duncan. 1984. The role of condensed tannins in the nutritional value of *Lotus pedunculatus* for sheep. 1. Voluntary intake. Brit. J. Nutr. 51:485–491.

Barry, T.N., and D.A. Forss. 1983. The condensed tannin content of vegetative *Lotus pedunculatus*, its regulation by fertiliser application, and effect upon protein solubility. J. Sci. Food Agric. 34:1047–1056.

Barry, T.N., and T.R. Manley. 1984. The role of condensed tannins in the nutritional value of *Lotus pedunculatus* for sheep. 2. Quantitative digestion of carbohydrates and proteins. Brit. J. Nutr. 51:493–504.

Barry, T.N., and T.R. Manley. 1986. Interrelationships between the concentrations of total condensed tannin, free condensed tannin and lignin in *Lotus* sp. and their possible consequences in ruminant nutrition. J. Sci. Food Agric. 37:248–254.

Barry, T.N., T.R. Manley, and S.J. Duncan. 1986b. The role of condensed tannins in the nutritional value of *Lotus pedunculatus* for sheep. 4. Sites of carbohydrate and protein digestion as influenced by dietary reactive tannin concentration. Brit. J. Nutr. 55:123–137.

Brooker, J.D., L.A. O'Donovan, I. Skene, K. Clarke, L. Blackall, and P. Muslera. 1994. *Streptococcus caprinus* sp. nov., a tannin-resistant ruminal bacterium from feral goats. Lett. Appl. Microbiol. 18:313–318.

Chiquette, J., K.-J. Cheng, J.W. Costerton, and L.P. Milligan. 1988. Effect of tannins on the digestibility of two isosynthetic strains of birdsfoot trefoil (*Lotus corniculatus* L.) using in vitro and in sacco techniques. Can. J. Anim. Sci. 68:571–760.

Cremin, J.D., Jr., J.K. Drackley, L.R. Hansen, D.E. Grum, J. Odle, and G.C. Fahey, Jr. 1995. Effects of glycine and bovine serum albumin on inhibition of propionate metabolism in ovine hepatocytes caused by reduced phenolic monomers. J. Anim. Sci. 73:3009–3021.

Degen, A.A., K. Becker, H.P.S. Makkar, and N. Borowy. 1995. *Acacia saligna* as a fodder tree for desert livestock and the interaction of its tannins with fibre fractions. J. Sci. Food Agric. 68:65–71.

Hagerman, A.E., C.T. Robbins, Y. Weerasuriya, T.C. Wilson, and C. McArthur. 1992. Tannin chemistry in relation to digestion. J. Range Manage. 45:57–62.

Horton, G.M.J., and D.A. Christensen. 1981. Nutritional value of black locust tree leaf meal (*Robinia pseudoacacia*) and alfalfa meal. Can. J. Anim. Sci. 61:503–506.

Jones, G.A., T.A. McAllister, A.D. Muir, and K.-J. Cheng. 1994. Effects of sainfoin (Onobrychis viciifolia Scop.) condensed tannins on growth and proteolysis by four strains of ruminal bacteria. Appl. Environ. Microbiol. 60:1374–1378.

Jung, H.C., and G.C. Fahey, Jr. 1983. Nutritional implications of phenolic monomers and lignin: a review. J. Anim. Sci. 57:206–219.

Khazaal, K., and E.R. Orskov. 1994. The in vitro gas production technique: An investigation on its potential use with insoluble polyvinylpyrrolidone for the assessment of phenolics-related antinutritive factors in browse species. Anim. Feed Sci. Tech. 47:305–320.

Kumar, R., and S. Vaithiyanathan. 1990. Occurrence, nutritional significance and effect on animal productivity of tannins in tree leaves. Anim. Feed Sci. Tech. 30:21–38.

Lowry, J.B., E.A. Sumpter, C.S. McSweeney, A.C. Schlink, and B. Bowden. 1993. Phenolic acids in the fibre of some tropical grasses, effect of feed quality, and their metabolism by sheep. Aust. J. Agr. Res. 44:1123–1133.

Mangan, J.L. 1988. Nutritional effects of tannins in animal feeds. Nutr. Res. Rev. 1:209–231.

McNabb, W.C., G.C. Waghorn, T.N. Barry, and I.D. Shelton. 1993. The effect of condensed tannins in *Lotus pedunculatus* on the digestion and metabolism of methionine, cystine and inorganic sulphur in sheep. Brit. J. Nutr. 70:647–661.

Mehansho, H., L.G. Butler, and D.M. Carlson. 1987. Dietary tannins and salivary proline-rich proteins: Interactions, induction and defense mechanisms. Annu. Rev. Nutr. 7:423–440.

Miller, S.M., J.D. Brooker, and L.L. Blackall. 1995. A feral goat rumen fluid inoculum improves nitrogen retention in sheep consuming a mulga (*Acacia aneura*) diet. Aust. J. Agric. Res. 46:1545–1553.

Miller, S.M., J.D. Brooker, A. Phillips, and L.L. Blackall. 1996. *Streptococcus caprinus* is ineffective as a rumen inoculum to improve digestion of mulga (*Acacia aneura*) by sheep. Aust. J. Agric. Res. 47:1323–1331.

Mole, S., L.G. Butler, and G. Iason. 1990. Defense against dietary tannin in herbivores: a survey for proline rich salivary proteins in mammals. Biochem. Systematics Ecol. 18:287–293.

Murdiati, T.B., C.S. McSweeney, and J.B. Lowry. 1992. Metabolism in sheep of gallic acid, tannic acid and hydrolysable tannin from *Terminalia oblongata*. Aust. J. Agric. Res. 43:1307–1319.

Nelson, K.E., A.N. Pell, P. Schofield, and S. Zinder. 1995. Isolation and characterization of an anaerobic ruminal bacterium capable of degrading hydrolyzable tannins. Appl. Environ. Microbiol. 61:3293–3298.

Niezen, J.H., T.N. Barry, P.R. Wilson, and G. Lane. 1992. Red urine from red deer grazed on pure red clover swards. N.Z. Vet. J. 40:164–167.

Niezen, J.H., T.S. Waghorn, W.A.G. Charleston, and G.C. Waghorn. 1995. Growth and gastrointestinal nematode parasitism in lambs grazing either lucerne (*Medicago sativa*) or sulla (*Hedysarum coronarium*) which contains condensed tannins. J. Agric. Sci. 125:281–289.

Nsimba-Lubaki, M., and W.J. Peumans. 1986. Seasonal fluctuations of lectins in barks of elderberry (*Sambucus nigra*) and black locust (*Robinia pseudoacacia*). Plant Physiol. 80:747–751.

Osawa, R.O. 1992. Tannin-protein complex–degrading enterobacteria isolated from the alimentary tracts of koalas and a selective medium for their enumeration. Appl. Environ. Microbiol. 58:1754–1759.

Osawa, R.O., and L.I. Sly. 1992. Occurrence of tannin-protein complex degrading *Streptoccus* sp. in feces of various animals. System. Appl. Microbiol. 15:144–147.

Reed, J.D. 1986. Relationships among soluble phenolics, insoluble proanthocyanidins and fiber in East African browse species. J. Range Manage. 39:5–7.

Reed, J.D. 1995. Nutritional toxicology of tannins and related polyphenols in forage legumes. J. Anim. Sci. 73:1516–1528.

Rittner, U., and J.D. Reed. 1992. Phenolics and in-vitro degradability of protein and fibre in West African browse. J. Sci. Food Agric. 58:21–28.

Tanner, G.J., A.E. Moore, and P.J. Larkin. 1994. Proanthocyanidins inhibit hydrolysis of leaf proteins by rumen microflora *in vitro*. Brit. J. Nutr. 71:947–958.

Terrill, T.H., G.B. Douglas, A.G. Foote, R.W. Purchas, G.F. Wilson, and T.N. Barry. 1992. Effect of condensed tannins upon body growth, wool growth and rumen metabolism in sheep grazing sulla (*Hedysarum coronarium)* and perennial pasture. J. Agric. Sci. 119:265–273.

Terrill, T.H., W.R. Windham, C.S. Hoveland, and H.E. Amos. 1989. Forage preservation method influences on tannin concentration, intake, and digestibility of sericea lespedeza by sheep. Agron. J. 81:435–439.

Topps, J.H. 1992. Potential, composition and use of legume shrubs and trees as fodders for livestock in the tropics. J. Agric. Sci. 118:1–8.

Van Hoven, W., and D. Furstenburg. 1992. The use of purified condensed tannin as a reference in determining its influence on rumen fermentation. Comp. Biochem. Physiol. 101A:381–385.

Waghorn, G.C., and I.D. Shelton. 1995. Effect of condensed tannins in *Lotus pedunculatus* on the nutritive value of ryegrass (*Lolium perenne*) fed to sheep. J. Agric. Sci. 125:291–297.

Waghorn, G.C., I.D. Shelton, and W.C. McNabb. 1994a. Effects of condensed tannins in *Lotus pedunculatus* on its nutritive value for sheep. 1. Non-nitrogenous aspects. J. Agric. Sci. 123:99–107.

Waghorn, G.C., I.D. Shelton, W.C. McNabb, and S.N. McCutcheon. 1994b. Effects of condensed tannins in *Lotus pedunculatus* on its nutritive value for sheep. 2. Nitrogenous aspects. J. Agric. Sci. 123:109–119.

Waghorn, G.C., M.J. Ulyatt, A. John, and M.T. Fisher. 1987. The effect of condensed tannins on the site of digestion of amino acids and other nutrients in sheep fed on *Lotus corniculatus* L. Brit. J. Nutr. 57:115–126.

Wang, Y., G.B. Douglas, G.C. Waghorn, T.N. Barry, and A.G. Foote. 1996a. Effect of condensed tannins in *Lotus corniculatus* upon lactation performance in ewes. J. Agric. Sci. 126:353–362.

Wang, Y., G.B. Douglas, G.C. Waghorn, T.N. Barry, A.G. Foote, and R.W. Purchas. 1996b. Effect of condensed tannins upon the performance of lambs grazing *Lotus corniculatus* and lucerne (*Medicago sativa*). J. Agric. Sci. 126:87–98.

PART IV

Plants Poisonous to Livestock

CHAPTER 12

Plants and Toxins Affecting the Gastrointestinal Tract and Liver

OAK POISONING

Poisoning of cattle from consumption of oak buds, leaves, twigs, and acorns occurs in many parts of North America and Europe. Oak poisoning is generally seasonal, caused by ingestion of buds and leaves in the spring and acorns in the fall. **Tannins**, such as tannic acid and its phenolic acid constituent, gallic acid, are the causative agents. The tannin content of leaves and acorns tends to be highest in the immature stages. An unusually heavy crop of acorns, such as occurred in Ohio (Sandusky *et al.*, 1977) and England (Dixon *et al.*, 1979) in 1976, often results in a large number of cases of oak poisoning. An outbreak of oak poisoning in cattle in northern California in 1985 was caused by an unusually late snowfall, which covered available forage and forced cattle to browse on buds and leaves of oaks (Spier *et al.*, 1987). Over 2700 cattle died in this incident. Surviving animals showed complete clinical recovery (Ostrowski *et al.*, 1989).

The initial **signs of oak poisoning** include anorexia, depression, clear watery nasal discharge, rumen stasis, excessive thirst, and frequent urination. Initial constipation is followed by the excretion of dark, thin, mucoid, and often bloody feces.

The principal lesions are gastritis and nephritis. The abomasum and small intestine are often inflamed and hemorrhagic. The major lesion of oak poisoning is necrosis of the renal tubules. Affected kidneys are pale and swollen. There is impaired kidney function, with elevated blood urea nitrogen. **Supplemental feeding** with a mixture containing calcium hydroxide has been recommended as a preventative measure. A mixture of 30% alfalfa, 54% cottonseed meal, 6% vegetable oil, and 10% calcium hydroxide has been recommended by Dollahite *et al.* (1966). The likely mechanism of action is binding of calcium hydroxide with tannins to form insoluble complexes that prevent their absorption (Murdiati *et al.*, 1990).

Gallotannins in oak are hydrolyzed in the rumen (Fig. 12–1) to gallic acid, pyrogallol, resorcinol and several other small molecular weight phenolic compounds (Shi, 1988; Tor *et al.*, 1996). These smaller molecules are absorbed, and react with tissue proteins, causing pathology. The kidney and liver lesions characteristic of oak poisoning are experimentally produced when tannic acid is administered directly into the abomasum, but not when given intraruminally (Zhu and Filippich, 1995; Zhu, 1995). This suggests that rumen metabolism detoxifies hydrolyzable tannins, and it is only those tannins that are released post-ruminally from plant material that produce pathology. Absorbed pyrogallol and resorcinol are excreted in the urine as glucuronides (Murdiati *et al.*, 1992).

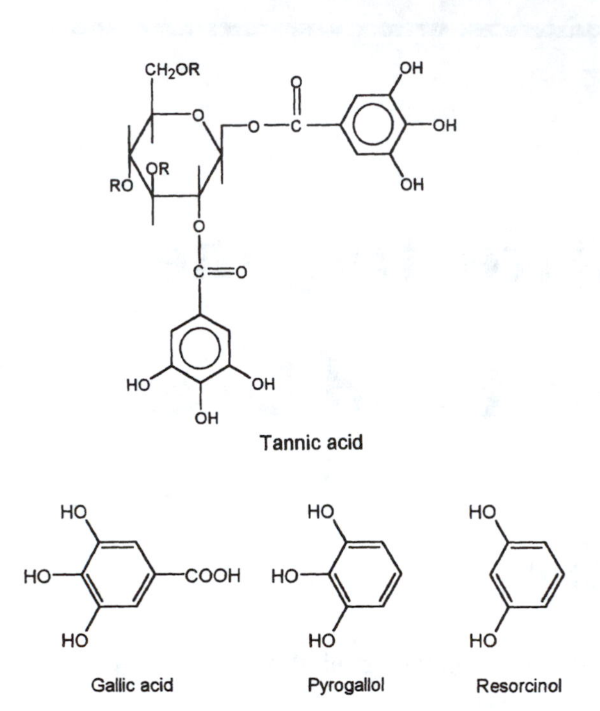

FIGURE 12–1 Tannic acid (R groups = gallic acid) is hydrolyzed in the rumen to monomeric phenolics such as pyrogallol and resorcinol, which are toxic when absorbed.

Goats can utilize oak browse productively. Davis *et al.* (1975) observed that the total grazing capacity of a gambel oak (*Quercus gambelii*) range was almost doubled by including goats in a mixed grazing scheme with cattle. Goats are also used for brush control. Feeding high levels of immature gambel oak to goats did not produce any toxicological reactions in a Utah study (Nastis and Malechek, 1981). These workers concluded that mature oak browse can contribute effectively to the nutrition of growing and lactating goats. Immature oak leaves, while not overtly toxic, had a low metabolizable energy content and were of low palatability. The resistance of goats to oak poisoning may be a result of secretion of tannin-binding salivary proteins, which is a defensive mechanism against tannins noted in many browsing species (see Chapter 6).

SOLANUM POISONING

Solanum spp. contain alkaloids and glycoalkaloids which are irritants and cholinesterase inhibitors. Alkaloids in potatoes (*S. tuberosum*) were described in Chapter 6. Other *Solanum* spp. include the nightshades, eggplant, the Jerusalem cherry (an ornamental plant) and a large number of wild spp. found throughout the world. These include *S. malacoxylon*, which contains **calcinogenic glycosides** (see Chapter 15). The flowers of *Solanum* spp. and related plants such as tomatoes are very similar in appearance (Fig. 12–2) and serve as good identifiers of *Solanum* spp.

A number of other *Solanum* spp. such as *S. fastigiatum*, *S. kwebense*, and *S. dimidiatum* cause **cerebellar degeneration** in cattle, characterized by progressive vacuolation and degeneration of Purkinje cells. The disorder is apparently a

FIGURE 12–2 Examples of *Solanum* spp. flowers. From left to right, tomato, potato, climbing nightshade and black nightshade.

lysosomal storage disease similar to that caused by *Swainsona* and *Astragalus* spp. The *Solanum*-induced disorder is discussed in Chapter 13.

Among the nightshades, **black nightshade** (*S. nigrum*) is probably of widest distribution. It is a common annual weed that grows from 1 to 3 feet in height, with white flowers and shiny black berries. It is a common garden weed and also invades forage crops such as new seedings of alfalfa, and cereals. Mortality has been reported for cattle, sheep, swine, horses, chickens, and ducks (Kingsbury, 1964) from consumption of the berries or from grazing the plant. Nightshade in stubble fields may be grazed in preference to the dry stubble. A cultivated vegetable plant, the garden huckleberry, is apparently a domesticated type of black nightshade. The berries are used for pies and are nontoxic. The wild and domestic plants do not interbreed.

Climbing nightshade (*Solanum dulcamara*) is a climbing or trailing perennial reaching 6 ft in length. The flowers are blue or purple, followed by large red berries. Climbing nightshade has been implicated in poisoning of livestock and children (Kingsbury, 1964). Silverleaf nightshade (*Solanum eleagnifolium*) is a common perennial weed in the U.S. southwestern states. Considerable losses of cattle from consumption of the plant have occurred (Kingsbury, 1964). Ripe berries produce moderate to severe poisoning when ingested at 0.1–0.3% of body weight.

The chemistry of **steroidal glycoalkaloids (SGA)** found in *Solanum* spp. has been reviewed by Van Gelder (1991) and Keeler *et al.* (1991). The SGA are divided into five groups, based upon the steroidal nucleus: solanidines, spirosolanes, epiminocholestanes, 3-aminospirostans and solanocapsine, occurring in plants as glycosides. Presumably mixtures of these SGA are found in the various *Solanum* spp. such as the nightshades. There is little data on toxicity of alkaloids in wild *Solanum* spp. Voss *et al.* (1993) fed dietary levels of up to 25% black nightshade berries to rats for 13 weeks, with no signs of toxicity except slight anemia. Cattle are more sensitive to SGA than sheep and goats (Buck *et al.*, 1960). The SGA are heat stable, as evidenced by the retention of toxicity in boiled green potatoes. With the lack of other information, Kingsbury's (1964) suggestion that intakes of green nightshade of 0.1–0.3% of body weight are potentially hazardous should be followed. Corn silage and other forages such as oat hay and first year alfalfa may be heavily contaminated, with nightshade comprising 50% or more of total dry matter. Such forage should be "test-fed" to a few animals to assure safety, or diluted with uncontaminated forage.

There is some evidence that SGA are teratogenic (see Chapter 14).

OXALATES

Various pasture and range plants are toxic because of their high oxalate content. In the U.S., the major oxalate-related livestock problem has been halogeton (*Halogeton glomeratus*) poisoning of range sheep. Large numbers of sheep have died, including numerous instances of hundreds and even a thousand or more sheep dying in a single outbreak. In Australia, soursob (*Oxalis pes-caprae*), an introduced plant from South Africa, causes extensive problems. In Australia and other parts of the tropics, certain tropical grasses such as setaria (*Setaria sphacelata*) and *Panicum* spp. (elephant grass, guinea grass) may contain toxic oxalate concentrations under some conditions and have been implicated in oxalate-induced problems in cattle and horses (see Chapter 11). Other oxalate-containing plants for which livestock problems have been suggested include redroot pigweed (*Amaranthus retroflexus,*) kochia (*Kochia scoparia*), and greasewood (*Sarcobatus vermiculatus*).

Oxalate is found in plants in two major forms. Some plants, such as soursob, have a cell sap pH of about 2, and the oxalate exists as salts of acid oxalate ($H_2CO_4^-$), such as acid potassium oxalate (Fig. 12–3). Other plants, such as halogeton, have a cell sap pH of about 6, and the oxalate exists as soluble sodium and insoluble calcium and magnesium oxalates. With the acid oxalate salts, both acute and chronic toxicities occur, while with halogeton, only acute toxicity is seen.

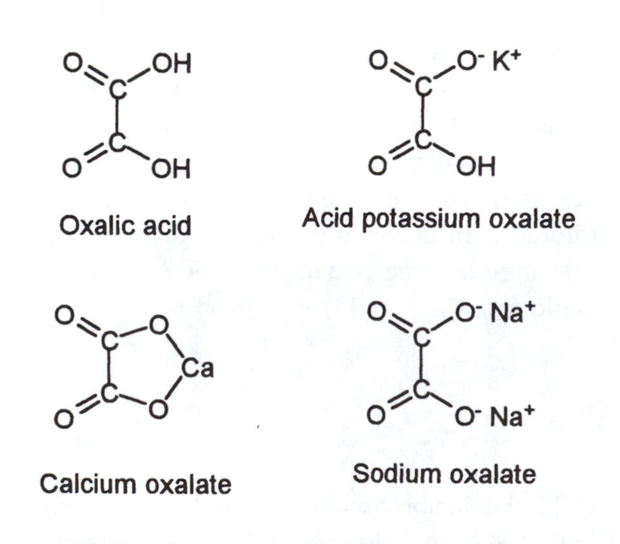

FIGURE 12–3 Examples of forms of oxalate in halogeton and other oxalate accumulators.

Halogeton is a branched annual herbaceous plant native to arid alkaline soils of Russia. It was accidentally introduced into the U.S. as a contaminant of agricultural products and was first collected and identified in 1934 in Nevada. Since that time, it has spread over more than 10 million acres of western rangelands, particularly in Nevada, Utah, and Idaho. Losses of sheep from halogeton consumption were suspected in the 1930s; in 1942, severe losses in Nevada prompted experimental studies which proved that halogeton is toxic. In the 1940s, the range sheep industry was still quite large in the western states, and therefore with a large number of animals exposed to the plant, some spectacular losses occurred. Numerous cases are documented where 500–1500 sheep died in a single day when a band was herded through halogeton-infested areas. Since that time, losses have been few, both because of increased awareness on the part of sheep producers of the toxicity of the plant and because the decline of the western range sheep industry has put fewer animals at risk.

Halogeton cannot compete with established perennial plants and therefore is found primarily in disturbed or barren soils and many winter ranges. It grows along roadsides, railroad tracks, and other disturbed areas. Livestock losses, including some cattle poisonings (Lincoln and Black, 1980), occur most frequently when hungry stock are unloaded in halogeton-infested areas or trailed along roads with stands of halogeton on the roadside (Fig. 12–4). The oxalate concentration is highest in fall and winter, and the plant is most likely to be consumed at this time after fall rains have softened the dry plant. Generally plants must contain at least 10% oxalate on a dry weight basis for toxicity to occur.

Signs of halogeton poisoning include labored breathing, depression, weakness, coma, and death. Some animals may have convulsions, and tetany may be seen. Animals with obvious symptoms may have serum calcium reduced by 20% or more, while severe hypocalcemia, with serum calcium only 20% of normal, occurs in animals that die (Fig. 12–5). Gross pathology

FIGURE 12–4 Halogeton growing along a road on a western range. Sheep are frequently poisoned when being moved along roads in such areas. (Courtesy of L.F. James)

FIGURE 12–5 Sheep losses from halogeton poisoning. Typically the animals die in a sleeping position, characteristic of effects of hypocalcemia. (Courtesy of L.F. James)

includes hemorrhages and edema of the rumen wall, swollen kidneys, and calcium oxalate crystals found in the rumen wall, kidney tubules, and other tissues. Rumen stasis also occurs.

The mode of action of oxalate in causing toxicity is not totally clear. Acute toxicity may be due to hypocalcemia, while uremia from kidney damage may contribute to chronic toxicity. Complicating the situation are the observations (James, 1978) that sheep made hypocalcemic by dialysis or EDTA infusion do not die, while blood calcium levels can be experimentally maintained in sheep poisoned by halogeton, and they still die. Hemorrhagic rumenitis and shock may be contributory factors. Rumen stasis may cause an increased pH of the rumen. Oxalate also inhibits several respiratory enzymes, including succinic dehydrogenase. Enzymes activated by calcium or magnesium may be inhibited by oxalate.

Dietary oxalate can be degraded by rumen microorganisms (Allison *et al.*, 1981). Ruminants adapted to diets with high oxalate content can tolerate oxalate levels that are lethal to nonadapted animals. Allison *et al.* (1981) demonstrated the presence of a rumen microorganism that degrades oxalate to CO_2 and formate and depends on oxalate for its growth. Since most vegetation contains a small amount of oxalate, residual populations of an oxalate-dependent organism could survive in the rumen. Alfalfa, for instance, has an appreciable portion of its calcium in the form of calcium oxalate (Ward *et al.*, 1979). In subsequent work, Allison *et al.* (1985) isolated an oxalate-degrading anaerobe from the rumen which they named *Oxalobacter formigenes*.

Prevention of halogeton poisoning depends primarily on livestock management. Hungry animals should not be exposed to abundant stands of halogeton, particularly if they have not had prior exposure to it. At least 4 days of oxalate exposure are needed for development of oxalate-degrading capacity of the rumen. Such preconditioning results in about a 30% increase in the level of oxalate required to kill sheep. The provision of calcium supplements has had some beneficial effect.

In the U.S., other plants besides halogeton have been implicated in livestock losses. **Greasewood** (*Sarcobatus vermiculatus*) is an erect spiny shrub that grows in alkaline soils on western ranges. The oxalate content of the leaves varies between 10 and 20% of the dry weight, reaching a maximum in late summer. Heavy losses of sheep have occurred from oxalate poisoning due to consumption of greasewood. It is regarded as a useful forage, and toxicity can be avoided with good animal management.

Redroot pigweed (*Amaranthus retroflexus*) causes perirenal edema and nephrosis in cattle and swine. Oxalates (Marshall *et al.*, 1967) have been suggested as one of the causative agents. Domesticated varieties of amaranthus are being developed as grain sources; they have growth-depressing properties which could be influenced by the fairly high oxalate content. Various *Chenopodiaceae* such as **lamb's-quarters** (*Chenopodium album*) accumulate oxalates. *Rumex* spp., the **sorrels and docks**, contain oxalates and sometimes cause oxalate toxicity (Panciera *et al.*, 1990). **Soursob** (*O. pes-caprae*) causes significant sheep losses in Australia. It contains acid potassium oxalate and causes chronic poisoning. Kidney damage due to formation of oxalate crystals is the major problem.

Oxalate is the toxic component of **rhubarb** (*Rheum rhaponeticum*) leaves. Rhubarb (pieplant) stems are a common item of the human diet, and it is frequently grown in home gardens. The leaves are poisonous to humans and livestock and have resulted in severe poisoning and death (Kingsbury, 1964).

Ward *et al.* (1979) demonstrated that about 20–30% of the calcium in **alfalfa** is in the form of oxalate and is unavailable to ruminants. While oxalate toxicity from alfalfa is not likely, the low availability of about one third of its calcium content should be considered in diet formulation.

The popular houseplant *Dieffenbachia sequine* (Fig. 12–6), or dumb cane, contains crystals of calcium oxalate that cause severe irritation of the mouth and throat if the plant is consumed. **Dieffenbachia** is a native of the West Indies, and according to Woodhouse (1983), it was used in torturing slaves. Woodhouse (1983) gives an interesting account of the occurrence of calcium oxalate in this plant. Found throughout the plant tissues are cells called **idioblasts**, shaped like double-ended microscopic lemons, which contain raphides or needle-shaped crystals of calcium oxalate. The crystals are slender, sharp, and packed together embedded in a gelatinous substance. If the tip of the idioblast is broken, juice from the plant or saliva enters and causes the gelatinous material to swell, increasing the internal pressure and expelling the needles. As Woodhouse (1983) describes,

FIGURE 12–6 Dumb cane (*Dieffenbachia sequine*) causes intense irritation of the buccal cavity due to calcium oxalate crystals.

> *They emerge like bullets one at a time, with sufficient force to cause the cell to recoil like a gun. This can be watched under a microscope, and goes on for many minutes. The result, then, of eating a piece of dieffenbachia leaf, is not a taste sensation, but simple pain.*

In addition, dieffenbachia contains a toxic protein which causes swelling of the mucous membranes of the mouth and throat, contributing to the severe irritation and pain.

SESQUITERPENE LACTONES

Various Compositae contain toxicants called **sesquiterpene lactones**. These include a variety of plants common on U.S. rangelands, such as bitterweed or bitter rubberweed (*Hymenoxys odorata*), pingue or Colorado rubberweed (*Hymenoxys richardsonii*) (Fig. 12–7), and the sneezeweeds (*Helenium* spp.), including orange sneezeweed (*H. hoopesii*), southeastern bitterweed (*H. amarum*), and bitterweed (*H. autumnale*). The sesquiterpene lactones are highly irritating to the nose, eyes, and gastrointestinal tract. Sheep and goats are the main livestock species affected, primarily because the plants are unpalatable and rarely consumed in toxic quantities by cattle and horses. Sneezeweed poisoning is often referred to as "spewing sickness" because of the characteristic vomiting seen. Affected sheep may have a green stain around the mouth and stand with an upturned head attempting to retain regurgitated material. Vomited material is often inhaled into the lungs, causing either death from inhalation pneumonia or permanent lung damage accompanied by chronic coughing. Primary lesions are gastrointestinal tract irritation, congestion of the liver and kidney, and pulmonary damage.

FIGURE 12–7 Pingue or Colorado rubberweed (*Hymenoxys richardsonii*).

A variety of sesquiterpene lactones have been identified in these species. A typical example is **hymenoxon**, found in *Hymenoxys odorata*. **Helenalin** is the major toxin in *Helenium autumnale* (Fig. 12–8).

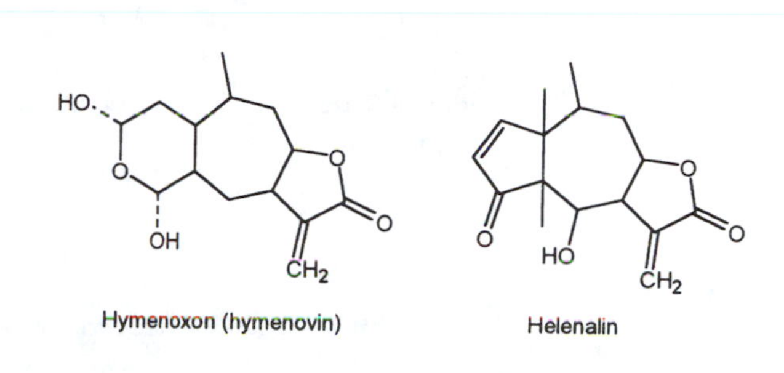

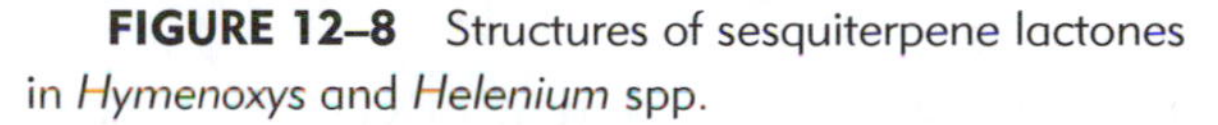

FIGURE 12–8 Structures of sesquiterpene lactones in *Hymenoxys* and *Helenium* spp.

Note that these compounds contain a seven-membered ring, a lactone (cyclic ester) structure, and an exocylic methylene group.

The loss of sheep and goats in Texas from *H. odorata* (bitterweed) has been estimated at several million dollars annually (Conner *et al.*, 1988). It is an annual weed that grows on disturbed soils and overgrazed pastures in the semiarid regions of the U.S. Southwest. Poisoning generally occurs from late December to early May when bitterweed may be the only green plant available, particularly during droughts. The oral LD_{50} of dried bitterweed in sheep is 2.9–8.5 g/kg body weight (Ueckert and Calhoun, 1988).

The toxicity of sesquiterpene lactones is due to binding of the exocyclic methylene group with tissue constituents, such as sulfhydryl groups and other nucleophilic components (Fig. 12–9). Hymenoxon and its metabolites appear to be excreted as glucuronides, (Hill *et al.*, 1980) and mercaptic acids (Terry *et al.*, 1983).

Kim and colleagues at Texas A&M University have shown protective effects of certain dietary additives against hymenoxon and bitterweed toxicity. Kim *et al.* (1974) reported that

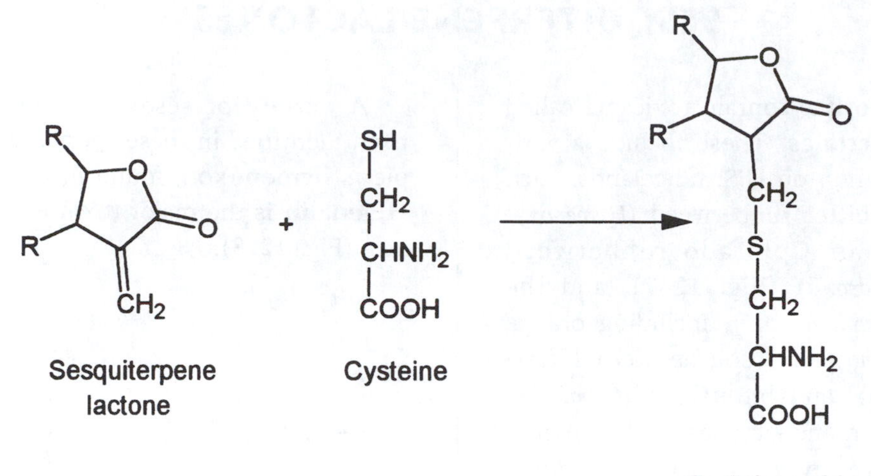

FIGURE 12–9 Reaction of sesquiterpene lactones with thiols.

simultaneous administration of cysteine with a lactone preparation from *H. odorata* gave up to 80% protection against an LD_{90} dose of the poisonous lactone in dogs. In later studies, Rowe *et al.* (1980) obtained similar results with sheep, with cysteine giving an increased survival rate of sheep injected with hymenoxon. Jones and Kim (1981) pretreated mice with microsomal enzyme inducers (phenobarbital, polychlorinated biphenyl) or an enzyme inhibitor (chloramphenicol) and found no effect on the toxicity of injected hymenoxon. This indicates that mixed function oxidases play little or no role in the detoxification of hymenoxon in mice. Prior administration of carbon tetrachloride did provide protection against hymenoxon; the LD_{50} was 241 mg/kg in controls and 630 mg/kg in carbon tetrachloride–treated mice. This suggests that metabolism of hymenoxon to a metabolite occurs, but not by the mixed function oxidase system. Carbon tetrachloride pretreatment would reduce the number of viable hepatocytes that could function in activation of the hymenoxon. Kim *et al.* (1981) also investigated the effects of dietary butylated hydroxyanisole (BHA), ethoxyquin, and methionine on the acute toxicity of hymenoxon in mice. The synthetic antioxidants BHA and ethoxyquin provided some protection against the toxicity of hymenoxon. Methionine also had protective activity and appeared to have synergistic effects with BHA. The role of methionine is probably to provide a precursor of cysteine or reduced glutathione. BHA increases the level of reduced glutathione and the activity of glutathione transferase. In a preliminary study with sheep (Kim *et al.*, 1981), the apparent LD_{50} of bitterweed was 4 and 8 g/kg for control and BHA-fed sheep, respectively, indicating some protective activity. Supplementary protein as soybean meal had a protective effect against bitterweed poisoning in sheep (Post and Bailey, 1992), whereas dietary urea increased bitterweed toxicity.

Losses of sheep to bitterweed can be reduced by grazing management techniques (Taylor and Ralphs, 1992), including using mixed grazing systems (sheep, cattle and goats) and rotational rather than continuous grazing. Short-term (about 7 days) grazing cycles alternating between bitterweed-infested and bitterweed-free pastures will prevent toxicosis in sheep.

SELENOAMINO ACIDS

Soils in many parts of the world contain high levels of selenium. Plants growing on these soils may accumulate toxic levels of the element, which occurs in the plant as selenoamino acids. In North America, high soil selenium levels are found in areas of the Great Plains (Fig. 12–10), including parts of Montana, North and South Dakota, Nebraska, Kansas, Oklahoma, Texas, Wyoming, Colorado, New Mexico, Idaho, Utah, Arizona, Nevada, and the provinces of western Canada. Because one of the principal effects of seleniferous compounds is irritation of mucous membranes, selenium-containing amino acids are discussed in this chapter, although they have post-absorptive effects as well, including hepatotoxicity.

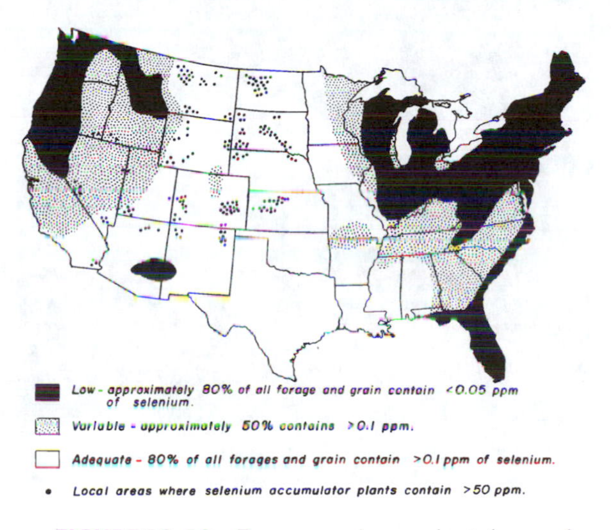

FIGURE 12–10 Forages and crop plants have selenium contents reflective of the soil selenium content. Areas of the U.S. where plants with toxic selenium concentrations are found are mainly in the Great Plains states.

Selenium poisoning, or selenosis, may be of several types, including acute toxicity and chronic forms known as blind staggers and alkali disease. **Acute toxicity** results from the consumption of plants containing very high selenium levels (several hundred ppm). These types of plants are called **selenium accumulators** or **selenium indicators**, as their growth is restricted to areas of seleniferous soils. They may contain extremely high selenium levels, of as much as 10,000 ppm on a dry weight basis. Examples of indicator plants include about 25 species of *Astragalus* (e.g., *A. bisulcatus*) and *Xylorrhiza* spp., the woody asters. These plants are generally unpalatable because of the offensive odors of the selenium-containing compounds, but are eaten when forage is sparse. Signs of acute toxicity include staggering, diarrhea, prostration, hemorrhage of internal organs, and abdominal pain.

Chronic toxicity has been suggested to be of two types: alkali disease and blind staggers, although it is now recognized that blind staggers is probably not a selenosis condition (O'Toole *et al.*, 1996). The **blind staggers syndrome** has been associated with the consumption of indicator or secondary selenium absorber plants of moderate (less than 200 ppm) selenium content. Secondary absorber plants are those which accumulate moderate (several hundred ppm) levels of selenium when growing on seleniferous soils, and their growth is not restricted to such soils. The blind staggers syndrome involves loss of appetite, impairment of vision, and wandering in circles, followed by weakness, recumbency, paralysis of the tongue and swallowing mechanisms, labored respiration, abdominal pain, grating of the teeth, respiratory failure, and death. Consumption of *A. bisulcatus* has been associated with the condition. These symptoms resemble those of locoism and cannot be duplicated by feeding high levels of inorganic selenium. Van Kampen and James (1978) fed *A. bisulcatus* with a high (180 ppm) selenium content to ewes and produced signs of locoism rather than of blind staggers. They suggested that the condition previously described as blind staggers due to chronic selenium intoxication may in fact be locoweed poisoning. In more recent work, Baker *et al.* (1989) found little difference in toxicity of the same dietary level of selenium as sodium selenite or selenium in *Astragalus* spp. Neurological effects were attributed to swainsonine (see Chapter 13) in the plants, rather than to selenium. Panter *et al.* (1996) found that selenium in *A. bisulcatus* caused more severe neurological signs in pigs than the same level of selenium in sodium selenate or seleno-DL-methionine, suggesting a contribution of swainsonine to the pathology observed.

O'Toole *et al.* (1996) made an exhaustive examination of research archives of the principal investigator who proposed the widely-accepted selenosis basis of blind staggers, and concluded that locoism and polioencephalomalacia (from consumption of high-sulfate alkali water) were the actual causes of the so-called blind staggers. The apparently erroneous implication of selenium in the blind staggers condition has resulted in widespread but unjustified public concern (e.g. Harris, 1991) about use of selenium as a feed additive and as an environmental toxicant (O'Toole *et al.*, 1996).

Alkali disease is caused by the chronic consumption of plants containing 5–50 ppm of selenium. To put these and previous levels mentioned in perspective, the selenium requirement of animals is about 0.1–0.5 ppm, while obvious deficiency symptoms (white muscle disease) occur when selenium levels are less than 0.05 ppm. Principal signs of chronic selenium toxicity include loss of the long hair from the mane and tail of horses (bobtail horses) (Fig. 12–11) and from the tail switch of cattle, loss of body hair from swine, sore feet with inflammation at the coronary band in horses, cattle, and swine, followed by hoof deformities (Fig. 12–11) and loss of condition, progressing to mortality. Decreased hatchability of chicken eggs due to embryonic malformation occurs. Impaired reproduction is characteristic of selenium toxicity. Sheep are more resistant to the reproductive effects than cattle (Panter *et al.*, 1995), suggesting that sheep may be more suitable than cattle on rangelands with seleniferous soils. Anemia, liver atrophy and cirrhosis, and hemosiderin deposits in tissues also occur. Alkali disease is associated with the consumption of normal crop and pasture species that have been grown in soils of moderate (2 ppm) selenium content.

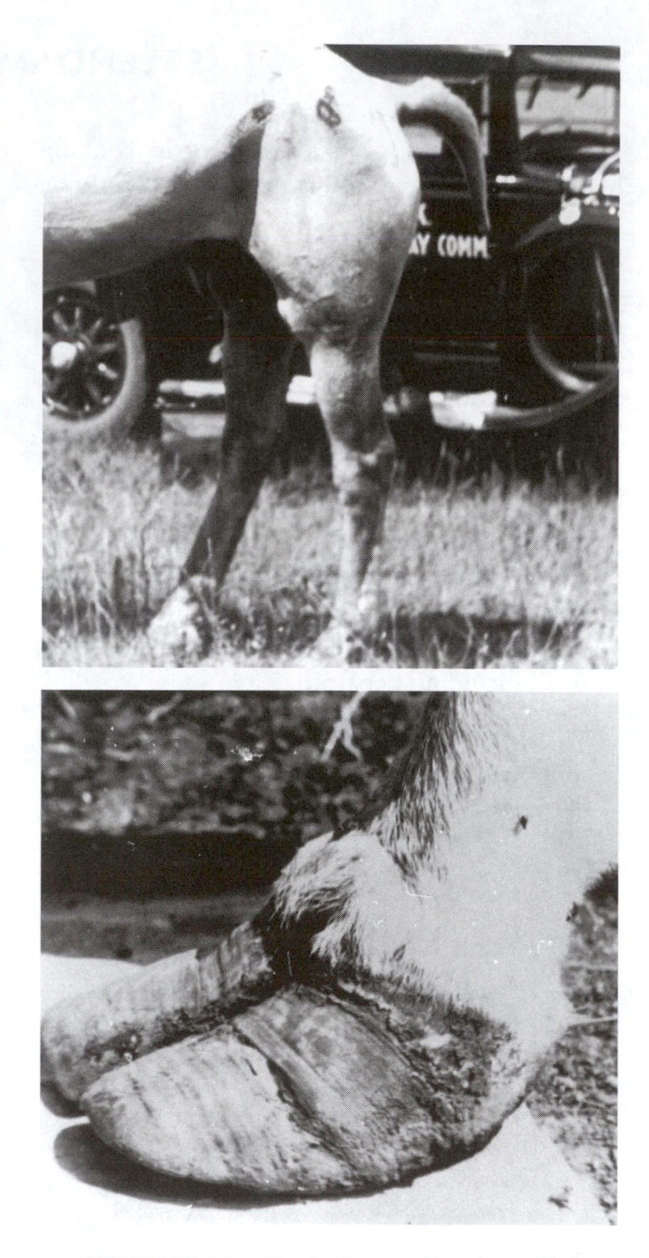

FIGURE 12–11 (Top): Chronic selenium toxicosis in a horse, showing emaciation and loss of long hair from the tail. (Bottom): Chronic selenium toxicosis in a bovine, showing elongated and cracked hooves. (Courtesy of O.E. Olson)

As many as 17 different selenocompounds have been identified in plants (Whanger, 1989). In the accumulator or primary indicator plants,

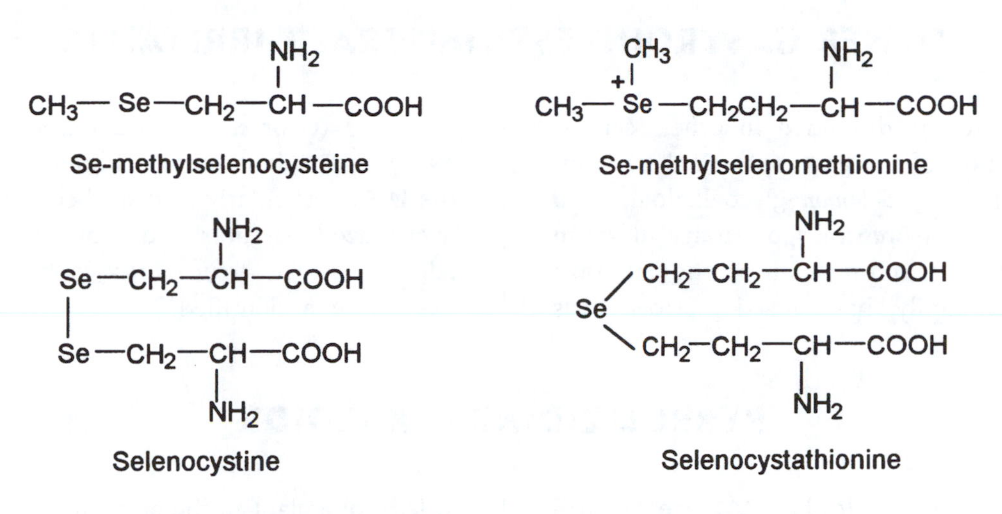

FIGURE 12–12 Structures of some selenoamino acids in plants.

the main selenocompounds are non-protein selenoamino acids such as Se-methylselenocysteine, selenocystathionine and γ-glutamyl-Se-methyl-selenocysteine (Fig. 12–12). The tolerance of these indicator plants for selenium is probably because the selenoamino acids are not incorporated into proteins, but instead are sequestered as non-physiological amino acids. In non-accumulator plants, selenomethionine and selenate are the major forms of the element.

The biochemical mode of action of selenium toxicity is not well understood. One of the main routes of excretion of excess selenium is exhalation via the lungs. Dimethyl- and trimethylselenide are synthesized and excreted in the urine and expired air. The breath of animals affected by selenium toxicosis has a strong garlic-like odor, attributed to the methylated selenium compounds.

An outbreak of selenium toxicity in migratory waterfowl occurred at the Kesterson Reservoir in California in 1984 (Fan *et al.*, 1988; Ohlendorf *et al.*, 1988). Severe embryonic defects occurred with very low reproductive success in nesting birds. This incident led to widespread criticism of the use of selenium as a feed additive (Harris, 1991), culminating in a reduction in the FDA-permitted level of added selenium in animal feeds from 0.3 ppm to 0.1 ppm (the approved level was subsequently restored to 0.3 ppm). However, the source of high levels of selenium in the reservoir water was from outcroppings of Cretaceous marine shale formations containing selenium salts which were leached into irrigation water, and subsequently concentrated by evaporation in the Kesterson Reservoir. Norman *et al.* (1992) found no evidence that supplementation of animals with selenium on ranches in northern California had any measurable effect on selenium concentrations in aquatic ecosystems.

The primary role of selenium in animal metabolism is as a component of glutathione peroxidase, which functions in detoxification of tissue peroxides (see Favism, Chapter 7). Selenium has also been identified (Behne *et al.*, 1990) as an essential component (as selenocysteine) of Type I iodothyronine 5-deiodinase, the enzyme that converts thyroxin (T4) to the metabolically active thyroid hormone, triiodothyronine (T3). Thus there are at least two selenium-dependent enzymes in animal metabolism.

OTHER GASTROINTESTINAL TRACT IRRITANTS

Various toxins discussed in other sections cause irritation of the gastrointestinal tract, including saponins, *Solanum* glycoalkaloids, and glucosinolates in *Brassica* spp. **Hoary alyssum** (*Berteroa incana*), a member of the mustard (Cruciferae) family, has caused gastroenteritis in horses (Geor *et al.*, 1992; Hovda and Rose, 1993). This plant occurs throughout much of the U.S., particularly in disturbed areas such as over-grazed pastures and drought-stressed alfalfa fields. The toxic entity in hoary alyssum has not been identified.

PYRROLIZIDINE ALKALOIDS

Pyrrolizidine alkaloids (PA) are hepatotoxins, causing irreversible liver damage. Several of the PA are carcinogens. A comprehensive review of the structure, analysis, physical and chemical properties, metabolism, toxicity and carcinogenicity of PA is that of Mattocks (1986). An earlier comprehensive review was compiled by Bull *et al.* (1968). The PA-containing plants are widespread throughout the world and have caused extensive poisonings of livestock, poultry and humans. Most PA-containing plants occur in three plant families, the Boraginaceae, Compositae and Leguminosae. Examples in the *Boraginacea* family include *Amsinckia intermedia* (tarweed), *Borago officinalis* (borage), *Cynoglossum officinale* (hound's tongue), *Echium plantagineum* (echium), *Heliotropium europaeum* (heliotrope) and *Symphytum officinale* (comfrey). Some of these, such as borage and comfrey, are commonly used in herbal preparations. In the **Compositae** family, the main PA-containing plants are *Senecio* spp. There are over 1200 *Senecio* spp. world-wide, with many of them containing toxic PA. Examples include tansy ragwort (*S. jacobaea*), common groundsel (*S. vulgaris*) and a variety of toxic rangeland spp. such as *S. longilobus*. There are over 250 *Senecio* spp. in South Africa (Kellerman *et al.*, 1988), and at least 128 spp. in Brazil (Habermehl *et al.*, 1988), of which *S. brasiliensis* is most often implicated in PA poisoning (Lombardo de Barros *et al.*, 1992). A number of toxic *Crotalaria* spp. occur in the family **Leguminosae**. In addition, at least one plant in the grass family, tall fescue, contains PA. However, tall fescue PA lack the 1,2 double bond in the pyrrolizidine nucleus that is necessary for hepatotoxicity. The significance of PA in fescue toxicosis is probably slight (see Chapter 10).

Some of the major toxicological problems associated with PA-containing plants will be discussed first, followed by consideration of PA metabolism.

Senecio Poisoning

The major livestock losses due to *Senecio* poisoning have been with *Senecio jacobaea*, commonly called **tansy ragwort** (Fig. 12–13). This

FIGURE 12–13 Flowers of tansy ragwort (*Senecio jacobaea*). These bright yellow daisy-like flowers are characteristic of *Senecio* spp.

plant is a native of the British Isles. It flourishes in humid temperate regions and has been inadvertently introduced to Western Europe, South Africa, Australia, New Zealand, the Pacific Northwest of the U.S., and eastern Canada (Ontario, Quebec, and Nova Scotia). Other *Senecio* species that have caused significant livestock losses are *S. vulgaris*, *S. longilobus*, *S. riddelli*, and *S. latifolius*.

Demonstration of the role of *Senecio* spp. and PAs in livestock poisoning occurred in the early 1900s, but livestock producers had suspected the plants of being poisonous much earlier. A problem in Wales called stomach staggers in the late 1700s was thought by farmers to be due to consumption of *S. jacobaea* (Bull *et al.*, 1968). Two major outbreaks of livestock poisonings occurred in the period from 1860 to 1900 in Nova Scotia and New Zealand. They were called Pictou disease and Winton disease, respectively. Cattle and horses developed liver cirrhosis, jaundice, and, in the case of horses, staggering, pressing the head against objects, and walking in a straight line regardless of obstructions. Livestock owners in Nova Scotia suspected ragwort consumption as the cause, but agricultural officials believed it to be a bacterial infection. For nearly 20 years, the Canadian government ordered slaughter of affected herds and disinfection of the premises. Finally, in 1906, feeding trials definitely established that Pictou disease was caused by *S. jacobaea*. This was largely due to the reports from New Zealand on the etiology of Winton disease. There, based on association of losses with the presence of ragwort in pastures and feeding trials with the plant, it was demonstrated that *S. jacobaea* was the causative agent. Similar conditions were noted in other parts of the world. In South Africa, Molteno horse disease was traced to the consumption of *S. latifolius* and other *Senecio* species. In the U.S., walking disease of horses in Nebraska, Colorado, and Wyoming was a considerable economic problem in the early 1900s; in 1929–1933, feeding trials established *S. riddelli* as the cause. Other *Senecio* poisonings have occurred, including losses of cattle to *S. longilobus* in Arizona and *S. vulgaris* in California and Oregon. In the 1960s, *S. jacobaea* became a significant problem in the Pacific Northwest. Several million dollars of livestock losses have occurred since that time. As in the case of the weed in Nova Scotia, the introduction of ragwort to Oregon was traced to contaminated straw ballast used in shipping.

Tansy ragwort is normally a biennial. In the Pacific Northwest, it germinates following the start of the autumn rains, and the seedlings develop into rosettes the following year. The next year, a flowering stalk appears in late spring, with characteristic compositae flowers (Fig. 12–13), and the plant sets seeds and dies. One plant can produce many thousands of seeds which are readily dispersed by wind. If the flowering stalk is cut or mown, the plant may become a perennial, living until it has had the opportunity to produce seeds. If ragwort is cut in the early flowering stage, the cut plant will still set viable seeds, so control mowing should be done before flowering. Tansy ragwort is unpalatable and is generally not voluntarily consumed by cattle and horses even if other pasture is scarce. Poisonings generally occur as a result of contamination of hay and silage, or in the spring when the leaves of the plant in a dense pasture are not readily differentiated by animals from grass and clovers, thus causing inadvertent consumption. All parts of the plant are toxic, with the flowers having about four times the PA concentration as the leaves. The stems are quite low in toxicity.

Senecio poisoning has been a significant human health problem in several parts of the world. In the 1920s "bread poisoning" of low-income Europeans in South Africa was caused by contamination of wheat with the flowers and seeds of *Senecio* spp, with many fatalities. In Africa and the West Indies, "bush tea" is prepared to combat various illnesses. Leaves from a variety of plants collected in the bush are steeped in water and the resulting tea used for medicinal purposes. *Crotalaria* and *Senecio* spp. are commonly used as leaf sources, with the result that the tea often contains PAs. Consumption of bush tea has been linked to veno-occlusive disease in which occlusion of the hepatic veins occurs. Huxtable (1989) reported an outbreak of PA poisoning in Arizona in which children died as a result of consuming herbal tea made from *S. longilobus*. Culvenor (1983) made a comprehensive evaluation of outbreaks of PA poisoning in humans and compared the estimated PA intakes with dose rates causing tumors in rats. In several instances, the PA intake in chronic poisonings in humans was comparable to a carcinogenic dose in rats. He suggested that long-term observation of survivors of PA poisonings be made to evaluate the carcinogenic potential in humans.

Tansy ragwort is commonly referred to as tansy in many areas. This is unfortunate because it is sometimes confused with *Tanacetum vulgare*, which is also commonly called tansy. Tanacetum is an herb that has been used in herbal medicines. An example of the possible consequences of confusing these two tansys is a letter to a farm paper (Oregon Farmer-Stockman, May 17, 1973) in which the following suggestion was made for helping to control tansy ragwort:

> *Tansy is an herb that has been used for years. Dried you can mix it with your tea and drink it hot or cold. A sprinkle of the dried tansy can be added to bread, rolls or even cookies. Tansy control is possible if all of us recognize this golden peril and attack it from every angle instead of waiting for someone to legislate it out of existence.*

It is hoped that this well-intentioned advice to consume tansy ragwort was not followed by anyone! If the full name is not used, it is preferable to refer to the plant as ragwort rather than as tansy to avoid this confusion. Another observation of this type was made by Burns (1972), who determined the effect of tansy ragwort consumption on the heart and pulmonary arteries of rats. He stated,

> *A disturbing feature that came to light during the course of the present study was the fact that I was able to purchase a large quantity of dried chopped Senecio jacobaea from a "Health Stores" in the Liverpool area. The material purchased was meant to be taken in the form of an infusion to cure a variety of ailments.*

Crotalaria Poisoning

The loss of large numbers of horses grazing on Missouri River bottomland in the late 1800s (Missouri bottom disease) was due to consumption of *Crotalaria sagittalis*. In the Pacific Northwest, walking disease occurred in horses grazing wheat stubble infested with *C. sagittalis*. A similar problem in South Africa was caused by consumption of *Crotalaria dura*. In the U.S., *Crotalaria spectabilis* was widely grown in the South as a green manure crop. Contamination of grain with seeds of *C. spectabilis* has resulted in numerous poisonings of swine and poultry. The problem is still a significant one in the southern states. In Western Australia, Kimberley horse disease, or "walkabout," was responsible for the death of large numbers of horses. *Crotalaria retusa* was identified as the causative plant. In the early 1970s, hundreds of people in India and Afghanistan died in an outbreak of veno-occlusive disease caused by contamination of grain with *Crotalaria* seeds (Mattocks, 1986). *Crotalaria* contamination of grains continues to be a significant animal and human health problem, particularly in developing countries. Routine spraying of grain fields for broad-leaved weeds in North America and Europe has virtually eliminated problems of grain contamination with toxic seeds in these areas.

Heliotropium and *Echium* Poisoning

There has been extensive poisoning of sheep and cattle in Australia due to consumption of *Heliotropium europaeum* (Fig. 12–14). The plant continues to be a major problem. Sheep are highly resistant to the hepatoxic effects of PAs, but exposure to PAs causes the liver to accumu-

FIGURE 12–14 The leaves and flowers of *Heliotropium europaeum*, a pyrrolizidine alkaloid–containing plant which causes extensive losses of sheep. (Courtesy of C.C.J. Culvenor)

late excessive copper levels. Death is frequently due to the hemolytic crisis of copper toxicity. Usually more than one grazing season on heliotrope is necessary for sheep losses to occur. Sheep are much more sensitive than other livestock to copper toxicity. *Echium plantagineum* (Fig. 12–15) grows extensively in Victoria, southern New South Wales, and South Australia. It is called Paterson's curse or Salvation Jane, depending on one's point of view. While it provides forage in the spring when other feed may be lacking, it contains PAs. Feeding trials

FIGURE 12–15 *Echium plantagineum*, a pyrrolizidine alkaloid–containing plant that is widespread in Australia. It is commonly referred to as Paterson's curse or Salvation Jane, depending on whether or not other forage is available.

with sheep (Culvenor *et al.*, 1984) have shown that they experience little detrimental effect from eating large quantities of dried echium. However, since echium and heliotrope frequently grow in the same areas, they may both contribute to poisoning problems. Prolonged grazing of echium may result in **chronic copper poisoning** of sheep (Seaman, 1985; Seaman and Dixon, 1989) with signs of severe jaundice and hemoglobinuria. The fat of affected animals is bright yellow, and the liver is enlarged, fibrotic, and bright yellow. The accumulation of copper is influenced by the dietary copper:molybdenum ratio. A high Cu:Mo ratio in excess of 100 promotes copper accumulation (Peterson *et al.*, 1992; Seaman and Dixon, 1989).

Other Plants with Pyrrolizidine Alkaloids

Amsinckia intermedia (tarweed, fiddleneck) is an annual weed growing in waste areas and grain fields of California and the Pacific Northwest. It contains a number of PAs, including intermedine, lycopsamine, echiumine, and sincamidine (Bull *et al.*, 1968). At one time, up until the 1940s, *Amsinckia* toxicity of livestock was quite common in the Pacific Northwest, affecting horses, swine, and cattle. The conditions were called "walking disease" in horses and "hard liver disease" in cattle and swine. Poisoning was due mainly to the presence of the weed seeds in grain and grain screenings fed to livestock. Following the widespread use of herbicides to control broad-leaved weeds in grain, the problems have disappeared.

Comfrey (*Symphytum officinale*) is a vigorous, deep-rooted perennial herb with large, coarse prickly leaves (Fig. 12–16). It has been promoted at various times as a forage crop for livestock and is also widely used as an herb with reputed medicinal properties. Comfrey contains at least eight PAs (Culvenor *et al.*, 1980), including echimidine, symphytine, lycopsamine, and intermedine. Administration of comfrey PAs to rats resulted in mortality and liver pathology characteristic of PA toxicosis. Hirono *et*

FIGURE 12–16 Comfrey is a succulent plant promoted as a forage crop and for medicinal properties. It contains toxic and carcinogenic pyrrolizidine alkaloids.

al. (1976) fed comfrey roots and leaves to rats and found that liver cancer was produced. Dietary levels of 0.5% root and 8% leaves produced hepatomas. The public health significance of comfrey PAs has not been fully established. It is a commonly used herb and is readily available in "health food" stores, although its sale has been banned in some countries (e.g. Canada). The comments of Culvenor *et al.* (1980) are pertinent:

> *The pyrrolizidine alkaloid content of Russian comfrey provides grounds for concern at the human consumption of this plant, especially by children because of the greater sensitivity to young animals of the effects of this type of alkaloid. The lack of reports of toxicity of this plant despite claims of dietary use over many years is not necessarily an indicator of safety. The effects of such alkaloids are cumulative and overt damage may be long delayed, thus preventing association with the plant cause.*

Even more succinctly, Tyler (1993) states "It is most unfortunate that this poisonous herb continues to be sold to uninformed consumers by organizations that are ostensibly interested in promoting the good health of their customers."

Another herb containing PA is borage (*Borago officinalis*); the PA is in low concentration and probably is not a significant health hazard (Larson *et al.*, 1984).

Cynoglossum officinale (**hound's tongue**) is an annual or biennial herbaceous plant (Fig. 12–17) containing PA, including heliosupine and echinatine (Pfister *et al.*, 1992). A native of Europe, it has become naturalized in various parts of the U.S. and Canada. The green plant has a distinctive unpleasant odor which discourages consumption, but is more palatable in hay, and poisoning may occur when contaminated hay is fed to livestock. Knight *et al.* (1984) reported classic signs of PA toxicity in horses fed contaminated hay in Colorado. The level of PA in the *C. officinale* was about 0.6–2.1% of the dry matter. Stegelmeier *et al.* (1996) also observed that horses are very susceptible to houndstongue toxicity. According to them, consumption of one plant a day for two weeks would be a toxic dose. Cattle are also poisoned by *C. officinale* (Baker *et al.*, 1991).

FIGURE 12–17 *Cynoglossum officinale* (hound's tongue) is naturalized in western North America.

Structure of Pyrrolizidine Alkaloids

Several hundred PAs have been identified and their structures determined (Mattocks, 1986). Of the hepatoxic PAs, most are esters of the bases retronecine and heliotridine (Fig. 12–18). These bases are called amino alcohols. Retronecine and heliotridine are diastereomers, with opposite configurations at C7. The PAs in toxic plants are of three general types: monoesters, noncyclic (open) diesters, and cyclic diesters. Examples of each are heliotrine, lasiocarpine, and monocrotaline (Fig. 12–19).

FIGURE 12–18 Pyrrolizidine bases.

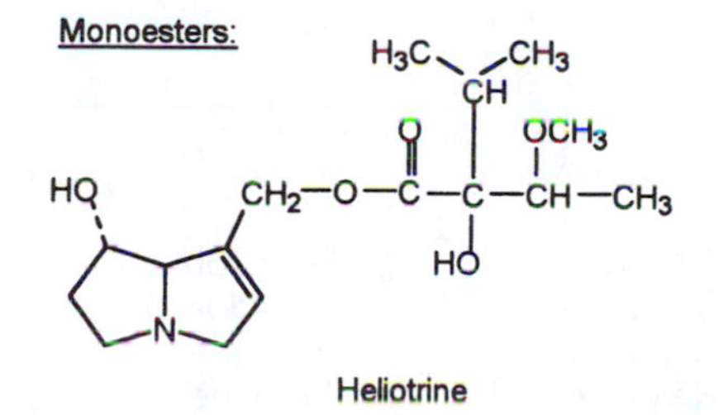

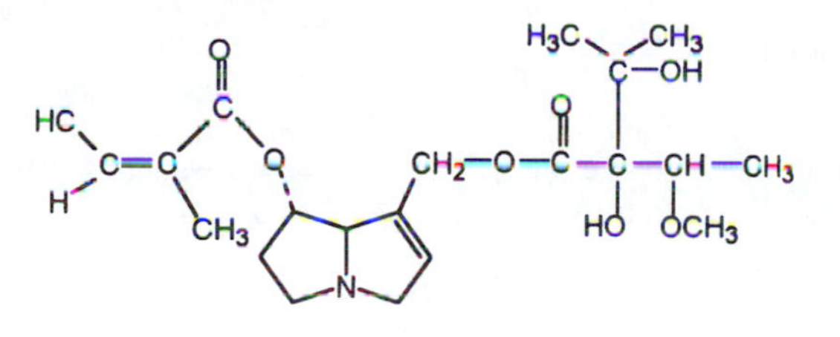

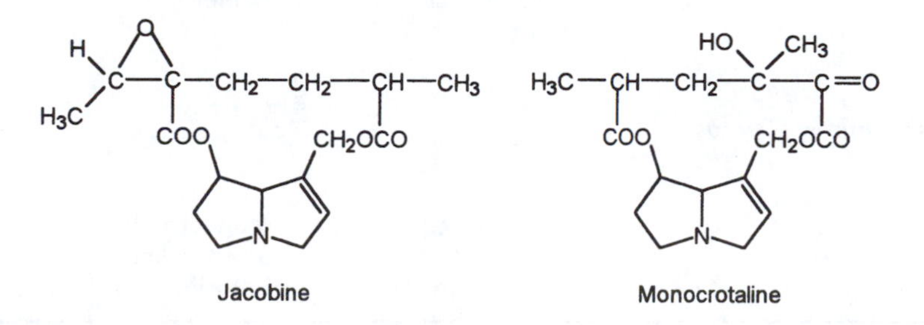

FIGURE 12–19 Examples of common pyrrolizidine alkaloids. The cyclic diesters tend to be most toxic, and monoesters the least toxic.

For toxicity, several structural features are necessary. There must be a 1,2 double bond, and there must be a branch in the ester group. The amino alcohols are not toxic. The acute toxicity is generally such that the cyclic diesters such as retrorsine are most toxic, the noncyclic diesters are of intermediate toxicity, and the monoesters are the least toxic. Esters of heliotridine are more toxic than retronecine esters; esters of these two bases are the main groups of hepatoxic PAs. The amino alcohols produce practically no pyrroles, which are the toxic metabolites formed in the liver. Diesters produce more pyrrole than corresponding monoesters. The PAs that are more easily hydrolyzed by esterases produce less pyrrole. Branching in the side chains hinders hydrolysis and therefore leads to greater pyrrole production. The PAs in several genera and species of toxic plants and their relative toxicity to rats (acute LD_{50}) are shown in Table 12–1. Representative structures are in Fig. 12–19. The PAs are also found in plants as N-oxides; these are of the same order of toxicity as the PAs.

TABLE 12–1 Distribution of Pyrrolizidine Alkaloids (PA) and Their Comparative Toxicities

Toxic plant	Amino alcohol	PA	LD_{50}[a]
Compositae			
Senecio jacobaea (tansy ragwort, stinking Willie)			
(British Isles, Western Europe, South Africa, Australia, New Zealand, North America)	Retronecine	Seneciphylline	77
	Retronecine	Senecionine	85
	Retronecine	Jacobine	77 (mouse)
	Retronecine	Jaconine	168 (female rat)
	Retronecine	Jacoline	NA[b]
	Retronecine	Jacozine	NA
Senecio vulgaris (common groundsel)			
(North America)	Retronecine	Senecionine	85
	Retronecine	Seneciphylline	77
	Retronecine	Retrorsine	38
Senecio longilobus (woolly or threadleaf groundsel)			
(North America)	Retronecine	Seneciphylline	77
	Retronecine	Retrorsine	38
	Retronecine	Riddelliine	105 (mouse)
Leguminosae			
Crotalaria spectabilis; C. retusa (rattlebox)			
(North America, Asia, Australia)	Retronecine	Monocrotaline	175
	Retronecine	Spectabiline	220
	Turneforcidine	Retusine	NA

(Continued)

TABLE 12–1 (Continued)

Toxic plant	Amino alcohol	PA	LD50[a]
Boraginaceae			
Heliotropium europaeum (heliotrope)			
(Australia)	Heliotridine	Heliotrine	300
	Heliotridine	Lasiocarpine	72
	Heliotridine	Europine	>1000
	Supinidine	Supinine	450
	Supinidine	Heleurine	140
Echium plantagineum (Paterson's curse, Salvation Jane)			
(Australia)	Retronecine	Echiumine	NA
	Retronecine	Echimidine	200
Amsinckia intermedia (tarweed)			
(U.S. Pacific Coast)	Retronecine	Intermedine	1500
	Retronecine	Echiumine	NA
	Retronecine	Lycopsamine	1500
Symphytum officinale (comfrey)			
(temperate regions)	Retronecine	Symphytine	300
	Retronecine	Echimidine	200
Cynoglossum officinale (hound's tongue)	Heliotridine	Heliosupine	60
	Heliotridine	Echinatine	350
Graminaea			
Festuca arundinacea (tall fescue)			
(temperate regions)	(no named amino alcohol for these alkaloids)	Loline	NA
		N-Acetyl loline	NA
		N-Formyl loline	NA

[a]All LD50 are for male rats unless otherwise indicated.

[b]NA = not available.

Toxicity, Metabolism, and Metabolic Effects

The toxic effects of PAs are due to their bioactivation in the liver to toxic metabolites called **pyrroles**, or dihydropyrrolizine (DHP) derivatives (Fig. 12–20). The pyrroles are chemically very reactive. Hepatoxic PAs have a 1,2 double bond which facilitates aromatization of the B ring by the hepatic mixed function oxidases. The pyrroles are powerful alkylating agents and react with tissue components. Pyrroles from diester PAs may act as bifunctional alkylating agents and can cross-link DNA, explaining the antimitotic effects of PAs (Hincks *et al.*, 1991; Kim *et al.*, 1993). Pyrroles may covalently bind with soluble nucleophiles such as glutathione and be excreted in the urine, or

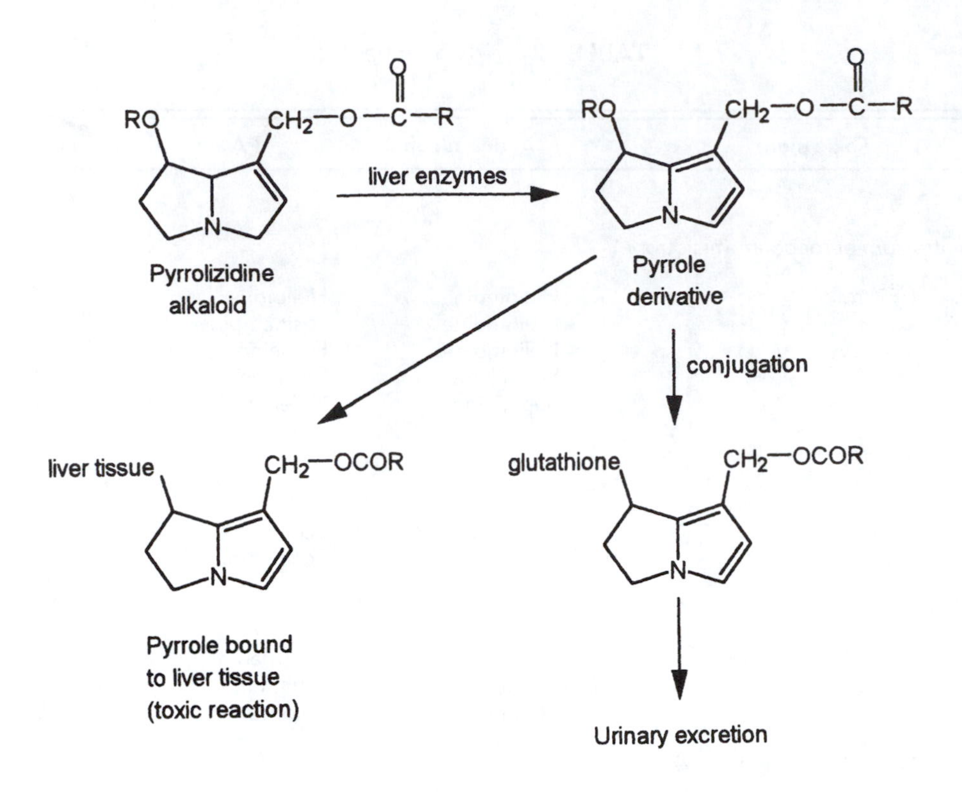

FIGURE 12–20 Hepatic metabolism of pyrrolizidine alkaloids.

they may form covalent bonds with DNA or liver enzymes. Small amounts of pyrrole may enter the blood and be transported to the lungs, causing pulmonary lesions. Besides formation of pyrroles, other possible fates of PAs include urinary excretion (with or without conjugation), hydrolysis of the ester groups by esterases with excretion of the acid and amino alcohol products, and formation and excretion of highly water-soluble N-oxides. Factors which tend to increase hepatotoxicity of PAs are those that maximize pyrrole production and minimize the other pathways (Mattocks, 1986). Factors favoring pyrrole production include a high degree of lipid solubility, a conformation favoring dehydrogenation rather than N-oxidation, and resistance to ester hydrolysis.

The **hepatoxic effects** of PAs are due to the reaction of pyrroles and/or other metabolites with tissue components, principally DNA. Typical hepatoxic signs are swelling of hepatocytes, centrilobular necrosis (the centrilobular region has microsomal enzyme activity, producing pyrroles), megalocytosis of the parenchymal cells (enlarged cells), karyomegaly (enlarged nuclei), fibrosis, bile duct proliferation, veno-occlusion, and loss of liver function. These effects are due to the alkylation of DNA with pyrroles, impairing cell division. The antimitotic effect results in a prevention of successful cell division. The specific stage of mitotic blockage is either in the latter half of the S phase or in early G_2 phase. The cells may be from 10 to 30 times normal size, and the DNA content may be 200 times normal. Megalocytosis is the result of an interaction between the antimitotic effect and the stimulus for regeneration. In most species affected by PA poisoning, the liver becomes extremely hard, fibrotic, and shrunken. In sheep, it tends to be mushy rather than hard. Liver weight as a percentage of body weight falls progressively in PA poisoning. Liver function is severely impaired. Because of decreased bile secretion, bilirubin levels in the blood rise, causing a typical jaundiced condition of the skin and mucous membranes. Other common **clinical signs** of PA toxicity include a rough unkempt appearance, diarrhea, prolapsed rectum, ascites, edema of tissues of the digestive tract, lassitude and dullness, photosensitization reactions, and abnormal behavior. In horses, neurological disturbances are seen, including "head pressing" against solid objects (Fig. 12–21) and walking in a straight line, regardless of obstacles or dangers in the path. Many deaths of horses are due to misadventure as a result of these symptoms. The clinical signs can be explained in terms of the loss of liver function.

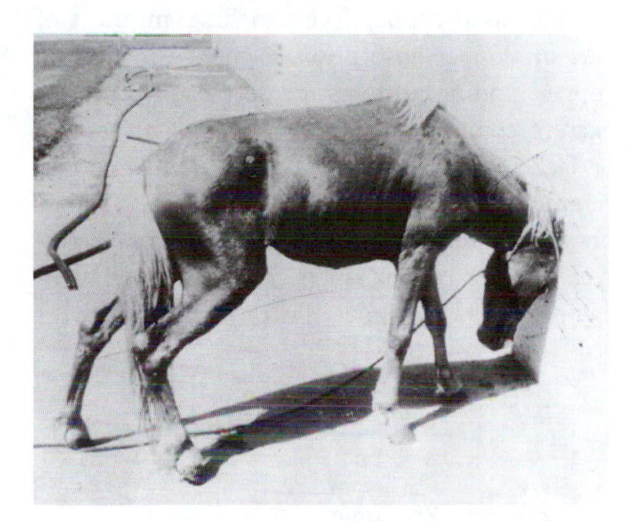

FIGURE 12–21 A horse exhibiting head pressing, a neurological sign associated with pyrrolizidine alkaloid toxicosis. This animal had consumed feed contaminated with *Senecio vulgaris*. (Courtesy of Humphrey Knight)

Due to decreased liver synthesis of plasma proteins, osmotic relationships between blood and other body fluids are disturbed, causing fluid movement from the blood to interstitial spaces (edema) and body cavities (ascites) (see Fig. 12–22). Cattle poisoned by PAs may have a bloated appearance, and there may be several liters of ascitic fluid in the abdominal cavity. Neurological signs in horses are due to elevated blood ammonia caused by decreased ability of the liver to convert ammonia to urea, with subsequent brain damage. Neurological signs and

FIGURE 12–22 Tansy ragwort toxicosis in rats. The intoxicated rat (right) shows pronounced ascites, compared to the control (left).

spongy changes in the brain, known as **hepatic encephalopathy**, are characteristic of the later stages of liver disease. Elevated blood ammonia causes an increased uptake of ammonia by the brain, where it reacts with α-ketoglutaric acid to form glutamine. The depletion of α-keto glutarate, which is an intermediate of the citric acid cycle, impairs ATP formation, leading to deficiency of ATP for normal brain metabolism.

With chronic PA toxicity, animals may look normal even after they have consumed a lethal dose. The liver damage is progressive, and death may occur months or even years after exposure to toxic plants containing PAs. A nutritional stress requiring greater liver function, such as the onset of lactation, may provoke signs of toxicity in animals that previously appeared normal (Synge and Stephen, 1993). Several serum enzymes have been proposed for **diagnosis** of PA toxicosis. These include γ-glutamyl transpeptidase (GGT), alkaline phosphatase, and glutamate dehydrogenase. While of some value for diagnostic purposes, changes in serum levels of these enzymes can be caused by other factors, and elevations in their levels due to PA poisoning are transitory (Craig *et al.*, 1991). The most useful screening test for subclinical PA intoxication appears to be serum GGT (Curran *et al.*, 1996).

Mattocks (1981) related the toxicity of PAs to the amount of covalently bound pyrrole in the liver. The liver pyrrole level 2 hr after administration of a dose of PAs is well correlated with the hepatoxic effect. Damage caused by pyrroles is not confined to the liver. Lesions in the lungs, heart, gastrointestinal tract, and kidneys have been observed. These may be due to extrahepatic metabolism of PAs to pyrroles, or a "spill-over" of pyrroles from the liver. The damage from *Senecio*, *Heliotropium*, and *Echium* is largely confined to the liver, while *Crotalaria* intoxication involves significant lung damage. Pyrroles that escape from the liver may bind with red blood cells by reaction with thiol groups on hemoglobin (Mattocks and Jukes, 1992). These pyrrolic thioethers are stable and may persist on the hemoglobin for long periods (Seawright *et al.*, 1991). The measurement of erythrocyte-bound pyrroles can be used for diagnosis of PA exposure long after the exposure has occurred (Seawright *et al.*, 1991; Winter *et al.*, 1993). Winter *et al.* (1993) used this method to verify PA poisoning of yaks in a remote area of the Himalayan kingdom of Bhutan.

Another effect of PA intoxication is anemia. This may be due to changes in copper and iron metabolism or may be a specific effect on hematopoesis. Swick *et al.* (1982a) demonstrated that rats poisoned with *S. jacobaea* had a severely impaired ability to incorporate ^{59}Fe into red blood cells. The spleen is greatly enlarged in PA poisoning. Deposition of hemosiderin in various tissues occurs. Further studies are needed to conclusively identify the mechanisms by which PAs cause anemia. Impaired protein synthesis in the liver would likely reduce hemoglobin synthesis.

A characteristic effect of PA poisoning in some animals is an increased liver copper concentration. In heliotrope poisoning of sheep, copper accumulates in the liver, even though forage copper is low, and leads to hemolytic crisis of copper toxicity. In sheep, the liver copper content may be 4–10 times normal values, while in rats, a 50- to 100-fold increase may occur (Fig. 12–23). Swick *et al.* (1982c) demonstrated that the nuclear and debris fractions of the liver cells are the main intracellular sites of copper accumulation. Possibilities to explain the increased copper levels include an effect of PAs on copper absorption, a change in hepatic copper-binding proteins, or decreased ability to excrete copper. It is probable that cholestasis, or lack of bile secretion, contributes to copper retention in the liver. Other hepatoxins, such as phomopsins involved in lupinosis, also cause increased liver copper, suggesting that liver damage per se is the primary factor involved. Deol *et*

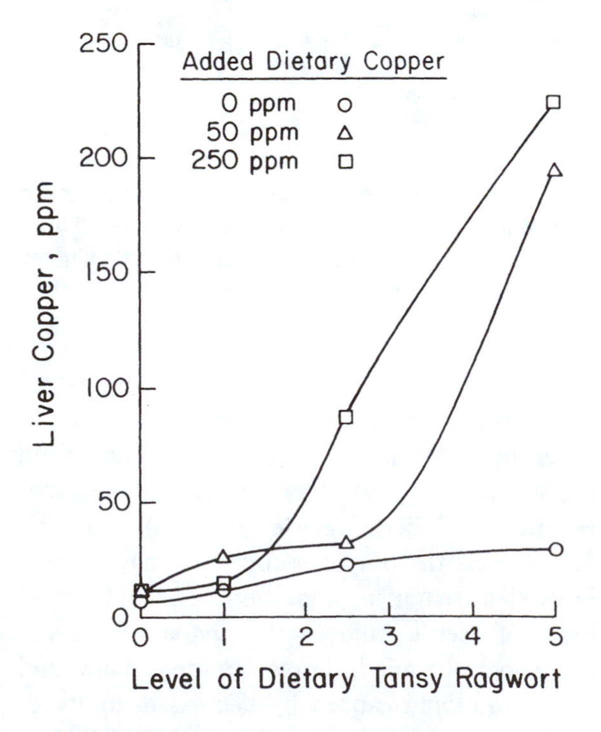

FIGURE 12–23 Effect of various dietary levels of tansy ragwort on liver copper levels in rats.

al. (1992) found that feeding heliotrope to sheep reduced the bile copper content without affecting bile flow, suggesting a specific effect on hepatic copper excretion.

Howell *et al.* (1991a,b) noted that dietary heliotrope and copper had synergistic toxic effects, especially if they were administered at the same time or if copper were given after heliotrope feeding ceased. Feeding heliotrope resulted in lowered liver concentrations of zinc and molybdenum, and increased selenium (Deol *et al.*, 1994). The increased selenium may be a response to liver necrosis, rather than a specific PA effect.

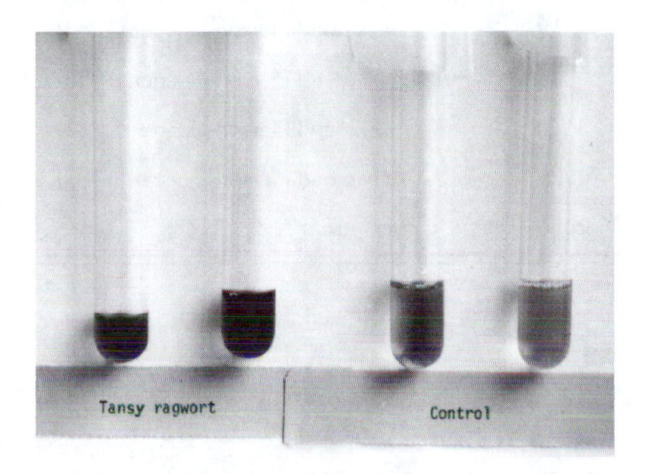

FIGURE 12–24 Erythrocytes from *Senecio*-intoxicated rats are susceptible to *in vitro* hemolysis, suggestive of vitamin E deficiency. *In vitro* erythrocyte hemolysis is a classic bioassay for vitamin E deficiency.

Blood and liver vitamin A levels are markedly reduced in PA toxicosis (Moghaddam and Cheeke, 1989; Huan *et al.*, 1992), probably because of impaired hepatic synthesis of retinol-binding protein (Huan *et al.*,1993). Erythrocytes from PA-intoxicated animals are susceptible to *in vitro* hemolysis (Fig. 12–24), suggesting possible effects of PA on vitamin E metabolism (Cheeke, 1989a,b). Reduced tissue vitamin E could contribute to PA pathology by reducing the liver's ability to detoxify oxidants.

Species Differences in Toxicity and Metabolism of PAs

Horses and cattle are the major livestock species poisoned by PAs, and they are of about equal susceptibility. In the case of *S. jacobaea* and *S. vulgaris,* a *Senecio* dry matter intake of 5–10% of body weight is lethal. Sheep, on the other hand, are very resistant to the hepatoxic effects of PAs. They can tolerate an intake of several times their body weight of *S. jacobaea*. Goats are slightly more susceptible than sheep. Hooper (1978) ranked the relative intakes of *S. jacobaea* to produce mortality as pigs, 1; chickens, 5; cattle and horses, 14; rats, 50; mice, 150; sheep and goats, 200. Japanese quail, gerbils, rabbits, and guinea pigs are also very resistant to effects of dietary PAs (Cheeke, 1988).

There are several possibilities to account for species differences in susceptibility to PA toxicity. These include lack of PA absorption, degradation of PAs in the rumen, lack of hepatic formation of pyrroles, and conjugation and excretion of PAs and pyrroles. Shull *et al.* (1976) reported that, in general, susceptible species had a high rate of liver pyrrole production, while resistant species had a low rate (Table 12–2). There were some exceptions to this, such as the rabbit, which has a high rate of pyrrole production, but is very resistant to dietary PAs (Pierson *et al.*, 1977). Evidence that hepatic metabolism is involved in some cases is that administration of inducing agents, such as phenobarbital, makes some resistant animals susceptible to PA poisoning. For example, guinea pigs are almost totally resistant to hepatoxic effects of monocrotaline. Prior administration of phenobarbital, which stimulates pyrrole production, results in guinea pigs becoming susceptible to monocrotaline toxicity. In contrast to their marked resistance to monocrotaline, guinea pigs are quite susceptible to *Senecio* PAs (Swick *et al.*, 1982b) such as jacobine (Chung and Buhler, 1995).

There are marked differences in **cytochrome P450 activities** among animal species (see

TABLE 12–2 Characterization of Animal Species by Susceptibility to Pyrrolizidine Alkaloid Toxicity and *In Vitro* Hepatic Pyrrole Production Rate

Species	Susceptibility to PA toxicosis	In vitro pyrrole production rate[a]	Lethal dose (as % of body weight)[b]	Reference
Cow	High	High	3.6	Cheeke *et al.* (1985)
Horse	High	High	7.3	Garrett *et al.* (1984)
Sheep	Low	Low	302	White *et al.* (1984)
Goat	Low	?	205	Goeger *et al.* (1982a)
Rat	High	High	21	Goeger *et al.* (1983)
Mouse	Intermediate	High	?	
Rabbit	Low	High	113	Pierson *et al.* (1977)
Guinea pig	Low	Low	119	Cheeke and Pierson-Goeger (1983)
Hamster	Low	High	338	Cheeke and Pierson-Goeger (1983)
Gerbil	Low	?	3640	Cheeke and Pierson-Goeger (1983)
Chicken	High	Low	39	Cheeke and Pierson-Goeger (1983)
Japanese quail	Low	Low	2450	Buckmaster *et al.* (1977)

[a]Adapted from Shull *et al.* (1976).

[b]Chronic lethal dose of *Senecio jacobaea*.

Chapter 3) and in hepatic metabolism of PA (Winter *et al.*, 1988; Chung *et al.*, 1995). For example, in guinea pigs, PA detoxification is largely accomplished by cytochrome P450 2B and flavin monooxygenase (FMO) while in the rat, cytochrome P450 3A2 is the major isozyme involved in PA bioactivation (Chung *et al.*, 1995a,b). In sheep and hamsters, cytochrome P4503A is the major isoform in the conversion of PA to pyrroles, while cytochrome P4502B has a minor role in these species (Huan, 1995). Thus PA activation and detoxification in various species are catalyzed by different constitutive and inducible cytochrome P450 isoforms.

The resistance of sheep and goats to PA toxicity in contrast to the susceptibility of cattle is of interest. Possibilities to account for their resistance are that PAs might be detoxified in the rumen, or that the sheep liver detoxifies pyrroles more efficiently than does cattle liver. Lanigan (1976) has shown that a microorganism (*Peptococcus heliotrinreductans*) isolated from sheep rumen contents is capable of metabolizing heliotrope PAs to nontoxic 1-methylene and 1-methylpyrrolizidine derivatives by reduction of the 1,2 double bond in the nucleus and cleavage of the esters (Fig. 12–25). Culvenor *et al.* (1976) found that 7α-angelyloxy-1-methylene pyrrolizidine from *in vitro* metabolism of lasiocarpine did not produce liver necrosis in rats or yield pyrroles when incubated with rat liver microsomes. The formation of 1-methylene derivatives requires metabolic hydrogen. The addition of methane inhibitors to the diet appears to increase the rumen formation of the nontoxic metabolites from heliotrope PAs

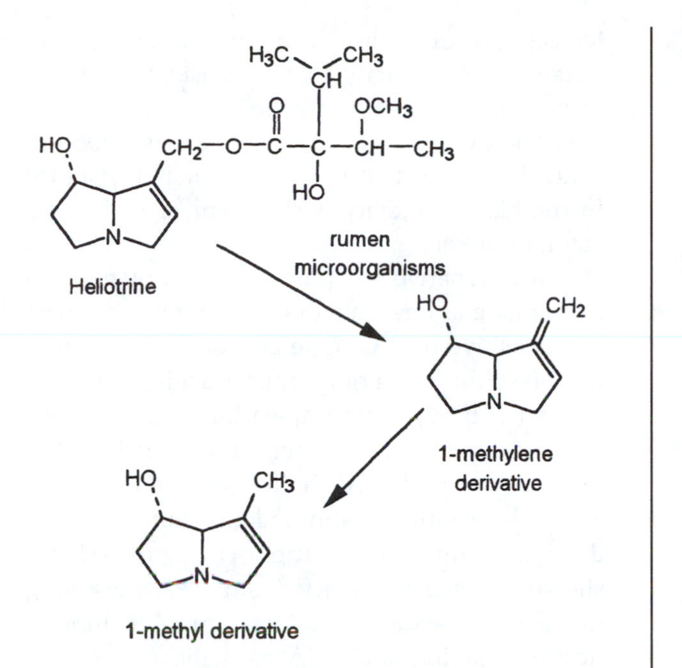

FIGURE 12–25 Rumen metabolism of heliotrine.

by providing more hydrogen (Lanigan *et al.*, 1978), but the resistance of sheep to heliotrope poisoning is not affected by continuous administration of antimethanogens (Peterson *et al.*, 1992). The resistance of sheep to *S. jacobaea* poisoning appears not to be due to **rumen metabolism** of PAs. Shull *et al.* (1976) and Swick *et al.* (1983a) found that incubation of *S. jacobaea* in sheep rumen fluid did not reduce its toxicity, and there was no evidence of formation of 1-methylene derivatives. The PAs in heliotrope are open esters of the base heliotridine, whereas those in *S. jacobaea* are all macrocyclic closed esters of retronecine. Because of steric hindrance, the closed esters are probably less readily converted to reduction metabolites. The resistance of sheep to *S. jacobaea* is probably due to a low rate of hepatic pyrrole production (Shull *et al.*, 1976; Winter *et al.*, 1988) and a high activity of detoxifying enzymes such as epoxide hydrolase (Swick *et al.*, 1983b) and N-oxidation and glutathione conjugation (Huan, 1995).

Some metabolism of *Senecio* PA in sheep, goat and bovine rumen fluid has been reported (Wachenheim *et al.*, 1992) with the suggestion that the resistance of sheep to PA toxicity may be related to ruminal detoxification of PA (Craig *et al.*, 1992). It is unlikely that sheep everywhere in the world have rumen microbes that detoxify PA whereas cattle do not. Sheep have been observed to be PA-resistant on all continents, whereas no resistant cattle occur anywhere. For example, in the remote Himalayan country of Bhutan, yaks (bovines) are PA susceptible while sheep are resistant (Winter *et al.*, 1992). Rumen bacteria are not specific to animal species, but are influenced primarily by diet. Sheep and goats have evolved as browsing animals exposed to many herbaceous plants containing toxins, whereas cattle and horses evolved on grasslands. Grasses have few chemical defenses. Thus on an evolutionary basis, sheep and goats have evolved superior hepatic detoxification mechanisms than grazers such as cattle. The PA-resistance of sheep and goats and other small herbivores (e.g. rabbits, guinea pigs, gerbils, hamsters, quail) is largely a reflection of hepatic detoxifying enzymes rather than gut microbes (Cheeke, 1994).

Milk Transfer of Pyrrolizidine Alkaloids

Dickinson and King (1978) demonstrated that PAs are found in the milk of dairy cattle and goats fed *S. jacobaea*. Goeger *et al.* (1982b) fed *S. jacobaea* to lactating goats and fed the milk to rats for a prolonged period. Very slight liver pathology was observed. They concluded that hazards to human health as a result of milk transfer of PAs were negligible. Candrian *et al.* (1991) demonstrated secretion of seniphylline in milk of cows dosed with this *Senecio* PA, but at levels unlikely to represent a health hazard. Most of the PA metabolites secreted in milk are water-soluble and non-toxic.

Prevention and Treatment of Pyrrolizidine Alkaloid Poisoning

Pyrrolizidine alkaloid poisoning of grazing animals in many areas can largely be eliminated through application of sound livestock management techniques. Plants containing PAs are unpalatable and not generally consumed in the presence of adequate feed. In the case of *S. jacobaea*, most poisonings occur when pastures are severely overgrazed and in early spring when other green forage is not available. The presence of *S. jacobaea* in a pasture is usually a reflection of poor management practices which have allowed invasion by the weed. In Australia, with harsh conditions and an extensive type of grazing system, exposure of livestock to PA-containing plants is often unavoidable. The potential toxicity problems associated with echium or heliotrope have to be balanced against the fact that often other feed is not available, hence the contradictory names Paterson's curse and Salvation Jane for *E. plantagineum*. Since it may take months or years for the lethal effects of PA ingestion to occur, grazing infested areas with animals destined for slaughter rather than with breeding stock has been recommended. Poisoning of swine and poultry generally is a result of contamination of grain with *Crotalaria* seeds. Spraying of grain fields with broad-leaf herbicides can readily eliminate this problem.

Senecio jacobaea infestations of pastures can be controlled with herbicides. However, areas where the weed is a problem are often of low productivity, and it covers extensive areas of rough ground, making spraying economically unfeasible. **Biological control** measures can be very successful. The cinnabar moth (*T. jacobaeae*) larvae feed only on plants containing PAs. In the Pacific Northwest, the cinnabar moth has had a useful role in helping to control ragwort by defoliation and prevention of seed production (Fig. 4–4). The flea beetle (*Longitarsus jacobaeae*) feeds on the roots of ragwort and, in western Oregon, has in conjunction with the cinnabar moth been an important management tool in controlling *S. jacobaea*. In Oregon and New Zealand, sheep have also been used as biological control agents. They are sufficiently resistant to PA toxicity that they can be used to control moderate ragwort infestations without likelihood of poisoning. Sheep are also used to control echium in Australia, as their resistance to the plant's toxicity is sufficient to make poisoning unlikely.

An alternative to controlling the plant is to manipulate animal metabolism to make animals less sensitive to PAs. One approach is the feeding of sulfur-containing amino acids to supply sulfhydryl groups for conjugation with pyrroles. Buckmaster *et al.* (1976) found that feeding cysteine to rats increased their tolerance to *S. jacobaea*. The synthetic antioxidants butylated hydroxyanisole (BHA) and ethoxyquin have shown protective activity against PA poisoning in mice (Miranda *et al.*, 1981a,b). A combination of cysteine and BHA as a dietary supplement has slightly increased the resistance of horses (Garrett *et al.*, 1984) and beef cattle (Cheeke *et al.*, 1985) to PA toxicity. Further evaluation of these supplements with cattle and horses is necessary to assess fully their potential as protective agents.

In Australia, copper toxicity in sheep grazing heliotrope is a major problem. The use of salt licks containing molybdenum, which increases copper excretion, should be examined as a potential protective regime. White *et al.* (1984) have studied interrelationships of copper and molybdenum with *S. jacobaea* toxicity and found that supplementary molybdenum tended to increase the susceptibility of sheep to PA toxicity.

The stability of PAs in hay and silage containing *S. alpinus* has been studied by Candrian *et al.* (1984). The PA content of hay remains constant over many months, while there is some degradation in silage. However, the reduction in PA levels during the ensilage process does not substantially reduce the toxicity of contaminated silage. Silage contaminated with more than about 5% *Senecio* plants is unsafe for cattle feeding. Clover and grass crops are sometimes contaminated with *S. jacobaea* and *S. vulgaris*, while corn may be contaminated with *S. vulgaris*.

LANTANA POISONING

Lantana cannara is an ornamental vine native to central America and Africa. It has escaped from cultivation in the southern U.S. (e.g. Florida, California) and is widely distributed as a weed of pastures, roadsides and rangelands in Australia, India and South America.

The toxicology of lantana poisoning has been thoroughly reviewed by Pass (1991). The following discussion is derived largely from this review.

Lantana contains a number of triterpenoid compounds, with the major toxic ones being **lantadenes A** and **B** (Fig. 12–26). These compounds are hepatotoxins that are metabolized by liver microsomal enzymes to active hepatoxic metabolites. Lantana toxins cause intrahepatic **cholestasis** (retention of bile in the liver) by inhibiting bile secretion by the hepatocytes. The lantadene metabolites damage the membranes of small bile ducts (canaliculi). The main consequences of the impaired bile flow are photosensitization, jaundice and rumen stasis. Photosensitization occurs because of the inhibited biliary excretion of phylloerythrin; jaundice is a result of impaired bilirubin excretion. Ruminal stasis, which is a prominent sign of lantana poisoning, is apparently due to a hepatoruminal reflex slowing ruminal motility, initiated by injured liver tissue.

Clinical signs of lantana poisoning include photosensitization, prominent jaundice of mucous membranes, anorexia and rumen stasis, dehydration and death. The liver is firm, enlarged, and yellowish in color. The gall bladder is grossly distended. The pronounced rumen stasis plays a role in the etiology of lantana poisoning, by retaining the toxins in the rumen, allowing their prolonged, continuous absorption. Removal of rumen contents usually results in rapid recovery. Kidney damage also occurs, and death may result from renal failure, particularly in chronic cases.

Lantana poisoning affects mainly ruminants, with numerous reports of poisoning of cattle,

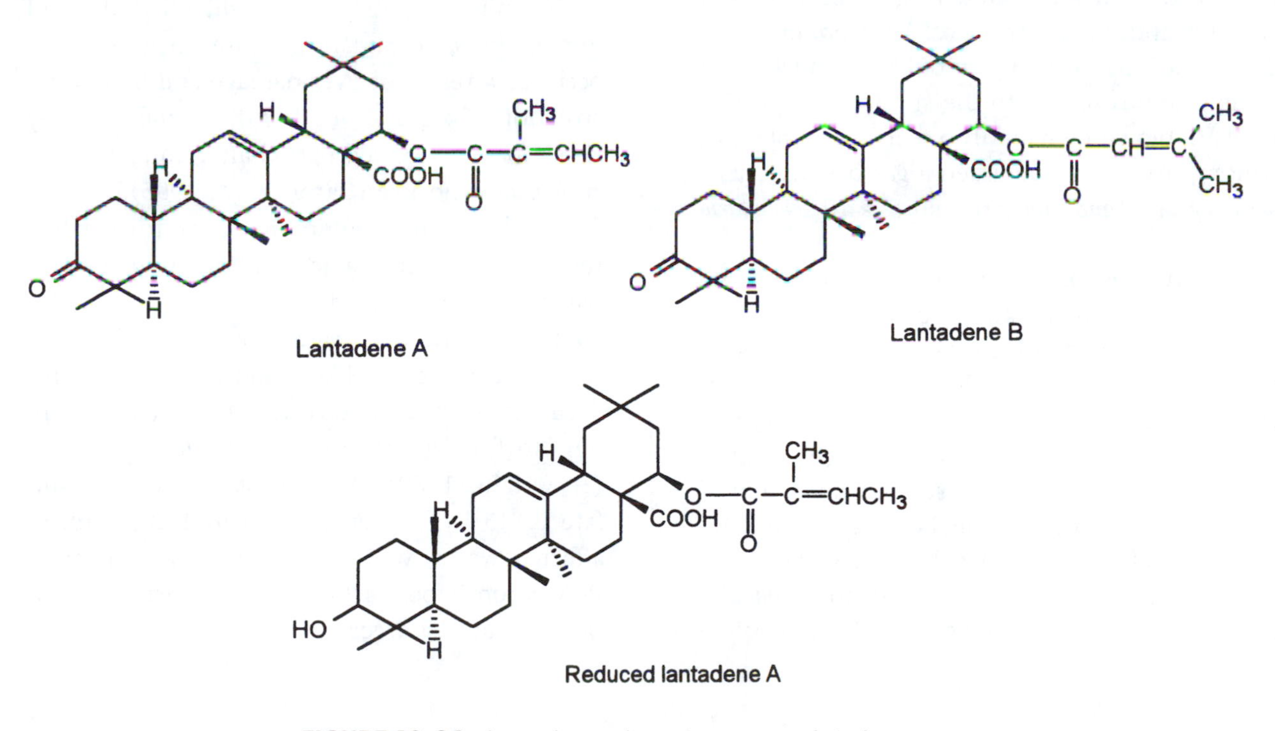

FIGURE 12–26 Lantadenes, the toxic compounds in *Lantana* spp.

sheep, goats and water buffalo. The acutely toxic dose of green lantana is about 1% of body weight. The toxin is not cumulative, so prevention of toxicity is facilitated by avoiding exposure of hungry animals to lantana. Activated charcoal at an oral dosage of 500 g for sheep and 2–2.5 kg for adult cattle is an effective treatment. Activated charcoal binds lantana toxins and reduces their absorption. Bentonite clay is also effective in treating lantana poisoning, at a dosage of 5 g/kg body weight (McKenzie, 1991).

Lantana poisoning of horses in Florida has been observed (Morton, 1994). Photosensitization, which the owners had mistakenly attributed to insect bites, was the major clinical sign.

Lantana is an attractive ornamental plant. It should not be used as a hedge in areas where livestock are kept, and care should be taken that hedge trimmings are not given to animals. This is good advice for all trimmings of hedges and ornamental plants.

BLUE-GREEN ALGAE

Blue-green algae of a number of species have caused livestock mortalities in many countries when animals consumed algae-infected water. Algae blooms occur in summer and autumn when stock watering ponds are low. Anaerobic bacteria in the bottom mud raise the soluble phosphorus and nitrogen levels and increase the availability of carbon dioxide. These factors, combined with long days, bright sunlight, and warm water, favor the proliferation of algae. The algal cells develop gas bubbles, causing the algae colonies to rise to the surface, after which they are blown by wind into dense algal blooms. Among the toxic blue-green algae are *Microcystis aeruginosa*, *Anabaena circinallis*, and *Nodularia spumingena*.

The two major classes of cyanobacterial toxins are hepatotoxins and neurotoxins (Beasley, 1991). The toxic principles of *Microcystis* spp. are hepatoxic cyclopeptides called **microcystins**. These are extremely toxic, with LD_{50} values in mice of about 0.056 mg/kg (Elleman *et al.*, 1978). Microcystins cause disorganization of the actin filaments of the hepatic cytoskeletal components, leading to cellular collapse. The toxin in the hepatotoxic mushroom *Amanita phalloides* acts in a similar way. Pale, swollen hemorrhagic livers, centrilobular necrosis, ascites and numerous small hemorrhages are observed at necropsy (Jackson *et al.*, 1984). Serum enzymes indicative of liver damage are elevated. Clinical signs include depression, dyspnea, paddling, prostration and muscle twitching in the terminal stages. Similar hepatoxic effects are seen with *Nodularia* intoxication.

In animals consuming a sublethal dose of microcystins, secondary photosensitization may occur as a result of liver damage and biliary obstruction (Carbis *et al.*, 1995). Mortalities may occur for a prolonged period after an apparently sublethal exposure (Carbis *et al.*, 1995).

Algal blooms tend to be most toxic when the cells are in the rapidly proliferating stage or when the bloom is disintegrating (Seawright, 1989). Pollution of water sources or fertilizer run-off increase the likelihood of algae blooms. Algae in ponds can be controlled by use of copper sulfate at about 1 kg/4,00,000 liters (Seawright, 1989). A spray of 70 g copper sulfate in 45 l of water, applied to 100 sq. m. of affected water, will provide this concentration. Prevention is best achieved by restricting pollution of water sources.

HEPATIC FATTY CIRRHOSIS

Hepatic fatty cirrhosis (HFC) or hard yellow liver syndrome has occurred for many years in cattle and sheep in west Texas (Helman *et al.*, 1993, 1995). Wild ruminants such as pronghorn are also affected. Animals show inital signs of unthriftness and poor performance, with signs of hepatic encephalopathy such as head pressing, depression and coma in terminal stages. The livers of affected animals are yellow and hard, representing fatty degeneration and cirrhosis, respectively. The cause of the condition is unknown. Mycotoxins or poisonous plants have been suggested as causative agents (Helman *et al.*, 1993). Over 120 species of plants have been tested without induction of signs of HFC (Helman *et al.*, 1993). Phomopsins produced by the fungus *Phomopsis leptostromiformis* might be involved (Helman *et al.*, 1993); the fungus has been identified on grasses in the areas where HFC occurs and the hepatic lesions are similar to those seen in lupinosis, a condition induced by phomopsins in lupins. However, the etiology of HFC remains uncertain.

ISOQUINOLINE ALKALOIDS

The Mexican poppy (*Argemone mexicana*) is an annual herb that grows in semi-arid areas, and favors disturbed sites such as roadsides and grain fields. The seeds occasionally cause toxicoses of livestock and poultry as contaminants of feed grains. Human toxicoses have also occurred by way of contaminated grains. The plant resembles a thistle in appearance. Its toxicity is due to isoquinoline alkaloids ((Fig. 12–27) such as dihydrosanguinarine, sanguinarine, berberine and coptisine (Fletcher *et al.*, 1993). Chickens fed grain contaminated with Mexican poppy seeds show edema of the lungs and heart (hydropericardium) and general subcutaneous edema (Norton and O'Rourke, 1980).

About 1% Mexican poppy seed in grains is the maximum tolerated level for pigs and poultry (Takken *et al.*, 1993). In humans, poisonings have occurred with consumption of contaminated grains, with clinical signs of dropsy (edema), congestive heart failure, liver disease and glaucoma (Dalvi, 1985). Sanguinarine is hepatoxic, and inhibits pyruvate oxidation and Na-K ATPase (Dalvi, 1985).

Sanguinarine is also found in the bloodroot plant, *Sanguinaria canadensis*. Bloodroot was used medicinally by Native Americans, as a corrosive agent to treat superficial neoplasms (Becci *et al.*, 1987).

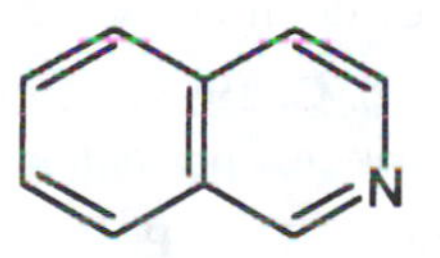

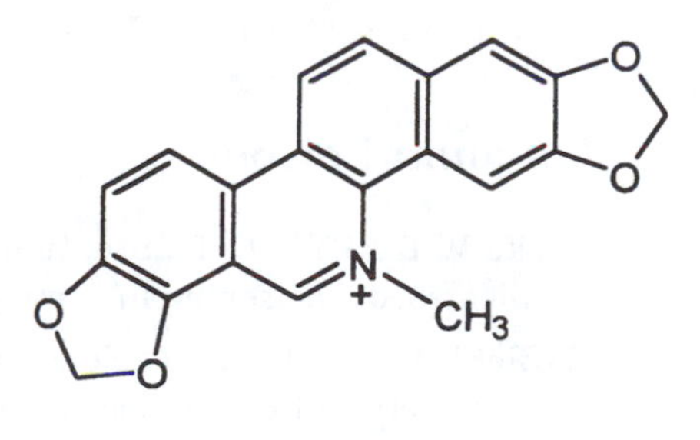

FIGURE 12–27 Isoquinoline alkaloids.

REFERENCES

Oak Poisoning

Davis, G.G., L.E. Bartel, and C.W. Cook. 1975. Control of gambel oak sprouts by goats. J. Range Manage. 28:216–218.

Dixon, P.M., E.A. McPherson, A.C. Rowland, and W. Maclennan. 1979. Acorn poisoning in cattle. Vet. Rec. 104:284–285.

Dollahite, J.W., G.T. Housholder, and B.J. Camp. 1966. Effect of calcium hydroxide on the toxicity of post oak (*Quercus stellata*) in calves. J. Am. Vet. Med. Assoc. 148:908–912.

Murdiati, T.B., C.S. McSweeney, and J.B. Lowry. 1992. Metabolism in sheep of gallic acid, tannic acid and hydrolysable tannin from *Terminalia oblongata*. Aust. J. Agric. Res. 43:1307–1319.

Murdiati, T.B., C.S. McSweeney, R.S.F. Campbell, and D.S. Stoltz. 1990. Prevention of hydrolysable tannin toxicity in goats fed *Clidemia hirta* by calcium hydroxide supplementation. J. Appl. Toxicol. 10:325–331.

Nastis, A.S., and J.C. Malechek. 1981. Digestion and utilization of nutrients in oak browse by goats. J. Anim. Sci. 53:283–290.

Ostrowski, S.R., B.P. Smith, S.J. Spier, B.B. Norman, and M.N. Oliver. 1989. Compensatory weight gain in steers recovered from oak bud toxicosis. J. Am. Vet. Med. Assoc. 195:481–484.

Sandusky, G.E., C.J. Fosnaugh, J.B. Smith, and R. Mohan. 1977. Oak poisoning of cattle in Ohio. J. Am Vet. Med. Assoc. 171:627–629.

Shi, Z.-C. 1988. Identification of the phenolic substances in bovine urine associated with oak leaf poisoning. Res. Vet. Sci. 45:152–155.

Spier, S.J., B.P. Smith, A.A. Seawright, B.B. Norman, S.R. Ostrowski, and M.N. Oliver. 1987. Oak toxicosis in cattle in northern California: Clinical and pathologic findings. J. Am. Vet. Med. Assoc. 191:958–964.

Tor, E.R., T.M. Francis, D.M. Holstege, and F.D. Galey. 1996. GC/MS determination of pyrogallol and gallic acid in biological matrices as diagnostic indicators of oak exposure. J. Agric. Food Chem. 44:1275–1279.

Zhu, J., and L.J. Filippich. 1995. Acute intra-abomasal toxicity of tannic acid in sheep. Vet. Human Toxicol. 37:50–54.

Solanum Poisoning

Buck, W.B., J.W. Dollahite, and T.J. Allen. 1960. *Solanum elaeagnifolium*, silver-leafed nightshade, poisoning in livestock. J. Am. Vet. Med. Assoc. 137:348–351.

Gaffield, W., and R.F. Keeler. 1996. Induction of terata in hamsters by solanidane alkaloids derived from *Solanum tuberosum*. Chem Res. Toxicol. 9:426–433.

Keeler, R.F., D.C. Baker, and W. Gaffield. 1991. Solanum alkaloids. pp. 607–636 in: R.P. Sharma and D.K. Salunkhe (Eds.). Mycotoxins and Phytoalexins. CRC Press, Boca Raton, FL.

Kingsbury, J.M. 1964. Poisonous Plants of the United States and Canada. Prentice-Hall, Englewood Cliffs, NJ.

Van Gelder, W.M.J. 1991. Steroidal glycoalkaloids in *Solanum*: Consequences for potato breeding and food safety. pp. 101–158 in: R.F. Keeler and A.T. Tu. 1991. Handbook of Natural Toxins. Vol. 6. Toxicology of Plant and Fungal Compounds. Marcel Dekker, Inc., New York.

Voss, K.A., W.J. Chamberlain, and L.H. Brennecke. 1993. Subchronic toxicity study of eastern black nightshade (*Solanum ptycanthum*) berries in Sprague–Dawley rats. J. Food Safety 13:91–97.

Oxalates

Allison, M.J., K.A. Dawson, W.R. Mayberry, and J.G. Foss. 1985. *Oxalobacter formigenes* gen. nov., sp. nov.: oxalate-degrading anaerobes that inhabit the gastrointestinal tract. Arch. Microbiol. 141:1–7.

Allison, M.J., H.M. Cook, and K.A. Dawson. 1981. Selection of oxalate-degrading rumen bacteria in continuous cultures. J. Anim. Sci. 53:810–816.

James, L.F. 1978. Oxalate poisoning in livestock. pp. 139–145 in: R.F. Keeler, K.R. Van Kampen and L.R. James (Eds.). Effects of Poisonous Plants on Livestock. Academic Press, NY.

Kingsbury, J.M. 1964. Poisonous Plants of the United States and Canada. Prentice–Hall, Englewood Cliffs, NJ.

Lincoln, S.D., and B. Black. 1980. Halogeton poisoning in range cattle. J. Am. Vet. Med. Assoc. 176:717–718.

Marshall, V.L., W.B. Buck, and G.L. Bell. 1967. Pigweed (*Amaranthus retroflexus*): An oxalate-containing plant. Am. J. Vet. Res. 28:888–889.

Panciera, R.J., T. Martin, G.E. Burrows, D.S. Taylor, and L.E. Rice. 1990. Acute oxalate poisoning attributable to ingestion of curly dock (*Rumex crispus*) in sheep. J. Am. Vet. Med. Assoc. 196:1981–1984.

Ward, G.M., L.H. Harbers, and J.J. Blaha. 1979. Calcium-containing crystals in alfalfa: Their fate in cattle. J. Dairy Sci. 62:715–722.

Woodhouse, E.D. 1983. Talking of dumb cane. Pac. Hortic. 44:47–48.

Sesquiterpene Lactones

Conner, J.R., J.L. Schuster, and E.M. Bailey, Jr. 1988. Impact of bitterweed on the economics of sheep production in the Texas Edwards Plateau. pp. 145–151 in: L.F. James, M.H. Ralphs, and D.B. Nielsen (Eds.). The Ecology and Economic Impact of Poisonous Plants on Livestock production. Westview Press, Boulder, CO.

Hill, D.W., E.M. Bailey, Jr., and B.J. Camp. 1980. Tissue distribution and disposition of hymenoxon. J. Agric. Food Chem. 28:1269–1273.

Jones, D.H., and H.L. Kim. 1981. Toxicity of hymenoxon in Swiss white mice following pretreatment with microsomal enzyme inducers, inhibitors and carbon tetrachloride. Res. Commun. Chem. Pathol. Pharm. 33:361–364.

Kim, H.L., A.C. Anderson, M.K. Terry, and E.M. Bailey, Jr. 1981. Protective effect of butylated hydroxyanisole on acute hymenoxon and bitterweed poisoning. Res. Commun. Chem. Pathol. Pharm. 33:365–368.

Kim, H.L., M. Szabuniewicz, L.D. Rowe, B.J. Camp, J.W. Dollahite, and C.H. Bridges. 1974. L-Cysteine, an antagonist to the toxic effects of an α-methylene-γ-lactone isolated from *Hymenoxys odorata* D.C. (Bitterweed). Res. Commun. Chem. Pathol. Pharm. 8:381–384.

Post, L.O., and E.M. Bailey, Jr. 1992. The effect of dietary supplements on chronic bitterweed (*Hymenoxys odorata*) poisoning in sheep. Vet. Human Tox. 34:209–213.

Rowe, L.D., H.L. Kim, and B.J. Camp. 1980. The antagonistic effect of L-cysteine in experimental hymenoxon intoxication in sheep. Am. J. Vet. Res. 41:484–486.

Taylor, C.A., Jr., and M.H. Ralphs. 1992. Reducing livestock losses from poisonous plants through grazing management. J. Range Manage. 45:9–12.

Terry, M.K., H.A. Williams, H.L. Kim, L.O. Post, and E.M. Bailey, Jr. 1983. Ovine urinary metabolites of hymenoxon, a toxic sesquiterpene lactone isolated from *Hymenoxys odorata* DC. J. Agric. Food Chem. 31:1208–1210.

Ueckert, D.N., and M.C. Calhoun. 1988. Ecology and toxicology of bitterweed (*Hymenoxys odorata*). pp. 131–143 in: L.F. James, M.H. Ralphs, and D.B. Nielsen (Eds.). The Ecology and Economic Impact of Poisonous Plants on Livestock Production. Westview Press, Boulder, CO.

Selenoamino Acids

Baker, D.C., L.F. James, W.J. Hartley, K.E. Panter, H.F. Maynard, and J. Pfister. 1989. Toxicosis in pigs fed selenium-accumulating *Astragalus* plant species or sodium selenate. Am. J. Vet. Res. 50:1396–1399.

Behne, D., A. Kyriakopoulos, H. Meinhold, and J. Kohrle. 1990. Identification of type I iodothyronine 5'-deiodinase as a selenoenzyme. Biochem. Biophys. Res. Comm. 173:1143–1149.

Fan, A.M., S.A. Book, R.R. Neutra, and D.M. Epstein. 1988. Selenium and human health implications in California's San Joaquin Valley. J. Tox. Environ. Health 23:539–559.

Harris, T. 1991. Death in the Marsh. Island Press, Washington, D.C.

Norman, B., G. Nader, M. Oliver, R. Delmas, D. Drake, and H. George. 1992. Effects of selenium supplementation in cattle on aquatic ecosystems in northern California. J. Am. Vet. Med. Assoc. 201:869–872.

Ohlendorf, H.M., A.W. Kilness, J.L. Simmons, R.K. Stroud, D.J. Hoffman, and J.F. Moore. 1988. Selenium toxicosis in wild aquatic birds. J. Tox. Environ. Health 24:67–92.

Ohlendorf, H.M., D.J. Hoffman, M.K. Saiki, and T.W. Aldrich. 1986. Embryonic mortality and abnormalities of aquatic birds: Apparent impacts of selenium from irrigation drainwater. The Science of the Total Environment 52:49–63.

O'Toole, D., M. Raisbeck, J.C. Case, and T.D. Whitson. 1996. Selenium-induced "Blind Staggers" and related myths. A commentary on the extent of historical livestock losses attributed to selenosis on western U.S. rangelands. Vet. Pathol. 33:104–116.

Panter, K.E., W.J. Hartley, L.F. James, H.F. Mayland, B.L. Stegelmeier, and P.O. Kechele. 1996. Comparative toxicity of selenium from seleno-DL-methionine, sodium selenate, and *Astragalus bisulcatus* in pigs. Fund. Appl. Toxicol. 32:217–223.

Panter, K.E., L.F. James, and H.F. Mayland. 1995. Reproductive response of ewes fed alfalfa pellets containing sodium selenate or *Astragalus bisulcatus* as a selenium source. Vet. Human Toxicol. 37:30–32.

Van Kampen, K.R., and L.F. James. 1978. Manifestations of intoxication by selenium-accumulating plants. pp. 135–138 in: R.F. Keeler, K.R. Van Kampen, and L.F. James (Editors), Effects of Poisonous Plants on Livestock. Academic Press, NY.

Whanger, P.D. 1989. Selenocompounds in plants and their effects on animals. pp. 141–167 in: P.R. Cheeke (Ed.). Toxicants of Plant Origin. Vol. III. Proteins and Amino Acids. CRC Press, Boca Raton, FL.

Other Gastrointestinal Tract Irritants

Geor, R.J., R.L. Becker, E.W. Kanara, L.R. Hovda, W.H. Sweeney, T.F. Winter, J.K. Rorick, G.R. Ruth, E. Hope, and M.J. Murphy. 1992. Toxicosis in horses after ingestion of hoary alyssum. J. Am. Vet. Med. Assoc. 201:63–67.

Hovda, L.R., and M.L. Rose. 1993. Hoary alyssum (*Berteroa incana*) toxicity in a herd of broodmare horses. Vet. Hum. Toxicol. 35:39–41.

Pyrrolizidine Alkaloids

Baker, D.C., J.A. Pfister, R.J. Molyneux, and P. Kechele. 1991. *Cynoglossum officinale* toxicity in calves. J. Comp. Path. 104:403–410.

Buckmaster, G.W., P.R. Cheeke, and L.R. Shull. 1976. Pyrrolizidine alkaloid poisoning in rats: Protective effects of dietary cysteine. J. Anim. Sci. 43:464–473.

Bull, L.B., C.C.J. Culvenor, and A.T. Dick. 1968. The Pyrrolizidine Alkaloids. Their Chemistry, Pathogenicity and Other Biological Properties. American Elsevier, New York.

Burns, J. 1972. The heart and pulmonary arteries in rats fed on *Senecio jacobaea*. J. Pathol. 106:187–194.

Candrian, U., U. Zweifel, J. Luthy, and C. Schlatter. 1991. Transfer of orally administered [^{3}H]seneciphylline into cow's milk. J. Agric. Food Chem. 39:930–933.

Candrian, U., J. Luthy, P. Schmid, C. Schlatter, and E. Gallasz. 1984. Stability of pyrrolizidine alkaloids in hay and silage. J. Agric. Food Chem. 32:935–937.

Cheeke, P.R. 1994. A review of the functional and evolutionary roles of the liver in the detoxification of poisonous plants, with special reference to pyrrolizidine alkaloids. Vet. Human Toxicol. 36:240–247.

Cheeke, P.R. 1989a. Nutritional implications of pyrrolizidine alkaloids as contaminants of foodstuffs. pp. 157–175 in: A.M. Rizk (Ed.). Poisonous Plant Contamination of Edible Plants. CRC Press, Boca Raton, FL.

Cheeke, P.R. 1989b. Pyrrolizidine alkaloid toxicity and metabolism in laboratory animals and livestock. pp. 1–22 in: P.R. Cheeke (Ed.). Toxicants of Plant Origin. Vol. I. Alkaloids. CRC Press, Boca Raton, FL.

Cheeke, P.R. 1988. Toxicity and metabolism of pyrrolizidine alkaloids. J. Anim. Sci. 66:2343–2350.

Cheeke, P.R., J.A. Schmitz, C.D. Lassen, and E.G. Pearson. 1985. Effects of dietary supplementation with ethoxyquin, magnesium oxide, methionine hydroxy analog and B vitamins on tansy ragwort (*Senecio jacobaea*) toxicosis in beef cattle. Am. J. Vet. Res. 46:2179–2183.

Chung, W.-G., and D.R. Buhler. 1995. Major factors for the susceptibility of guinea pig to the pyrrolizidine alkaloid jacobine. Drug Metab. Dispos. 23:1263–1267.

Chung, W.-G., C.L. Miranda, and D.R. Buhler. 1995. A cytochrome P4502B form is the major bioactivation enzyme for the pyrrolizidine alkaloid senecionine in guinea pig. Xenobiotica 25:929–939.

Craig, A.M., C.J. Latham, L.L. Blythe, W.B. Schmotzer, and O.A. O'Connor. 1992. Metabolism of toxic pyrrolizidine alkaloids from tansy ragwort (*Senecio jacobaea*) in ovine ruminal fluid under anaerobic conditions. Appl. Environ. Microbiol. 58:2730–2736.

Craig, A.M., E.G. Pearson, C. Meyer, and J.A. Schmitz. 1991. Serum liver enzyme and histopathologic changes in calves with chronic and chronic-delayed *Senecio jacobaea* toxicosis. Am. J. Vet. Res. 52:1969–1978.

Culvenor, C.C.J. 1983. Estimated intakes of pyrrolizidine alkaloids by humans. A comparison with dose rates causing tumors in rats. J. Tox. Environ. Health 11:625–635.

Culvenor, C.C.J., M.V. Jago, J.E. Peterson, L.W. Smith, A.L. Payne, D.G. Campbell, J.A. Edgar, and J.L. Frahn. 1984. Toxicity of *Echium plantagineum* (Paterson's Curse). I. Marginal toxic effects in Merino wethers from long-term feeding. Aust. J. Agric. Res. 35:293–304.

Culvenor, C.C.J., M. Clarke, J.A. Edgar, J.L. Frahn, M.V. Jago, J.E. Peterson, and L.W. Smith. 1980. Structure and toxicity of the alkaloids of Russian comfrey (*Symphytum* x *uplandicum* Nyman), a medicinal herb and item of the human diet. Experientia 36:377–379.

Culvenor, C.C.J., J.A. Edgar, M.V. Jago, A. Outleridge, J.E. Peterson, and L.W. Smith. 1976. Hepato- and pneumotoxicity of pyrrolizidine alkaloids and derivatives in relation to molecular structure. Chem.-Biol. Interact. 12:299–324.

Curran, J.M., R.J. Sutherland, and R.L. Peet. 1996. A screening test for subclinical liver disease in horses affected by pyrrolizidine alkaloid toxicosis. Aust. Vet. J. 74:236–240.

Deol, H.S., J. McC. Howell, and P.R. Dorling. 1994. Effect of the ingestion of heliotrope and copper on the concentration of zinc, selenium and molybdenum in the liver of sheep. J. Comp. Path. 110:303–307.

Deol, H.S., J. McC. Howell, P.R. Dorling, and H.W. Symonds. 1992. The effect of copper and heliotrope on the composition of bile in sheep. Res. Vet. Sci. 53:324–330.

Dickson, J.O., and R.R. King. 1978. The transfer of pyrrolizidine alkaloids from *Senecio jacobaea* into the milk of lactating cows and goats. pp. 201–208 in: R.F. Keeler, K.R. Van Kampen, and L.F. James (Editors). Effects of Poisonous Plants on Livestock. Academic Press, NY.

Garrett, B.J., D.W. Holtan, P.R. Cheeke, J.A. Schmitz, and Q.R. Rogers. 1984. Effects of dietary supplementation with butylated hydroxyanisole, cysteine and B vitamins on tansy ragwort (*Senecio jacobaea*) toxicosis in ponies. Am. J. Vet. Res. 45:459–464.

Goeger, D.E., P.R. Cheeke, J.A. Schmitz, and D.R. Buhler. 1982a. Toxicity of tansy ragwort (*Senecio jacobaea*) to goats. Am. J. Vet. Res. 43:252–254.

Goeger, D.E., P.R. Cheeke, J.A. Schmitz, and D.R. Buhler. 1982b. Effect of feeding milk from goats fed tansy ragwort (*Senecio jacobaea*) to rats and calves. Am. J. Vet. Res. 43:1631–1633.

Griffin, D.S., and H.J. Segall. 1989. Effects of the pyrrolizidine alkaloid senecionine and the alkenals *trans*-4-OH-hexenal and *trans*-2-hexenal on intracellular calcium compartmentation in isolated hepatocytes. Biochem. Pharm. 38:391–397.

Habermehl, G.G., W. Martz, C.H. Tokarnia, J. Dobereiner, and M.C. Mendez. 1988. Livestock poisoning in South America by species of the *Senecio* plant. Toxicon 26:275–286.

Hincks, J.R., H.-Y. Kim, H.J. Segall, R.J. Molyneux, F.R. Stermitz, and R.A. Coulombe, Jr. 1991. DNA cross-linking in mammalian cells by pyrrolizidine alkaloids: Structure-activity relationships. Tox. Appl. Pharm. 111:90–98.

Hirono, I., H. Mori, and M. Haga. 1978. Carcinogenic activity of *Symphytum officinale*. J. Natl. Cancer Inst. 61:865–868.

Hooper, P.T. 1978. Pyrrolizidine alkaloid poisoning—pathology with particular reference to differences in animal and plant species. In: R.F. Keeler, K.R. Van Kampen, and L.F. James (Editors). Effects of Poisonous Plants on Livestock. Academic Press, NY.

Howell, J. McC., H.S. Deol, and P.R. Dorling. 1991a. Experimental copper and *Heliotropium europaeum* intoxication in sheep: Clinical syndromes and trace element concentrations. Aust. J. Agric. Res. 42:979–92.

Howell, J.McC., H.S. Deol, P.R. Dorling, and J.B. Thomas. 1991b. Experimental copper and heliotrope intoxication in sheep: Morphological changes. J. Comp. Path. 105:49–74.

Huan, J. 1995. Species differences in bioactivation and detoxification of pyrrolizidine (*Senecionine*) alkaloids. Ph.D. thesis, Oregon State University.

Huan, J., P.R. Cheeke, and W.S. Blaner. 1993. Modification of vitamin A metabolism by dietary pyrrolizidine (*Senecio*) alkaloids in rats. Proc. West. Sec. Am. Soc. Anim. Sci. 44:275–278.

Huan, J., P.R. Cheeke, R.R. Lowry, H.S. Nakaue, S.P. Snyder, and P.D. Whanger. 1992. Dietary pyrrolizidine (*Senecio*) alkaloids and tissue distribution of copper and vitamin A in broiler chickens. Toxicol. Lett. 62:139–153.

Kellerman, T.S., J.A.W. Coetzer, and T.W. Naude. 1988. Plant Poisonings and Mycotoxicoses of Livestock in Southern Africa. Oxford University Press, Cape Town.

Kim, H.-Y., F.R. Stermitz, R.J. Molyneux, D.W. Wilson, D. Taylor, and R.A. Coulombe, Jr. 1993. Structural influences on pyrrolizidine alkaloid–induced cytopathology. Toxicol. Appl. Pharm. 122:61–69.

Knight, A.P., C.V. Kimberling, F.R. Stermitz, and M.R. Roby. 1984. *Cynoglossum officinale* (hounds' tongue)—a cause of pyrrolizidine alkaloid poisoning in horses. J. Am. Vet. Med. Assoc. 185:647–650.

Lanigan, G.W., A.L. Payne, and J.E. Peterson. 1978. Anti-methanogenic drugs and *Heliotropium europaeum* poisoning in penned sheep. Aust. J. Agric. Res. 29:1281–1292.

Lanigan, G.W. 1976. *Peptococcus heliotrinreducans*, sp. nov., a cytochrome-producing anaerobe which metabolizes pyrrolizidine alkaloids. J. Gen. Microbiol. 94:1–10.

Larson, K.M., M.R. Roby, and F.R. Stermitz. 1984. Unsaturated pyrrolizidines from borage (*Borago officinalis*), a common garden herb. J. Nat. Prod. 47:747.

Lombardo de Barros, C.S., D. Driemeier, C. Pilati, S.S. Barros, and L.M.L. Castilhos. 1992. Senecio spp. poisoning in cattle in southern Brazil. Vet. Hum. Toxicol. 34:241–246.

Mattocks, A.R. 1986. Chemistry and Toxicology of Pyrrolizidine Alkaloids. Academic Press, San Diego, CA.

Mattocks, A.R. 1981. Relation of structural features to pyrrolic metabolites in livers of rats given pyrrolizidine alkaloids and derivatives. Chem.-Biol. Interact. 35:301–310.

Mattocks, A.R., and R. Jukes. 1992. Detection of sulphur-conjugated pyrrolic metabolites in blood and fresh or fixed liver tissue from rats given a variety of toxic pyrrolizidine alkaloids. Toxicol. Lett. 63:47–55.

Miranda, C.L., R.L. Reed, P.R. Cheeke, and D.R. Buhler. 1981a. Protective effects of butylated hydroxyanisole against the acute toxicity of monocrotaline in mice. Toxicol. Appl. Pharm. 59:424–430.

Miranda, C.L., H.M. Carpenter, P.R. Cheeke, and D.R. Buhler. 1981b. Effect of ethoxyquin on the toxicity of the pyrrolizidine alkaloid monocrotaline and on hepatic drug metabolism in mice. Chem.-Biol. Interact. 37:95–107.

Moghaddam, M.F., and Cheeke, P.R. 1989. Effects of dietary pyrrolizidine (*Senecio*) alkaloids on vitamin A metabolism in rats. Toxicol. Lett. 45:149–156.

Molyneux, R.J., A.E. Johnson, J.D. Olsen, and D.C. Baker. 1991. Toxicity of pyrrolizidine alkaloids from Riddell groundsel (*Senecio riddellii*) to cattle. Am. J. Vet. Res. 52:146–151.

Peterson, J.E., A.L. Payne, and C.C.J. Culvenor. 1992. *Heliotropium europaeum* poisoning of sheep with low liver copper concentrations and the preventive efficacy of cobalt and antimethanogen. Aust. Vet. J. 69:51–56.

Pfister, J.A., R.J. Molyneux, and D.C. Baker. 1992. Pyrrolizidine alkaloid content of houndstongue (*Cynoglossum officinale* L.). J. Range Manage. 45:254–256.

Pierson, M.L., P.R. Cheeke, and E.O. Dickinson. 1977. Resistance of the rabbit to dietary pyrrolizidine (*Senecio*) alkaloid. Res. Commun. Chem. Pathol. Pharm. 16:561–564.

Seaman, J.T. 1985. Hepatogenous chronic copper poisoning in sheep associated with grazing *Echium plantagineum*. Aust. Vet. J. 62:247.

Seaman, J.T., W.S. Turvey, S.J. Ottaway, R.J. Dixon, and A.R. Gilmour. 1989. Investigations into the toxicity of *Echium plantagineum* in sheep. 1. Field grazing experiments. Aust. Vet. J. 66:279–285.

Seaman, J.T., and R.J. Dixon. 1989. Investigations into the toxicity of *Echium plantagineum* in sheep. 2. Pen feeding experiments. Aust. Vet. J. 66:286–292.

Seawright, A.A., J. Hrdlicka, J.D. Wright, D.R. Kerr, A.R. Mattocks, and R. Jukes. 1991. The identification of hepatoxic pyrrolizidine alkaloid exposure in horses by the demonstration of sulphur-bound pyrrolic metabolites on their hemoglobin. Vet. Hum. Toxicol. 33:286–287.

Shull, L.R., G.W. Buckmaster, and P.R. Cheeke. 1976. Factors influencing pyrrolizidine (*Senecio*) alkaloid metabolism: Species, liver sulfhydryls and rumen fermentation. J. Anim. Sci. 43:1247–1253.

Stegelmeier, B.L., D.R. Gardner, L.F. James, and R.J. Molyneux. 1996. Pyrrole detection and the pathologic progression of *Cynoglossum officinale* (houndstongue) poisoning in horses. J. Vet. Diagn. Invest. 8:81–90.

Swick, R.A., P.R. Cheeke, H.S. Ramsdell, and D.R. Buhler. 1983a. Effect of sheep rumen fermentation and methane inhibition on the toxicity of *Senecio jacobaea*. J. Anim. Sci. 56:645–651.

Swick, R.A., C.L. Miranda, P.R. Cheeke, and D.R. Buhler. 1983b. Effect of phenobarbital on toxicity of pyrrolizidine (*Senecio*) alkaloids in sheep. J. Anim. Sci. 56:887–894.

Swick, R.A., P.R. Cheeke, C.L. Miranda, and D.R. Buhler. 1982a. The effect of consumption of the pyrrolizidine alkaloid–containing plant *Senecio jacobaea* on iron and copper metabolism in the rat. J. Toxicol. Environ. Health 10:757–768.

Swick, R.A., P.R. Cheeke, D.E. Goeger, and D.R. Buhler. 1982b. Effect of dietary *Senecio jacobaea* and injected *Senecio* alkaloids and monocrotaline on guinea pigs. J. Anim. Sci. 55:1411–1416.

Swick, R.A., P.R. Cheeke, and D.R. Buhler. 1982c. Subcellular distribution of hepatic copper, zinc and iron and serum ceruloplasmin in rats intoxicated by oral pyrrolizidine (*Senecio*) alkaloids. J. Anim. Sci. 55:1425–1430.

Synge, B.A., and F.B. Stephen. 1993. Delayed ragwort poisoning associated with lactation stress in cows. Vet. Record 132:327.

Tyler, V.E. 1993. The Honest Herbal. A Sensible Guide to the Use of Herbs and Related Remedies. The Haworth Press, Binghamton, N.Y.

Wachenheim, D.E., L.L. Blythe, and A.M. Craig. 1992. Characterization of rumen bacterial pyrrolizidine alkaloid biotransformation in ruminants of various species. Vet. Human Toxicol. 34:513–517.

White, R.D., R.A. Swick, and P.R. Cheeke. 1984. Effects of dietary copper and molybdenum on tansy ragwort (*Senecio jacobaea*) toxicity in sheep. Am. J. Vet. Res. 45:159–161.

Winter, H., A.A. Seawright, J. Hrdlicka, A.R. Mattocks, R. Jukes, K. Wangdi, and K.B. Gurung. 1993. Pyrrolizidine alkaloid poisoning of yaks: diagnosis of pyrrolizidine alkaloid exposure by the demonstration of sulphur-conjugated pyrrolic metabolites of the alkaloid in circulating haemoglobin. Aust. Vet. J. 70:312–313.

Winter, H., A.A. Seawright, J. Hrdlicka, U. Tshewang, and B.J. Gurung. 1992. Pyrrolizidine alkaloid poisoning of yaks (*Bos grunniens*) and confirmation by recovery of pyrrolic metabolites from formalin-fixed liver tissue. Res. Vet. Sci. 52:187–194.

Winter, C.K., H.J. Segall, and A.D. Jones. 1988. Species differences in the hepatic microsomal metabolism of the pyrrolizidine alkaloid senecionine. Comp. Biochem. Physiol. 90:429–433.

Lantana Poisoning

McKenzie, R.A. 1991. Bentonite as therapy for *Lantana camara* poisoning of cattle. Aust. Vet. J. 68:146–148.

Morton, J.F. 1994. Lantana, or red sage (*Lantana camara* L., [Verbenaceae]), notorious weed and popular garden flower; some cases of poisoning in Florida. Econ. Bot. 48:259–270.

Pass, M.A. 1991. Poisoning of livestock by *Lantana* plants. pp. 297–311 in: R.F. Keeler and A.T. Tu (Eds.). Toxicology of Plant and Fungal Compounds. Vol. 6. Marcel Dekker, Inc., New York.

Blue-Green Algae

Beasley, V.R., R.A. Lovell, A.M. Dahlem, W.M. Haschek, and S.B. Hooser. 1991. Cyclic peptide hepatotoxins from Cyanobacteria. pp. 459–495 in: R.F. Keeler and A.T. Tu (Eds.). Toxicology of Plant and Fungal Compounds. Vol. 6. Handbook of Natural Toxins. Marcel Dekker, Inc., New York.

Carbis, C.R., D.L. Waldron, G.F. Mitchell, J.W. Anderson, and I. McCauley. 1995. Recovery of hepatic function and latent mortalities in sheep exposed to the blue-green alga *Microcystis aeruginosa*. Vet. Record 137:12–15.

Elleman, T.C., I.R. Falconer, A.R.B. Jackson, and M.T. Runnegar. 1978. Isolation, characterization and pathology of the toxin from a *Microcystis aeruginosa* (=*Anacystis cyanea*) bloom. Aust. J. Biol. Sci. 31:209–218.

Jackson, A.R.B., A. McInnes, I.R. Falconer, and M.T.C. Runnegar. 1984. Clinical and pathological changes in sheep experimentally poisoned by the blue-green algae *Microcystis aeruginosa*. Vet. Pathol. 21:102–113.

Seawright, A.A. 1989. Animal Health in Australia. Vol. 2. Chemical and Plant Poisons. Australian Government Publishing Service, Canberra.

Hepatic Fatty Cirrhosis

Helman, R.G., L.G. Adams, and C.H. Bridges. 1995. The lesions of hepatic fatty cirrhosis in sheep. Vet. Pathol. 32:635–640.

Helman, R.G., L.G. Adams, and C.H. Bridges. 1993. Hepatic fatty cirrhosis in ruminants from western Texas. J. Am. Vet. Med. Assoc. 202:129–132.

Isoquinoline Alkaloids

Becci, P.J., H. Schwartz, H. Barnes, and G.L. Southard. 1987. Short-term toxicity studies of sanguinarine and of two alkaloid extracts of *Sanguinaria canadensis* L. J. Toxicol. Environ. Health 20:199–208.

Dalvi, R.R. 1985. Sanguinarine: its potential as a liver toxic alkaloid present in the seeds of *Argemone mexicana*. Experientia 41:77–78.

Fletcher, M.T., G. Takken, B.J. Blaney, and V. Alberts. 1993. Isoquinoline alkaloids and keto-fatty acids of *Argemone ochroleuca* and *A. mexicana* (Mexican poppy) seed. I. An assay method and factors affecting their concentration. Aust. J. Agric. Res. 44:265–275.

Norton, J.H., and P.K. O'Rourke. 1980. Oedema disease in chickens caused by Mexican poppy (*Argemone mexicana*) seed. Aust. Vet. J. 56:187–189.

Takken, G., M.T. Fletcher, and B.J. Blaney. 1993. Isoquinoline alkaloids and keto-fatty acids of *Argemone ochroleuca* and *A. mexicana* (Mexican poppy) seed. II. Concentrations tolerated by pigs. Aust. J. Agric. Res. 44:277–285.

CHAPTER 13

Neurotoxins, Cardiac/Pulmonary Toxins, and Nephrotoxins

NEUROTOXINS

The Hemlocks: *Conium* and *Cicuta* spp.

Poison hemlock (*Conium maculatum*) and water hemlock (*Cicuta* spp., e.g. *C. maculata*) are common poisonous plants found throughout North America. They belong to the **Umbelliferae** family, which includes such common vegetables as carrots, celery, and parsnip, as well as many wild plants or weeds including poison and water hemlock, wild carrot (Queen Anne's lace), wild parsnip, water celery and water parsnip. The Umbelliferae have their flowers in umbrella-shaped clusters (umbels). They have many features in common which often makes it difficult for the casual observer to conclusively

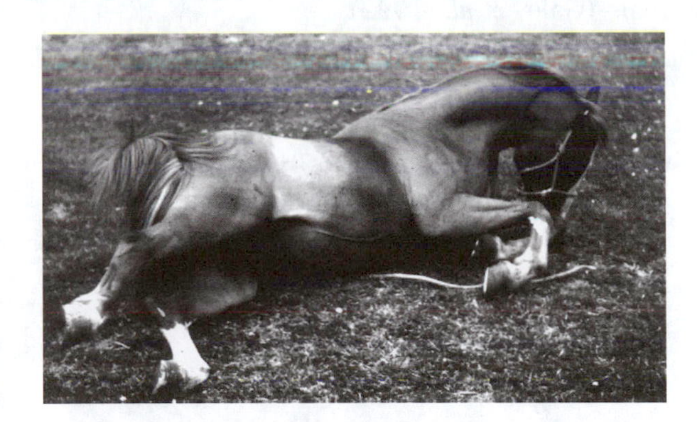

FIGURE 13–1 A horse poisoned by water hemlock (*Cicuta maculata*). The toxin in this plant, cicutoxin, causes violent convulsions. (Courtesy of M.E. Fowler)

identify them. For example, water hemlock, poison hemlock, wild carrot and water celery are very similar in appearance, and often grow in similar habitats.

Poison Hemlock (*Conium maculatum*)

Poison hemlock, a native of Europe, is distributed throughout the U.S., especially in the Pacific Northwest, and Northcentral and Northeast regions. It is a vigorous plant, growing 6–10 ft high along roadsides, ditches, cattle yards, and the edges of grain and alfalfa fields. It resembles both wild carrot (*Daucus carota*) and water hemlock in appearance (Fig. 13–2.) It has a fleshy, white tap root, whereas water hemlock has a thickened, branching root stalk much like a dahlia tuber in appearance. The lower portion of the stem of water hemlock is divided into chambers which contain the toxins. These features can be used to distinguish the two plants. Poison hemlock has a characteristic unpleasant mouse urine–like odor, which is detectable when one is near the plant or crushes a leaf or stem. This odor is associated with the piperidine alkaloids which the plant contains. The odor can also be detected in the gut contents, breath and urine of *Conium*-poisoned animals (Galey *et al.*, 1992).

FIGURE 13–2 Poison hemlock (*Conium maculatum*). (Courtesy of W. C. Krueger)

Poison hemlock contains eight **piperidine alkaloids** (Panter *et al.*, 1988a); the two major ones are coniine and α-coniceine. α-coniceine predominates in the vegetative growth while coniine is the major alkaloid in the seed. Coniine has the distinction of being the first alkaloid to be chemically synthesized, in 1886 (Kingsbury, 1964). Major conium alkaloids are shown in Fig. 13–3.

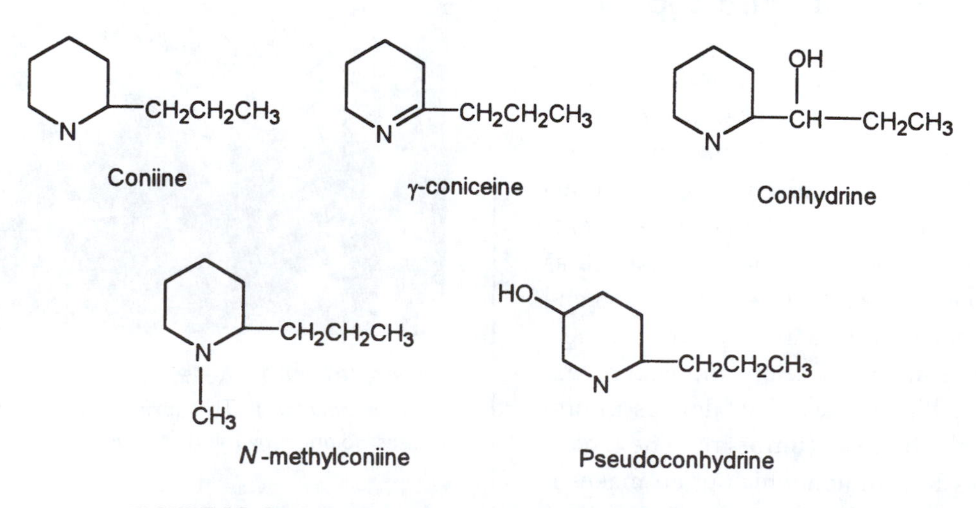

FIGURE 13–3 Piperidine alkaloids in *Conium* spp. (poison hemlock.)

The conium alkaloids have two major effects: they cause acute effects on the nervous system, and they are teratogenic agents. The three major *Conium* alkaloids, coniine, α-coniceine and N-methyl coniine, cause paralysis of the muscles, by acting as neuromuscular junction blocking agents (Panter and Keeler, 1989). The symptoms of **acute poisoning**, in order of their appearance, are nervousness, trembling, incoordination, dilation of the pupils, weakened heartbeat, coldness of the extremities, coma, and death through respiratory failure. Keeler *et al.* (1980) have reported the signs of sublethal coniine toxicity in cows, ewes, and mares. The alkaloid was given orally. Within 30–40 min after dosing in mares, and about 1.5–2 hr after dosing in cows and sheep, the animals became nervous, followed by tremors and ataxia. These signs lasted 4–5 hr in mares and 6–7 hr in cows and sheep. Cows were most sensitive, with horses intermediate, and sheep quite resistant to toxicity. Cows had severe toxic signs with a dose of 3.3 mg coniine per kilogram body weight, mares had severe signs at 15.5 mg/kg, while ewes showed only moderate toxic signs when given 44 mg/kg. Keeler *et al.* (1980) believe that the difference of about 10 times in susceptibility of cows and sheep to coniine is due to liver metabolism, because a similar difference in toxicity is seen when conium is administered either orally or by injection. They reported that when progressively larger daily doses of coniine were given by injection, about 16 mg/kg killed cows, whereas 240 mg/kg was the lethal dose for sheep.

Concentration of the alkaloids is highest in the seeds. In the past, poisoning of swine and poultry by poison hemlock seeds in grain was fairly common. The use of herbicides in grain fields has virtually eliminated contamination of grain with conium seeds. In terms of acute toxicity, poison hemlock is least toxic to pigs, moderately toxic to sheep and most toxic to cows. The teratogenic effects (to be discussed below) are most severe in pigs and cattle, and least in sheep (Panter *et al.*, 1988a).

Conium foliage has an unpleasant odor and is not readily consumed by livestock unless other forage is scarce. It is more likely to be consumed when it is a contaminant of hay or silage. Alfalfa fields are often contaminated with *Conium*. Green-chop, silage and hay from such fields could be toxic. The alkaloids are volatile, so hay would tend to be less toxic. However, Galey *et al.* (1992) have observed *Conium* poisoning in cattle fed *Conium*-contaminated alfalfa hay.

Panter *et al.* (1985a,b) reported that a single oral dose of 1 g/kg body weight of *C. maculatum* seed or 8 g/kg of fresh plant (6- to 10-in. stage) was lethal to swine (gilts). The material was consumed avidly by the animals, and there was evidence of a craving for and possible addiction to the plant. Daily sublethal doses were given to gilts for the study of **teratogenic effects**. About 140 g of seed and 1000 g of fresh plant per pig per day were used. About 15 min after dosing, the third eyelid began moving across the eye, causing complete blindness by 45 min. The blindness lasted about 15–30 min. The animals became weak and docile for about 1–2 hr. Trembling and weakness were observed. Both the seed and plant foliage were teratogenic. When fed in the 30- to 45-day period of gestation, *Conium* caused cleft palate. In the 43- to 53-day period, arthrogryposis occurred, along with severe flexure of fetlock joints, scoliosis, and hydrocephalus. When *Conium* was given in the 51- to 61-day period of gestation, limb malformations similar in type but less severe than in the 43- to 53-day group were observed. Edmonds et al. (1972) also noted that *C. maculatum* causes fetal deformities in swine.

In the western U.S., fetal deformities in cattle have been quite common. **Crooked calf disease** has been attributed primarily to maternal consumption of lupins at a particular stage of gestation. Implication of poison hemlock as a causative agent was reported by Keeler (1974). Several crooked calves were found in herds that were not exposed to any known teratogenic plants. Poison hemlock was widely distributed in the pastures, suggesting it as a possibility.

Because one of the conium alkaloids, coniine, was available commercially, Keeler and associates administered coniine to cows during the 55–75 days of gestation and observed skeletal deformities similar to those seen in the natural outbreak. Later studies (Keeler and Balls, 1978) showed that administration of the whole plant to cows also caused crooked calf disease. Poison hemlock was given orally to cows daily during days 45–75 of gestation. Daily doses of green plant material of 410–840 g resulted in deformed calves. Cows forced to breathe fresh poison hemlock for about 20 hr/day had normal calves, but did show moderate signs of toxicity, demonstrating that the volatile alkaloids are toxic by inhalation. These workers further studied the structural features of the conium alkaloids that produced teratogenic effects. Five piperidine alkaloids in *C. maculatum* were identified (Fig. 13–3). Over 98% of the alkaloid in fresh plant was α-coniceine, while it was less than 20% in dried plant material. A third of the alkaloid in dry plant was coniine. A variety of analogs of coniine were tested for teratogenic activity. Introduction of a double bond between C-1 and the nitrogen did not alter activity. Both the length of the side chain and the degree of unsaturation of the ring influenced activity. The fully unsaturated ring analog of coniine was not teratogenic. Compounds with a side chain of less than three carbons did not have activity. Piperidine alkaloids with a side chain α to the nitrogen, at least propyl in length, and with a saturated ring are candidates for having teratogenic activity.

In the study of Keeler *et al.* (1980) in which coniine was administered to pregnant cows, mares, and ewes, only the cows had deformed off-spring. The foal fetus may be resistant because the mares showed severe toxic signs and had normal foals. It was concluded that the sheep fetus is definitely resistant to the teratogenic effects of coniine because the large doses given over a wide period (12–65 days) of gestation did not cause fetal defects. However, feeding pre-flowering vegetative *Conium* to ewes induced toxicity to the ewes and fetal defects including excessive carpal flexure in the lambs (Panter *et al.*, 1988b). The likely alkaloid involved was α-coniceine, which is the main piperidine alkaloid in pre-flowering *Conium.*

Frank and Reed (1987) reported a case of poison hemlock toxicosis in a flock of range turkeys. The birds showed neurological signs such as excessive salivation, tremors, paralysis and mortality. *Conium* seeds were found in the digestive tract. Similar signs were induced in a subsequent study, with administration of coniine to turkeys (Frank and Reed, 1990).

The early Greeks used the juice of poison hemlock to prepare a hemlock potion used to execute criminals. The philosopher Socrates was put to death for not believing in the correct gods. An account of his execution is provided by Panter and Keeler (1989). After consuming a cup of hemlock, Socrates described a heaviness and numbness in his feet and legs. He continued in conversation until the feelings of coldness and numbness reached his waist, when he spoke for the last time. He then peacefully expired.

Since Biblical times, there have been reports of humans suffering poisoning after consuming meat of the European migratory quail (*Coturnix coturnix coturnix*). Kennedy and Grivetti (1980) reviewed historical evidence linking **quail poisoning** (coturnism) with birds that had consumed poison hemlock. It had been hypothesized that the quail are resistant to poison hemlock toxicity, but by accumulating the toxins in their tissues, they are toxic to humans. Kennedy and Grivetti (1980) concluded that *Conium* is not involved in quail poisoning. They determined that quail are in fact poisoned by ingestion of small amounts of *Conium* seed, and noted that coturnism occurs during periods when *Conium* seeds are not produced. Frank and Reed (1990) also determined that quail are very sensitive to *Conium* toxicity.

Water Hemlock (*Cicuta*)

Various *Cicuta* spp. grow throughout North America. They are considered the most violently

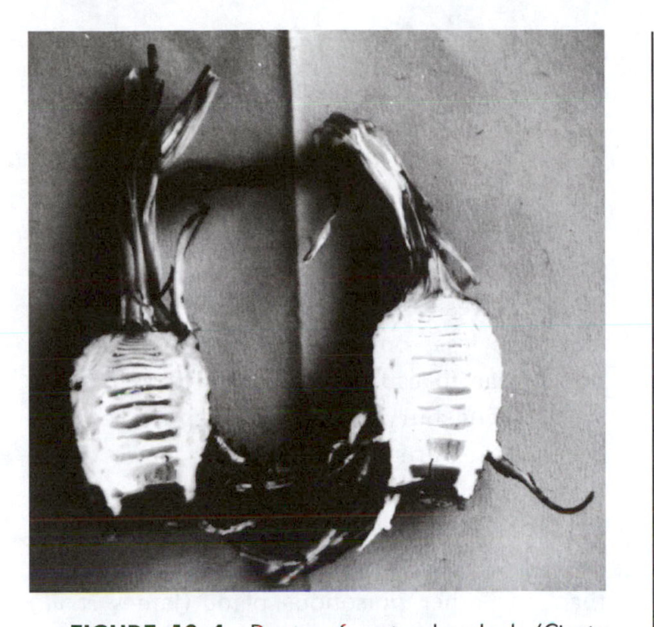

FIGURE 13–4 Roots of water hemlock (*Cicuta maculata*), showing the characteristic chambers where the toxin is located.

poisonous plants of the North Temperate Zone (Kingsbury, 1964). Water hemlock is the common name usually applied to *Cicuta maculata*. It is similar in appearance to numerous other Umbelliferae, including poison hemlock (*Conium maculatum*) and wild carrot (*Daucus carota*). It is found *only* in swampy or wet habitats, such as along streams and in swampy areas and marshes. The plant may grow from 5 to 10 ft tall with a jointed, hollow stem. A very characteristic feature of water hemlock is that it has a thickened storage organ at the base of the stem which is divided into chambers (Fig. 13–4). This chambered rootstock contains a yellowish oily liquid with the characteristic pungent odor of raw parsnip. The liquid is the poisonous principle, which is also found to a lesser extent in the lower portions of the stem. The toxin is called **cicutoxin**, a highly unsaturated higher alcohol (Fig. 13–5).

Many people have died from the consumption of water hemlock. Cases are reported sporadically when fishermen, hikers, and others in wilderness areas mistake it for an edible plant such as wild parsnip. Children may mistake the plant for an edible one. Cicutoxin acts directly on the central nervous system and is a violent convulsant (Knutsen and Paszkowski, 1984). The description by Kingsbury (1964) conveys its violent nature:

> *Symptoms appear within 15 minutes to more than an hour, but usually within about a half-hour after ingestion of a lethal dose. Excessive salivation is first noted. This is quickly followed by tremors and then by spasmodic convulsions interspersed intermittently with periods of relaxation. The convulsions are extremely violent; head and neck are thrown rigidly back, legs may flex as though running, and clamping or chewing motions of the jaw and grinding of the teeth occur. Abdominal pain is evident. In some cases the tongue is chewed to shreds; in others, teeth have been broken in an unsuccessful attempt to pry the mouth open to administer treatment.*

Early signs of poisoning occur within 10–15 minutes after ingestion of a lethal dose of *Cicuta*

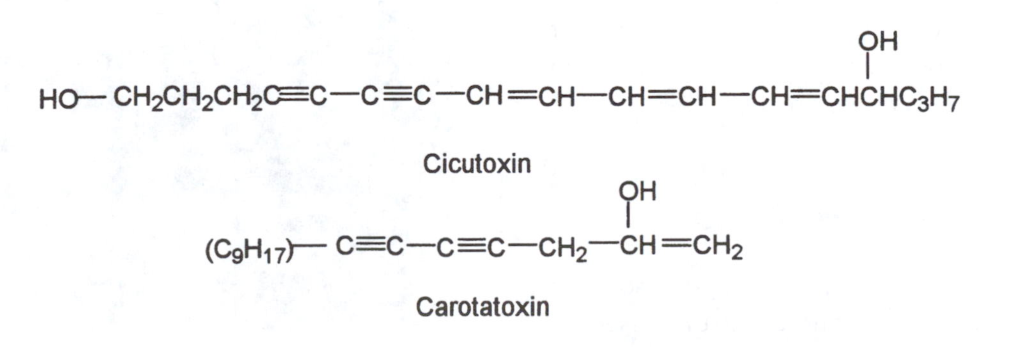

FIGURE 13–5 Cicutoxin is a highly unsaturated alcohol. It is violently toxic. Interestingly, caratotoxin in carrots has a similar structure but is of very low toxicity.

(Panter *et al.*, 1988a). Excessive salivation, frothing, nervousness, stumbling and incoordination are the first signs, followed by tremors, muscular weakness and convulsive seizures, with intermittent periods of relaxation. Each successive relaxation period between convulsions is shorter, the subject becomes exhausted and weak, with death due to anoxia during a prolonged convulsive episode (Panter *et al.*, 1988a). If the seizures are prevented by administration of barbituates, death is prevented. Panter *et al.* (1996) gavaged sheep with water hemlock tubers, and found that if sodium pentobarbital was administered intravenously at the first sign of seizures, further seizures were prevented, with a rapid and complete recovery. Panter *et al.* (1996) determined that 1.2–2.7 g fresh water hemlock tuber per kg body weight was a lethal dose for sheep.

Most losses of livestock occur in early spring when the new growth appears before other forage is available. When the ground is soft, the rootstock may be pulled up and consumed along with the tops (Smith and Lewis, 1987). The plants usually grow in small patches, which can be eliminated manually or with chemical sprays. Both the roots and stems of water hemlock float on water. In some areas, spring floods wash out *Cicuta* plants into streams, where they may collect in masses at log jams or other obstructions in the stream. If this is near a livestock watering point, losses of animals may occur when they consume the plants.

Garden carrots and celery contain a similar acetylinic alcohol, which is much less toxic than cicutoxin, called carotatoxin (Fig. 13–5).

Lupinus spp.

Toxicity of *Lupinus* spp. (lupins) associated with quinolizidine alkaloids is discussed in Chapter 11. Teratogenic effects of lupins are covered in Chapter 14.

Larkspur (*Delphinium* spp.)

Toxic **polycyclic diterpene alkaloids** occur in larkspurs (*Delphinium* spp.) and in *Aconitum* spp. (aconite or monkshood). These genera are similar in appearance, habitat preference, and toxic effects. Various species of *Aconitum* are found in western range areas of the U.S. They are much less common than larkspurs and do not appear to cause significant losses of livestock (Kingsbury, 1964). Larkspurs, on the other hand, are responsible for large cattle losses. Since before 1900, and to the present, larkspurs have been considered to cause more cattle losses on western rangelands in the U.S. than any other poisonous plant (James *et al.*, 1992; Ralphs *et al.*, 1988).

The larkspurs are divided arbitrarily into two groups: low larkspurs and tall larkspurs. This refers specifically to their growth habits; low

FIGURE 13–6 A single specimen of tall larkspur on a mountain range. (Courtesy of A.P. Appleby)

larkspurs are usually under 3 ft in height, while tall larkspurs average from 3 to 6 ft or more. The low larkspurs are found at lower elevations while tall larkspurs are found in high mountain regions (Fig. 13–6). *Delphinium barbeyi, D. occidentale, D. glaucum, D. glaucescens* and *D. trolliifolium* are principal species of tall larkspur, while *D. andersonii, D. nuttallianum,* and *D. menziesii* are major low larkspur types. The **low larkspurs** tend to grow at lower elevations where moisture is quite limiting in the summer; they begin growth very early in the spring, sometimes growing through snow. Cattle losses occur at this time when animals are seeking out new green herbage. The **tall larkspurs** grow at high elevations in deep soils where moisture is abundant. They are found in open meadows or under dense tree cover such as aspen and conifer groves. They are climax species and increase with improving range conditions. They are quite palatable to cattle.

Cattle on western U.S. rangelands are commonly moved from low elevations in the spring to high mountain ranges in the summer. They are exposed to a progression of various spp. of low and tall larkspurs as they make this annual migration. Calves are often poisoned in the mountain ranges. They tend to remain in groves of conifers or aspen with a "babysitting cow" while most of the cow herd is out grazing in meadows. The calves may graze on small patches of tall larkspur in the tree groves and because of their small size, can easily ingest a lethal dose. Periods of gluttonous larkspur consumption by cows often occur during summer storms (Ralphs *et al.*, 1994). These authors suggest that changes in barometric pressure, temperature and plant chemistry are involved in altering the grazing behavior of cattle. For as yet undefined reasons, cattle sometimes engage in gluttonous bouts of larkspur feeding, resulting in toxicosis.

Larkspur toxicity symptoms are rarely observed in range livestock because of the quick-acting effects of the alkaloids; generally, when first seen, the animals are dead (Fig. 13–7). Rapid bloating occurs as the animals often fall down with their head pointing downhill, resulting in an accumulation of rumen gases. Cattle have been experimentally poisoned (Pfister *et al.*, 1994a), and the toxic signs are those of impairment of the nervous system. Signs include uneasiness, stiffness of gait, a straddled stance, with the hindlimbs far apart, followed by collapse. Signs of nausea, abdominal pain, weakness, and involuntary muscle twitching are observed. Death occurs from respiratory paralysis. Sheep are less susceptible than cattle; it takes about four times as much larkspur per kilogram body weight to kill sheep as cattle. The lethal dose in cattle is about 17 g of green plant per kilogram body weight, or about 0.5% of body weight.

FIGURE 13–7 A typical example of larkspur poisoning. The aspen grove in the background, on a high mountain range, is typical larkspur habitat. (Courtesy of J.D. Olsen)

The toxicity of various larkspur spp. varies, depending upon their alkaloid profile and total alkaloid content. *D. barbeyi* is the most toxic species (Manners *et al.*, 1993; Pfister *et al.*, 1994b). Alkaloid levels vary markedly from site to site and year to year. Larkspur consumption by cattle also varies greatly, with virtually no consumption in some years and high intakes in others, with no clear associations of intake with alkaloid levels, availability of other forage, etc. Pfister *et al.* (1988) proposed a "toxic window" for cattle losses. During early growth, the toxicity is high but intake is low. After flowering, the plant is palatable but alkaloid content is low. The "toxic window" when poisoning is most likely to occur is during the flowering stage, when both palatability and alkaloid content are high (Fig. 13–8). There is no consistent relationship between alkaloid content and apparent palatability of larkspur to cattle and sheep (Pfister *et al.*, 1996).

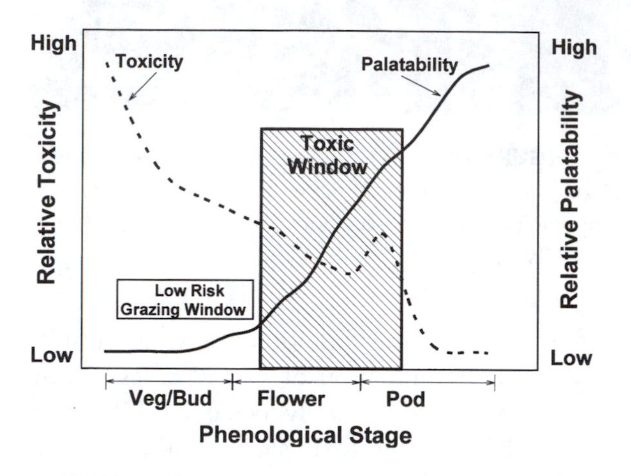

FIGURE 13–8 The "toxic window" of larkspur poisoning, when palatability and alkaloid content of larkspur during the flowering period are likely to result in larkspur poisoning of cattle. (Courtesy of J.A. Pfister)

Because sheep are quite resistant to larkspur poisoning, they may in some situations be useful in controlling the plant and reducing cattle losses. Ralphs and Olsen (1992) demonstrated that sheep grazing could be used to reduce consumption of larkspur by cattle. Herding sheep into patches of larkspur, with trampling and soiling of the plants, helps to reduce subsequent larkspur consumption by cattle (Ralphs *et al.*, 1991). Sheep consume larkspur when it is in the flowering stage, but may be unwilling to feed on the early growth. Even under heavy sheep grazing, most larkspur plants survive and set seed each year.

Larkspurs contain complex diterpene alkaloids. Alkaloids in poisonous rangeland larkspurs have been reviewed by Manners *et al.* (1992, 1993, 1995). Over 40 specific alkaloids have been isolated from toxic larkspurs. **Methyllycaconitine** (Fig. 13–9) is the most

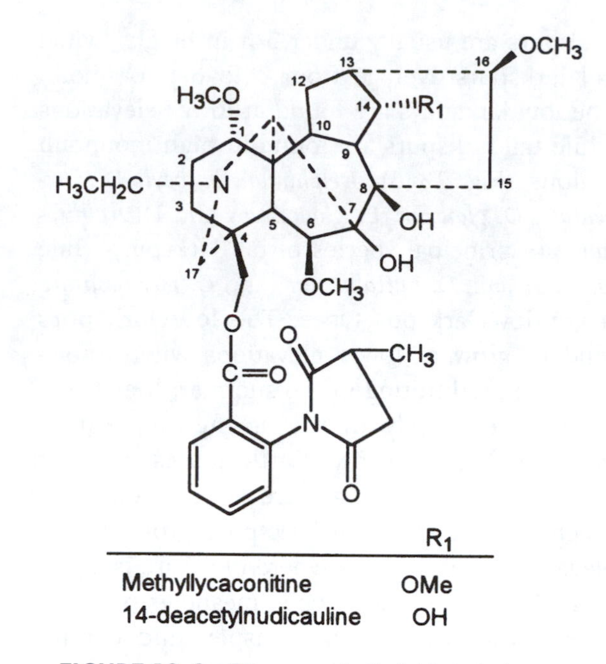

FIGURE 13–9 Diterpenoid alkaloids in larkspurs.

toxic and widely distributed diterpene alkaloid in the toxic larkspurs (Manners *et al.*, 1993, 1995). The larkspur alkaloids are neuromuscular junction blocking agents; they block post-synaptic acetylcholine receptors in the central and peripheral nervous systems (Pfister and Manners, 1995), resulting in clinical signs of neurological disturbance and death from respiratory failure. The alkaloids are found in highest concentrations in the early plant growth, in growing tips and new leaf growth. In general, the plants are most toxic in the spring and lose toxicity throughout the growing season, particularly after flowering. Losses can be reduced by managing cattle so that they are not allowed to graze larkspur-infested areas until after the plants have flowered.

There is little information on the metabolism of larkspur alkaloids. They are stable in sheep and cattle rumen fluid, with no evidence of **rumen detoxification** (Majak, 1993; Siemion *et al.*, 1992).

Several means of reducing larkspur poisoning have been investigated. As previously mentioned, prior grazing of larkspur-infested areas with sheep before cattle grazing may reduce losses. For many years, **mineral supplements** have been marketed with the claim that they will reduce cattle losses. However, no positive effects of mineral supplementation have ever been verified experimentally (Pfister and Manners, 1991, 1995).

Under experimental conditions, the drug **physostigmine** is effective in treating larkspur poisoning (Nation *et al.*, 1982; Pfister *et al.*, 1994c). This drug increases acetylcholine concentrations at neuromuscular junction receptor sites. However, use of this treatment would be impractical under field conditions, because of the rapid course of larkspur intoxication. Carbachol is a long-acting drug that activates cholinergic receptors. However, administration of carbachol does not reduce susceptibility of cattle to larkspur toxicosis (Pfister and Manners, 1995).

An interesting approach to control of larkspur losses is to train cattle not to eat the plant. This approach has been reviewed by Lane *et al.* (1990) and Ralphs (1992). In this procedure, cattle are fed larkspur and at the same time administered a dose of lithium chloride. This drug stimulates the emetic center, causing the animal to feel nauseous. It then associates the feelings of illness with the larkspur recently consumed. This process is known as **aversive conditioning**. While cattle can be successfully rendered aversive to larkspur by this technique, the aversiveness can be lost if averted cattle observe other cattle consuming larkspur. The grazing behavior of livestock is strongly influenced by what peers are consuming. The practicality of aversive conditioning under field conditions is not established.

The most promising management technique to prevent cattle losses appears to be spot spraying of larkspur patches with herbicides, followed by efforts to revegetate with other types of plants. Conversion of a larkspur-dominated plant community to one dominated by grass improves the range for livestock and wild animals and the quality of a watershed. It must also be recognized that larkspurs are wildflowers with aesthetic value to hikers and other users of rangelands. Thus, on public lands, livestock management to control losses is preferred over herbicide treatments, in accordance with multiple use concepts. Opponents of public lands grazing in the U.S. have claimed to have distributed larkspur seeds on rangelands in an effort to discourage cattle ranching.

The Locoweeds (*Astragalus* and *Oxytropis* spp.) and Other Plants with Indolizidine Alkaloids

Locoism, caused by the consumption of certain *Astragalus* and *Oxytropis* spp. by cattle, horses, and sheep on western rangelands, has been recognized for over 75 years as being very similar physiologically to *Swainsona* spp. (poison pea) poisoning of livestock in Western Australia. The toxic factor(s) in both cases remained elusive until researchers at Murdoch University in Western Australia identified the indolizidine alkaloid **swainsonine** as the causative agent (Dorling *et al.*, 1980).

There are over 50 species of ***Swainsona*** in Australia, with toxicity problems occurring mainly in Western Australia and the Northern Territory (Fig. 13–10). These plants are found primarily in unimproved range areas where extensive livestock production with year-round

FIGURE 13–10 A *Swainsona* spp. from Australia. The toxic indolizidine alkaloid swainsonine was first isolated from *Swainsona*.

grazing is conducted. The plants are perennials; following rains, they are often the dominant forage available. Locoweeds in the U.S. are mainly ***Astragalus*** spp. and a few ***Oxytropis*** spp. (Fig. 13–11). There are at least 372 *Astragalus* spp. in the U.S., and most are toxic. Some contain poisonous nitrocompounds (e.g, miserotoxin), while others accumulate toxic levels of selenium. A third group of about 13 species causes locoism. The symptoms of locoweed and *Swainsona* poisoning are similar and will be discussed collectively.

FIGURE 13–11 *Oxytropis serecia*, a locoweed. (Courtesy of L.F. James)

Consumption of the plants for a few weeks to a month is necessary before obvious signs of intoxication are seen. The principal signs are those of nervous system impairment. These include dullness and depression, excitement when disturbed (loco means crazy in Spanish), loss of sense of direction and of herding instinct in sheep, and **habituation** to the plants, or an apparent craving for them. Affected animals seek out locoweed. A true addiction is not involved, but animals may acquire a preference (habituation) for locoweed (Ralphs *et al.*, 1990, 1991), and consume it in preference to other available forage. Cattle may learn to eat locoweed by observing others eating it (Ralphs *et al.*, 1994), indicating a role of social facilitation in grazing behavior. Removal of cattle that begin to eat locoweeds may help to prevent a locoweed problem involving the entire herd (Ralphs *et al.*, 1994). Cattle graze locoweeds primarily in early spring when there is a shortage of other forage. Even those habituated to locoweed will switch to grass when it becomes abundant. According to Ralphs *et al.* (1993), the surest means to prevent locoweed poisoning in cattle is to have locoweed-free pastures available for early spring grazing.

Up to 60% of ewes on locoweed-infested range may abort, while abortion is common in cows consuming *Swainsona*. With locoweed, **congenital malformations** in lambs and calves are seen. These include a permanent flexure of the carpal joints and contracted tendons (Fig. 13–12). These deformities are observed when ewes ingest locoweed at almost any period of gestation. When locoweed is consumed in the period of 90–120 days of pregnancy by ewes, fetal abnormalities such as enlarged heart, spleen, and thyroid may be observed. Fetal edema and lesions of the placenta may be seen. Lambs born to ewes fed locoweed are developmentally impaired, lacking a nursing instinct and the ability to seek their mothers (Pfister *et al.*, 1992). Limb deformities in foals (Fig. 13–12) from mares consuming locoweed have been reported (McIlwraith and James, 1982). The teratogenic effects are related to swainsonine-induced defects in glycoprotein synthesis and organization in the fetus (Warren *et al.*, 1989).

Locoweed-poisoned animals seem to be susceptible to infections such as pneumonia, footrot, and pink eye, suggesting possible impairment of the **immune system** (Sharma *et al.*, 1984). Sharma and his associates reported that sheep fed locoweed had decreases in total leukocytes and lymphocytes, with an indication of

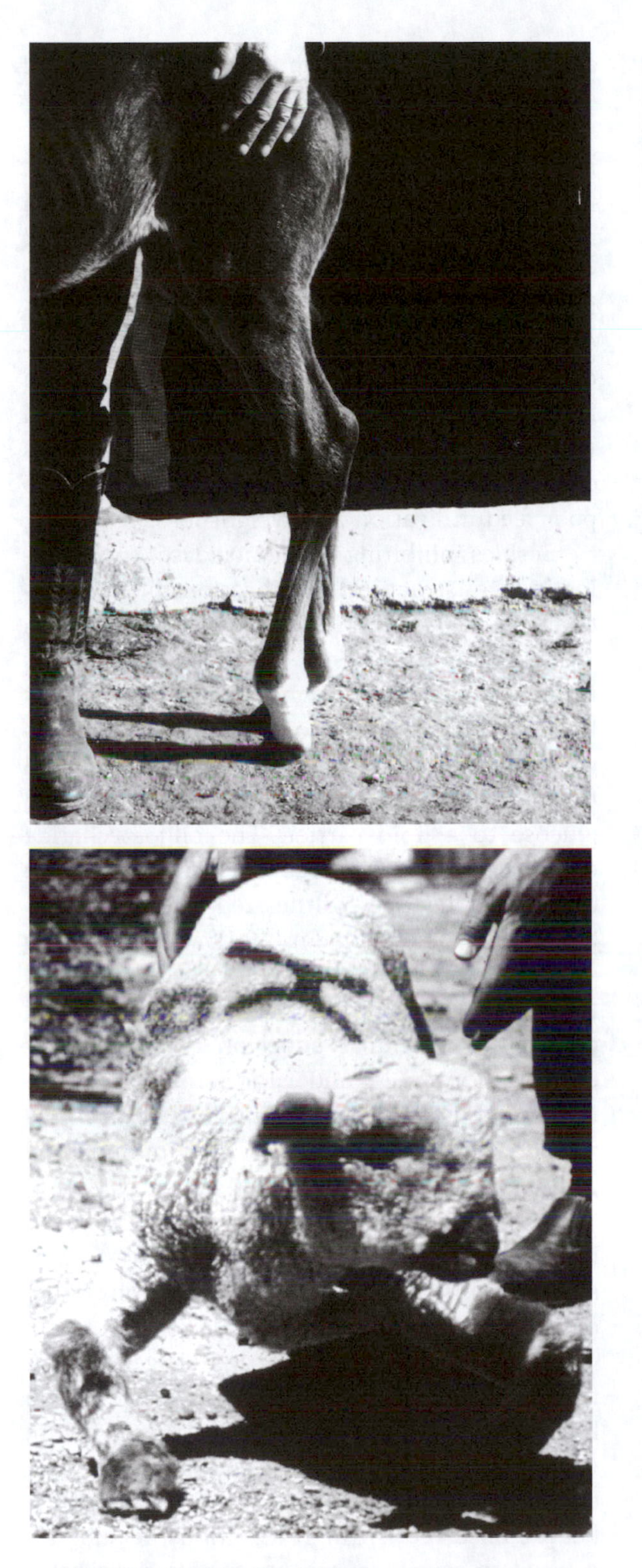

FIGURE 13–12 (Top): Flexure deformity in a foal from a mare that consumed locoweed. (Courtesy of C.W. McIlwraith) (Bottom): Lamb from a ewe fed locoweed (*Astragalus lentiginosus*). (Courtesy of L.F. James)

a selective effect on cell-mediated immune responses (see also Chapter 16).

Locoweed consumption may result in mortality of affected animals. Signs of toxicity in cattle and sheep are seen after consumption of about 90% of body weight of locoweed (Kingsbury, 1964), or about 2 months of exposure. Death may occur after an intake of about 300% of body weight of locoweed. Horses are most susceptible, with intake of about 30% of body weight being a lethal dose. Affected horses are listless and unaware of their surroundings, but when excited, become wild and unmanageable. Horses in the early stages of loco intoxication are dangerous to ride. Deaths of affected livestock may occur from misadventure, such as tumbling over cliffs and banks.

Signs of toxicity from ***Swainsona* poisoning** are similar to those from North American locoweeds. Habituation to *Swainsona* is also seen. With both locoweeds and *Swainsona*, some differences in neurological signs between sheep and cattle are observed. Cattle have a low head carriage, a stiff clumsy gait, head shaking, and staring eyes. Sheep have a high head carriage, and a high-stepping stiff gait, and also show head tremors, staring eyes, disturbed vision and incoordination.

The predominant microscopic lesions are cytoplasmic vacuoles in organs and nerve tissue that have been identified as swollen lysosomes. Dorling *et al.* (1978) demonstrated that the lysosomes contained high levels of oligosaccharides composed largely of mannose. The lesions and composition of the lysosomes were strikingly similar to those observed in **mannosidosis** in humans and cattle, which is a lysosome storage disease due to a genetic deficiency of lysosomal α-mannosidase. Subsequent studies by these workers (Dorling *et al.*, 1980) showed that *Swainsona* contains **indolizidine alkaloids** that inhibit α-mannosidase. The major alkaloid involved is **swainsonine** (Fig. 13–13).

The α-mannosidase acts upon the mannose residues appearing at the terminal nonreducing end of a chain of sugars. Other glycosidases remove their respective sugars from oligosaccha-

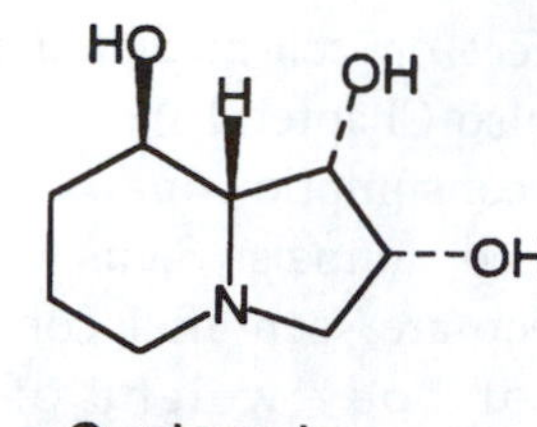

FIGURE 13–13 Swainsonine, the toxic indolizidine alkaloid in locoweeds.

rides. If α-mannosidase is inhibited, other glycosidases remove their respective sugars until a mannose residue is reached, and then hydrolysis of the carbohydrate stops. The undigested oligosaccharides accumulate within the lysosomes. The number of lysosomes in the affected cells increases to accommodate the increasing quantity of oligosaccharide, which eventually disrupts cellular function (Huxtable and Dorling, 1982). The effect of indolizidine alkaloid on α-mannosidase is reversible, and the accumulation of stored carbohydrate diminishes when alkaloid is no longer consumed. The vacuoles will largely disappear in 10–12 days, so that animals showing early clinical signs generally recover completely. Advanced clinical signs are irreversible, because of axon degeneration from the crowding of perikaryon with storage vacuoles. Although vacuolation occurs in most tissues, it is the central nervous system that is most sensitive to mannose accumulation, with functional impairment and degenerative changes. After the Australian work was reported, indolizidine alkaloids were shown to be the active compounds in locoweeds (Molyneux and James, 1982). Although the locoweeds contain several indolizidine alkaloids, swainsonine appears to be the principal cause of toxicity.

Swainsonine is rapidly absorbed, leading to rapid **mannosidase inhibition**. Mannosidase activity returns to normal within six days after removal of locoweeds from the diet (Stegelmeier *et al.*, 1995a). Swainsonine is apparently not detoxified prior to excretion, as urine from sheep fed locoweed is toxic to rats (James and Van Kampen, 1976) and milk from cows fed *Astragalus* produced locoism in kittens, lambs and calves (James and Hartley, 1977).

A characteristic feature of swainsonine intoxication is failure to thrive. Pritchard *et al.* (1990) observed that swainsonine causes a marked impairment of growth as a result of suppression of appetite. Water intake is also impaired. Emaciation and wasting in chronic locoweed poisoning may be in part a consequence of neurologic damage, resulting in difficulty in finding, prehending and masticating food (Stegelmeier *et al.*, 1995b). Inhibited synthesis of intestinal glycosidases may also contribute to poor feed utilization and weight loss.

Besides inhibiting α-mannosidase, swainsonine induces structural modifications of glycoproteins, which contributes to clinical signs (Pritchard *et al.*, 1990). Swainsonine inhibits Golgi mannosidase II, which is involved in N-linked **glycoprotein processing**. N-linked oligosaccharides are synthesized via the sequential transfer of N-acetyl-glucosamine, mannose and glucose to a lipid carrier. The oligosaccharide chain is then transferred to growing peptide chains as they are synthesized on membrane-bound ribosomes (Elbein, 1991). The resulting alterations in the synthesis, processing and transport of glycoproteins result in altered and dysfunctional cellular adhesion molecules, circulating hormones, and various membrane receptors, thus explaining clinical signs such as abnormal endocrine, reproductive, immune and gastrointestinal function (Stegelmeier *et al.*, 1995b). The toxicology of swainsonine has been reviewed in detail by James *et al.* (1989). Swainsonine has potential medical applications as an anti-cancer agent (James *et al.*, 1989).

Cattle grazing high-elevation rangelands infested with the locoweed *Oxytropis sericea* have a high incidence of congestive right heart failure (**high mountain disease**). Calves are most susceptible. Swelling under the jaw and of the brisket are observed. At high elevations, feeding locoweed to calves caused congestive right heart failure, while at lower elevations, locoweed poisoning occurred (James *et al.*, 1986, 1991). Because swainsonine influences the vascular system, its effects are probably intensified by high elevations.

Supplementation of animals fed locoweed with **mineral supplements** did not reduce toxicity (Bachman *et al.*, 1992; Pulsipher *et al.*, 1994), despite commercial claims made for such products. Although consumption of locoweed by cattle does result in reduced liver copper concentrations, toxic effects of locoweed would occur before copper deficiency would be induced, and copper supplementation seems unlikely to alter the course of locoweed intoxication (Galyean *et al.*, 1996).

Other toxic indolizidine alkaloids include **slaframine** in moldy red clover (see Chapter 10) and **castanospermine**, which occurs in the leaves, seeds, and bark of *Castanospermum australe* ((Moreton Bay chesnut), a tree native to eastern Australia. Livestock poisonings occur during dry periods when forage is scarce and animals consume large quantities of the seeds. Severe diarrhea is the most consistent sign of toxicity, accompanied by general debilitation. Severe inflammation of the digestive tract occurs (McKenzie *et al.*, 1988). Inhibition of digestive enzymes may contribute to these effects (Pan *et al.*, 1993).

A comprehensive review of glycosidase inhibitors and their mode of action in oligosaccharide and glycoprotein processing is that of Elbein (1991).

Nitrotoxins in *Astragalus* spp.

In addition to swainsonine-containing *Astragalus* spp., others are **selenium accumulators** (see Table 13–1). Selenium toxicity is discussed in Chapter 12.

A third type of toxin found in *Astragalus* spp. is nitroglycosides. Simple aliphatic compounds containing a nitrite moiety (**nitrotoxins**) are found in several *Leguminosae* of agricultural importance. These include various *Astragalus* spp., crown vetch (*Coronilla varia*), trefoils (*Lotus* spp.), and *Indigofera spicata*, a tropical forage.

TABLE 13–1 Classification of Common *Astragalus* spp. According to Type of Toxin

Selenium accumulators	Indolizidine alkaloids (locoweeds)	Nitro-containing glycosides[a]
A. bisulcatus	*A. argillophilus*	*A. atropubescens* (3-NPOH)
A. racemosus	*A. bisulcatus*	*A. canadensis* (3-NPA)
A. pectinatus	*A. earlei*	*A. cibaria* (3-NPA)
A. pattersonii	*A. lentiginosus*	*A. convallarius* (3-NPOH)
	A. mollissimus	*A. diversifolius* (3-NPOH)
	A. nothoxys	*A. falcatus* (3-NPA)
	A. pubentissumus	*A. miser* (3-NPOH)
	A. thurberi	*A. pterocarpus* (3-NPOH)
	A. wootonii	
	Oxytropis lambertii *Oxytropis sericea* (= *O. lambertii* var. *sericea*)	

[a]The glycosides are metabolized in ruminants to 3-NPOH or 3-NPA as indicated for each species in parentheses.

The nitrocompounds in *Astragalus* spp. are of the most significance in livestock production.

Of the at least 372 *Astragalus* spp. in North America, about 263 are poisonous to livestock because of the nitro-containing glycosides they contain. About 13 species cause locoism, while about 25 species may accumulate toxic levels of selenium. Examples of some of these are listed in Table 13–1. A number of *Astragalus* spp. are excellent forages containing no toxic entities. An example is cicer milk vetch (*A. cicer*) which is grown as an irrigated and dryland hay and pasture crop in the Great Plains and western states of the U.S. In the Middle East (i.e. Pakistan, Iran, Turkey), wild *Astragalus* spp. are tapped for their latex, which is used to make **gum tragacanth**, a pharmaceutical and food processing additive.

Timber milk vetch (*A. miser*) is a prominent example of a nitro-containing species (Fig. 13–14). The toxic principles are glycosides containing a simple aliphatic moiety with a nitrite group. The glycoside in milk vetch is called **miserotoxin**, which contains β-D-glucose and 3-nitropropanol (3-NPOH). Some *Astragalus* spp. contain glycosides with 3-NPOH, while three others contain 3-nitropropionic acid (3-NPA). The two nitrocompounds do not occur together in the same species.

FIGURE 13–14 A specimen of *Astragalus miser* (timber milk vetch). (Courtesy of M.C. Williams)

Both acute and chronic toxicity occur in livestock that consume *Astragalus* spp. containing nitroglycosides. **Acute poisoning** is characterized by weakness, convulsions, frequent urination, rapid beating of the heart, labored and rasping breathing, coma, and death. Cyanosis of the oral cavity and nasal passages is evident, associated with methemoglobinemia. **Chronic toxicity** is often referred to as "cracker heels" because the fetlocks knuckle over and general weakness in the hindquarters causes the hooves to rub against each other (Fig. 13–15). There is loss of neural control of the extremities, so the animal staggers or weaves when walking. Blindness or impaired vision may occur because of degeneration of the optic nerve. Livestock rarely recover from *Astragalus* intoxication, but may linger for several months after being removed from the toxic forage. James *et al.* (1980) performed feeding trials with various *Astragalus* spp. and provide a good summary of toxic doses and signs of toxicity. Williams *et al.* (1979) described in detail an outbreak of emory milk vetch poisoning of cattle and sheep in New Mexico.

FIGURE 13–15 Chronic toxicity signs due to ingestion of *Astragalus emoryanus* (Emory milk vetch). The cow displays the knuckled-over fetlocks, or "cracker heels," and general weakness of the hindquarters. (Courtesy of M.C. Williams)

$$\text{glucose}-O-CH_2-CH_2-CH_2-NO_2$$

Miserotoxin

$$HOOC-CH_2-CH_2-NO_2$$

3-nitropropionic acid

$$HO-CH_2-CH_2-CH_2-NO_2$$

3-nitro-l-propanol

FIGURE 13–16 Miserotoxin is one of the most important toxic nitroglycosides. It can be hydrolyzed to both 3-NPA and 3-NPOH.

The toxic principles are glycosides of 3-NPOH or 3-NPA (Fig. 13–16). Glycosides containing 3-NPOH, such as miserotoxin, are hydrolyzed to release 3-NPOH in the rumen, whereas in nonruminants, 3-NPA is produced. The 3-NPOH is rapidly absorbed from the rumen and is toxic, whereas 3-NPA is absorbed more slowly and appears to be degraded in the rumen to nontoxic compounds. Majak and Clark (1980) and Anderson *et al.* (1993) have studied the ruminal metabolism of aliphatic nitrocompounds. Miserotoxin is rapidly hydrolyzed by rumen microorganisms; the 3-NPOH released can be degraded slowly, but at a much slower rate than the hydrolysis of the glycosidic bond. Per unit of NO_2 absorbed, 3-NPOH is much more toxic than 3-NPA. However, *Astragalus* spp. containing 3-NPA often have several times as much glycoside as those with 3-NPOH, so the toxicity of the plants is often similar. Nitrotoxin degrading ruminal microbes rapidly adapt to nitrocompounds in the diet with enhanced detoxification rates (Majak, 1992; Anderson *et al.*, 1993). The toxic dose of *A. miser* is about 4.8 g/kg body weight in cattle, which is equivalent to about 25 mg NO_2 per kilogram (about 100 mg NO_2 per kilogram is lethal). Sheep can tolerate 3–4 times as much per kilogram as cattle.

The metabolic basis of the toxic effect is not entirely clear. The acute toxicity is related to, but not exclusively caused by, **methemoglobinemia** in which hemoglobin is oxidized by nitrite:

$$\underset{\text{ferrous}}{\text{Hemoglobin-}Fe^{2+}} + NO_2^- \longrightarrow \underset{\text{ferric}}{\text{Methemoglobin-}Fe^{3+}}$$

Methemoglobin is incapable of carrying oxygen. Death usually occurs in acute *Astragalus* toxicity when methemoglobin exceeds 20% of the total hemoglobin. This degree of methemoglobinemia is not lethal when induced by nitrite administration alone. Therefore, an additional factor, such as a metabolite formed when NO_2 is released from 3-NPOH or 3-NPA, seems to be involved. Majak *et al.* (1981) studied the metabolism of intravenously administered 3-NPOH in cattle. They obtained evidence that 3-NPOH is metabolized at a site other than the circulatory system to yield NO_2^- and an unidentified metabolite(s) that may be involved in the intoxication. Nitrite is released into the blood, oxidizing hemoglobin to methemoglobin. Hemoglobin is subsequently regenerated by methemoglobin reductase, and the NO_2^- is converted to nitrate and excreted.

In **chronic toxicity**, levels of methemoglobin of 4–6% of total hemoglobin are observed. The mechanisms by which the nitrocompounds cause chronic toxicity are not conclusively known (Majak and Pass, 1989; Maricle *et al.*, 1996). Alston *et al.* (1977) have shown that 3-NPA is an inhibitor of the tricarboxylic acid cycle enzyme succinic dehydrogenase. The 3-NPA may bind to FAD, a co-factor for succinic dehydrogenase, making the FAD unreactive. In the process, nitrite is given off. The 3-NPA is structurally similar to succinate and may be acting as a competitive inhibitor. This reaction with FAD could explain both the metabolic effects and the occurrence of methemoglobinemia. Ludolph *et al.* (1991, 1992) propose that 3-NPA is a neurotoxin which causes selective degeneration of striatal and hippocampal regions of the brain. These investigators demonstrated that 3-NPA is the active neurotoxin in a neurological disorder of humans in China, associated with the consumption of mildewed sugar cane. Sugar cane is harvested and stored for consumption as a confectionary during celebration of the new year. Ludolph (1991, 1992) reviewed the **sugar cane toxicosis** condition; the 3-NPA is a metabolite of *Arthrinium* fungi that grow on the sugar cane. Moldy sugar cane poisoning includes signs of vomiting, diarrhea, and neurologic symptoms such as seizures, involuntary muscle contractions and coma (Ming, 1995).

The *Astragalus* nitrocompounds occur primarily in the leaves and reach their highest concentrations during pod formation. The levels drop rapidly when the leaves begin to dry. Herbicides cause the leaves to bleach and lose their activity. The toxins are stable in dried green specimens; nitrocompounds have been measured in herbarium specimens collected over 100 years ago (Williams and James, 1978). Williams (1981b) has examined 1690 spp. of Old World and South American *Astragalus* for alipatic nitrocompounds, using small (20 mg) samples of herbarium specimens. Nitrocompounds were detected in 12% of the Old World and 45% of the South American species, including some specimens collected between 1822 and 1836. Because *Astragalus* spp. are often considered for seeding on mine spoils, and reclamation of disturbed sites following oil, gas, and coal extraction on public grazing lands in the western states, it is important that introduced species be screened to be certain they are nontoxic. Sicklepod milk vetch (*A. falcatus*) was introduced to western ranges and subsequently was found to be highly toxic to sheep and cattle (Williams *et al.*, 1976).

Bees have been poisoned by foraging on *Astragalus* spp. James *et al.* (1978) demonstrated that the pollen of *A. lentiginosus* was toxic to mice, so presumably the toxins would be present to affect bees. Majak *et al.* (1980) noted several outbreaks of high mortality of bees foraging on timber milk vetch *A. miser*). The nectar was found to contain miserotoxin. Feeding trials with caged bees demonstrated the toxicity of the glycoside, with signs of incoordination, weakness, and inability to fly, followed by death. The possibility of toxicity of the honey, produced by bees feeding on *Astragalus* spp., to other organisms was not examined. The toxicity of *Astragalus* to bees is unusual in that comparatively few plants produce nectar or pollen which is poisonous to honeybees (Majak *et al.*, 1980).

Crown vetch (*C. varia*) is well adapted to the U.S. Northeast and Midwest (Fig. 13–17). It is grown to a limited extent as a forage crop. It contains nitroglycosides (Gustine, 1979) such as coronarian (Fig. 13–18). Crown vetch is not toxic to ruminants. Coronarian is hydrolyzed in the rumen, releasing 3-NPA. Gustine *et al.* (1977) demonstrated that 3-NPA is degraded in rumen fluid to nontoxic metabolites. They established that 3-NPA can be completely detoxi-

FIGURE 13–17 Crown vetch (*Coronilla varia*) is a forage plant grown in the U.S. Northeast and Midwest. It contains nitroglycosides, such as coronarian.

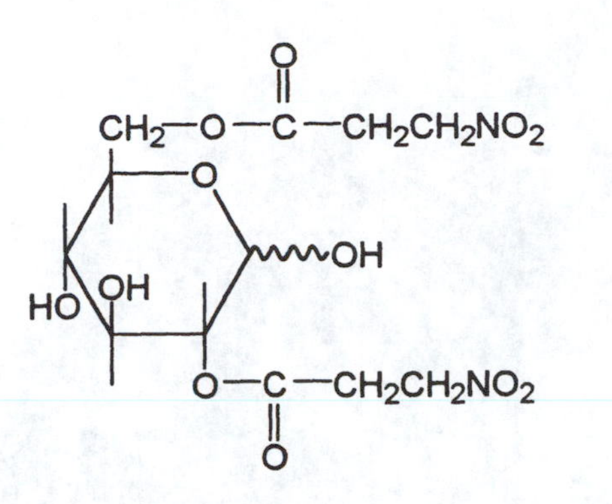

FIGURE 13–18 Coronarian, the nitroglycoside in crown vetch.

fied in the rumen if its concentration does not exceed 1 mg/ml of rumen fluid. Such a level is unlikely to be surpassed as a result of consumption of crown vetch.

Nonruminants can be poisoned by crown vetch. Symptoms of 3-NPA toxicity include growth depression, ataxia, posterior paralysis, and death (Shenk *et al.*, 1976). The toxicity may be due to a combination of methemoglobinemia and inhibition of succinate dehydrogenase (Alston *et al.*, 1977).

Neurotoxic *Solanum* spp.

The glycoalkaloids in potatoes and nightshades are cholinesterase inhibitors, causing neurological signs. Solanum glycoalkaloids in potatoes are discussed in Chapter 6. Nightshades are discussed in Chapter 12. A neurological problem associated with consumption of certain *Solanum* spp. will be discussed here.

Several *Solanum* spp. cause **cerebellar degeneration** in cattle of a similar nature to that induced by *Swainsona* and *Astragalus* spp. The condition was first described by Pienaar *et al.* (1976) in South Africa, in cattle consuming foliage of *S. kwebense*. The disorder is characterized by recurrent seizures with loss of equilibrium, extension of the head and thoracic limbs, opisthotonus, rapid eye movements, and falling to the side or backward (Fig. 13–19). The main pathological signs are vacuolation, degeneration, and loss of Purkinje cells, with cytoplasmic inclusion bodies similar to those of induced lysosomal storage diseases such as locoism. It is usually necessary to frighten or disturb affected animals to elicit symptoms (Kellerman *et al.*, 1988). Riet-Correa *et al.* (1983) described a similar condition in Brazil in cattle consuming *S. fastigiatum* and *S. bonariensis*. Similar signs and lesions were described for intoxication of cattle by *S. dimidiatum* in the southwestern U.S. (Menzies *et al.*, 1979). Zambrano *et al.* (1985) investigated rats, guinea pigs, rabbits and sheep for their susceptibility to *S. fastigiatum* poisoning. Only sheep were affected, with signs similar to those of cattle.

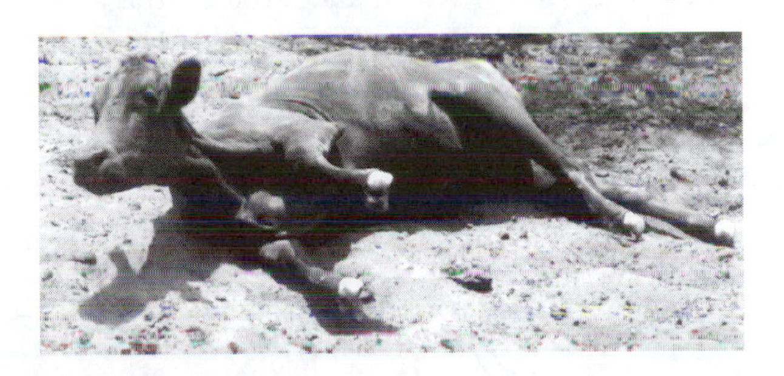

FIGURE 13–19 A bovine in South Africa showing signs of *Solanum kwebense* toxicity.

The toxic entity in *Solanum* spp. causing cerebellar degeneration has not been conclusively identified. Molyneux *et al.* (1994) suggested that **calystegins** (polyhydroxy alkaloids similar in structure to indolizidine alkaloids such as swainsonine) may be the active components. They identified calystegins in *S. dimidiatum* and *S. kwebense*, and demonstrated that calystegins are glycosidase inhibitors similar in action to swainsonine. Molyneux *et al.* (1994) suggested that the neurological signs and cerebellar degeneration associated with the *Solanum* species is an induced lysosomal storage disease, analagous to locoism, caused by calystegin inhibition of β-glucosidase. Properties of calystegins have been reviewed by Molyneux *et al.* (1993).

A locomotor disorder in sheep in Australia, with degeneration of spinal cord white matter, has been linked to consumption of *Solanum esuriale* (Dunster and McKenzie, 1987).

Datura spp.

Jimsonweed, or thorn apple (*Datura stramonium*), is a common inhabitant of waste areas, barnyards, and edges of cultivated fields in much of the U.S. It is a large (3–5 ft tall) coarse annual. The fruit is an erect spiny capsule, while the flowers are erect, up to 4 in long, and white (Fig. 13–20). It contains **tropane alkaloids** (Fig. 13–21) which affect the central nervous system, with the major alkaloids being **atropine** and **scopolamine** (Friedman and Levin, 1989). The tropane alkaloids are divided into two main groups: the solanaceous and the coca alkaloids. Examples of the solanaceous alkaloids are hyoscyamine (atropine) and its epoxide derivative, scopolamine (hyoscine). **Cocaine** is the most notorious of the coca alkaloids.

FIGURE 13–20 Jimsonweed (*Datura stramonium*) showing the spiny seed pods.

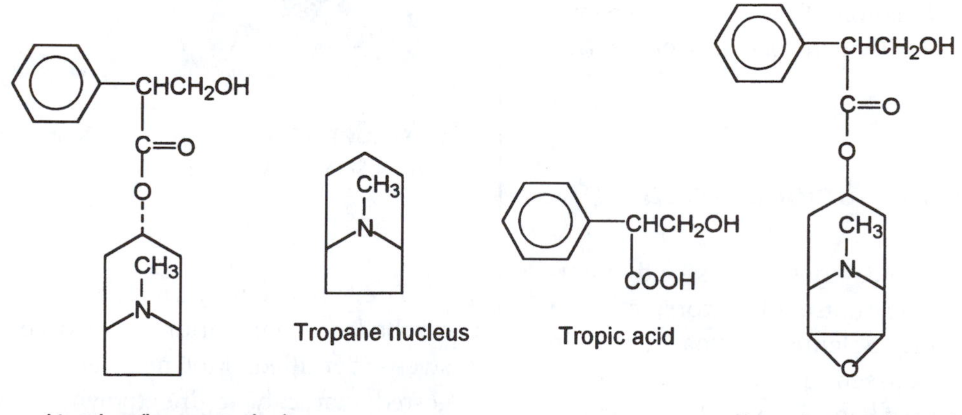

FIGURE 13–21 Tropane alkaloids are composed of the tropane nucleus and a tropic acid side chain.

Jimsonweed is unpalatable to livestock, although reports of poisoning of most classes of livestock are in the literature (Kingsbury, 1964). Poisonings have resulted from grazing fresh material, from contamination of hay and silage, and from ingestion of the seeds. Symptoms in both humans and livestock include intense thirst, disturbed vision, delirium, and violent behavior. Human poisonings result from sucking the nectar from flowers or consuming the seeds, and from contamination of beans and grains. In 1676 a mass poisoning of soldiers in

Jamestown, Virginia, occurred, giving rise to the common names "jamestown weed" and "jimsonweed."

Leipold *et al.* (1973) reported an outbreak of **arthrogryposis** (permanent joint flexure) in newborn pigs that appeared to be associated with consumption of jimsonweed by the pregnant sows. However, Keeler (1981) fed a high level (maintaining chronic signs of intoxication) of jimsonweed to pregnant sows, and no arthrogryposis was observed in the offspring, suggesting that jimsonweed is not teratogenic in pigs. Worthington *et al.* (1981) investigated the toxicity of *Datura* seeds to pigs as a result of a legal action stemming from *Datura*-seed contamination of pig feed. They found that an alkaloid intake of 2.2 mg/kg body weight from seeds containing 0.2–0.6% alkaloid was tolerated with little or no effect. The seeds were very unpalatable, and it was not possible to produce severe toxicity signs because of feed rejection. A high proportion of the whole seed passed through the gut undigested. They concluded that it is difficult if not impossible to kill pigs by feeding *Datura* seeds because of rejection of levels high enough to cause toxicity.

In a study with cattle, Nelson *et al.* (1982) fed diets with various levels of jimsonweed seed containing 0.26% atropine and 0.55% scopolamine. They concluded that death of cattle from *Datura* poisoning is unlikely because rumen atony and anorexia limit intake of the contaminated feed below lethal levels. They suggested that the toxic dose is 2.49 mg atropine and 0.5 mg scopolamine per kilogram body weight, or about 107 seeds per kilogram body weight.

Williams and Scott (1984) noted that contamination of corn with 0.5% *Datura* seeds caused a variety of signs in horses, including anorexia, weight loss, rapid heart and respiration rates, dilation of the pupils, excessive thirst, diarrhea, and excessive urination. Removal of the offending feed resulted in recovery of the affected animals.

Day and Dilworth (1984) examined the toxicity of jimsonweed seed to broiler chicks. About 1% jimsonweed seed was the maximum level that could be used without detrimental effects. Higher levels (3 and 6%) caused pronounced growth depression. Thus, the major effects of jimsonweed contamination of grain or grain screenings appears to be a depression of feed intake.

Chewing or smoking jimsonweed seeds is a common type of **substance abuse** (Dugan *et al.*, 1989; Guharoy and Barajas, 1991). The effects include altered perception of the environment, visual hallucinations, dilation of the pupils and increase in heart rate.

The tropane alkaloids are metabolized by hydrolysis of the ester linkage of the side chain to the tropane ring. A variety of metabolites are produced from scopolamine, including phenolic derivatives and glucuronides, with marked species differences in metabolism (Wada *et al.*, 1991, 1994). A good review of the metabolism of tropane alkaloids is provided by Scheline (1991).

Another plant containing tropane alkaloids is ***Atropa belladonna*** (belladonna, deadly nightshade). Despite its sinister name, it rarely causes livestock poisoning problems, largely because it is unpalatable to animals (Kingsbury, 1964). Human poisonings have occurred with belladonna contamination of herbal products, such as comfrey tea adulterated with belladonna leaves (Tyler, 1993). Schneider *et al.* (1996) reported a case of poisoning of a family who mistook belladonna berries for edible fruit, and consumed a pie made from the berries. Delerium, hallucinations and coma occurred, from which the individuals recovered. Henbane (*Hyoscyamus niger*) is another common weed containing tropane alkaloids. Because of its unpalatability, livestock poisonings are rare. It has been used as an herb for medicinal purposes since ancient times.

Tobacco (*Nicotiana* spp.)

Nicotine is a highly toxic alkaloid produced by *Nicotiana* spp., the wild and cultivated tobaccos. Livestock are exposed to nicotine through the consumption of tobacco or through exposure to nicotine-containing insecticides such as Black Leaf 40. Wild tobaccos (*N. attenuata* and *N. trigonophylla*) are erect herbaceous annuals found in dry sandy desert soils of the western U.S. Poisonings of sheep, cattle, and horses from consumption of these plants have been reported (Kingsbury, 1964). Kingsbury (1964) also mentions a report of a family that used wild tobacco as a boiled green; one fatality resulted. Numerous cases of livestock poisonings from consumption of cultivated tobacco (*N. tabacum*) are known. Horses have died from consumption of tobacco leaves, while pigs have been fatally poisoned when they broke into a tobacco field (Kingsbury, 1964). Nicotine is the presumed toxic agent.

Symptoms of **nicotine toxicity** include excitement, shaking and twitching, rapid respiration, staggering, weakness and prostration, followed by coma, paralysis, and death. Death is due to respiratory paralysis. Nicotine blocks autonomic ganglia and neuromuscular junctions. Large doses cause a descending paralysis of the central nervous system.

Nicotine has been implicated as a **teratogen**. Bush and Crowe (1989) reviewed a number of reports on congenital abnormalities in swine caused by consumption of tobacco stalks. A total of 15 farms in Kentucky experienced epidemics of congenital limb deformities (arthrogryposis) involving 246 sows and 1148 abnormal pigs. The pigs had a persistent flexure or contracture of a limb joint (Fig. 13–22). The affected bones were thickened, curved, or twisted, with a misalignment of joints, resulting in an inability of the pigs to flex or extend their limbs. The pregnant sows had access to freshly discarded tobacco stalks. The stalks were very large, with an abundance of succulent pith, resembling celery. The sows shredded the stalks and consumed the pith. Experimental feeding of the stalks confirmed the teratogenic effect, but it was not conclusively shown that nicotine was the active principle. Later studies, reviewed by Panter *et al.* (1992), have shown that nicotine is not teratogenic and that the active compound is probably **anabasine**, a piperidine-pyridine alkaloid (Fig. 13–23).

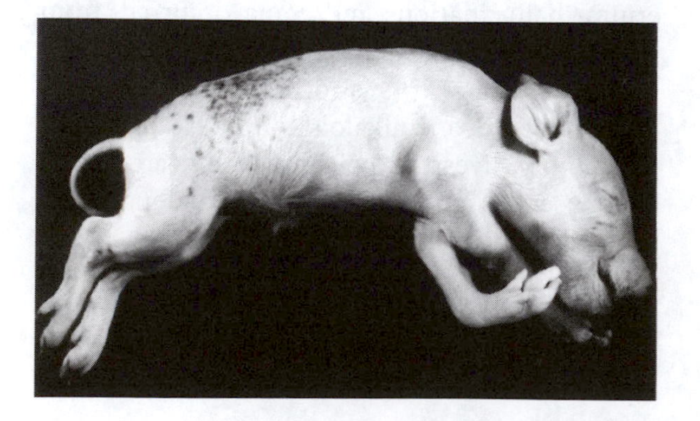

FIGURE 13–22 Arthrogryposis in a pig from a sow which consumed tobacco stalks during gestation. (Courtesy of M. Ward Crowe)

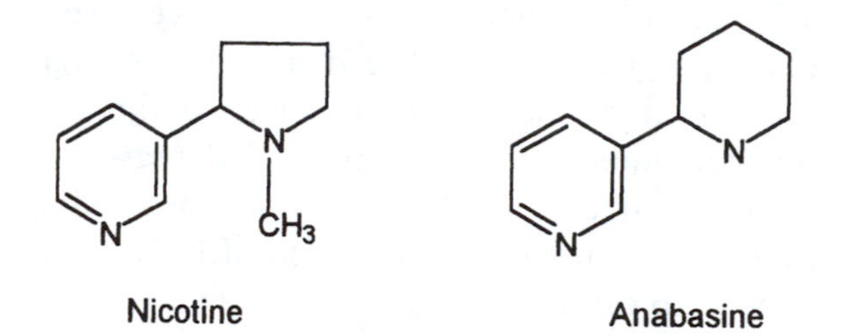

FIGURE 13–23 Piperidine alkaloids in tobacco.

Nicotiana glauca (**tree tobacco**) is an evergreen shrub or small tree commonly found in low-elevation areas in California and Arizona. Consumption of the plant results in teratogenic effects in cattle, pigs and sheep (Panter *et al.*, 1992). Calves from cows gavaged with dried *N. glauca* had arthrogryposis of the forelimbs or curvature of the spine, with the deformities clinically indistinguishable from those caused by maternal ingestion of lupin or *C. maculatum*. Fetal defects in sheep included carpal flexure, lordosis, and cleft palate. The defects in pigs given *N. glauca* were similar to those in pigs from dams that consumed tobacco stalks, except that palate closure defects noted with *N. glauca* consumption are not seen following maternal consumption of tobacco. About 99% of the total alkaloid in *N. glauca* is **anabasine**; this alkaloid is also found in tobacco. It meets the criteria for teratogenicity of coniine analogs established by Keeler and Balls (1978) and is the probable teratogen in both *N. glauca* and tobacco. It has been shown to be teratogenic in swine (Keeler *et al.*, 1984). As with *Lupinus* and *Conium* alkaloids, the teratogenic effects of *Nicotiana* alkaloids are due to their inhibition of fetal movement (Bunch *et al.*, 1992).

A sudden death syndrome of cattle in California was attributed to poisoning by *N. glauca* (Plumlee *et al.*, 1993). Death was from respiratory paralysis.

Yellow Star Thistle (*Centaurea solstitialis*)

The prolonged consumption by horses of yellow star thistle (*Centaurea solstitialis*) or Russian knapweed (*C. repens*) causes a nervous disorder called **equine nigropallidal encephalomalacia (ENE)**. Yellow star thistle causes significant ENE problems in northern California, with a lesser incidence in southern California and southern Oregon. It has also been reported in Argentina and Australia. Russian knapweed has been linked to ENE incidence in several western states, including Colorado, Utah, and Washington (Cordy, 1978). In northern California, two annual peaks of ENE occur: June–July, and October–November. Generally, yellow star thistle grows in weedy horse paddocks or fallow fields.

Toxicity of *Centaurea* species seems restricted to horses. A variety of other domestic and laboratory animals have been fed the plants with no ill effects (Cordy, 1978). Clinical signs of toxicity occur abruptly in horses, characterized by drowsiness, difficulty in eating and drinking, and aimless walking with the muzzle to the ground, or else total inactivity. The animals have particular difficulty in eating and drinking, apparently due to impairment of neural activity of the fifth, seventh, and twelfth cranial nerves. Lesions are typically found in any of four sites in the brain: the globus pallidus and substantia nigra of the left and right sides. The lesions are basically foci of necrotic tissue. Cordy (1978) has suggested that ENE is caused by disruption of the dopaminergic nigrostriatal pathway. The initial toxicity signs are due to release of dopamine from the nigrostriatal nerve endings, and the later signs may reflect dopamine deficiency. Affected horses generally die of starvation or dehydration.

The toxic principle of *Centaurea* species has not been conclusively identified. Sesquiterpene lactones may be the neurotoxin constituent (Wang *et al.*, 1991). The lethal dose of green plant material is about 1.8–2.5 kg/100 kg body weight/day for *C. repens* and 2.3–2.6 kg for *C. solstitialis*.

White Snakeroot

White snakeroot (*Eupatorium rugosum*) (Fig. 13–24) is a showy herbaceous perennial that grows over much of eastern North America. It generally grows in low, moist areas, near streams, and in open woods. It may form dense stands after an area is logged. It grows in late summer and early autumn, reaching 3–4 ft in height, with a white composite-type flower. During dry weather when other forage becomes scarce, livestock are likely to move into wooded areas and graze the plant. It often stays green and succulent late in the fall because it is protected from frost in its woodland habitat. After consuming white snakeroot for several days, livestock become depressed and develop a condition referred to as **trembles**, in which there are muscle tremors around the neck, shoulders, and legs. Affected animals stand in a hunched position. Often there is the odor of acetone on the breath, labored breathing, urinary incontinence, and constipation. In horses, there may be partial throat paralysis. Affected animals may recover, though there may be a long period of inappetence and muscular weakness. In severe poisoning, prostration and death occur. Conges-

FIGURE 13–24 White snakeroot (*Eupatorium rugosum*). (Courtesy of S.S. Nicholson)

tion and fatty degeneration of the liver and kidneys are noted. Myocardial problems, including ascites of the pericardial sac and massive degeneration of the myocardium, have been noted in horses (Olson *et al.*, 1984). Centrilobular necrosis of the liver and elevated serum enzymes, such as SGOT, CPK, LDH (see Chapter 2), also occur. Oral administration of activated charcoal, at 1 g/kg body weight, shows promise as a treatment (W.B. Buck, 1982, Personal communication, University of Illinois, Urbana). White snakeroot causes significant livestock losses in the Midwest. In states such as Illinois, it is one of the major poisonous plants.

White snakeroot provides a classic example of a milk-transferred toxin. The toxin can be transferred in cow's milk to humans, producing a condition called **milk sickness**. Symptoms are weakness, nausea, prostration, ketosis, delirium, coma, and death. In the pioneer days, it sometimes reached epidemic proportions, and whole villages were abandoned because of milk sickness. It is said to be responsible for the death of Abraham Lincoln's mother. For many years, the cause was not known, until in the early 1900s it was shown conclusively to be due to a toxin in white snakeroot.

The toxic entity in white snakeroot has apparently not been conclusively identified. A ketone called tremetone has been isolated from the plant, but according to Beier and Norman (1990) and Beier *et al.* (1987), it is not the toxic agent. Beier *et al.* (1987) determined that the toxin is likely activated to toxic metabolites by cytochrome P450 activity.

The disease was associated with pioneer living conditions, when most families had their own cow that grazed on newly cleared land, along stream banks, and so on, where *Eupatorium* would often be found. People consumed milk and butter from their own cow, and if it happened to be grazing on white snakeroot, the toxin was transferred to the consumers. Milk sickness has gradually disappeared from the American scene as more intensive dairying, with milk pooled from many different herds, occurred. However, with the resurgence of small-scale farming, often associated with lack of spraying for weed control, outbreaks of white snakeroot toxicity may occur. Stotts (1984) reported such an incident in which a calf receiving milk from a family cow developed trembles.

Eupatorium adenophorum or Crofton weed is known to cause respiratory disease in horses, characterized by coughing, rapid heaving respiration, and lung lesions (O'Sullivan, 1979). The condition has occurred in Australia and Hawaii. The toxic agent has not been identified.

Rayless goldenrod or jimmyweed (*Haplopappus heterophyllus*) is a tall, erect perennial plant common on dry rangelands in southern Colorado, Texas, New Mexico, and Arizona. It is usually found along irrigation canals, ditches, and in river valleys. Significant losses of range livestock have occurred in various parts of the southwestern U.S. as a result of consumption of this plant. The toxicity signs are as in white snakeroot poisoning, with trembles a predominant sign. Ingestion of 1–1.5% of body weight of the green plant over 1–3 weeks may cause toxicity in horses, cattle, and sheep.

Blind Grass (*Stypandra imbricata*)

Stypandra imbricata, a perennial grasslike plant of the lily family which grows in Western Australia, is commonly referred to as blind grass. It has been responsible for numerous losses of sheep; animals either die after an acute illness with signs of neurological disturbance, or survive, but are permanently blind. Studies with sheep and goats (Main *et al.*, 1981; Whittington *et al.*, 1988) and with rats (Huxtable *et al.*, 1980) have shown that the lesions associated with blind grass toxicity are confined to the nervous system, with vacuolation of myelin in the brain and spinal cord. The optic nerves suffer a severe loss of myelinated axons, with a progressive disintegration until they are reduced to thin strands of scar tissue (Fig. 13–25).

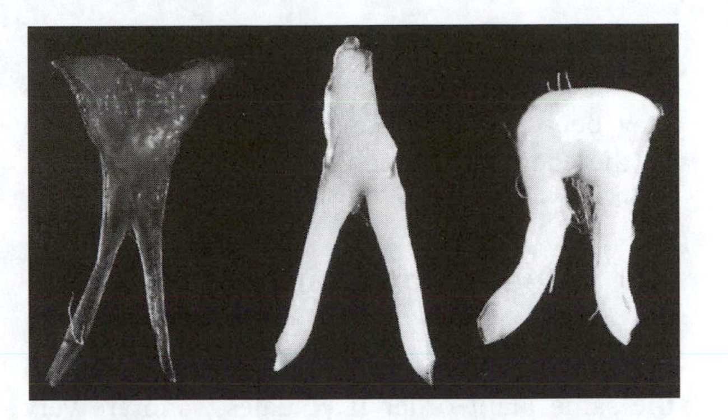

FIGURE 13–25 Optic nerves from a control rat (center), a rat acutely intoxicated with *Stypandra imbricata* (right), and a rat 12 weeks after acute intoxication (left). Note the swelling with acute intoxication (right) followed by severe atrophy (left). (Courtesy of P.R. Dorling)

The toxic agent is a hydroquinone, **stypandrol** (Huxtable *et al.*, 1989). It is a dimer of dianellidin and undergoes oxidation to the quinone, dianellinone (Fig. 13–26). Dorling *et al.* (1992) suggested that stypandrol acts to generate free radicals in the myelin which damage stabilizing protein bridges, leading to enlarged

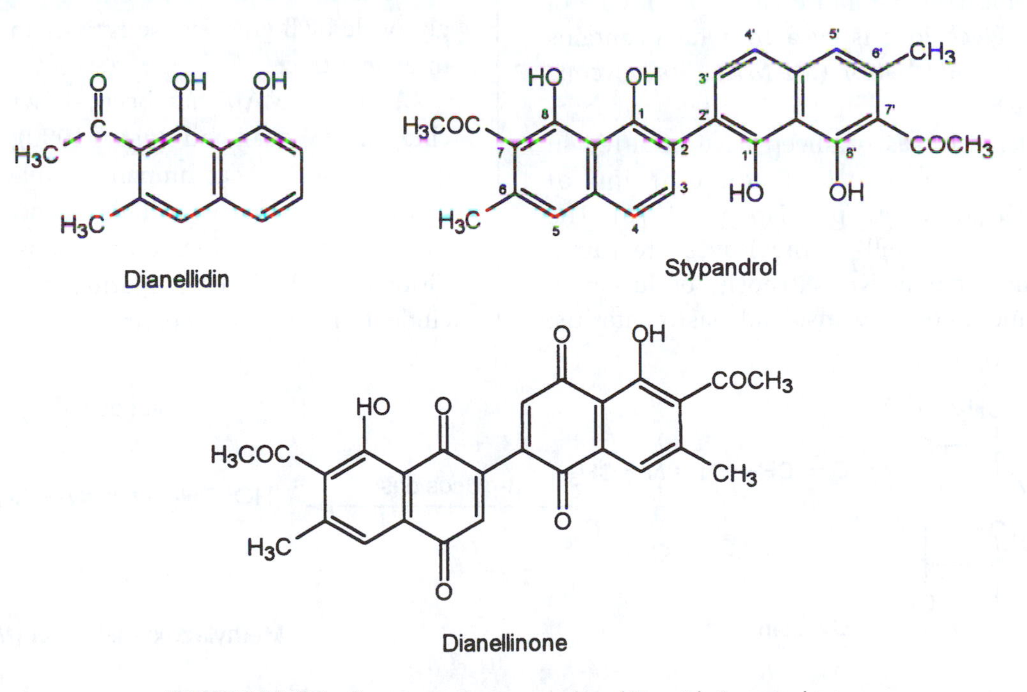

FIGURE 13–26 Structure of stypandrol and its oxidation product.

extracellular spaces that become edematous. Destruction of the axons of the optic nerves may be a consequence of compression caused by the edematous myelin.

Neurotoxins in Cycads

Cycas and *Macrozamia* spp. (cycads) belong to the same plant order (Cycadales), and are very similar in appearance and botanical characteristics. Cycads (Fig. 13–27) are ancient palm-like plants that were widely distributed in the Mesozoic (100–200 million years ago) period. They are found in tropical and subtropical areas. Livestock poisonings from consumption of cycads are important in Australia, where *Cycas* and *Macrozamia* spp. are found in tropical grazing areas. Cycads survive adverse conditions, such as drought and fire. The roots, seeds, and stems contain high levels of starch. They are consumed by grazing animals and by humans in some tropical areas.

The cycads contain glycosides that cause hepatic and gastrointestinal diseases in livestock (Hooper, 1978) and that have been shown to be carcinogenic (Wogan and Busby, 1980). One of the main glycosides is **cycasin**, which contains **methylazoxymethanol (MAM)** as its aglycone (Fig. 13–28).

Extensive losses of sheep have occurred in Australia as a result of consumption of *Macrozamia* and *Cycas* spp. (Hooper, 1978). The seeds (nuts) as well as the leaves are eaten. Symptoms include liver cirrhosis, occlusion of central and hepatic veins, and gastroenteritis.

FIGURE 13–27 A specimen of *Cycas media*, a cycad common in Western Australia. (Courtesy of P.R. Dorling)

The causative agent is MAM, released from the glycoside by β-glucosidase activity in the gastrointestinal tract.

Although MAM has been shown to be carcinogenic (Wogan and Busby, 1980), there is no direct evidence that human populations which consume cycads have a higher cancer incidence than populations which do not consume it. Traditional methods of preparing cycad flour include fermentation, heating, water extraction,

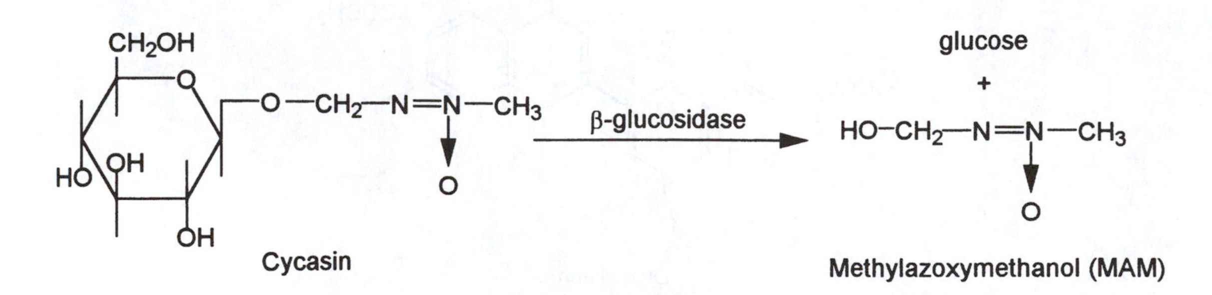

FIGURE 13–28 Cycasin and its carcinogenic aglycone, MAM.

and sun-drying, which seem to destroy the carcinogenic activity (Beck, 1992).

An interesting historical note is that Captain Cook's crew consumed seeds of *Cycas media* when their ship ran aground at Cooktown, Australia. Cook wrote in his diary that the "sailors were violently ill, both upwards and downwards"! (A.E. Bell, Kew Gardens, personal communication, May, 1984). Another description, by Cook, cited by Mills *et al.* (1996), is that clinical symptoms were profound, causing "ungovernable movements" and with such severe vomiting "that there was hardly any difference between us and death."

Reddy *et al.* (1982) reported that butylated hydroxyanisole (BHA) when administered to mice protected against the acute toxicity of MAM. This protection was associated with increased hepatic levels of cytochromes P450 and b5, and a reduction in necrotic changes in the liver. The MAM is a colon-specific carcinogen in mice.

Cattle consuming these plants develop a neural condition in which the hind legs become paralyzed because of axon degeneration in the central nervous system. The causative agent(s) has not been identified, but does not seem to be MAM (Hooper, 1978). Cycads contain a neurotoxic amino acid, β-methylamino-L-alanine (BMAA), which has been implicated in **Guam disease** (see Chapter 11). Charlton *et al.* (1992) concluded that BMAA in cycads was not the causative agent of hind limb ataxia.

Stringhalt

Stringhalt is a disease of horses characterized by an exaggerated flexion and delayed distention of one or both hocks during forward movement (Cahill *et al.*, 1985). The condition has occurred throughout the world, mainly in Australia, New Zealand and North America. Stringhalt in horses on pasture is often referred to as Australian stringhalt. A common feature of outbreaks of this disorder is that it occurs in late summer or autumn, usually following a dry summer with sparse forage (Robertson-Smith *et al.*, 1985). Although the cause has not been conclusively identified, generally the plant *Hypochaeris radicata* (flatweed, false dandelion, **summer dandelion**) is abundantly present in pastures where stringhalt occurs (Cahill *et al.*, 1985; Galey *et al.*, 1991). However, attempts at reproducing the condition by administering the plant to horses have not been successful (Cahill *et al.*, 1985). Stringhalt is a neurological condition, with evidence of axonal degeneration. Animals generally recover over a period of several months when removed from the offending pasture (Cahill *et al.*, 1985), although recovery can take as long as two years (Huntington *et al.*, 1991). These authors reported that administration of the drug phenytoin reduced the severity of Australian stringhalt gait abnormalities, probably through a tranquillizer effect. Excitement exaggerates the clinical symptoms.

β-phenethylamine Neurotoxins

Acacia berlandieri is a leguminous shrub that grows on rangelands in the U.S. Southwest. During periods of drought, sheep and goats browsing on this plant may develop locomotor ataxia, commonly referred to as "guajillo wobbles" or "limberleg." This condition is caused by the neurotoxic β-phenethylamine compounds (Fig. 13–29) that occur in the plant (Smith, 1977.) These include N-methyl-β-phenethylamine, tyramine, N-methyltyramine and hordenine (Pemberton *et al.*, 1993; Forbes *et al.*, 1994). Some of these compounds are also involved in *Phalaris* toxicity (see Chapter 11),

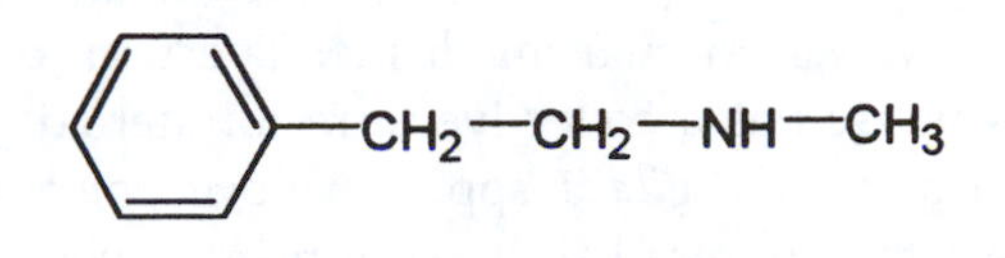

***N*-methyl-β–phenethylamine**

FIGURE 13–29 Structure of N-methyl-β-phenethylamine, a toxic constituent of *Acacia berlandieri*.

probably accounting for similar neurotoxic effects with both conditions.

Forbes *et al.* (1994) and Vera-Avila *et al.* (1996) suggested that β-phenethylamines suppress release of luteinizing hormone (LH), with potential adverse effects on reproduction. Forbes *et al.* (1995) observed that with severe defoliation of *Acacia berlandieri,* as typically occurs with heavy browsing activity of sheep and goats under drought conditions, the regrowth has increased β-phenethylamine concentrations. Thus both animal performance and reproduction may be negatively impacted when livestock are forced to rely on *Acacia berlandieri* when other forage is of limited availability.

CARDIAC, PULMONARY, AND RENAL TOXINS

Cardiac Glycosides (Cardenolides)

Cardenolides are steroids which have a lactone group attached to the steroid nucleus (Fig. 13–32). Cardenolides occur in plants as glycosides, and are commonly referred to as cardiac glycosides. Their mode of action as heart poisons is an inhibition of Na^+, K^+-ATPase in cardiac muscle. The elevated Na^+ leads to an increased accumulation of Ca via Na-Ca exchange. The increased Ca concentration causes a more forceful contraction of the myocardium (Joubert, 1989). The heart rate is slowed but the individual contractions are more forceful.

Cardiac glycosides occur in the common **garden foxglove** (*Digitalis purpurea*), a native of Europe (Fig. 13–30). Foxglove has naturalized in the U.S. Pacific Northwest and is found along roadsides, in logged-off areas, and in pastures. Foxgloves contain a number of cardiac glycosides which strengthen the force of contraction of the heart and prolong the duration of the diastolic phase. Digitalis drugs such as digitonin are derived from *D. purpurea,* and are used extensively in human medicine. The term **digitalis** is used as a collective term for steroid glycosides from *Digitalis* spp. Foxglove poisoning of livestock is rare because the plant is not normally consumed. Signs of toxicity include gastric disturbances, drowsiness, irregular heartbeat and pulse, tremors, and convulsions. Kingsbury (1964) cites early work showing fresh foxglove leaves to be highly toxic with only a few hundredths percent of body weight being a lethal dose.

FIGURE 13–30 Foxglove (*Digitalis purpurea*) contains cardiac glycosides. It is a common roadside weed of the Pacific Northwest.

Oleander (*Nerium oleander*) is an introduced ornamental shrub widely grown in California and other southern areas of the U.S. It has large showy blossoms. Several cardiac glycosides from oleander have been isolated (Tor *et al.*, 1996) and have similar pharmacological effects as digitalis glycosides. The major one is oleandrin (Galey *et al.*, 1996). Kingsbury (1964) has reviewed studies on the toxicity of oleander. It is extremely toxic to livestock and humans, with all plant parts, green or dry, being toxic. A single leaf can be lethal to humans. People have been poisoned when meat or frankfurters have been skewered on oleander branches for cooking. An intake of leaves at 0.005% of body weight is lethal to cattle and horses. Symptoms of toxicity are similar to those of foxglove poisoning. Oleander is unpalatable, and is consumed only when inadvertently mixed with other forage such as lawn clippings.

The garden ornamental **lily-of-the-valley** (*Convallaria majalis*) contains cardiac glycosides. Livestock poisoning from this plant occurs occasionally. Moxley *et al.* (1989) reported the poisoning of a dog which consumed lily-of-the-valley leaves.

Various species of **milkweed** (***Asclepias* spp.**) grow throughout North America (Fig. 13–31). They are perennial herbs with a thick milky latex juice. Some may have commercial potential as sources of latex and hydrocarbons (Nielsen *et al.*, 1977). They are of two major groups, narrow leaved and broad leaved. Some, such as the woolly pod or California milkweed (*A. eriocarpa*) and *A. labriformis*, which occurs in Utah, are hazards to range animals. As little as 10–20 g of these milkweeds will kill a sheep (Benson *et al.*, 1979; Seiber *et al.*, 1983; Ogden *et al.*, 1992a). They are unpalatable and are consumed only in times of drought when the range is overgrazed or when contaminated hay is fed.

Signs of milkweed toxicity in livestock include a profuse depression accompanied by staggering. Following collapse, there is labored respiration, elevated temperature, and dilation of the pupils. Pathological lesions include congestion of the lungs and kidneys and irritation

FIGURE 13–31 Plants of the broad-leaved milkweed *Asclepias syriaca*.

of the intestinal mucosa (Ogden *et al.*, 1992a,b). The cardiac glycosides in milkweeds have been reviewed by Seiber *et al.* (1983).

The milkweed cardenolides are structurally similar to oubain, a drug used in clinical medicine. Milkweed cardenolides have structures similar to the general structure shown in Fig. 13–32. Some of them contain one sugar moiety attached to the steroid nucleus through bonding with hydroxyl groups on both C-2 and C-3 positions of the aglycone, while others have one site of glycosidation, generally at C-3.

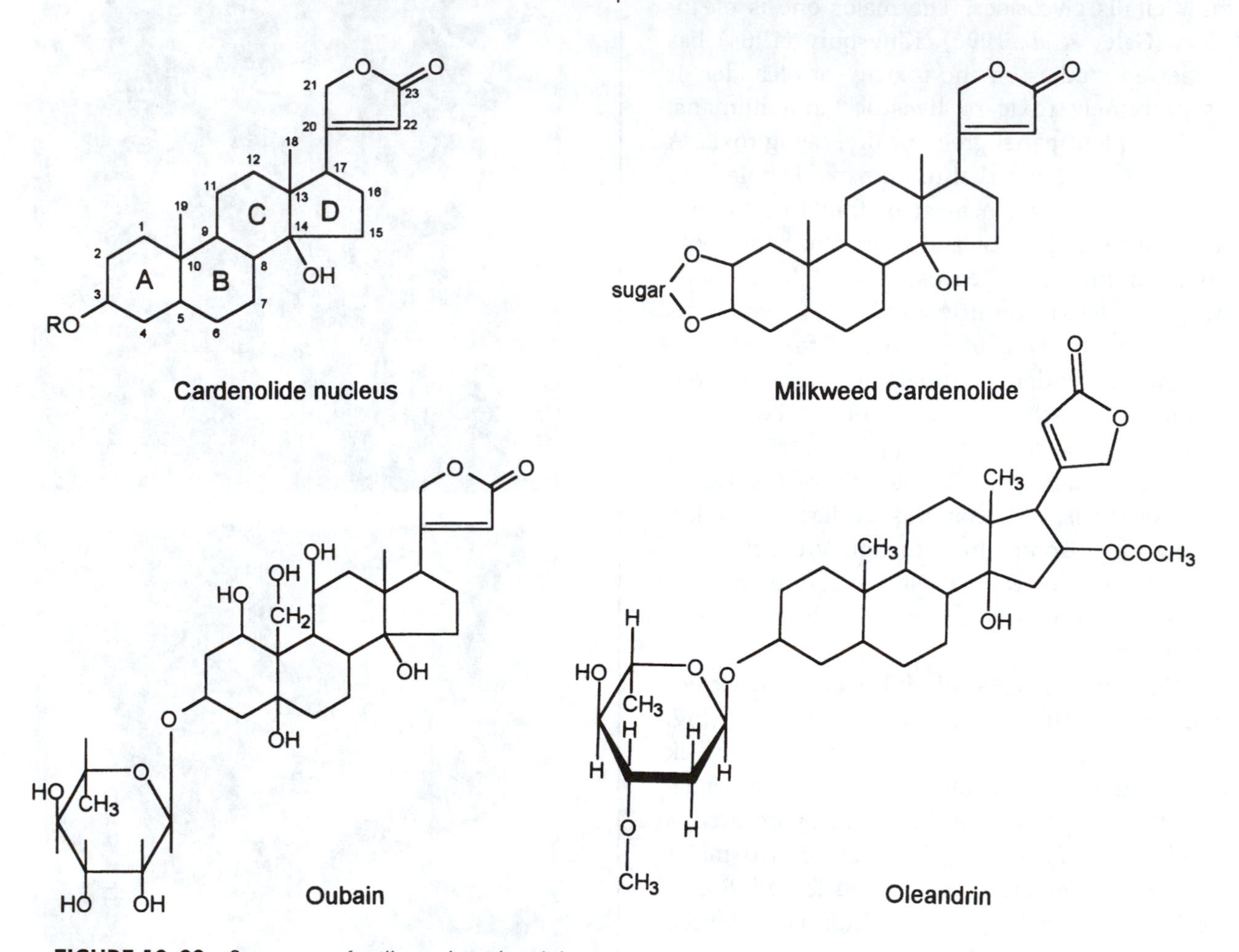

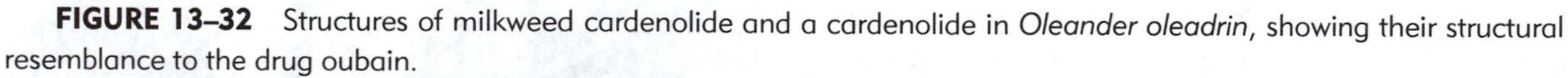
FIGURE 13–32 Structures of milkweed cardenolide and a cardenolide in *Oleander oleadrin*, showing their structural resemblance to the drug oubain.

An interesting relationship is that of milkweeds and the **monarch butterfly** (Harborne, 1993). The larvae of this butterfly feed on milkweed (Fig. 13–33) and accumulate the cardenolides. If the level of toxin is sufficiently high, it causes birds which consume the caterpillar, pupa, or adult to vomit. This response may be learned by the predators, and they subsequently avoid the monarch butterfly. An extension of this relationship is that other butterflies such as the Viceroy which do not feed on milkweed have a nearly identical color pattern (mimicry), so predators avoid these nontoxic insects also.

Bufadienolides are cardiac glycosides very similar to cardenolides, differing only in the structure of the C-17 substituent on the D ring.

FIGURE 13–33 A monarch butterfly caterpillar feeding on leaves of *Asclepias*. The larvae accumulate cardiac glycosides, making them unpalatable to predators. (Courtesy of J.N. Seiber)

They are found in a variety of South African plant species (Kellerman *et al.*, 1988). In South Africa, *Morea polystachya* or blue tulp causes livestock poisoning due to its cardiac glycosides (Joubert and Schultz, 1982). Clinical signs include gastrointestinal, cardiac, and neuromuscular disturbances. Gastrointestinal signs include rumen stasis, bloat, and diarrhea; cardiac effects involve bradycardia, tachycardia, and arrhythmia, while neuromuscular effects are hypersensitivity, paresis, ataxia, paralysis, dyspnea, and drooling. Joubert and Schultz (1982) reported that an oral administration of a mixture of activated charcoal (5 g/kg body weight) and potassium chloride (1 g/kg body weight) protected against blue tulp toxicity in sheep and cattle.

South African cattle raisers have for many years followed a management practice of predosing cattle orally with blue tulp before turning them into tulp-infested pastures, believing that the prior exposure lessened the likelihood of poisoning. However, Strydom and Joubert (1983) and Joubert (1983) found no beneficial effects of predosing cattle with the plant in protecting against subsequent toxicity.

Anderson *et al.* (1983) in South Africa reported that *Kalanchoe lanceolata* contains three bufanienolides, causing a disease in sheep (krimpsiekte) with neurological signs typical of cardiac glycoside poisoning. Williams and Smith (1984) found that several *Kalanchoe* spp. are toxic to chicks, causing neurological signs. They also cite an incident involving a rabbit which consumed three-fourths of a leaf of *K. daigremontiana* and was severely poisoned. The widespread use of *Kalanchoe* spp. as garden and house plants and their apparent toxicity indicate that care should be taken to ensure that the plants are not accessible to children or herbivorous pets.

Cardiac glycoside poisoning of livestock can be successfully treated if toxicosis is not advanced. The objectives of treatment are to reduce the absorption of toxins, promote toxin excretion, ameliorate the effects on the heart, redydrate animals suffering from diarrhea, and restore ruminal activity and appetite (McKenzie and Dunster, 1987). This may be accomplished by oral administration of activated charcoal with electrolytes to reduce glycoside absorption and promote excretion; injection of propanolol (an adrenergic blocking agent) to treat tachycardia (rapid heart rate); and injection of atropine to treat atrioventricular block. Acidification of the rumen has been found effective in treating oleander poisoning (Rezakhani and Maham, 1992).

Veratrum and *Zigadenus* Toxicity

Veratrum californicum (false hellebore) is a tall, coarse-leaved plant that grows in moist habitats of western North America. Over 50 steroid alkaloids have been identified in *Veratrum* spp. The **veratrum alkaloids** have pharmacological effects with marked hypotensive properties. Lowering of blood pressure is caused by dilation of arterioles with constriction in venous vascular beds and a slowing of heart rate. The alkaloids have been used for medicinal purposes. Sheep sometimes consume toxic quantities of *Veratrum*; signs of intoxication include salivation, prostration, depressed heart action, weakness, and dyspnea. Of particular interest with *Veratrum* spp. is their teratogenic effects. This aspect will be discussed in Chapter 14.

Death camas (*Zigadenus* spp.) contains veratrum-type alkaloids. Death camas resembles a wild onion in appearance, with a small bulb and grasslike leaves (Fig. 13–34). Various spp. grow over the western U.S. Death camas is one of the major poisonous plants affecting range sheep. The plant begins growth in early spring and may be consumed because of the lack of other forage. All parts of the plant are toxic. The toxic bulbs may be pulled up and eaten when the ground is wet. Individual herd losses of 500–2000 sheep have occurred (Kingsbury, 1964). The lethal dose varies from about 0.6 to 6% of body weight. Death camas contains several steroid alkaloids, including zygacine (Fig. 13–35).

FIGURE 13–34 Death camas (*Zigadenus paniculatus*). (Courtesy of W.C. Krueger).

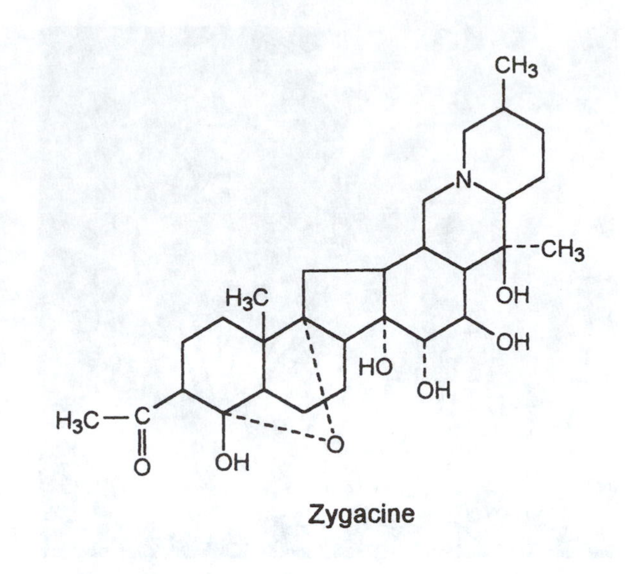

FIGURE 13–35 Zygacine, a steroid alkaloid in death camas.

The zygacine content of death camas leaves increases with plant maturity (Majak *et al.*, 1992). Panter *et al.* (1987) described an incident in which 250 sheep died from death camas poisoning. Clinical signs included muscular weakness, ataxia, trembling and prostration. Heart action becomes slow and weak, with a marked drop in blood pressure. Lesions are those of severe pulmonary congestion. Wild herbivores such as mule deer are apparently more tolerant of death camas than cattle and sheep, as they have been observed to consume the leaves without ill effect (Longland and Clements, 1995).

Yew (*Taxus*) Poisoning

Yews (*Taxus* spp.) (Fig. 13–36) are small or shrub-like evergreen trees, commonly grown as ornamental shrubs and hedge plants. Common ornamental types include the English and Japanese yews. In western North America, the Pacific or Western yew (*T. breviofolia*) is found in the understory of old growth forests. In the early 1990's, there was widespread interest in the Pacific yew as a source of taxol, which has

FIGURE 13–36 Yews (*Taxus* spp.) are conifers, with fleshy fruits rather than cones.

anti-tumor activity (Rowinsky *et al.*, 1990). **Taxol** is one of several diterpenoid taxanes in yews (Kelsey and Vance, 1992; Witherup *et al.*, 1990). It prevents cell division without affecting DNA, RNA or protein synthesis, by binding to microtubules (Rowinsky *et al.*, 1990). Microtubules function as components of the mitotic spindle, as well as being involved in many cell functions including maintenance of cell shape, and intracellular transport. Taxol is extremely potent in disrupting the microtubular structure of cancer cells (Rowinsky *et al.*, 1990). Clinical trials with taxol are in progress (1997).

The foliage of yew trees is toxic to livestock. The usual mode of exposure is the feeding of hedge trimmings or plant clippings to animals. For example, Panter *et al.* (1993) described the death of 35 cattle from **yew poisoning**. Clippings of English yew had been discarded into a pasture containing 43 cattle. Within 4 hours of ingestion of yew foliage, most of the 35 animals had died. Clinical signs were lethargy, recumbency, dyspnea, jugular pulsation and distention, and rapid death. The major lesions were hemorrhagic areas on the right ventricular myocardium and right atrium, and pulmonary congestion and edema. The mode of action of taxol apparently involves calcium and sodium inhibiting properties in heart muscle, similar to those of other antiarrhythmic drugs, leading to cardiac arrest (Panter *et al.*, 1993). Veatch *et al.* (1988) reported the death of a pony from yew poisoning within 15 minutes after the onset of clinical signs. The lethal dose of green yew foliage to ruminants is about 0.5% of body weight (Panter *et al.*, 1993).

Yew poisoning is an example of the hazard of disposing of garden prunings and clippings by feeding them to animals. Many garden ornamental plants are toxic, and the clippings often become more palatable when wilted.

Rhododendron Toxicity

A number of plants in the Ericaceae (heath) family are toxic, including laurel (*Kalmia* and *Ledum* spp.), and rhododendrons. The toxic principles in rhododendrons are diterpenoid compounds called **grayanotoxins**. They have neurotoxic and cardiac activity. Signs of intoxication include bradycardia, cardiac arrhythmia, blurred vision, paralysis, and convulsions (Meier and Hemmick, 1992; Onat *et al.*, 1991).

In the Meditteranean region, so-called "mad honey" is quite common. It is produced by bees foraging on rhododendron nectar. When consumed by humans, bradycardia and convulsions occur. According to legend, Roman troops were sometimes fed mad honey by local people, and then slain when they were intoxicated.

Although data on livestock are lacking, it would be advisable to prevent access to ornamental rhododendrons and laurels. Prunings and clippings of these and other ornamentals should not be fed to livestock.

Avocado Toxicity

The fruit and leaves of avocado (*Persea americana*) are toxic. **Avocado poisoning** results in lung edema and myocardial degeneration (Grant *et al.*, 1991; Sani *et al.*, 1991; Stadler et al., 1991). Hargis *et al.* (1989) demonstrated that mashed avocado fruit is highly toxic to caged birds such as canaries and budgerigars. Severe respiratory distress and rapid death occurred, with ascites and edema of heart and lung tissue.

Similar signs of acute toxicity, with respiratory and cardiac distress, were noted in sheep and goats fed avocado leaves (Grant *et al.*, 1991). Burger *et al.* (1994a,b) reported on the poisoning of a flock of ostriches that were introduced into an avocado orchard in South Africa. The birds ingested avocado fruit and leaves, with numerous mortalities due to cardiac failure. Horses have been poisoned by avocado consumption (McKenzie and Brown, 1991). Avocado also inhibits lactation (see Chapter 14.)

The toxicity of avocado raises the obvious question of its safety in the human diet. The situation is comparable to that of many other toxicants in foods, in that the dose makes the poison. It is unlikely that people would consume enough avocado to experience toxicity. Nevertheless, the cardiotoxic properties of avocado are such that the toxic principle(s) and its pharmacology should be established, to adequately assess the hazards, if any, of human consumption of avocado.

Galegine Toxicity

Galegine (Fig. 13–37) is a guanidine compound (containing a guanidine group; another guanidine compound is the amino acid arginine). Galegine is the toxic entity of several poisonous plants, including *Galega officinalis, Verbesina encelioides* and *Schoenus asperocarpus*.

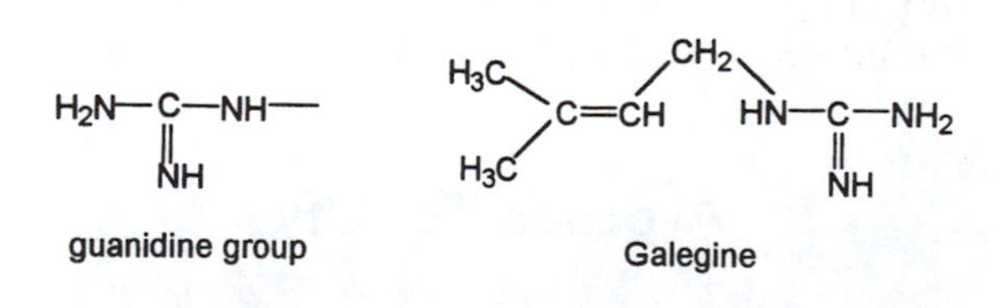

FIGURE 13–37 Galegine contains a guanidine group.

***Galega officinalis* (goatsrue)** is a perennial legume quite similar in appearance to alfalfa. A native of Europe, it has been introduced into various temperate parts of the world. In the United States, goatsrue was introduced and tested as a potential forage at Utah State University. It was deemed unsuitable for forage use because of its toxicity. However, it escaped from test plots and now infests several thousand acres in Utah (Keeler *et al.*, 1992). Eradication efforts are successfully reducing its distribution, with total eradication the long-term goal. The plant is unpalatable to livestock, but occasionally is consumed in sufficient quantities to cause toxicity. Clinical signs include labored breathing, a foaming nasal discharge, and severe hydrothorax (ascites of pleural cavity) and pulmonary edema (Keeler *et al.*, 1986, 1988, 1992). **Galegine**, the toxic component, lowers blood pressure and has a direct effect on pulmonary vascular permeability, leading to the hydrothorax condition (Huxtable *et al.*, 1993). In sheep, there is a large variation in individual susceptibility to *G. officinalis* poisoning (Keeler *et al.*, 1988). In some animals, 5 g/kg B.W. of dried plant was a toxic dose, whereas in other animals 24 g/kg had no effect. Another species, *G. orientalis*, is used as a forage (Moller and Hostrup, 1996), and is presumably nontoxic.

Verbesina enceloides is a Compositae native to North America, and naturalized in Australia. Although it is widely distributed in western North America (Keeler *et al.*, 1992), it has not been reported to cause livestock poisonings on this continent. However, it has caused extensive losses in Australia (Oelrichs *et al.*, 1985), with signs similar to those of *G. officinalis* toxicity. *V. enceloides* contains galegine as its toxic entity. Concentrations of galegine in the plant were 0.8% in the U.S. (Keeler *et al.*, 1992) and about three times higher in Australian plants (Oelrichs *et al.*, 1985), compared to a level of 0.46% in *G. officinalis* (Keeler *et al.*, 1992).

Galegine is the toxic component of *Schoenus asperocarpus* (poison sedge), a poisonous plant in Western Australia (Huxtable *et al.*, 1993). Signs of severe respiratory distress, hydrothorax and pulmonary edema occur in sheep grazing the plant. Similar signs were induced by administration of galegine isolated from the plant.

Other Cardiac, Pulmonary, and Renal Toxins

Gossypol (see Chapter 7) in cottonseed products cause heart lesions and death by cardiac failure. **Gousiekte** (quick disease) is a toxicosis of ruminants in South Africa, characterized by acute heart failure following consumption of certain *Pachystigma* and *Pavetta* spp. These are shrubs or small trees. After ingestion of a lethal dose of the plants involved, there is a latency period of 4–6 weeks, after which the animals suddenly die with no premonitory signs (Kellerman *et al.*, 1988). An earlier report, cited by Fourie *et al.* (1994) states: "Apparently healthy animals, which are feeding up to the last minute, are suddenly seized by a fit, jump, and drop dead on the ground". Pathological changes include signs of congestive heart failure, with degeneration of cardiac muscles and replacement fibrosis (Kellerman *et al.*, 1988). The toxic compounds in the plants have not been identified (Kellerman *et al.*, 1988).

The pyrrolizidine alkaloids (PA) can be **lung toxins**. This is especially true of **monocrotaline**, the major PA in *Crotalaria spectabilis*. Dietary consumption of *Crotalaria* seeds or administration of monocrotaline results in a pulmonary vascular syndrome characterized by pulmonary hypertension and edema (Wilson *et al.*, 1992). Reactive metabolites of monocrotaline produced in the liver are transported to the lungs in association with red blood cells (Pan *et al.*, 1991). The PA are discussed in Chapter 12. Acute bovine pulmonary emphysema associated with 3-methyl indole is discussed in Chapter 9.

Kidney damage occurs with several plant toxicoses such as oak and oxalate (Chapter 12) poisoning. Pyrrolizidine alkaloid toxicosis may involve kidney damage, especially in swine. Nephrosis occurs in *Lantana* poisoning (Chapter 12) and with vetch toxicity (Chapter 11). Ochratoxins in moldy grains (Chapter 5) are potent nephrotoxins. *Narthecium ossifragum* (bog asphodel), an herb growing in northern Europe, is nephrotoxic to cattle (Flaoyen *et al.*, 1993, 1994, 1995). This plant contains saponins (see Chapter 11) but it is not known if they are the cause of the nephrotoxicity.

REFERENCES

The Hemlocks: *Conium* and *Cicuta* spp.

Edmonds, L.D., L.A. Selby, and A.A. Case. 1972. Poisoning and congenital malformation associated with consumption of poison hemlock by sows. J. Am. Vet. Med. Assoc. 160:1319–1324.

Frank, A.A., and W.M. Reed. 1990. Comparative toxicity of coniine, an alkaloid of *Conium maculatum* (poison hemlock), in chickens, quails, and turkeys. Avian Dis. 34:433–437.

Frank, A.A., and W.M. Reed. 1987. *Conium maculatum* (poison hemlock) toxicosis in a flock of range turkeys. Avian Dis. 31:386–388.

Galey, F.D., D.M. Holstege, and E.G. Fisher. 1992. Toxicosis in dairy cattle exposed to poison hemlock (*Conium maculatum*) in hay: isolation of *Conium* alkaloids in plants, hay and urine. J. Vet. Diagn. Invest. 4:60–64.

Keeler, R.F. 1974. Coniine, a teratogenic principle from *Conium maculatum*, producing congenital malformations in calves. Clin. Toxicol. 7:195–206.

Keeler, R.F., and L.D. Balls. 1978. Teratogenic effects in cattle of *Conium maculatum* and *Conium* alkaloids and analogs. Clin. Toxicol. 12:49–64.

Keeler, R.F., L.D. Balls, J.L. Shupe, and M.W. Crowe. 1980. Tertogenicity and toxicity of coniine in cows, ewes and mares. Cornell Vet. 70, 19–26.

Kennedy, B.W., and L.E. Grivetti. 1980. Toxic quail: A cultural-ecological investigation of coturnism. Ecol. Food Nutr. 9:15–42.

Kingsbury, J.M. 1964. Poisonous Plants of the United States and Canada. Prentice–Hall, Englewood Cliffs, NJ.

Knutsen, O.H., and P. Paszkowski. 1984. New aspects in the treatment of water hemlock poisoning. Clin. Toxicol. 22:157–166.

Panter, K.E., and R.F. Keeler. 1989. Piperidine alkaloids of poison hemlock (*Conium masculatum*). pp. 109–132 in: P.R. Cheeke (Ed). Toxicants of Plant Origin, Vol. I. Alkaloids. CRC Press, Boca Raton, FL.

Panter, K.E., D.C. Baker, and P.O. Kechel. 1996. Water hemlock (*Cicuta douglassii*) toxicoses in sheep: Pathological description and prevention of lesions and death. J. Vet. Diag. Invest. 8: in press.

Panter, K.E., R.F. Keeler, and D.C. Baker. 1988a. Toxicoses in livestock from the hemlocks (*Conium* and *Cicuta* spp.) J. Anim. Sci. 66:2407–2413.

Panter, K.E., T.D. Bunch, and R.F. Keeler. 1988b. Maternal and fetal toxicity of poison hemlock (*Conium maculatum*) in sheep. Am. J. Vet. Res. 49:281–283.

Panter, K.E., R.F. Keeler, and W.B. Buck. 1985a. Induction of cleft palate in newborn pigs by maternal ingestion of poison hemlock (*Conium maculatum*). Amer. J. Vet. Res. 46:1368–1371.

Panter, K.E., R.F. Keeler, and W.B. Buck. 1985b. Congenital skeletal malformations induced by maternal ingestion of *Conium maculatum* (poison hemlock) in newborn pigs. Am J. Vet. Res. 46:2064–2066.

Smith, R.A., and D. Lewis. 1987. Cicuta toxicosis in cattle: Case history and simplified analytical method. Vet. Human Toxicol. 29:240–241.

Larkspur (*Delphinium* spp.)

James, L.F., D.B. Nielsen, and K.E. Panter. 1992. Impact of poisonous plants on the livestock industry. J. Range Manage. 45:3–8.

Kingsbury, J.M. 1964. Poisonous Plants of the United States and Canada. Prentice–Hall, Englewood Cliffs, NJ.

Lane, M.A., M.H. Ralphs, J.D. Olsen, F.D. Provenza, and J.A. Pfister. 1990. Conditioned taste aversion: potential for reducing cattle loss to larkspur. J. Range Manage. 43:127–131.

Majak, W. 1993. Alkaloid levels in a species of low larkspur and their stability in rumen fluid. J. Range Manage. 46:100–103.

Manners, G.D., K.E. Panter, and S.W. Pelletier. 1995. Structure-activity relationships of norditerpenoid alkaloids occurring in toxic larkspur (*Delphinium*) species. J. Natural Products 58:863–869.

Manners, G.D., K.E. Panter, M.H. Ralphs, J.A. Pfister, J.D. Olsen, and L.F. James. 1993. Toxicity and chemical phenology of norditerpenoid alkaloids in the tall larkspurs (*Delphinium* species). J. Agric. Food Chem. 41:96–100.

Manners, G.D., J.A. Pfister, M.H. Ralphs, K.E. Panter, and J.D. Olsen. 1992. Larkspur chemistry: Toxic alkaloids in tall larkspurs. J. Range Manage. 45:63–67.

Nation, P.N., M.H. Benn, S.H. Roth, and J.L. Wilkens. 1982. Clinical signs and studies of the site of action of purified larkspur alkaloid, methyllycaconitine, administered parenterally to calves. Can. Vet. J. 23:264–266.

Pfister, J.A., and G.D. Manners. 1995. Effects of carbachol administration in cattle grazing tall larkspur–infested range. J. Range Manage. 48:343–349.

Pfister, J.A., and G.D. Manners. 1991. Mineral salt supplementation of cattle grazing tall larkspur–infested rangeland during drought. J. Range Manage. 44:105–111.

Pfister, J.A., G.D. Manners, D.R. Gardner, K.W. Price, and M.H. Ralphs. 1996. Influence of alkaloid concentration on acceptability of tall larkspur (*Delphinium* spp.) to cattle and sheep. J. Chem. Ecol. 22:1147–1168.

Pfister, J.A., K.E. Panter, and G.D. Manners. 1994a. Effective dose in cattle of toxic alkaloids from tall larkspur (Delphinium barbeyi). Vet. Human Toxicol. 36:10–11.

Pfister, J.A., G.D. Manners, D.R. Gardner, and M.H. Ralphs. 1994b. Toxic alkaloid levels in tall larkspur (*Delphinium barbeyi*) in western Colorado. J. Range Manage. 47:355–358.

Pfister, J.A., K.E. Panter, G.D. Manners, and C.D. Cheney. 1994c. Reversal of tall larkspur (Delphinium barbeyi) poisoning in cattle with physostigmine. Vet. Human Toxicol. 36:511–513.

Pfister, J.A., G.D. Manners, M.H. Ralphs, Z.X. Hong, and M.A. Lane. 1988. Effects of phenology, site, and rumen fill on tall larkspur consumption by cattle. J. Range Manage. 41:509–514.

Ralphs, M.H. 1992. Continued food aversion: Training livestock to avoid eating poisonous plants. J. Range Manage. 45:46–51.

Ralphs, M.H., and J.D. Olsen. 1992. Prior grazing by sheep reduces waxy larkspur consumption by cattle: An observation. J. Range Manage. 45:136–139.

Ralphs, M.H., D.T. Jensen, J.A. Pfister, D.B. Nielsen, and L.F. James. 1994. Storms influence cattle to graze larkspur: An observation. J. Range Manage. 47:275–278.

Ralphs, M.H., J.E. Bowns, and G.D. Manners. 1991. Utilization of larkspur by sheep. J. Range Manage. 44:619–622.

Ralphs, M.H., J.D. Olsen, J.A. Pfister, and G.D. Manners. 1988. Plant–animal interactions in larkspur poisoning in cattle. J. Anim. Sci. 66:2334–2342.

Siemion, R.S., M.F. Raisbeck, J.W. Waggoner, M.A. Tidwell, and D.A. Sanchez. 1992. In vitro ruminal metabolism of larkspur alkaloids. Vet. Hum. Toxicol. 34:206–208.

Locoweeds and Indolizidine Alkaloids

Bachman, S.E., M.L. Galyean, G.S. Smith, D.M. Hallford, and J.D. Graham. 1992. Early aspects of locoweed toxicosis and evaluation of a mineral supplement or clinoptilolite as dietary treatments. J. Anim. Sci. 70:3125–3132.

Dorling, P.R., C.R. Huxtable, and S.M. Colegate. 1980. Inhibition of lysosomal α-mannosidase by swainsonine, an indolizidine alkaloid isolated from *Swainsona canescens*. Biochem. J. 191:649–651.

Dorling, P.R., C.R. Huxtable, and P. Vogel. 1978. Lysosomal storage in *Swainsona* spp. toxicosis: An induced mannosidosis. Neuropathol. Appl. Neurobiol. 4:285–295.

Elbein, A.D. 1991. Glycosidase inhibitors: inhibitors of N-linked oligosaccharide processing. FASEB J. 5:3055–3063.

Galyean, M.L., M.H. Ralphs, M.N. Reif, J.D. Graham, and W.E. Braselton, Jr. 1996. Effects of previous grazing treatment and consumption of locoweed on liver mineral concentrations in beef steers. J. Anim. Sci. 74:827–833.

Huxtable, C.R., and P.R. Dorling. 1982. Poisoning of livestock by *Swainsona* spp.: Current status. Aust. Vet. J. 59:50–53.

James, L.F., and W.J. Hartley. 1977. Effects of milk from animals fed locoweed on kittens, calves and lambs. Am. J. Vet. Res. 38:1263–1265.

James, L.F., and K.R. Van Kampen. 1976. Effects of locoweed toxin on rats. Am. J. Vet. Res. 37:845–846.

James, L.F., K.E. Panter, H.P. Broquist, and W.J. Hartley. 1991. Swainsonine-induced high mountain disease in calves. Vet. Hum. Toxicol. 33:217–219.

James, L.F., A.D. Elbein, R.J. Molyneux, and C.D. Warren (Eds.). 1989. Swainsonine and Related Glycosidase Inhibitors. Iowa State University Press, Ames.

James, L.F., W.J. Hartley, D. Nielsen, S. Allen, and K.E. Panter. 1986. Locoweed (*Oxytropis sericea*) poisoning and congestive heart failure in cattle. J. Amer. Vet. Med. Assn. 189:1549–1556.

Kingsbury, J.M. 1964. Poisonous Plants of the United States and Canada. Prentice–Hall, Englewood Cliffs, N.J.

McIlwraith, C.W., and L.F. James. 1982. Limb deformities in foals associated with ingestion of locoweed by mares. J. Am. Vet. Med. Assoc. 181:255–258.

McKenzie, R.A., K.G. Reichmann, C.K. Dimmock, P.J. Dunster, and J.O. Twist. 1988. The toxicity of *Castanospermum australe* seeds for cattle. Aust. Vet. J. 65:165–167.

Molyneux, R.J., and L.R. James. 1982. Loco intoxication: Indolizidine alkaloids of spotted locoweed (*Astragalus lentiginosus*). Science 216:190–191.

Pan, Y.T., J. Ghidoni, and A.D. Elbein. 1993. The effects of castanospermine and swainsonine on the activity and synthesis of intestinal sucrase. Arch. Biochem. Biophys. 303:134–144.

Pfister, J.A., C.D. Cheney, and F.D. Provenza. 1992. Behavioral toxicology of livestock ingesting poisonous plants. J. Range Manage. 45:30–36.

Pritchard, D.H., C.R.R. Huxtable, and P.R. Dorling. 1990. Swainsonine toxicosis suppresses appetite and retards growth in weanling rats. Res. Vet. Sci. 48:228–230.

Pulsipher, G.D., M.L. Galyean, D.M. Hallford, G.S. Smith, and D.E. Kiehl. 1994. Effects of graded levels of bentonite on serum clinical profiles, metabolic hormones, and serum swainsonine concentrations in lambs fed locoweed (*Oxytropis sericea*). J. Anim. Sci. 72:1561–1569.

Ralphs, M.H., D. Graham, and L.F. James. 1994. Social facilitation influences cattle to graze locoweed. J. Range Manage. 47:123–126.

Ralphs, M.H., D. Graham, R.J. Molyneux, and L.F. James. 1993. Seasonal grazing of locoweeds by cattle in northeastern New Mexico. J. Range Manage. 46:416–420.

Ralphs, M.H., K.E. Panter, and L.F. James. 1991. Grazing behavior and forage preference of sheep with chronic locoweed toxicosis suggest no addiction. J. Range Manage. 44:208–209.

Ralphs, M.H., K.E. Panter, and L.F. James. 1990. Feed preferences and habituation of sheep poisoned by locoweed. J. Anim. Sci. 68:1354–1362.

Sharma, R.P., L.F. James, and R.J. Molyneux. 1984. Effect of repeated locoweed feeding on peripheral lymphocytic function and plasma proteins in sheep. Am. J. Vet. Res. 45:2090–2093.

Stegelmeier, B.L., L.F. James, K.E. Panter, and R.J. Molyneux. 1995a. Serum swainsonine concentration and α-mannosidase activity in cattle and sheep ingesting *Oxytropis sericea* and *Astragalus lentiginosus* (locoweeds). Am. J. Vet. Res. 56:149–154.

Stegelmeier, B.L., R.J. Molyneux, A.D. Elbein, and L.F. James. 1995b. The lesions of locoweed (*Astragalus mollissimus*), swainsonine, and castanospermine in rats. Vet. Pathol. 32:289–298.

Warren, C.D., B. Bugge, P.F. Daniel, K.B. Linsley, D.K. Daniels, L.F. James, and R.W. Jeanloz. 1989. Locoweed toxicosis in sheep: oligosaccharides accumulated in fetal and maternal tissues. pp. 344–359 in: L.F. James, A.D. Elbein, R.J. Molyneux, and C.D. Warren. Swainsonine and Related Glycosidase Inhibitors. Iowa State Univ. Press, Ames.

Nitrotoxins in *Astragalus* spp.

Alston, T.A., L. Mela, and H.J. Bright. 1977. 3-Nitro-propionate, the toxic substance of *Indigofera*, is a suicide inactivator of succinate dehydrogenase. Proc. Natl. Acad. Sci. U.S.A. 74:3767–3771.

Anderson, R.C., M.A. Rasmussen, and M.J. Allison. 1993. Metabolism of the plant toxins nitropropionic acid and nitropropanol by ruminal microorganisms. Appl. Environ. Microbiol. 59:3056–3061.

Gustine, D.L. 1979. Aliphatic nitro compounds in crown vetch: A review. Crop Sci. 19:197–203.

Gustine, D.L., B.G. Moyer, P.J. Wangness, and J.S. Shenk. 1977. Ruminal metabolism of 3-nitropropanoyl-D-glucopyranoses from crown vetch. J. Anim. Sci. 44:1107–1111.

James, L.F., W.J. Hartley, M.C. Williams, and K.R. Van Kampen. 1980. Field and experimental studies in cattle and sheep poisoned by nitro-bearing *Astragalus* or their toxins. Am. J. Vet. Res. 41:377–382.

James, L.F., W. Foote, W. Nye, and W.J. Hartley. 1978. Effects of feeding *Oxytropis* and *Astragalus* pollen to mice and *Astragalus* seeds to rats. Am. J. Vet. Res. 39:711–712.

Ludolph, A.C., M. Seelig, A. Ludolph, P. Novitt, C.N. Allen, P.S. Spencer, and M.I. Sabri. 1992. 3-Nitropropionic acid decreases cellular energy levels and causes neuronal degeneration in cortical explants. Neurodegeneration 1:155–161.

Ludolph, A.C., F. He, P.S. Spencer, J. Hammerstad, and M. Sabri. 1991. 3-Nitropropionic acid—exogenous animal neurotoxin and possible human striatal toxin. Can. J. Neurol. Sci. 18:492–498.

Majak, W. 1992. Further enhancement of 3-nitropropanol detoxification by ruminal bacteria in cattle. Can. J. Anim. Sci. 72:863–870.

Majak, W., and M.A. Pass. 1989. Aliphatic nitrocompounds. pp. 143–159 in: P.R. Cheeke (Ed.). Toxicants of Plant Origin. II. Glycosides. CRC Press, Boca Raton, FL.

Majak, W., and L.J. Clark. 1980. Metabolism of aliphatic nitro compounds in bovine rumen fluid. Can. J. Anim. Sci. 60:699–708.

Majak, W., T. Udenberg, R.E. McDiarmid, and H. Douwes. 1981. Toxicity and metabolic effects of intravenously administered 3-nitropropanol in cattle. Can. J. Anim. Sci. 61:639–647.

Majak, W., R. Neufeld, and J. Corner. 1980. Toxicity of *Astragalus miser* V. *serotinus* to the honeybee. J. Apicult. Res. 19:196–199.

Maricle, B., J. Tobey, W. Majak, and J.W. Hall. 1996. Evaluation of clinicopathological parameters in cattle grazing timber milkvetch. Can. Vet. J. 37:153–156.

Ming, L. 1995. Moldy sugarcane poisoning—a case report with a brief review. Clin. Toxicol. 33:363–367.

Shenk, J.S., P.J. Wanganess, R.M. Leach, D.L. Gustine, J.L. Gobble, and R.F. Barnes. 1976. Relationship between β-nitropropionic acid content of crown vetch and toxicity in nonruminant animals. J. Anim. Sci. 42:616–621.

Williams, M.C. 1981a. Nitro compounds in *Indigofera* species. Agron. J. 73:434–436.

Williams, M.C. 1981b. Nitro compounds in foreign species of *Astragalus*. Weed Sci. 29:261–269.

Williams, M.C., and L.F. James. 1978. Livestock poisoning from nitro-bearing *Astragalus*. pp. 379–389 in: R.F. Keeler, K.R. Van Kampen, and L.F. James (Eds.). Effect of Poisonous Plants on Livestock. Academic Press, NY.

Williams, M.C., L.F. James, and B.O. Bond. 1979. Emory milkvetch (*Astragalus emoryanus*) poisoning in chicks, sheep and cattle. Am. J. Vet. Res. 40:403–406.

Williams, M.C., L.F. James, and A.T. Bleak. 1976. Toxicity of introduced nitro-containing *Astragalus* to sheep, cattle and chicks. J. Range Manage. 29:30–32.

Neurotoxic *Solanum* spp.

Dunster, P.J., and R.A. McKenzie. Does *Solanum esuriale* cause humpyback in sheep? Aust. Vet J. 64:119–120.

Kellerman, T.S., J.A.W. Coetzer, and T.W. Naude. 1988. Plant Poisonings and Mycotoxicoses of Livestock in Southern Africa. Oxford University Press, Capetown.

Menzies, J.S., C.H. Bridges, and E.M. Bailey. 1979. A neurological disease of cattle associated with *Solanum dimidiatum*. Southwest. Vet. 32:45–49.

Molyneux, R.J., L.F. James, M.H. Ralphs, J.A. Pfister, K.E. Panter, and R.J. Nash. 1994. Polyhydroxyalkaloid glycosidase inhibitors from poisonous plants of global distribution: Analysis and identification. pp. 107–112 in: S.M. Colegate and P.R. Dorling (Eds.). Plant-Associated Toxins. Agricultural, Phytochemical and Ecological Aspects. CAB International, Wallingford, United Kingdom.

Molyneux, R.J., Y.T. Pan, A. Goldmann, D.A. Tepfer, and A.D. Elbein. 1993. Calstegins, a novel class of alkaloid glycosidase inhibitors. Arch. Biochem. Biophys. 304:81–88.

Pienaar, J.G., T.S. Kellerman, P.A. Basson, W.L. Jenkins, and J. Vahrmeijer. 1976. Maldronksiekte in cattle. A neuropathy caused by *Solanum kwebense*. Onderstepoort J. Vet. Res. 43:67–74.

Riet-Correa, F., M.D.C. Mendez, A.L. Schield, B.A. Summers, and J.A. Oliveira. 1983. Intoxication by *Solanum fastigiatum* var. *fastigiatum* as a cause of cerebellar degeneration in cattle. Cornell Vet. 73:240–256.

Zambrano, M.D.S., F. Riet-Correa, A.L. Schild, and M.D.C. Mendez. 1985. Intoxication by *Solanum fastigiatum* var. *fastigiatum*: evolution and reversibility of the lesions in cattle and susceptibility of sheep, rabbits, guinea pigs and rats. Pesq. Vet. Bras. 5:133–141.

Datura spp. and Tropane Alkaloids

Day, E.J., and B.C. Dilworth. 1984. Toxicity of jimson weed seed and cocoa shell meal to broilers. Poult. Sci. 63:466–468.

Dugan, G.M., M.R. Gumbmann, and M. Friedman. 1989. Toxicological evaluation of jimson weed (*Datura stramonium*) seed. Food Chem. Toxic. 27:501–510.

Friedman, M., and C.E. Levin. 1989. Composition of jimson weed (*Datura stramonium*) seeds. J. Agric. Food Chem. 37:998–1005.

Guharoy, S.R., and M. Barajas. 1991. Atropine intoxication from the ingestion and smoking of jimson weed (*Datura stramonium*). Vet. Hum. Toxicol. 33:588–589.

Keeler, R.F. 1981. Absence of arthrogryposis in newborn Hampshire pigs from sows ingesting toxic levels of jimsonweed during gestation. Vet. Hum. Toxicol. 23:413–415.

Kingsbury, J.M. 1964. Poisonous Plants of the United States and Canada. Prentice–Hall, Englewood Cliffs, N.J.

Leipold, H.W., F.W. Oehme, and J.E. Cook. 1973. Congenital arthrogryposis associated with ingestion of jimsonweed by pregnant sows. J. Am. Vet. Med. Assoc. 162:1059–1060.

Nelson, P.D., H.D. Mercer, H.W. Essig, and J.P. Minyard. 1982. Jimson weed toxicity in cattle. Vet. Hum. Toxicol. 24:321–325.

Scheline, R.R. 1991. CRC Handbook of Mammalian Metabolism of Plant Compounds. CRC Press, Boca Raton, FL.

Schneider, F., P. Lutun, P. Kintz, D. Astruc, F. Flesch, and J.-D. Tempe. 1996. Plasma and urine concentrations of atropine after the ingestion of cooked deadly nightshade berries. Clin. Toxicol. 32:113–117.

Tyler, V.E. 1993. The Honest Herbal: A Sensible Guide to the Use of Herbs and Related Remedies. Haworth Press, New York.

Wada, S., T. Shimizudani, H. Yamada, K. Oguri, and H. Yoshimura. 1994. Sulphotransferase-dependent dehydration of atropine and scopolamine in guinea pig. Xenobiotica 24:853–861.

Wada, S., T. Yoshimitsu, N. Koga, H. Yamada, K. Oguri, and H. Yoshimura. 1991. Metabolism *in vivo* of the tropane alkaloid, scopolamine, in several mammalian species. Xenobiotica 21:1289–1300.

Williams, S., and P. Scott. 1984. The toxicity of *Datura stramonium* (Thorn apple) to horses. N.Z. Vet. J. 32:47.

Worthington, T.R., E.P. Nelson, and M.J. Bryant. 1981. Toxicity of thornapple (*Datura stramonium* L.) seeds to the pig. Vet. Rec. 108:208–211.

Nicotiana Toxicity

Bunch, T.D., K.E. Panter, and L.F. James. 1992. Ultrasound studies of the effects of certain poisonous plants on uterine function and fetal development in livestock. J. Anim. Sci. 70:1639–1643.

Bush, L.P., and M.W. Crowe. 1989. *Nicotiana* Alkaloids. pp. 87–107 in: P.R. Cheeke (Ed.). Toxicants of plant Origin, Vol. I. CRC Press, Boca Raton, FL.

Keeler, R.F., and L.D. Balls. 1978. Teratogenic effects in cattle of *Conium maculatum* and *Conium* alkaloids and analogs. Clin. Toxicol. 12:49–64.

Keeler, R.F., M.W. Crowe, and E.A. Lambert. 1984. Teratogenicity in swine of the tobacco alkaloid anabasine isolated from *Nicotiana glauca*. Teratology 30:61–69.

Kingsbury, J.M. 1964. Poisonous Plants of the United States and Canada. Prentice–Hall, Englewood Cliffs, N.J.

Panter, K.P., R.F. Keeler, L.F. James, and T.D. Bunch. 1992. Impact of plant toxins on fetal and neonatal development: a review. J. Range Manage. 45:52–57.

Plumlee, K.H., D.M. Holstege, P.C. Blanchard, K.M. Fiser, and F.D. Galey. 1993. *Nicotiana glauca* toxicosis of cattle. J. Vet. Diagn. Invest. 5:498–499.

Yellow Star Thistle

Cordy, D.R. 1978. *Centaurea* species and equine nigropallidal encephalomalacia. pp. 327–336 in: R.F. Keeler, K.R. Van Kamen, and L.F. James (Eds.). Effects of Poisonous Plants on Livestock. Academic Press, N.Y.

Wang, Y., M. Hamburger, C.H.K. Cheng, B. Costall, R.J. Naylor, P. Jenner, and K. Hostettmann. 1991. Neurotoxic sesquiterpenoids from the yellow star thistle *Centaurea solstitialis* L. (Asteraceae). Helvetica Chimica Acta 74:117–123.

White Snakeroot (*Eupatorium* spp.)

Beier, R.C., and J.O. Norman. 1990. The toxic factor in white snakeroot: identity, analysis and prevention. Human Vet. Toxicol. 32(Suppl.):81–88.

Beier, R.C., J.O. Norman, T.R. Irvin, and D.A. Witzel. 1987. Microsomal activation of constituents of white snakeroot (*Eupatorium rugosum* Houtt) to form toxic products. Am. J. Vet. Res. 48:583–585.

O'Sullivan, B.M. 1979. Crofton weed (*Eupatorium adenophorum*) toxicity in horses. Aust. Vet. J. 55:19–21.

Olson, C.T., W.C. Keller, D.F. Gerken, and S.M. Reed. 1984. Suspected tremetol poisoning in horses. J. Am. Vet. Med. Assoc. 185:1001–1003.

Stotts, R. 1984. White snakeroot toxicity in dairy cattle. VM/SAC, Vet. Med. Small An. Clin. 79:118–120.

Stypandra (Blind Grass) Toxicity

Dorling, P.R., S.M. Colegate, and C.R. Huxtable. 1992. The molecular mechanism of stypandrol poisoning. pp. 469–473 in: L.F. James, R.F. Keeler, E.M. Bailey, Jr., P.R. Cheeke, and M.P. Hegarty (Eds.). Poisonous Plants. Proceedings of the Third International Symposium. Iowa State University Press, Ames.

Huxtable, C.R., S.M. Colegate, and P.R. Dorling. 1989. Stypandrol and *Stypandra* toxicosis. pp. 83–94 in: P.R. Cheeke (Ed.). Toxicants of Plant Origin, Vol. IV. Phenolics. CRC Press, Boca Raton, FL.

Huxtable, C.R., P.R. Dorling, and D.H. Slatter. 1980. Myelin oedema, optic neuropathy and retinopathy in experimental *Stypandra imbricata* toxicosis. Neuropathol. Appl. Neurobiol. 6:221–232.

Main, D.C., D.H. Slatter, C.R. Huxtable, I.C. Constable, and P.R. Dorling. 1981. *Stypandra imbricata* (blindgrass) toxicosis in goats and sheep—clinical and pathologic findings in 4 field cases. Aust. Vet. J. 57:132–135.

Whittington, R.J., Searson, J.E., Whittaker, S.J., and J.R.W. Glastonbury. 1988. Blindness in goats following ingestion of *Stypandra glauca*. Aust. Vet. J. 65:176–181.

Cycads

Beck, W. 1992. Aboriginal preparation of *Cycas* seeds in Australia. Econ. Bot. 46:133–147.

Charlton, T.S., A.M. Marini, S.P. Markey, K. Norstog, and M.W. Duncan. 1992. Quantification of the neurotoxin 2-amino-3-(methylamino)-propanoic acid (BMAA) in cycadales. Phytochem. 31:3429–3432.

Hooper, P.T. 1978. Cycad poisoning in Australia—Etiology and pathology. pp. 337–347 in: R.F. Keeler, K.R. Van Kampen, and L.F. James (Eds.). Effects of Poisonous Plants on Livestock. Academic Press, N.Y.

Mills, J.N., M.J. Lawley, and J. Thomas. 1996. Macrozamia toxicosis in a dog. Aust. Vet. J. 73:69–72.

Reddy, B.S., K. Furuya, D. Hanson, J. Dibello, and B. Berke. 1982. EFfect of dietary butylated hydroxyanisole on methlazoxymethanolacetate-induced toxicity in mice. Food Chem. Toxicol. 20:853–860.

Tustin, R.C. 1983. Notes on the toxicity and carcinogenicity of some South African cycad species with special reference to that of *Encephalartos lanatus*. J. S. Afr. Vet. Assoc. 54:33–42.

Wogan, G.N., and W.F. Busby. 1980. Cycasin. pp. 350–353 in: I.E. Liener (Ed.). Toxic Constituents of Plant Foodstuffs. Academic Press, N.Y.

Stringhalt

Cahill, J.I., B.E. Goulden, and H.G. Pearce. 1985. A review and some observations on stringhalt. N. Z. Vet. J. 33:101–104.

Galey, F.D., P.J. Hullinger, and J. McCaskill. 1991. Toxic plant case reports. Outbreaks of stringhalt in northern California. Vet. Hum. Toxicol. 33:176–177.

Huntington, P.J., S. Seneque, R.F. Slocombe, L.B. Jeffcott, A. McLean, and A.R. Luff. 1991. Use of phenytoin to treat horses with Australian stringhalt. Aust. Vet. J. 68:221–224.

Robertson-Smith, R.G., L.B. Jeffcott, S.C.E. Friend, and L.M. Badcoe. 1985. An unusual incidence of neurological disease affecting horses during a drought. Aust. Vet. J. 62:6–12.

β-phenethylene Neurotoxins

Forbes, T.D.A., I.J. Pemberton, G.R. Smith, and C.M. Hensarling. 1995. Seasonal variation of two phenolic amines in *Acacia berlandieri*. J. Arid Environ. 30:403–415.

Forbes, T.D.A., B.B. Carpenter, R.D. Randel, and D.R. Tolleson. 1994. Effects of phenolic monoamines on release of luteinizing hormone stimulated by gonadotropin-releasing hormone and on plasma adrenocorticotropic hormone, norepinephrine, and cortisol concentrations in wethers. J. Anim. Sci. 72:464–469.

Pemberton, I.J., G.R. Smith, T.D.A. Forbes, and C.M. Hensarling. 1993. Technical note: An improved method for extraction and quantification of toxic phenethylamines from *Acacia berlandieri*. J. Anim. Sci. 71:467–470.

Smith, T.S. 1977. Phenethylamine and related compounds in plants. Phytochemistry 16:9–18.

Vera-Avila, H.R., T.D.A. Forbes, and R.D. Randel. 1996. Plant phenolic amines: Potential effects on sympathoadrenal medullary, hypothalamic-pituitary-adrenal, and hypothalamic-pituitary-gonadal function in ruminants. Domestic Anim. Endocrinol. 13:285–296.

Cardiac Glycosides

Anderson, L.A.P., R.A. Schultz, L.P.J. Joubert, L. Prozesky, T.S. Kellerman, G.L. Erasmus, and J. Procos. 1983. Krimpiekte and acute cardiac glycoside poisoning in sheep caused by bufadienolides from the plant *Kalanchoe lanceolata* Forsk. Onderstepoort J. Vet. Res. 50:295–300.

Benson, J.M., J.N. Seiber, C.V. Bagley, R.F. Keeler, A.E. Johnson, and S. Young. 1979. Effects on sheep of the milkweeds *Asclepias eriocarpa* and *A. labriformis* and of cardiac glycoside-containing derivative material. Toxicon 17:155–165.

Galey, F.D., D.M. Holstege, K.H. Plumlee, E. Tor, B. Johnson, M.L. Anderson, P.C. Blanchard, and F. Brown. 1996. Diagnosis of oleander poisoning in livestock. J. Vet. Diagn. Invest. (in press).

Harborne, J.B. 1993. Introduction to Ecological Biochemistry. Academic Press, San Diego.

Joubert, J.P.J. 1989. Cardiac glycosides. pp. 61–96 in: P.R. Cheeke (Ed.). Toxicants of Plant Origin. Vol. II. Glycosides. CRC Press, Boca Raton, FL.

Joubert, J.P.J. 1983. Attempted prevention and treatment of *Geigeria filifolia* Mattf. poisoning (vermeersiekte) in sheep. J.S. Afr. Vet. Assoc. 54:255–258.

Joubert, J.P.J., and R.A. Schultz. 1982. The treatment of *Morea polystachya* (Thunb.) Ker-Gawl (cardiac glycoside) poisoning in sheep and cattle with activated charcoal and potassium chloride. J. S. Afr. Vet. Assoc. 53:249–253.

Kellerman, T.S., J.A.W. Coetzer, and T.W. Naude. 1988. Plant Poisonings and Mycotoxicoses of Livestock in Southern Africa. Oxford University Press, Capetown.

Kingsbury, J.M. 1964. Poisonous Plants of the United States and Canada. Prentice-Hall, Englewood Cliffs, N.J.

McKenzie, R.A., and P.J. Dunster. 1987. Curing experimental *Bryophyllum tubiflorum* poisoning of cattle with activated carbon, electrolyte replacement solution and antiarrhythmic drugs. Aust. Vet. J. 64:211–214.

Moxley, R.A., N.R. Schneider, D.H. Steinegger, and M.P. Carlson. 1989. Apparent toxicosis associated with lily-of-the-valley (*Convallaria majalis*) ingestion in a dog. J. Am. Vet. Med. Assoc. 195:485–487.

Nielsen, D.E., H. Nishimura, J.W. Otos, and M. Calvin. 1977. Plant crops as a source of fuel and hydrocarbon-like materials. Science 198:942–944.

Ogden, L., G.E. Burrows, R.J. Tyrl, and R.W. Ely. 1992a. Experimental intoxication in sheep by *Asclepias*. pp. 495–499 in: L.F. James, R.F. Keeler, E.M. Bailey, Jr., P.R. Cheeke, and M.P. Hegarty (Eds.). Poisonous Plants. Proc. of the Third Internat. Symposium. Iowa State University Press, Ames.

Ogden, L., G.E. Burrows, R.J. Tyrl, and S.L. Gorham. 1992b. Comparison of *Asclepias* species based on their toxic effects in chickens. pp. 500–505 in: L.F. James, R.F. Keeler, E.M. Bailey, Jr., P.R. Cheeke, nd M.P. Hegarty (Eds.). Poisonous Plants. Proc. of the Third Internat. Symposium. Iowa State University Press, Ames.

Rezakhani, A., and M. Maham. 1992. Oleander poisoning in cattle of the Fars province, Iran. Vet. Human Tox. 34:549.

Seiber, J.N., S.M. Lee, and J.M. Benson. 1983. Cardiac glycosides (cardenolides) in species of *Asclepias* (Asclepiadaceae). pp. 43–83 in: R.F. Keeler and A.T. Tu (Eds.). Handbook of Natural Toxins. Plant and Fungal Toxins. Marcel Dekker, Inc., New York.

Strydom, J.A., and J.P.J. Joubert. 1983. The effect of predosing *Homeria pallida* Bak. to cattle to prevent tulp poisoning. J. S. Afr. Vet. Assoc. 54:201–203.

Tor, E.R., D.M. Holstege, and F.D. Galey. 1996. Determination of oleander glycosides in biological matrices by high-performance liquid chromatography. J. Agric. Food Chem. 44:2716–2719.

Williams, M.C., and M.C. Smith. 1984. Toxicity of *Kalanchoe* spp. to chicks. Am. J. Vet. Res. 45:53–546.

Veratrum and *Zygadenus* spp.

Kingsbury, J.M. 1964. Poisonous Plants of the United States and Canada. Prentice–Hall, Englewood Cliffs, New Jersey.

Longland, W.S., and C. Clements. 1995. Consumption of a toxic plant (*Zigadenus paniculatus*) by mule deer. Great Basin Naturalist 55:188–191.

Majak, W., R.E. McDiarmid, W. Cristofoli, F. Sun, and M. Benn. 1992. Content of zygacine in *Zygadenus venenosus* at different stages of growth. Phytochemistry 31:3417–3418.

Panter, K.E., M.H. Ralphs, R.A. Smart, and B. Duelke. 1987. Death camas poisoning in sheep: A case report. Vet. Human Toxicol. 29:45–48.

Yew (*Taxus*) Toxicity

Kelsey, R.G., and N.C. Vance. 1992. Taxol and cephalomannine concentrations in the foliage and bark of shade-grown and sun-exposed *Taxus breviofolia* trees. J. Natural Products 55:912–917.

Panter, K.E., R.J. Molyneux, R.A. Smart, L. Mitchell, and S. Hansen. 1993. English yew poisoning in 43 cattle. J. Am. Vet. Med. Assoc. 202:1476–1477.

Rowinsky, E.K., L.A. Cazenave, and R.C. Donehower. 1990. Taxol: A novel investigational antimicrotubule agent. J. National Cancer Inst. 82:1247–1259.

Veatch, J.K., F.M. Reid, and G.A. Kennedy. 1988. Differentiating yew poisoning from other toxicoses. Vet. Med. March, 298–300.

Witherup, K.M., S.A. Look, M.W. Stasko, T.J. Ghiorzi, and G.M. Muschik. 1990. *Taxus* spp. needles contain amounts of taxol comparable to the bark of *Taxus brevifolia*: Analysis and isolation. J. Natural Products 53:1249–1255.

Rhododendron

Meier, K.H., and R.S. Hemmick. 1992. Bradycardia and complete heart block after ingestion of rhododendron tea. Vet. Human Tox. 34:351 (Abst.)

Onat, F., B.C. Yegen, R. Lawrence, A. Oktay, and S. Oktay. 1991. Site of action of grayanotoxins in mad honey in rats. J. Appl. Tox. 11:199–201.

Avocado Toxicity

Burger, W.P., T.W. Naude, I.B.J. Van Rensburg, C.J. Botha, and A.C.E. Pienaar. 1994a. Cardiomyopathy in ostriches (*Struthio camelus*) due to avocado (*Persea americana* var. *guatemalensis*) intoxication. J. S. Afr. Vet. Assoc. 65:113–118.

Burger, W.P., T.W. Naude, I.B.J. Van Rensburg, C.J. Botha, and A.C.E. Pienaar. 1994b. Avocado poisoning in ostriches. pp. 546–551 in: S.M. Colegate and P.R. Dorling (Eds.). Plant-associated Toxins. Agricultural, Phytochemical and Ecological Aspects. CAB International, Wallingford, Oxon, UK.

Grant, R., P.A. Basson, H.H. Booker, B. Hofherr, and M. Anthonissen. 1991. Cardiomyopathy caused by avocado (*Persea americana* Mill.) leaves. J.S. Afr. Vet. Assoc. 62:21–22.

Hargis, A.M., E. Sauber, S. Casteel, and D. Eitner. 1989. Avocado (*Persea americana*) intoxication in caged birds. J. Amer. Vet. Med. Assoc. 194:64–66.

McKenzie, R.A., and O.P. Brown. 1991. Avocado (*Persea americana*) poisoning of horses. Aust. Vet. J. 68:77–78.

Sani, Y., R.B. Atwell, and A.A. Seawright. 1991. The cardiotoxicology of avocado leaves. Aust. Vet. J. 68:150–151.

Galegine-Containing Plants

Huxtable, C.R., P.R. Dorling, and S.M. Colegate. 1993. Identification of galegine, an isoprenyl guanidine, as the toxic principle of *Schoenus asperocarpus* (poison sedge). Aust. Vet. J. 70:169–172.

Keeler, R.F., D.C. Baker, and K.E. Panter. 1992. Concentration of galegine in *Verbesina encelioides* and *Galega officinalis* and the toxic and pathologic effects induced by the plants. J. Environ. Path. Toxicol. Oncol. 11:75–81.

Keeler, R.F., D.C. Baker, and J.O. Evans. 1988. Individual animal susceptibility and its relationship to induced adaptation or tolerance in sheep to *Galega officinalis* L. Vet. Human Toxicol. 30:420–423.

Keeler, R.F., A.E. Johnson, L.D. Stuart, and J.O. Evans. 1986. Toxicosis from and possible adaptation to *Galega officinalis* in sheep and the relationship to *Verbesina encelioides* toxicosis. Vet. Human Toxicol. 28:309–315.

Moller, E., and S.B. Hostrup. 1996. Digestibility and feeding value of fodder galega (*Galega orientalis* Lam.). Acta Agric. Scand., Sect. A, Animal Sci. 46:97–104.

Oelrichs, P.B., P.J. Vallely, J.K. MacLeod, and I.A.S. Lewis. 1985. Chemistry and toxic effects of *Verbesina encelioides*. pp. 479–483 in: A.A. Seawright, M.P. Hegarty, L.F. James and R.F. Keeler (Eds.). Plant Toxicology. Queensland Poisonous Plants Committee, Yeerongpilly, Australia.

Other Heart, Lung, and Renal Toxins

Flaoyen, A., M. Binde, B. Bratberg, B. Djonne, M. Fjolstad, H. Gronstol, H. Hassan, P.G. Mantle, T. Landsverk, J. Schonheit, and M.H. Tonnesen. 1995. Nephrotoxicity of *Narthecium ossifragum* in cattle in Norway. Vet. Rec. 137:259–263.

Flaoyen, A., B. Bratberg, M. Fjolstad, H. Gronstol, T. Landsverk, and J. Schonheit. 1994. A pasture-related nephrotoxicosis of cattle in Norway: clinical signs and pathological findings. pp. 557–560 in: S.M. Colegate and, P.R. Dorling (Eds.). Plant-Associated Toxins. Agricultural, Phytochemical and Ecological Aspects. CAB International, Wallingford, Oxon, UK.

Flaoyen, A., M. Binde, B. Djonne, H. Gronstol, H. Hassan, and P.G. Mantle. 1993. A pasture-related nephrotoxicosis in Norway; epidemiology and aetiology. N. Z. Vet. J. 41:221–222.

Fourie, N., G.L. Erasmus, L. Prozesky, and R.A. Schultz. 1994. Gousiekte, an important plant-induced cardiotoxicosis of ruminants in southern Africa caused by certain members of the Rubiaceae. pp. 529–533 in: S.M. Colegate and P.R. Dorling (Eds.). Plant-Associated Toxins. Agricultural, Phytochemical and Ecological Aspects. CAB International, Wallingford, UK.

Kellerman, T.S., J.A.W. Coetzer, and T.W. Naude. 1988. Plant Poisonings and Mycotoxicoses of Livestock in Southern Africa. Oxford University Press, Cape Town.

Pan, L.C., M.W. Lame, D. Morin, D.W. Wilson, and H.J. Segall. 1991. Red blood cells augment transport of reactive metabolites of monocrotaline from liver to lung in isolated and tandem liver and lung preparations. Tox. Appl. Pharm. 110:336–346.

Wilson, D.W., H.J. Segall, L.C. Pan, M.W. Lame, J.E. Estep, and D. Morin. 1992. Mechanisms and pathology of monocrotaline pulmonary toxicity. Crit. Rev. Toxicol. 22:307–325.

CHAPTER 14

Reproductive Toxins

Numerous toxins in plants cause reproductive problems. In most cases, these effects have been discussed in previous chapters, where a discussion of reproductive effects along with other responses was appropriate. In such instances, the toxins are only briefly mentioned in this chapter, with the reader directed to the appropriate chapter for more information.

The general effects of poisonous plants on reproduction in animals is reviewed by James *et al.* (1992) and Panter *et al.* (1992).

PHYTOESTROGENS

Isoflavones with estrogenic activity are common in various legume forages, with those in subterranean clover being the most prominent in causing impairment of livestock reproduction. Phytoestrogens in legume forages are discussed in Chapter 11, and those in soybeans in Chapter 7.

Zearalenone, a mycotoxin produced by *Fusarium* fungi, is estrogenic. It is discussed in Chapter 5.

INHIBITORS OF MALE FERTILITY

Gossypol in cottonseed inhibits spermatogenesis in many species, including humans. Gossypol-induced infertility is discussed in Chapter 7. Consumption of snakeweeds (*Gutierrezia* spp.) inhibits fertility of male rats, with increased numbers of abnormal sperm (see Saponins, Chapter 11).

TERATOGENS IN POISONOUS PLANTS

Veratrum Alkaloids

Veratrum californicum (false hellebore) (Fig. 14–1) contains **steroid alkaloids** which cause both acute toxicity (see Chapter 13) and teratogenic effects. Investigators at the USDA Poisonous Plant Laboratory demonstrated that consumption of this plant by pregnant ewes on the fourteenth day of pregnancy resulted in production of "cyclops" or "monkey-face" lambs. This condition had occurred in western range areas, particularly in Idaho, in epidemic proportions. In southwest Idaho, the incidence of the condition varied from less than 1% to nearly 25% of the lambs born in a given flock of sheep. The lamb deformities vary from slight deformities of the upper jaw to a complete cyclops condition, with one centrally placed eye (Fig. 14–2).

FIGURE 14–1 Leaves and flowers of *Veratrum californicum*, which contains teratogenic steroid alkaloids.

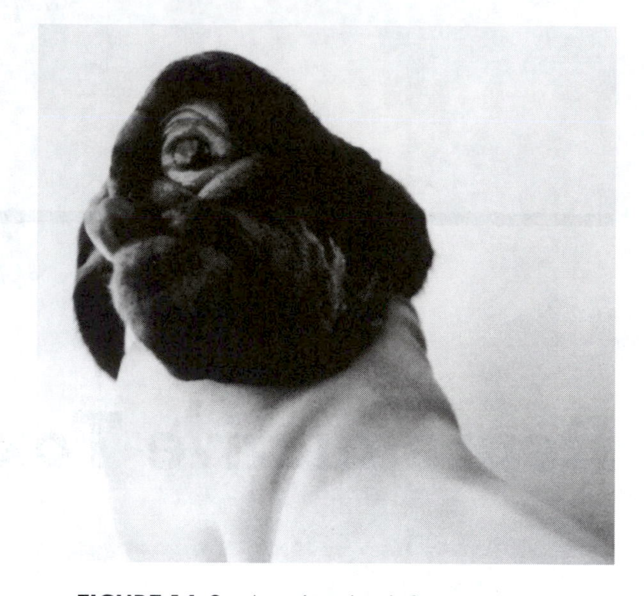

FIGURE 14–2 A cyclops lamb from a ewe that had consumed *Veratrum californicum* in early gestation. (Courtesy of R.F. Keeler)

A ewe carrying a deformed lamb(s) due to *Veratrum* may often have a prolonged gestation period. The deformed lamb continues to live and grow in utero and may reach a weight of 20–30 lb. Eventually both the lamb and ewe die.

When the *Veratrum* is consumed by the ewe on days 28–30 of pregnancy, the lamb may have

FIGURE 14–3 Shortened metatarsal and metacarpal bones in a lamb born to a ewe that consumed *Veratrum californicum*. (Courtesy of R.F. Keeler)

FIGURE 14–4 Deformity in a sheep caused by maternal consumption of *Veratrum californicum*. (Courtesy of M.E. Fowler)

shortened metatarsal and metacarpal bones, resulting in an abnormal squat appearance (Fig. 14–3). A cleft lip and palate (Fig. 14–4) may also occur (Keeler and Stuart, 1987).

Other animals such as cattle and goats are affected by the teratogenic properties of veratrum when it is administered experimentally. Field cases are seen only with sheep. Prevention of cylops lambs can be accomplished by waiting until after the first frost before breeding ewes; *Veratrum* is killed by frost and loses its toxicity. The teratogenic effects of *Veratrum* have been reviewed by Panter *et al.* (1992).

Veratrum spp. contain over 50 individual **steroid alkaloids**. Veratramine is one found in high concentration. Five of the alkaloids found in largest concentrations in *V. californicum* are shown in Fig. 14–5.

The compounds that are active in producing teratogenicity are cylopamine, cycloposine, and jervine. Other *Veratrum* alkaloids are not teratogenic. The fused rings with an oxide bridge appear necessary for activity. Also, C-5, C-6 unsaturation is a key structural factor in *Vera-*

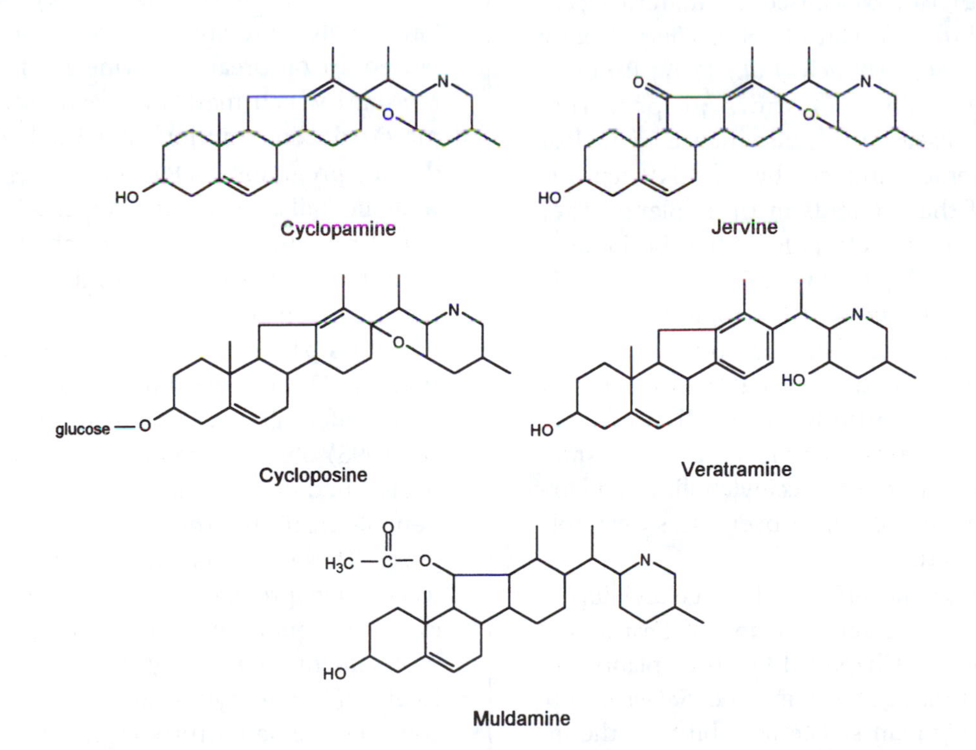

FIGURE 14–5 Structures of five steroidal alkaloids from *Veratrum californicum*.

trum-induced teratogenicity (Gaffield and Keeler, 1993).

According to Panter *et al.* (1994), there are two modes of action of *Veratrum* alkaloids in their teratogenic action. The craniofacial deformities occur with disturbances in embryo development during the neural tube stage. *Veratrum* alkaloids interfere with catecholamine-secreting cells in embryonic neuroepithelium; the inhibition of catecholamine release disrupts the normal migration and development of the embryonic cells. The *Veratrum* effects on bone development are due to an interference with cartilage metabolism. Effects on chondrocyte development are also implicated in the tracheal stenosis (narrowing or stricture of trachea) condition observed in neonatal lambs from ewes consuming *Veratrum* during days 27–33 of gestation (Keeler *et al.*, 1985).

Other Teratogens

Congenital skeletal malformations (**crooked calf disease**) can be caused by maternal consumption during gestation of *Lupinus, Conium* and *Nicotiana* spp. (Bunch *et al.*, 1992; Panter *et al.*, 1990a,b). These authors propose that crooked calf disease is a consequence of the loss of fetal movement induced by the CNS-depressing effect of the alkaloids in these plants. They also suggest that **cleft palate** may be induced by the same lack of fetal activity, resulting in mechanical obstruction of the tongue. Panter *et al.* (1994) conclude that the contracture type skeletal defects and cleft palate induced by piperidine and quinolizidine alkaloids are caused by neuromuscular blocking action similar to that of curare or succinylcholine. The exposure must be prolonged over the susceptible period of gestation.

The teratogenic effects of rangeland lupins are discussed in Chapter 11 and of *Conium* and *Nicotiana* spp. in Chapter 13. Other plants implicated in teratogenesis include *Solanum* spp. and cyanogenic plants. The possibility of the involvement of **potato glycoalkaloids** as teratogens in humans is discussed in Chapter 6. The steroidal *Solanum* glycoalkaloids are very similar in chemical structure to teratogenic *Veratrum* alkaloids, so teratogenic effects might be expected. Various weedy *Solanum* spp. have been shown to induce birth defects. For example, *S. dimidiatum* (potato-weed), a toxic plant in the south-western U.S., was demonstrated to induce a neural tube defect in hamsters (Casteel *et al.*, 1989). *S. eleagnifolium* and *S. dulcamara* were also observed to be teratogenic in hamsters (Keeler *et al.*, 1991). The terata induced by *Solanum* spp. were mainly exencephaly (brain outside of skull) and encephalocele (hernia of brain). According to Keeler *et al.* (1991), the "real world" hazard of teratogenic *Solanum* alkaloids cannot yet be stated.

Various ***Astragalus*** spp. are teratogenic; they are discussed in Chapter 13. Swainsonine is not teratogenic; thus another constituent of locoweeds must be responsible for the teratogenic activity (Panter *et al.*, 1994), although there is evidence (see Chapter 13) that swainsonine may affect glycoprotein synthesis in the fetus.

There are indications that **cyanogens** may have teratogenic effects. Selby *et al.* (1971) reported an outbreak of swine malformations in Missouri which may have been associated with maternal consumption of wild black cherries during pregnancy. Deformed pigs were born with no tails, very small external sex organs, and limb deformities. Pritchard and Voss (1967) observed fetal deformities (ankylosis of the joints) in foals from mares grazing on Sudan grass pasture, while more recently, Seaman *et al.* (1981) reported limb deformities in calves from heifers grazing on Sudan grass. Bradley *et al.* (1995) observed contracted limbs (arthrogryposis) in lambs born to ewes that exhibited neurological disorders while grazing sudan grass. These reports, indicating teratogenic effects when pregnant animals were exposed to cyanogen-containing plants, suggest a possible involvement of these glycosides as teratogens. Keeler (1984) hypothesized that this effect could be mediated through hypoxia induced by cyanide.

Beef calves born with generalized joint laxity and dwarfism have been noted sporadically in North America and Australia. The condition is often called **acorn calves** because of its occurrence on oak-dominated rangelands of California. However, the condition is not reproduced by feeding acorns, but is apparently feed-related. Ribble *et al.* (1989) introduced the term **congenital joint laxity and dwarfism (CJLD)** to describe affected calves (Fig. 14–6). The condition was associated with the feeding of grass and clover silage as the sole winter feed for pregnant beef cows. Supplementation with hay and grain eliminated the problem. CJLD has been reported from Australia (Hawkins, 1994; Peet and Creeper, 1994), accompanied by a high incidence of retained placenta. No causative agents have been conclusively identified, although mycotoxins are a possibility. Another condition, **congenital spinal stenosis (CSS)** of newborn calves, shares some similarity with CJLD (Ribble *et al.*, 1993). Both conditions result from exposure of pregnant cows to a causative agent in the winter feeding period, and in both conditions the fetal bone growth plate is affected. In CJLD, there is impeded endochondral ossification (Ribble *et al.*, 1989), while with CSS there is premature growth plate closure (Ribble *et al.*, 1993). The CSS condition is associated with the feeding of moldy straw, suggesting mycotoxin involvement.

FIGURE 14–6 An "acorn calf," showing congenital joint laxity and dwarfism (CJLD). The calf shows shortened long bones, with overdistention of the distal joints, and slight doming of the cranium. (Courtesy of C.D. Hawkins)

PONDEROSA PINE NEEDLE ABORTION

Ponderosa pine (*Pinus ponderosa*) needle abortion is a common problem in the rangeland areas of the western U.S. and Canada. The effect of pine needle consumption is actually premature parturition rather than abortion, but the condition is generally referred to as pine needle abortion (Fig. 14–7). In addition to the abortion loss, affected cows have an increased incidence of retained placenta and impaired breeding performance. Cows in the last trimester of pregnancy are susceptible, and the problem is generally observed only in the winter and spring when other forage is scarce. Green needles, needles from logging slash, or dried needles that have fallen from trees are all potentially toxic. Pine needles are not normally consumed; when

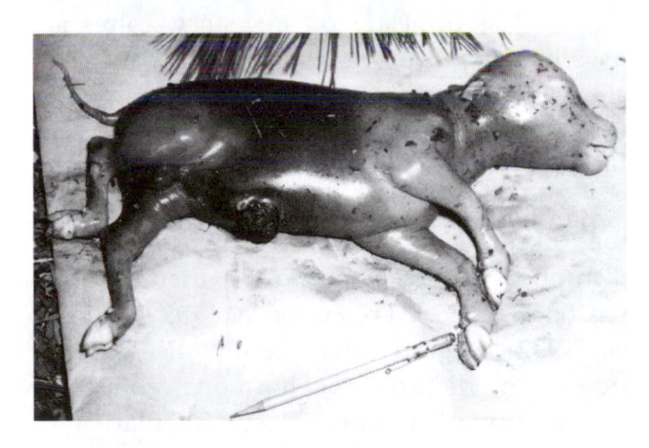

FIGURE 14–7 A case of pine needle abortion caused by maternal consumption of Ponderosa pine needles. (Courtesy of W. Majak)

the problem occurs, there are generally some environmental factors that cause cows to consume them. These include winter storms that force cattle to seek shelter under pine trees when grass and other forage is in short supply, sudden access to pine needles, feeding of poor-quality hay, or boredom. Pine needle consumption and risk of abortion are greatly diminished during mild winter weather (Pfister and Adams, 1993). All parts of the tree are toxic; the branch tips and bark are more potent than the needles (Panter *et al.*, 1990). A minimum intake of about 0.68 kg of needles per cow per day for several days induces abortion (Short *et al.*, 1992). Dietary supplements such as mineral mixtures are not effective in preventing losses (Short *et al.*, 1994). Short *et al.* (1994) found that feeding corn silage to cows exposed to pine needles prevented abortion, because the silage-fed cows did not eat pine needles. Abortions may occur within 48 hr of pine needle consumption and may be observed as much as 2 weeks after needles are eaten. Affected cows become depressed and dull in appearance, and edema of the genitalia and udder occurs prior to abortion. There may be a bloody discharge from the vulva. Cows may have a toxemia prior to and after the abortion. The aborted fetuses show autolysis, indicating fetal death in utero, necrosis of the kidney tubules and pulmonary congestion. When cows ingest pine needles during the last trimester of pregnancy, calves are usually born alive (Panter *et al.*, 1992). Calves aborted after day 250 of gestation may survive with extra care; the closer to normal parturition, the more likely a calf will survive.

The physiological mechanism involved in pine needle–induced abortion appears to be a disruption of blood flow to the uterus (Christenson *et al.*, 1992a,b; Ford *et al.*, 1992). Ford *et al.* (1992) suggested that the pine needle abortion factor disturbs the hormonal control of **uterine arterial blood flow**. Regulation of uterine arterial smooth muscle tone results from the coordinated interaction of α_2-adrenergic receptors, potential sensitive Ca^{++} channels (PSC) and catechol estrogens (Christenson *et al.*, 1993). Catechol estrogens are vasoactive metabolites of estrogen which prevent α_2-adrenergic receptor–mediated changes in blood vessel diameter by inhibiting Ca^{++} uptake through PSC (Christenson *et al.*, 1993). The abortifacient agent in pine needles inhibits the ability of catechol estrogen to block PSC activity. It may be a PSC agonist, inhibit catechol estrogen synthesis, or prevent catechol estrogen from interacting with PSC (Christenson *et al.*, 1993). The result of these effects is a profound decrease in blood flow to the uterus with the subsequent stress to the calf causing the normal cascade of events associated with parturition (Christenson *et al.*, 1993; Short *et al.*, 1995a; Stegelmeier *et al.*, 1996).

Hormonal changes are induced by pine needle consumption. Short *et al.* (1989) observed that feeding pine needles to pregnant cows induced a premature normal rise in cortisol and estradiol-17β associated with parturition. An abrupt rise in serum progesterone occurred after the start of pine needle feeding which then decreased sharply for 7–8 days before parturition. In cows, progesterone declines shortly before parturition, removing the progesterone block of estrogen, thus allowing the stimulatory effect of estrogen on uterine contractions. Jensen *et al.* (1989) also noted a reduction in serum progesterone in cows fed pine needles. They attributed this response to luteal cell necrosis.

Pine needle abortion seems to affect only bovines such as cattle and bison. Short *et al.* (1992) fed ponderosa pine needles to cattle, bison, sheep and goats, and induced abortion only in the cattle and bison. Although sheep do not abort after eating pine needles, they do have a greater incidence of dead lambs (Short *et al.*, 1995b), and have a similar plasma level of vasoconstrictive activity (as measured by a uterine tissue perfusion bioassay) as cattle. Short *et al.* (1995b) further demonstrated that transfer into cattle of rumen contents of sheep fed pine needles did not affect abortifacient action, suggesting that differences between sheep and cattle in abortion are not due to ruminal metabolism.

Further evidence that sheep ruminal action does not inactivate the abortion-inducing agent is the similar plasma levels of vasoconstrictive factor in sheep and cattle fed pine needles (Short *et al.*, 1995b). Deer may consume high levels of ponderosa pine needles without aborting (Short *et al.*, 1992). Pine needle–induced abortion does not seem to occur in wild bison, because they apparently do not consume pine foliage. Their ability to obtain grass during storms and heavy snowfalls is greater than is the case with cattle.

Numerous investigators have attempted to identify the abortion-inducing factor(s) in ponderosa pine needles. Research was hampered by the lack of a laboratory animal model, so that purified fractions had to be tested with pregnant cows. James *et al.* (1994) reported that the active fraction was extracted from pine needles with methylene chloride. Analysis of this fraction (Gardner *et al.*, 1994) revealed four abietane diterpene acids (isopimaric, dehydroabietic, abietic and neoabietic acids) and two labdane diterpene acids (imbricataloic and isocupressic acids). Pregnant cows gavaged with a mixture of the abietane diterpene acids did not abort but toxicity occurred with high doses, with signs of depression, anorexia, rumen stasis, bloat, respiratory distress, peripheral neuropathy and death. These signs had also been observed in cows fed ponderosa pine branch tips (Panter *et al.*, 1990). When pregnant cows were gavaged with isocupressic acid, the animals aborted. Thus it appears that the major abortifacient constituent in ponderosa pine is **isocupressic acid** (Fig. 14–8), while the abietane diterpene acids have toxic effects, including nephrotoxic and neurologic actions (Stegelmeier *et al.*, 1996). Much of the isocupressic acid in ponderosa pine needles occurs as acetyl and succinyl esters. These are hydrolyzed in the rumen, releasing free isocupressic acid (Gardner *et al.*, 1996). Al-Mahmoud *et al.* (1995) isolated some vasoactive lipids from pine needles which may be involved as well. Gardner *et al.* (1994) suggested that the toxic effects (depression, decreased appetite) of the abietane acids might expedite the abortion-inducing effects of isocupressic acid. Consumption of ponderosa pine needles also adversely affects digestive processes in the rumen (Adams *et al.*, 1992). Pfister *et al.* (1992) observed that ponderosa pine needles in the diet reduce protein and energy digestibility and are toxic to cellulolytic rumen microbes. This is not surprising, because pine resins function as antimicrobial agents, protecting the tree against fungal and bacterial invasion (Himejima, 1992).

Although abortion problems are primarily associated with ponderosa pine, there is evidence that some other conifers may have similar properties. Knowles and Dewes (1980) reported abortions associated with consumption by cattle of *Pinus radiata* and *Cupressus macrocarpa* (a cypress). Because of the widespread use of radiata pine in agroforestry systems, its abortion-inducing properties should be further investigated. *Cupressus macrocarpa* has been shown to contain levels of isocupressic acid comparable

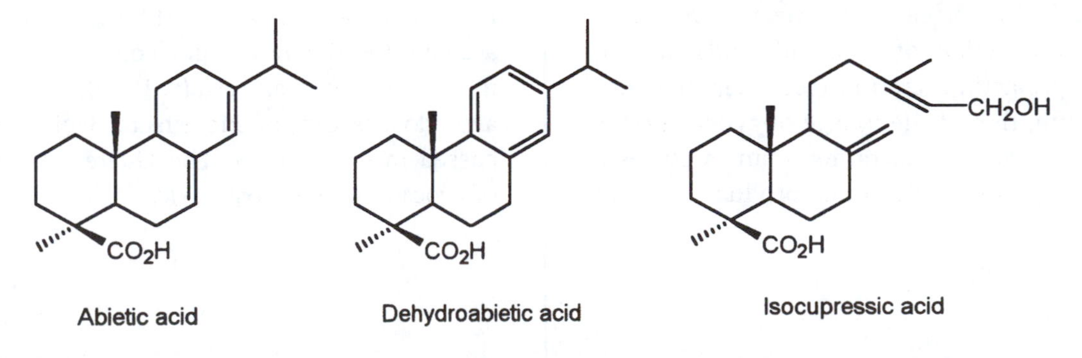

FIGURE 14–8 Diterpenes in Ponderosa pine needles. Isocupressic acid is the major abortifacient in pine needles.

to those in Ponderosa pine (Parton *et al.*, 1996). Western juniper (*Juniperus* spp.) foliage induced abortion when given by gavage to sheep (Johnson *et al.*, 1976). The red pine, *P. resinosa,* does not have abortifacient activity in cattle (Short *et al.*, 1996).

OTHER ABORTIFACIENT-CONTAINING PLANTS

Snakeweeds (*Gutierrezia microcephala*) and *G. sarothrae* are perennial shrub-like plants which occur on semi-arid rangelands of the southwestern U.S. and Mexico. Consumption of snakeweed by livestock during pregnancy may cause abortion or birth of premature, weak offspring (Edrington *et al.*, 1993a,b). Administration of snakeweed to rats caused increased embryonic mortality (Edrington *et al.*, 1993b), but the mechanisms involved and the toxic agent(s) were not identified. Undernourished animals are particularly sensitive to reproductive impairment by snakeweed (Smith *et al.*, 1994). Saponins in snakeweed have been implicated as the abortifacients, but recent studies do not support their involvement. Diterpene acids have been identified in snakeweed (Roitman *et al.*, 1994), which is of interest because diterpene acids are the abortifacients in Ponderosa pine.

Mistletoe is a parasitic plant which forms a dense bushy growth on trees, especially oaks. American mistletoes are of the genus *Phoradendron* while European mistletoes are *Viscum* spp. According to folk lore, mistletoe is poisonous and has medicinal properties (Locock, 1986). The active constituents are not fully known, but include lectins. Kingsbury (1964) cites examples of cattle poisoned by consuming mistletoe. Anecdotal reports of cattle abortions associated with mistletoe consumption suggest possible abortifacient activity, particularly as mistletoe extracts have been used pharmacologically as powerful stimulants of smooth muscle contraction, especially of the intestine and uterus. A tea brewed from the berries of American mistletoe (*Phoradendron flavescens*) has a long history of use as an abortifacient in folk-lore medicine (Spiller *et al.*, 1996). Scientific evidence of its effectiveness is lacking, and mortality has occurred from ingestion of mistletoe tea (Spiller *et al.*, 1996).

AGALACTIA-INDUCING FACTORS IN PLANTS

Ergot alkaloids in endophyte-infected tall fescue cause agalactia (lack of milk secretion), particularly in horses. This effect is mediated through the action of ergot alkaloids in suppressing prolactin secretion (see Chapter 10).

Consumption of the foliage of **avocado** trees (*Persea americana*) by lactating animals causes a marked suppression of milk production, with damage to the microcirculation in the mammary gland, followed by necrosis of the secretory epithelium (Craigmill *et al.*, 1989). These lesions account for the rapid and complete failure of milk secretion (Sani *et al.*, 1994). In addition, avocado has cardiotoxic effects, which are discussed in Chapter 13. The Guatemalan variety of avocado is the most toxic.

REFERENCES

General Reproductive Toxins

James, L.F., K.E. Panter, D.B. Nielsen and R.J. Molyneux. 1992. The effect of natural toxins on reproduction in livestock. J. Anim. Sci. 70:1573–1579.

Panter, K.E., R.F. Keeler, L.F. James and T.D. Bunch. 1992. Impact of plant toxins on fetal and neonatal development: A review. J. Range Manage. 45:52–57.

Teratogens

Bunch, T.D., K.E. Panter and L.F. James. 1992. Ultrasound studies of the effects of certain poisonous plants on uterine function and fetal development in livestock. J. Anim. Sci. 70:1639–1643.

Bradley, G.A., H.C. Metcalf, C. Reggiardo, T.H. Noon, E.J. Bicknell, F. Lozano-Alarcon, R.E. Reed and M.W. Riggs. 1995. Neuroaxonal degeneration in sheep grazing *Sorghum* pastures. J. Vet. Diagn. Invest. 7:229–236.

Casteel, S.W., E.M. Bailey, B.J. Camp, C.H. Bridges, E. Hejtmancik and D.N. Ueckert. 1989. The developmental toxicity of an isolate from the plant *Solanum dimidiatum* (potato-weed) in Syrian golden hamsters. Toxicon 27:757–762.

Gaffield, W., and R.F. Keeler. 1993. Implication of C-5, C-6 unsaturation as a key structural factor in steroidal alkaloid–induced mammalian teratogenesis. Experentia 49:922–924.

Hawkins, C.D. 1994. "Acorn" calves and retained placentae following grazing on sandplain lupins (*Lupinus cosentinii*). pp. 351–356 in: S.M. Colegate and P.R. Dorling (Eds.). Plant-associated Toxins. Agricultural, Phytochemical and Ecological Aspects. CAB International, Wallingford, UK.

Keeler, R.F. 1984. Teratogens in plants. J. Anim. Sci. 58:1029–1039.

Keeler, R.F., and L.D. Stuart. 1987. The nature of congenital limb defects induced in lambs by maternal ingestion of *Veratrum californicum*. Clin. Toxicol. 25:273–286.

Keeler, R.F., D.C. Baker and W. Gaffield. 1991. Teratogenic *Solanum* species and the responsible teratogens. pp. 83–99 in: R.F. Keeler and A.T. Tu (Eds.). Handbook of Natural Toxins. Vol. 6. Toxicology of plant and fungal compounds. Marcel Dekker, Inc., New York.

Keeler, R.F., S. Young and R. Smart. 1985. Congenital tracheal stenosis in lambs induced by maternal ingestion of *Veratrum californicum*. Teratology 31:83–88.

Menzies, J.S., C.H. Bridges and E.M. Bailey, Jr. 1979. A neurological disease of cattle associated with *Solanum dimidiatum*. Southwestern Vet. 32:45–49.

Panter, K.E., L.F. James, D.R. Gardner and R.J. Molyneux. 1994. The effects of poisonous plants on embryonic and fetal development in livestock. pp. 325–332 in: S.M. Colegate and P.R. Darling (Eds.). Plant-associated Toxins. Agricultural, Phytochemical and Ecological Aspects. CAB International, Wallingford, UK.

Panter, K.E., R.F. Keeler, L.F. James and T.D. Bunch. 1992. Impact of plant toxins on fetal and neonatal development: A review. J. Range Manage. 45:52–57.

Panter, K.E., R.F. Keeler, T.D. Bunch and R.J. Callan. 1990a. Congenital skeletal malformations and cleft palate induced in goats by ingestion of *Lupinus, Conium* and *Nicotiana* species. Toxicon 28:1377–1385.

Panter, K.E., R.F. Keeler, T.D. Bunch, D.V. Sisson and R.J. Callan. 1990b. Multiple congenital contractures (MCC) and cleft palate induced in goats by ingestion of piperidine alkaloid–containing plants. Reduction in fetal movement as the probable cause. Clin. Toxicol. 28:69–83.

Peet, R.L., and J. Creeper. 1994. Congenital joint laxity and dwarfism in calves. Aust. Vet. J. 71:58.

Pritchard, J.T., and J.L. Voss. 1967. Fetal ankylosis in horses associated with hybrid Sudangrass pasture. J. Am. Vet. Med. Assoc. 150:871–873.

Ribble, C.S., E.D. Janzen and C.E. Doige. 1993. Congenital spinal stenosis and dam mortality associated with feeding moldy cereal straw. Can. Vet. J. 34:221–225.

Ribble, C.S., E.D. Janzen and J.G. Proulx. 1989. Congenital joint laxity and dwarfism: A feed-associated congenital anomaly of beef calves in Canada. Can. Vet. J. 30:331–338.

Seaman, J.T., M.G. Smeal and J.C. Wright. 1981. The possible association of a sorghum (*Sorghum sudanese*) hybrid as a cause of developmental defects in calves. Aust. Vet. J. 57:351–352.

Selby, L.A., R.W. Menges, E.C. Houser, R.E. Flatt and A.C. Case. 1971. An outbreak of swine malformations associated with wild black cherry, *Prunus serotina*. Arch. Environ. Health 22:496–501.

Abortifacients

Adams, D.C., J.A. Pfister, R.E. Short, R.G. Cates, B.W. Knapp and R.D. Wiedmeier. 1992. Pine needle effects on in vivo and in vitro digestibility of crested wheatgrass. J. Range Manage. 45:249–253.

Al-Mahmoud, M.S., S.P. Ford, R.E. Short, D.B. Farley, L. Christenson and J.P.N. Rosazza. 1995. Isolation and characterization of vasoactive lipids from the needles of *Pinus ponderosa*. J. Agric. Food Chem. 43:2154–2161.

Christenson, L.K., R.E. Short, D.B. Farley and S.P. Ford. 1993. Effects of ingestion of pine needles (*Pinus ponderosa*) by late-pregnant beef cows on potential sensitive Ca^{2+} channel activity of caruncular arteries. J. Reprod. Fert. 98:301–306.

Christenson, L.K., R.E. Short, J.P. Rosazza and S.P. Ford. 1992a. Specific effects of blood plasma from beef cows fed pine needles during late pregnancy on increasing tone of caruncular arteries in vitro. J. Anim. Sci. 70:525–530.

Christenson, L.K., R.E. Short and S.P. Ford. 1992b. Effects of ingestion of Ponderosa pine needles by late-pregnant cows on uterine blood flow and steroid secretion. J. Anim. Sci. 70:531–537.

Edrington, T.S., G.I. Flores-Rodriguez, G.S. Smith and D.M. Hallford. 1993a. Effect of ingested snakeweed (*Gutierrezia microcephala*) foliage on reproduction, semen quality, and serum clinical profiles of male rats. J. Anim. Sci. 71:1520–1525.

Edrington, T.S., G.S. Smith, T.T. Ross, D.M. Hallford, M.D. Samford and J.P. Thilsted. 1993b. Embryonic mortality in Sprague–Dawley rats induced by snakeweed (*Gutierrezia microcephala*). J. Anim. Sci. 71:2193–2198.

Ford, S.P., L.K. Christenson, J.P. Rosazza, and R.E. Short. 1992. Effects of Ponderosa pine needle ingestion on uterine vascular function in late-gestation beef cows. J. Anim. Sci. 70:1609–1614.

Gardner, D.R., K.E. Panter, R.J. Molyneux, L.F. James, and B.L. Stegelmeier. 1996. Abortifacient activity in beef cattle of acetyl- and succinylisocupressic acid from ponderosa pine. J. Agric. Food Chem. 44:3257–3261.

Gardner, D.R., R.J. Molyneux, L.F. James, K.E. Panter and B.L. Stegelmeier. 1994. Ponderosa pine needle–induced abortion in beef cattle: Identification of isocupressic acid as the principal active compound. J. Agric. Food Chem. 42:756–761.

Himejima, M., K.R. Hobson, T. Otsuka, D.L. Wood and I. Kubo. 1992. Antimicrobial terpenes from oleoresin of Ponderosa pine tree *Pinus ponderosa*: A defense mechanism against microbial invasion. J. Chem. Ecol. 18:1809–1817.

James, L.F., R.J. Molyneux, K.E. Panter, D.R. Gardner and B.L. Stegelmeier. 1994. Effect of feeding Ponderosa pine needle extracts and their residues to pregnant cattle. Cornell Vet. 84:33–39.

James, L.F., R.E. Short, K.E. Panter, R.J. Molyneux, L.D. Stuart and R.A. Bellows. 1989. Pine needle abortion in cattle: A review and report of 1973–1984 research. Cornell Vet. 79:39–52.

Jensen, R., A.C. Pier, C.C. Kaltenbach, W.J. Murdoch, V.M. Becerra, K.W. Mills and J.L. Robinson. 1989. Evaluation of histopathologic and physiologic changes in cows having premature births after consuming Ponderosa pine needles. Am. J. Vet. Res. 50:285–289.

Johnson, A.E., L.F. James and J. Spillett. 1976. The abortifacient and toxic effects of big sagebrush (*Artemisia tridentata*) and juniper (*Juniperus osteosperma*) on domestic sheep. J. Range Manage. 29:278–280.

Kingsbury, J.M. 1964. Poisonous Plants of the United States and Canada. Prentice–Hall, Englewood Cliffs, NJ.

Knowles, R.L., and H.F. Dewes. 1980. *Pinus radiata* implicated in abortion. N. Z. Vet. J. 28:103.

Locock, R.A. 1986. Mistletoe. Can. Pharm. J. 119:125–127.

Panter, K.E., L.F. James and R.J. Molyneux. 1992. Ponderosa pine needle-induced parturition in cattle. J. Anim. Sci. 70:1604–1608.

Panter, K.E., L.F. James, R.J. Molyneux, R.E. Short and D.V. Sisson. 1990. Premature bovine parturition induced by Ponderosa pine: Effects of pine needles, bark and branch tips. Cornell Vet. 80:329–338.

Parton, K., D. Gardner and N.B. Williamson. 1996. Isocupressic acid, an abortifacient component of *Cupressus macrocarpa*. N. Z. Vet. J. 44:109–111.

Pfister, J.A., and D.C. Adams. 1993. Factors influencing pine needle consumption by grazing cattle during winter. J. Range Manage. 46:394–398.

Pfister, J.A., D.C. Adams, R.D. Wiedmeier, and R.G. Cates. 1992. Adverse effects of pine needles on aspects of digestive performance in cattle. J. Range Manage. 45:528–533.

Roitman, J.N., L.F. James and K.E. Panter. 1994. Constituents of broom snakeweed (*Gutierrezia sarothrae*), an abortifacient rangeland plant. pp. 345–350 in: S.M. Colegate and P.R. Dorling (Eds.). Plant-associated Toxins. Agricultural, Phytochemical and Ecological Aspects. CAB International, Wallingford, UK.

Short, R.E., S.P. Ford, J.P.N. Rosazza, D.B. Farley, J.A. Klavons, and J.B. Hall. 1996. Effects of feeding pine needles and pine needle components to late pregnant cattle. Proc. West. Sec. Am. Soc. Anim. Sci. 47:193–195.

Short, R.E., R.B. Staigmiller, R.A. Bellows and S.P. Ford. 1995a. Endocrine responses in cows fed Ponderosa pine needles and the effects of stress, corpus luteum regression, progestin, and ketoprofen. J. Anim. Sci. 73:198–205.

Short, R.E., S.P. Ford, E.E. Grings and S.L. Kronberg. 1995b. Abortifacient response and plasma vasoconstrictive activity after feeding needles from Ponderosa pine trees to cattle and sheep. J. Anim. Sci. 73:2102–2104.

Short, R.E., R.A. Bellows, R.B. Staigmiller and S.P. Ford. 1994. Pine needle abortion in cattle: Effects of diet variables on consumption of pine needles and parturition response. J. Anim. Sci. 72:805–810.

Short, R.E., L.F. James, K.E. Panter, R.B. Staigmiller, R.A. Bellows, J. Malcolm and S.P. Ford. 1992. Effects of feeding Ponderosa pine needles during pregnancy: Comparative studies with bison, cattle, goats, and sheep. J. Anim. Sci. 70:3498–3504.

Short, R.E., L.F. James, R.B. Staigmiller and K.E. Panter. 1989. Pine needle abortion in cattle: Associated changes in serum cortisol, estradiol and progesterone. Cornell Vet. 79:53–60.

Smith, G.S., T.T. Ross, D.M. Hallford, J.P. Thilsted, E.C. Staley, J.A. Greenberg and R.J. Miller. 1994. Toxicology of snakeweeds (*Gutierrezia sarothrae* and *G. microcephala*. Proc. West. Sec. Am. Soc. Anim. Sci. 45:98–102.

Spiller, H.A., D.B. Willias, S.E. Gorman and J. Sanftleban. 1996. Retrospective study of mistletoe ingestion. Clin. Toxicol. 34:405–408.

Stegelmeier, B.L., D.R. Gardner, L.F. James, K.E. Panter and R.J. Molyneux. 1996. The toxic and abortifacient effects of ponderosa pine. Vet. Pathol. 33:22–28.

Agalactia-Inducing Factors

Craigmill, A.L., A.A. Seawright, T. Mattila and A.J. Frost. 1989. Pathological changes in the mammary gland and biochemical changes in milk of the goat following oral dosing with leaf of the avocado (*Persea americana*). Aust. Vet. J. 66:206–211.

Sani, Y., A.A. Seawright, J.C. Ng, G.O'Brien and P.B. Oelrichs. 1994. The toxicity of avocado leaves (*Persea americana*) for the heart and lactating mammary gland of the mouse. pp. 552–556 in: S.M. Colegate and P.R. Dorling (Eds.). Plant-associatd toxins. Agricultural, Phytochemical and Ecological Aspects. CAB International, Wallingford, Oxon, UK.

CHAPTER 15

Carcinogens and Metabolic Inhibitors

In this chapter, naturally-occurring carcinogens will be discussed, as well as toxins which affect cellular metabolism. In some cases (e.g. bracken fern), both carcinogens and metabolic inhibitors are present in the same plants.

CARCINOGENS IN PLANTS

Bracken Fern

Bracken fern (*Pteridium aquilinum*) (Fig. 15–1) is one of the world's most abundant and widely distributed plants. It grows from the equator to the northern reaches of Europe, and occurs on all continents except Antarctica. It is an ancient plant; fossil records of bracken spores and fronds indicate its presence on Earth for more than two million years (Fenwick, 1988; Rymer, 1976). Bracken fern reproduces by spores and underground rhizomes. It favors lightly shaded areas or cleared land. The clearing of forests by humans over the past 5000 years has greatly expanded the habitat for bracken. For example, large areas of England, Scotland and Wales are virtual monocultures of bracken. Bracken fern very effectively suppresses other vegetation, by shading, by the build-up of dead biomass when the tops die back each autumn, and the secretion of allelopathic substances. Individual bracken plants may be as much as 1000 years old and have a root system diameter of about 400 m (Sheffield *et al.*, 1989).

Bracken fern has been extensively used by humans, for a variety of purposes. The young plants (fiddleheads) and the rhizomes have been consumed as food in many cultures. Bracken is considered a delicacy in many Asian countries and is widely consumed in Japan,

FIGURE 15–1 Bracken fern (*Pteridium aquilinum*).

China and Korea. Until the development of industrialized agriculture, dried bracken was commonly used as a bedding or litter material for livestock. Numerous other traditional uses for bracken are described by Rymer (1976).

There are numerous livestock toxicoses associated with bracken fern, with the following being the most significant:

1. induced thiamin deficiency
2. acute hemorrhagic disease
3. retina degeneration and blindness
4. urinary tract cancer (enzootic hematuria)
5. upper alimentary tract carcinoma

These conditions are caused by one or more of several toxic entities in bracken, including thiaminase, ptaquiloside, pterosides and pterosins. Bracken also contains tannins and cyanogenic glycosides (e.g. prunasin). Bracken toxicoses have been observed in all parts of the world, and particularly in Britain, western Europe, Turkey, China, Japan, Australia, New Zealand, and much of North and South America. Bracken fern is the only higher plant known to cause natural incidences of cancer in animals.

Bracken-Induced Thiamin Deficiency

The **thiaminase** enzyme splits thiamin, a B vitamin, rendering it inactive. The enzyme is found in a variety of sources, including the viscera of certain fish such as carp, in bracken (*Pteridium aquilinum*), horsetail (*Equisetum arvense*), and nardoo (*Marsilea drummondii*), an Australian fern.

Thiamin deficiency induced by consumption of bracken fern mainly affects nonruminants such as horses and swine. Ruminants are rarely affected by the thiaminase activity, because of the abundant thiamin synthesis by rumen microbes. Thiamin deficiency has been induced in sheep experimentally by feeding 15–25% dried rhizome, but such a situation is unlikely to occur under practical conditions. In North America, bracken poisoning of horses and pigs was more common in pioneer times, when land was being cleared for farming, and was often heavily infested with bracken. Horses were exposed to bracken in hay, and when the dried plant was used for bedding. Pigs were often used to clear land by their rooting activities, and consumed bracken rhizomes. Horses have been the major animals affected by bracken-induced thiamin deficiency. Ingestion of hay containing more than 20% bracken produces toxicity signs in about a month. Signs of thiamin deficiency include anorexia (loss of appetite), ataxia, opisthotonus (head retraction), convulsions, and death. Clinical signs are an elevated level of pyruvate in the blood, cardiac irregularity, and decreased blood thiamin. Response to thiamin administration is dramatic, with a complete reversal of symptoms within a short time. Metabolically, thiamin is involved as a cofactor in decarboxylation reactions. These include the conversion of pyruvate to acetyl-CoA, and oxidation of α-ketoglutarate to succinyl-CoA in the citric acid cycle. As is apparent from Fig. 15–2, thiamin deficiency results in impaired pyruvate utilization. Therefore, as pyruvic acid is formed via glycolysis, it accumulates, and the blood pyruvate level rises. The animal suffers from

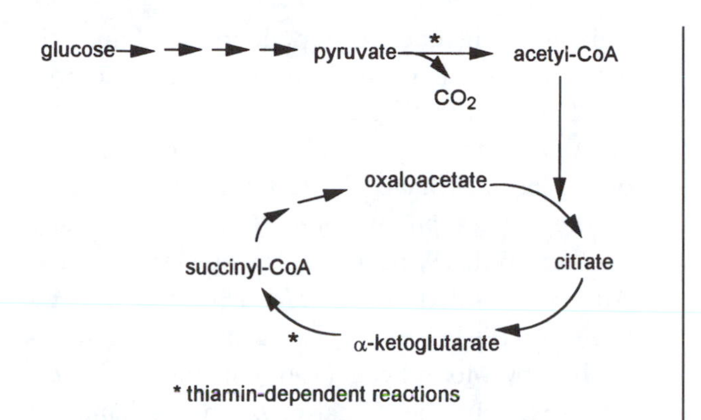

FIGURE 15–2 Role of thiamin in cellular metabolism.

impaired energy metabolism and a cellular shortage of ATP. The elevated pyruvate may affect central nervous system function.

The action of the thiaminase enzyme is to split the thiamin molecule apart, separating its pyrimidine and thiazole rings (Fig. 15–3). A cosubstrate, usually an amine or a sulfhydryl-containing compound such as proline or cysteine, is required. The pyrimidine analog produced when thiamin is cleaved can also be a thiamin antagonist, depending upon the structure of the cosubstrate, which increases the severity of the thiamin deficiency. The thiaminase activity of bracken is highest in the rhizomes, and the activity shows a characteristic seasonal variation.

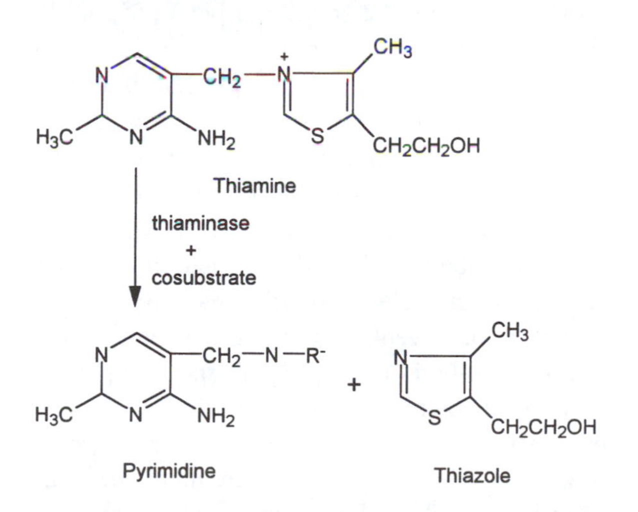

FIGURE 15–3 Action of thiaminase, splitting thiamin into its pyrimidine and thiazole rings, thus inactivating the vitamin.

In the rhizomes, activity is highest during the summer, while in the fronds it is very high during early growth (fiddlehead stage) and decreases progressively during the growing season.

Horsetail (*E. arvense*) is a common thiaminase-containing weed in moist areas of the U.S. and Canada (Fig. 15–4). Cases of poisoning of horses in North America have been documented; hay containing 20% or more horsetail may produce symptoms of thiamin deficiency in horses in 2–5 weeks.

FIGURE 15–4 *Equisetum* spp. or horsetails contain thiaminase activity.

Various fish such as carp contain thiaminase activity. Thiamin deficiency may occur in mink and other fur animals when large amounts of raw fish are used as feed. This condition is called *Chastek's paralysis*, named for the Minnesota fox farmer who first noted the problem. Mink develop classic signs of thiamin deficiency and recover dramatically when thiamin is injected. Thiamin deficiency has occurred in dogs fed large amounts of raw carp (Houston and Hulland, 1988).

Nardoo (*M. drummondii*) is an Australian fern (Fig. 15–5) that grows in damp areas such as water courses. It often grows following periods of flooding or water logging. Extensive losses of sheep with typical symptoms of thiamin deficiency have occurred. Thiamin deficiency in ru-

FIGURE 15–5 Nardoo, an Australian fern with high thiaminase activity. Note the "nuts" (spore capsules) at the base of the plant.

minants is referred to as **polioencephalomalacia**, characterized by depression, incoordination, convulsions, and cerebrocortical necrosis (Fig. 15–6). Nardoo contains thiaminase levels up to 100 times those of bracken fern.

FIGURE 15–6 A sheep with polioencephalomalacia, resulting from a thiamin deficiency induced from consumption of nardoo, showing the characteristic head retraction (opisthotonus). (Courtesy of B. Chick)

Thiamin deficiency and polioencephalomalacia are widespread in New South Wales, Victoria, and South Australia.

As an interesting sidelight, the consumption of nardoo may have played a role in the deaths of the Australian explorers Robert Burke and William Wills, whose tragic expedition across Australia ended in their deaths at Cooper's Creek in 1861. Their epic journey has been described by Moorehead (1963) in his book *Cooper's Creek*. In their final days on the banks of Cooper's Creek, the explorers subsisted largely on the seeds and leaves of nardoo, which grew abundantly in the area and was a major food of the aborigines. In his diary, a few days before his death, Wills wrote

> *my dear Father, these are probably the last lines you will ever get from me. We are on the point of starvation, not so much from absolute want of food, but want of nutriment in what we can get—I am weaker than ever altho' I have a good appetite and relish the nardoo much but it seems to give us no nutriment—but starvation on nardoo is by no means very unpleasant, but for the weakness one feels, and the utter inability to move oneself, for as far as appetite is concerned, it gives me the greatest satisfaction —.*

These symptoms of weakness are consistent with a thiamin deficiency (beri-beri) induced by the high thiaminase activity of nardoo. This conclusion was also reached by Bergin (1981) who recreated the Burke and Wills expedition. Burke and Wills exhausted their supply of pork, their main dietary source of thiamin, and so would have had a low thiamin status. Symptoms described in their diary, including edema, wasting of muscles, altered sensitivity to cold and pain, and weakness in the legs, are indicative of beri-beri. During April and May, when nardoo was being consumed, its toxicity is highest. Bergin (1981) concluded, "— nardoo would obviously have been lethal to men suffering from beri-beri!" The aborigines, who consumed large amounts of nardoo, avoided poisoning by cooking the plant, thus destroying thiaminase.

Another Australian plant with a high thiaminase activity is rock fern (*Cheilanthes sieberi*), which is widely distributed in coastal areas and subtropical regions of Australia. It also appears to contain a bone marrow toxin similar to that found in bracken (Everist, 1981).

Acute Hemorrhagic Disease

Bracken poisoning in cattle occurs after animals have consumed significant quantities of the fern. This may occur when other feed is scarce, as in early spring when the bracken "fiddleheads" are emerging. Conversely, when there is abundant lush, succulent forage, cattle may consume bracken as a source of roughage. The palatable young green fronds are about five times as toxic as the mature fronds, and the rhizomes are highly toxic. **Bracken poisoning** involves severe damage to the bone marrow, resembling radiation damage, with a consequent loss of cellular blood components, leading to severe leukopenia and thrombocytopenia. Hemorrhaging occurs, with blood in the feces, bleeding from the nose and vagina, and bleeding from the membranes of the eyes and mouth. A high fever (107° –109° F) develops in the terminal stages. On postmortem, much hemorrhaging in the stomach, intestines, lungs, and heart can be seen. Cattle are the most susceptible; sheep are poisoned occasionally, while nonruminants such as horses are highly resistant (Fenwick, 1988).

The cause of acute bracken poisoning has not been conclusively determined. The compound ptaquiloside is suggested as the likely toxin (Hirono *et al.*, 1984a).

Retina Degeneration and Blindness

"Bright blindness" characterized by stenosis (narrowing) of blood vessels in the eye and progressive retinal atrophy, occurs in sheep consuming bracken fern (Fenwick, 1988). The causative agent has not been identified. The syndrome has been produced experimentally by feeding a diet of 50% bracken to sheep for 63 weeks (Fenwick, 1988).

Enzootic Hematuria

Enzootic hematuria (red water disease), characterized by red urine, results from the production of bladder cancer in ruminants consuming bracken on a chronic basis. The red urine is caused by bleeding from multiple bladder tumors (Fig. 15–7). This condition has been reported in cattle in many parts of the world. It is a serious problem in many parts of China (Xu, 1992). The carcinogen in bracken is a glycoside, **ptaquiloside** (Hirono *et al.*, 1984a,b; Saito *et al.*, 1989). In addition to ptaquiloside, bracken contains pterosides, which are glycosides of pterosins such as **pterosin B** (Fig. 15–8). The content

FIGURE 15–7 Bladder tumors inside the bladder of a bovine poisoned by bracken fern. (Courtesy of C.H. Tokarnia)

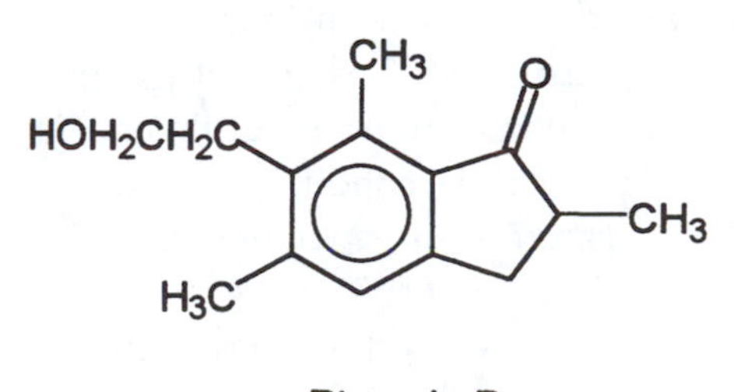

FIGURE 15–8 Pterosin B, the aglycone of pterosides, which are glycosides in bracken.

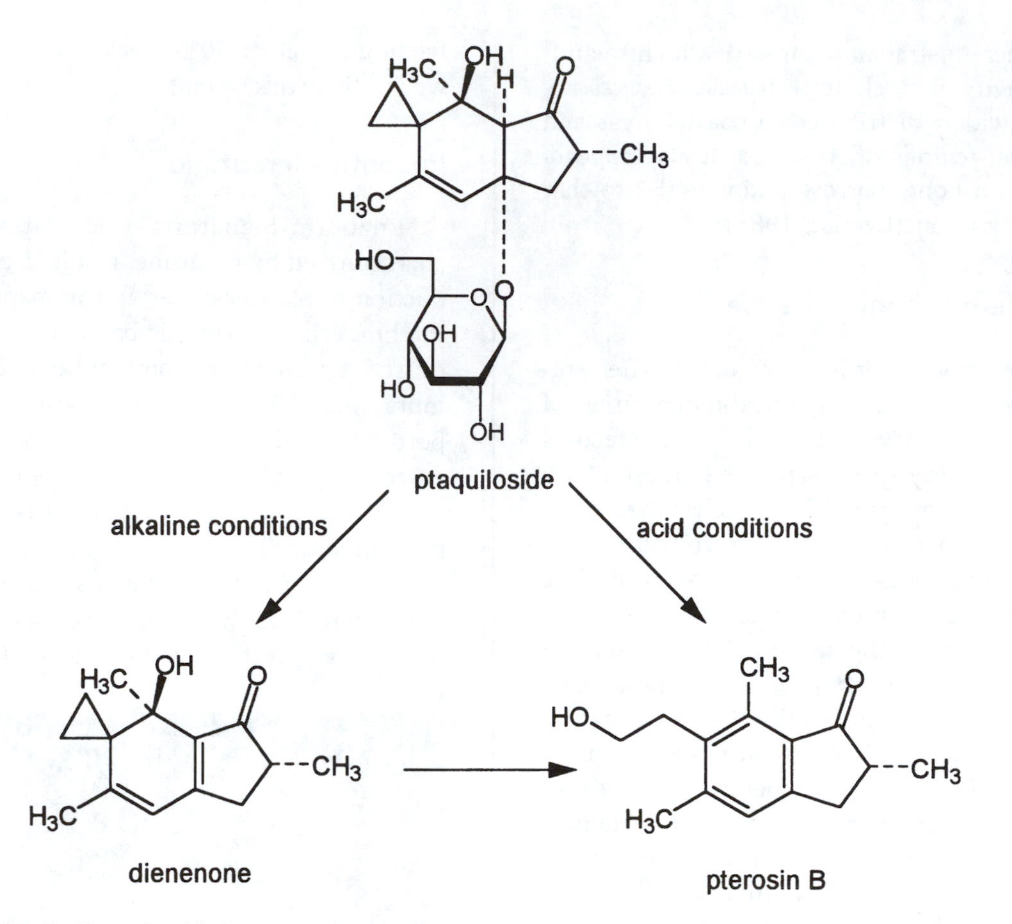

FIGURE 15–9 Ptaquiloside, the carcinogenic glycoside in bracken fern, is activated under alkaline conditions to the active carcinogen, a dienenone.

of pterosides is higher in rhizomes than in fronds, while the fronds are higher than rhizomes in ptaquiloside (Saito *et al.*, 1989). Smith *et al.* (1988) also noted marked differences in carcinogenicity and ptaquiloside contents of bracken from different geographic sites. Fenwick (1988) provides a thorough review of the pterosins and pterosides in bracken (14 pterosins and 4 pterosides). Toxicity of pterosins has not been demonstrated. Ptaquiloside is activated under alkaline conditions (Fig. 15–9) to the active carcinogen, a dienenone (Fenwick, 1988). Smith *et al.* (1988) suggested that because herbivores tend to have an alkaline urine, ptaquiloside is activated in the bladder, leading to tumor formation in the bladder wall, while in rats, the ileum is the predominant site of tumors. Under alkaline conditions, the ptaquiloside glycoside is converted to the aglycone by removal of the glucose. The aglycone intermediate can form DNA adducts (Smith *et al.*, 1994), leading to tumor initiation.

Ptaquiloside has also been isolated from rock fern (*Cheilanthes sieberi*) in Australia and New Zealand (Smith *et al.*, 1989).

Upper Alimentary Tract Carcinoma

In addition to urinary tract cancer, tumors also occur in the nasopharynx, esophagus and forestomachs of cattle consuming bracken (Fenwick, 1988). Intestinal (ileal) cancer occurs in rats fed bracken (Hirono *et al.*, 1987). In cattle, there is evidence that there is an interaction of bracken with the bovine papilloma virus (Jarrett *et al.*, 1978; Moura *et al.*, 1988). Bovine papilloma virus type 4 (BPV-4) infects the mucous epithelium of the upper alimentary tract, inducing epithelial papillomas. Consumption of bracken fern can result in the papillomas progressing to malignancy. Pennie and Campo (1992) suggested that the flavonoid compound **quercetin** in bracken is the cocarcinogen which in conjunction with BPV-4 leads to the development of cancer. Pamuku *et al.* (1980) have reported that quercetin in bracken had a role in the carcinogenicity of the plant. It is not known if ptaquiloside is involved in the upper alimentary tract carcinoma condition.

Cyanogenic Glycosides and Tannins in Bracken

Bracken contains **prunasin**, a cyanogenic glycoside. It apparently functions to deter herbivory. Cooper-Driver and Swain (1976) and Cooper-Driver *et al.* (1977) observed that about 98% of bracken plants exposed to sheep and deer herbivory in a London park were cyanogenic. They were rarely browsed upon, but the acyanogenic plants were heavily grazed by the animals. **Condensed tannins** also occur in bracken. Cooper-Driver *et al.* (1977) found that there were two periods when herbivory was most strongly inhibited. These corresponded to peaks in cyanogenesis in the young fronds and in late summer when tannin concentrations were highest.

Bracken Toxins and Human Health

Bracken fern is a potential human health hazard. The carcinogenic activity of bracken is highest at the crozier stage (Alonso-Amelot *et al.*, 1992). The young fiddleheads or croziers are consumed as a delicacy in Japan, China and Korea (Fig. 15–10). It has been suggested that the high incidence of stomach cancer in Japan could be partially due to consumption of bracken. In Japan, bracken is processed by boiling in alkaline solution (wood ash or sodium bicarbonate), or pickled in salt (Hirono, 1989). Tumor incidence in rats fed a diet with unprocessed bracken was 78.5%, while with wood ash, sodium bicarbonate or sodium chloride treated bracken, tumor incidences were 25, 10 and 4.7% (Hirono, 1989). Thus while the processing treatments markedly reduced carcinogenicity, the processed bracken was still carcinogenic.

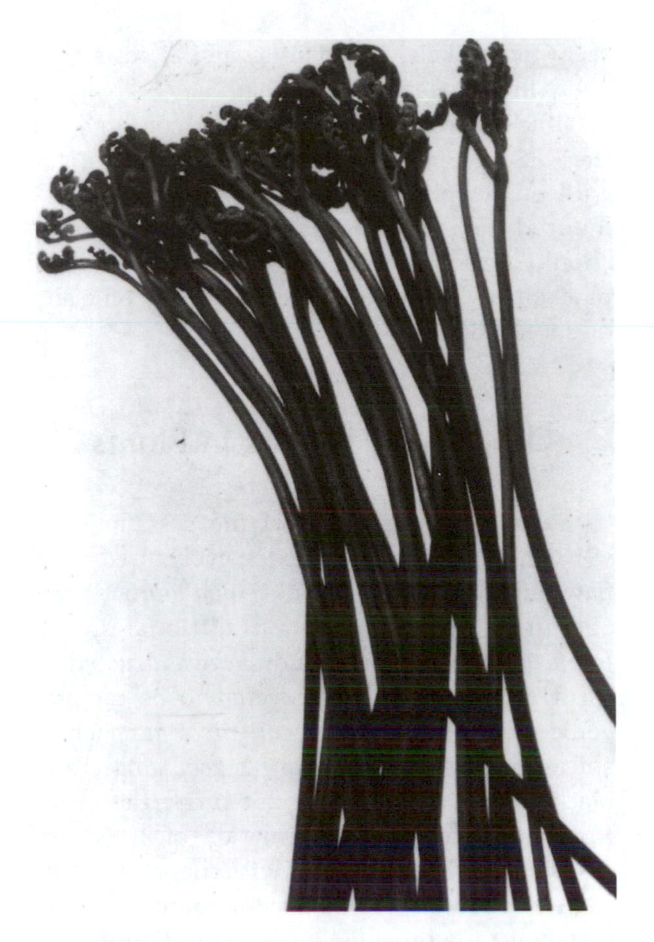

FIGURE 15–10 Fresh young bracken fronds prepared for market in Tokyo, Japan, where bracken is consumed in large amounts by humans. (Courtesy of Iwao Hirono)

Transfer of the bracken carcinogen into milk has been demonstrated, as reviewed by Hopkins (1990) and Alonso-Amelot (1992). The human health hazard is likely to be very slight, except perhaps in cases where a family dairy animal is exposed to high bracken intakes. This would seem most likely with dairy goats, as cows generally find bracken unpalatable.

Bracken croziers are sometimes consumed in the U.S. by people interested in foraging for wild edible plants. In view of the well-documented carcinogenic activity of bracken, it should never be eaten. The consumption of bracken in Asian countries is being officially discouraged.

Other Carcinogens in Plants

Safrole, a constituent of the sassafras tree (*Sassafras albidum*), was extensively used as a flavoring agent in soft drinks such as root beer. Sassafras tea was once a traditional spring tonic. Its use as a food additive was banned in the U.S. after it was discovered to be carcinogenic for rats. Safrole occurs in a number of spices, including cinnamon, mace, ginger and black pepper. The carcinogenic properties of safrole are linked to the formation of 1-hydroxy safrole which may react with hepatic DNA, RNA and proteins. A detailed account of safrole metabolism is provided by Scheline (1991).

Other carcinogens in plants include some of the **pyrrolizidine alkaloids** (see Chapter 12) and glycosides in *Cycad* spp. (see Chapter 13). **Mycotoxins** such as aflatoxin and fumonisins (see Chapter 5) are carcinogens. Cocarcinogens (tumor promoters) include **phorbol esters** (see Chapter 16) and **cyclopropenoid fatty acids** (see Chapter 7). Weber and Hecker (1978) hypothesized that a high rate of esophageal cancer in people inhabiting the Caribbean island of Curacao may be associated with the consumption of a "bush tea" made from the leaves of *Croton flavens*. The leaves of this plant contain tumor-promoting **phorbol esters** such as 12-o-tetradecanoyl-phorbol-13-acetate (TPA), the strongest known tumor promoter (see Chapter 16). In China, in areas with a high incidence of nasopharyngeal cancer, there is also a high utilization of phorbol ester–containing plant materials such as tung oil and croton oil, which are used in herbal preparations (Kinghorn, 1991). **Indole-3-carbinol** in *Brassica* spp. can function both as a tumor promoter and an anticarcinogenic substance, depending upon when it is administered relative to the intake of carcinogens (see Chapter 7).

Anticarcinogens in Plants

There are numerous substances in plants which have been demonstrated to have anticancer properties, leading to general recommendations that the human diet should contain an abundant quantity of fruits and vegetables (Block *et al.*, 1992; Bailey and Williams, 1993; Stavric, 1994; Elson and Yu, 1994). Anticancer agents may act by blocking the initiation phase of carcinogenesis, inhibiting the promotion phase, or having a direct tumorstatic effect (Elson and Yu, 1994). For example, **d-limonene**, a constituent of citrus oils, has all three of these activities (Elson and Yu, 1994).

Many of the compounds with anticancer activity are **phenolics**. The roles of phenolic compounds in human health have been reviewed by Ho *et al.* (1992) and Huang *et al.* (1992). Phenolics such as ellagic acid, chlorogenic acid, quercetin, rutin and catechin, are some of the most common. They are widely distributed among common fruits and vegetables (Stavric *et al.*, 1994). Green tea, widely consumed in Asia, has substantial anticancer activity (Huang *et al.*, 1992).

The flavonoids, a class of plant phenolics, are abundant in many fruits (Cook and Samman, 1996). **Flavonoids** in tea, wine and fruit juices have anti-cancer effects which may be a result of antioxidant activity (Hertog *et al.*, 1993; Wiseman, 1996). Wines, both red and white (Vinson and Hontz, 1995) are excellent sources of antioxidant phenolics, which may have roles in preventing both cancer (Hertog *et al.*, 1993) and coronary heart disease (Frankel *et al.*, 1995; Teissedre *et al.*, 1996). Green tea contains antioxidant phenols which appear to protect against heart disease (Vinson *et al.*, 1995). The possible modes of action of phenolics in

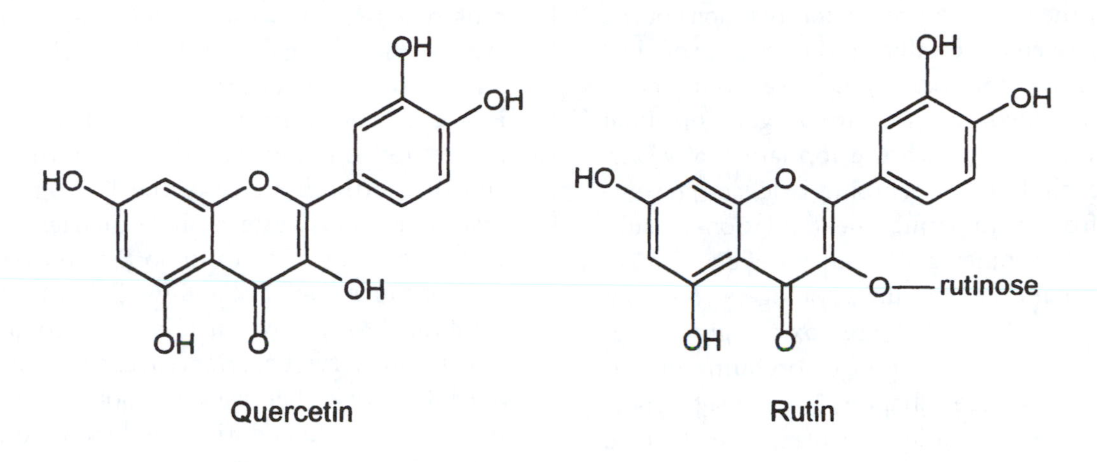

FIGURE 15–11 Quercetin and rutin are phenolics in green tea and red wine. They appear to have beneficial effects on human health.

preventing human diseases have been reviewed by Formica and Regelson (1995). Flavonoids such as **quercetin** and **rutin** (Fig. 15–11) appear to modify the biosynthesis of **leukotriene** and **prostaglandin** eicosanoids from arachidonic acid. An imbalance in leukotriene homeostasis can result in inflammatory responses such as asthma, arthritis and inflammatory bowel disease. An imbalance in prostaglandin synthesis can provoke cardiovascular and renal disease. Selective inhibition of eicosanoid biosynthesis by flavonoids could favorably influence the balance of these compounds. Flavonoids influence the activity of DNA and RNA polymerases and reverse transcriptase, and enhance the antiviral activity of interferon and tumor necrosis factor. These effects could be involved in anti-cancer activity (Formica and Regelson, 1995).

Other components of foods for which anticancer activity has been reported include the **phytoestrogens** in soybeans (Stavric *et al.*, 1994) and the **allylsulfide** compounds in onions and garlic (Wattenburg, 1990). Components of cruciferous vegetables having anti-carcinogenic effects, such as sulforaphane and indole-3-carbinol, were discussed in the section on rapeseed in Chapter 7.

Finally, a ubiquitous component of plants, **chlorophyll**, appears to have anti-cancer activity. Both chlorophyll and its metabolite **chlorophyllin** have protective effects against cancer (Breinholt *et al.*, 1995). The porphyrin nucleus is the active moiety (Hayatsu *et al.*, 1993). Thus the consumption of abundant amounts of green leafy vegetables would seem advisable because of the variety of beneficial phytochemicals they contain, including chlorophyll.

METABOLIC INHIBITORS IN PLANTS

Various metabolic pathways in animal tissue are inhibited by toxicants in plants. This discussion pertains mainly to pathways of energy metabolism.

Cyanogens

Cyanide is a potent inhibitor of cytochrome oxidase. Cytochrome oxidase is the terminal enzyme in the oxidation-reduction reactions of the respiratory chain, catalyzing the transfer of electrons arising from oxidation of substrate molecules by dehydrogenases to oxygen, the final electron acceptor, with the formation of water. Cyanide binds to cytochrome oxidase, preventing it from completing the oxidation of substrates, thus causing a cellular deficit of ATP. Death occurs as a result. Cyanogens occur in virtually all plants, but are toxicologically significant to livestock in forage sorghums and related species (see Chapter 11), linseed meal (see Chapter 7), and cassava products (see Chapter 6).

Wild cherries (*Prunus* spp.) such as the **chokecherry** have caused extensive cattle losses from cyanide poisoning (Fig. 15–12). The chokecherry leaves are highest in cyanogen content in the spring and early summer. The hydrolysis of the glycosides requires water. It has been observed that mortality often occurs soon after drinking. The water promotes quick release of cyanide from previously ingested dry cherry leaves. **Arrow grass** (*Triglochin* spp.), which grows in damp areas, marshes, and sloughs over much of North America, has caused losses of livestock because it contains cyanogenic glycosides (Majak *et al.*, 1980a). **Saskatoon serviceberry** (*Amelanchier alnifolia*) is a cyanogen-containing shrub that is widely distributed on rangelands of western Canada and the northern states of the U.S. Frequently saskatoons and western chokecherries grow in the same areas and are important browse species for range livestock and wildlife. Majak *et al.* (1980b, 1981) have studied the toxicity of saskatoon serviceberries to cattle. They are much less toxic than western chokecherries, but could be hazardous during the bloom period.

FIGURE 15–12 Leaves and flowers of chokecherry (*Prunus virginiana*). (Courtesy of B.R. LeaMaster and U.S. Sheep Experiment Station)

White clover (*Trifolium repens*) contains moderate amounts of linamarin (Vickery *et al.*, 1987). It does not generally cause livestock problems due to its cyanogenic activity. In midsummer, it often becomes unpalatable and animals will refuse to graze it. This may be due to bitterness associated with the cyanogens. Plant-breeding studies have shown that the cyanogen content of white clover can be easily modified by selection to produce low-cyanogen types. However, the cyanogens are not a problem of sufficient magnitude to warrant a serious effort to reduce them, and they have a role in protecting the clover from slugs, snails and insects. The major livestock problem associated with white clover is bloat due to bloat-producing soluble proteins (see Chapter 7).

Hypoglycin

Blighia sapida is one of the plants that Captain Bligh brought to the West Indies on his second voyage, after he survived the mutiny on the Bounty. The plant is named after him. It is a small tree, native to Africa, that is cultivated in southern Florida and the tropics for its fruit, which is edible when ripe (Fig. 15–13). The fruit is called ackee in Jamaica and isin in Nigeria. Ackee is known as the national fruit of Jamaica and is popularized in calypso songs. The unripe fruit has a high level of a toxic amino acid, β-methylenecyclopropyl alanine, known as hypoglycin A (Fig. 15–14). It is also found as an

FIGURE 15–13 Fruit and leaves of ackee (*Blighia sapida*), a fruit widely used in Jamaica. Consumption of immature ackee fruit causes a condition called "vomiting sickness." (Courtesy of J.F. Morton)

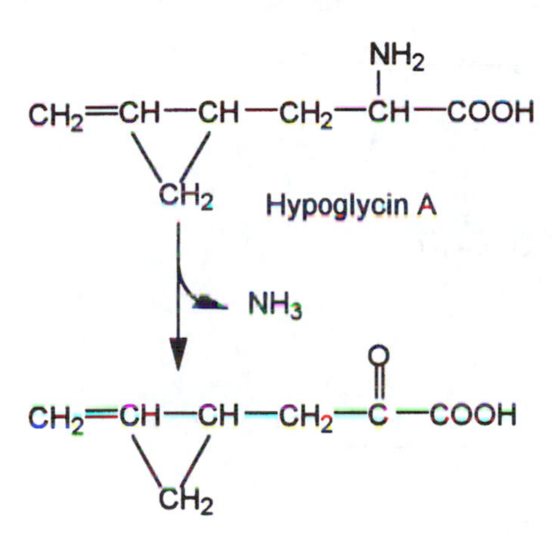

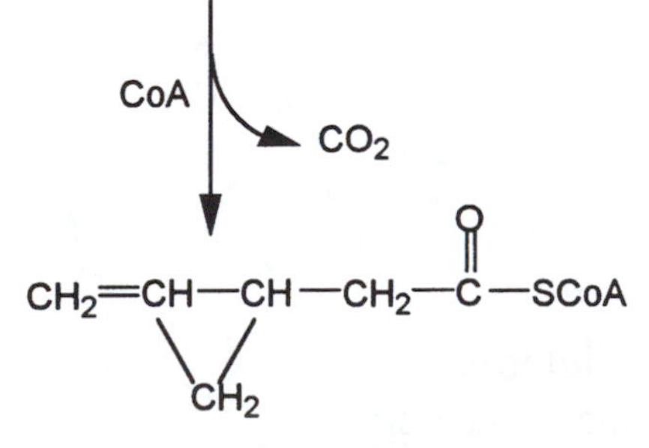

FIGURE 15–14 Metabolism of hypoglycin A.

α-glutamyl dipeptide conjugate, hypoglycin B. The hypoglycin A content of unripe fruit can exceed 1000 ppm, which declines to an undetectable level when the fruit is fully ripe (Brown *et al.*, 1992).

When consumed by humans, unripe ackee can cause a condition known as **vomiting sickness**. It occurs in malnourished people who have consumed the unripe fruit. Its onset is sudden with violent vomiting, followed by convulsions, coma, and death, usually all in a period of 12 hr or less. The principal clinical sign is severe hypoglycemia, with a blood glucose level as low as 20 mg/100 ml blood, compared to a normal value of about 100 mg/100 ml. Hypoglycin is metabolized in a similar manner as the branched chain amino acids. It is deaminated to β-methylenecyclopropyl pyruvate, which then undergoes oxidative decarboxylation to β-methylenecyclopropyl acetyl-CoA (Fig. 15–14). This compound inhibits transfer of long-chain fatty acid CoA residues to carnitine, blocking the process of β-oxidation. This impairs gluconeogenesis, so total depletion of stored glycogen occurs and severe hypoglycemia follows. The vomiting syndrome may be a consequence of inhibition of leucine metabolism due to structural antagonism by the chemically similar hypoglycin. Isovalaric and α-methylbutyric acid are formed; since short-chain branched amino acids are depressants of the central nervous system, the vomiting syndrome could be a result of these compounds.

The biochemistry of hypoglycin A toxicity has been reviewed in depth by Kean (1989).

Vitamin Antagonists

Several natural toxicants exert their effects by inhibiting some aspect of vitamin metabolism. Hepatoxic agents such as aflatoxin and pyrrolizidine alkaloids have inhibitory effects on **vitamin A** and **vitamin E** tissue levels and metabolism, as discussed in Chapter 5 (aflatoxin) and Chapter 12 (pyrrolizidine alkaloids). **Vitamin K** deficiency is induced by dicumarol in moldy sweet clover (see Chapter 11). **Vitamin**

B_6 (pyridoxine)-antagonists are found in flax seed and linseed meal (see Chapter 7). **Pyridoxine antagonists** are also responsible for the toxicity of the seeds of *Albizia versicolor* and *A. tanganyicensis* in Africa (Gummow *et al.*, 1992). ***Albizia*** contains 4-methoxy pyridone, which is a structural antagonist of pyridoxine. Outbreaks of "albiziosis" occur in cattle in South Africa when strong winds blow the seed pods to the ground. Poisoned animals exhibit severe neurologic dysfunction, including convulsions, and markedly elevated body temperature. Treatment with pyridoxine results in dramatic recovery (Gummow *et al.*, 1992).

Some plants contain glycosides of 1,25-dihydroxycholecalciferol (1,25-OHD_3), the active

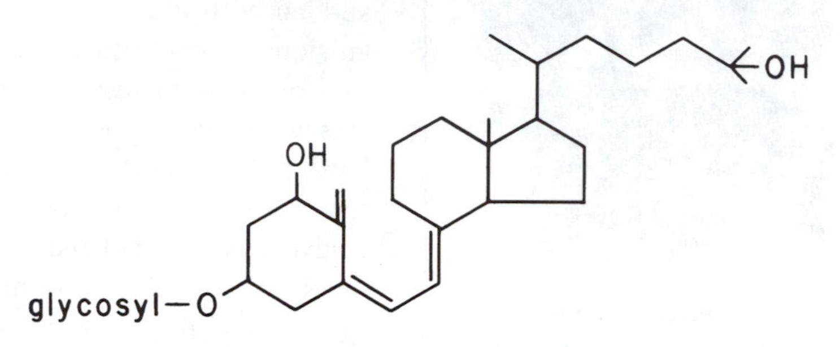

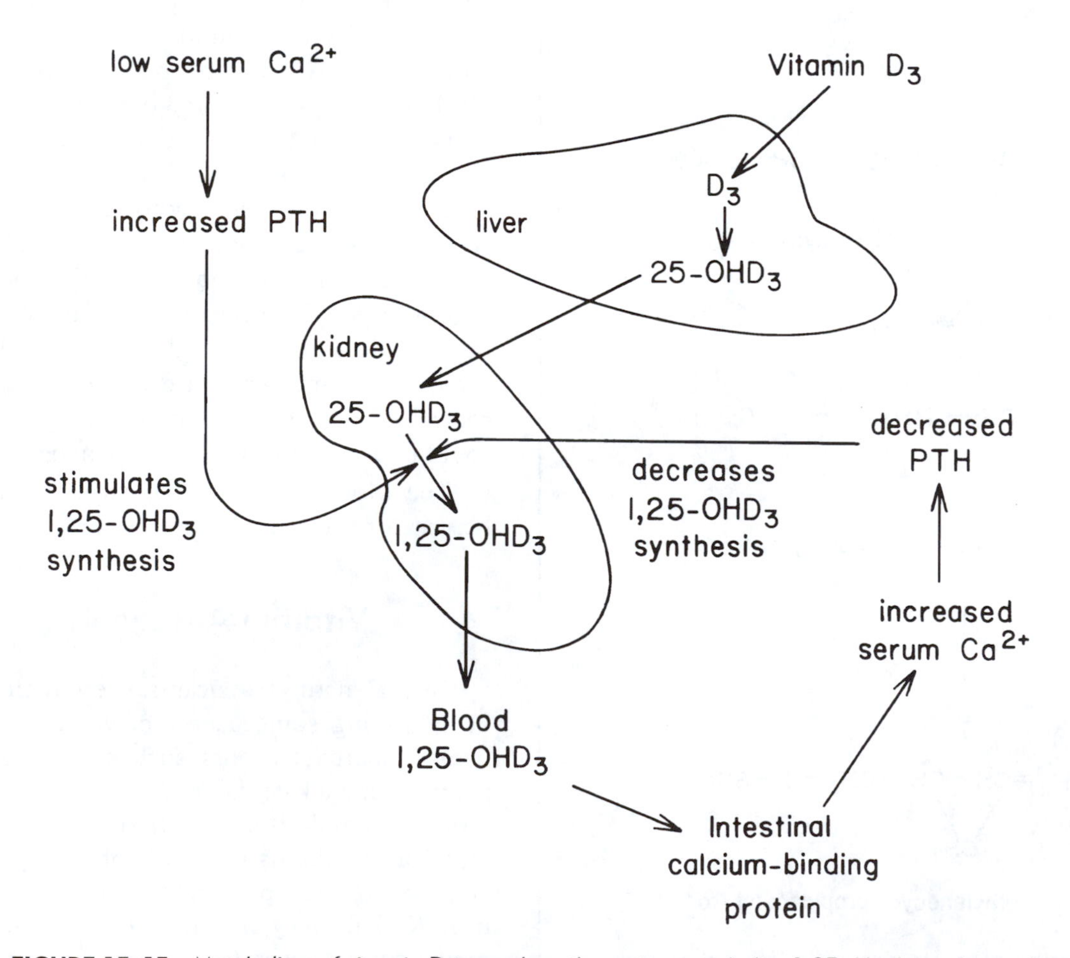

FIGURE 15–15 Metabolism of vitamin D to produce the active metabolite 1,25-dihydroxy vitamin D_3.

metabolite of **vitamin D**. Consumption of these plants by grazing animals causes an induced vitamin D toxicity manifested by calcinosis, the deposition of calcium in the soft tissues. Three plants have been implicated in **calcinosis**. These are *Solanum malacoxylon, Cestrum diurnum*, and *Trisetum flavescens*, which have caused calcinosis in grazing animals in South America (Argentina and Brazil), Florida, and the alpine region of Germany and Austria, respectively. The mode of action of the calcinogenic glycosides can best be perceived after a brief consideration of the metabolism and mode of action of vitamin D.

The primary function of vitamin D is the regulation of calcium (and secondarily, phosphorus) absorption. A metabolite of vitamin D, 1,25 dihydroxy cholecalciferol (1,25-OHD_3), regulates the synthesis and activity of calcium-binding protein (CaBP) in the intestinal mucosa, which transports calcium from the intestine to the blood.

The level of 1,25-OHD_3 is governed by the serum calcium level. If the serum calcium falls, the parathyroid hormone (PTH) secretion is increased, which increases formation of 1,25-OHD_3, stimulating calcium absorption to bring the serum calcium level back to normal. These relationships are shown in Fig. 15–15.

Researchers at Cornell University (Wasserman, 1978) have demonstrated the presence of glycosides of 1,25-OHD_3 in *S. malacoxylon* and *C. diurnum*. The plant compounds seem to act in an identical manner with the vitamin D metabolite formed in animal tissue. Thus, consumption of the plants overrides the regulatory role of 1,25-OHD_3 in animals in which its synthesis is governed by PTH, which in turn is controlled by the serum Ca level. The result is excessive calcium absorption and the deposition of the extra calcium in the soft tissues. Symptoms observed in grazing animals consuming these calcinogenic plants include progressive weight loss, lameness and stiffness of limbs (Fig. 15–16), arching of the back, hypercalcemia and hyperphosphatemia, and calcification of the cardiovascular system, tendons, ligaments, lungs, diaphragm, and kidney.

FIGURE 15–16 A 3-year-old Angus bull showing signs of *Cestrum diurnum* intoxication. A solid stand of *C. diurnum* is in the background. (Courtesy of L. Krook and The Cornell Veterinarian)

The activity of *S. malocoxylon* is about 1.3 x 10^5 IU vitamin D_3 equivalent per kilogram of dried leaf, while for *C. diurnum* the activity is about 3 x 10^3 IU D_3 per kilogram dried leaf, or about one-tenth that of *S. malacoxylon* (Wasserman, 1978). A level of 1.5–3% dietary *S. malacoxylon* will induce calcinosis.

The glycoside is hydrolyzed by microbial enzymes in the rumen (Weissenberg, 1989), which might make calcinogenic plants more toxic to ruminants than to nonruminants. Calcinogenic plants can be used as sources of vitamin D in animal nutrition, but probably this application is impractical under most circumstances.

The functions of calcinogenic glycosides in plants are unclear but they may participate in plant calcium metabolism, possibly by stimulating calmodulin synthesis (Weissenberg, 1989).

Carboxyatractyloside

Cocklebur (*Xanthium* spp.) is a coarse herbaceous annual weed found in many parts of the world. In the U.S., *X. strumarium* is the species involved in livestock poisoning. Significant losses of animals occur in the U.S, Australia, and South Africa. The cocklebur has a fruit containing two seeds surrounded by a spiny capsule. One seed germinates the first growing season and the other the following year. Cocklebur often grows in areas under water for extensive periods and that dry out during the summer. These conditions are found along streams or along the shores of shallow farm ponds. It also grows in pastures and fields. Only the seedlings in the cotyledon stage are poisonous (Fig. 15–17). As the first true leaves develop, toxicity is rapidly lost. The seeds also contain the toxin.

FIGURE 15–17 Cocklebur (*Xanthium strumarium*) seedlings which are the source of toxicity problems with cocklebur. (Courtesy of M.E. Fowler)

In the U.S., pigs seem to be the livestock most frequently poisoned. Signs of toxicity include depression, reluctance to move, a hunched posture, nausea, vomiting, weakness and prostration, dyspnea, opisthotonus, paddling of the limbs and convulsions when recumbent, coma, and death. Severe **hypoglycemia** occurs, with blood glucose levels going from a normal of about 100 mg/100 ml to levels as low as 16 mg/100 ml. The principal gross lesions seen are related to increased vascular permeability. These include edema of the gallbladder wall and ascites of the peritoneal cavity. There is evidence of gastrointestinal tract irritation. Acute centrilobular liver necrosis occurs. Martin *et al.* (1986) described cocklebur poisoning in cattle, with pathologic signs similar to those described above. The minimum lethal dose of cocklebur seeds is about 0.3% of body weight. Witte *et al.* (1990) reported an incidence of cocklebur toxicosis in cattle which were fed hay contaminated with mature cocklebur plants. Clinical signs ranged from acute death to hyperexcitability, blindness, tense musculature, spastic gaits, recumbancy and convulsions. When the mature plants are consumed, the spiney seed burs or capsules can cause gastrointestinal obstruction and irritation.

Cole *et al.* (1980) and Stuart *et al.* (1981) demonstrated that the toxic agent is **carboxyatractyloside** (Fig. 15–18), a glycoside which had previously been isolated from cocklebur and shown to be hypoglycemic (Craig *et al.*, 1976). Carboxyatractyloside causes uncoupling of oxidative phosphorylation, which probably contributes to its hypoglycemic effect.

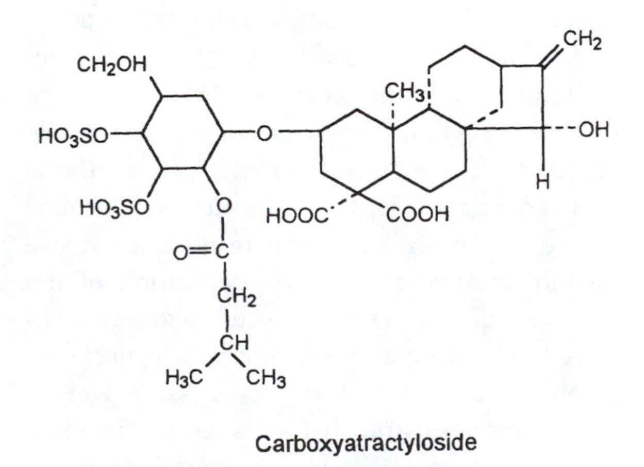

FIGURE 15–18 Structure of carboxytractyloside.

Hatch *et al.* (1982) have used various enzyme inducers and inhibitors to attempt to elucidate the mechanisms of metabolism of the cocklebur glycoside. The use of compounds such as phenobarbital to stimulate cytochrome P450-dependent enzymes did not alter toxicity. Phenylbutazone, which apparently induces synthesis of a non-cytochrome P450-dependent detoxification enzyme, did reduce the toxic effects. Glutathione precursors or blockers did not affect toxicity. Further studies are necessary to completely identify the metabolic pathways involved in carboxyatractyloside metabolism.

Cutler and Cole (1983) demonstrated that carboxyatractyloside is a plant growth inhibitor, and hypothesized that a higher content of the compound in one of the two seeds in the cocklebur seed bur may account for the delayed dormancy of one of the seeds. More detail on the metabolism of carboxytractyloside is provided by Cole *et al.* (1989).

Fluoroacetate (1080)

Sodium monofluoroacetate is compound 1080, which has been widely used as a poison to kill coyotes in the U.S. and other vertebrate pests elsewhere. Certain plants, particularly in Australia and South Africa, contain toxic levels of 1080. It is a very potent toxin, inhibiting aconitate hydratase, one of the tricarboxylic acid (TCA) cycle enzymes. Cellular respiration is thus brought to a halt, and death occurs from cellular lack of ATP.

Fluoroacetate occurs in a variety of Australian plants of the family Leguminosae. These include various species of *Acacia*, *Gastrolobium*, and *Oxylobium*. They are shrubby browse plants which may be consumed by livestock. In most cases they are quite palatable. Because of the high toxicity of 1080, only a few leaves (about 30 g wet weight) can kill a sheep. Brazilian and South African plants containing 1080 are known as well. The main South African 1080-containing plant poisonous to livestock is *Dichapetalum cymosum* (poison leaf).

The toxic **mode of action of fluoroacetate** involves an inhibition of the TCA cycle. Fluoroacetate can substitute for acetate, and be converted to fluorocitrate in the first step of the TCA cycle. However, the fluorocitrate competitively inhibits aconitate dehydrogenase, thus blocking the TCA cycle at the citrate stage (Fig. 15–19). This results in accumulation of citrate in the tissues, energy deprivation of the cells, and death (Twigg and King, 1991). Fluoroacetate may also inhibit citrate transport through mitochondrial membranes. Fluorocitrate forms a thiol-ester bond with the sulfhydryl groups of two enzymes in the mitochondrial membrane. These enzymes function in the transfer of citrate across the membrane.

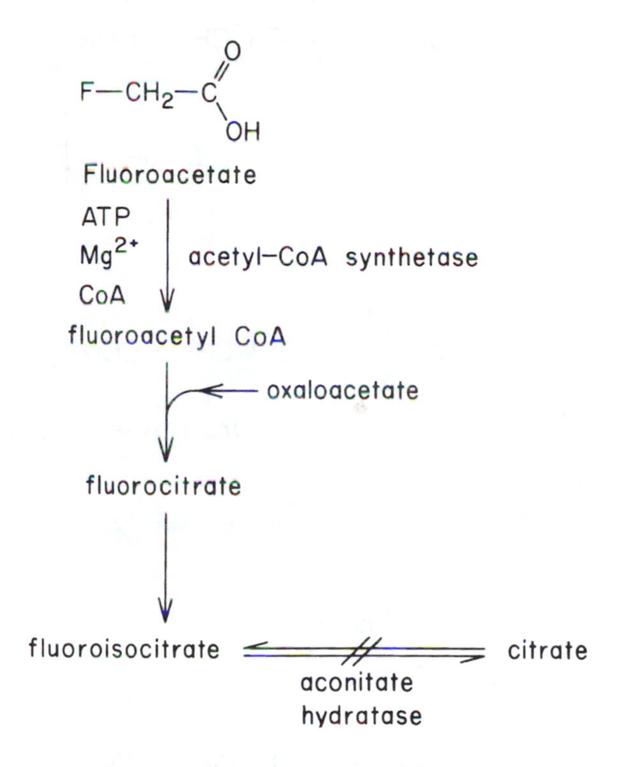

FIGURE 15–19 Metabolism of fluoroacetate.

Mead *et al.* (1979), in their study of western Australian mammals resistant to 1080, found that detoxification of fluoroacetate resulted in an elevation of fluoride in the blood. They implicated a glutathione-dependent reaction, shown in Fig. 15–20. Defluorination occurs mainly in the liver.

It is interesting that while fluoroacetate is very toxic, fluoropropionate is nontoxic. Higher fluoro fatty acids are toxic if they have an even number of carbons, and nontoxic if they have an odd number. This can be explained on the basis of the products of β-oxidation (Fig. 15–21).

The latent period between consumption of fluoroacetate and signs of poisoning is about 3 hours (Twigg and King, 1991). Toxicity signs

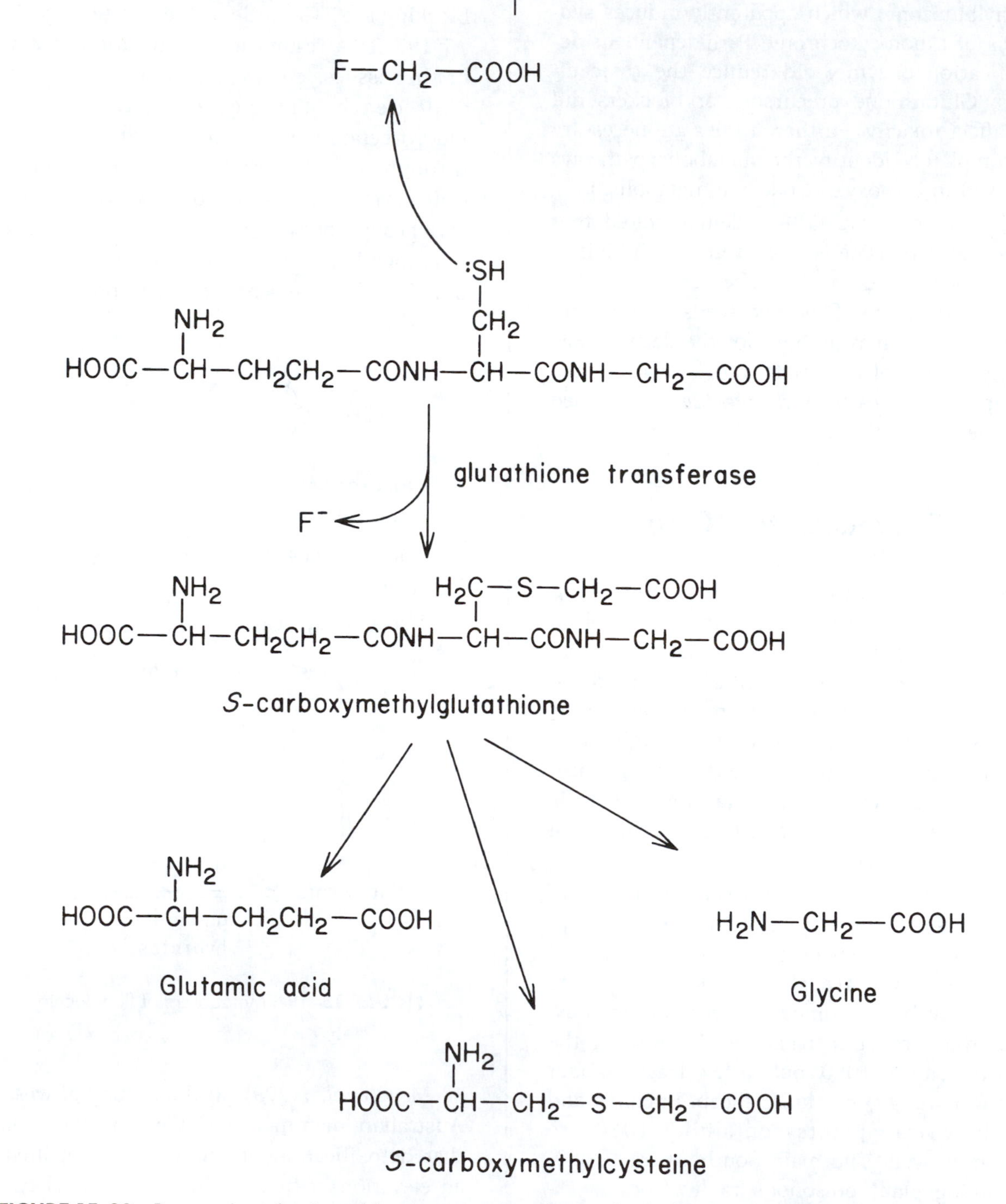

FIGURE 15–20 Proposed mechanism of fluoroacetate detoxification by the Western Australian brush-tailed possum.

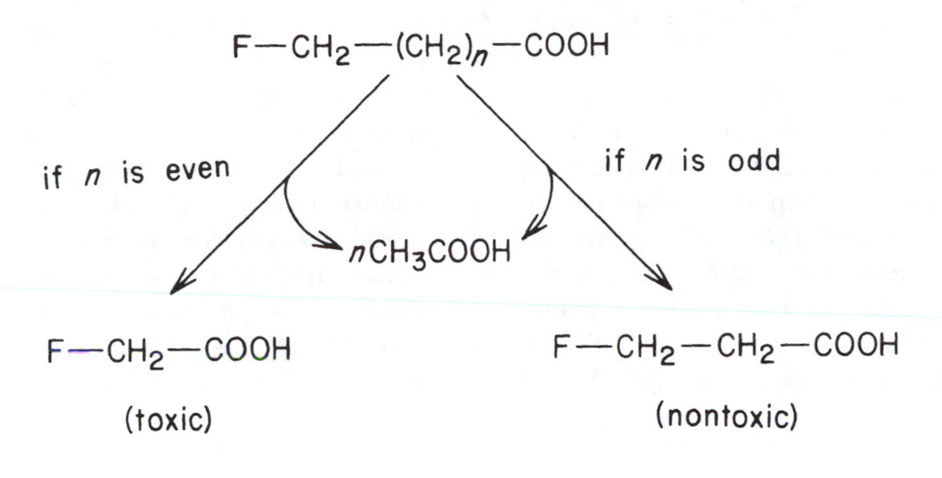

FIGURE 15–21 β-oxidation of higher fluoro fatty acids.

include neurological changes, ataxia, respiratory distress and muscle tremors.

Because fluoroacetate blocks the TCA cycle, glucose metabolism is impaired and hyperglycemia occurs. In ruminants, cardiac failure occurs, with evidence of damage to the myocardium apparent on necropsy. Other findings include general cyanosis of the mucous membranes and other tissues, and the liver and kidney are dark and congested. There is no antidote to 1080 poisoning, although under experimental conditions, provision of metabolic sources of acetate such as acetamide may decrease toxicity by competition with fluoroacetate for metabolic conversion to citrate (or fluorocitrate). The lethal oral dose of 1080 for livestock is about 0.3–1.75 mg per kg body weight. Dogs and other canines are highly sensitive, while birds are more tolerant than mammals. The emu is especially resistant to 1080 (Twigg *et al.*, 1988).

Fluoroacetate poisoning provides some good examples of **coevolution**. In Australia, 1080-containing plants are primarily in Western Australia. Many native herbivores in Western Australia are highly resistant to 1080, and can safely consume leaves of fluoroacetate-containing plants. For example, the LD_{50} for 1080 in the brush-tailed possum from Western Australia is about 100 mg/kg, whereas for the same species in eastern Australia the LD_{50} is 0.68 mg/kg, about a 150-fold difference (Mead *et al.*, 1979). Twigg and King (1991) provide a good review of coevolution of Australian animals and 1080-containing plants. They suggest that the main selection pressure for tolerance to fluoroacetate has been selection against reduced fertility. The high demand for glutathione for 1080 detoxification, combined with inhibited ATP production, reduces fertility (Twigg and King, 1991). Evolution of tolerance to 1080 has also occurred in Africa and South America, the other continents where 1080-containing plants are found (Twigg and King, 1991).

Extensive studies on the tolerance of native Australian animals to 1080 have been conducted (McIroy, 1986a,b), for purposes of wildlife management and preservation of endangered species. Some small native animals, such as rat kangaroos and rock wallabies, are threatened with extinction from predation by introduced predators such as cats and foxes. In many situations, these alien predators can be controlled with 1080-containing baits to which the native animals are resistant. The resistance of native species to 1080 has been assessed by LD_{50} values. As a humane alternative to LD_{50} determination, sensitivity to 1080 can be assessed by measuring plasma citrate following administration of a non-lethal dose of 1080. Those species or individuals with the greatest tolerance to

1080 have the lowest plasma citrate levels (Twigg and King, 1991).

South African researchers (Meyer *et al.*, 1990) have isolated a *Pseudomonas* bacterium from the tissues of the fluoroacetate-containing plant *D. cymosum*, which defluoronates the fluoroacetate molecule. They are attempting to introduce the gene that encodes for the defluoronating enzyme into a rumen bacterium, with the potential result of making ruminants resistant to fluoroacetate toxicity (Meyer and van Rooyen, 1994). Similar work is being conducted in Australia (Gregg and Sharpe, 1991). These workers have modified the rumen bacterium *Butyrivbrio fibrosolvens* by inserting a gene isolated from a soil bacterium that can degrade fluoroacetate (Gregg *et al.*, 1994). The present strain of engineered rumen bacteria does not confer satisfactory levels of 1080 resistance to sheep (C. Cooper, University of New England, Armidale, N.S.W., personal communication, July 10, 1995).

REFERENCES

Bracken Fern

Alonso-Amelot, M.E., M. Perez-Mena, M.P. Calcagno, R. Jaimes-Espinoza and U. Castillo. 1992. Ontogenic variation of biologically active metabolites of *Pteridium aquilinum* (L. Kuhn), pterosins A and B, and ptaquiloside in a bracken population of the tropical Andes. J. Chem. Ecol. 18:1405–1420.

Bergin, T. 1981. In the Steps of Burke and Wills. Australian Broadcasting Commission, Sydney, N.S.W.

Cooper-Driver, G.A., and T. Swain. 1976. Cyanogenic polymorphism in bracken in relation to herbivore predation. Nature 260:604.

Cooper-Driver, G., S. Finch and T. Swain. 1977. Seasonal variation in secondary plant compounds in relation to the palatability of *Pteridium aquilinum*. Biochem. Syst. Ecol. 5:177–183.

Everist, S.L. 1981. Poisonous Plants of Australia. Angus & Robertson Publishers, Sydney.

Fenwick, G.R. 1988. Bracken (*Pteridium aquilinum*)—toxic effects and toxic constituents. J. Sci. Food Agric. 46:147–173.

Hirono, I. 1989. Carcinogenic bracken glycosides. pp. 239–251 in: P.R. Cheeke (Ed.) Toxicants of Plant Origin. CRC Press, Boca Raton, FL.

Hirono, I., H. Ogino, M. Fujimoto, K. Yamada, Y. Yoshida, M. Ikagawa and M. Okumura. 1987. Induction of tumors in ACI rats given a diet containing ptaquiloside, a bracken carcinogen. J. Nat. Cancer Inst. 79:1143–1149.

Hirono, I., Y. Kono, K. Takahashi, K. Yamada, H. Niwa, M. Ojika, H. Kigoshi, K. Niiyama and Y. Uosaki. 1984a. Reproduction of acute bracken poisoning in a calf with ptaquiloside, a bracken constituent. Vet. Rec. 115:375–378.

Hirono, I., K. Yamada, H. Niwa, Y. Shizuri, M. Ojika, S. Hosaka, T. Yamaji, K. Wakamatsu, H. Kigoshi, K. Niiyama and Y. Uosaki. 1984b. Separation of carcinogenic fraction of bracken fern. Cancer Letters 21:239–246.

Hopkins, A. 1990. Bracken (*Pteridium aquilinum*): Its distribution and animal health implications. Brit. Vet. J. 146:316–326.

Houston, D.M., and T.J. Hulland. 1988. Thiamine deficiency in a team of sled dogs. Can. Vet. J. 29:383–385.

Jarrett, W.F.H., P.E. McNeil, W.T.R. Grimshaw, I.E. Selman and W.I.M. McIntyre. 1978. High incidence area of cattle cancer with a possible interaction between an environmental carcinogen and a papilloma virus. Nature 274:215–217.

Moorehead, A. 1963. Cooper's Creek. Harper & Row, N.Y.

Moura, J.W., R.C. Stocco dos Santos, M.L.Z. Dagli, J.L.D'Angelino, E.H. Birgel and W. Becak. 1988. Chromosome aberrations in cattle raised on bracken fern pasture. Experientia 44:785–788.

Pamucku, A.M., S. Yalciner, J.F. Hatcher and G.T. Bryan. 1980. Quercetin: A rat intestinal and bladder carcinogen present in bracken fern. Cancer Res. 40:3466–3472.

Pennie, W.D., and M. Saveria Campo. 1992. Synergism between bovine papillomavirus type 4 and the flavonoid quercetin in cell transformation in vitro. Virology 190:861–865.

Rymer, L. 1976. The history and ethnobotany of bracken. Bot. J. Linnean Soc. 73:151–176.

Saito, K., T. Nagao, M. Matoba, K. Koyama, S. Natori, T. Murakami and Y. Saiki. 1989. Chemical assay of ptaquiloside, the carcinogen of *Pteridium aquilinum*, and the distribution of related compounds in the pteridaceae. Phytochem. 28:1605–1611.

Sheffield, E., P.G. Wolf, and C.H. Haufler. 1989. How big is a bracken plant? Weed Res. 29:455–460.

Smith, B.L., G. Shaw, A. Prakash, and A.A. Seawright. 1994. Studies on DNA adduct formation by ptaquiloside, the carcinogen of bracken ferns (*Pteridium* spp.). pp. 167–172 in: S.M. Colegate and P.R. Dorling (Eds.) Plant-associated Toxins. Agricultural, Phytochemical and Ecological Aspects. CAB International, Wallingford, UK.

Smith, B.L., P.P. Embling, D.R. Lauren, M.P. Agnew, A.D. Ross and P.L. Greentree. 1989. Carcinogen in rock fern (*Cheilanthes sieberi*) from New Zealand and Australia. Aust. Vet. J. 66:154–155.

Smith, B.L., P.P. Embling, M.P. Agnew, D.R. Lauren, and P.T. Holland. 1988. Carcinogenicity of bracken fern (*Pteridium esculentum*) in New Zealand. N. Z. Vet. J. 36:56–58.

Sunderman, F.M. 1987. Bracken poisoning in sheep. Aust. Vet. J. 64:25–26.

Xu, L.R. 1992. Bracken poisoning and enzootic haematuria in cattle in China. Res. Vet. Sci. 53:116–121.

Other Carcinogens

Cook, N.C., and S. Samman. 1996. Flavonoids—Chemistry, metabolism, cardioprotective effects, and dietary sources. Nutritional Biochem. 7:66–76.

Kinghorn, A.D. 1991. New techniques for the isolation and identification of phorbol esters and structurally related diterpenes. pp. 217–242 in: R.F. Keeler and A.T. Tu (Eds.). Handbook of Natural Toxins. Toxicology of Plant and Fungal Compounds. Marcel Dekker, Inc., New York.

Scheline, R.R. 1991. Handbook of Mammalian Metabolism of Plant Compounds. CRC Press, Boca Raton, FL.

Teissedre, P.L., E.N. Frankel, A.L. Waterhouse, H. Peleg and J.B. German. 1996. Inhibition of *in vitro* human LDL oxidation by phenolic antioxidants from grapes and wines. J. Sci. Food Agric. 70:55–61.

Weber, J., and E. Hecker. 1978. Cocarcinogens of the diterpene ester type from *Croton flavens* L. and esophageal cancer in Curacao. Experientia 34:679–822.

Wiseman, H. 1996. Dietary influences on membrane function: Importance in protection against oxidative damage and disease. Nutritional Biochem. 7:2–15.

Anticarcinogens

Bailey, G.S., and D.E. Williams. 1993. Potential mechanisms for food-related carcinogens and anticarcinogens. Food Tech.(Feb.):105–118.

Block, G., B. Patterson and A. Subar. 1992. Fruit, vegetables, and cancer prevention: A review of the epidemiological evidence. Nutrition and Cancer 18:1–29.

Breinholt, V., J. Hendricks, C. Pereira, D. Arbogast, and G. Bailey. 1995. Dietary chlorophyllin is a potent inhibitor of aflatoxin B_1 hepatocarcinogenesis in rainbow trout. Cancer Res. 55:57–62.

Elson, C.E., and S.G. Yu. 1994. The chemoprevention of cancer by mevalonate-derived constituents of fruits and vegetables. J. Nutr. 124:607–614.

Formica, J.V., and W. Regelson. 1995. Review of the biology of quercetin and related bioflavonoids. Fd. Chem. Toxic. 33:1061–1080.

Frankel, E.N., A.L. Waterhouse and P.L. Teissedre. 1995. Principal phenolic phytochemicals in selected California wines and their antioxidant activity in inhibiting oxidation of human low-density lipoproteins. J. Agric. Food Chem. 43:890–894.

Hayatsu, H., T. Negishi, S. Arimoto and T. Hayatsu. 1993. Porphyrins as potential inhibitors against exposure to carcinogens and mutagens. Mut. Res. 290:79–85.

Hertog, M.G.L., P.C.H. Hollman, and B. van de Putte. 1993. Content of potentially anticarcinogenic flavonoids of tea infusions, wines, and fruit juices. J. Agric. Food Chem. 41:1242–1246.

Ho, C.-T., C.Y. Lee and M.-T. Huang (Eds.). 1992. Phenolic Compounds in Food and Their Effects on Health. I. Analysis, Occurrence and Chemistry. Am. Chem. Soc., Washington, D.C.

Huang, M.-T., C.-T. Ho and C.Y. Lee (Eds.). Phenolic Compounds in Food and Their Effects on Health. II. Antioxidants and Cancer Prevention. Am. Chem. Soc., Washington, D.C.

Stavric, B. 1994. Antimutagens and anticarcinogens in foods. Fd. Chem. Toxic. 32:79–90.

Vinson, J.A., and B.A. Hontz. 1995. Phenol antioxidant index: Comparative antioxidant effectiveness of red and white wines. J. Agric. Food Chem. 43:401–403.

Vinson, J.A., Y.A. Dabbagh, M.M. Serry and J. Jang. 1995. Plant flavonoids, especially tea flavonols, are powerful antioxidants using an *in vitro* oxidation model for heart disease. J. Agric. Food Chem. 43:2800–2802.

Wattenberg, L.W. 1990. Inhibition of carcinogenesis by minor anutrient constituents of the diet. Proc. Nutr. Soc. 49:173–183.

Cyanogens

Majak, W., R.E. McDiarmid and J.W. Hall. 1981. The cyanide potential of saskatoon serviceberry (*Amelanchier alnifolia*) and chokecherry (*Prunus virginiana*). Can. J. Anim. Sci. 61:681–686.

Majak, W., R.E. McDiarmid, J.W. Hall and A.L. Van Ryswyk. 1980a. Seasonal variation in the cyanide potential of arrowgrass (*Triglochin mariima*). Can. J. Plant Sci. 60:1235–1241.

Majak, W., T. Ubenberg, L.J. Clark and A. McLean. 1980b. Toxicity of saskatoon serviceberry to cattle. Can. Vet. J. 21:74–76.

Vickery, P.J., J.L. Wheeler and C. Mulcahy. 1987. Factors affecting the hydrogen cyanide potential of white clover (*Trifolium repens* L.) Aust. J. Agric. Res. 38:1053–1059.

Hypoglycin

Brown, M., R.P. Bates, C. McGowan and J.A. Cornell. 1992. Influence of fruit maturity on the hypoglycin A level in ackee (*Blighia sapida*). J. Food Safety 12:167–177.

Kean, E.A. 1989. Hypoglycin. pp. 229–262 in: P.R. Cheeke (Ed.). Toxicants of Plant Origin. Vol. III. Proteins and Amino Acids. CRC Press, Boca Raton, FL.

Vitamin Antagonists

Gummow, B., S.S. Bastianello, L. Labuschagne and G.L. Erasmus. 1992. Experimental *Albizia versicolor* poisoning in sheep and its successful treatment with pyridoxine hydrochloride. Onderstepoort J. Vet. Res. 59:111–118.

Krook, L., R.H. Wasserman, J.H. Shively, A.H. Tashjian, T.D. Brokken and J.F. Morton. 1975a. Hypercalcemia and calcinosis in Florida horses: Implication of the shrub, *Cestrum diurnum*, as the causative agent. Cornell Vet. 65:26–56.

Krook, L., R.H. Wasserman, K. McEntee, T.D. Brokken and M.B. Teigland. 1975b. *Cestrum diurnum* poisoning in Florida cattle. Cornell Vet. 65:557–575.

Wasserman, R.H. 1978. The nature and mechanism of action of the calcinogenic principle of *Solanum malacoxylon* and *Cestrum diurnum*, and a comment on *Trisetum flavescens*. pp. 545–553 in: R.F. Keeler, K.R. Van Kampen and L.F. James (Eds.). Academic Press, NY.

Weissenberg, M. 1989. Calcinogenic glycosides. pp. 201–238. in: P.R. Cheeke (Ed.). Toxicants of Plant Origin, Vol. II. CRC Press, Boca Raton, FL.

Carboxytractyloside

Cole, R.J., H.C. Cutler and B.P. Stuart. 1989. Carboxytractyloside. pp. 253–263 in: P.R. Cheeke (Ed.). Toxicants of Plant Origin. Vol. 2. Glycosides. CRC Press, Boca Raton, FL.

Cole, R.J., B.P. Stuart, J.A. Lansden and R.H. Cox. 1980. Isolation and redefinition of the toxic agent from cocklebur (*Xanthium strumarium*). J. Agric. Food Chem. 28:1330–1332.

Craig, J.C., M.L. Mole, S. Billets and F. El-Feraly. 1976. Isolation and identification of the hypoglycemic agent, carboxyatractylate, from *Xanthium strumarium*. Phytochemistry 15:1178.

Cutler, H.G., and R.J. Cole. 1983. Carboxytractyloside—a compound from *Xanthium strumarium* and *Atractylis gummifera* with plant growth inhibiting properties—the probable inhibitor. J. Nat. Prod. 46:609–613.

Hatch, R.C., A.V. Jain, R. Weiss and J.D. Clark. 1982. Toxicologic study of carboxyatractyloside (active principle in cocklebur, *Xanthium strumarium*) in rats treated with enzyme inducers and inhibitors and glutathione precursor and depletor. Am. J. Vet. Res. 43:111–116.

Martin, T., E.L. Stair and L. Dawson. 1986. Cocklebur poisoning in cattle. J. Am. Vet. Med. Assoc. 189:562–563.

Stuart, B.P., R.J. Cole and H.S. Gosser. 1981. Cocklebur (*Xanthium strumarium*. L. var. *strumarium*) intoxication in swine: Review and redefinition of the toxic principle. Vet. Pathol. 18:368–383.

Witte, S.T., G.D. Osweiler, H.M. Stahr and G. Mobley. 1990. Cocklebur toxicosis in cattle associated with the consumption of mature *Xanthium strumarium*. J. Vet. Diagn. Invest. 2:263–267.

Fluoroacetate

Gregg, K., and H. Sharpe. 1991. Enhancement of rumen microbial detoxification by gene transfer. pp. 719–735 in: T. Tsuda, Y. Sasaki and R. Kawashima (Eds.). Physiological Aspects of Digestion and Metabolism in Ruminants. Academic Press, San Diego.

Gregg, K., C.L. Cooper, D.J. Schafer, H. Sharpe, C.E. Beard, G. Allen and J. Xu. 1994. Detoxification of the plant toxin fluoroacetate by a genetically modified rumen bacterium. Bio/Technology 12:1361–1365.

McIlroy, J.C. 1986. The sensitivity of Australian animals to 1080 poison. IX. Comparisons between the major troups of animals, and the potential danger non-target species face from 1080-poisoning campaigns. Aust. Wildl. Res. 13:39–48.

McIlroy, J.C. 1982a. The sensitivity of Australian animals to 1080 poison. III. Marsupial and eutherian herbivores. Aust. Wildl. J. 9:487–504.

McIlroy, J.C. 1982b. The sensitivity of Australian animals to 1080 poison. IV. Native and introduced rodents. Aust. Wildl. J. 9:505–517.

Mead, R.J., A.J. Oliver and D.R. King. 1979. Metabolism and defluorination of fluoroacetate in the brush-tailed possum (*Trichosurus vulpecula*). Aust. J. Biol. Sci. 32:15–26.

Meyer, J.J.M., and S.W. van Rooyen. 1994. Fluoroacetate metabolism by a bacterium from *Dichapetalum braunii*. pp. 457–461 in: S.M. Colegate and P.R. Dorling (Eds.). Plant-Associated Toxins. Agricultural, Phytochemical and Ecological Aspects. CAB International, Wallingford, UK.

Meyer, J.J.M., N. Grobbelaar and P.L Steyn. 1990. Fluoroacetate-metabolizing pseudomonad isolated from *Dichapetalum cymosum*. Appl. Environ. Micro. 56:2152–2155.

Twigg, L.E., and D.R. King. 1991. The impact of fluoroacetate-bearing vegetation on native Australian fauna: a review. OIKOS 61:412–430.

Twigg, L.E., D.R. King, H.M. Davis, D.A. Saunders and R.J. Mead. 1988. Tolerance to, and metabolism of, fluoroacetate in the emu. Aust. Wildl. Res. 15:239–247.

CHAPTER 16

Irritants, Dermatitic Agents, and Miscellaneous Toxins

CONTACT DERMATITIS FROM PLANTS

Poison Oak/Ivy Dermatitis

In North America, major sources of contact dermatitis are three members of the genus *Toxicodendron*: poison oak (*T. diversilobum*), poison ivy (*T. radicans*) and poison sumac (*T. vernix*). **Poison oak** occurs mainly in western North America from British Columbia to northern Mexico. It grows as a shrub in open or lightly wooded areas, and may grow upright on trees as a vine. Each leaf is composed of three leaflets, which turn a bright red in the autumn. The leaves have a shiny appearance. **Poison ivy** grows in all areas of the U.S. east of the Cascade mountains, the Great Basin and the Mojave Desert. **Poison sumac** occurs mainly east of the Mississippi River, and grows only in swamps and other wetlands.

These *Toxicodendron* spp. contain dermatitic oleoresins collectively called **urushiols**. The toxins occur in canals inside the leaves, stems and roots; they are released to the surface when the plant tissue is crushed or bruised. Urushiols consist of a catechol nucleus with a hydrocarbon side chain (usually 17 carbons long, with one, two or three double bonds (Fig. 16–1). Urushiols are absorbed through the skin, and the catechol nucleus is oxidized to a quinone form. The quinones react with proteins in the

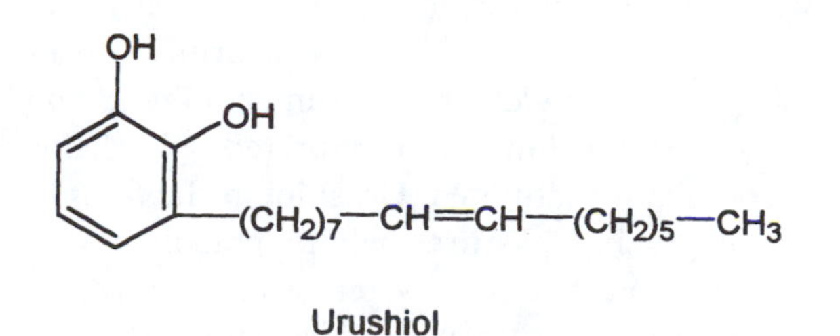

FIGURE 16–1 Urushiol, the toxin in poison oak and poison ivy.

skin. The immune system responds to the quinone-protein complex as an antigen. Thus the dermatitis is an antigen-antibody response, not to the urushiol, but to the urushiol quinone-protein complex (Baer, 1983; Dunn *et al.*, 1986). The mechanisms of the immune response to urushiols are discussed comprehensively by Epstein (1991).

Contact dermatitis from exposure to these plants is a major public health problem in the United States. It is a particular problem for fire fighters, foresters, hikers, and others who spend considerable time in the outdoors. Sensitivity to urushiols is increased by repeated skin contact with the plants. People vary considerably in their susceptibility to development of an allergenic response. Interestingly, if the compounds are introduced to the body for the first time by the oral route, immune tolerance is produced (Baer, 1983). Use of oral preparations of urushiol for induction of tolerance to poison oak and poison ivy shows promise (Epstein, 1991).

Livestock are generally not affected by *Toxicodendron* spp., apparently because of protection by hair. Urushiols must make direct contact with bare skin to induce an allergenic response. Goats are often used as biological control agents for poison oak (Kouakou *et al.*, 1992). Consumption of milk from goats browsing on poison oak is, according to folk legend, a means of increasing one's tolerance to poison oak (Kouakou *et al.*, 1992). However, Kouakou *et al.* (1992) could not detect urushiols in milk from goats fed poison oak. These authors could not account for most of the urushiol consumed by the goats; about 10% of the ingested urushiol was excreted in the feces, and the remainder could not be recovered in milk or urine. Urushiol is not degraded in the rumen (Tzeng and Brown, 1994), but the unsaturated side chains become more saturated. Urushiol in the feces of cattle grazing pastures where poison oak is common is a hazard to veterinarians doing rectal palpations for pregnancy determination.

Several other members of the Anacardiaceae family besides *Toxicodendrum* spp. contain urushiols or similar compounds, including the **cashew** nut tree (*Anacardium occidentale*), the **mango** tree (*Mangifera indica*) and the **lacquer tree** (*Toxicodendrum vernicifluum*) of China and Japan. The toxins are found in the shell of the cashew, and not in the nut, and the exocarp of the mango, and not in the fruit itself. Urushiols oxidize in air to produce black pigments. The sap of the lacquer tree is used to make the black lacquer used on many decorative Japanese products. According to military anecdotes during the American military occupation of Japan after World War II, American soldiers experienced rashes on their forearms and horseshoe-shaped rashes on their buttocks. Apparently, trace amounts of urushiol in the lacquer on wooden bar counters and toilet seats were sufficient to cause allergenic responses in previously sensitized individuals who had been exposed to poison ivy or oak. Japanese people were not affected because they never had the initial exposure to concentrated urushiol in poison ivy to sensitize their immune systems (D.L. Brown, Cornell University, personal communication).

Plants with Stinging Hairs

Stinging nettles (*Urtica* spp.) can cause intense skin irritation. When the leaves are brushed by a person or animal, leaf hairs penetrate the skin and break off. A syringe-like action of bulb cells in the hair injects the toxins into the skin (Epstein, 1991). The irritants are a mixture of histamine and similar amines. According to Epstein (1991), the stinging nettle *U. dioica* "causes fierce urticaria, at times causes collapse, and can fell animals as large as a horse." Horses exposed to stinging nettles may exhibit "nettle rash" as well as neurological signs such as ataxia, and muscle weakness with difficulty in walking (Bathe, 1994). Recovery occurs in a few hours. Treatment with sedatives and analgesics reduces animal discomfort. Stinging nettles contain phospholipids which act as platelet-activating factors; Antonopoulou *et al.* (1996) suggest that these substances are involved in the release of secondary mediators in the affected tissue, causing a persistent skin irritation.

Stinging trees (*Dendrocnide* spp.) occur in tropical rainforests in Australia. Contact with the leaves with penetration of the skin by the stinging hairs causes intense pain. The pain may persist for several days. Horses have been driven to frenzy and self-destruction when they have been exposed to the plant (Everist, 1981). The sting occurs when the skin is pierced by long needle-like hairs on the leaves. The hairs contain the poison, which is injected directly into the skin. The toxin has been isolated from *D. moroides*, and named moroidin (Oelrichs and Williams, 1992). It is a bicyclic octapeptide. It is probably bioactivated in the skin, because there is a consistent lag time of 15–20 seconds before the pain is experienced.

Miscellaneous Irritants in Plants

Phorbol esters are common constituents of plants in the Euphorbiaceae and Thymelaceaceae families. *Euphorbia* spp. (spurge family) often contain strongly irritant latex containing phorbol esters. Croton oil (*Croton tiglium*) and tung oil (*Aleurites fordii*) have severe skin irritant and purgative properties. *Pimelea* spp. in Australia cause livestock poisonings (Seawright, 1989). Clinical signs of phorbol ester toxicosis include severe hemorrhagic gastrointestinal necrosis, diarrhea, abdominal pain and death. Contact dermatitis with exposure to irritant sap or latex may cause severe skin inflammation, loss of hair, and swelling around the eyes and mouth (Kinghorn, 1991). Phorbol esters have tumor-promoting effects (cocarcinogens) and are widely used in cancer research. Phorbol esters activate **protein kinase C** (PKC), an enzyme with an important role in signal transduction for several hormones (Hong *et al.*, 1995).

Poisoning of cattle by *Pimelea* spp. is caused by phorbol esters such as simplexin (Fig. 16–2). Pulmonary disease occurs because of sustained constriction of the pulmonary venous system, caused by activation of PKC in the smooth muscles of the pulmonary veins.

On the Carribbean island of Curacao, roots and leaves of *Croton flavens* are chewed and also brewed into tea. An exceptionally high rate of esophageal cancer on Curacao has been linked to the phorbol esters consumed via *Croton flavens* (Weber and Hecker, 1978). The phorbol esters may cause chronic irritation of the esophagous, and act as cocarcinogens to promote the carcinogenic action of other as-yet unidentified toxins.

Leafy spurge (*Euphorbia esula*) is an important noxious weed of western U.S. rangelands. Although not considered poisonous, it has important implications in livestock production.

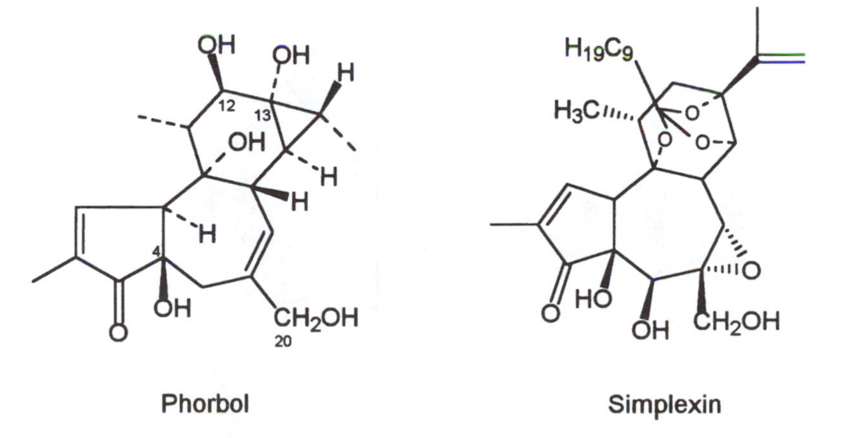

FIGURE 16–2 Phorbol and an example of a phorbol ester, simplexin, found in toxic *Pimelea* spp.

Leafy spurge is an aggressive perennial plant which displaces other vegetation and forms solid stands. It is unpalatable to most types of livestock and wildlife, and thus reduces biodiversity, wildlife habitat and carrying capacity of rangeland for livestock. Sheep and goats are somewhat effective as biological control agents. Goats are more willing to consume the plant than are other livestock species (Walker *et al.*, 1994). Cattle develop a conditioned aversion to leafy spurge after initial consumption (Kronberg *et al.*, 1993). The aversive compounds are believed to be **diterpenoid ingenol esters** (Kronberg *et al.*, 1995). These compounds are activators of protein kinase C, which induces feed aversion by activation of the hypothalamic-pituitary-adrenal axis (Kronberg *et al.*, 1995).

Buttercups (*Ranunculus* spp.) are common pasture weeds in North America, Europe, South Africa, and Australia. They generally grow in wet soils and marshy areas. Pastures heavily infested with buttercups are often characterized by acid soil, and poor fertilization and pasture management practices. Common species in North America are *Ranunculus acris* (tall field buttercup) and *Ranunculus repens* (creeping buttercup). Both are common across the northern U.S. and southern Canada.

Buttercups contain a glycoside, **ranunculin** (Fig. 16–3), which upon crushing of the plant tissue is enzymatically converted to a yellow volatile oil, **protoanemonin** (Bai *et al.*, 1996). Protoanemonin is unstable and either polymerizes to non-toxic anemonin or is volatilized. Therefore, the dried plant, as in hay, is nontoxic. Protoanemonin is an irritant and may cause blisters on the lips and irritation of the mouth and digestive tract, producing salivation, abdominal pain, and diarrhea. Buttercups are unpalatable and will not be consumed unless

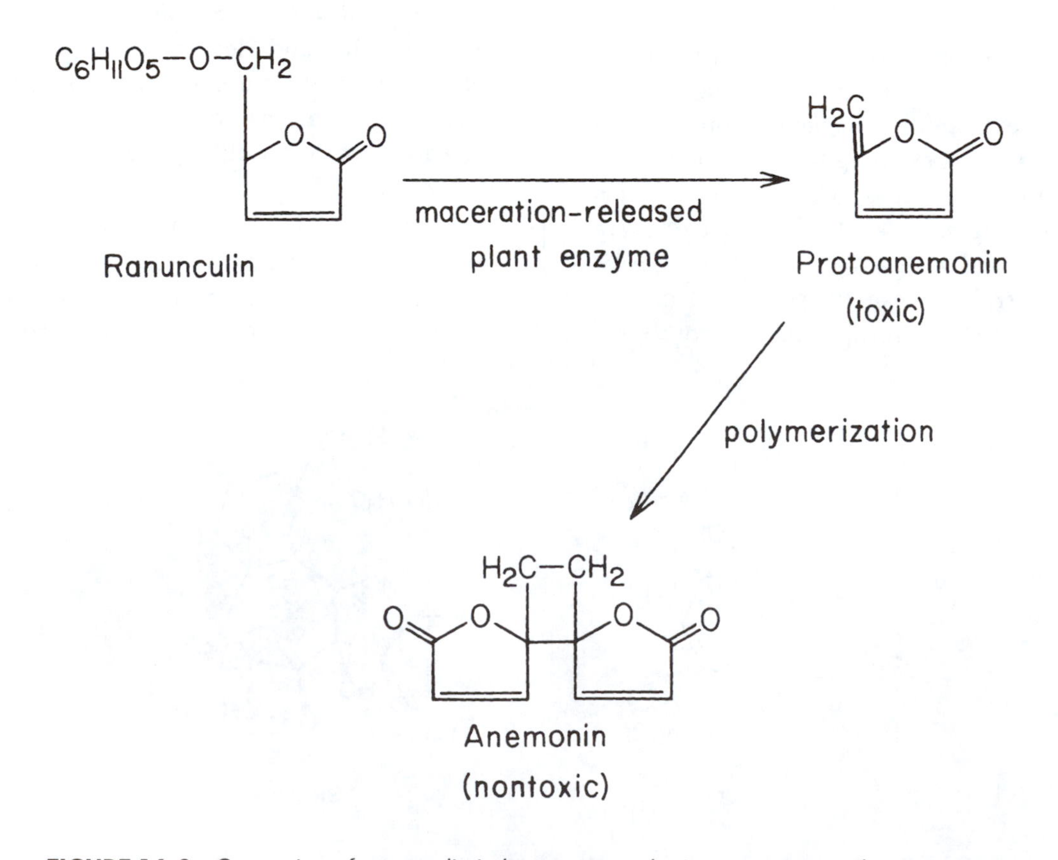

FIGURE 16–3 Conversion of ranunculin in buttercups to the irritant compound, protoanemonin.

other feed is sparse. While a number of poisonings in North America have occurred in the past (Kingsbury, 1964), buttercup is not a significant problem now, probably due to improved pasture management, including liming, fertilization, and seeding with improved pasture species. Therrien *et al.* (1962) and Hidiroglou and Knutti (1963) conducted feeding trials with sheep and cattle and found that even when high levels of buttercup (*R. acris*) were fed, there was no indication of toxicological problems. Photosensitization and liver damage induced by buttercup ingestion in cattle has been reported by Kelch *et al.* (1992). In general, it appears that any hazards associated with consumption of common buttercup are slight.

Bur buttercup (*Ceratocephalus testiculatus*) is an annual weed that infests large areas of several western states. It is of Mediterranean origin and was first identified in the U.S. in Utah in 1932. It is not a true buttercup (*Ranunculus* spp.), but closely resembles the buttercups in appearance. Bur buttercup frequents roadsides, heavily grazed areas, old sheep bed grounds, and other disturbed areas. It grows from 1 to 5 in. in height and forms dense stands. It has small, yellow buttercup-like flowers, with characteristic burlike clusters of fruit. The plant has been shown to be toxic to sheep. Olsen *et al.* (1983) reported an incident in Utah in which about 150 of a band of 800 ewes brought into lambing pastures died from bur buttercup poisoning. The sheep were hungry when unloaded into the pastures, which contained about 50% bur buttercup and 50% cheatgrass (*Bromus tectorum*). A few hours after grazing the plant, the sheep developed watery diarrhea, weakness, and labored breathing. Necropsy revealed edema of the rumen wall, hemorrhage on the inside of the left ventricle of the heart, congestion of the lungs, liver, and kidneys, and fluid accumulation in the thoracic and abdominal cavities. Toxicity of the bur buttercup was confirmed by an LD_{50} test with sheep gavaged with green plant material. About 10.9 g of green bur buttercup per kilogram of body weight was the lethal dose, with identical toxicity signs as observed in the field outbreak. Bur buttercup is unpalatable and a lethal dose is not likely to be consumed under normal grazing conditions. However, when hungry sheep are put into an areas heavily infested with the plant, significant stock losses can be expected. Nachman and Olsen (1983) demonstrated that bur buttercup contains ranunculin. The early flower stage had the highest concentration of the toxin.

Mayweed, mayweed chamomile, false chamomile or **dog fennel** (*Anthemis cotula*) is a common weed of barnyards and over-grazed pastures, particularly those that have suffered trampling damage when wet. Dog fennel contains sesquiterpene lactone irritants that cause allergic contract dermatitis in humans. Although unpalatable, dog fennel does not seem to be toxic to livestock.

PLANTS CAUSING MECHANICAL INJURY

The seeds of some grasses and weeds have arrow-like or fish hook–like properties in that they penetrate the skin and cannot readily be pulled back out. They may cause painful wounds and abscesses. Grasses with these characteristics are often called speargrass or ripgut. Pastures containing these types of plants should be mechanically clipped if possible to prevent seed head formation. Barley has abrasive seed awns; feeding ensiled whole crop barley to cattle may result in mouth lesions due to penetration by awns (Karren *et al.*, 1994).

BLISTER BEETLE TOXICOSIS

A sometimes fatal intoxication of horses and occasionally other livestock is caused by the consumption of hay (mainly alfalfa) containing blister beetles (*Epicauta* spp.). Blister beetles (Fig. 16–4) produce a terpenoid toxin, cantharidin (Spanish fly):

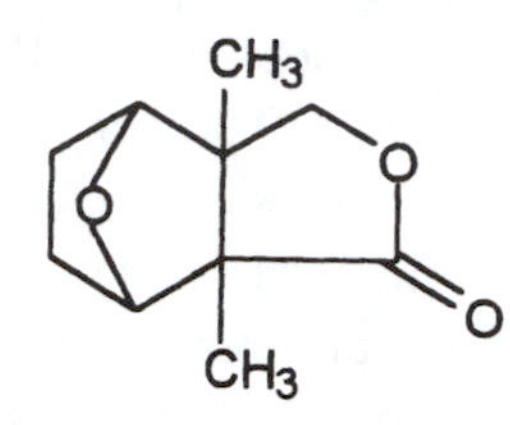

Cantharidin

Cantharidin is an irritant causing severe colic, gastroenteritis, shock, lethargy, frequent urination, cystitis, profuse salivation, renal dysfunction and death (Beasley *et al.*, 1983; Ray *et al.*, 1989). Blister beetle poisoning is a significant problem in much of the U.S.; for example, it is the most frequently diagnosed equine toxicosis at the Texas A and M University Veterinary Diagnostic Laboratory (Ray *et al.*, 1989). The lethal dose of cantharidin is less than 1 mg per kg body weight. The typical etiology of the condition is that horses are fed alfalfa hay that has been contaminated with blister beetles during the harvesting process. Adult blister beetles feed on alfalfa leaves and particularly the flowers. Thus alfalfa in full bloom when cut is commonly implicated in toxicoses, especially if the hay has been crimped or conditioned, which kills the beetles and leaves them in the hay.

Treatment for blister beetle poisoning is "empiric and generally unrewarding" (Ray *et al.*, 1989). Fluids and electrolytes may combat shock, increase urinary excretion and reduce renal damage by cantharidin (Beasley *et al.*, 1983).

FIGURE 16–4 Close-up of a blister beetle on alfalfa (left), and blister beetles feeding on alfalfa (right). (Left photo courtesy of M. Rice; right photo courtesy of S. Blodgett)

Diagnosis may involve inspection of hay for presence of beetles, and measurement of cantharidin in urine and gastric samples (Ray *et al.*, 1989).

Although most blister beetle toxicoses involve horses, chickens have been poisoned when they have eaten the beetles in poultry house litter (Penrith and Naude, 1996).

PHOTOSENSITIZATION

Photosensitization is dermatitis caused by compounds in the blood reacting with ultra-violet light at the skin surface, producing free radicals which react with proteins in the dermal tissue. The substances which react with light are called **photodynamic agents**. They absorb the energy of a photon, and are raised to a higher, unstable energy state. In this excited state, they react with other molecules, including proteins and cell membranes in the epidermis. They may also react with oxygen, producing the superoxide radical and hydrogen peroxide, which can also cause tissue damage. Photosensitization affects white or lightly pigmented skin; pigments in dark skin absorb photons without producing reactive compounds. Photosensitization is classified as either primary or secondary. In **primary photosensitization**, the photodynamic agents occur as such in plants, and are absorbed into the blood and react with light at the skin surface. Only a few plants, such as *Hypericum* and buckwheat, are important causes of primary photosensitization. **Secondary (hepatogenous) photosensizitation** is secondary to liver damage. It is caused by phylloerythrin, a photodynamic metabolite of chlorophyll. Chlorophyll is metabolized by gut microbes, with the removal of the hydrocarbon side chain, forming phylloerythrin. Although both chlorophyll and phylloerythrin are photodynamic, only the latter, being a smaller molecule, is absorbed. Normally, the absorbed phylloerythrin is removed from the blood by the liver, and excreted in the bile. With liver damage or biliary obstruction, it cannot be excreted, and thus enters the general circulation. Thus secondary photosensitization is a common occurrence with liver damage, and is intensified when the consumption of chlorophyll from green plants is high. For example, outbreaks of photosensitization may occur when animals are turned out on lush green pasture, following a wintering period on dry pasture or non-green hay. The increased sunlight exposure that is associated with good pasture growing conditions is also a contributing factor.

Primary Photosensitization Agents

Hypericin. The ingestion of St. Johnswort, *Hypericum perforatum*, by grazing animals may cause the development of photosensitization reactions. There are reports of this condition from Europe, Australia, New Zealand, South America, and the U.S. St. Johnswort is an erect perennial herb, 1–3 ft tall, with yellow flowers (Fig. 16–5). It bears a superficial resemblance (mainly the yellow flowers) to tansy ragwort (*Senecio jacobaea*), with which it is sometimes confused by casual observers. The leaves of St. Johnswort contain numerous small dots, just visible to the naked eye, that are translucent pigment granules. The photodynamic agent is a conjugated quinone derivative of naphthodianthrone called hypericin (Fig. 16–6). Hypericin occurs in the clear pigment granules. The leaves give the appearance of having many pin-pricks; hence, the species name. Hypericin functions in the plant as a light-activated insecticidal agent (Arnason *et al.*, 1992).

All types of grazing livestock are susceptible to *Hypericum* toxicity, although the plant is unpalatable to cattle and horses. Sheep and goats consume it more readily. Sheep are estimated to be intoxicated by a St. Johnswort intake of 4%

FIGURE 16–5 Leaves and flowers of St. Johnswort.

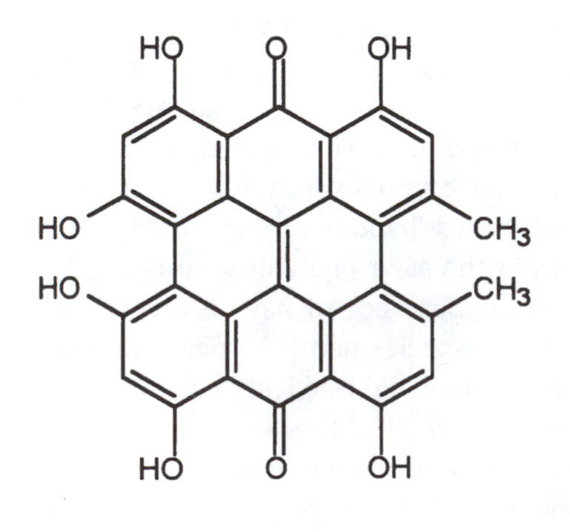

FIGURE 16–6 Structure of hypericin.

of their body weight, and cattle by 1% (Southwell and Campbell, 1991). The leaves and flowers are the plant parts with greatest hypericin content. Signs of toxicity include photophobia (reluctance to be in sunlight), reddening of the skin (erythema) with serous oozing and skin necrosis (Fig. 16–7). Animals may die of starvation or infection of the infected areas. Affected animals may have intense itching and discomfort, and become demented and frantic. It is primarily light-skinned animals that are affected. Kingsbury (1964) cites a case of *Hypericum* poisoning in Holstein cattle, in which the entire herd was found "with the white skin hanging in rags and the dark skin soft and supple as a glove." Newly shorn sheep are especially susceptible.

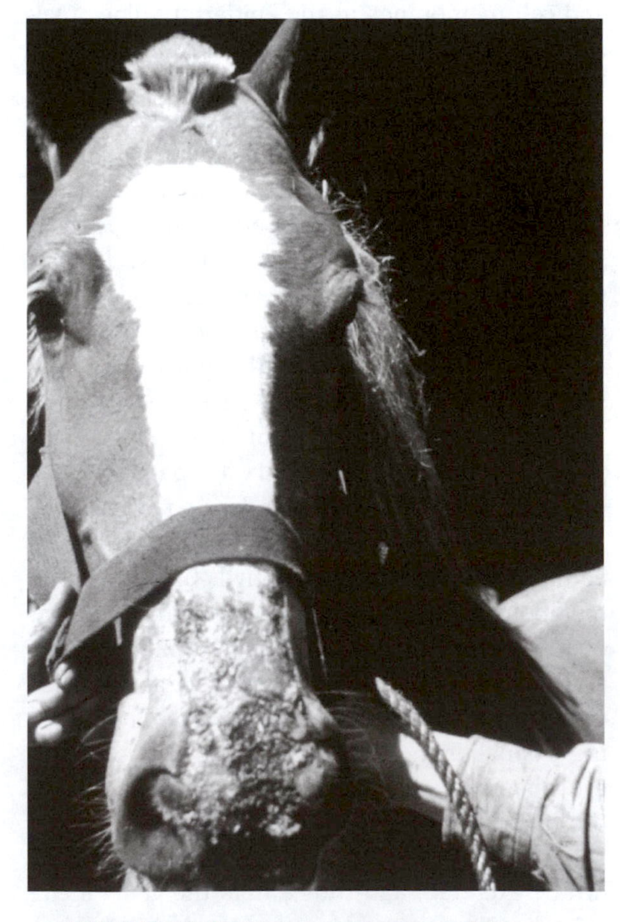

FIGURE 16–7 Photosensitization exhibited by a horse due to ingestion of St. Johnswort (*Hypericum perforatum*).

Other clinical signs include red cell hemolysis, which is a consequence of free radical damage to the erythrocyte membranes (Kako *et al.*, 1993). The mode of action of hypericin involves formation of singlet oxygen and free radicals.

St. Johnswort has traditionally been used in herbal medicine. Its common name is apparently derived from its use in religious rites in celebration of the June 24 birthday of John the Baptist. The suffix "wort" was used in the names of medicinal plants. Interest in medicinal uses of St. Johnswort has intensified with reports that hypericin has antiviral activity (Lopez-Bazzocchi *et al.*, 1991), with anti-HIV (human immunodeficiency virus) properties. Thus it could conceivably have medical use in the treatment of AIDS (Lenard *et al.*, 1993).

St. Johnswort has been controlled in the western U.S. (northern California, Pacific Northwest) and in Australia with a biological control agent, the beetle *Chrysolina quadrigemina*, which defoliates the plant (Fig. 4–9).

Fagopyrin

Buckwheat (*Fagopyrum esculentum*) is sometimes grown as a grain (see Chapter 6) and forage crop. Both the seeds and foliage contain a conjugated quinone called fagopyrin. Fagopyrin is hypericin (Fig. 16–6) with a $C_6H_{11}ON$ side chain attached to each methyl group. Buckwheat seed and forage may cause photosensitization (fagopyrism) when consumed by light-skinned animals. Mulholland and Coombe (1979) in Australia reported an outbreak of fagopyrism in sheep grazing buckwheat stubble. About one third of the flock was affected. It is advisable not to graze light-skinned animals on buckwheat forage.

Furocoumarins

Furocoumarins are compounds with a furan ring fused with a coumarin nucleus. Most are derivatives of psoralen (Fig. 16–8).

Furocoumarins are primary photosensitizing agents. They have been used for several thousand years in India and Egypt to treat skin depigmentation (leukoderma) by application to the skin with subsequent exposure to sunlight. Besides being natural components of many plants, they are in some cases **phytoalexins**; that is, they are produced when some plants are infested with fungal pathogens. Celery and parsnips are plants which may elaborate furocoumarins as phytoalexins. Most of the psoralen-containing plants causing photosensitization problems are members of the Umbelliferae family (the carrot family).

In livestock the major toxicological problem associated with furocoumarins is photosensitization. Two furocoumarin-containing plants in the U.S. that cause livestock problems are bishop's weed (*Ammi majus*) and spring parsley (*Cymopterus watsonii*). **Bishop's weed** grows in coastal regions of the southern U.S. and has caused severe periodic outbreaks of photosensitization in Texas. In cattle, severe blistering and peeling of light-skinned animals occurs as well as clouding of the cornea, which may produce blindness (Dollahite *et al.*, 1978). Photosensitization with erythema and blistering on the beak, feet, and eye lesions have occurred in poultry fed grain containing *Ammi* seeds.

Spring parsley grows on rangelands in Utah and Nevada; it is one of the first plants to begin growing in early spring. Sheep grazing on spring parsley suffer severe photosensitization, with the udder and teats so sensitive from erythema and blistering that the ewes may refuse to nurse their lambs. Losses of up to 25% of the lambs from starvation have occurred in some range sheep flocks exposed to the plant. Similar effects are seen in cattle. The main losses are of lambs and calves that the dams refuse to nurse. Losses are avoided mainly by keeping sheep and cattle off infested ranges in early spring until other plants begin growth. The furocoumarins responsible for the condition are xanthotoxin and bergapten.

Dutchman's breeches (*Thamnosma texana*) and *Thamnosma montana* are perennial weeds of the western and southwestern U.S. Oertli *et al.* (1983) identified at least nine psoralens, including xanthotoxin and bergapten, in *T. texana*. These workers induced photosensitization reactions in sheep by administering the plant (9–12 g/kg body weight/day) and reported an incident of photosensitization in cattle grazing a *T. texana*–infested pasture.

Another condition due to furocoumarins is **celery dermatitis** (Schlatter *et al.*, 1991). Outbreaks of this condition sometimes occur in workers handling the plant, with development of erythema and blistering on the hands and forearms. The causative agents are xanthotoxin and trisoralen, phytoalexins which celery produces in response to "pink rot" infections with *Sclerotinia sclerotiorum*.

The mode of dermatitic action of furanocoumarins such as psoralen involves photoactivation of the molecule to an excited state. The activated metabolites bind to DNA bases, with the formation of DNA adducts. Specific receptor proteins for psoralens have been identified in the membranes of responsive cells (Laskin, 1994). Activated psoralens share a number of similar effects with phorbol esters, including a modulation of epidermal growth factor (Laskin, 1994).

Coevolution of psoralens and the parsnip webworm have been extensively studied by Berenbaum (1991). Structural evolution in psoralen structure (linear vs. angular furanocoumarins) (Fig. 16–8) have been accompanied by enzymatic changes in the parsnip webworm, al-

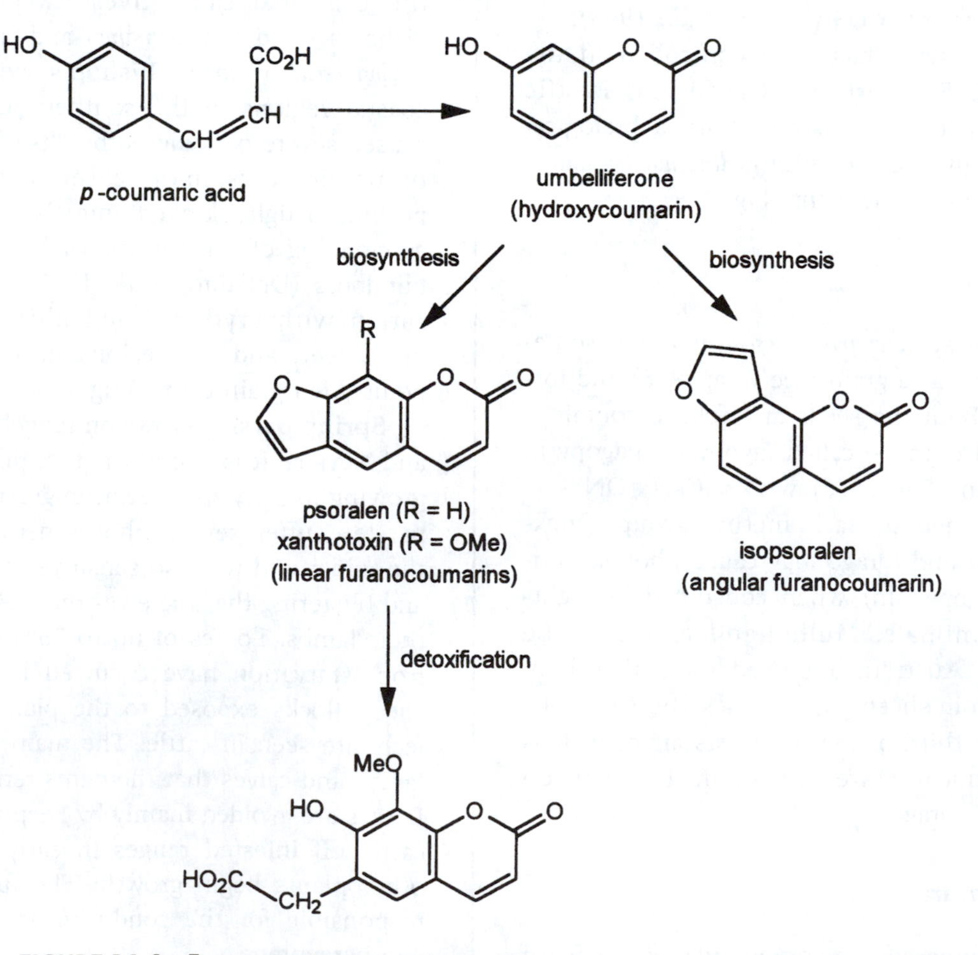

FIGURE 16–8 Furocoumarins are synthesized from a phenolic acid, p-coumaric acid. Plants in the Umbelliferae family synthesize these compounds with increasing complexity (e.g. angular furanocoumarins) as a coevolutionary response to insect herbivory.

lowing some populations to detoxify the modified psoralens. Population dynamics involving the insect and psoralen-containing wild parsnip support the coevolutionary concept (Berenbaum *et al.*, 1986).

Although metabolites of furanocoumarins are excreted in milk and can be transferred to the egg in laying hens, no significant hazard to human health appears to exist from these routes of exposure (Pangilinan *et al.*, 1992).

Secondary Photosensitization

Secondary photosensitization is a consequence of liver damage, as described earlier. Some of the more significant syndromes of this type include "geeldikkop" (yellow head disease), an important problem in South Africa associated with consumption of *Tribulus* (see Chapter 11). Similar conditions occur with tropical grasses such as *Panicum* spp. (see Chapter 11). In both of these cases, biliary crystals of insoluble calcium salts of saponin glucuronides are implicated, as previously described. Other previously-described secondary photosensitization conditions include those associated with lantana (Chapter 12), alsike clover (Chapter 11) and mycotoxicoses such as lupinosis (Chapter 7) and facial eczema (Chapter 10).

It is not clear why secondary photosensitization is more prominent with some liver dysfunctions than with others. For example, *Senecio* poisoning causes severe liver injury, but rarely results in photosensitization.

Tetradymia-Artemisia Poisoning

Tetradymia canescens (spineless horsebrush) and *T. glabrata* (little leaf horsebrush, spring rabbit brush, coal oil brush) are densely branched woody shrubs of the dry desert and sagebrush ranges of the western U.S. *Tetradymia glabrata* is one of the first plants to become green in the spring and one of the major feeds available when sheep are moved from winter to summer range. Thousands of sheep have died as a result of consumption of horsebrush. Cattle do not seem to be affected. Sheep may die from acute liver dysfunction or as a result of photosensitization. *Tetradymia* contains toxins that cause liver damage, with centrolobular necrosis and fatty degeneration (Jennings *et al.*, 1978). If acute toxicity occurs, symptoms observed are anorexia, depression, twitching, incoordination, a rapid weak pulse, dyspnea, prostration, coma, and death. If the liver damage is less severe, secondary photosensitization may occur. The photosensitization condition in sheep consuming *Tetradymia* spp. is commonly called **bighead.** A day or so after consumption of the plant, redness of the skin about the head accompanied by uneasiness and itching develops. The tissues swell as a result of edema. When the edema is severe, the head may become greatly enlarged (Fig. 16–9). Secondary infections and blindness may occur.

FIGURE 16–9 A sheep with "bighead" (right) caused by *Tetradymia*-induced photosensitization. (Courtesy of M.E. Fowler)

Johnson (1978) has conducted a number of feeding experiments with *T. canescens* and *T. glabrata* and was unable to routinely induce toxicity. He concluded that a number of predisposing factors are involved. One of these is prior consumption of *Artemisia* (sagebrush) spp., such as *A. nova* (black sagebrush). In field experiments, exposing sheep first to sagebrush and then to *Tetradymia* effectively induced bighead in range sheep, but not in farm-reared sheep

(Johnson, 1978), implicating additional predisposing conditions. Another factor that could be involved is the amount of green plant matter consumed, as this would influence the amount of chlorophyll available to produce phylloerythrin.

Jennings *et al.* (1978) conducted studies to identify the toxic factor(s) in *Tetradymia* spp. They isolated several furanosesquiterpenes (furanoeremophilanes) that appear to be the toxic agents. The major one was given the name **tetradymol** (Fig. 16–10).

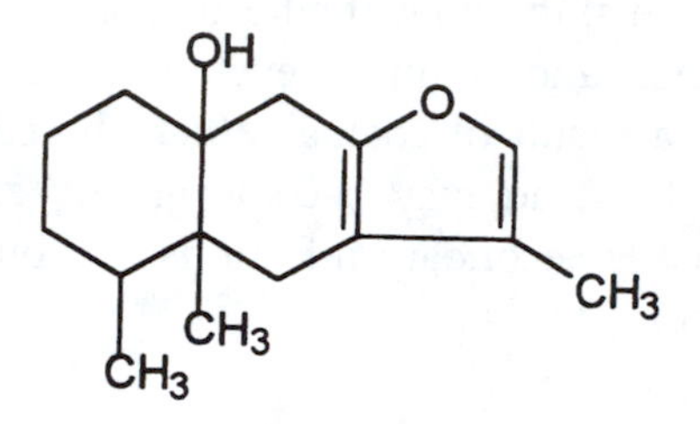

FIGURE 16–10 Structure of tetradymol.

Administration of tetetradymol to laboratory animals revealed that the liver was the only affected organ, with centrolobular necrosis observed. Studies with microsomal mixed function oxidase inducers and inhibitors indicated that tetradymol is bioactivated in the liver to at least two different metabolites. These metabolites appear to disrupt energy metabolism by uncoupling oxidative phosphorylation from electron transport, producing a cellular ATP deficiency (Jennings *et al.*, 1978).

The mode of action of black sagebrush in potentiating *Tetradymia* toxicity is not understood. *Artemisia* spp. are recognized as being potentially toxic (Kingsbury, 1964); **sesquiterpene lactones** are almost universal constituents of *Artemisia* (Herz, 1978) and are a possibility to account for the *Artemisia–Tetradymia* interaction. Sagebrush also contains **monoterpenes** or essential oils. Welch and Pederson (1981) demonstrated large differences in *in vitro* digestibility of a number of sagebrush species, but there was no apparent correlation with monoterpene content. Sagebrush species vary widely in palatability to grazing animals (Sheehy and Winward, 1981); the constituents influencing preference by animals such as sheep and deer which browse on sagebrush have not been identified. Welch *et al.* (1983) reported that the monoterpenoid concentration in big sagebrush was not correlated with mule deer feeding preferences.

Tansymustard Poisoning

Tansymustard (*Descurainia pinnata*) and flixweed (*D. sophia*) are annual weeds of the Brassica family, found throughout the semi-arid regions of western North America. Tansymustard has been circumstantially linked to photosensitization in cattle, but the condition has not been induced in feeding trials when cattle were fed the plant (Pfister *et al.*, 1989, 1990). In the south-western U.S., a condition known as **wooden tongue** or "paralyzed tongue" occurs in cattle consuming tansymustard. It is characterized by partial or complete blindness, an inability to use the tongue or swallow, leading to emaciation and death. A toxic entity in *Descurainia* spp. has not yet been identified.

Forage-Induced Photosensitization

Alfalfa hay has periodically induced secondary photosensitization in livestock (Casteel *et al.*, 1995; Scruggs and Blue, 1994). Causative agents, which may be mycotoxins, have not been identified. House *et al.* (1996) reported photosensitization of dairy cattle fed alfalfa silage. They suggested that during fermentation, chlorophyll was degraded to photodynamic compounds such as pheophorbide and pheophytin, which act as primary photosensitizing agents when absorbed. Various other legume forages, such as alsike clover and cicer milkvetch (*Astragalus cicer*) (see Chapter 11), have caused photosensitization (Marten *et al.*, 1990). Birdsfoot trefoil (*Lotus corniculatus*) has occasionally caused photosensitization in sheep (Stafford *et al.*, 1995). It was not determined if the condensed tannins in trefoil had a role in the photosensitivity. Forage-induced photosensitization may occur with forages that are not visibly moldy and present no visual evidence that they might induce problems.

Moldy Straw–Induced Photosensitization

Bagley *et al.* (1983) described a photosensitization condition in cattle in Utah which appeared to be associated with the consumption of moldy straw. It is commonly referred to as "straw-pile disease." In the dry-land farming and ranching areas of the western U.S., it is a common practice to use a straw catcher on wheat combines to periodically discharge the straw and chaff in large piles. Wintering beef cattle are turned into the fields and may consume the straw throughout the winter. A heavy infestation of mold is often evident in the top several inches of the piles. Straw-pile disease is associated with this type of wintering program. Affected cattle show the usual clinical signs of photosensitization, including ulcerated lesions on nonpigmented skin. The eyes and skin around the eyes are particularly affected, as are the muzzle, underside of the jaw, the teats, and the lower limbs. These areas have the shortest hair coat and receive both direct light and light reflected from snow. Evidence of liver damage, including elevated serum γ-glutamyl transpeptidase and glutamic-oxaloacetic transaminase, are observed. The affected cattle recover when removed from access to the straw piles, although a recurrence can occur when they are placed on green spring pasture, indicating sufficient liver damage to interfere with biliary excretion of pylloerythrin. The presumed etiology is mycotoxin-induced hepatic damage, leading to secondary photosensitization. The causative fungi have not been identified.

IMMUNOTOXINS

There is relatively little information available on the effects of natural toxins and poisonous plants on the immune system. Most low molecular weight natural toxicants are unable to provoke an immune response, because they lack the molecular features recognized by the receptors on the surface of cells of the immune system (Edgar, 1994).

Panciera *et al.* (1992) suggested that **hairy vetch poisoning** (see Chapter 11) may involve an immunologically-mediated response producing a granulomatous inflammatory disease. **Swainsonine** in *Astragalus* spp. is an immunomodulator, which may stimulate the release or binding of endogenous lymphocyte mitogenic factors (White *et al.*, 1989). *Astragalus*-poisoned animals seem to be of increased susceptibility to infections, such as pneumonia, footrot and pink eye, suggesting possible impairment of the immune system (Sharma *et al.*, 1984). Sharma and associates reported that sheep fed locoweed had decreases in total leucocytes and lymphocytes, with an indication of a selective effect on cell-mediated immune responses. Stegelmeier *et al.* (1995) reviewed the effects of swainsonine on immune function. Because of its inhibitory effect on glycoprotein processing, swainsonine and locoweed intoxication can both promote and impair various immune functions, including lymphocyte proliferation, lymphocyte cytotoxicity, lymphocyte expression of interleukin receptors and bone marrow immunocyte formation (Stegelmeier *et al.*, 1995). Aflatoxin (see Chapter 5) has immunosuppressive activity, interfering with the development of acquired immunity and impairment of resistance to infection.

Numerous efforts to prepare **vaccines against natural toxins** have been made, usually unsuccessfully (Edgar, 1994). The procedure generally followed is to attach the natural toxin to a carrier protein to produce an immunogenic toxicant–protein conjugate. The conjugate acts as an antigen determinant capable of binding to complementary Ig receptors on the surface of resting B lymphocytes. This leads to activation of B cells with toxicant-specific Ig receptors; these proliferate and become plasma cells secreting anti-toxicant antibodies (Edgar, 1994). Even though a high level of circulating anti-toxicant antibodies has been achieved in a number of studies, this has not resulted in protection against poisoning.

MYOTOXINS

Muscle degeneration is a relatively unusual consequence of plant poisonings. Cardiac myopathy is noted in some cases (e.g. gossypol toxicity) but there are only a few cases of natural toxins causing skeletal muscle degeneration.

One plant which induces skeletal muscle pathology is *Thermopsis montana* (false lupin, **mountain thermopsis**). It is widely distributed on rangelands of the western United States. It is an erect perennial legume that grows 1–2 feet in height. Cattle poisoned by mountain thermopsis become recumbant, and upon necropsy show varying degrees of skeletal muscle degeneration (Keeler *et al.*, 1986; Baker and Keeler, 1989). A dosage of 0.6–2.75 g of dried plant per kg body weight caused severe toxic signs (depression, anorexia, arched back, swollen eyelids) and prolonged recumbency (Keeler *et al.*, 1986). Subsequent work demonstrated that administration of a **quinolizidine alkaloid** extract from *T. montana* induced the same lesions as did the whole plant (Keeler and Baker, 1990). The quinolizidine alkaloids N-methylcytisine, cytisine, 5,6-dehydrolupanine, thermopsine and anagyrine were isolated from the alkaloid extract. Acute poisoning of children has occurred from the consumption of as few as 6 seeds of *Thermopsis* spp. (Spoerke *et al.*, 1988). Primary symptoms are nausea, vomiting and headache. Other plants containing quinolizidine alkaloids include lupins (see Chapter 11) and laburnum or golden chain tree (*Laburnum anagyroides*). **Laburnum** is a small ornamental tree grown in parks and gardens. It has long hanging racemes of yellow flowers followed by pods containing small seeds which contain the quinolizidine alkaloid cytisine. Children are sometimes poisoned from consumption of the seeds. The tree is not recommended as an ornamental in areas such as public parks where children may play. **Scotch broom** (*Cytisus scoparius*) is a leguminous weedy shrub native to Europe, and naturalized throughout temperate areas of the world. It contains the quinolizidine alkaloids sparteine and isosparteine. Toxicity of the plant is very low, and in many areas Scotch broom is used as a forage for sheep and goats.

Myopathy occurs in the condition **lupinosis**, a mycotoxicosis in sheep and cattle grazing lupin stubble (see Chapter 7). Although lupins contain quinolizidine alkaloids, they are not believed to be involved in lupinosis.

Cassia spp. (**cassia**, coffeeweed, sicklepod, coffee senna) are tropical legumes widely naturalized throughout the southeastern United States, particularly in the sandy soil areas of the coastal plains. *C. occidentalis* and *C. obtusifolia* are common weeds in corn, sorghum and soybean fields. The grains, especially sorghum, may be contaminated with cassia seeds. Sorghum and cassia seeds are of similar size and density. Intoxication of livestock from consumption of cassia-contaminated grains causes extensive degenerative skeletal myopathy and sometimes cardiomyopathy (Colvin *et al.*, 1986). Clinical signs include ataxia, neuromuscular dysfunction and recumbancy. Dietary levels of about 2% cassia seeds are toxic to swine and poultry (Colvin *et al.*, 1986; Flunker *et al.*, 1989). Intoxication of grazing animals also occurs. Signs include progressive muscular dysfunction, with generalized muscle weakness and recumbancy (Schmitz and Denton, 1977; Barth *et al.*, 1994). The myotoxic principle in *Cassia* spp. has not been identified (Rowe, 1991). Interestingly, the administration of vitamin E and selenium has adverse effects on *Cassia* poisoning (O'Hara *et al.*, 1969), even though deficiencies of these nutrients cause myodegeneration (white muscle disease).

REFERENCES

Poison Oak/Ivy Dermatitis

Baer, H. 1983. Allergic contact dermatitis from plants. pp. 421–442 in: R.F. Keeler and A.T. Tu (Eds.). Handbook of Natural Toxins. Vol. 1. Plant and Fungal Toxins. Marcel Dekker, Inc., New York.

Dunn, I.S., D.J. Liberato, N. Castagnoli, Jr., and V.S. Byers. 1986. Influence of chemical reactivity of urushiol-type haptens on sensitization and the induction of tolerance. Cellular Immunology 97:189–196.

Epstein, W.L. 1991. Cutaneous responses to plant toxins. pp. 611–634 in: R.F. Keeler and A.T. Tu (Eds.). Handbook of Natural Toxins. Vol. 6. Toxicology of Plant and Fungal Toxins. Marcel Dekker, Inc., New York.

Kouakou, B., D. Rampersad, E. Rodriguez and D.L. Brown. 1992. Dairy goats used to clear poison oak do not transfer toxicant to milk. Cal. Agric. 46:4–6.

Tzeng, C.C., and D.L. Brown. 1994. Effects of in vitro rumen fluid incubation and pepsin HCl treatment on poison oak urushiol. J. Anim. Sci. 72(Suppl. 1):105(Abst.).

Plants with Stinging Hairs

Antonopoulou, S., C.A. Demopoulos, and N.K. Andrikopoulos. 1996. Lipid separation from *Urtica dioica:* Existence of platelet-activating factor. J. Agric. Food Chem. 44:3052–3056.

Bathe, A.P. 1994. An unusual manifestation of nettle rash in three horses. Vet. Rec. 13[illegible]:11–12.

Epstein, W.L. 1991. Cutaneous responses to plant toxins. pp. 611–634 in: R.F. Keeler and A.T. Tu (Eds.). Handbook of Natural Toxins. Vol. 6. Toxicology of Plant and Fungal Toxins. Marcel Dekker, Inc., New York.

Everist, S.L. 1981. Poisonous Plants of Australia. Angus and Robertson Publishers, Sydney.

Oelrichs, P.B., and D.H. Williams. 1992. Isolation and identification of the pain-producing peptide moroidin from *Dendrocnide moroides* (Laportea). pp. 556–560 in: L.F. James, R.F. Keeler, E.M. Bailey, Jr., P.R. Cheeke and M.P. Hegarty (Eds.). Poisonous Plants. Proceedings of the Third International Symposium. Iowa State Univ. Press, Ames.

Phorbol Esters

Hong, D.H., J. Huan, B. Ou, J. Yeh, T.C. Saido, P.R. Cheeke and N.E. Forsberg. 1995. Protein kinase C isoforms in muscle cells and their regulation by phorbol ester and calpain. Biochimica et Biophysica Acta 1267:45–54.

Kinghorn, A.D. 1991. New techniques for the isolation and identification of phorbol esters and structurally related diterpenes. pp. 217–242 in: R.F. Keeler and A.T. Tu (Eds.). Handbook of Natural Toxins. Vol. 6. Toxicology of Plant and Fungal Compounds. Marcel Dekker, Inc., New York.

Seawright, A.A. 1989. Animal Health in Australia. Vol. 2. Chemical and Plant Poisons. Australian Govt. Publ. Service, Canberra.

Weber, J., and E. Hecker. 1978. Cocarcinogens of the diterpene ester type from *Croton flavens* L. and esophageal cancer in Curacao. Experientia 34:679–682.

Leafy Spurge

Kronberg, S.L., W.C. Lynch, C.D. Cheney and J.W. Walker. 1995. Potential aversive compounds in leafy spurge for ruminants and rats. J. Chem. Ecol. 21:1387–1399.

Kronberg, S.L., R.B. Muntifering, E.L. Ayers and C.B. Marlow. 1993. Cattle avoidance of leafy spurge: A case of conditioned aversion. J. Range Manage. 46:364–366.

Walker, J.W., S.L. Kronberg, S.L. Al-Rowaily and N.E. West. 1994. Comparison of sheep and goat preferences for leafy spurge. J. Range Manage. 47:429–434.

Buttercup Toxicity

Bai, Y., M.H. Benn, W. Majak and R. McDiarmid. 1996. Extraction and HPLC determination of ranunculin in species of the buttercup family. J. Agric. Food Chem. 44:2235–2238.

Hidiroglou, M., and H.J. Knutti. 1963. The effects of green tall buttercup in roughage on the growth and health of beef cattle and sheep. Can. J. Anim. Sci. 43:68–71.

Kelch, W.J., L.A. Kerr, H.S. Adair and G.D. Boyd. 1992. Suspected buttercup (*Ranunculus bulbosus*) toxicosis with secondary photosensitization in a Charolais heifer. Vet. Hum. Toxicol. 34:238–239.

Kingsbury, J.M. 1964. Poisonous Plants of the United States and Canada. Prentice-Hall, Englewood Cliffs, NJ.

Nachman, R.J., and J.D. Olson. 1983. Ranunculin. A toxic constituent of the poisonous range plant bur buttercup (*Ceratocephalus testiculatus.*) J. Agric. Food Chem. 31:1358–1360.

Olson, J.D., T.E. Anderson, J.C. Murphy and G. Madsen. 1983. Bur buttercup poisoning of sheep. J. Am. Vet. Med. Assoc. 183:538–543.

Therrien, H.P., M. Hidiroglou and L.A. Charette. 1962. The toxicity of tall buttercup (*Ranunculus acris* L.) to cattle. Can. J. Anim. Sci. 42:123–124.

Plants Causing Mechanical Injury

Karren, D.B., L.A. Goonewardene and J.A. Bradley. 1994. The effect of feed type on mouth lesions in slaughter cattle. Can. J. Anim. Sci. 74:571–573.

Blister Beetle Toxicosis

Beasley, V.R., G.A. Wolf, D.C. Fischer, A.C. Ray and W.C. Edwards. 1983. Cantharidin toxicosis in horses. J. Am. Vet. Med. Assoc. 182:283–284.

Blodgett, S.L., J.E. Carrel and R.A. Higgins. 1992. Cantharidin contamination of alfalfa hay. J. Medical Entomol. 29:700–703.

Penrith, M.-L., and T.W. Naude. 1996. Mortality in chickens associated with blister beetle consumption. J. S. Afr. Vet. Assoc. 67:97–99.

Ray, A.C., A.L.G. Kyle, M.J. Murphy and J.C. Reagor. 1989. Etiologic agents, incidence, and improved diagnostic methods of cantharidin toxicosis in horses. Am. J. Vet. Res. 50:187–191.

Primary Photosensitization: Hypericin and Fagopyrin

Arnason, J.T., B.J.R. Philogene and G.H.N. Towers. 1992. Phototoxins in Plant–Insect Interactions. pp. 317–341 in: G.A. Rosenthal and M.R. Berenbaum (Eds.). Herbivores. Their Interactions with Secondary Plant Metabolites. Vol. II. Ecological and Evolutionary Processes. Academic Press, San Diego.

Kako, M.D.N., I. I. Al-Sultan and A.N. Saleem. 1993. Studies of sheep experimentally poisoned with *Hypericum perforatum*. Vet. Hum. Toxicol. 35:298–300.

Kingsbury, J.M. 1964. Poisonous Plants of the United States and Canada. Prentice–Hall, Englewood Cliffs, NJ.

Lenard, J., A. Rabson and R. Vanderoef. 1993. Photodynamic inactivation of infectivity of human immunodeficiency virus and other enveloped viruses using hypericin and rose bengal: Inhibition of fusion and syncytia formation. Proc. Natl. Acad. Sci. USA 90:158–162.

Lopez-Bazzocchi, I., J.B. Hudson and G.H.N. Towers. 1991. Antiviral activity of the photoactive plant pigment hypericin. Photochem. Photobiol. 54:95–98.

Mulholland, J.G., and J.B. Coombe. 1979. A comparison of the forage value for sheep of buckwheat and sorghum stubbles grown on the Southern Tablelands of New South Wales. Aust. J. Exp. Agric. Anim. Husb. 19:297–302.

Southwell, I.A., and M.H. Campbell. 1991. Hypericin content variation in *Hypericum perforatum* in Australia. Phytochemistry 30:475–478.

Primary Photosensitization: Furocoumarins

Berenbaum, M.R. 1991. Coumarins. pp. 221–249 in: G.A. Rosenthal and M.R. Berenbaum (Eds.). Herbivores. Their Interactions with Secondary Plant Metabolites. Vol. 1. The Chemical Participants. Academic Press, San Diego.

Berenbaum, M.R., A.R. Zangerl and J.K. Nitao. 1986. Constraints on chemical coevolution: wild parsnips and the parsnip webworm. Evolution 40:1215–1228.

Dollahite, J.W., R.L. Younger and G.O. Hoffman. 1978. Photosensitization in cattle and sheep caused by feeding *Ammi majus* (Greater *Ammi*, Bishop's weed). Am. J. Vet. Res. 39:193–197.

Laskin, J.D. 1994. Cellular and molecular mechanisms in photochemical sensitization: Studies on the mechanism of action of psoralens. Fd. Chem. Toxic. 32:119–127.

Oertli, E.H., L.D. Rowe, S.L. Slovering, G.W. Ivie and E.M. Bailey. 1983. Phototoxic effect of *Thamnosma texana* (Dutchman's breeches) in sheep. Am. J. Vet. Res. 44:1126–1129.

Pangilinan, N.C., G.W. Ivie, B.A. Clement, R.C. Beier and M. Uwayjan. 1992. Fate of [^{14}C]xanthotoxin (8-methoxypsoralen) in laying hens and a lactating goat. J. Chem. Ecol. 18:253–270.

Schlatter, J., B. Zimmerli, R. Dick, R. Panizzon and Ch. Schlatter. 1991. Dietary intake and risk assessment of phototoxic furocoumarins in humans. Fd. Chem. Toxic. 29:523–530.

Secondary Photosensitization

Bagley, C.V., J.B. McKinnon and C.S. Asay. 1983. Photosensitization associated with exposure of cattle to moldy straw. J. Am. Vet. Med. Assoc. 183:802.

Casteel, S.W., G.E. Rottinghaus, G.C. Johnson and D.T. Wicklow. 1995. Liver disease in cattle induced by consumption of moldy hay. Vet. Human Toxicol. 37:248–251.

Herz, W. 1978. Sesquiterpene lactones from livestock poisons. pp. 487–497 in: R.F. Keller, K.R. Van Kampen and L.F. James (Eds.). Effects of Poisonous Plants on Livestock. Academic Press, NY.

House, J.K., L.W. George, K.L. Oslund, F.D. Galey, A.W. Stannard, and L.M. Koch. 1996. Primary photosensitization related to ingestion of alfalfa silage by cattle. J. Am. Vet. Med. Assoc. 209:1604–1607.

Jennings, P.W., S.K. Reeder, J.C. Hurley, J.E. Robbins, S.K. Holian, A. Holian, P. Lee, J.A.S. Pribanic and M.W. Hull. 1978. Toxic constituents and hepatotoxicity of the plant *Tetradymia glabrata* (Asteroceae). pp. 217–228 in: R.F. Keeler, K.R. Van Kampen and L.F. James (Eds.). Effects of Poisonous Plants on Livestock. Academic Press, NY.

Johnson, A.E. 1978. Tetradymia toxicity—a new look at an old problem. pp. 209–216 in: R.F. Keeler, K.R. Van Kampen and L.F. James (Eds.). Effects of Poisonous Plants on Livestock. Academic Press, NY.

Kingsbury, J.M. 1964. Poisonous Plants of the United States and Canada. Prentice–Hall, Englewood Cliffs, NJ.

Marten, G.C., R.M. Jordan and E.A. Ristau. 1990. Performance and adverse response of sheep during grazing of four legumes. Crop Sci. 30:860–866.

Pfister, J.A., J.R. Lacey, D.C. Baker, L.F. James and R. Brownson. 1990. Is tansymustard causing photosensitization of cattle in Montana? Rangelands 12:170–172.

Pfister, J.A., D.C. Baker, J.R. Lacey and R. Brownson. 1989. Photosensitization of cattle in Montana: Is *Descurainia pinnata* the culprit? Vet. Hum. Toxicol. 31:225–227.

Scruggs, D.W., and G.K. Blue. 1994. Toxic hepatopathy and photosensitization in cattle fed moldy alfalfa hay. J. Am. Vet. Med. Assoc. 204:264–266.

Sheehy, D.P., and A.H. Winward. 1981. Relative palatability of seven *Artemisia* taxa to mule deer and sheep. J. Range Manage. 34:397–399.

Stafford, K.J., D.M. West, M.R. Alley, and G.C. Waghorn. 1995. Suspected photosensitisation in lambs grazing birdsfoot trefoil (*Lotus corniculatus*). N. Z. Vet. J. 43:114–117.

Welch, B.L., E.D. McArthur and J.N. Davis. 1983. Mule deer preference and monoterpenoids (essential oils). J. Range Manage. 36:485–487.

Immunotoxins

Edgar, J.A. 1994. Vaccination against poisoning diseases. pp. 421–426 in: S.M. Colegate and P.R. Dorling (Eds.). Plant-Associated Toxins. Agricultural, Phytochemical and Ecological Aspects. CAB International, Wallingford, UK.

Panciera, R.J., D.A. Mosier and J.W. Ritchey. 1992. Hairy vetch (*Vicia villosa* Roth) poisoning in cattle: update and experimental induction of disease. J. Vet. Diagn. Invest. 4:318–325.

Sharma, R.P., L.F. James and R.J. Molyneux. 1984. Effect of repeated locoweed feeding on peripheral lymphocytic function and plasma proteins in sheep. Am. J. Vet. Res. 45:2090–2093.

Stegelmeier, B.L., R.J. Molyneux, A.D. Elbein and L.F. James. 1995. The lesions of locoweed (*Astragalus mollissimus*), swainsonine, and castanospermine in rats. Vet. Pathol. 32:289–298.

White, S.L., M.J. Humphries, R.J. Molyneux and K. Olden. 1989. Swainsonine: a new immunomodulator which enhances murine lymphoproliferation and natural killer activity. pp. 425–444 in: L.F. James, A.D. Elbein, R.J. Molyneux, and C.D. Warren (Eds.). Swainsonine and Related Glycosidase Inhibitors. Iowa State University Press, Ames.

Myotoxins

Baker, D.C., and R.F. Keeler. 1989. *Thermopsis montana*–induced myopathy in calves. J. Am. Vet. Med. Assoc. 194:1269–1272.

Barth, A.T., G.D. Kommers, M.S. Salles, F. Wouters, and C.S. Lombardo de Barros. 1994. Coffee senna (*Senna occidentalis*) poisoning in cattle in Brazil. Vet. Human Toxicol. 36:541–545.

Colvin, B.M., L.R. Harrison, L.T. Sangster and H.S. Gosser. 1986. *Cassia occidentalis* toxicosis in growing pigs. J. Am. Vet. Med. Assoc. 189:423–426.

Flunker, L.K., B.L. Damron and S.F. Sundlof. 1989. Response of white leghorn hens to various dietary levels of *Cassia obtusifolia* and nutrient fortification as a means of alleviating depressed performance. Poult. Sci. 68:909–913.

Keeler, R.F., and D.C. Baker. 1990. Myopathy in cattle induced by alkaloid extracts from *Thermopsis montana*, *Laburnum anagyroides* and a *Lupinus* sp. J. Comp. Path. 103:169–182.

Keeler, R.F., A.E. Johnson and (the late) R.L. Chase. 1986. Toxicity of *Thermopsis montana* in cattle. Cornell Vet. 76:115–127.

O'Hara, P.J., K.R. Pierce and W.K. Read. 1969. Degenerative myopathy associated with the ingestion of *Cassia occidentalis* L.: Clinical and pathological features of the experimentally induced disease. Am. J. Vet. Res. 30:2173–2180.

Rowe, L.D. 1991. *Cassia*-induced myopathy. pp. 335–351 in: R.F. Keeler and A.T. Tu (Eds.). Handbook of Natural Toxins. Toxicology of Plant and Fungal Compounds, Marcel Dekker, Inc., New York.

Schmitz, D.G., and J.H. Denton. 1977. Senna bean toxicity in cattle. Southwest. Vet. 30:165–170.

Spoerke, D.G., M.M. Murphy, K.M. Wruk and B.H. Rumack. 1988. Five cases of *Thermopsis* poisoning. Clin. Toxicol. 26:397–406.

INDEX

Index

B

D

M

N

Q

R